THIRD EDITION

BASIC HUMAN ANATOMY

Of related interest from
The Benjamin/Cummings Series in the Life Sciences

Anatomy and Physiology

R.B. Chase
The Bassett Atlas of Human Anatomy (1989)

S.W. Langjahr and R.D. Brister
Human Anatomy: A Computerized Review and Coloring Atlas (1985)

E.N. Marieb
The A&P Coloring Workbook: A Complete Study Guide to Anatomy and Physiology,
third edition (1991)

E.N. Marieb
Essentials of Human Anatomy and Physiology, third edition (1991)

E.N. Marieb
Human Anatomy and Physiology (1989)

E.N. Marieb
Human Anatomy and Physiology Laboratory Manuals, Brief Version, second edition (1987)

E.N. Marieb
Human Anatomy and Physiology Laboratory Manuals: Cat and Pig Versions,
third edition (1989)

E.B. Mason
Human Physiology (1983)

General Biology

N.A. Campbell
Biology, second edition (1990)

Microbiology and Immunology

I.E. Alcamo
Fundamentals of Microbiology, third edition (1991)

G.J. Tortora, B.R. Funke, and C.L. Case
Microbiology: An Introduction, third edition (1989)

P.L. VanDemark and B.L. Batzing
The Microbes: An Introduction to Their Nature and Importance (1987)

THIRD EDITION

BASIC HUMAN ANATOMY

ALEXANDER P. SPENCE

State University of New York College at Cortland

Illustrations by Fran Milner

THE BENJAMIN/CUMMINGS PUBLISHING COMPANY, INC.

Redwood City, California • Fort Collins, Colorado • Menlo Park, California
Reading, Massachusetts • New York • Don Mills, Ontario • Wokingham, U.K.
Amsterdam • Bonn • Sydney • Singapore • Tokyo • Madrid • San Juan

To Torry, who makes it all worthwhile.

Sponsoring Editor: *Melinda Adams*
Production Supervisor: *Anne Friedman*
Book Designer and Cover Concept: *John Edeen*
Artists: *Fran Milner, Carol Verbeeck (Figures 9.2 and 21.9), and Linda McVay (Figures 3.2–3.8, 3.12–3.23)*

Magnifications listed with the microscope symbol indicate the final enlargement of a photomicrograph. Magnifications listed in legends indicate the enlargement at which the photomicrograph was shot, with no adjustment for final printed size.

Library of Congress Cataloging-in-Publication Data

Spence, Alexander P., 1929–
  Basic human anatomy / Alexander P. Spence: illustrations by Fran Milner. — 3rd ed.
      p.   cm.
    Includes index.
    ISBN 0-8053-8860-5
    1. Human anatomy.   I. Title.
    [DNLM: 1. Anatomy.   QS 4 S744b]
QM23.2.S68   1990
611—dc20
DNLM/DLC
for Library of Congress   90-921
                             CIP

  5 6 7 8 9 10 - VH - 95 94 93

The Benjamin/Cummings Publishing Company, Inc.
390 Bridge Parkway
Redwood City, CA 94065

About the Cover

Shown on the cover is *Studies for the Libyan Sibyl* by Michaelangelo (1475–1564) by courtesy of The Metropolitan Museum of Art (Purchase, 1924, Joseph Pulitzer Bequest). (24.197.2)

Preface

When a textbook has been as well received as have the first two editions of *Basic Human Anatomy*, an author is hesitant to make major changes in a revision in fear of upsetting what seems to be an acceptable balance of coverage. At the same time, there are corrections to be made, updating to be done, and improved descriptions or explanations to be included. Therefore, in line with the suggestions I have received from the reviewers, my revisions for this third edition are relatively conservative.

My goals for the third edition remain the same as for the first two editions; to write a text for a one-term course in human anatomy that is accurate and up-to-date, yet suitable for a wide range of students whose interests may vary between biology, physical education, allied health, health sciences, and liberal arts. The very positive feedback that I have received from users concerning this book indicates that it is meeting the needs of such a variety of students. I am deeply indebted to the many instructors who have been supportive of my work over the first two editions, and hope that they will find the third edition equally satisfactory.

In the past several years textbooks have become increasingly colorful. Color can be very useful in distinguishing structures from one another, but if overused can distract and even confuse. I have, therefore, resisted the trend and continue to use color only when it serves an instructional purpose. I feel that this allows the figures to remain clear and uncluttered. The one place we have expanded the use of color in this edition is in the nervous system. Yellow is a standard color for nerves, and its use should remove any possibility of confusion in the figures.

I have continued to update terminology to conform with the fifth edition of *Nomina Anatomica* whenever such a change contributes to the clarity of discussion. The new terms are generally more descriptive than the old terms, which often include a person's name. I have generally left the old terms in parentheses to assist the reader in the transition.

SPECIAL FEATURES

This new edition retains several special features that were well received in the first two editions because they enhance the text's usefulness as a teaching tool and help to increase student interest in the human body. An exciting new feature, a set of Regional Anatomy Reference Plates, has been incorporated into this edition as well. The special features include:

- Regional Anatomy Reference Plates

  New to this edition, the Regional Anatomy Reference Plates consist of full color labeled photographs of dissected cadavers from the Bassett photo collection. The photographs will be valuable to anyone wishing to study regional anatomy, especially to those with access to a cadaver in the laboratory. The Reference Plates have been strategically placed early in the text, in the muscular system chapter, to increase their usefulness both in lecture and in lab.

- Integration of Embryology

  I have found that structural relationships within the body are best understood when students have some knowledge of the embryonic development of the structures involved. For this reason, the discussion of each system begins with a brief consideration of the development of the system in the embryo. These discussions are self-contained and, if the instructor chooses, may be omitted without detracting from the remainder of the chapter.

- Surface Anatomy

  When students can visualize structures, their retention and understanding of anatomical relationships is improved. One of the best ways to visualize anatomical relationships is to have students locate structures on their own bodies. The chapter on surface anatomy (Chapter 8), placed after the skeletal and muscular system chapters, will facilitate this understanding. The surface anatomy chapter provides an excellent review of many of the structures identified in previous chapters.

- Special Topic Boxes

  Brief essays on topics of special interest have been included in several chapters. Accompanied by a photograph, these special topics were selected for student interest as a welcome break from the heavily factual material in the text.

- ## Conditions of Clinical Significance

  While my emphasis throughout the text is on normal human anatomy, brief discussions of diseases, dysfunctions, and aging are included where they serve to enhance and reinforce an understanding of normal human anatomy. Because of the urgent need to better understand aging in order to provide care to the rapidly increasing numbers of older persons, I have expanded the discussions of age-related changes and included them in consideration of all body systems. Discussions of conditions of clinical significance have been placed in separate display boxes following the discussion of each system. This special treatment allows the instructor to emphasize or de-emphasize these conditions according to the objectives of his or her particular course.

- ## Frontiers in Health Boxes

  To stimulate the reader's interest, *Frontiers in Health* boxes which present a short discussion of a new medical treatment are included in several chapters. These treatments have been selected primarily because they illustrate anatomical relationships.

PEDAGOGICAL DEVICES

This new edition makes use of numerous pedagogical aids designed to increase readability and enhance the study of anatomy. These in-text learning aids include:

- ## Learning Objectives

  Each chapter opens with a set of learning objectives. These objectives identify the most important concepts found in the chapter.

- ## Chapter Contents

  Each chapter opens with a list of chapter contents that outline the major topics in the chapter.

- ## Marginal Figure and Table References

  Cited as one of the most useful pedagogical devices unique to this text, every figure and table is referenced in the margin. This allows immediate recognition of where each piece of art is discussed in the text. This careful integration of art and text, with easily identifiable referencing of all figures and tables, makes this text a superior teaching and learning tool.

- ## Tables

  The text makes use of numerous tables to organize the large amount of information that must be mastered in a Human Anatomy course.

- ## Study Outline with Page References

  Thorough chapter summaries in a study outline format with page references will help students review the material covered in each chapter.

- ## Self Quizzes

  Incorporating the benefits of a study guide into the text itself, self quizzes appear at the end of every chapter and allow students to test their own mastery of the chapter's contents.

- ## Self Quiz Answers

  Found in Appendix 1, answers to the chapter self quizzes are listed along with the page reference of where the material can be located in the text. Page referencing greatly facilitates the studying and learning of Human Anatomy with this text.

- ## Word Roots, Prefixes, Suffixes, and Combining Forms

  Learning a new vocabulary is often cited as one of the more difficult challenges facing students in a Human Anatomy course. Found in Appendix 2, this guide will be a valuable aid to students throughout the entire course.

- ## Units of the Metric System

  Appendix 3 provides metric/English conversion constants. Because many measurements use metric units, this appendix is a valuable reference for the student of Human Anatomy.

- ## Glossary and Index

  At the end of this book, the student will find an extensive glossary that includes both the definition and phonetic pronunciation of each term. In addition, the Index is comprehensive and contains many cross-references.

SUPPLEMENTS

- ## Instructor's Guide with Test Items

  Written by Leann Blem of Virginia Commonwealth University and Robin Lenn of Sacramento City College, the Instructor's Manual is completely revised for this edition. With two different tests per chapter, it provides roughly 1100 multiple choice questions in all. Many supplemental objective type questions have been added to each chapter test as well. It also includes two final exams of approximately 300 multiple choice questions each. An updated audiovisual reference guide is also included to enrich your lectures and laboratories. (Code 38861)

- ## Computerized Testbank

  For your convenience, this computerized test bank offers a flexible, time-saving method of creating challenging and original tests. It allows you to create tests quickly and easily, format tests to personal specifications, and edit and add questions at will. This Testbank is available to adopters of the text for the Macintosh and IBM family of computers.

- ## Transparency Acetates

  A set of color overhead transparencies is available to adopters of the text. With illustrations unique to this set, these transparencies feature the most important figures from the text. (Code 38862)

- ## Human Musculature Videotape

  Produced by Rose Leigh Vines and Allan Hinderstein of California State University at Sacramento, this 23 minute videotape offers students an anatomical tour of the muscles in the human body. It illustrates and reviews 33 important muscles in a prosected cadaver, explaining the origin, insertion, and action of each muscle. The videotape is not only an inexpensive alternative to teaching students with a cadaver, it is also a dramatic improvement over other abstract models and techniques. (Code 30106)

- ## Coloring Workbook

  A truly unique study guide, *The A&P Coloring Workbook* by Elaine Marieb can be ordered through the college bookstore. It is the first and only book to combine the best features of a study guide with an entertaining and instructive coloring book. (Code 34806)

- ## The Bassett Atlas of Human Anatomy

  Authored by Robert A. Chase, M.D., Professor of Anatomy at Stanford University, *The Bassett Atlas of Human Anatomy* offers an extraordinary collection of full-color dissection photos. An inexpensive alternative to the costly collections now available, it will be extremely useful to your students in their laboratories. (Code 30118)

- ## Human Dissection Slides

  A complimentary collection of slides, made up of the full set of 86 dissection photographs from *The Bassett Atlas of Human Anatomy*, is available to qualified adopters. (Code 30119)

- ## Computer Assisted Instruction

  Benjamin/Cummings offers two innovative software programs to assist students in understanding Human Anatomy. One is *Human Anatomy: A Computerized Review and Coloring Atlas* (Code 35810) by Stephen W. Langjahr and Robert D. Brister. This software package is written for the Apple family of computers and consists of a disk for the instructor and workbook for students. Also available to adopters of *Basic Human Anatomy* is a complimentary Macintosh program using Hypercard. Written by Dr. Marvin Branstrom of Canada College, *The A&P Tutor* offers an animated review of important anatomical concepts. For details on either package or *The Benjamin/Cummings Guide to A&P software*, please contact your local representative or call 1-800/950-BOOK, ext. 756.

ACKNOWLEDGMENTS

The success of any textbook depends to a large extent upon the strengths of its reviewers. The reviewers for the first two editions of Basic Human Anatomy were outstanding, and my good fortune continues into the third edition. While I received numerous suggestions and comments from instructors who were using the text, I am particularly indebted to: Dr. Shirley Bayer of Purdue University, Dr. Annabelle Cohen of City University of New York, Dr. Larry Ganion of Ball State University, Dr. John Hein of the University of Wisconsin, Dr. Robert Laird of the University of Central Florida, and Dr. Sherwin Mizell of Indiana University. Their detailed comments contributed greatly to the improvement of this edition.

Finally, I would like to acknowledge all the help, direction, and encouragement that I received from the dedicated people at Benjamin/Cummings. I am especially indebted to—Connie Spatz, executive editor; Melinda Adams, sponsoring editor; Anne Friedman, production supervisor; and Mark Thomas Childs, editorial assistant.

Alexander P. Spence
Department of Biological Sciences
State University of New York College at Cortland
Cortland, New York 13045

Brief Contents

Detailed Contents

5 The Skeletal System   93

6 Articulations   153

7 The Muscular System   181

8 Surface Anatomy   247

9 The Circulatory System: The Blood   265

20 The Digestive System   531

19 The Respiratory System   511

LEARNING OBJECTIVES

After completing this chapter, you should be able to:

- Describe several subdivisions of the study of anatomy.

- Name four types of tissues produced by the three embryonic cell layers.

- Name the ten major organ systems in the human body.

- Describe the locations of structures, using common directional and regional terms.

- Describe the anatomical position.

- Name the planes and cavities of the body.

- Distinguish between the parietal and the visceral membranes of the ventral body cavities.

- Describe the mesenteries of the abdominopelvic cavity.

CHAPTER CONTENTS

FIELDS OF ANATOMY

PHYSIOLOGY

BASIC STRUCTURAL LEVELS

ANATOMICAL TERMINOLOGY

BODY POSITIONS

DIRECTIONAL TERMS

REGIONAL TERMS

BODY PLANES

BODY CAVITIES

MEMBRANES OF THE VENTRAL BODY CAVITIES

Anatomical Orientation

Anatomy is the study of the structure of an organism and the relationships of its parts. This book presents a basic, but comprehensive, coverage of *human anatomy*. The term *anatomy* is derived from the Greek words meaning "apart" and "to cut." As this derivation indicates, anatomy is based largely on dissection of the body. Nevertheless, some of the more recent fields of anatomy involve the use of machines that provide valuable supplements to dissection.

FIELDS OF ANATOMY

The study of anatomy involves examination of the general structures of the body **(gross anatomy)** as well as those structures that can be seen only with the aid of a microscope **(microscopic anatomy).** Gross anatomy can be studied by regions, such as the head, neck, thorax, abdomen, pelvis, or limbs. This approach, referred to as **regional anatomy,** is often used in dissection, in which all structures in a region are studied simultaneously. However, for our purposes the study of anatomy by organ systems that perform common functions **(systemic anatomy)** is most beneficial and this book uses that approach. Microscopic anatomy includes the study of cells *(cytology)* and the study of tissues *(histology)*. When anatomy is studied under the extremely high magnifications possible with the electron microscope, it is referred to as *fine structure* or *ultrastructure*. **Developmental anatomy,** another subdivision of anatomy, focuses on the development of the body from the fertilized egg to the adult form. Developmental anatomy includes *embryology*, which is limited to prenatal development.

Radiographic anatomy is particularly valuable in the diagnosis of disorders and injuries. Until recent years, the only tools available for this field were the X-ray machine and the fluoroscopic machine. Both instruments produce *roentgen rays* (X rays), which pass through the structure under examination and either expose an X-ray film or illuminate a fluorescent screen. When an exposed film is developed, the result is a photographic image called a *roentgenogram*, popularly known as an X ray. Although roentgenograms are diagnostically useful, they are somewhat limited in that the three-dimensional relationships of the body's parts are lost on the film, where the body image is compressed into a flat image. Moreover, small differences in the densities of tissues are not always detectable on a roentgenogram.

In the 1970s, technological advances made *tomography* possible, thus expanding radiographic anatomy. Tomography involves radiographing a specific level in the body while blurring structures above and below this level. This effect is achieved by rotating the X-ray tube and film around the selected level while exposing it repeatedly to X rays. The technique, in effect, produces a cross section of the body at the selected level (Figure 1.1). Since the **F1.1** rotation of the X-ray tube and film are coordinated by computer, this technique is referred to as *computed tomography (CT scan)* or *computed axial tomography (CAT scan)*. A CT scan not only shows the three-dimensional relationships of the body parts but produces better differentiation between tissues than is possible by the use of X rays.

The recent development of a complex X-ray machine called the *dynamic spatial reconstructor (DSR)* represents a significant advance in radiographic

3

Anterior

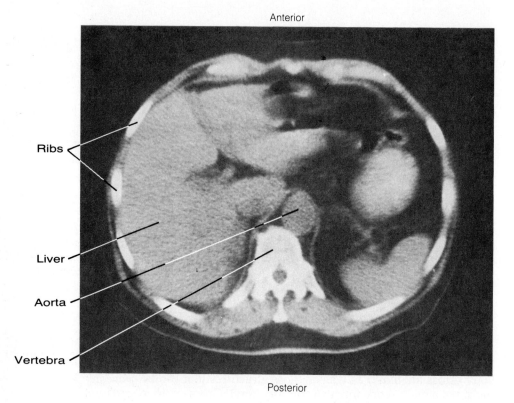

Ribs

Liver

Aorta

Vertebra

Posterior

Figure 1.1
CT scan from the mid-trunk region. The liver forms the opaque area on the left. Note the ribs (white) in the body wall and the vertebra at the bottom. The aorta can be observed against the upper right side of the vertebra, passing through the aortic hiatus of the diaphragm. (For a fuller explanation of CT scans, see Box 5.2.)

anatomy. The DSR uses multiple X-ray tubes that rapidly revolve around the patient, producing thousands of cross sections within a time span of a few seconds. This is much faster than a CT scanner, which would produce only a single section during the same time span. The DSR produces moving, three-dimensional images of an organ that can be rotated and tipped in any direction, allowing all sides of the organ to be viewed (Figure 1.2). A CT scan, in contrast, produces only cross-sectional slices of the body. In addition, the image produced by the DSR can be "sliced open" on the screen so that the interior of an organ can be viewed.

Nuclear magnetic resonance (NMR) is another recent advance in radiographic anatomy. NMR has been used primarily as a research instrument, but it is beginning to be used for diagnostic purposes. It has the advantage of using nonionizing radiation, which is less damaging to cells than are X rays. In NMR the patient is placed in a body-size chamber within a large magnet. The magnetic field causes the hydrogen nuclei, as well as certain other nuclei, throughout the body to become aligned. By adjusting the magnetic energy generated by the magnet it is possible to detect the amount of energy absorbed by the various nuclei. This information, when fed into a computer, can be used to plot the distribution of the nuclei and thus provides images of the body organs.

Positron-emission tomography (PET) is a specialized radiographic anatomy process that is useful in producing images that provide information concerning organ functioning. While CT scans produce images that show what an organ looks like, PET, which thus far has only been used to view the brain, produces images that give a measure of the activity of the cells within the organ. During a PET examination, radioactive glucose is injected into the patient's bloodstream. The glucose concentrates in the cells of the most metabolically active regions of the brain. The patient's head is inserted into the PET machine, where detectors absorb the radiation and send signals to a computer. The computer combines the signals into a video image, with bright spots showing where the radioactive glucose has accumulated, thus providing a functional map of the patient's brain.

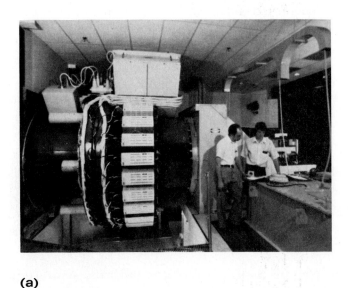

(a)

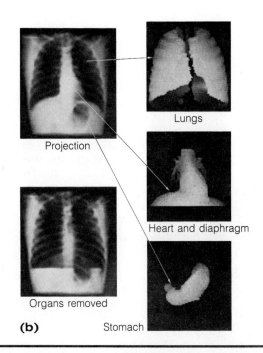

Projection

Lungs

Heart and diaphragm

Organs removed

(b)

Stomach

Figure 1.2
(a) Dynamic spatial reconstructor. (b) Three-dimensional images of lungs, diaphragm, and stomach produced by a DSR.

The use of *ultrasonography* provides another means of obtaining images of the body organs. In this procedure, sound waves are sent into the body and the echoes they make when they strike various tissues are used to define the boundaries of organs. Since sound waves appear to have no significant harmful side effects on cells, ultrasonography can be safely used to produce images of unborn fetuses. It can also produce images of movement, such as bloodflow within a vessel.

PHYSIOLOGY

The study of body functions is called **physiology.** Physiology explains, in chemical and physical terms, how the body and its parts work. Although this text focuses primarily on anatomy, structure is generally related to function, and an awareness of the relationship between structure and function will help make the study of anatomy more meaningful. The text thus includes enough physiology to explain the anatomy of a particular system or structure.

BASIC STRUCTURAL LEVELS

There are four structural levels in the body: *cells, tissues, organs,* and *systems.* Each level of body structure has specific functions that contribute not only to the structure itself but also to the general well-being of the entire body.

Cells

At the most basic structural level, the body is composed of **cells.** Even at the cellular level there are structural differences and these cellular differences are closely related to the physiology of each type of cell. We consider these cellular differences in Chapter 3.

Tissues

Only the simplest animals are able to exist as single cells. In most animals, including humans, groups of similar cells join together to form **tissues.** In the early embryo, where tissue formation first occurs, similar cells group together into three layers: the **ectoderm,** which forms both the outer covering of the

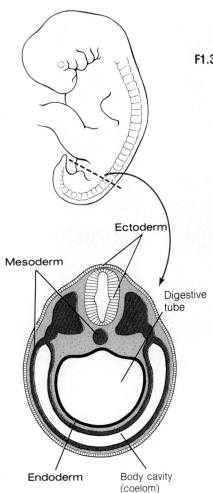

Figure 1.3
Schematic cross section of an embryo showing the location of endoderm, mesoderm, and ectoderm. The dotted line on the embryo indicates the site of the cross section.

Ectoderm

Mesoderm

Digestive tube

Endoderm   Body cavity (coelom)

F1.3 body and the nervous tissue; the **endoderm,** which forms the inner lining of the digestive tube and its associated structures; and the **mesoderm,** the layer located between the ectoderm and endoderm tissues, which forms the skeleton and the muscles of the body (Figure 1.3). The three embryonic cell layers—the ectoderm, endoderm, and mesoderm—give rise to four types of tissue. These tissues are briefly discussed here and are studied in detail in Chapter 3.

EPITHELIAL TISSUES   These tissues cover the surface of the body and line the various body cavities, ducts, and vessels. The epithelial tissue that forms the outer protective layer of the body, the **epidermis,** is derived from embryonic ectoderm. The rest of the epithelial tissues originate from either the mesoderm or the endoderm of the embryo.

MUSCULAR TISSUES   Muscular tissues are composed of specialized cells that are capable of contracting and thereby decreasing in length. These tissues move the skeleton, propel the blood throughout the body, and aid in digestion by moving food through the digestive tract. As will be explained in Chapter 7, there are three types of muscle tissues: *skeletal, cardiac,* and *visceral.* Each type is derived from embryonic mesoderm.

NERVOUS TISSUES   Nervous tissues, which form the brain, spinal cord, and nerves, consist of cells with long protoplasmic extensions. These nerve cells, or **neurons,** transmit messages throughout the body. They originate from the ectodermal layer of the embryo.

CONNECTIVE TISSUES   Several types of cells are involved in the formation of various connective tissues. These tissues, most of which are derived from mesoderm, are used for support (bones and cartilage), for the attachment of other tissues (tendons, ligaments, and fasciae), or for other specialized functions (for example, blood).

Organs

Tissue formation allows the body to perform more involved physiological activities than would be possible for individual cells. Physiological processes that are even more complex are possible when two or more tissues combine to form an **organ.** The stomach, for instance, is lined with epithelial tissue and its walls are formed by muscular tissue. These tissues are held together as a discrete structure by various connective tissues and are innervated by nervous tissue. Each of these types of tissues contributes in a specific manner to the functioning of the stomach, and without such tissue combinations it would not be possible to process large particles of complex foods. This same principle holds true for all organs, each serving as a specialized physiologic center for the body.

Systems

The ability of the organs to function for the general well-being of the body is enhanced by the fact that certain organs work together as a system, each organ in the system performing a specific part of a general body function. For example, one shared function is obtaining energy from food for use throughout the body: food is prepared and partially digested in the mouth; it is transported by the esophagus to the stomach, where it is further prepared and digested; it is absorbed into the blood vessels through the walls of the intestines; and finally, the residue that is not absorbed is eliminated from the body through the rectum. When the various organs and systems work together to maintain a stable internal environment, a state of *homeostasis* exists.

Organs that function cooperatively to accomplish a common purpose (such as the digestion and absorption of food) are said to be a part of a body **system.** There are ten major systems in the human body: *integumentary, skeletal, muscular, nervous, endocrine, circulatory, respiratory, digestive, urinary,* and **Table 1.1** *reproductive.* The structure of each of these body systems is listed in Table 1.1 and discussed in later chapters.

Table 1.1   Organ Systems of the Body

System	Major Components	Representative Functions
Integumentary	Skin and associated structures such as hair and nails	Protects internal body structures against injury and foreign substances; prevents fluid loss (dehydration); important in temperature regulation
Skeletal	Bones	Supports and protects soft tissues and organs
Muscular	Skeletal muscles	Moves body and its parts
Nervous	Brain, spinal cord, nerves, special sense organs	Controls and integrates body activities; responsible for "higher functions" such as thought and abstract reasoning
Endocrine	Hormone-secreting glands such as the pituitary, thyroid, parathyroid, adrenals, pancreas, and gonads	Controls and integrates body activities; function closely allied with that of the nervous system
Circulatory	Heart, blood and lymphatic vessels, blood, lymph	Links internal and external environments of the body; transports materials between different cells and tissues
Respiratory	Nose, trachea, lungs	Transfers oxygen from the atmosphere to the blood and carbon dioxide from the blood to the atmosphere
Digestive	Mouth, esophagus, stomach, small intestine, large intestine; accessory structures include salivary glands, pancreas, liver, gallbladder	Supplies body with substances (food materials) from which energy for activity is derived and from which components for synthesis of required substances are obtained
Urinary	Kidneys, ureters, urinary bladder, urethra	Eliminates variety of metabolic end products such as urea; conserves or excretes water and other substances as required
Reproductive	Male: seminal vesicles, testes, prostate gland, bulbourethral glands, penis, associated ducts	Produces male gametes (sperm); provides method for introducing sperm into the female
	Female: ovaries, uterine tubes, uterus, vagina, mammary glands	Produces female gametes (ova); provides proper environment for development of fertilized ovum

ANATOMICAL TERMINOLOGY

Every branch of science has developed its own special terminology, and anatomy is no exception. You will read many terms in this text that are new to you. The terms are hardly new, however, as a sizable number of them originated centuries ago and have Greek or Latin backgrounds. Although these terms may appear formidable, they are quite descriptive if you understand their roots. For example, *ilio* refers to the hip bone (ilium), and *costal* refers to the ribs. Therefore, the *iliocostalis* muscle is clearly a muscle that passes from the hip bone to the rib cage.

A knowledge of prefixes and suffixes is also helpful in understanding anatomical terms. For example, the prefix *endo-* means "within" and is used in many anatomical terms, including the following:

endocardium (*cardium* refers to the heart) the inner lining of the heart

endocarditis (*itis* refers to an inflammation) an inflammation of the inner lining of the heart

endochondral (*chondral* refers to cartilage) development within cartilage (for example, endochondral bone)

endometrium (*metra* refers to the uterus) the inner lining of the uterus

The meanings of other prefixes, suffixes, and roots will be explained as they are introduced in the text. For a more thorough list that will prove helpful in understanding unfamiliar words, see Appendix 3 at the back of the book.

BODY POSITIONS

While studying the detailed description of each body structure, you must also understand the positional relationships between the various body structures. For this reason, it is essential that you become familiar with the terms that are used to describe these relationships.

If the body is lying horizontally with the face downward, it is in the *prone position*. If the body is on its back, with the face upward, it is in the *supine position*. The relationships of the various body structures to each other are different in these positions. Therefore, in order to communicate effectively concerning human anatomy, the body must always be assumed to be in a standard position so that the structural relationships are clear and consistent.

F1.4a This standard position is referred to as the **anatomical position** (Figure 1.4*a*). In this position, the body is erect, with the feet together. The upper limbs hang at the side, with the palms of the hands facing forward, the fingers extended, and the thumbs pointing away from the body. With the hands in this position, the bones of the hands and fingers are exposed and therefore their relationships are easily described. Moreover, when the palms are facing forward, the bones of the forearm are uncrossed. *Unless it is stated otherwise, all anatomical descriptions refer to a body that is in the anatomical position.*

DIRECTIONAL TERMS

The terms used to denote direction come in pairs, each indicating an opposite
F1.4b, F1.4c direction (Figure 1.4*b* and *c*). **Anterior (ventral)** refers to the front, whereas its opposite, **posterior (dorsal)**, refers to the back. **Superior (cranial)** means toward the head; **inferior (caudal)** means away from the head. Notice that these terms are used differently for humans than when they refer to four-legged animals. Since humans stand upright, this alters the direction indicated by these terms. In this text we are only interested in their meaning as they relate
Table 1.2 to humans. The directional pairs are listed in Table 1.2.

REGIONAL TERMS

In addition to directional terms, there are several frequently used terms that
F1.4a refer only to special areas of the body (Figure 1.4*a*):

cervical refers to the neck

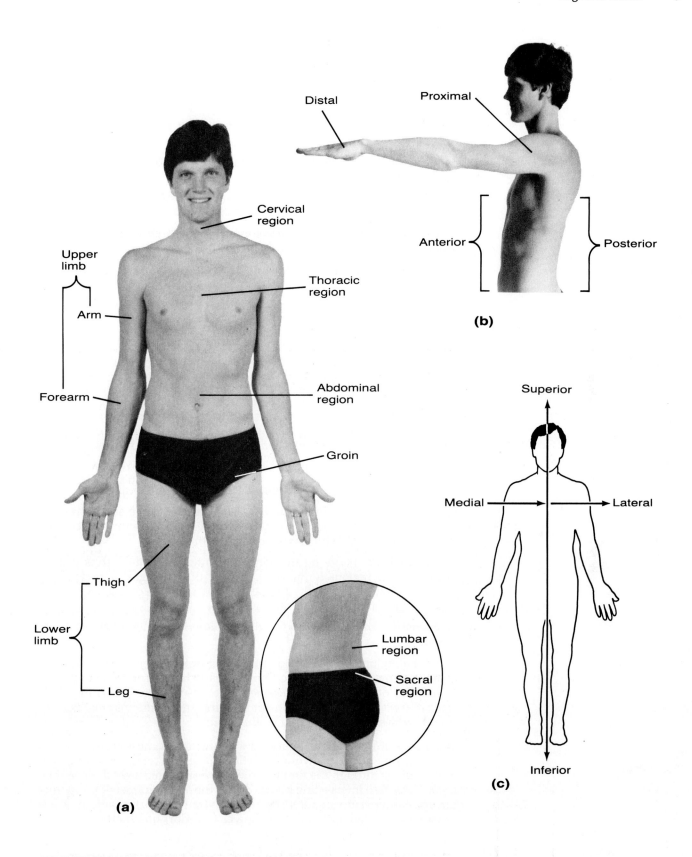

Figure 1.4
(a) Anatomical position
and regions of the body.
(b, c) Directional terms.

thoracic   the portion of the body between the neck and the abdomen that is commonly referred to as the chest (thorax)

lumbar   the portion of the back between the thorax and the pelvis

sacral   the lower portion of the back, just superior to the buttocks

plantar   the sole of the foot; the top of the foot is the *dorsal* surface

palmar   the anterior surface of the hand; the posterior surface of the hand is the *dorsal* surface

axilla (armpit)   the depression on the inferior surface of the attachment of the upper limb and the body trunk

groin (inguinal region)   the junction of the thigh with the abdominal wall

arm   the portion of the upper limb between the shoulder and the elbow

forearm   the portion of the upper limb between the elbow and the wrist

thigh   the portion of the lower limb between the hip and the knee

leg   the portion of the lower limb between the knee and the ankle

To make it easier to describe the location of the organs of the abdomen, the abdominal cavity is divided into nine regions: first by two vertical lines that bisect the clavicles (collarbones); and then by two horizontal lines, one that passes along the lower edge of the rib cage and another that runs across the upper edges of the hip bones (iliac crests) (Figure 1.5*a*). These regions are:

F1.5a

Umbilical   located centrally, surrounding the *umbilicus* (navel)

Lumbar   the regions to the right and left of the umbilical region

Epigastric   (*epi-* means on or above; *gastric* refers to the stomach) the midline region superior to the umbilical region. As the name implies, most of the stomach is located in this region

Hypochondriac   (*hypo-* means beneath or under; *chondral* refers to cartilage)   the regions to the right and left of the epigastric region. The name indicates that the hypochondriac regions are located beneath the cartilage of the rib cage

Hypogastric   the midline region directly inferior to the umbilical region

Iliac   the regions on either side of the hypogastric region. The name is derived from the iliac (hip) bones, which form the lateral boundaries of the regions. These areas are also referred to as the *inguinal regions* because their lower margins end at the inguinal ligament, which follows the fold of the groin

In practice, it is more common to divide the abdominopelvic cavity into four quadrants by means of two intersecting planes: an imaginary horizontal plane that passes through the umbilicus and a vertical midsagittal plane (Figure 1.5*b*). These two intersecting planes divide the abdominopelvic cavity into a **right upper (superior) quadrant**; a **right lower (inferior) quadrant**; a **left upper (superior) quadrant**; and a **left lower (inferior) quadrant**.

F1.5b

BODY PLANES

In the study of anatomy, it is useful to visualize the body as cut or sectioned into various planes of reference (Figure 1.6). A **sagittal plane** is a longitudinal section that divides the body or any of its parts into right and left portions. If

F1.6

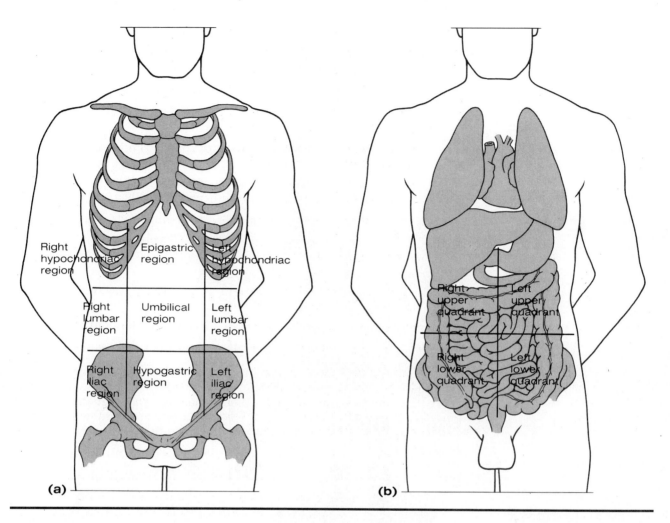

Right hypochondriac region

Epigastric region

Left hypochondriac region

Right lumbar region

Umbilical region

Left lumbar region

Right iliac region

Hypogastric region

Left iliac region

(a)

Right upper quadrant

Left upper quadrant

Right lower quadrant

Left lower quadrant

(b)

the section passes through the midline of the body, it is referred to as a **median sagittal (midsagittal) section.** Such a section divides the body into equal right and left *halves.* Sagittal sections other than the median sagittal section are often referred to as **parasagittal sections.** Parasagittal sections divide the body into *unequal* right and left portions. The **frontal (coronal) plane** is also a longitudinal section, but it runs at right angles to the sagittal plane, dividing the body into anterior and posterior portions. A **transverse plane (cross section** or **horizontal section)** divides the body or any of its parts into superior and inferior portions.

BODY CAVITIES

The body contains two main cavities: the **dorsal (posterior) cavity** and the **ventral (anterior) cavity** (Figure 1.7*a*). Each of these cavities is lined with membranes and contains a small amount of fluid surrounding the organs that fill the cavities.   **F1.7a**

The dorsal cavity has two subdivisions: the **cranial cavity,** which houses the brain, and the **spinal (vertebral) cavity,** which contains the spinal cord. The spinal cavity communicates with the cranial cavity through the *foramen magnum,* a large opening in the base of the skull. The membranes associated with these dorsal cavities are examined in greater detail in Chapter 14. For now, it is sufficient to call the membranes that cover the brain and spinal cord the *meninges.* The fluid found in these dorsal cavities is the *cerebrospinal fluid.* It too is considered in more detail with the nervous system.

Figure 1.5
(a) Abdominal regions. The top horizontal line passes along the lower edge of the rib cage. The lower horizontal line passes across the upper margins of the hip bones. The vertical lines pass through the midpoints of the clavicles and the inguinal ligaments.
(b) The abdominal wall and abdominopelvic cavity subdivided into four quadrants.

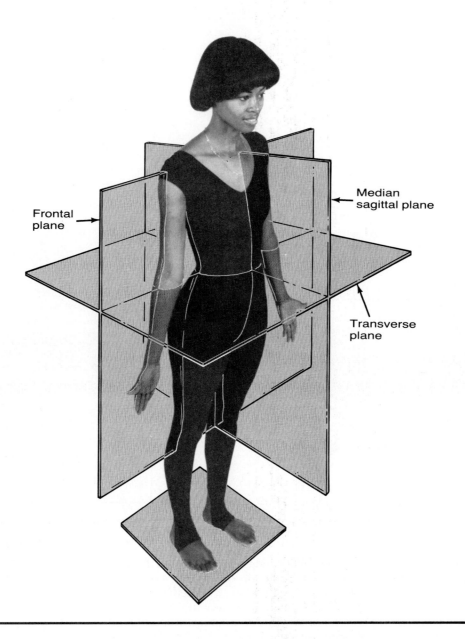

Figure 1.6
Body planes.

The ventral cavity of the body also has two subdivisions. It is separated by a muscle called the *diaphragm* into an upper **thoracic cavity** and a lower **abdominopelvic (peritoneal) cavity.** Each of these cavities is, in turn, further subdivided. The thoracic cavity is divided into a **pericardial cavity,** which surrounds the heart, and right and left **pleural cavities,** each of which encom-

F1.7b passes a lung (Figure 1.7*b*). The portion of the thoracic cavity between the two pleural cavities is called the **mediastinum.** The heart, trachea, esophagus, thymus gland, and several major blood vessels are also located within, or pass through, the mediastinum.

The abdominopelvic cavity, which is the second subdivision of the ventral body cavity, is divided for descriptive purposes into a superior **abdominal cavity** and an inferior **pelvic cavity (true pelvis)** by an imaginary oblique plane that passes from the superior margin of the **symphysis pubis** anteriorly

F1.8 and the **sacral promontory** posteriorly (Figure 1.8). The circumference of this plane is called the **pelvic brim.** The pelvic cavity is completely surrounded by the bones of the pelvis. In contrast, the lower portion of the abdominal cavity is bounded posteriorly by the flat hip bones, but its anterior wall is formed by the abdominal wall. This expanded region, which is located just above the

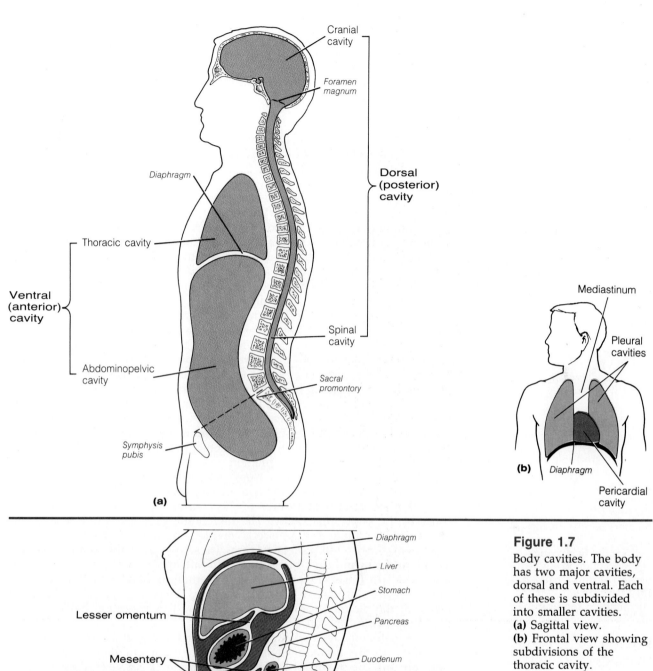

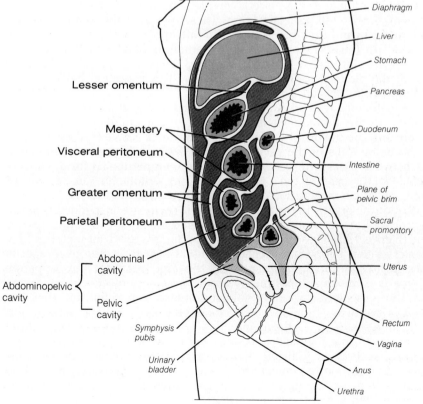

Figure 1.7
Body cavities. The body
has two major cavities,
dorsal and ventral. Each
of these is subdivided
into smaller cavities.
(a) Sagittal view.
(b) Frontal view showing
subdivisions of the
thoracic cavity.

Figure 1.8
Sagittal section of the
body showing membrane
relationships of the
abdominopelvic
(peritoneal) cavity.

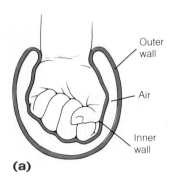

(a)

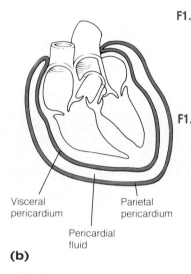

Visceral
pericardium

Parietal
pericardium

Pericardial
fluid

(b)

Figure 1.9
Membrane relationships
in the ventral body
cavities. **(a)** Schematic
representation using a
fist thrust into a balloon.
(b) Membranes
surrounding the heart.

pelvic brim, is the **false pelvis.** The abdominal cavity contains the stomach, spleen, liver, gallbladder, pancreas, and the small and large intestines. The pelvic cavity contains the lower part of the digestive system (rectum), the urinary bladder, and, in the female, the internal reproductive organs.

MEMBRANES OF THE VENTRAL BODY CAVITIES

To understand the membranes associated with the ventral body cavities, imagine that you have thrust your fist into an inflated balloon and have **F1.9a** pushed in one side of the balloon (Figure 1.9a). Now suppose that your fist is an organ. Notice that the wall of the balloon that has been pushed in lies close against your fist. The outer balloon wall remains separated from the inner wall that covers your fist because of the air in the balloon. The membranes of the ventral body cavities have the same relationship, except that they are kept separated by fluid rather than air.

In the pericardial cavity, the heart (rather than your fist) pushes in one **F1.9b** side of the membranous sac (rather than a balloon) (Figure 1.9b). The membrane that lies closely adherent to the heart is the **visceral pericardium.** The membrane that covers both the heart and the visceral pericardium is the **parietal pericardium** (*parietal* refers to the walls of a body cavity). The parietal pericardium is separated from the visceral pericardium by **pericardial fluid,** which is secreted by the cells of the pericardial membranes. In other words, the outer wall of the pericardial cavity is *lined* with parietal pericardium and the heart is *covered* with visceral pericardium. However, both of these are simply different regions of the same membrane. The amount of pericardial fluid that separates these two membranes is very small—just enough to reduce friction and maintain the tissues in a healthy state. If the membranes of the heart become inflamed, the condition is known as *pericarditis.*

The membrane relationships of the pleural cavities are quite similar to those of the pericardial cavity. The membrane that lies tightly against the surface of the lungs is the **visceral pleura.** The outer walls of the pleural cavities are lined with **parietal pleura.** These two membranes are separated by the **pleural fluid** that they secrete. An inflammation of these membranes may result in the secretion of excessive amounts of pleural fluid into the pleural cavity. Prolonged inflammation may cause the visceral and parietal layers of the pleura to adhere to each other. This condition, which makes breathing painful, is called *pleurisy.*

The membrane relationships within the abdominopelvic cavity are similar to those of the thoracic cavity, but here the membrane is called the **peritoneum.** The organs of this cavity are covered with **visceral peritoneum** and the outer walls of the cavity are lined with **parietal peritoneum.** The space between these two membranes is filled with **peritoneal fluid,** which is secreted by the cells of the peritoneum. The peritoneum can become inflamed, causing a very serious condition called *peritonitis.*

Most of the organs in the abdominopelvic cavity are suspended from the posterior wall of the cavity by a double-layered membrane of parietal perito- **F1.8** neum (Figure 1.8). These membrane supports are called **mesenteries.** The mesenteries that support particular organs or structures are given specific names such as mesocolon (mesentery of the large intestine), mesoappendix (mesentery of the appendix), mesovarium (mesentery of the ovary), and so forth. The mesenteries not only hold the organs in position, but they also provide a pathway through which blood vessels, lymphatic vessels, and nerves can reach them. As the membranes that form the mesenteries continue over the organ that they suspend, they become visceral membranes. Some structures, such as the kidneys, do not hang into the cavity on mesenteries. Instead, they are located outside the cavity, between the body wall and the **F1.10** parietal peritoneum (Figure 1.10). These structures are *retroperitoneal*—that is, they are located *behind* the peritoneum.

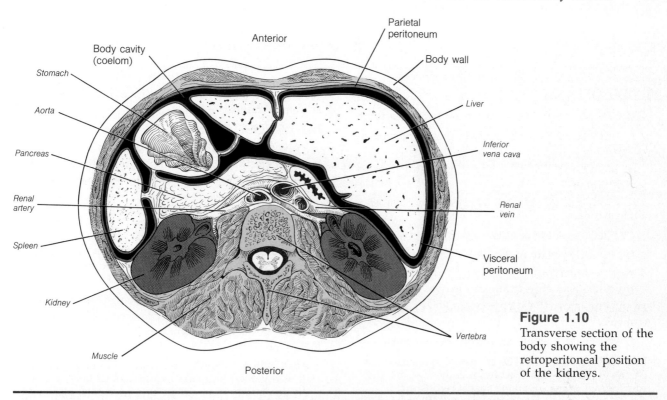

Figure 1.10
Transverse section of the body showing the retroperitoneal position of the kidneys.

Table 1.2 Directional Terms

Term	Definition	Example
Anterior (ventral)	Situated in front of; the front of the body	The chest is on the anterior surface of the body.
Posterior (dorsal)	Situated in back of; the back of the body	The buttocks are on the posterior surface of the body.
Superior (cranial)	Toward the head; relatively higher in position	The eyebrows are superior to the eyes.
Inferior (caudal)	Away from the head; relatively lower in position	The mouth is inferior to the nose.
Medial	Toward the midline of the body	The breast is medial to the armpit.
Lateral	Away from the midline of the body	The hip is on the lateral surface of the body.
Proximal	Closer to any point of reference, such as the attached end of a limb, the origin of a structure, or the center of the body	The arm is proximal to the forearm.
Distal	Farther from any point of reference, such as the attached end of a limb, the origin of a structure, or the center of the body	The hand is distal to the wrist.
Superficial (external)	Located close to or on the body surface	The skin is superficial to the muscles.
Deep (internal)	Located further beneath the body surface than superficial structures	The muscles are deep to the skin.

STUDY OUTLINE

FIELDS OF ANATOMY pp. 3–5

GROSS ANATOMY study of general body structures.

REGIONAL ANATOMY study of gross structures by region.

SYSTEMIC ANATOMY study of gross structures by organ systems.

MICROSCOPIC ANATOMY study of structures with a microscope.

CYTOLOGY study of cells.

HISTOLOGY study of tissues.

FINE STRUCTURE (ULTRASTRUCTURE) study of structures with an electron microscope.

DEVELOPMENTAL ANATOMY study of body development.

EMBRYOLOGY study of prenatal development.

RADIOGRAPHIC ANATOMY use of specialized machines to study body structure; roentgenograms, fluoroscopy, CT scans, dynamic spatial reconstructor (DSR), nuclear magnetic resonance (NMR), positron-emission tomography (PET), ultrasonography.

PHYSIOLOGY study of body functions. p. 5

BASIC STRUCTURAL LEVELS there are four structural levels in body: cells, tissues, organs, and systems. pp. 5–6

CELLS simplest structural level of body.

TISSUES three embryonic cell layers—ectoderm, endoderm, and mesoderm—give rise to four types of tissues.

EPITHELIAL TISSUES cover body surface, line body cavities, ducts, vessels; develop from embryonic ectoderm, mesoderm, or endoderm.

MUSCULAR TISSUES move skeleton, pump blood, move food through digestive tract; develop from embryonic mesoderm.

NERVOUS TISSUES form brain, spinal cord, nerves; develop from embryonic ectoderm.

CONNECTIVE TISSUES used for support, attachment of other tissues; develop from embryonic mesoderm.

ORGANS tissues combine to form organs, e.g., the stomach.

SYSTEMS

HOMEOSTASIS organs function for general well-being of entire body, maintaining stable internal environment.

MAJOR SYSTEMS ten major organ systems make up body: integumentary, skeletal, muscular, nervous, endocrine, circulatory, respiratory, digestive, urinary, reproductive.

ANATOMICAL TERMINOLOGY knowing word roots, prefixes, and suffixes is helpful in understanding anatomical terms. pp. 7–8

BODY POSITIONS p. 8

ANATOMICAL POSITION is achieved when body stands erect, feet together, upper limbs hang at sides, palms forward, fingers extended, thumbs pointing away from body.

PRONE POSITION lying face down.

SUPINE POSITION lying face up.

DIRECTIONAL TERMS denote direction in pairs of opposites; for example, medial (toward midline of body) and lateral (away from midline of body).
p. 8

REGIONAL TERMS refer to special areas, for example, cervical (neck), thoracic (chest), plantar (sole of foot). pp. 8–10

ABDOMINAL CAVITY is divided into nine regions by two vertical lines and two horizontal lines; is also divided into four quadrants by a horizontal plane and a vertical plane.

BODY PLANES pp. 10–11

SAGITTAL PLANE is longitudinal section that divides the body, or its parts, into right and left portions.

FRONTAL PLANE is also a longitudinal section, but runs at right angles to sagittal plane, dividing body into anterior and posterior portions.

TRANSVERSE PLANE divides body or its parts into superior and inferior portions.

BODY CAVITIES the body contains two main cavities. pp. 11–14

DORSAL (POSTERIOR) CAVITY has two subdivisions.

CRANIAL CAVITY houses brain.

SPINAL (VERTEBRAL) CAVITY contains spinal cord.

VENTRAL (ANTERIOR) CAVITY also has two subdivisions.

THORACIC CAVITY divided into:

Pericardial Cavity around heart.

Pleural Cavities right and left, around lungs.

ABDOMINOPELVIC CAVITY divided into:

Abdominal Cavity

Pelvic Cavity

MEMBRANES OF THE VENTRAL BODY CAVITIES both thoracic and abdominopelvic cavities have visceral membranes that cover organs and parietal membranes that line outer walls of cavities. The space between the two membrane regions is filled with fluid. p. 14

SELF-QUIZ

1. Histology is a subdivision of: (a) gross anatomy; (b) developmental anatomy; (c) microscopic anatomy.

2. Each different function of each body structure is usually for the well-being of the structure itself rather than for the general well-being of the entire body. True or False?

3. Nervous tissue is derived from the embryonic: (a) ectoderm; (b) endoderm; (c) mesoderm.

4. Muscular tissue is derived from the embryonic: (a) ectoderm; (b) endoderm; (c) mesoderm.

5. The component of the word *endocarditis* that signifies "inflammation" is: (a) *endo*; (b) *card*; (c) *itis*.

6. Match the following types of tissue with the appropriate lettered descriptions.

Epithelial tissues	(a)	Derived from embryonic mesoderm.
Muscular tissues	(b)	Derived for the most part from embryonic mesoderm.
Nervous tissues		
Connective tissues	(c)	Derived from embryonic ectoderm, mesoderm, and endoderm.
	(d)	Forms the brain, spinal cord, and nerves.
	(e)	Cover the surface of the body and line the body cavities.
	(f)	Derived from embryonic ectoderm.
	(g)	Aid in digestion by moving food through the digestive tract.
	(h)	Used for support or attachment.

7. Match the following body position and directional terms with the appropriate description.

Prone position	(a)	Toward the head.
Anterior (ventral)	(b)	Away from the midline of the body.
Posterior (dorsal)	(c)	Lying face up.
Anatomical position	(d)	Away from the head.
	(e)	Toward the attached end of a limb.
Superior (cranial)	(f)	Sole of the foot.
	(g)	Front.
Medial	(h)	Standing erect with feet together, arms at side, palms forward.
Supine position		
Inferior (caudal)		
Lateral	(i)	Away from the attached end of a limb.
Proximal		
Cervical	(j)	Part of the back between the thorax and pelvis.
Plantar		
Distal	(k)	Lying face down.
Lumbar	(l)	Back.
Palmar	(m)	Neck.
	(n)	Toward the midline of the body.
	(o)	Anterior surface of the hands.

8. Which of the following directional terms are paired correctly? (a) superficial and deep; (b) medial and distal; (c) proximal and lateral.

9. The midline region superior to the umbilical region is called the epigastric, and it contains most of the stomach. True or false?

10. The midline region directly inferior to the umbilical region is the: (a) hypochondriac; (b) lumbar; (c) hypogastric.

11. All sagittal sections other than the median sagittal section are: (a) intersagittal; (b) parasagittal; (c) intrasagittal.

12. The dorsal body cavity contains the brain and spinal cord. True or false?

13. The fluid associated with the body's dorsal cavities is the cerebrospinal fluid. True or false?

14. The portion of the thoracic cavity between the two pleural cavities is called the: (a) mediastinum; (b) meninges; (c) foramen magnum.

15. If the membranes of the heart become inflamed, the condition is known as: (a) pericarditis; (b) peritonitis; (c) pleuritis.

16. Which one of the following is *not* located in the abdominal cavity? (a) spleen; (b) liver; (c) lungs.

17. Which one of the following is *not* located in the pelvic cavity? (a) urinary bladder; (b) liver; (c) female internal reproductive organs.

18. Most of the organs in the abdominal cavity are suspended from the anterior wall of the cavity by a single-layered membrane of parietal peritoneum. True or false?

19. The mesenteries: (a) support the kidneys; (b) are mostly associated with the dorsal activities; (c) are double-layered membranes.

LEARNING OBJECTIVES

After completing this chapter, you should be able to:

- Describe the major structural models of the plasma membrane.

- Describe the active and passive processes by which substances move across the plasma membrane.

- Describe the nuclear envelope.

- Distinguish between m-RNA, r-RNA, and t-RNA.

- Name the cytoplasmic organelles and describe the structure and function of each.

- Distinguish between rough and smooth endoplasmic reticulum.

- Distinguish between mitosis and meiosis, and describe the events that occur during each phase of both division processes.

CHAPTER CONTENTS

CELLULAR COMPONENTS

EXTRACELLULAR MATERIALS

CELL DIVISION

EFFECTS OF AGING ON CELLS

The Cell

It has been well established that all organisms, from bacteria to humans, consist of cellular units. Although more complex animals are organized into tissues, organs, and systems, the basic cellular nature of these structures is evident upon microscopic examination. In humans and other higher organisms, cells have become specialized both anatomically and physiologically. Muscle cells, for example, have a well-developed property of *contractility* (the ability to move or contract), whereas nerve cells are specialized for *conductivity* (the ability to transmit impulses). Other cells may exhibit highly developed properties of *metabolism* (the ability to process foods, obtain energy, and synthesize products), *irritability* (the capacity to respond to stimuli), or *reproduction* (the ability to duplicate themselves). It is important to remember, however, that these properties are present to some degree in all cells and as such may be regarded as general cell characteristics. Even single-celled organisms, such as amebas, are able to carry out all the basic activities characteristic of living organisms.

CELLULAR COMPONENTS

During the early years of cell study, it was believed that cells consisted of a uniform material called **protoplasm** and that all the properties characteristic of life were inherent in this material. As more sophisticated instruments and techniques were developed, it was found that cellular structure was far from uniform. A central body called the **nucleus** was observed within the cell. Moreover, the nucleus appeared to be surrounded by a medium called **cytoplasm.** Further study revealed that the cytoplasm itself contained numerous structures collectively called **organelles** (small organs) and chemical substances, such as glycogen granules or lipid droplets, that are called **inclusions.** With the advent of the electron microscope, it was possible to determine more precisely the structures of the various organelles and to examine closely the limiting boundary of the cell—the **plasma membrane** (or *cell membrane*) (Figure 2.1). Furthermore, improved physiological and biochemical techniques have provided new information about the functioning of the various cellular structures. **F2.1**

Plasma Membrane

Even before the plasma membrane was observed microscopically, its existence was inferred from experiments that dealt with the ability of various substances to enter cells. When researchers found that not all materials were able to enter cells with equal ease, they began to suspect the existence of some sort of selective barrier at the cell's surface. This barrier was found to retard free diffusion. Some substances moved across the barrier much more slowly than they would diffuse in water. Moreover, this barrier, the plasma membrane, was found to be highly selective—that is, it affected the entry of some molecules into the cell much more than others.

Viewed with an electron microscope, the plasma membrane appears as a trilaminar (three-layered) structure approximately 80 to 100 angstroms (Å) thick (1 Å = 1×10^{-7} mm) (Figure 2.2). Until recent years, the three layers **F2.2** have been interpreted as corresponding to a central layer of phospholipid

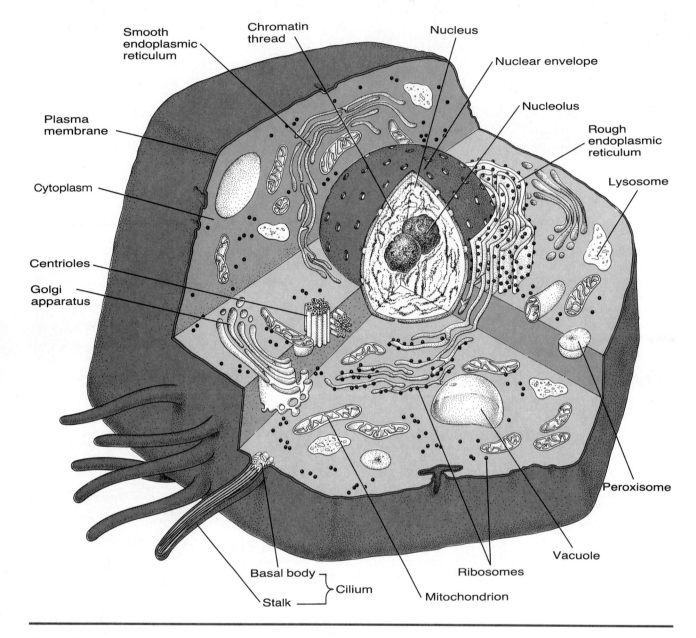

Figure 2.1
A "typical" animal cell showing subcellular organelles. There is probably no actual cell that can be considered "typical" in all respects.

surrounded by two protein layers. More recent evidence has resulted in proposals of several alternative molecular arrangements for the plasma membrane. One popular alternative suggests that the plasma membrane consists of a bimolecular layer of lipid (particularly phospholipid and cholesterol) with various proteins either attached to the membrane surface (*peripheral proteins*) or embedded in the lipid (*integral proteins*) in an asymmetrical pattern (Figure
F2.3 2.3). Some groups of proteins extend completely through the membrane, forming aqueous channels or pores that connect the interior of the cell with the external environment. Carbohydrates are often linked to lipid and protein molecules at the extracellular surface of the membrane. Many cells have a thin
F2.2 external coating of polysaccharides called the *glycocalyx* (Figure 2.2). This coat is thought to have an important role in maintaining adhesiveness between cells, and may be involved in regulating the cell's uptake of substances.

Movement of Materials Across the Plasma Membrane

All materials that enter or leave a cell must pass through the plasma membrane or through the pores in the plasma membrane. Thus the properties of

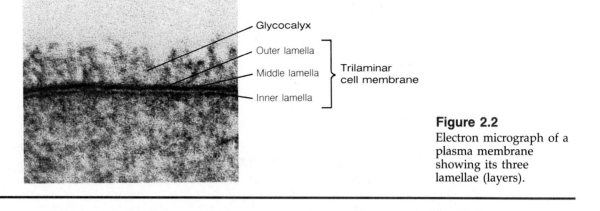

Figure 2.2
Electron micrograph of a plasma membrane showing its three lamellae (layers).

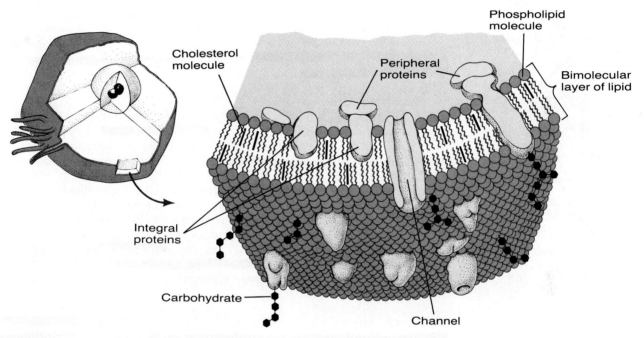

Figure 2.3
The composition of the plasma membrane.

the membrane (as well as the properties of the penetrating molecules) are important in determining the relative ease with which substances will be able to enter or leave a cell. If the plasma membrane were completely impermeable, the cell could not survive, since no substances could enter or leave the cell and the cell is not self-sustaining—it requires certain ions and molecules from its environment in order to function and must eliminate waste products. If the plasma membrane were completely permeable, however, it would allow the passage of everything and could not even retain the components of the cell. Actually, the plasma membrane permits certain ions or molecules to enter or leave the cell but restricts the movements of others. It does so on the basis of several molecular characteristics, including size, electrical charge, and lipid solubility. Thus the plasma membrane is said to be **semipermeable** or **selectively permeable.**

Substances can pass through the plasma membrane by means of energy or force supplied from outside the cell or by the cell itself. When the energy or force is supplied by the extracellular environment rather than by the cell, the process is referred to as a **passive process.** When the cell supplies the energy for transport, the process is called an **active process.** In considering the different ways in which materials may enter or leave cells, remember that a substance may enter or leave a cell by several different routes.

Passive Processes

There are five passive transport processes by which substances move across the plasma membrane: diffusion, facilitated diffusion, osmosis, dialysis, and filtration.

DIFFUSION **Diffusion** is the movement of a substance from a region of higher concentration of the substance to a region where it is less concentrated. It occurs because of collisions between atoms or molecules, which are always in motion due to their kinetic energy. Diffusion continues until the molecules are evenly distributed—at this point, *equilibrium* has been reached. Diffusion does not require the presence of a membrane; however, many compounds that are soluble in lipids (fats) move with relative ease through the plasma membrane by simple diffusion. Other water-soluble compounds that are not soluble in lipids generally have difficulty moving through the plasma membrane by simple diffusion—if they are able to do so at all. Thus the lipid portion of the membrane is thought to be a principal barrier to substances that are not soluble in lipids and it hinders their entering or leaving the cell by diffusion.

FACILITATED DIFFUSION Some substances are able to diffuse through the plasma membrane even though they are insoluble in lipids. They are thought to do this by **facilitated diffusion.** In this process the lipid-insoluble substance is bound to a *carrier molecule* on one side of the plasma membrane. In this bound form the substance is soluble in the lipids of the membrane and can move across the membrane. Once across the membrane, the transported molecule is released from the carrier and leaves the membrane. In facilitated diffusion the carrier is believed to move material equally well in either direction, into or out of the cell. However, the net movement of facilitated diffusion, as with any diffusion, is from a region of high concentration of a substance to a region of low concentration of the substance. Facilitated diffusion is therefore considered to be a passive process in which no cellular energy is used.

OSMOSIS **Osmosis** is a passive process by which water moves across the semipermeable plasma membrane from an area of higher water concentration to an area of lower water concentration. The membrane's selective permeability is important in osmosis as it allows water to cross the membrane while the movement of substances in solution in the water may be restricted by the membrane. In this manner, osmosis is responsible for the net movement of water into or out of cells.

DIALYSIS Because the plasma membrane is selectively permeable, it is capable of separating substances that are in solution but have differing diffusibilities through the membrane. As a result, water and small molecules dissolved in the water (that is, crystalloids) pass through the membrane (by osmosis and diffusion) whereas larger molecules, such as protein molecules, may not be allowed to pass. This process of selectively separating substances in solution is called **dialysis.**

The principle of dialysis is used in the artificial kidney, where blood is passed through a membranous tube that is immersed in a solution that has a specific ionic balance. The membranous tube has properties similar to those of the plasma membrane. Since most waste products are small crystalloids, they can be removed from the blood while vital protein molecules are unable to pass through the tube and are thus retained in the blood. Those crystalloids required by the body may also be retained in the blood if they are present in the bathing fluid in the same concentration as in the blood. Thus there will be no concentration difference of these molecules on either side of the membrane and, therefore, no net diffusion.

FILTRATION **Filtration** is the passage of solvents, such as water, and various dissolved substances through a membrane as the result of *mechanical forces* such as gravity or hydrostatic pressure (for example, blood pressure). The direction of movement is from the region of higher pressure to the region of lower pressure. Since large molecules dispersed in the solvent may not be able to pass through the membrane, filtration removes them from the solvent. Filtration is considered a passive process because the force that causes the movement originates outside the cell.

Within the body, filtration is important in the kidney. The blood pressure in the capillaries of the kidney forces the fluid portion of the blood and small dissolved substances such as salts and glucose out of the blood vessels, but blood cells and protein molecules that are too large to leave the vessels remain in the blood.

The movement of substances across the plasma membrane by these passive processes is greatly influenced by the presence of pores or channels in the membrane. These pores are generally considered to be less than 10 Å in diameter; therefore only the smaller molecules can enter or leave the cell through them. In addition, the movement of ions (charged atoms) through the pores is highly specific, and certain ions pass most easily through certain pores. For example, sodium ions pass easily through some pores and potassium ions pass easily through other pores.

Active Processes

Active processes are those methods of transport in which the expenditure of energy by the cell is necessary in order for certain substances to enter or leave the cell. The active processes include active transport, endocytosis, and exocytosis.

ACTIVE TRANSPORT This type of transport is not only able to move substances across plasma membranes that might not otherwise pass, it is also able to move the substances from regions of *low* concentration to regions of *high* concentration, *against normal concentration* or *diffusion gradients*. Thus, active transport systems are able to accumulate substances on one side of a membrane at concentrations many times greater than those on the opposite side of the membrane. This activity requires the input of cellular energy and depends on the metabolic processes of the living cell. Active transport also makes it possible for many polar molecules that are insoluble in lipids and too large to fit through the pores in the plasma membrane to enter a cell.

Several theories have been proposed to explain how substances are moved across the plasma membrane by active transport. One theory, called the **carrier mechanism,** proposes that a protein carrier molecule that is free within the plasma membrane binds on one side of the membrane to a molecule of a substance to be transported (Figure 2.4). As a result of the expenditure of metabolic energy by the cell, the entire carrier protein rotates across the membrane, delivers its bound transported molecule to the other side, and releases it. The carrier molecule then rotates back across the membrane and picks up another molecule of the substance to be transported. **F2.4**

ENDOCYTOSIS (PHAGOCYTOSIS AND PINOCYTOSIS) Some cells are known to form sacs or vesicles that enclose small portions of the external environment. In this process, a cell surrounds a part of the environment with a section of its plasma membrane. This section then separates from the plasma membrane itself and moves into the interior of the cell. The general term for this process is **endocytosis.** There are three types of endocytosis: phagocytosis, pinocytosis, and receptor-mediated endocytosis.

Phagocytosis ("cell eating") involves surrounding solid materials located external to the cell by projections of cytoplasm called pseudopodia (Figure 2.5). When the pseudopodia join, the solid materials are enclosed within a **F2.5**

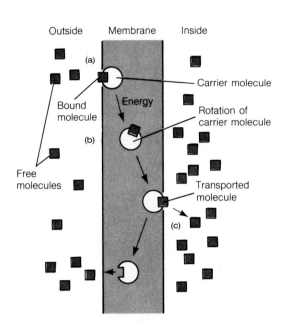

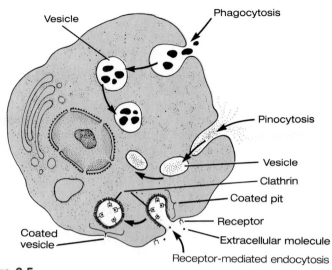

Figure 2.5
The process of endocytosis showing phagocytosis of solid materials, pinocytosis of liquids, and receptor-mediated endocytosis of specific molecules.

Figure 2.4
The carrier mechanism of active transport. **(a)** The substance to be transported across the plasma membrane attaches to a binding site on a carrier molecule on the membrane. **(b)** The carrier molecule rotates across the plasma membrane, carrying the substance with it. **(c)** The substance is released into the interior of the cell. After releasing the transported molecule, the carrier molecule returns to the side of the membrane from which it began.

vesicle formed from the plasma membrane. As the vesicle moves through the cytoplasm of the cell it generally becomes fused with a specialized cellular organelle called a lysosome, releasing digestive enzymes contained within the lysosome. The enzymes, in turn, digest the ingested materials.

The process of phagocytosis provides the body with an important defense mechanism. Certain cells of the body (for example, some types of white blood cells) are able to phagocytose (engulf) and subsequently destroy many potentially harmful foreign materials, such as bacteria, that enter the body.

In **pinocytosis** ("cell drinking") a small amount of extracellular fluid, plus any substances dissolved in the fluid, flow into tiny invaginations of the plasma membrane. The outer regions of the invaginated membrane fuse, engulfing the extracellular fluid in a vesicle. Pinocytosis occurs in most cells, but only certain cells are capable of phagocytosis.

In contrast to phagocytosis and pinocytosis, both of which are nonspecific ingestion processes, **receptor-mediated endocytosis** involves the ingestion only of specific molecules. Located on the plasma membrane are proteins that serve as receptors for specific molecules by binding them to the plasma membrane. An invagination called a *coated pit* forms in the plasma membrane at the site where the extracellular material becomes attached to the receptor. With continued invagination the pit forms into a *coated vesicle* which completely surrounds the receptor/extracellular material complex. The pit and the vesicle are referred to as being coated because they have a coating of the protein clathrin on their cytoplasmic surfaces.

Some hormones, low-density lipoproteins, amino acids, and vitamins enter cells via receptor-mediated endocytosis. When the coated vesicle combines with a lysosome the extracellular material is broken down or released into the cytoplasm and the membranes and their receptors return to the plasma membrane and may be used again.

EXOCYTOSIS Substances that are contained in membrane-bounded vesicles (or *vacuoles*) located inside of cells may leave the cells by a process called
F2.6 **exocytosis** (Figure 2.6). In this process, a membrane-bounded vesicle within a cell fuses with the plasma membrane and the contents of the vesicle are released from the cell. Apart from releasing material from the cell, the process of exocytosis may also generate additional plasma membrane when the membranous vesicles fuse with the plasma membrane. Exocytosis provides a means by which various cell products and secretions may leave the cells in which they are produced.

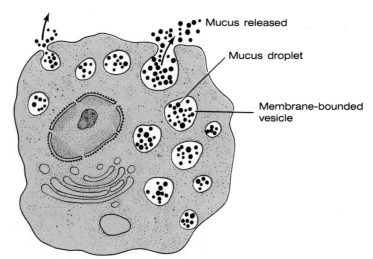

Figure 2.6
The process of
exocytosis.

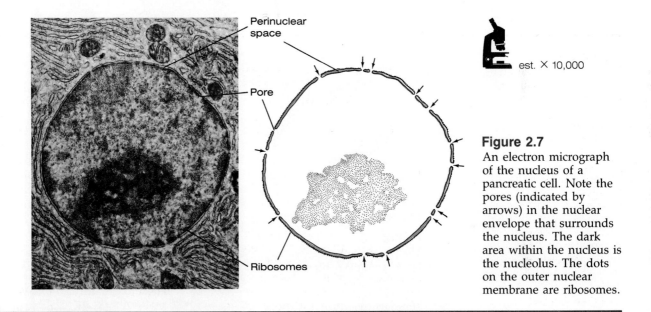

est. × 10,000

Figure 2.7
An electron micrograph
of the nucleus of a
pancreatic cell. Note the
pores (indicated by
arrows) in the nuclear
envelope that surrounds
the nucleus. The dark
area within the nucleus is
the nucleolus. The dots
on the outer nuclear
membrane are ribosomes.

The Nucleus

The **nucleus** of a cell is a large organelle that contains the genetic material and
manufactures molecules (ribosomal-RNA, transfer-RNA, messenger-RNA)
that control the synthetic activities of other organelles in the cytoplasm. Most
cells contain a single nucleus. Some cells contain two or more nuclei, however; and mature red blood cells and blood platelets do not contain any nucleus. Cells that lack a nucleus are unable to synthesize proteins or undergo
cell division.

The nucleus of a cell that is not undergoing division is surrounded by a
nuclear envelope consisting of two parallel membranes separated by a **peri-
nuclear space** (Figure 2.7). The outer membrane—that is, the side toward the
cytoplasm—often has ribosomes attached to it. (The function of ribosomes is
discussed later in this chapter.) The nuclear envelope is about 400 Å thick.
Numerous octagonal **nuclear pores** are present where the inner and outer
membranes of the nuclear envelope join together. These pores range from 300
to 1000 Å in diameter. While the pores are thought to allow passage of substances between the nucleus and the cytoplasm, their small diameters indicate that they do not allow free passage of large substances. In fact, the nuclear pores may be even more restrictive than their diameters would indicate,

F2.7

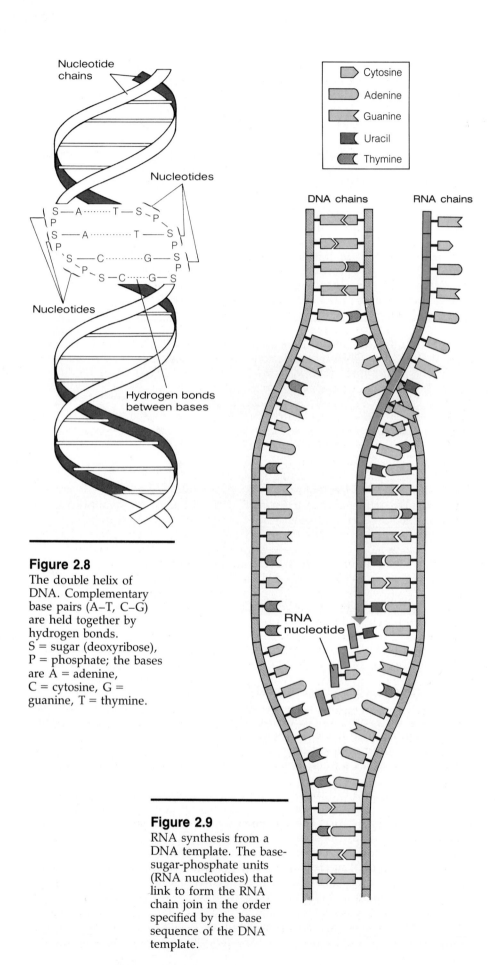

Nucleotide chains

Nucleotides

Nucleotides

Hydrogen bonds between bases

Figure 2.8
The double helix of DNA. Complementary base pairs (A–T, C–G) are held together by hydrogen bonds.
S = sugar (deoxyribose), P = phosphate; the bases are A = adenine, C = cytosine, G = guanine, T = thymine.

Cytosine
Adenine
Guanine
Uracil
Thymine

DNA chains   RNA chains

RNA nucleotide

Figure 2.9
RNA synthesis from a DNA template. The base-sugar-phosphate units (RNA nucleotides) that link to form the RNA chain join in the order specified by the base sequence of the DNA template.

since electron micrographs suggest that the openings of many pores are partially or completely closed by thin diaphragms. The outer layer of the nuclear envelope is often continuous with the membranes of the endoplasmic reticulum, a series of canals that pass throughout the cytoplasm (Figure 2.1). The **F2.1** nuclear envelope breaks down during cell division and reforms after the completion of division.

Between periods of active cell division, the cell's genetic material is found in the nucleus in the form of **chromatin** threads or granules that consist of **deoxyribonucleic acid (DNA)** combined with protein. During active cell division, the chromatin threads or granules become tightly coiled and visible as structures called *chromosomes*. The nucleus also contains one or more structures known as **nucleoli** (little nuclei). Nucleoli are not bounded by membranes and consist mostly of **ribonucleic acid (RNA)** and protein and rather small amounts of DNA. It is believed that the nucleoli are the sites of synthesis (on DNA) of a specific type of RNA known as **ribosomal RNA (r-RNA)** and that r-RNA combines with protein in the nucleolar area. The r-RNA–protein complexes are then believed to migrate out of the nucleus and into the cytoplasm, where they become organized into structures called *ribosomes.*

The DNA in the nucleus is capable of replicating itself for purposes of cellular reproduction and it serves as a template for the assemblage of molecules of RNA. The DNA molecule consists of two chains of base-sugar-phosphate units called **nucleotides** (Figure 2.8). The two chains are joined by hy- **F2.8** drogen bonds between the bases of the different chains; and the bases of the different chains always pair with one another in a predictable fashion. When DNA serves as a template for RNA synthesis, sections of the two linked chains of DNA separate from one another for some distance, exposing part of the base sequence of one of the chains. RNA nucleotides then assemble along the single DNA chains (Figure 2.9). **F2.9**

One type of RNA that is produced from nuclear DNA is **messenger RNA (m-RNA).** Within its base sequence, messenger RNA contains coded information derived from DNA that can be used to direct the synthesis of proteins by the cell. This protein synthesis occurs in the cytoplasm on ribosomes. To reach the ribosomes, m-RNA passes through the nuclear membrane and enters the cytoplasm. Its coded message, derived from DNA, specifies the sequence in which different amino acids are to be joined to form specific polypeptides or proteins. The code is contained in the base sequence of the m-RNA and, therefore, ultimately in the base sequence of the DNA that served as the template for the synthesis of the m-RNA molecule. Each sequence of three bases has been found to constitute a **codon** or "word" of the code (Figure 2.10). In a similar manner, nuclear DNA serves as the template **F2.10** for the production of **transfer RNA (t-RNA).** Transfer RNA leaves the nucleus and transports amino acids in the cytoplasm of the cell to messenger RNA, which assembles the amino acids into proteins. The method by which this is accomplished is discussed in the section on ribosomes.

The DNA molecules that comprise the hereditary genetic material of a cell act via messenger RNA to specify the particular amino acid sequences of the various polypeptides or proteins that are synthesized by the cell. As a result of this activity of the DNA hereditary material, we can define a gene biochemically or physiologically as a portion of DNA that specifies the amino acid sequence of one polypeptide or protein.

Once polypeptides or proteins are synthesized according to the specifications of DNA, they may act as enzymes, as hormones, or in other ways to accomplish the work of a cell. DNA (and its associated protein) regulates cellular function by determining which polypeptides or proteins will be synthesized.

Although every cell (except the reproductive cells) contains the full genetic complement of DNA (hereditary material), all regions of the DNA are not active at all times in all cells. In some cells, one DNA region may be active in specifying the synthesis of a particular polypeptide or protein. In another cell, this DNA region may be inactive but another region may be active in specifying the synthesis of a different polypeptide or protein. Even within the

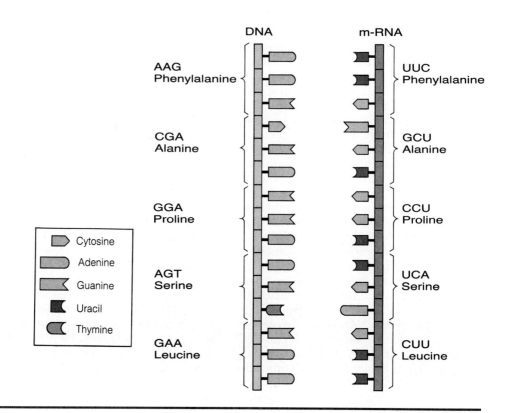

Figure 2.10
The codes of DNA and m-RNA. Each three-base sequence (codon) specifies a given amino acid. Note that a three-base sequence of m-RNA that specifies a given amino acid is complementary to the three base sequence of DNA that specifies the same amino acid.

same cell, a specific region of DNA may not be active at all times. Thus DNA plays a central role in directing the activities of the cell by specifying the particular polypeptides or proteins that will be synthesized by a cell and, further, *when* they will be synthesized.

Cytoplasm

Cytoplasm is a thick fluid located between the plasma membrane and the nuclear membrane. It receives substances from the external environment of the cell as well as from the nucleus of the cell. Many chemical reactions occur within the cytoplasm, and it contains numerous specialized structures called *organelles* that function to provide energy to the cell, synthesize new molecules, transport molecules within the cell, provide structural support to the cell, or facilitate the excretion of waste products from the cell.

Cytoplasmic Organelles

RIBOSOMES **Ribosomes** are small particles in the cytoplasm that are composed of ribosomal RNA (r-RNA) and protein. A single ribosome consists of two subunits of unequal size, each subunit containing r-RNA and protein. Ribosomes are from 120 to 150 Å in diameter. They may be found free in the cytoplasm or attached to membranes, including the membranes of the nuclear F2.1 envelope and the endoplasmic reticulum (Figure 2.1). Whether they are free or attached, ribosomes are generally found in clusters called **polysomes** or **polyribosomes.**

Ribosomes are the sites where new proteins are synthesized in a cell. It has been suggested that the free ribosomes synthesize proteins that are used by the cell for its own needs, such as replication, whereas the ribosomes attached to membranes synthesize proteins that will be secreted by the cell and used elsewhere in the body. Although cells that produce large amounts of protein that are exported out of the cell do have more attached ribosomes and those whose protein remains in the cell do tend to have more free ribosomes, the functions of free and attached ribosomes are probably not as distinct as this scheme would suggest.

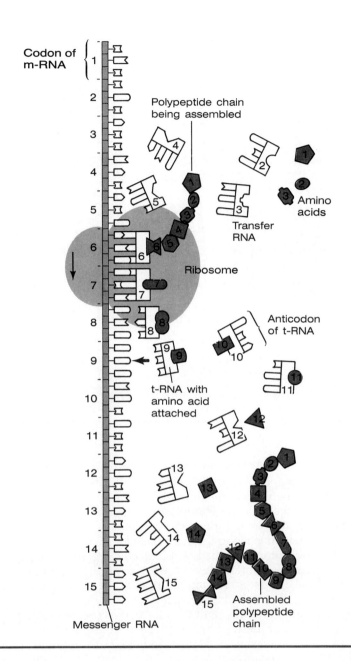

Codon of m-RNA

Polypeptide chain being assembled

Amino acids

Transfer RNA

Ribosome

Anticodon of t-RNA

t-RNA with amino acid attached

Messenger RNA

Assembled polypeptide chain

Figure 2.11
A ribosome reading the code of m-RNA and assembling a polypeptide or protein chain of amino acids. Molecules of t-RNA bring amino acids to the ribosomes for incorporation into the growing chain.

Proteins are synthesized at the ribosomes by linking amino acids together in the sequence specified by messenger RNA (and ultimately DNA) to form polypeptides or proteins (Figure 2.11). In this process, an m-RNA chain that F2.11 carries the coded instructions from nuclear DNA for synthesizing a specific polypeptide or protein enters the cytoplasm and becomes attached to a ribosome. The ribosome then travels along the m-RNA chain and "reads" the m-RNA code. Often, several ribosomes may be attached to a single m-RNA, each reading the m-RNA code and forming an amino acid chain that will become the polypeptide or protein specified by the m-RNA molecule.

The amino acids that will be incorporated into the growing polypeptide chain float free in the cytoplasm until they are brought to the ribosome by another type of RNA, called *transfer RNA (t-RNA)*. It is thought that there is at least one different t-RNA for each different amino acid. A specific t-RNA contains within its sequence of bases a region called the **anticodon** that enables it to recognize the proper codon on the m-RNA to which it can attach. This allows the amino acid it carries to be inserted into the growing polypeptide chain at the proper position. When the ribosome has read the entire

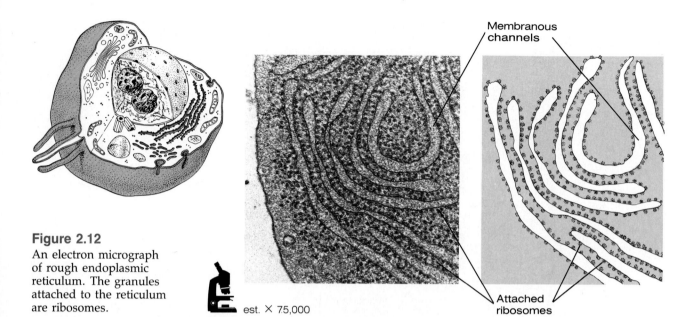

Membranous channels

Attached ribosomes

est. × 75,000

Figure 2.12
An electron micrograph of rough endoplasmic reticulum. The granules attached to the reticulum are ribosomes.

m-RNA molecule, it releases the newly formed polypeptide or protein and the m-RNA. The ribosome is then able to read other m-RNA molecules and form new proteins, while the m-RNA that is released can be used by other ribosomes.

ENDOPLASMIC RETICULUM   A membranous network of tubular or sac-like channels called the **endoplasmic reticulum** (*endo-* = within; *plasm* = cytoplasm; *reticulum* = network) runs throughout much of the cytoplasm of a cell (Figure 2.12). The walls of the endoplasmic reticulum contain enzymes associated with the synthesis of various substances.

F2.12

Ribosomes are often attached to the endoplasmic reticulum. Endoplasmic reticulum that has ribosomes attached is called *rough (granular) endoplasmic reticulum.* Rough endoplasmic reticulum is generally in the form of flattened saccules called *cisterna* and is often continuous with the nuclear membrane. The enzymes in the wall of rough endoplasmic reticulum are thought to be involved in the synthesis of lipids, and the attached ribosomes manufacture proteins. Protein that is synthesized by the attached ribosomes may pass into the channels of the endoplasmic reticulum and travel throughout the cell. Other portions of the reticulum can pinch off, giving rise to membrane-bounded vesicles that contain protein.

Endoplasmic reticulum that does not have ribosomes attached to it is called *smooth (agranular) endoplasmic reticulum.* Smooth endoplasmic reticulum is generally in the form of a complex network of anastomosing tubules, with few, if any, cisterna present. The enzymes in the wall of smooth endoplasmic reticulum are believed to be involved with the synthesis of steroids as well as lipids. In some cells (liver, for example), the enzymes of smooth endoplasmic reticulum function to detoxify poisons.

GOLGI APPARATUS   The **Golgi apparatus** or **Golgi complex** consists of stacks of flattened membranous sacs located in the cytoplasm of many cells, frequently near the nucleus (Figure 2.1, Figure 2.13). Arrayed around the margins of the flattened sacs are vesicles of various sizes. It is believed that the structures of the Golgi apparatus originate from the membranes of the endoplasmic reticulum.

F2.1, F2.13

The Golgi apparatus is particularly evident in cells during periods of increased secretory activity. In glandular cells the Golgi apparatus serves as the site for the accumulation and concentration of the cell's secretory products. In cells whose secretions are protein, the protein is synthesized in the ribosomes

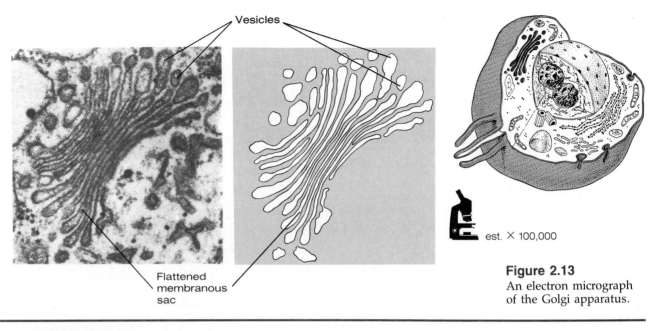

Vesicles

Flattened
membranous
sac

est. × 100,000

Figure 2.13
An electron micrograph
of the Golgi apparatus.

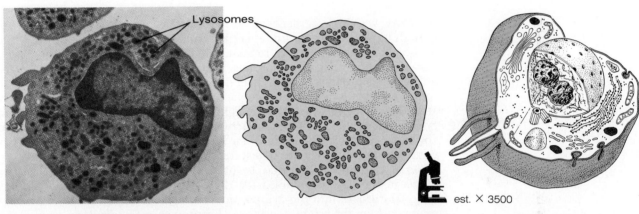

Lysosomes

est. × 3500

of rough endoplasmic reticulum. The protein is then transported within the tubules of the endoplasmic reticulum to the sacs of the Golgi apparatus, where it is concentrated and stored. Eventually, the stored protein is released by the Golgi apparatus in the form of vesicles called *secretory granules* that move toward the surface of the cell. Before being released from the Golgi apparatus, the secretory granules acquire an enveloping membrane that is capable of fusing with the plasma membrane—thus releasing the protein secretion from the cell by the process of exocytosis. Apart from releasing material from the cell, this process may also generate additional plasma membrane when the membranous vesicles fuse with the plasma membrane.

The Golgi apparatus also plays an important role in those cells whose secretions are glycoproteins (combinations of carbohydrate and protein). The membranes of the Golgi apparatus contain the enzymes necessary to synthesize carbohydrates and couple them to protein. In cells that secrete glycoproteins, the protein that is synthesized by the ribosomes of rough endoplasmic reticulum is coupled with the carbohydrate of the Golgi apparatus after entering the sacs of the Golgi apparatus. The resulting glycoprotein is then released as secretory granules that move to the plasma membrane and exit the cell the same way that other Golgi secretions are released—via exocytosis.

LYSOSOMES   **Lysosomes** (*lyso-* = dissolution; *soma* = body) are membrane-bounded cytoplasmic structures that appear granular during inactivity but assume the appearance of vesicles when active (Figure 2.14). Lysosomes, like

Figure 2.14
An electron micrograph
of lysosomes in a white
blood cell.

F2.14

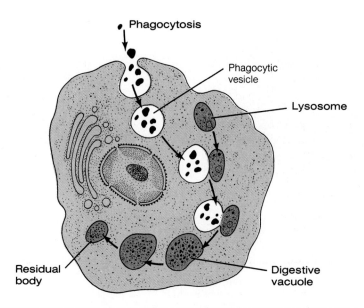

Figure 2.15
Enzymatic digestion of phagocytosed substances by lysosomes.

secretory vesicles, are believed to originate from the Golgi apparatus, although in certain cells or under certain conditions they may be derived from portions of the endoplasmic reticulum that break away and give rise to protein-containing vesicles. Lysosomes contain strong enzymes that are capable of digesting proteins, lipids, certain carbohydrates, DNA, and RNA. Lysosomes are able to form digestive vacuoles (also called *secondary lysosomes*) by attaching to vesicles that have been formed in the cell as the result of phago-

F2.15   cytosis (Figure 2.15). During attachment, the membranes of the lysosomes and the phagocytic vesicles fuse, the lysosomes release their enzymes into the vacuoles, and the enzymes digest the phagocytosed material, converting the material to products that may be used by the cell. Undigested material remains in the vacuoles (which are then called *residual bodies*), and digested material diffuses into the cytoplasm. The residual bodies are thought to be removed from the cell by exocytosis. Lysosomes are particularly abundant in those white blood cells whose principal activity is the phagocytosis of foreign materials in the body.

The complete role of the lysosome is still not clear, but it involves more than the digestion of phagocytosed materials. Parts of a cell itself may sometimes appear within a lysosomal vacuole (which is then called an *autophagic vacuole*) and be broken down. This response, which may occur during starvation, allows the cell to use part of its own substance as an energy source without doing itself irreparable harm. After a cell has been severely injured or has died, lysosomal membranes may rupture. The released enzymes then digest the material of the cell itself. Destruction of cells by their lysosomes also seems to play an important role in normal embryonic development and in the regression of the mother's mammary glands when her infant is no longer nursing. In both cases there is an excess of cells that must be gotten rid of. The destruction of apparently normal cells by their lysosomes seems to perform a vital role in these cases of cell death.

PEROXISOMES   **Peroxisomes,** which are also called microbodies, are membrane-bounded cytoplasmic structures that appear to be very similar to lysosomes in microscopic cell sections, although they may actually be larger, ir-

F2.1   regularly shaped structures (Figure 2.1). Peroxisomes differ from lysosomes by virtue of the fact that lysosomes contain digestive enzymes, whereas peroxisomes contain a variety of powerful oxidative enzymes that are used to obtain energy from molecules, as well as enzymes that increase the rate of the oxidative decomposition of hydrogen peroxide into water and oxygen. This is

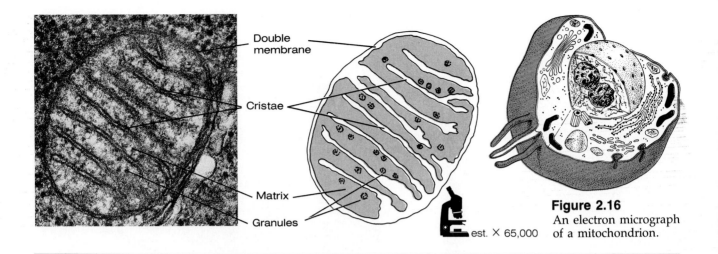

Double
membrane

Cristae

Matrix

Granules

est. × 65,000

Figure 2.16
An electron micrograph
of a mitochondrion.

an important function, as an accumulation of hydrogen peroxide can damage the cell.

MITOCHONDRIA **Mitochondria** (*mito* = thread; *chondros* = granule) are rod-shaped or filamentous cytoplasmic organelles that are bounded by a double membrane (Figure 2.1, Figure 2.16). The outer membrane is smooth and it **F2.1, F2.16** surrounds the mitochondrion itself. The inner membrane folds at intervals into the central portion of the mitochondrion, forming partial partitions known as *cristae*. A semisolid substance called the *matrix* fills the interior of each mitochondrion, lying between the cristae. In the matrix there are ribosomes and often small *granules* that serve as binding sites for various positively charged ions (such as calcium). These ions are required for the functioning of enzymes located in the mitochondria. Enzymes located on the inner membrane of the mitochondria are involved in the generation of adenosine triphosphate (ATP)—a vital molecule that provides metabolic energy for cellular activities. Since mitochondria play a major role in the generation of metabolic energy, they have come to be called the powerhouses of the cell. Mitochondria have their own supply of genetic material in the form of DNA and are capable of self-duplication. As might be expected, more mitochondria seem to be present in very active cells than in cells that are less active metabolically.

CYTOSKELETON Within the cytoplasm of cells is a complex network of threadlike structures referred to as the **cytoskeleton** (cell skeleton). The cytoskeleton helps to maintain cell shape and the position of various organelles within the cytoplasm. It is also involved in the transport of substances within the cell, cellular movement, and cell division.

The cytoskeleton is composed of thin filaments called *microtubules, intermediate filaments,* and *microfilaments.* These components appear to be interconnected with a three-dimensional network of thin bridges, forming a complex called the **microtrabecular lattice** (Figure 2.17). **F2.17**

Microtubules are very small, hollow, unbranched, cylindrical tubules (Figure 2.18). They are not bounded by a membrane, and are composed of **F2.18** subunits of a protein called *tubulin,* as well as various other proteins known as microtubule-associated proteins. Microtubules form a supporting framework within a cell; thus they are thought to play a role in the development and maintenance of cell shape. It has also been suggested that microtubules may serve as conducting channels for the movement of substances within the cell; but this function has not been well established. Microtubules form structures, called **spindle fibers,** that are evident in a cell during cell division, and they are believed to be similar to the tubules of cilia and flagella. Microtubules sliding past one another seem to be involved in the movements of cilia and flagella.

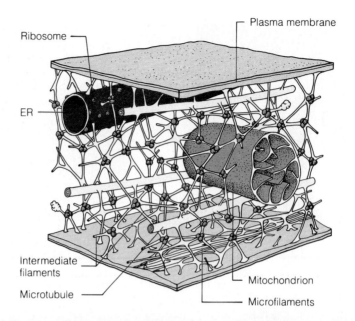

Figure 2.17
Three-dimensional
reconstruction of the
cytoskeleton.

Intermediate filaments, which have a smaller diameter than microtubules, are believed to provide structural support for the cytoskeletal framework. Because some intermediate filaments form a ring around the nucleus with branches extending outward through the cytoplasm and other intermediate filaments extend into the pores of the nuclear envelope, they are thought to help maintain the position of the nucleus within the cell, as well as being involved in the transport of materials into and out of the nucleus. Five classes of intermediate fibers have been identified, based on the presence of different structural proteins.

F2.19   **Microfilaments** are very small fibrils which, in contrast to microtubules, are not hollow and have a much smaller diameter (Figure 2.19). They generally occur in bundles and other groupings rather than singly. There are several different subclasses of microfilaments, some of which appear to be associated with the contractile activities of cells that lead to such phenomena as cell locomotion or modifications of cell shape. Muscle cells in particular exhibit a highly developed and organized array of microfilaments.

CILIA, FLAGELLA, AND BASAL BODIES   Many cells have one or more thin, cylindrical extensions projecting from their surfaces. These structures can move substances over the surface of the cell, or they can move the entire cell through a liquid medium. If these extensions are short and numerous,
F2.1   they are called **cilia** (Figure 2.1). If they are longer and not as numerous, they are called **flagella.** Both cilia and flagella exhibit the same basic organizational pattern, and both are believed to arise from cytoplasmic structures called **basal bodies.** Basal bodies are also believed to play a role in coordinating the movement of cilia and flagella.

Cilia and flagella consist of a membranous sheath that encloses a series of microtubules. The sheath is continuous with the plasma membrane and the microtubules are arranged in a characteristic circular pattern of nine groups of tubules with two tubules per group. Two additional tubules are located in the
F2.20a   center of this pattern (Figure 2.20*a*). Basal bodies also exhibit a characteristic pattern of nine groups of microtubules in a circle, but they have three tubules
F2.20b   per group and do not have the central tubules (Figure 2.20*b*).

CENTRIOLES   Cytoplasmic structures called **centrioles** are found near the nucleus of the cell, in a region called the *centrosome* or *microtubule-organizing center.* Centrioles are involved in cell division. They normally occur in pairs, with each member of the pair oriented at right angles to the other. Centrioles

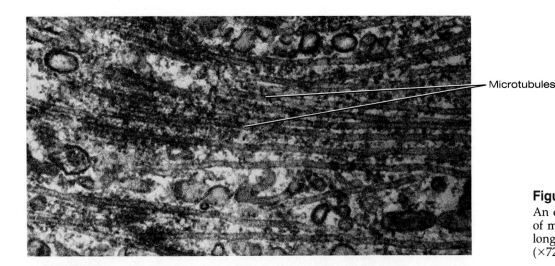

Microtubules

Figure 2.18
An electron micrograph of microtubules in longitudinal section (×72,000).

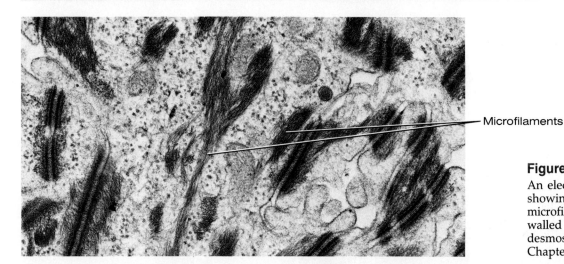

Microfilaments

Figure 2.19
An electron micrograph showing bundles of microfilaments. The dark-walled structures are desmosomes (see Chapter 3).

are cylindrical in shape, and contain a series of microtubules that are arranged in the same pattern as the microtubules of basal bodies (Figure 2.1, Figure 2.20*b* and *c*). In fact, basal bodies and centrioles are believed to have common F2.1, F2.20b, c origins and they may even be identical structures.

Centrioles are self-duplicating organelles. During cell division, two pairs of centrioles can be seen, each pair consisting of two cylinders lying at right angles to one another. As cell division proceeds, one pair of centrioles moves to one end or pole of the cell while the other pair moves to the opposite end or pole. During cell division, thin microtubules can be seen to radiate in all directions from the vicinity of each pair of centrioles. Some of these microtubules, which are called *spindle fibers,* are in the form of an organized system, known as a **spindle,** that is located between the two pairs of centrioles. Spindle fibers are important in the redistribution of chromosomes during cell division. Other microtubules, called *astral fibrils,* radiate blindly from the centriole regions and form what is called an *aster* about each centriole pair.

Inclusions

In addition to organelles, which we have just discussed, the cytoplasm of cells also contains a wide variety of chemical substances that are collectively referred to as **inclusions.** The hemoglobin molecules of red blood cells, which transport oxygen and carbon dioxide, are inclusions. So are particles of the pigment melanin, which is found in some cells of the eyes, skin, and hair. A

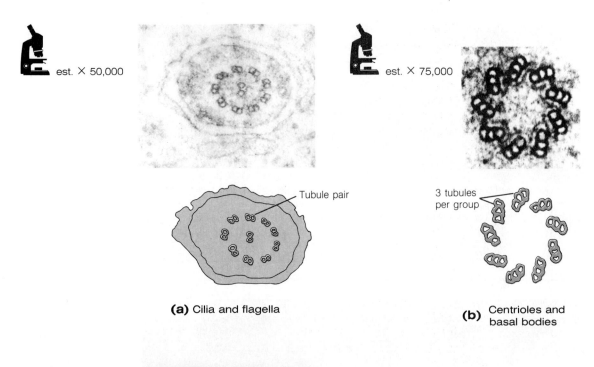

est. × 50,000

Tubule pair

(a) Cilia and flagella

est. × 75,000

3 tubules
per group

(b) Centrioles and
basal bodies

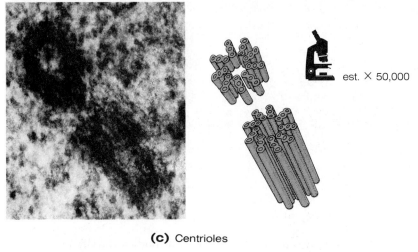

est. × 50,000

(c) Centrioles

Figure 2.20
(a) Cross-sectional view of typical 9 + 2 arrangement of microtubules in cilia and flagella. **(b)** Cross-sectional view of a typical arrangement of microtubules in basal bodies and centrioles. Note that there are three tubules per group and no tubules in the center. **(c)** Centrioles.

number of metabolically important substances are also found in cells as inclusions. For example, the polysaccharide glycogen, which is a storage form of carbohydrate, is particularly evident in liver and muscle cells, and fats are stored in cells of adipose tissue. When a cellular inclusion is a liquid that could mix with the cytoplasm of the cell, the inclusion may be surrounded by a membrane that forms a structure called a *vacuole*.

EXTRACELLULAR MATERIALS

Many body substances are found outside of cells rather than within them. These are collectively called **extracellular materials.** Extracellular materials include the body fluids and the framework or matrix in which many cells are embedded. Many extracellular materials are products of the cells themselves. Among these are chondroitin sulfate, which is a jellylike substance found in bone, cartilage, and heart valves; and hyaluronic acid, which is a viscous fluid-like substance present in a number of tissues. A variety of fibrous materials such as the proteins collagen and elastin also occurs extracellularly. Connective tissue is particularly rich in extracellular materials.

CELL DIVISION

Some highly specialized cells, such as muscle and nerve cells, do not undergo cell division once they have differentiated (attained their mature form). Other cells, however, such as liver, intestinal, bone marrow, and epidermal cells, retain the capacity to divide and reproduce themselves. The processes by which these cells divide involve several basic events. The first event is the replication of genetic material in the nucleus of the cell. A second is the redistribution of the replicated genetic material into two new nuclei. The process of redistributing the genetic material into two new nuclei, each with the same number and kind of chromosomes as the original nucleus, is called **mitosis,** or **karyokinesis.** A third event is the division of the cytoplasm into two new cells (called daughter cells), each with its own nucleus. This process is called **cytokinesis.** Usually cytokinesis occurs immediately following the completion of mitosis.

Interphase

The period between active cell divisions is called **interphase.** During interphase, the DNA molecules that comprise the genetic material of the cell (and protein associated with DNA) appear only as indistinct chromatin threads or granules in the nucleus. One of the principal events that occurs during interphase is *DNA replication*—that is, DNA molecules serve as templates for the replication of additional DNA molecules (Figure 2.21). In this process, the two **F2.21** chains of a DNA molecule separate from one another for some distance and individual base-sugar-phosphate units (DNA nucleotides) attach to the exposed template DNA chains according to the complementary base-pairing pattern. That is, an adenine-containing nucleotide is incorporated into the new DNA chain wherever the template chain has thymine, and vice versa; a cytosine-containing nucleotide is incorporated wherever the template DNA has guanine, and vice versa. The individual base-sugar-phosphate units themselves become linked together by enzymes called DNA polymerases to form a new DNA chain that is the complement of the original template DNA chain. The newly replicated chain of DNA remains attached to the original template DNA, thus forming a complete DNA molecule exactly like the original. Thus each new two-chain DNA consists of one chain from the original DNA that acted as a template and one newly synthesized chain. By the end of interphase, the cell therefore contains twice the amount of DNA it contained when it entered interphase.

Because the activities that occur in the cell change as the cell approaches mitosis, interphase is sometimes divided into three phases: G_1 (first gap), S (DNA synthesis), and G_2 (second gap).

1. The **G_1 phase** immediately follows the completion of cell division. During this phase there is active synthesis of protein and RNA, the nucleus and the cytoplasm enlarge, and there is increased pinocytotic activity by the cell.

2. The **S phase** follows the G_1 phase. Pinocytotic activity decreases during this phase; however, the most notable event is the synthesis of DNA molecules.

3. The **G_2 phase** follows the completion of DNA synthesis. During this phase the metabolic activities of the cell decrease as changes occur in preparation for mitosis (which is sometimes called the *M* **phase).**

Mitosis

For descriptive purposes, mitosis (nuclear division) is divided into four stages: *prophase, metaphase, anaphase,* and *telophase.* It must be emphasized, however, that mitosis is a continuous event and not a series of discrete steps.

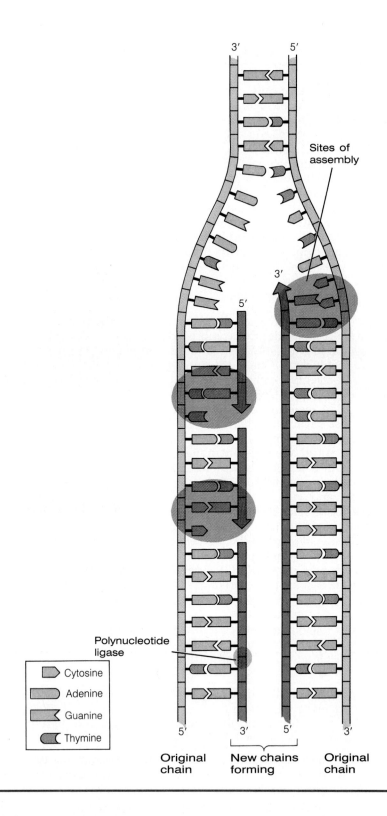

Figure 2.21
Replication of DNA. The original two chains of a DNA molecule separate, each chain then serving as the template for the assembly of a new complementary chain. This results in two DNA molecules, each exactly like the original.

Sites of assembly

Polynucleotide ligase

◁ Cytosine
◁ Adenine
◁ Guanine
◁ Thymine

Original chain   New chains forming   Original chain

Prophase

F2.22 The initial phase of mitosis is called **prophase** (Figure 2.22*b*). Early in prophase two pairs of centrioles are present. During prophase, the centrioles begin to move toward opposite poles of the cell. The nucleolus disappears and the nuclear envelope begins to disintegrate. Also during prophase, the chromatin threads or granules of DNA and protein become tightly coiled and
F2.23 visible as **chromosomes** (Figure 2.23). Each chromosome is made up of two separate strands called **chromatids** that are joined together at one point by a

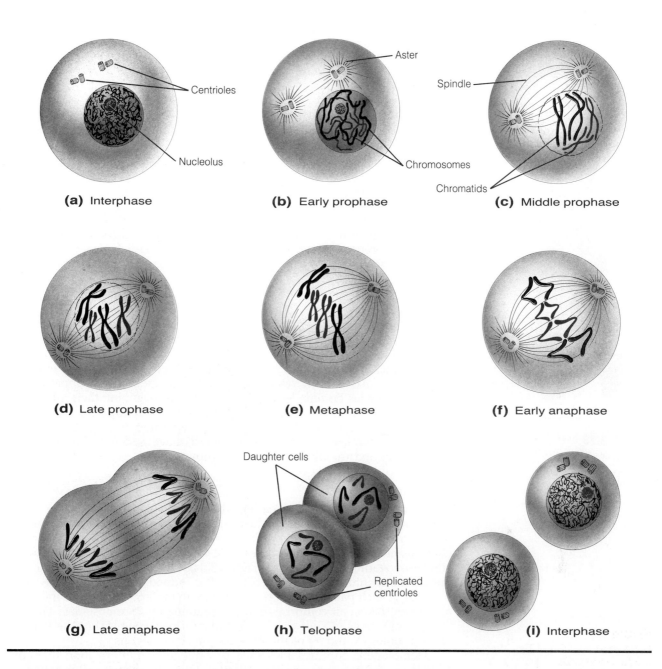

(a) Interphase

(b) Early prophase

(c) Middle prophase

(d) Late prophase

(e) Metaphase

(f) Early anaphase

(g) Late anaphase

(h) Telophase

(i) Interphase

structure called a **centromere.** Actually, each chromatid is a complete chromosome composed of a double-stranded DNA molecule that was replicated during interphase.

By the end of prophase (Figure 2.22*d*), the centrioles have nearly reached **F2.22d** opposite poles of the cell and the chromosomes have moved toward a position at the middle or equator of the cell, halfway between the two centriole pairs. Some of the microtubules that radiate from the centriole regions end blindly. These are known as *astral fibrils,* and they form an aster about each centriole pair. Other microtubules, called *spindle fibers,* form a *spindle apparatus* between one centriole pair and the other. The spindle apparatus is composed of at least two kinds of spindle fibers (which are actually microtubules). One kind of spindle fiber *(continuous microtubules)* extends from one pole of the cell to the other. A second kind *(chromosomal microtubules)* attaches to a chromosome and terminates near one pole of the spindle apparatus.

Metaphase

By the beginning of **metaphase** (Figure 2.22*e*), the nuclear envelope has disap- **F2.22e** peared completely. As metaphase proceeds, the chromosomes line up along

Figure 2.22
Interphase and the phases of mitosis. See the text for a detailed discussion.

Chromatids (each includes a two chain DNA molecule)

Centromere

Figure 2.23
A chromosome as it appears during late prophase.

the central or equatorial plate of the cell and the chromatids of each chromosome become attached by their centromeres to spindle fibers. At the conclusion of this stage, the centromeres divide and each of the former chromatids becomes a separate, single-stranded chromosome. In determining the number of chromosomes, one counts the number of centromeres and not the number of strands. Therefore, during prophase a chromosome consists of two chromatids, but following metaphase each chromatid is considered to be a chromosome.

Anaphase

F2.22f, g   During **anaphase** (Figure 2.22*f* and *g*), the single-stranded chromosomes separate and move toward opposite poles of the cell. By the end of anaphase, the single-stranded chromosomes (each of which is composed of a complete, two-chain DNA molecule) have reached the poles of the cell. When mitosis and cytokinesis occur together, the beginning of cytokinesis is generally evident during anaphase as an inward pinching of the plasma membrane at the equatorial region.

Telophase

F2.22h   In **telophase** (Figure 2.22*h*), a new nuclear envelope forms, presumably from the endoplasmic reticulum, and the nucleolus reappears. If mitosis and cytokinesis are occurring together, cytokinesis is complete when the continued inward pinching of the plasma membrane has separated the cell into two *daughter cells.* Also during telophase, the spindle apparatus disappears, and the chromosomes become less distinct, gradually assuming their interphase appearance of chromatin threads. In many cells, the centrioles are replicated during this phase, but in other cells they are replicated during interphase. By the end of telophase, the two daughter cells have assumed the interphase appearance, and the division cycle is complete.

Meiosis

There are 46 chromosomes in each of the human *somatic cells* (all cells except the reproductive cells). Two of these are **sex chromosomes** (two X chromosomes in females; one X and one Y chromosome in males). The remaining 44 chromosomes are called **autosomes.** The 44 autosomes consist of 22 pairs of F2.24   similar-appearing chromosomes (Figure 2.24). One member of each pair contains genetic information derived from the individual's mother. The other member of each pair contains genetic information derived from the individual's father. Each pair comprises a set of **homologous chromosomes.** The two sex chromosomes (XX) of the female are also homologous, but the two sex chromosomes (X and Y) of the male are not homologous.

Homologous chromosomes each possess genetic information that controls the same functions or characteristics. Often the genetic information on one chromosome of a homologous pair takes precedence over the corresponding genetic information on the other chromosome of the pair. If the genetic information derived from the individual's father takes precedence over the genetic information derived from the individual's mother, the paternal functions or characteristics will be *dominant* and the individual will display the paternal functions or characteristics. In such a case the maternal functions or characteristics are said to be *recessive.* If the maternal genetic information takes precedence over the corresponding paternal genetic information the individual will display the maternal functions or characteristics. If neither the paternal nor the maternal genetic information for a particular function or characteristic is dominant, the individual may display some intermediate function or characteristic.

The 46 chromosomes of human somatic cells actually consist of two 23-chromosome sets (22 autosomes and 1 sex chromosome per set), with one set derived from the individual's father and one from the individual's mother. Thus the *gametes* (reproductive cells) of the male (the sperm from the testes) and female (the ova from the ovaries) each contain only 23 chromosomes.

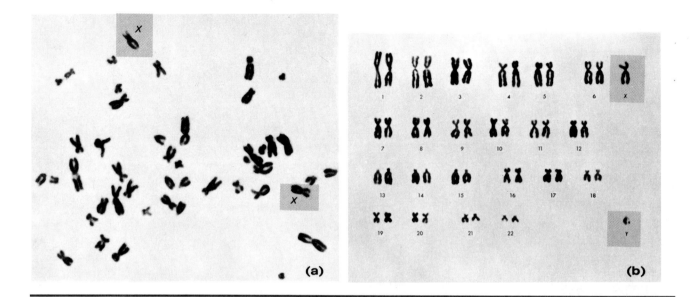

Figure 2.24
Human chromosomes.
(a) Chromosomes of a
female, with X sex
chromosomes indicated.
(b) Chromosomes of a
male, with homologous
chromosomes arranged in
pairs. X and Y are sex
chromosomes.

When a sperm fertilizes an ovum, each gamete contributes 23 chromosomes, thereby establishing the full 46 chromosomes in the new individual.

Cells with two complete sets of chromosomes (46 chromosomes) are known as **diploid cells.** The formation of gametes, however, must result in the formation of cells that have only one set of chromosomes (23 chromosomes rather than 46). Such cells are called **haploid cells.** Haploid cells are not produced by the normal processes of mitosis. A second type of cell division, a reduction division known as **meiosis,** is responsible for the production of haploid reproductive cells.

Two successive division sequences occur in meiosis (Figure 2.25). In the first sequence (Figure 2.25*a–f*), prophase occurs essentially as in mitosis except that the 46 chromosomes do not all move separately toward the spindle fibers. Rather, homologous chromosomes pair **(synapse)** with one another (as do the nonhomologous sex chromosomes of the male). The paired chromosomes move toward the spindle fibers as two-chromosome units. During metaphase, the chromatids of the synapsed chromosomes do not uncouple. At anaphase, the paired homologous chromosomes simply separate, with one member of the pair moving to one pole of the cell and the other member moving to the opposite pole. Thus, 23 double-stranded chromosomes are moved to each pole during the first division sequence of meiosis, rather than the 46 single-stranded chromosomes that are moved to each pole in mitosis. The remaining events of first anaphase and first telophase occur as in mitosis (with cytokinesis), resulting in two daughter cells that each contain only 23 chromosomes, but the chromosomes are double-stranded rather than single-stranded. Following the first division sequence of meiosis, a short period called **interkinesis** occurs (Figure 2.25*g*). During this period, which is similar to the interphase period between mitotic divisions, the 23 double-stranded chromosomes of the daughter cells do not duplicate themselves.

Following the interkinesis period, the second division sequence of meiosis occurs (Figure 2.25*h–k*). Each of the two 23-chromosome daughter cells undergoes a typical *mitotic* division with cytokinesis. The result of this second meiotic sequence is four haploid cells, each of which contains 23 *single-stranded* chromosomes (Figure 2.25*l*). It is these haploid cells that ultimately differentiate into sperm in the male testes or ova in the female ovaries. However, the second meiotic division of the ovum is completed in the uterine tube, and only if it is fertilized by a sperm.

Meiosis allows a great deal of genetic diversity in the makeup of sperm and ova. The chromosomes that originally came from the individual's male parent, for example, generally do not all line up toward one centriole pair, nor do those from the individual's female parent all line up toward the other

F2.25

F2.25g

F2.25h–k

F2.25l

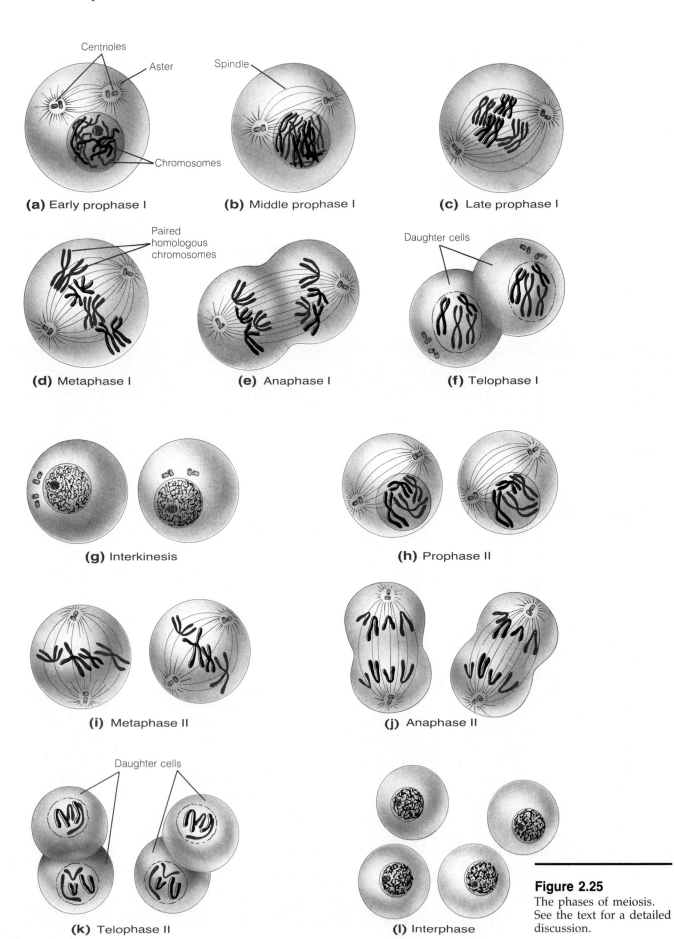

(a) Early prophase I

(b) Middle prophase I

(c) Late prophase I

(d) Metaphase I

(e) Anaphase I

(f) Telophase I

(g) Interkinesis

(h) Prophase II

(i) Metaphase II

(j) Anaphase II

(k) Telophase II

(l) Interphase

Figure 2.25
The phases of meiosis.
See the text for a detailed
discussion.

FRONTIERS IN HEALTH

Doctoring Our Genes

Kevin is 21 years old. Unlike other young men his age, who have graduated from college and are busy starting families and new jobs, Kevin sits at home under his mother's care. With his hands strapped to his wheelchair or bed, Kevin passes his hours watching television. He has never gone to school and he knows practically nothing of the world outside his home. At times Kevin shakes out of control, and his mind drives him toward self-destruction. During one of his fits, he may bite a hand with enough force to amputate a finger; hence the straps.

Sad as it may sound, Kevin has virtually no hope of leaving his bed or wheelchair; no hope of throwing off the straps that bind his hands every moment of his life. Kevin, like 2000 other people in the United States, is afflicted with a rare genetic disorder called Lesch-Nyhan syndrome. This disease, which is detected in about 200 children every year, has turned Kevin's body into a living prison.

A single defective gene has made the difference between a normal life and a life of agony for this young man. The defective gene synthesizes an enzyme, HPRT (hypoxanthine guanine phosphoribosyl transferase), that plays a critical role in the body's metabolism. In its absence, uric acid builds up in the body's tissues, causing the symptoms seen in Kevin, as well as gout and recurrent kidney stones. New drugs can help reduce the buildup of uric acid, and permit victims a somewhat longer life. Most victims now live well into their 20s, whereas previously they generally died in early childhood. However, these drugs do nothing to block the episodes of self-destructiveness.

Curing diseases of this nature has previously seemed impossible, because the only way to reverse such a disease would be to replace the defective genes with normal ones, a task well beyond the scope of medicine—until recently, that is.

New advances in genetic engineering may someday provide a cure for Kevin and thousands of other victims of single-gene defects. Many medical scientists believe that normal genes can be administered to victims of such defects. To do this it is necessary first to isolate the gene in question. Once it is isolated, new copies of the gene are made, usually in a host bacterium. These copies then can be inserted into the genetic material of special viruses that infect human cells but do not cause any undesirable effects. These viruses, carrying the inserted genes, can be injected into the body. If all goes as planned, the viruses will infect body cells and the DNA of the injected gene will be incorporated into the genetic material of the cell. The defective gene remains—it was not functioning anyway—but the new gene becomes active, producing the missing enzymes.

The prospects of gene therapy are exciting, because there are at least 1600 single-gene defects known to medical science. However, a major obstacle to be overcome is perfecting the insertion steps; that is, getting the gene from the virus into the cells of the body, or at least into the cells that require the normal gene.

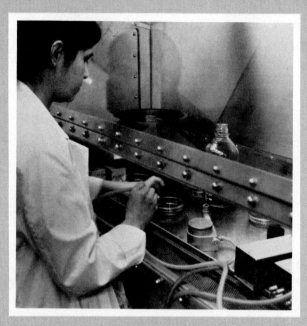

Cell tissue experiment at cancer research lab.

Encouraging results have come from a team of scientists headed by University of California, San Diego geneticist Dr. Theodore Friedman and Dr. Inder Verma of the Salk Institute. After years of tedious work, they have succeeded in transferring the HPRT gene to cultured white blood cells taken from victims of Lesch-Nyhan syndrome. The transplanted genes have raised the enzyme levels to one-quarter of the normal levels. The researchers plan to reinject these white blood cells into Lesch-Nyhan syndrome patients with the hope that the genes will be transferred to other body cells. Even though they are producing only a fraction of the normal levels of HPRT, the transplanted genes may provide enough enzyme to reverse the disease.

It is questionable, however, whether the genes will transfer from the white blood cells to the rest of the body, and it has been suggested that a better option would be to attempt to get the genes directly into every cell of the body, especially those cells of the brain where many of the symptoms originate. Such procedures raise serious questions. For example, how can researchers be certain that the gene will insert properly? What if it interferes with another gene and causes a harmful condition, such as cancer, or turns out to be lethal?

Many medical researchers believe that the possibility of opening the door to a healthy, productive life for the thousands of victims who suffer from genetic diseases far outweighs the risks involved. The future for people suffering from genetic disorders thus appears a bit more hopeful.

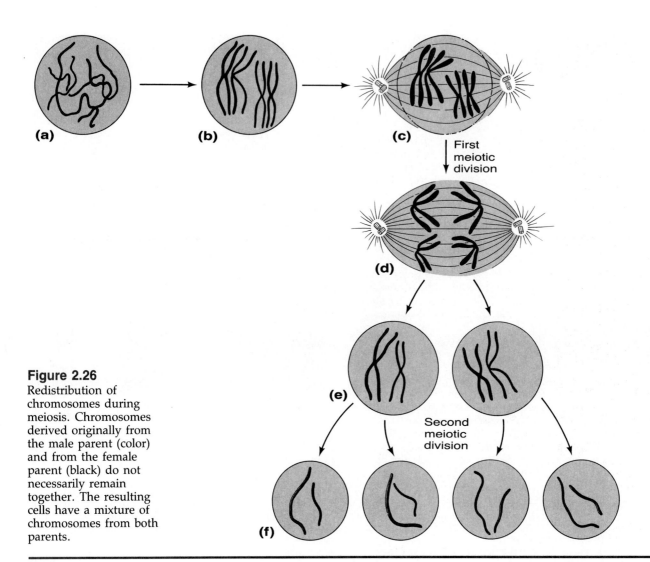

Figure 2.26
Redistribution of chromosomes during meiosis. Chromosomes derived originally from the male parent (color) and from the female parent (black) do not necessarily remain together. The resulting cells have a mixture of chromosomes from both parents.

F2.26

F2.27

centriole pair during the synapsis of chromosomes that occurs in the first division sequence of meiosis (Figure 2.26). Rather, a mixture of positions is assumed so that the resulting daughter cells each receive some chromosomes derived from the individual's male parent and some from the female parent, in an apparently random assortment.

Further genetic diversity may result from the phenomenon of **crossing over,** which takes place occasionally during the first stage of meiosis, while the chromosomes are synapsed (Figure 2.27). In this process the chromatids of synapsed chromosomes may break and the two free chromatid fragments then exchange places with one another, so that when the breaks are repaired, the fragment originating from the chromatid of one of the synapsed chromosomes is attached to the chromatid of the other chromosome, and vice versa. When the synapsed chromosomes separate and daughter cells ultimately are produced, the daughter cells can have a different genetic composition than they would have if crossing over had not occurred. Thus, the gametes that are ultimately produced from different cells will differ from one another genetically depending on the particular chromosomal redistribution that occurs during meiosis.

EFFECTS OF AGING ON CELLS

Aging is the result of numerous progressive structural and functional changes that cause us to be less able to adequately maintain homeostatic conditions

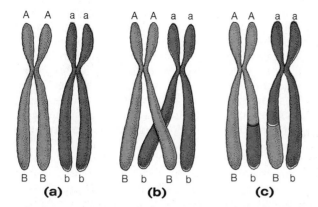

A A   a a   A A A a   A A   a a

B B   b b   B b B b   B b   B b
(a)   **(b)**   **(c)**

Figure 2.27
The process of crossing over.

within our bodies, and thus we may become more vulnerable to disease as we grow older. We gradually become aware of the aging changes because we are no longer able to do some things that we could do when we were younger, or we develop dysfunctions that seem to be more common in older persons. However, we are not, as a rule, aware of the many cellular changes that may be directly responsible for the observable alterations in structure and function. We conclude our study of the cell by considering some of the cellular changes that have been attributed to the aging process.

A number of theories have been proposed to explain what causes the body to age. Some suggest that aging occurs at the organismic level. One such theory proposes that with age changes occur in the immune system that disrupt its ability to recognize the body's own cells. As a consequence, it is thought that in older persons antibodies may be produced that attack the body cells. Such autoimmune responses would destroy normal body proteins. Another organismic theory suggests that aging is programmed by a region of the brain called the hypothalamus. The hypothalamus controls the secretion of hormones by the pituitary, thyroid, and adrenal glands, as well as the testes and ovaries. This theory proposes that with aging the control exerted by the hypothalamus over these glands is altered, thereby producing wide-ranging bodily changes.

Other theories consider cellular changes to be the main cause of aging. These theories are of particular interest to us in this chapter. Some of these theories propose that aging changes are programmed into the cells, and the cells will die after a predetermined life span. Experiments have shown that cells grown in culture (outside the body) will divide only a certain number of times, and then they die. When older cells are cultured they undergo fewer divisions before dying than do younger cells. These findings support the concept that each cell is genetically programmed to divide only a certain number of times, and older cells have fewer divisions remaining before they die.

Another cellular theory of aging proposes that with age cross-links form between cellular proteins, thereby altering the structure and functioning of the proteins. The cellular proteins that are thought to be most often affected by cross-linking are enzymes.

Still another cellular theory suggests that, with time, there may be a gradual accumulation of chemical compounds called free radicals. Free radicals are oxygen molecules that contain an unpaired electron. Free radicals react with fats and other substances within the cell and, if allowed to accumulate, are thought to alter the plasma membrane as well as the membranes surrounding mitochondria and lysosomes. They may even damage chromosomes. Free radicals are produced by certain foods as well as by radiation. They are generally kept at harmless levels within the cell by the enzymes contained in peroxisomes. The suggestion is that with aging these enzymes may become overwhelmed, thus allowing free radicals to reach harmful levels within the cell.

Whatever the cause, there are changes in cell structure and functioning with age. Some of the more documented cellular changes include an increase in the incidence of damage to DNA (genes), perhaps due in part to a decline in the activities of the repair process that normally replaces damaged DNA. There is also some evidence that an inhibitor substance may be produced in the cytoplasm of older cells and diffuse into the nucleus, where it inhibits DNA formation. A common cytoplasmic change in aging cells is the accumulation of deposits of a substance called lipofuscin. The presence of lipofuscin in older cells is so consistent that it is referred to as the "age-pigment." It is thought that these accumulations may affect cell functioning and thus contribute to aging. Many older cells have a decreased number of mitochondria and there may be a reduction of folds (cristae) on the inner membrane of the mitochondria. Because mitochondria are the main sources of metabolic energy, any alterations they undergo may be correlated with general age-related changes. Lysosomes are thought to undergo changes in older cells that reduce their efficiency, causing them to be less able to handle the digestive and waste-removal needs of the cells. Accumulation of wastes within cells may affect cellular functioning and thus contribute to age-related changes.

The fastest growing segment of the population in the United States is the group over 85 years old. This group is increasing steadily from 0.5% of the population in 1960 to a projected 4.5% by 2040. Because of this increase, the study of aging has become a high priority for the medical profession as well as for all levels of government. These elderly people will be present in large enough numbers to severely strain the health care facilities and the social services facilities. In order that you might have some appreciation of the biological changes that commonly occur within the body during aging we consider the effects of aging on each body system in the chapters in which those systems are discussed.

STUDY OUTLINE

CELLULAR COMPONENTS nucleus surrounded by cytoplasm, which contains numerous structures called organelles. Limiting boundary of cell is plasma membrane. **pp. 19—36**

PLASMA MEMBRANE a highly selective barrier that affects movement of substances into or out of cells. Membrane is lipid bilayer with protein interspersed; pores in membrane connect cell interior with outside. Materials that enter or leave cell must pass through membrane or pores.

MOVEMENT OF MATERIALS ACROSS PLASMA MEMBRANE

PASSIVE PROCESSES energy or forces necessary for movement across plasma membrane supplied by extracellular environment.

Diffusion movement from region of higher concentration to region of lower concentration due to collisions between atoms or molecules.

Facilitated Diffusion molecules unable to penetrate membrane by themselves bind to carrier molecules on one side of membrane and cross it in the bound form. Movement is with concentration gradient and does not require cellular energy.

Osmosis movement of water across a semipermeable membrane from area of higher water concentration to area of lower water concentration.

Dialysis selective separation of substances in solution by plasma membrane. Used in artificial kidneys.

Filtration mechanical forces (pressures) move substances across plasma membrane.

ACTIVE PROCESSES energy necessary for movement across plasma membrane supplied by cell.

Active Transport molecules moved across plasma membrane against concentration gradient by carrier compounds.

Endocytosis plasma membrane forms small vesicles that enclose substances outside cell; vesicles then move into cell.
1. Phagocytosis: vesicles enclose solid materials.
2. Pinocytosis: vesicles enclose liquids.
3. Receptor-mediated endocytosis: plasma membrane receptors bind specific molecules.

Exocytosis expulsion of material from cell by means of membrane-bounded vesicles.

THE NUCLEUS
1. Contains genetic material of cell in form of chromosomes; consists of DNA combined with protein.
2. Contains nucleoli, which consist of RNA, protein, some DNA.
3. DNA replicates itself and also serves as template for assemblage of RNA.
4. r-RNA becomes part of ribosomes in cytoplasm.
5. m-RNA contains coded information, derived from DNA; used to direct protein synthesis at ribosomes.
6. Codon sequence of m-RNA specifies amino acid sequence of polypeptide or protein.

7. Polypeptides or proteins synthesized according to DNA specifications may act as enzymes or hormones.

CYTOPLASM thick fluid located between plasma membrane and nuclear membrane; site of many chemical reactions and specialized organelles.

CYTOPLASMIC ORGANELLES

Ribosomes sites of formation of polypeptides or proteins; consist of r-RNA and protein.

Endoplasmic Reticulum membranous network of tubular or saclike channels. Two types: smooth or rough. Walls contain enzymes that play a role in fatty acid and steroid synthesis; proteins synthesized by attached ribosomes can enter channels, and protein-containing vesicles can separate off.

Golgi Apparatus flattened membranous sacs with which vesicles from endoplasmic reticulum can fuse. A site of protein synthesis.

Lysosomes membrane-bounded structures that are believed to originate from Golgi apparatus; contain digestive enzymes that act on proteins, lipids, certain carbohydrates, DNA, and RNA.

Peroxisomes contain variety of powerful oxidative enzymes.

Mitochondria bounded by double membrane; inner membrane has folds. Involved in generation of metabolic energy for cell activities.

CYTOSKELETON an intricate network of filamentous structures.

Microtubules small, hollow, unbranched, cylindrical tubules that may function as supporting framework of cells or as conducting channels.

Intermediate Filaments may be involved in maintaining the position of the nucleus and in transporting materials into and out of the nucleus.

Microfilaments occur in bundles or other groupings; may be associated with contractile activities involved with cell movement.

Cilia, Flagella, and Basal Bodies cilia and flagella are motile extensions from plasma membrane; help move substances over cell surface or help move entire cell about; both believed to arise from basal bodies. Cilia, flagella, and basal bodies contain characteristic patterns of microtubules.

Centrioles are similar in structure to basal bodies; involved in cell division.

INCLUSIONS are chemical substances in cells, such as hemoglobin and melanin.

EXTRACELLULAR MATERIALS include body fluids and extracellular matrix in which many cells are embedded. Connective tissues are particularly rich in extracellular materials. **p. 36**

CELL DIVISION several basic events involved, including replication of genetic material in nucleus, redistribution of this material into two new nuclei, and division of cytoplasm into two new cells, each with its own nucleus. **pp. 37–44**

INTERPHASE period between active cell divisions; DNA synthesis occurs.

MITOSIS is continuous event, although four phases are observable.

PROPHASE chromosomes become visible; centrioles move toward opposite poles of cell; spindle forms.

METAPHASE centromeres divide, and each chromatid becomes separate, single-stranded chromosome.

ANAPHASE single-stranded chromosomes separate and move to opposite poles of cell.

TELOPHASE chromosomes assume interphase appearance as indistinct chromatin threads; spindle disappears; cytokinesis, if it occurs, is completed.

MEIOSIS a reduction-division process responsible for production of haploid reproductive cells; involves two successive division sequences.

FIRST SEQUENCE

Prophase essentially, same as in mitosis, except that homologous chromosomes pair and move toward spindle fibers as two-chromosome units.

Metaphase chromatids of synapsed chromosomes do not uncouple.

Anaphase paired chromosomes move apart; one member of pair moves to one pole of cell and other member moves to opposite pole.

SECOND SEQUENCE each 23-chromosome daughter cell undergoes mitosis with cytokinesis. Result is four haploid cells, each with 23 single-stranded chromosomes. Haploid cells differentiate into gametes.

EFFECTS OF AGING ON CELLS cellular age-related changes include reductions in DNA synthesis and repair, the formation of lipofuscin granules in the cytoplasm, decreased numbers of mitochondria, and declining lysosomal efficiency. **p. 44–46**

SELF-QUIZ

1. Many nonpolar compounds that are soluble in lipids move relatively easily through the plasma membrane by: (a) bulk flow; (b) active transport; (c) simple diffusion.

2. Match the following terms associated with the movement of substances across the plasma membrane with the appropriate description.

Active transport
Exocytosis
Facilitated
  diffusion
Osmosis
Dialysis
Filtration

(a) Passive process by which molecules unable to pass through plasma membrane by themselves are able to do so with the concentration gradient while bound to a carrier compound.

(b) Movement of water from area of higher water concentration to area of lower water concentration.

(c) Movement of molecules across plasma membrane against the concentration gradient by means of carrier compounds.

(d) Expulsion of material from within a cell by means of membrane-bounded vesicles.

(e) Selective separation of substances in solution by the plasma membrane.

(f) Movement of substances across plasma membrane by means of external forces.

3. The term for a cell's taking in of material from the external environment by means of membrane vesicles is: (a) exocytosis; (b) endocytosis; (c) cytokinesis.

4. Ribosomes are formed from r-RNA–protein complexes that are believed to migrate out of the nucleus and into the cytoplasm. True or false?

5. Ribosomes generally are found as individual units in the cytoplasm. True or false?

6. Those organelles whose walls contain enzymes associated with fatty acid and steroid synthesis are: (a) endoplasmic reticulum; (b) ribosomes; (c) mitochondria.

7. The endoplasmic reticulum that has ribosomes attached to it is called smooth endoplasmic reticulum. True or false?

8. Those organelles that contain strong digestive enzymes capable of digesting proteins, lipids, DNA, and RNA are: (a) ribosomes; (b) microtubules; (c) lysosomes.

9. Those organelles that are involved in the generation of metabolic energy for cellular activities are: (a) Golgi apparatus; (b) mitochondria; (c) lysosomes.

10. Thin, short, and numerous extensions that may move substances over the surface of a cell are known as: (a) cilia; (b) flagella; (c) basal bodies.

11. Match the following cytoplasmic organelles with their related function.

Ribosomes
Golgi apparatus
Peroxisomes
Basal bodies
Microfilaments

(a) These organelles contain powerful oxidative enzymes.

(b) Cilia and flagella are thought to arise from these organelles.

(c) These organelles are associated with various forms of contractile activities of cells.

(d) The site of polypeptide and protein synthesis.

(e) These vesicular structures are especially evident in cells during periods of increased secretory activity.

12. These organelles may function as a supporting framework within cells: (a) microtubules; (b) centrioles; (c) peroxisomes.

13. The skin pigment melanin is classed as an "inclusion." True or false?

14. Mitosis and cytokinesis always occur simultaneously. True or false?

15. The DNA molecules that comprise the genetic material of the cell appear as indistinct chromatin threads or granules within the nucleus during: (a) metaphase; (b) anaphase; (c) interphase.

16. Match the following items associated with meiosis with the appropriate item.

Autosomes
XY
Homologous
  chromosomes
46
Diploid
23
Haploid
Interkinesis
Crossing over

(a) XX

(b) The stage between the first and second meiotic divisions.

(c) Cells with two complete sets of chromosomes are said to be in this condition.

(d) This process promotes genetic diversity.

(e) This cell condition is characteristic of gametes.

(f) The total number of chromosomes in a human somatic cell.

(g) Male sex chromosomes.

(h) Chromosome classification that does not include sex chromosomes.

(i) The number of chromosomes contained in each of the human male and female gametes.

LEARNING OBJECTIVES

After completing this chapter, you should be able to:

- Name the four primary tissues, and cite one example of each.

- Describe the specializations by which adjacent cells may be attached to one another.

- List three means of classifying epithelial tissue, and cite at least one example of each tissue.

- Describe the shape that characterizes each of these cell types: squamous, cuboidal, and columnar.

- Distinguish between simple epithelium and stratified epithelium.

- Distinguish between exocrine glands and endocrine glands.

- Classify three types of glands by mode of secretion, and describe how each type functions.

- List the types of connective tissue, and state at least one function of each.

- Cite several structural differences between bone and cartilage.

- Name the three main types of muscle tissue, and describe the appearance of the cells of each type.

- Cite two kinds of tissues that cannot undergo regeneration in a mature person, and two kinds that can.

CHAPTER CONTENTS

EPITHELIAL TISSUES

CONNECTIVE TISSUES

MUSCLE TISSUE

NERVOUS TISSUE

TISSUE REPAIR

Tissues

3

In the previous chapter you learned that the body is composed of millions of cells and that each cell contains various organelles, which carry on a number of physiological processes. It is important to realize, however, that it is more common for groups of cells to cooperate for the benefit of the organism as a whole, rather than simply for their own individual needs. Groups of cells that are similar in structure, function, and embryonic origin and that are bound together with varying amounts of intercellular material are referred to as **tissues.** There are four primary tissues in the body: *epithelial, connective, muscular*, and *nervous*. Since it is these four tissues that join together to form the organs of the body, an understanding of the structure and function of each tissue type contributes to our understanding of the organ systems.

EPITHELIAL TISSUES

Epithelial tissues are formed of closely joined cells with only a minimum of intercellular material between them. Epithelial cells are always underlain by connective tissue, to which they are attached by a thin layer called the **basement membrane.** The basement membrane consists of two layers; one layer, called the *basal lamina*, is composed of collagen and glycoproteins, and is a product of the epithelial cells; the deeper layer is composed of *reticular fibers* that develop from the connective tissue. Epithelia may develop from either the ectoderm, endoderm, or mesoderm of the embryo.

Epithelia are, by definition, sheets of cells that cover body surfaces and line body cavities. In general, they cover most of the free surfaces of the body, both internal and external. For instance, they form the outer layer of the skin, the lining of the digestive tube, the linings of the ventral body cavities, the lining of the blood vessels, and those glandular ducts and tubules that develop from epithelial linings or coverings. In addition, some epithelial tissues are incorporated into the various glands, where they serve as the functional part of the gland. With such a variety of locations, it is not surprising that epithelial tissues have diverse functions. The epidermis of the skin, for instance, is a *protective* layer that forms a barrier between the organism and its external environment, whereas the linings of the internal body organs are involved in the *absorption* of materials into the body, the *excretion* of waste products, and the *secretion* of special products into the cavities. Regardless of their functions, none of the epithelial tissues contain blood vessels (i.e., they are *avascular*). Blood vessels within the connective tissues underlying the epithelial tissues supply nutrients to, and remove wastes from, the epithelial cells.

Specializations of Epithelial Cell Surfaces

The portion of epithelial cells that forms the surface of the body or lines the cavities and lumina (interior space) of the various tubes in the body is referred to as the **free surface.** The free surfaces of the epithelial cells that line the blood vessels are smooth. The electron microscope shows that other epithelial cells have their free surfaces folded into tiny protoplasmic projections called **microvilli** (Figure 3.1). Because microvilli greatly increase the area of the free F3.1

51

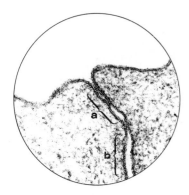

(a) Tight junction

(b) Intermediate junction

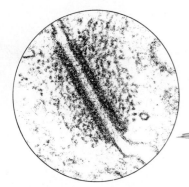

(c) Desmosome

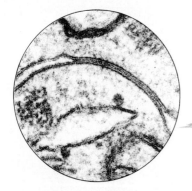

(d) Gap junction

Figure 3.1
Typical epithelial cell as seen with an electron microscope. On the right is a highly magnified junctional complex consisting of a tight junction, an intermediate junction, and a desmosome. Parts **(a)** through **(c)** are photomicrographs of the components of a junctional complex. Part **(d)** is a gap junction.

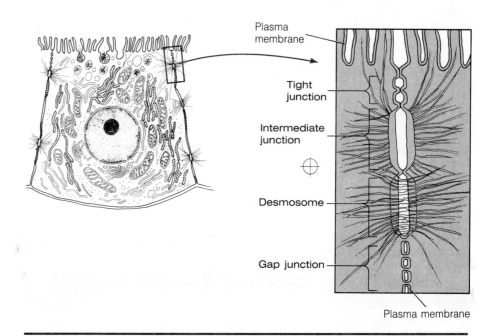

surface, they are especially abundant in locations where absorption is the main activity, as in the lining of the digestive tract. Before the electron microscope made it possible to view their structure clearly, these dense groups of microvilli were identified as *striated borders* or *brush borders*. Microvilli line some surfaces that are not absorptive, and their function is less well understood there. They may serve to anchor mucus to the cell surface. Microvilli are not motile. In some locations, the free surfaces of epithelial cells are modified by the presence of **cilia.** Most cilia are motile and move rhythmically, thus serving to propel materials along the epithelial surface.

Specializations for Cell Attachments

We have noted that one characteristic of epithelial tissues is that their cells are situated close together, with little intercellular material between them. In fact, adjacent epithelial cells are generally joined together into coherent sheets. It is these cellular junctions that make epithelial tissues so well suited to cover the body surfaces and to line the body cavities.

Before the advent of the electron microscope, epithelial cells were thought to be held together by combinations of cement and bridges between adjacent cells and by several structures that showed up under the light microscope as dark spots on the cell boundaries. These spots were called desmosomes or terminal bars, depending on their location. The electron microscope has clarified the structures of these cellular junctions. In columnar and some cuboidal epithelium, the intercellular relationships, as revealed by the electron microscope, are referred to as **junctional complexes.**

Junctional Complexes

F3.1 Every junctional complex generally has three distinct components—a *tight junction*, an *intermediate junction*, and a *desmosome*—all of which are located on the lateral cell boundaries (Figure 3.1).

TIGHT JUNCTIONS   **Tight junctions (zonula occludens)** are located just below the free surface of the epithelium. In this region protein molecules in the outer layers of the cell membranes of adjacent cells fuse in several places, leaving intercellular separations between the sites of membrane fusion. Tight junctions extend around the outer margin of the cell like a belt. Tight junctions not only connect adjacent cells but, because they obliterate the intercellular spaces, they also restrict the movement of substances through the epithelial membrane via the intercellular spaces.

INTERMEDIATE JUNCTIONS   **Intermediate junctions (zonula adherens)** are located just below tight junctions. In the intermediate junction the cell membranes of adjacent cells are not modified and are separated by a gap of about 200 Å. There is, however, a mat of filaments located against the inner layer of the cell membrane of each cell. Like the tight junction, the intermediate junction extends like a belt around each cell.

DESMOSOMES   The third component of a typical junctional complex is the **desmosome** or **macula adherens.** Each desmosome is an individual point of cell attachment rather than a beltlike zone. The cell membranes remain about 200 Å apart in a desmosome, and the inner layer of each cell membrane is thickened. Cytoplasmic filaments that form part of the cytoskeleton of the cells attach to the thickened cell membranes. Indistinct filaments of glycoprotein run across the intercellular space between adjacent desmosomes. It is thought that these filaments are related to the cell-to-cell binding at these points. Desmosomes can occur anywhere around the periphery of an epithelial cell. Where the cell membrane contacts connective tissue, as along the basal lamina, half desmosomes **(hemidesmosomes)** are sometimes found. These are often seen in stratified squamous epithelium, such as is found in the outer layer of the skin. While desmosomes are particularly prevalent in epithilia they are also found in other tissues, such as cardiac muscle and muscles of the uterus.

Gap Junctions

Apart from the components of the junctional complex is another intercellular specialization called the **gap junction** or **communicating junction.** In this junction the cell membranes of adjacent cells are separate but very close together—only approximately 20 Å apart. This extremely narrow gap is bridged by small tubular channels formed by proteins that protrude through the lipid bilayer of the cell membrane and extend into the intercellular space, where they join with similar channels from the adjacent cells. These channels, called *connexons,* directly link the cytoplasm of adjacent cells. Gap junctions are sites at which small molecules and ions can pass from one cell to another, and they play an important role in the transmission of electrical activity between cells. Unlike junctional complexes, gap junctions are not restricted to epithelia. They are also found connecting the cells of smooth muscle, cardiac muscle, some nerve cells, and bone-forming cells located within compact bones.

Classification of Epithelia

Epithelial tissues are generally classified on the basis of the *number and arrangement of cell layers* within the tissue and the *shape of the cells at the free surface* of the tissue.

According to Cell Layers

If an epithelium is formed by a single layer of cells, all of which are in contact with the basal lamina, it is called **simple epithelium.** If it has two or more layers of cells, and only the deepest layer is in contact with the basal lamina, it is termed **stratified epithelium.** If the tissue appears to consist of several layers but is actually formed of a single layer with all cells touching the basal lamina, it is **pseudostratified epithelium** (*pseudo* = false). This false impression of stratification occurs because some cells are shorter than others and the taller cells overlap the short ones, preventing them from reaching the free surface of the tissue.

According to Cell Shape

The cells that form the free surface of epithelial tissues are of three shapes. **Squamous cells** are flat and thin. **Cuboidal cells** are about as tall as they are wide and therefore appear almost square in vertical section. **Columnar cells** are taller than they are wide and appear rectangular in vertical section. Epithelia can be named according to which of these cell types form their free surfaces.

×1250

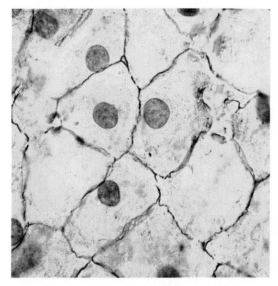

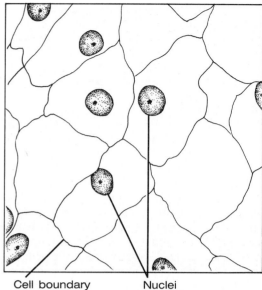

Cell boundary   Nuclei

Figure 3.2
Simple squamous
epithelium (surface
view).

×1250

Squamous cells

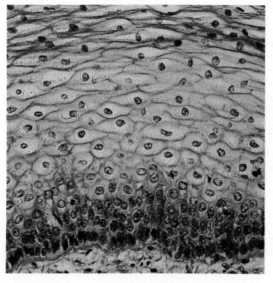

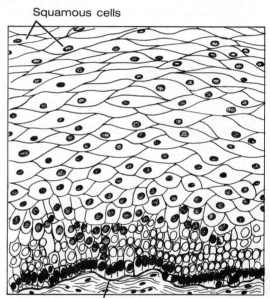

Basal lamina

Figure 3.3
Stratified squamous
epithelium.

General Classification

The general classification of epithelial tissues takes into consideration both the shape of the cells that form the free surface and the number of layers of cells in the tissue.

✓
F3.2   SIMPLE SQUAMOUS EPITHELIUM   **Simple squamous epithelium** is formed of a single layer of squamous cells (Figure 3.2). Since this thin sheet does not form a very effective barrier, substances can move easily across it. And the flat cells do not contain enough cytoplasmic inclusions to aid secretion or absorption.

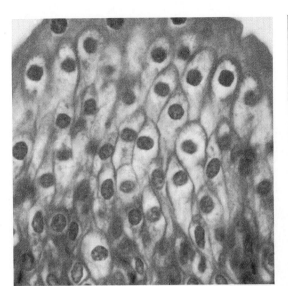
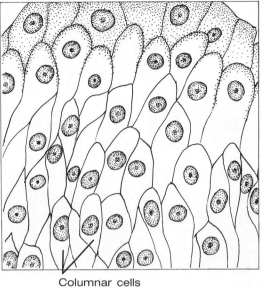

×3100

Columnar cells

Figure 3.4
Transitional epithelium.

In general, simple squamous epithelium is found in regions where diffusion and filtration occur. Specifically, it lines the heart and the blood vessels and is the only barrier that separates the blood in capillaries from the tissue fluid. The simple squamous lining of the vascular system is called **endothelium.** Similarly, it lines the air sacs (alveoli) of the lungs, where it separates the air from the tissue fluid, and lines the surfaces of body cavities. The simple squamous lining of the body cavities is called **mesothelium.** Simple squamous epithelium also forms the glomerular capsules of the kidneys, the sites where substances are filtered from the blood to form urine.

✓ STRATIFIED SQUAMOUS EPITHELIUM   As the name implies, **stratified squamous epithelium** (Figure 3.3) is composed of many layers, the precise number of which varies in different locations. The deeper cells, close to the basal lamina, tend to be cuboidal, but those against the surface are typical squamous cells. The deeper cells undergo mitosis and thus increase in number. These newly formed cells are pushed toward the surface, where they replace the older surface cells that are being continually sloughed off. Because of the ability of stratified squamous epithelium to replace the cells of the superficial layers, this tissue is capable of compensating for the loss of cells due to such actions as abrasion. Thus, stratified squamous epithelium forms a protective layer on the body surface as the epidermis of the skin, and also in areas that are subjected to friction, such as the linings of the mouth, pharynx, esophagus, anus, and vagina.

F3.3

TRANSITIONAL EPITHELIUM   **Transitional epithelium** is a specialized stratified tissue that lines the urinary bladder and a few other hollow organs (Figure 3.4). The surface cells of transitional epithelium vary between cuboidal and squamous, depending upon whether the bladder is empty or expanded. The deepest cells are columnar and they are overlaid by many layers of cells when the bladder is empty. When the bladder is full and its walls are stretched, the cells flatten and slide over each other, leaving only three or four strata (layers) between the deepest layer and the free surface. In this condition, the surface cells are flattened, like squamous cells. This specialized tissue allows an organ to expand with only minimal resistance from the tissue, thus lessening the chance of the organ rupturing, and reducing the discomfort that occurs as the organ becomes full.

F3.4

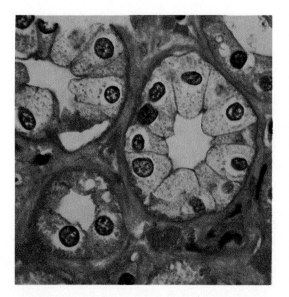
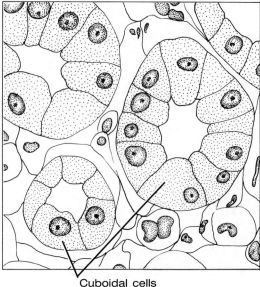

Figure 3.5
Simple cuboidal
epithelium.

Cuboidal cells

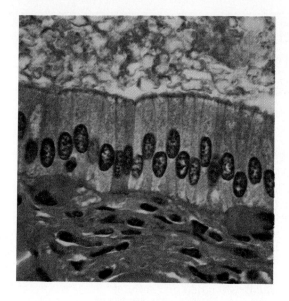
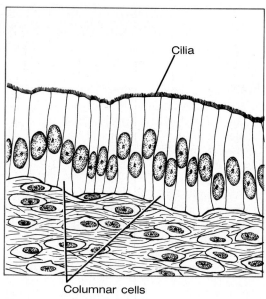

Cilia

Figure 3.6
Simple columnar
epithelium.

Columnar cells

✓   SIMPLE CUBOIDAL EPITHELIUM   Cuboidal cells are generally found as a
F3.5 single layer of cells that form **simple cuboidal epithelium** (Figure 3.5). Only
rarely are cuboidal cells found in layers (stratified). Some cuboidal cells are
capable of forming secretions and consequently are found in glands such as
the thyroid, the sweat glands, and the salivary glands. Cuboidal cells also
form the ducts of glands and parts of the kidney tubules as well as the outer
cell layer covering the ovary.

✓   SIMPLE COLUMNAR EPITHELIUM   Columnar cells, with their large
amounts of cytoplasm and cytoplasmic organelles, can perform rather complex chemical reactions and are therefore found in regions where secretion
F3.6 and absorption occur. **Simple columnar epithelium** (Figure 3.6) lines the digestive tube from the stomach to the anal canal. It also forms the ducts of
many glands. Cilia are present on the free surfaces of the columnar cells that
form the membranes lining the bronchi of the lungs, the nasal cavity, the
oviducts, and scattered regions of the uterus. The presence of cilia causes
these tissues to be classified as **simple columnar ciliated epithelium.**

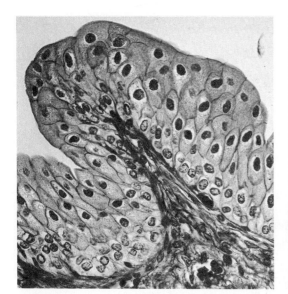

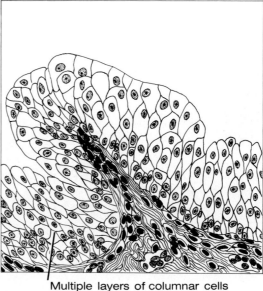

×340

Multiple layers of columnar cells

Figure 3.7
Stratified columnar epithelium.

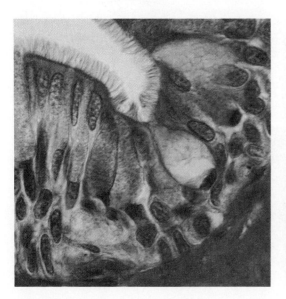

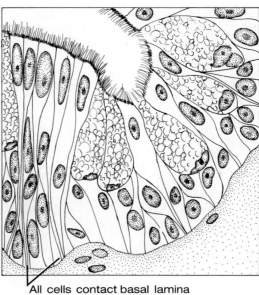

×3100

Figure 3.8
Pseudostratified columnar ciliated epithelium.

All cells contact basal lamina

STRATIFIED COLUMNAR EPITHELIUM   There are only a few locations where columnar tissue is truly stratified. In these tissues, the cells next to the basal lamina are small and rounded, and the columnar cells that form the free surface of the tissue do not contact the basal lamina (Figure 3.7). **Stratified columnar epithelium** appears on the epiglottis, in parts of the pharynx and anal canal, and in the male urethra.   F3.7

PSEUDOSTRATIFIED COLUMNAR EPITHELIUM   In **pseudostratified columnar epithelium,** which is much more common than the truly stratified columnar epithelium, all the cells contact the basal lamina but some are shorter than others and do not reach the free surface. The nuclei are also found at different levels within the cells, which adds to the impression of stratification. This epithelium is found in the large ducts of some glands, such as the parotid gland, and in regions of the male urethra. The free surface of pseudostratified tissue often has cilia, and the tissue is then called **pseudostratified columnar ciliated epithelium** (Figure 3.8). These ciliated tissues line   F3.8

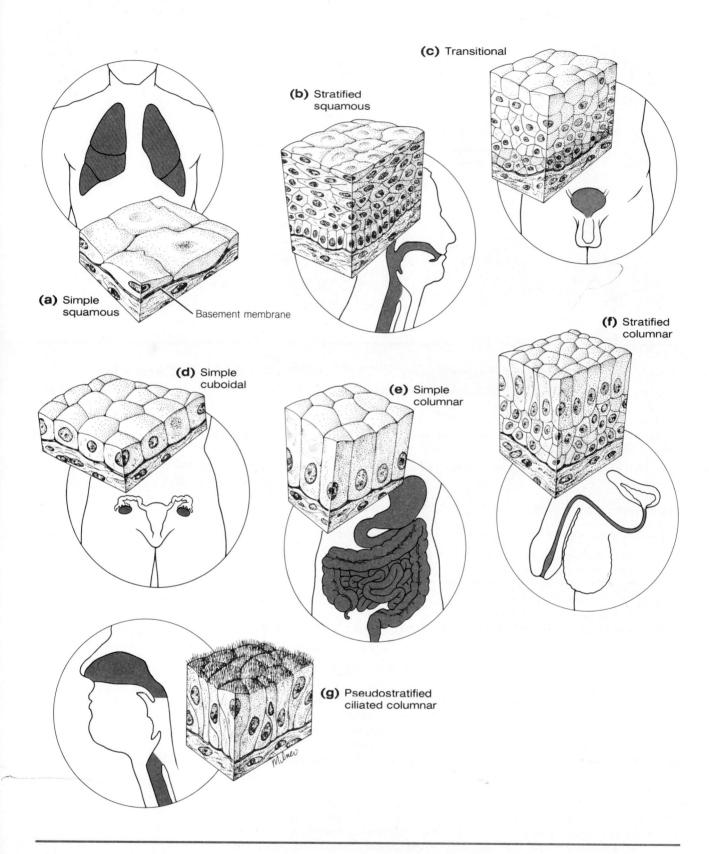

Figure 3.9
Classification of epithelial tissues according to cell layers and shape.

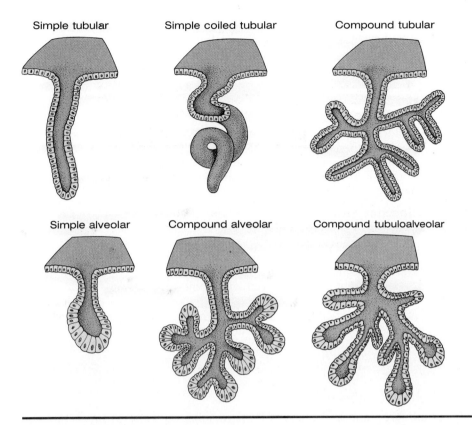

Simple tubular   Simple coiled tubular   Compound tubular

Simple alveolar   Compound alveolar   Compound tubuloalveolar

Figure 3.10
Classification of exocrine glands according to structure.

the mucous membranes of the respiratory passageways and the auditory (eustachian) tubes.

A summary of the classification of epithelial tissues according to cell layers and shape, as well as a typical location for each type of tissue, is given in Figure 3.9.   **F3.9**

Glandular Epithelium

Most glands of the body are composed of epithelial cells that produce a specific secretion (such as sweat, milk, a hormone, or an enzyme) or excrete certain waste products (such as bile pigments). The mucus-secreting **goblet cells** of the respiratory and digestive tracts are examples of *individual cells* that function as glands. Most glands, however, are *multicellular*—that is, they are formed of clusters of cells.

Embryologically, all glands originate from an epithelium. And most glands retain their connection with the epithelium—a connection that serves as a duct through which the secretions of the gland are carried to a particular site. Such glands are called **exocrine glands.** Some glands, however, lose their connection with the epithelium and empty their secretions directly into the blood. These are the **endocrine glands** and their secretions are **hormones.**

The multicellular exocrine glands are classified according to (1) their structure and (2) the manner in which they produce their secretions. Structurally, the ducts of the glands may be *unbranched* or *branched.* Glands whose ducts do not branch are called *simple glands;* glands whose ducts branch repeatedly are called *compound glands.* Simple and compound glands can be further subdivided according to whether their secreting portions are (1) *tubular,* (2) composed of small sacs (*alveoli* or *acini*), or (3) a combination of blindly ending tubules and alveoli (*tubuloalveolar*) (Figure 3.10). Compound tubuloalveolar glands are the most common type of exocrine gland, being present in the pancreas, prostate, salivary glands, and mammary glands.   **F3.10**

When glands are classified according to their *mode of secretion* there are three different types: merocrine, holocrine, and apocrine (Figure 3.11).   **F3.11**

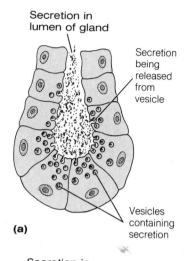

Secretion in lumen of gland

Secretion being released from vesicle

Vesicles containing secretion

(a)

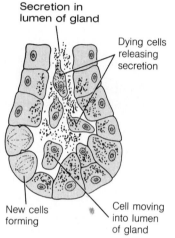

Secretion in lumen of gland

Dying cells releasing secretion

New cells forming

Cell moving into lumen of gland

(b)

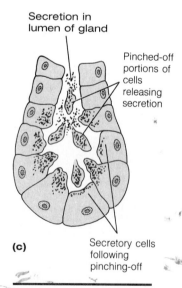

Secretion in lumen of gland

Pinched-off portions of cells releasing secretion

Secretory cells following pinching-off

(c)

Figure 3.11

Classification of glands according to mode of secretion: **(a)** merocrine, **(b)** holocrine, **(c)** apocrine.

1. **Merocrine glands** produce secretions that do not accumulate significantly in the gland cells. Rather, the secretions pass through the cell membrane within membranous vesicles by the process of exocytosis. In merocrine glands there is no destruction of glandular cells during secretion. The pancreas, the salivary glands, and most sweat glands are merocrine glands. In fact, most exocrine glands are of this type.

2. **Holocrine glands** accumulate their secretions in their cells and discharge them only when the cells rupture and die. New cells then form to replace those that have died. About the only example of holocrine glands are the sebaceous (oil) glands of the skin.

3. **Apocrine glands** produce secretions that accumulate toward the outer ends of the gland cells. The secretions are released and a small amount of cytoplasm is lost when these ends pinch off. But rather than dying, as happens in holocrine glands, the cell is only slightly damaged and it repeats the accumulation of secretory products. Some sweat glands are apocrine glands and the mammary glands are generally considered to be apocrine glands, although they are actually mixed glands as some of their secretion is of a merocrine type.

Epithelial Membranes

There are two important types of body membranes—*mucous membranes* and *serous membranes*—that, while not actually epithelial tissues, are composed of an epithelial layer on their free surface and an underlying connective tissue layer. For this reason they are often referred to as *epithelial membranes;* we will consider them here with the epithelial tissues.

MUCOUS MEMBRANES   **Mucous membranes** are moist epithelial membranes forming the linings of the digestive, respiratory, urinary, and reproductive tracts—all of which open to the exterior of the body. Their free surfaces vary in type, but they are usually either stratified squamous epithelium (mouth and esophagus) or simple columnar epithelium (stomach and intestine).

Mucous membranes are absorptive and secretory, making them particularly well suited to function in digestion and respiration. For example, food substances are absorbed into the body through the mucous membranes of the digestive tract. These membranes also secrete **mucin,** a viscous fluid that moistens the free surface of the membranes and lubricates food as it passes through the digestive tract.

Most mucous membranes have a layer of loose connective tissue—the **lamina propria**—deep to the innermost epithelial layer. Beneath the lamina propria there is often a thin layer of smooth muscle called the **muscularis mucosae.** Because mucous membranes are composed of several types of tissues, they are often considered to be simple organs.

SEROUS MEMBRANES   The ventral body cavities, which are not open to the exterior of the body, are lined with **serous membranes.** Serous membranes form both the parietal and the visceral portions of the pleura, the pericardium, and the peritoneum. The serous membranes consist of a thin layer of loose connective tissue covered by a surface layer of simple squamous epithelium called **mesothelium.** Mesothelium is derived from embryonic mesenchyme (mesoderm). The cells of the mesothelium secrete a clear, watery fluid, called **serous fluid,** that keeps the membranes moist. Serous membranes, like mucous membranes, are composed of more than one type of tissue and may therefore be considered simple organs.

CONNECTIVE TISSUES

Connective tissues vary considerably in form as well as function. Some serve as the framework upon which epithelial cells cluster to form organs; others bind various tissues and organs together, supporting them in their proper locations; some contain the media (tissue fluid) through which nutrients and wastes pass while traveling between blood and body cells; others serve as storage sites for excess food materials in the form of fat; and still others form the rigid skeletal framework of the body.

As you have seen, epithelial tissues are formed of closely packed cells that have very little material (intercellular matrix) between adjoining cells. In contrast, connective tissues are characterized by abundant intercellular matrix that surrounds relatively few cells.

Several types of cells are associated with the connective tissues, but fibroblasts and macrophages are the most common. **Fibroblasts** are spindle-shaped cells that form the various fibers characteristic of connective tissues. Fibroblasts that are in a resting or less active phase are called **fibrocytes.** **Macrophages,** which are generally not as abundant as fibroblasts, are active phagocytes. The macrophages may be *fixed* (attached to the connective tissue fibers) or they may remain *free* and be capable of moving throughout the matrix of the connective tissue. Both fixed and free macrophages engulf foreign matter and dead or dying cells, but free macrophages have the advantage of being able to move to the affected area. Furthermore, during an infection, fixed macrophages can detach and become free cells.

The activities of macrophages are so important in protecting the body against invasion by microorganisms that they are often referred to collectively as the **macrophage system,** even though they do not form a discrete system and are distributed widely and rather randomly throughout the body. For instance, macrophages are found in many tissues throughout the body, including the loose connective tissue, lymphatic tissues, mesenteries of the digestive tract, bone marrow, spleen, adrenal gland, and pituitary gland. In some locations, macrophages are given special names—such as *Kupffer cells* that line the blood sinusoids of the liver, the *dust cells* of the lungs, the *histiocytes* of loose connective tissue, and the *microglia* of the central nervous system. Whereas all these cells have a similar function—phagocytosis— macrophages in specific tissues or organs are often selective as to what they ingest. The macrophages of the spleen and the liver, for example, are particularly active in breaking down aging red blood cells. Macrophages are also involved in the activities of the body's immune system. These selective processes, in addition to their phagocytic role, make the cells of the macrophage system a major defense mechanism of the body. Because macrophages develop from a type of blood cell called a monocyte that has a single, unlobed nucleus, they are also referred to as the **mononuclear phagocytic system.**

Intercellular Material

The **intercellular material (matrix)** of connective tissue is formed of *ground substance* and *fibers*. The ground substance is a homogeneous product of the connective tissue cells that it surrounds. It is composed of tissue fluid and proteins joined with polysaccharides (*proteoglycans*). The polysaccharides are in the form of long chains called *glycosaminoglycans* (*GAGs*). Because of the ability of the polysaccharides to interact with water, the ground substance varies in consistency from fluid to a semisolid gel. One long, unbranched polysaccharide chain that is present in most connective tissues is *hyaluronic acid*. In order for dissolved substances to pass between cells and blood capillaries they must diffuse through the ground substance. The fibers, which are also produced by the connective tissue cells (fibroblasts), are found in varying amounts within the ground substance. There are three types of fibers: *collagenous, elastic,* and *reticular.*

×1250

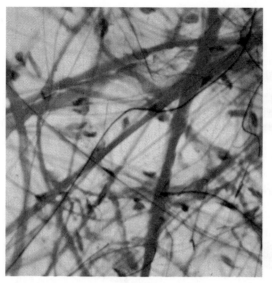

Reticular   Fibroblast   Collagenous
fibers   fibers

Figure 3.12
Typical composition of
loose connective tissue.

×950

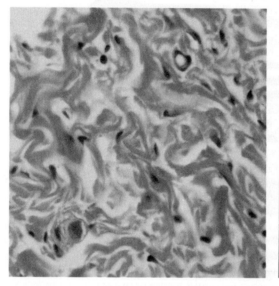

Fibroblasts   Collagenous
                   fibers

Figure 3.13
Dense irregular
connective tissue.

Collagenous Fibers

Collagenous fibers, the most abundant type, appear as wavy bands under the microscope. Each fiber is made up of bundles of smaller fibrils. Collagenous fibers are very strong and inelastic and are composed primarily of the protein *collagen*. Collagenous fibers that are closely packed have a white color and are sometimes referred to as white fibers.

Elastic Fibers

Elastic fibers are long, threadlike branching fibers that often form interwoven networks. Their main protein, called *elastin*, gives the fibers the capacity of

returning to their original lengths after being stretched. They therefore function to give resilience to connective tissue. Collagenous fibers are always found in the same tissue with elastic fibers. The collagenous fibers are capable of only a limited amount of stretch, after which they prevent further stretching. When the tension on the tissue is reduced, the elastic fibers return the connective tissue back to its normal length. Because of this capability, elastic fibers are found in abundance in tissues that routinely undergo stretching, such as the walls of blood vessels and the lungs. Large masses of elastic fibers have a slightly yellow color.

Reticular Fibers

Reticular fibers are short and very thin. They branch freely, forming a tight network called a **reticulum.** These fibers often form a gland's internal framework *(stroma),* to which the epithelial cells that make up the bulk of the gland are attached. Reticular fibers also join connective tissues to other types of tissue. Reticular fibers are inelastic and composed primarily of a type of collagen called *reticulin.* In fact, reticular fibers are believed to be quite similar to collagenous fibers, but are considerably thinner.

Types of Connective Tissue

Connective tissues are classified according to the nature of the ground substance and the types and organization of the fibers in the ground substance.

Loose Connective Tissue

Because the unorganized arrangement of the fibers in **loose connective tissue** leaves many spaces between them, it is also called **areolar connective tissue** *(areolar = space).* Most of the fibers in loose connective tissue are collagenous, but elastic and reticular fibers are also present (Figure 3.12). Loose **F3.12** connective tissue contains several different types of cells, but fibroblasts and macrophages are the most common. It is the most widespread connective tissue of the body, being used to (1) attach the skin to the underlying tissue *(subcutaneous tissue),* (2) fill the spaces between the various organs and thus hold them in place, and (3) surround and support the blood vessels. Because of the large spaces between cells and fibers, loose connective tissue contains a large amount of intercellular fluid (tissue fluid), which enters the spaces from capillaries. Tissue fluid is used to carry nutrients to, and waste products away from, the cells. If excessive fluid accumulates in these spaces, the affected area becomes swollen—a condition called *edema.* The presence of macrophages in loose connective tissue provides the body with a widespread defense against microorganisms.

Dense Irregular Connective Tissue

The dense connective tissues are distinguished by an abundance of collagen fibers, which provide the capacity to resist exceptional degrees of tension. **Dense irregular connective tissue** (Figure 3.13) is essentially a dense areolar **F3.13** tissue that contains all the same elements as loose connective tissue but has fewer cells and more numerous collagenous fibers. The fibers are closely interwoven, forming a compact tissue with fewer spaces. Because the tensions that need to be resisted by these tissues come from all directions, the fibers are oriented randomly. The dermis layer of the skin, the fibrous coverings of cartilage (perichondrium), bone (periosteum), and nerves (perineurium), and the strong fibrous capsules that surround the liver, kidneys, spleen, and some other organs are all formed of dense irregular connective tissue.

Dense Regular Connective Tissue

Dense regular connective tissue (Figure 3.14) is characterized by a predomi- **F3.14** nance of collagenous fibers that are tightly packed in parallel bundles. In this tissue the tensions to be resisted come from a single direction, parallel to the orientation of the fibers. Because of the prevalence of collagenous fibers, this tissue is sometimes referred to as *white fibrous connective tissue.* The only cells present are fibroblasts, which are located between the fiber bundles. The

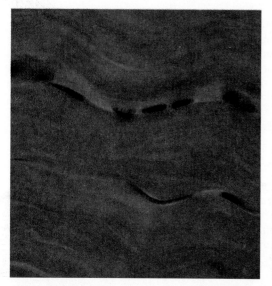

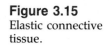

×1250

Collagenous fibers

Fibroblasts

Figure 3.14
Dense regular
connective
tissue.

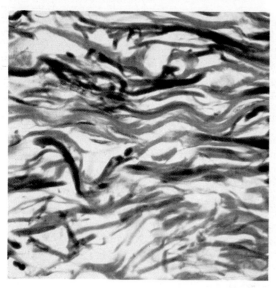

×550

Elastic fibers   Collagenous fibers

Figure 3.15
Elastic connective
tissue.

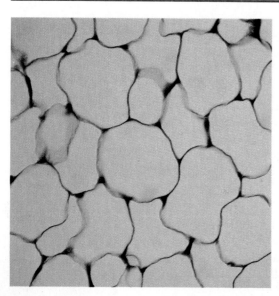

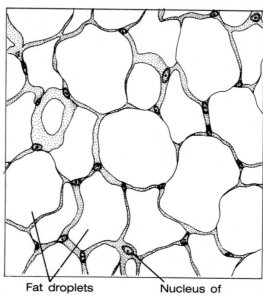

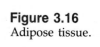

×1250

Fat droplets   Nucleus of
                         fat cell

Figure 3.16
Adipose tissue.

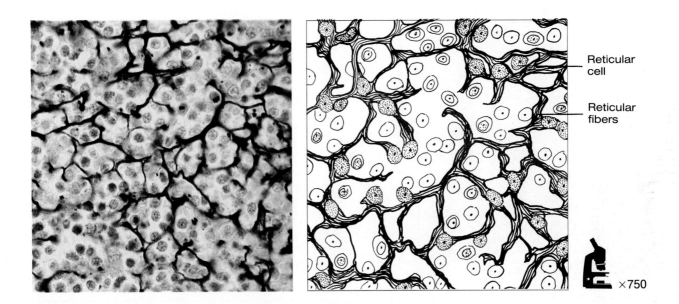

Reticular cell

Reticular fibers

×750

Figure 3.17
Reticular connective tissue.

abundance of fibers gives this tissue great strength. It forms the tendons of muscles, the ligaments of joints (which also contain some elastic fibers), and various fibrous membranes, such as fascia and aponeuroses. **Fascia** surrounds the organs and the muscles; **aponeuroses** are broad sheets that function as thin tendons, attaching muscles to other structures.

Elastic Connective Tissue

In contrast to dense connective tissues, **elastic connective tissue** (Figure 3.15) contains more elastic fibers than collagenous fibers. Although this tissue is quite strong, it allows some stretching—which dense tissues do not. Elastic tissue is found in the walls of arteries, in the trachea and bronchi, and in the vocal cords, as well as in the walls of some hollow organs.

Adipose Tissue

Adipose tissue (Figure 3.16) is essentially composed of fat cells dispersed in loose connective tissue. Each cell contains a large droplet of fat that squeezes and flattens the nucleus and forces the cytoplasm into a thin ring around the cell's periphery. Adipose tissue serves as a storage site for fats, and it pads and protects certain regions of the body.

Reticular Connective Tissue

Reticular connective tissue resembles loose connective tissue in that it contains considerable ground substance located between a network of interlacing fibers (Figure 3.17). However, the fibers in this tissue are primarily reticular fibers, which are thinner than the collagenous fibers that predominate in loose connective tissue.

While reticular fibers are commonly present in other types of connective tissues, the distribution of reticular connective tissue is rather limited. The tissue forms a delicate internal framework (*stroma*) in the liver, spleen, lymph nodes, and other organs. The glandular cells of the organs are anchored to the reticular framework. Reticular connective tissue also binds some muscle cells together and is found in the basal lamina of epithelial tissues.

Cartilage

Cartilage is a specialized fibrous connective tissue that contains numerous collagenous fibers embedded in a firm matrix whose ground substance is formed of various proteoglycans, chondroitin sulfate, and hyaluronic acid. The connective tissues that we have studied up to this point all have fluid or, at most, semisolid matrices. With a firm matrix, as is present in cartilage, the

F3.15

F3.16

F3.17

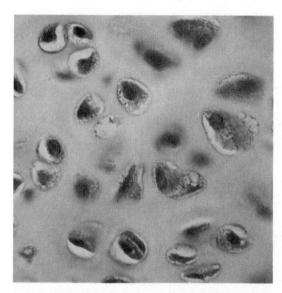

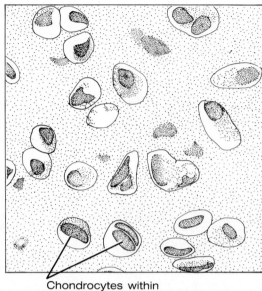

×1250

Figure 3.18
Hyaline cartilage.

Chondrocytes within
lacunae

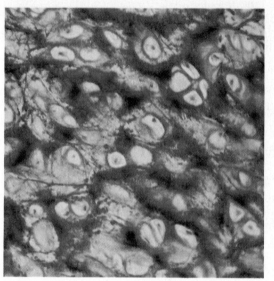

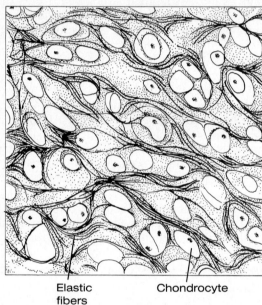

×170

Figure 3.19
Elastic cartilage.

Elastic
fibers

Chondrocyte

connective tissue is able to function as a structural support. At the same time, the presence of fibers in the matrix imparts a certain amount of flexibility to cartilage.

The fibers and matrix of cartilage are formed by cells called **chondroblasts.** Each chondroblast becomes surrounded by fibers and matrix that it produces. As a result, the cartilage-forming cells eventually occupy small spaces called **lacunae.** When cartilage formation is complete, the chondroblast produces only enough matrix to maintain the cartilage. These mature cells are then called **chondrocytes.** The matrix is avascular—that is, it contains no blood vessels. The only blood supply to cartilage is provided by blood vessels found in the inner layer of the **perichondrium.** The perichondrium is a membrane of dense irregular connective tissue that covers the external surfaces of all cartilaginous structures (with the exception of the articular cartilages of joints) and is vitally important in the growth of cartilage. Because there is no direct blood supply to the matrix of cartilage, the nourishment of chondrocytes depends on the diffusion of nutrients through the ma-

×450

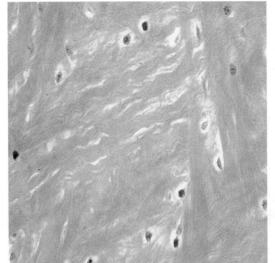

Figure 3.20
Fibrocartilage.

Collagenous
fibers

Chondrocyte

trix from capillaries located in the perichondrium or from the synovial fluid of joint cavities. Similarly, waste materials must diffuse from the chondrocytes to the vascular perichondrium.

Cartilage is especially prevalent in the embryo, but it also forms many adult structures. It is divided into three types according to variations in its fibrous structure: *hyaline, elastic,* and *fibrocartilage.*

HYALINE CARTILAGE Cartilage that contains many closely packed collagenous fibers dispersed throughout the matrix is called **hyaline cartilage** (Figure 3.18). Since the fibers have the same refractive index and staining properties **F3.18** as the matrix, they are not distinguishable by ordinary microscopic examination. Hyaline cartilage is semitransparent and is smooth and firm, but flexible. This type of cartilage, the most abundant in the body, is found primarily in places where strong support is needed but some flexibility is desirable. Hyaline cartilage forms most of the embryonic skeleton, which is gradually replaced by bone. It also makes up the costal cartilages, which attach the ribs to the sternum (breastbone) and allow the thorax to expand during respiration. The cartilage rings of the trachea are also hyaline cartilage, as are the articular cartilages on the ends of bones where two bones meet to form a movable joint. These joint cartilages provide smooth, moist surfaces that permit body movement with a minimum of friction.

ELASTIC CARTILAGE Some body structures must furnish firm but elastic support, and this is the function of **elastic cartilage** (Figure 3.19). Elastic carti- **F3.19** lage contains collagenous fibers like hyaline cartilage, but the fibers are not so closely packed. Moreover, elastic cartilage contains a generous network of elastic fibers, which are stainable and therefore show up under the microscope. This cartilage forms the external ear, the epiglottis, and the auditory tubes.

FIBROCARTILAGE The construction of **fibrocartilage** (Figure 3.20) differs **F3.20** from that of hyaline cartilage in that its collagenous fibers are arranged in thick, parallel bundles that give the matrix a coarse appearance. Actually, fibrocartilage resembles dense regular connective tissue. Because the fibers are not compacted as much as those in hyaline cartilage, fibrocartilage is slightly compressible—which makes it beneficial in regions that support the body weight or that must withstand heavy pressure. It occurs in the intervertebral discs, which provide cushions between the vertebrae; the articular

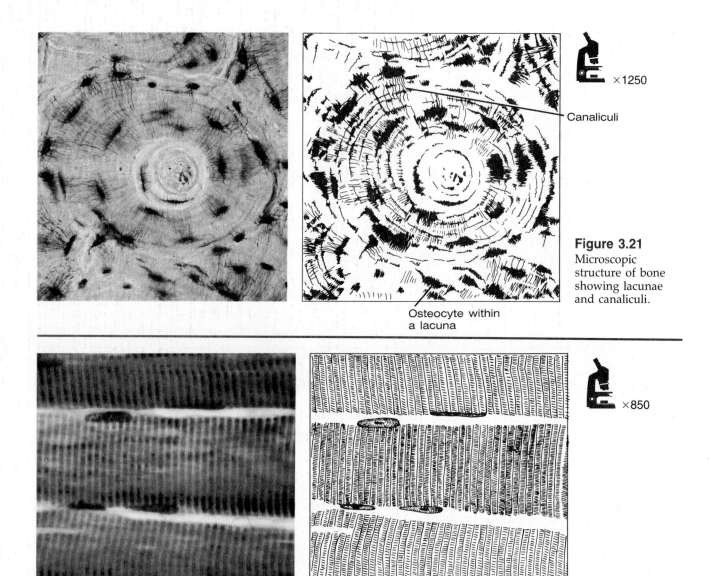

×1250

Canaliculi

Figure 3.21
Microscopic
structure of bone
showing lacunae
and canaliculi.

Osteocyte within
a lacuna

×850

Figure 3.22
Skeletal (striated)
muscle tissue.

Nucleus   Myofibrils

discs, which are located in the knee joint; and the pad of the pubic symphysis, which creates a partially movable joint between the two sides of the pelvis.

Bone

F3.21 Because it has become mineralized, the matrix of **bone** (Figure 3.21) is even harder than the matrix of cartilage. Like cartilage, bone contains collagenous fibers, but its rigidity and strength are greatly increased because of the deposition of inorganic salts among the fibers by cells called **osteoblasts.** There are two other structural differences between bone and cartilage: (1) bone is well supplied by blood vessels throughout its matrix and (2) its **lacunae** are interconnected by very small canals called **canaliculi.** Bone forms the major portion of the adult skeleton. Its structure is considered in greater detail in Chapter 5, where the skeletal system is discussed.

Blood

Because the cells of blood are interspersed in abundant matrix it is often considered to be a type of connective tissue. The matrix of blood, which

×3100

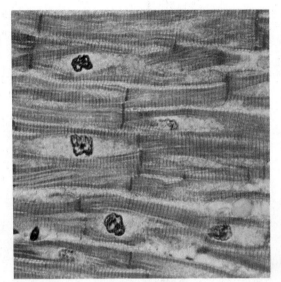

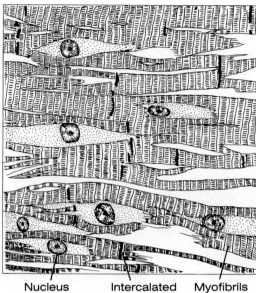

Nucleus   Intercalated   Myofibrils
                      disc

Figure 3.23
Cardiac muscle
tissue.

surrounds the blood cells, is a fluid called **plasma.** Both the cells and the fluid matrix of blood are discussed in Chapter 9.

MUSCLE TISSUE

The long, thin cells of **muscle tissue** are called **fibers.** It is important to realize that muscle cells are living cells and are in no way similar to the fibers of connective tissue. Muscle fibers are highly contractile. There are three structurally different types of muscle tissue: *skeletal, cardiac,* and *smooth.*

Skeletal Muscle

Skeletal muscle (Figure 3.22) is attached to various bones of the skeleton. The **F3.22** cells of skeletal muscle are long and cylindrical—in fact, some skeletal muscle cells are thought to extend the entire length of the muscle. Running longitudinally throughout the skeletal muscle cells are regularly ordered threadlike arrays of proteins called **myofibrils.** Transverse light and dark bands that alternate along the myofibrils give skeletal muscle cells a characteristic *striated* appearance. Each skeletal muscle cell is multinucleate—that is, it has more than one nucleus. These nuclei are located on the periphery of the cell, just inside the cell membrane.

Cardiac Muscle

Cardiac muscle (Figure 3.23) forms the wall of the heart. The cells of cardiac **F3.23** muscle, unlike those of skeletal muscle, form branching networks throughout the tissue. Where adjoining cells meet end to end, their junctions form structures called *intercalated discs* that are visible under the microscope and are unique to cardiac muscle. Cardiac muscle cells contain myofibrils that are arranged in a pattern similar to that of skeletal muscle and give cardiac muscle cells the same *striated* appearance. The nuclei of cardiac muscle cells, in contrast to those of skeletal muscle, are centrally located.

Smooth Muscle

Smooth muscle (Figure 3.24) is so named because its cells do not have the **F3.24** striated appearance of skeletal and cardiac muscle cells. Smooth muscle is also called *visceral muscle* because it is located in the walls of hollow internal structures, such as ducts, blood vessels, and the digestive tract, as well as in nu-

FRONTIERS IN HEALTH

Can Aging Be Delayed?

The oldest living things on earth are trees, which can live thousands of years. The life-span of animals is much shorter. The tortoise holds the record for longevity, living to be 200 years old. Humans are the only other animals that can live to be over 100, and few of us live that long. Disease and old age cause most of us to die in our 70s. Could more of us live to be 100?

The most common measure of longevity is life expectancy at birth—that is, the number of years the average person can expect to live. Of course, not everyone will live that long. Some will die younger; others will die older.

Life expectancy has increased significantly in the past 85 years. In 1900 the average white American female could expect to live only 50 years; today she can expect to live 78.3 years. The prospects for males have increased correspondingly. In 1900 the average white American male could expect to live 47 years; today he can expect to live 70.9 years.

These numbers are deceptive, however. They would seem to indicate that we are living longer—that is, that the human life-span has been extended and the aging process delayed. In reality, though, this is not the case. What has happened to produce these numbers is that there has been a reduction in infant deaths, so more people are living past the first year of life. This makes it appear that the aging problem is being solved, but this is unfortunately not the case.

To illustrate this point, consider an example: suppose ten people are born in a given year on a small island. If five of these people die in the first year and the other five live until they are 70 years old, the average life expectancy on the island is 35 years. Now suppose a doctor moves onto the island and is able to reduce the death rate so that only one child out of every ten dies, while the rest of the people continue to live to be 70. The average life expectancy is now 63 years. In this example, the island's residents have not conquered aging, and life-span has not really increased. What has raised the average life expectancy is that more children live past the dangerous first year of life.

This is what is happening in the United States. The gain in life-span is largely due to the significant drop in infant death that has occurred since the early 1900s (from more than 100 per thousand to about 12 per thousand), thanks to improved hygiene and advances in medical practice.

Increased infant survival is not the only factor producing increased longevity. Another way of producing a longer life is to delay death in later life. Improvements in diagnosis and treatment of heart disease have been one of the most important contributors to extended life.

Although such improvements come closer to actually increasing our life-span, they have done nothing to alter the aging process itself. Most medical scientists agree that although it has become possible to maximize the longevity potential of individual people, there has been no significant increase in life-span—that is, no actual slowing of the aging process.

Two young boys suffering from progeria, a disease that causes premature aging.

The search for ways to delay the aging process has been a frustrating one for medical scientists, because no one actually knows what causes aging. There are a number of theories, none of which is completely satisfactory. Some researchers suggest that aging results from a buildup of toxic substances in the body's tissues. These toxins may be the by-products of enzymatic reactions or irradiation, both natural and human produced. Other medical scientists think that cells are preprogrammed to divide only a certain number of times. When they have divided that many times, they die. Still others suggest that aging is largely the result of gradual physiological deterioration of the immune system.

One of the newest discoveries concerning aging is a protein called stomatin. Stomatin was first detected in cultured fibroblasts that had stopped dividing. Since then it has also been found in other cells that do not divide, such as skeletal muscle cells and certain cells of taste buds. It is thought that stomatin may actually cause a cell to stop dividing. If this is the case, then the gene that controls the production of stomatin may affect cellular aging.

The discovery of stomatin could be a significant step in the control of the aging process. Suppose, for instance, that the stomatin gene can be inactivated. Tissues may then be able to continue to regenerate beyond the normal 70-odd years, and humans, like trees, could conceivably live for many, many years.

Despite extensive research, however, there is no indication that we can slow down, much less prevent, aging. The death rate in the aged and the rate of infant death can be reduced by advances in medicine, but the basic process of cellular aging is still beyond our understanding.

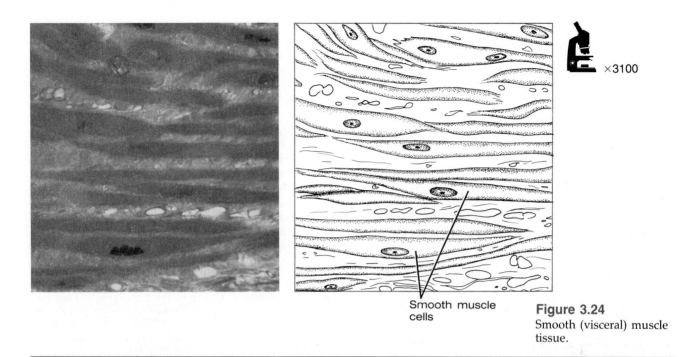

Smooth muscle
cells

Figure 3.24
Smooth (visceral) muscle
tissue.

×3100

merous other locations. Each cell of smooth muscle is shaped like a long spindle, with each end of the cell tapering to a point. A smooth muscle cell contains a single, centrally located nucleus. Muscular tissue is considered further in Chapter 7.

NERVOUS TISSUE

Nervous tissue is composed of **neurons**—highly specialized cells capable of receiving and transmitting impulses very rapidly—plus supportive cells, including **neuroglia** and **Schwann cells.** The structure of the neuron is adaptive to its function. Each neuron consists of a cell body with two or more thin cytoplasmic extensions. Because of the anatomical arrangement of nervous tissue, certain of these cytoplasmic extensions transmit impulses toward the cell body of the neuron while others carry impulses away from the cell body— either to another neuron or to a specific structure. Nervous tissue is studied in greater detail with the nervous system in Chapter 13.

TISSUE REPAIR

In the embryo, the cells of all tissues are capable of dividing by mitosis, thus enabling the tissues to grow and to repair damage. As the body continues to develop following birth, however, the ability of the cells of certain tissues to divide is greatly reduced or lost completely. Thus the capability of postembryonic tissues to undergo growth and repair depends on the tissue involved. Cells of nervous tissue and muscle tissue generally become mitotically inactive once the tissues have completed their development. In contrast, cells of the epithelial tissues—including those of the skin, digestive tract, respiratory tract, urogenital tract, and various glands and organs—remain mitotically active and are thus capable of undergoing repair. Fibroblasts also retain the capacity to divide; therefore, like epithelial tissue, connective tissue is able to undergo repair.

STUDY OUTLINE

EPITHELIAL TISSUES occur as coverings for most free surfaces of the body—internal and external. **pp. 51–60**

SPECIALIZATIONS OF EPITHELIAL CELL SURFACES

MICROVILLI are tiny protoplasmic projections that increase the area of the free surface; not highly motile.

CILIA are motile processes that move rhythmically to propel materials along the epithelial surface.

SPECIALIZATIONS FOR CELL ATTACHMENT

JUNCTIONAL COMPLEXES attach adjacent cells in columnar and some cuboidal epithelia. Each junctional complex consists of a *tight junction (zonula occludens)*, an *intermediate junction (zonula adherens)*, and a *desmosome (macula adherens)*.

GAP JUNCTIONS hold adjacent cells together in muscle and nerve tissue as well as in epithelia.

CLASSIFICATION OF EPITHELIA epithelia can be classified according to (1) the number and arrangement of cell layers and (2) the shape of cells on the free surface.

ACCORDING TO CELL LAYERS

Simple epithelium has a single layer of cells, all contacting basal lamina.

Stratified epithelium has two or more layers, only deepest layer contacting basal lamina.

Pseudostratified epithelium appears multilayered, but is actually only a single layer with all cells touching basal lamina.

ACCORDING TO CELL SHAPE

Squamous epithelium cells are flat and thin.

Cuboidal epithelium cells are about as tall as they are wide; appear almost square in vertical section.

Columnar epithelium cells are taller than they are wide; appear rectangular in vertical section.

GENERAL CLASSIFICATION is based both on the shape of cells that form free surface and on the number of layers.

Simple squamous epithelium a single layer of squamous cells; found in regions where diffusion and filtration occur; includes endothelium that lines the blood vessels and mesothelium that lines the body cavities.

Stratified squamous epithelium consists of multiple layers; forms protective layer on body surface (epidermis) and other sites of abrasion.

Transitional epithelium specialized stratified tissue lining urinary bladder and certain other hollow organs.

Simple cuboidal epithelium generally a single layer of cuboidal cells; occurs in many glands, such as the salivary glands.

Simple columnar epithelium single layer of columnar cells; capable of performing complex reactions, such as secretion and absorption; lines digestive tube.

Stratified columnar epithelium truly stratified columnar tissue; found on epiglottis and parts of pharynx, anus, and urethra.

Pseudostratified columnar epithelium columnar tissue in which all cells contact basal lamina but not all reach the free surface; found in large ducts of some glands, such as parotid; is often ciliated, as in respiratory tract.

GLANDULAR EPITHELIUM consists of cells that secrete various substances or excrete wastes such as bile pigments.

Gland types
1. *Merocrine glands:* produce secretions that do not accumulate in the gland cell; secrete with no glandular destruction; most common type.
2. *Holocrine glands:* accumulate secretions in their cells, discharging the secretions only when cells rupture and die.
3. *Apocrine glands:* accumulate secretions in outer ends of gland cells; released when end of cell pinches off.

EPITHELIAL MEMBRANES composed of a surface layer of epithelial cells underlain with loose connective tissue.

Mucous membranes absorptive and secretory tissue that line digestive, respiratory, urinary, and reproductive tracts; surface layer of stratified squamous or simple columnar epithelium.

Serous membranes line central body cavities; surface layer of simple squamous epithelium called mesothelium.

CONNECTIVE TISSUES vary considerably in form and function; all have abundant intercellular matrix; serve as internal framework of organs; bind tissues and organs together for support; some provide media through which nutrients and wastes pass between blood and body cells; serve as food storage sites; form rigid skeletal framework of body. **pp. 61–69**

INTERCELLULAR MATERIAL

GROUND SUBSTANCE varies from fluid to gel; produced by connective tissue cells.

MACROPHAGES active phagocytes; found throughout body; comprise the macrophage system.

FIBROBLASTS form various fibers characteristic of connective tissues.

Collagenous fibers inelastic fibers composed of bundles of strong fibrils.

Elastic fibers long, threadlike, branching fibers that often form interwoven networks.

Reticular fibers short, thin fibers that branch freely, forming tight, inelastic networks.

TYPES OF CONNECTIVE TISSUES

LOOSE CONNECTIVE TISSUE has unorganized fiber arrangement; is most widespread type. Used to attach skin to underlying tissue; fill spaces between organs, holding them in place; surround and support blood vessels.

DENSE IRREGULAR CONNECTIVE TISSUE dense areolar tissue but contains more randomly oriented collagenous fibers than loose tissue does; forms dermis layer of skin.

DENSE REGULAR CONNECTIVE TISSUE very strong because of numerous parallel collagenous fibers in tightly packed bundles; forms tendons of muscles, ligaments of joints.

ELASTIC CONNECTIVE TISSUE allows some stretching; found for instance, in walls of arteries and trachea.

ADIPOSE TISSUE consists of fat cells dispersed in loose connective tissue; serves as storage site for fats and also pads certain body regions.

RETICULAR CONNECTIVE TISSUE consists of interlacing reticular fibers; serves as the framework for gland cells in many organs.

CARTILAGE has firm matrix; functions as structural support but is somewhat flexible; divided into three types:

Hyaline cartilage most abundant type; contains tightly packed collagenous fibers; provides strong support, as in costal cartilage and rings of trachea.

Elastic cartilage contains elastic fibers; furnishes firm but elastic support, as in outer ear and epiglottis.

Fibrocartilage tough and slightly compressible; found in intervertebral discs of vertebral column and articular discs of knee joint.

BONE has great strength and rigidity provided by inorganic salts; osteoblasts located in lacunae; forms major part of adult skeleton.

BLOOD the matrix that surrounds blood cells is fluid plasma.

MUSCLE TISSUE living cells of muscle tissue are called fibers. **pp. 69–71**

SKELETAL MUSCLE attached to skeleton and moves it.

CARDIAC MUSCLE forms walls of heart.

SMOOTH MUSCLE located in walls of hollow internal structures (organs, ducts, and blood vessels).

NERVOUS TISSUE composed of neurons—highly specialized cells capable of receiving and transmitting impulses; also contains supportive cells called neuroglia and Schwann cells. **p. 71**

TISSUE REPAIR embryonic cells of all tissues are capable of mitosis. Some postembryonic tissues are also able to grow and repair damage. With continued development, cells of certain tissues are rendered incapable of mitosis, or mitotic activity is greatly reduced. **p. 71**

SELF-QUIZ

1. In which one of the following are the cell membranes of adjacent cells closest together? (a) desmosomes; (b) zonula occludens; (c) zonula adherens.

2. Which cell junction seems to allow for the passage of ions and small molecules? (a) gap junction; (b) zonula occludens; (c) zonula adherens.

3. Epithelial tissues are generally classified on the basis of the number and arrangement of cell layers within the tissue, the shape of the cells on the free surface, and/or the location or function of certain epithelia. True or false?

4. Epithelia with two or more layers of cells, with only the deepest layer in contact with the basal lamina, are: (a) stratified epithelia; (b) simple epithelia; (c) pseudostratified epithelia.

5. Specialized tissue that allows for the expansion of an organ with only minimal resistance from the tissue is composed of: (a) simple cuboidal epithelium; (b) transitional epithelium; (c) stratified squamous epithelium.

6. Mucous membranes are particularly well suited to function in digestion and respiration. True or false?

7. Match the following terms associated with epithelial cells with the appropriate description.

Squamous cells	(a)	Cells taller than they are wide.
Stratified epithelium	(b)	A single layer of squamous cells.
Simple columnar	(c)	Cells that are flat and thin.
Columnar cells	(d)	Cells about as tall as they are wide.
Cuboidal cells	(e)	A single layer of epithelial cells all in contact with the basal lamina.
Simple squamous epithelium	(f)	Epithelium with two or more layers of cells, only the deepest layer in contact with the basal lamina.

8. Which one of these epithelial tissue types has ciliated cells that line the mucous membranes of the respiratory passageways? (a) simple columnar; (b) stratified columnar; (c) pseudostratified columnar.

9. Match the following items with the appropriate description.

   Mucous
     membranes
   Lamina propria
   Serous
     membranes
   Mucin
   Mesothelium
   Endothelium
   Goblet cells

   (a) Tissue lining the wall of the heart.
   (b) Tissue whose cells secrete serous fluid.
   (c) Loose connective tissue that is part of the mucous membranes.
   (d) Absorptive and secretory tissue lining the respiratory tract.
   (e) Function as glands and secrete mucus.
   (f) Tissue lining the ventral body cavity.
   (g) Viscous fluid that lubricates mucous membranes.

10. These glands accumulate their secretions and release them only when the individual cells that store the secretions rupture and die: (a) merocrine; (b) holocrine; (c) apocrine.

11. Match the following terms associated with connective tissues with the appropriate description.

    Fibroblasts
    Macrophages
    Ground
      substance
    Fibrocytes
    Fibers
    Collagenous
    Elastic
    Reticular

    (a) Fiber cells in a resting phase.
    (b) Fibers that form the stroma of glands.
    (c) Structures that occur in the ground substance.
    (d) Homogeneous product of the connective tissue cells that it surrounds.
    (e) Long, threadlike, branching fibers yellow in color.
    (f) Active phagocytes of loose connective tissue.
    (g) Cells that form various fibers of connective tissues.
    (h) Most common type of fiber; composed of bundles of strong fibrils.

12. Loose connective tissue is the most common type and forms the dermis layer of the skin. True or false?

13. This connective tissue forms the tendons of muscles and the ligaments of joints: (a) dense irregular; (b) areolar; (c) dense regular.

14. This connective tissue contains numerous collagenous fibers embedded in a firm matrix: (a) cartilage; (b) elastic; (c) adipose.

15. The original fibers and matrix of cartilage are formed by: (a) chondrocytes; (b) osteoblasts; (c) chondroblasts.

16. The cartilage that forms the intervertebral discs is termed: (a) hyalin; (b) elastic; (c) fibrocartilage.

17. Although all the cells of the macrophage system have a similar function (phagocytosis), macrophages in specific tissues or organs are often selective as to what they ingest. True or false?

18. Intercalated discs are found in cardiac, skeletal, and smooth muscle tissues alike. True or false?

19. With continued development, the ability of the cells of certain tissues to divide is greatly reduced or is completely lost. True or false?

20. Which one of the following contains cells that remain mitotically active and thus capable of undergoing repair during a human's life span? (a) muscle tissue; (b) nervous tissue; (c) epithelial tissue.

LEARNING OBJECTIVES

After completing this chapter, you should be able to:

- Name the four body structures that make up the integumentary system.

- Name the two layers that form the skin, and describe the composition of each.

- Describe the process by which an epithelial cell becomes cornified.

- Describe the factors responsible for skin color.

- Name and describe the two layers that compose the dermis.

- Distinguish between eccrine sweat glands and apocrine sweat glands, and cite one example of the latter.

- List five functions of the integument.

- Describe the rule of nines.

CHAPTER CONTENTS

The Integumentary System

4

In Chapter 3 we discussed the organization of individual cells into tissues and noted that when two or more tissues join together, as in the formation of serous membranes and mucous membranes, an organ is formed. This chapter describes another combination of tissues into a simple organ—the **skin.** Although the skin is not often viewed as an organ, it is, in fact, one of the larger organs of the body in terms of surface area and weight. The skin and its accessory structures—hair, nails, and glands—comprise the **integumentary system.**

The skin forms the entire external covering of the body. It is continuous with, but differs structurally from, the mucous membranes lining the respiratory, digestive, and urogenital systems at their external openings (mouth, nose, anus, urethra, and vagina). The skin is composed of two main layers: (1) a surface layer of closely packed epithelial cells, the *epidermis,* and (2) a deeper layer of dense irregular connective tissue, the *dermis.* The dermis is connected to the underlying fascia of the muscles by a layer of loose connective tissue called the *hypodermis.* In many areas, fat is deposited in the loose connective tissue, thus forming adipose tissue. The hypodermis connects the skin and the underlying fascia of the muscles loosely, which allows muscles to contract without pulling on the skin. In some areas, where muscles do not lie beneath the skin, there is only a small amount of hypodermis and the integument is more tightly attached. On the shins, for example, the skin is connected directly to the membrane (periosteum) that covers the bone.

EPIDERMIS

The **epidermis** (Figure 4.1) develops from the single layer of surface ectoderm F4.1 of the embryo. By the time of birth, it consists of several layers of squamous cells that form a stratified squamous epithelium. The cells of the epidermis only live for about one month; therefore, the epidermis is constantly regenerating itself. The epidermis is generally quite thin—less than 0.12 mm over most of the body—but it is considerably thicker in areas subjected to constant pressure or friction, such as the soles of the feet and the palms of the hands. Continued pressure at one location causes the epidermis to thicken into calluses and corns.

The most common cell-type found in the epidermis is the **keratinocyte.** As the keratinocytes, which are connected to one another by desmosomes, are pushed toward the surface of the skin by the production of new cells in the deeper layers of the epidermis they begin to produce a fibrous protein called *keratin.* Keratin accumulates within the cytoplasm, and by the time the keratinocytes have reached the free surface of the skin they have died and their cytoplasm has been almost completely replaced by keratin. Thus, the superficial layers of the epidermis serve as a thin protective shield composed of keratinized cells.

Epidermal Layers

Where the epidermis is thick, it is possible to identify five layers, or **strata** of keratinocytes. The innermost layer is the *stratum basale.* This layer is overlaid

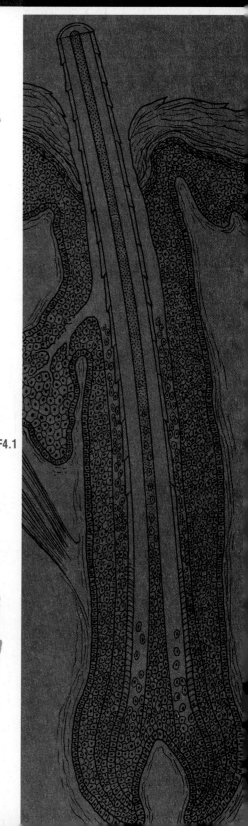

77

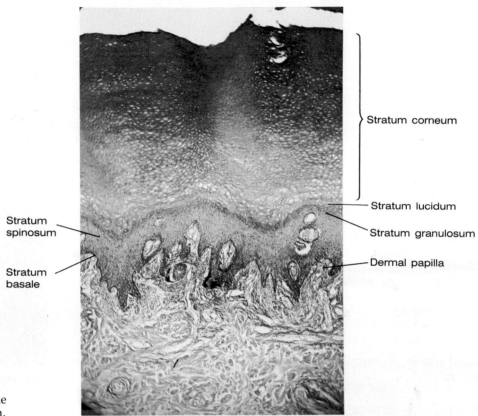

Figure 4.1
Photomicrograph of the epidermis of thick skin.

by the *stratum spinosum,* the *stratum granulosum,* the *stratum lucidum,* and the *stratum corneum,* in that order. In regions where the epidermis is thin, the stratum lucidum is often absent.

Stratum Basale

F4.1, F4.2b

The **stratum basale** is the deepest layer of the epidermis. As the name indicates, it lies directly on the dermis (Figure 4.1, Figure 4.2*b*). It is within this layer that mitosis occurs, furnishing cells to replace those that are lost from the more superficial strata of the epidermis. For this reason, this layer is also called the *stratum germinativum.* The cells of the stratum basale are attached to one another by desmosomes and contain bundles of microfibrils called tonofibrils in their cytoplasm. The cells in the deepest layer of the stratum basale, which contacts the dermis, are columnar cells. Most mitoses take place in this deep layer. Specialized cells called *melanocytes* are located in the stratum basale. The melanocytes produce a pigment called melanin, which provides protection against ultraviolet radiation.

Stratum Spinosum

Superficial to the basal layer, the cells become somewhat flattened and polyhedral in shape. Under the microscope there appear to be cytoplasmic extensions that connect adjacent cells. Because of these extensions, the layers of cells just superficial to the stratum basale are referred to as the **stratum spinosum** (*spine* = projection).

Stratum Granulosum

F4.1, F4.2b

The cells of the **stratum granulosum** (*granulosum* = granular) are flattened and are arranged in about three layers just superficial to the stratum spinosum (Figure 4.1, Figure 4.2*b*). This stratum derives its name from the presence of granules of **keratohyalin** in the cytoplasm of its cells. As the granules increase in size, the nucleus disintegrates, with the result that the outermost cells of the stratum granulosum are dead.

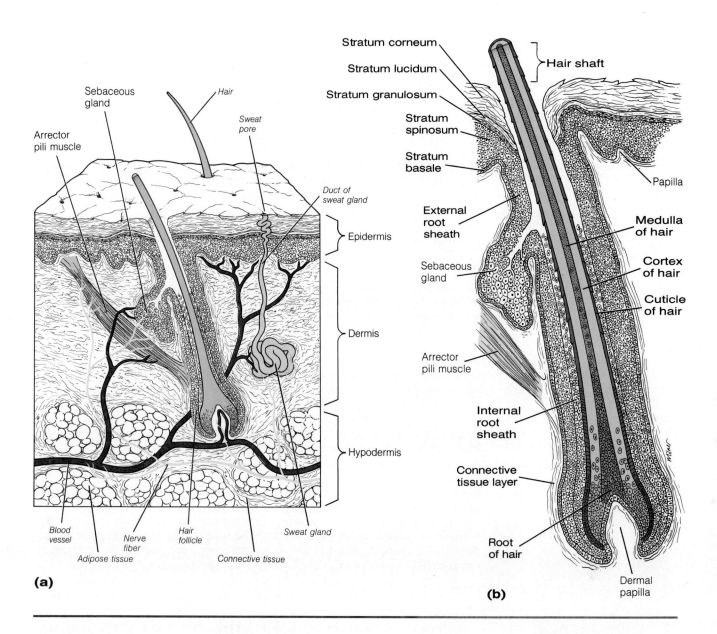

(a)

(b)

Figure 4.2
(a) Structure of the epidermis and dermis layers of the skin and the hypodermis. **(b)** Vertical section through a hair follicle showing the structure of a hair and the relationship of a sebaceous gland and arrector pili muscle to the follicle.

Stratum Lucidum

The **stratum lucidum** (*lucid* = clear) is a clear band superficial to the stratum granulosum (Figure 4.1, Figure 4.2*b*). It consists of several layers of flattened, closely packed cells, most of which have only indistinct outer boundaries and have lost all their cytoplasmic inclusions except for keratin fibrils and some droplets of a substance called **eleidin.** Eleidin is transformed into keratin as the cells of the stratum lucidum become part of the outer stratum corneum. The stratum lucidum is most prominent in areas of thick skin and is absent in some locations.

F4.1, F4.2b

Stratum Corneum

The **stratum corneum** (*cornu* = horn) is the most superficial layer of the epidermis. It is formed of varying numbers of layers of flat, closely packed, dead cells (Figure 4.1, Figure 4.2*b*). Since their cytoplasm has been largely replaced by keratin, these dead cells are referred to as being *cornified,* or *horny.* The cornified cells form a covering over the entire body surface that not only protects the body against invasion by substances in the external environment but also helps to restrict the loss of body water. The cells of the stratum corneum are held together by modified desmosomes that include dense extracellular material. The outermost layers of the stratum corneum are constantly

F4.1, F4.2b

being lost as the result of abrasion—for example, by friction with clothing. Nevertheless, the lost cells are constantly being replaced by cells from the deeper layers of the epidermis. When subjected to repeated trauma the stratum corneum becomes thickened, forming calluses in severe instances.

Nourishment of the Epidermis

As is typical of all epithelia, there are no blood vessels in the epidermis, although the underlying dermis is well vascularized. As a result, the only method by which cells of the epidermis can obtain nourishment is by way of diffusion from the capillary beds of the dermis. This method is sufficient for those cells closest to the dermis, but as cells divide and some are forced toward the body surface—and thus farther from the source of nourishment—they die. Their cytoplasm gradually becomes replaced with keratin, thus forming the structures typical of the outer layers of the epidermis.

Skin Color

Skin color is determined primarily by the presence and distribution of a dark pigment called **melanin.** Melanin is produced by cells called **melanocytes,** which migrate into the epidermis and transfer the pigment to cells of the stratum basale. There are no great differences in the number of melanocytes found in the skin of the various human races. Differences in skin color are due primarily to the amount of melanin produced by the cells and its distribution. Dark-skinned people have appreciable amounts of melanin in all layers of the epidermis. In light-skinned races, relatively little melanin is distributed among the layers of the epidermis, except in heavily pigmented areas such as the nipples of the breasts. The presence of the yellow pigment **carotene** in the strata of the epidermis, in combination with melanin, produces the yellowish hue that is typical of Oriental people. Skin color is also influenced by a reddish hue that results from the blood vessels of the dermis being visible through the epidermis. Whereas the amount of carotene in the skin is relatively constant for each person, a change in the amount of blood in the capillaries of the dermis—or in the amount of oxygen carried in the blood of these capillaries—can cause intense temporary changes in skin color. For example, *blushing* is caused by an expansion of the capillary bed, whereas the bluish skin that is characteristic of *cyanosis* results from a decrease in the amount of oxygenated hemoglobin in the blood in the capillaries.

DERMIS

F4.2a Lying deep to the stratum basale is a layer of dense irregular connective tissue called the **dermis** (Figure 4.2*a*). In contrast to the epidermis, the dermis develops from the mesoderm of the embryo, as do the muscles and skeleton. The dermis contains some elastic and reticular fibers, as well as many collagenous fibers, and is well supplied with blood vessels, lymph vessels, and nerves. It also contains specialized glands and sense organs. The thickness of the dermis varies in different locations, but it averages about 2 mm. It is composed of two indistinctly separated layers: the *papillary layer* and the *reticular layer*.

Papillary Layer

The outer **papillary layer** fits closely against the basal layer of the stratum basale. This layer is so named because it has many **dermal papillae** (projections) that protrude into the epidermal region. On the palms and the soles, these papillae are in the form of curving parallel ridges that cause the overlying epidermis to form the characteristic fingerprint and footprint patterns. Many papillae contain capillary loops; others contain specialized sensory receptors that react to external stimuli such as temperature and pressure changes. These receptors are described in Chapter 13.

FRONTIERS IN HEALTH

New Hope for Burn Victims

When 6-year-old Glen and his 5-year-old brother Jamie sneaked into an empty house near their home one day they had no idea that their mischief—and misfortune—would make medical history. Joined by a friend, the boys entered the house, where they found cans of paint. They pried open the lids and began splattering paint on the walls and floor and, inadvertently, on themselves. When the fun was over, the boys began to clean themselves with a solvent that was stored with the cans of paint. For some unknown reason, one of the boys struck a match.

The room instantly exploded in fire. The boys raced out of the house in flames. Within a few days of the accident, the friend died. Jamie and Glen, both severely burned over 80% of their bodies, were rushed to Boston. There, plastic surgeon Dr. G. Gregory Gallico III began the long task of stabilizing the boys' condition and replacing the skin that had been burned.

Because of the severity of the burns, Dr. Gallico tried a new technique developed by Dr. Howard Green of the Harvard Medical School. He took postage-stamp-sized pieces of skin from the boys' armpits and groins—about the only skin left intact on their bodies—diced it, and used an enzyme to separate the epidermal cells. The cells were then cultured in special flasks. After several weeks, the skin cells had multiplied thousands of times.

The cultured epidermal cells were next spread over thin gauze pads the size of playing cards and sewn into place on the boys' bodies. In 4 weeks the patches had grown together and had developed the full thickness of normal epidermis. Below the epidermis a thin dermal layer had developed, consisting only of blood vessels, fibroblasts, and collagen fibers. Elastic fibers, which give normal skin its flexibility, were lacking.

Eventually more than 50% of the boys' bodies was covered by new skin. The remaining burned areas either healed on their own or received skin transplants from less severely burned areas that had healed by themselves.

This new epidermal culture and graft technique could help save the lives of 10,000 or more of the 100,000 people who are hospitalized in the United States each year with severe burns. Still, there are some drawbacks to the procedure. While new skin is being grown, burned areas are covered with skin from cadavers, and gauze and antiseptic ointments are applied to help reduce fluid loss and prevent infection. These substances all have potential problems associated with their use.

Medical researchers have therefore sought new ways to speed the regrowth of skin. One of the most promising methods is the use of artificial skin. Researchers at the Massachusetts Institute of Technology have developed an artificial skin that can be applied to newly burned areas to help reduce fluid loss, prevent infection, and promote faster recovery.

The artificial skin consists of two layers that resemble the body's natural epidermal and dermal coverings. The artificial dermis is a spongy layer made of collagen fibers extracted from cowhide and another substance

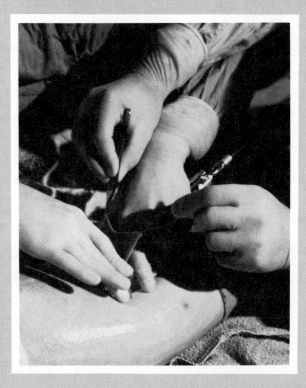

Dermatone used to harvest skin for split-thickness autograft at the Brooke Burn Center.

extracted from the cartilaginous skeleton of sharks. The artificial epidermal layer is made of plastic.

In severely burned patients, physicians clean away the burned flesh down to the muscle fascia. The artificial skin is then sewn into place. Within a short time fibroblasts and blood vessels invade the artificial dermis, and over a period of several months, new collagen fibers produced by the patient replace the fibers of the artificial skin.

The plastic epidermis is peeled off the dermis within a few weeks, after the dermis has revascularized and shown signs of regrowth. A new epidermis is then reconstructed, using some of the patient's own epidermal cells. Grafted in thin sheets onto the vascularized dermis, the epidermal cells gradually replace the plastic layer.

Artificial skin has been used on patients with moderate burns covering up to 95% of their body surface. In severely burned patients, physicians have successfully used artificial skin to cover 60% of the body.

Artificial skin is relatively easy to produce and can be sterilized and stored at room temperature. Therefore, it can be available for immediate grafting. So far, medical researchers have found no evidence of rejection and virtually no infection associated with the use of artificial skin. Moreover, the new skin grows in without the severe scarring that accompanies more conventional burn treatment.

Reticular Layer

The deeper **reticular layer** of the dermis consists of dense bundles of collagenous fibers that run in various directions, thus forming a *reticulum*. The fibers are continuous with the fibers of the hypodermis. When properly treated, this layer in the dermis of cows becomes leather.

HYPODERMIS

F4.2a The **hypodermis** (*hypo* = beneath) is not part of the skin, but it is important because it attaches the skin to the underlying structures (Figure 4.2*a*). This tissue is also referred to as *subcutaneous tissue* or *superficial fascia*. As we have noted, the hypodermis is composed of loose connective tissue, often having fat cells deposited among its fibers. In some regions, such as over the abdomen and the buttocks, the accumulation of fat in subcutaneous tissue can become quite extensive. The hypodermis is well supplied by blood vessels and nerve endings.

GLANDS OF THE SKIN

Two types of glands have a widespread distribution in the skin: sweat glands and sebaceous glands. In addition, the ceruminous (wax) glands of the external ear canal, the ciliary and meibomian glands of the upper eyelids, and the mammary glands are specialized skin glands. The skin glands begin their development as solid downgrowths from the embryonic ectoderm. The downgrowing cords become hollow tubes as they continue to grow and extend into the dermis, forming the skin glands and their associated ducts.

Sweat Glands

F4.2a The **sweat glands,** which are also called **sudoriferous glands** (*sudor* = sweat), are distributed over most of the body surface (Figure 4.2*a*). In only a few places, such as the lips, the nipples, and portions of the skin of the genital organs, are they absent. The typical sweat glands—the *eccrine sweat glands*—are merocrine glands, each in the form of a simple tubule that becomes coiled within the dermis. Stimulation of the sympathetic nerves to these glands causes them to secrete a watery solution of sodium chloride, with traces of urea, sulfates, and phosphates. The amount of sweat secreted depends on such factors as the environmental temperature and humidity, the amount of muscular activity, and various conditions that cause stress. Sweat glands that are located in the axilla, around the anus, on the scrotum, and on the labia majora of the female external genitalia are unusually large and extend into the subcutaneous tissue. Glands in these locations often empty into a hair follicle rather than directly onto the surface of the skin. These large glands are *apocrine sweat glands*—that is, part of the cytoplasm of the secreting cells is included in the secretion, which is thicker and more complex than true sweat. In the female these glands periodically become enlarged and hyperactive in conjunction with the menstrual cycle. The **ceruminous glands,** which produce "wax" **(cerumen)** in the ear canal, are also apocrine glands that are considered to be modified sweat glands.

Sebaceous Glands

F4.2a Most **sebaceous glands** develop from, and empty their secretions into, hair follicles (Figure 4.2*a*). Their secretion **(sebum)** is an oily substance that is rich in lipids. It travels along the shaft of the hair to the surface of the skin. Sebum not only oils the skin and the hair, preventing them from drying, but also contains substances that are toxic to certain bacteria. The sebaceous glands, which are known to be stimulated by the presence of sex hormones (mainly testosterone), are particularly active during adolescence. If their secretion accumulates within the duct of the gland, it forms a white pimple. This blocked

Internal
root sheath

External
root sheath

Connective
tissue layer

Blood
vessel

Medulla

Cortex

Cuticle

Figure 4.3
Scanning electron
micrograph of a hair
within a hair follicle
(×3845). (From *Tissues
and Organs: A Text-Atlas
of Scanning Electron
Microscopy* by Richard G.
Kessel and Randy H.
Kardon. W. H. Freeman
and Company. Copyright
© 1979.)

sebum may become oxidized, darken, and form a "blackhead." Most hairless
regions of the body, such as the palms of the hands and the soles of the feet,
lack sebaceous glands. However, some areas that lack hair, such as the lips,
the glans penis, and the labia minora, do have sebaceous glands. In these
regions, the glands empty their secretions directly onto the surface of the
epidermis.

Structurally, the typical sebaceous glands are of the simple alveolar type,
although some are compound alveolar (for instance, the *meibomian glands* of
the upper eyelids). Functionally, all sebaceous glands are holocrine glands.

HAIR

Although **hair** is most obvious on the head and in the axillary and pubic
regions, it is also present—though much less conspicuously—over most of
the body. The only hairless skin is on the lips, the palms of the hands, the
soles of the feet, the nipples, and parts of the external genitalia. Hair grows as
the result of the mitotic activity of epidermal cells at the bottom of a **hair
follicle** (Figure 4.2*b*). The follicles extend from the epidermis into the dermis. **F4.2b**
The outermost layer of the follicle, the **external root sheath,** is a downgrowth
of the epidermis. From the bottom of the follicle up to the level of the sebaceous glands the follicles are lined by the **internal root sheath,** which consists
of several layers of keratinized cells. Covering the follicle is an outer layer of
connective tissue that develops from the dermis. A portion of the dermis
protrudes into the bottom of each follicle, forming a **papilla.** Papillae contain
blood capillaries, which nourish the overlying follicle cells and permit them to
undergo repeated mitosis. Emptying into each hair follicle is one or more
sebaceous glands, whose secretions help to soften the hair.

Each hair is essentially a column of keratinized cells. The mitotically active
cells that cover the papilla are called the **matrix.** The part of the hair that is
below the surface of the skin, within the hair follicle, is the **root.** The **shaft** of
the hair develops from the cells of the matrix; the free end of the shaft extends
beyond the surface of the skin. The **medulla,** the central core of the hair shaft,
consists of loosely connected horny cells with air spaces between them (Figure 4.3). The **cortex,** which surrounds the medulla, is formed of tightly com- **F4.3**
pressed keratinized cells. Outside the cortex is a **cuticle** of very hard keratin-

ized cells. Straight hairs are cylindrical or oval; curly hairs are somewhat flattened.

The color of hair is determined primarily by the pigment melanin. Melanin, which ranges from black to brown, to yellow, is formed by melanocytes in the follicle and becomes located in the cortex and medulla of each hair. As people become older their hair tends to gradually become gray. This process is due to a decrease in the amount of pigments present, possibly as a result of a decline in the level of a specific enzyme that is necessary for the production of melanin. In the complete absence of pigments the hair appears white.

Hair follicles exhibit cyclic activity, having active periods alternating with periods of inactivity. During the active periods cells in the matrix of a follicle undergo mitosis, pushing the older cells upward and causing the hair to elongate. The cells die and become keratinized as they are pushed farther from the nourishment provided by the blood vessels in the papillae. The keratinized cells are incorporated into the hair.

During inactive periods, when the matrix cells are not undergoing mitosis, the root of a hair becomes detached from the matrix and the hair gradually moves up the follicle. The detached hair may be pulled from the follicle by brushing or combing, or it may remain within the follicle until the next active period, when the new hair produced by matrix cells pushes it out.

Hair follicles in different parts of the body follow different patterns of cyclic activity. For instance, in the scalp individual follicles can remain active and cause their hairs to elongate continuously for several years before becoming inactive for a period of months. In other regions of the body the follicles may be active for only a few months before entering an inactive phase. Cutting or shaving hair does not affect the cyclic activity of the follicles, and therefore has no effect on the growth of hair.

Because of the repeated formation of new hairs during the active periods, normal hair loss does not generally lead to baldness. Baldness is a genetic trait that requires the presence of the male hormones—the androgens—for the hereditary tendency to become effective. For that reason, baldness is more common in males than in females, who may have inherited the trait but lack the androgens necessary to activate it.

The hair follicles are generally at an oblique angle to the surface of the skin, as are the hairs themselves. Running diagonally from the connective-tissue covering of each follicle to the papillary layer of the dermis is a smooth **F4.2a** muscle: the **arrector pili** (Figure 4.2a). Contraction of this muscle pulls the follicle and causes the hair to "stand up"—that is, to be perpendicular to the skin surface—and causes the skin to bulge in front of the follicle, producing the "goose pimples" that form in response to cold or to frightening situations. In animals whose bodies are heavily covered with hairs, the erection of hairs traps air between them and the body surface, thus producing an insulating effect and reducing the loss of body heat. This response is probably of little importance in humans, on whom body hair is generally quite sparse.

NAILS

On the dorsal surfaces of the distal phalanges of the fingers and toes, the outer two epidermal layers—the strata corneum and lucidum—are heavily **F4.4** cornified, forming the nails (Figure 4.4). The **nail bed,** upon which the nail rests, is formed by the stratum basale and the stratum spinosum. These strata are thickened under the proximal end of the nail, forming a half-moon shaped whitish area called the **lunula** (*luna* = moon) that is visible through the nail. The region of thickened strata is called the **nail matrix.** It is within the nail matrix that mitosis occurs, pushing forward the previously formed cells that have cornified and thus causing the nail to grow. At the proximal end of the nail a narrow fold of epidermis extends onto the free surface, forming the **eponychium** (cuticle). Below the free edge of the nail the stratum corneum is thickened and is called the **hyponychium.** The nails generally have a pink coloration because the capillary network beneath them is visible through the cornified cells.

FUNCTIONS OF THE INTEGUMENTARY SYSTEM

Having an understanding of the structure of the skin and its associated organs should make it easier for you to understand the various functions that the skin performs. These functions can be grouped into several categories: protection, regulation of body temperature, excretion, sensation, and production of vitamin D. Because of its role in such activities, the skin plays a major role in maintaining the internal homeostasis of the body and ensuring the continued normal activity of individual cells.

Protection

The skin serves as a barrier that prevents microorganisms and other substances from invading the body. Because of the multiple layers of keratinized cells in the epidermis it forms a physical barrier that blocks the diffusion of water and water-soluble substances either into or out of the body. The barrier is not absolute, however, and some substances—such as those that are soluble in lipids—are able to pass through the skin. Because of the presence of melanin, the cells in the deeper layers of the skin are protected against ultraviolet damage. Another form of protection is provided by the acidity of the secretions of the skin glands. Because of this, the surface of the skin is coated with a thin liquid film that tends to be acidic (pH 4 to pH 6.9). This acid film acts as an antiseptic layer and retards the growth of microorganisms that are always present in large numbers on the surface of the skin. Yet another form of protection results from the presence of phagocytic cells called *macrophages* in the dermis and *Langerhans' cells* in the epidermis. These specialized cells are both thought to activate the immune system against foreign substances.

Body Temperature Regulation

Even under conditions of high environmental temperature or during exercise, the body temperature remains almost normal, in part because considerable heat is lost through the skin. As body temperature begins to rise, the arterioles in the dermis dilate, bringing a greater volume of blood to the body surface and thus allowing more of the internal heat to be lost to the environment. At the same time, the body surface may become wet because of increased secretory activity by the sweat glands. The evaporation of this sweat further facilitates the loss of body heat. In a similar manner, under cold conditions, body heat can be conserved by the constriction of dermal arterioles. This constriction reduces the amount of blood that flows toward the body surfaces so that less heat will be lost to the external environment.

Blood flow to the skin can vary tremendously. Under ordinary circumstances, cutaneous blood flow is approximately 400 ml per minute. However, under extreme conditions, up to 2500 ml of blood may flow through the vessels of the skin each minute. A unique arrangement of blood vessels in the skin makes this large blood flow possible (Figure 4.5). In addition to the usual **F4.5** capillary beds, there are extensive **subcutaneous venous plexuses** (*plexus* = network) that are capable of holding large volumes of blood. These plexuses are located close enough to the surface to allow heat to pass from the blood to the surface and thus to be lost from the body. Moreover, in some locations it is possible for blood to flow into the venous plexuses directly from the arteries

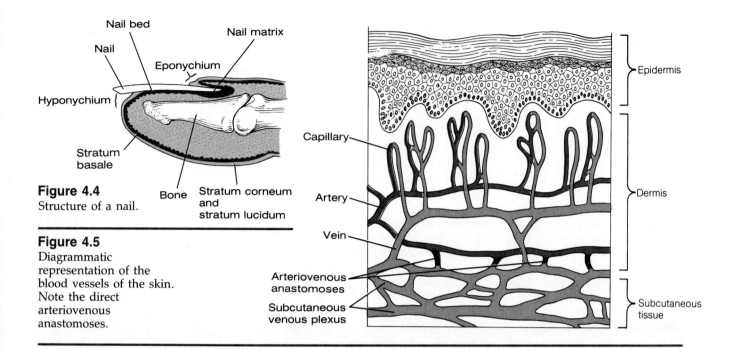

Figure 4.4
Structure of a nail.

Figure 4.5
Diagrammatic
representation of the
blood vessels of the skin.
Note the direct
arteriovenous
anastomoses.

by means of **arteriovenous anastomoses.** The walls of these arteriovenous shunts are muscular and they constrict under stimulation by the sympathetic nervous system. This constriction decreases the flow of blood into the venous plexuses and reduces the loss of body heat.

Excretion

In addition to its cooling effect, the secretion of sweat functions, to a limited extent, as a means of excretion. Small amounts of nitrogenous waste products and sodium chloride leave the body via the sweat. Both the volume and the composition of sweat vary according to the changing needs of the body.

Sensation

Because of the presence of nerve endings and specialized receptors, the skin provides the body with much information concerning the external environment. Events such as temperature change, light touch, pressure, and painful trauma all stimulate integumentary receptors. These receptors, in turn, alert the central nervous system of the particular event, thus enabling appropriate action to be taken. This action might be simple and automatic, such as withdrawing the hand from a flame, or it might require a more complicated act, such as deciding that a warmer coat should be worn.

Vitamin D Production

The skin is also involved in the specialized function of producing vitamin D. In the presence of sunlight or ultraviolet radiation, one of the sterols (7-dehydrocholesterol) found in the skin is altered in such a way that it forms vitamin D_3 (cholecalciferol). After being metabolically transformed, vitamin D_3 assists in the absorption of calcium and phosphate from ingested food. Vitamin D_3 is therefore important in maintaining optimum levels of calcium and phosphate in the body, thus facilitating the normal growth of bones and their repair following a fracture.

CONDITIONS OF CLINICAL SIGNIFICANCE
The Integumentary System

The importance of the skin in preventing invasion of the body by microorganisms is apparent from the abundance of these organisms that are normally present, even on healthy skin, and yet that cause no bodily harm unless the epidermis is damaged, allowing them to enter the body. The secretions of the sweat glands and the sebaceous glands provide ample nutrients as well as a favorable environment in which these microorganisms can thrive. The **fungi** that cause *athlete's foot* are often present on the soles of the feet and between the toes without causing harm. Then, due to some change in their environment, the fungi rapidly proliferate and cause the disease condition. There are also some **yeasts** that live harmlessly on healthy skin. By far, the most abundant organisms on the skin are **bacteria.** These include the rod-shaped forms and the spherical cocci. Most of the cocci are harmless, but one species, *Staphylococcus aureus*, can cause pimples, boils, and other more serious infections. However, even these powerful disease-producing microorganisms, which are normally present on certain areas of the skin, do not cause skin diseases unless the epidermis is penetrated.

The number of bacteria that are present on the skin varies in different regions of the body as well as from person to person. The largest bacterial populations are found on the face and neck, and in the axilla and the groin. Reported population densities range from 2.41 million bacteria per square centimeter of epidermis in the male axilla to 314 bacteria per square centimeter on the back. There are many recognized diseases of the integumentary system, but we will consider only a few of the more common pathologies.

Acne

This inflammatory disease is caused by a rod-shaped bacterium *(Corynebacterium acnes)*. These microorganisms provoke excessive secretion by the sebaceous glands, which, in turn, causes the formation of pimples, blackheads, and dandruff. Acne is most prevalent during puberty because of the hormonal changes that occur during that period. After several years, usually, the skin becomes adapted to the higher levels of sex hormones and the condition disappears.

Warts

The common wart is the result of a viral invasion of the skin. This condition is most common in adolescents and young adults. Warts are often found in groups because they are capable of spreading to adjacent areas. Warts that occur on the sole of the foot—*plantar warts*—are particularly painful because they are almost constantly subjected to pressure.

Dermatitis and Eczema

These are general terms that refer to many inflammatory skin conditions. Also included in this category are nonspecific allergic responses of the skin to many different substances.

Psoriasis

This fairly common condition is characterized by small reddish-brown elevations and patches that are covered by layers of silvery scales. When the patches are scraped away, bleeding occurs from minute points that correspond to the tops of the papillae of the dermis. Tiny abscesses form under the stratum corneum, producing an exudate. The cause of psoriasis is unknown.

Impetigo

Impetigo is a highly contagious skin infection that is most common in children. It results from the invasion of the epidermis by various strains of *Staphylococci* and *Streptococci bacteria*. Pus-filled sacs *(pustules)* form beneath the stratum corneum, causing inflammation and swelling. The pustules then rupture and form a crust.

Moles

Moles, which are elevations of the skin that are generally pigmented, are very common. Almost everyone has at least one mole, and the average person has about 20 at various locations on the body. Moles are considered to be congenital, although they often do not appear until adulthood. It has been suggested that their appearance may be stimulated by steroid hormones. Most moles are *benign*—that is, they do not develop into tumors. They grow slowly over a period of time, remain stable for a long period, and then gradually diminish in size (atrophy). A few, however, may become *malignant*, or capable of spreading to other parts of the body. This change is generally indicated by an increase in size and pigmentation, a reddening of the skin around the mole, and itching.

Herpes Simplex

This condition is commonly called a *fever blister* or *cold sore*. It occurs when a virus that has been dormant in a spinal nerve travels along the processes of the nerve cells and becomes active on the skin and mucous membranes. The active viruses cause clusters of watery blisters to form. The blisters generally occur on the lips or the external genitalia. They are often associated with any disease that causes an elevated body temperature.

Herpes Zoster (Shingles)

Like a cold sore, shingles results from the invasion of the body by a virus that remains dormant in the spinal nerves, generally in the thoracic region. Once the virus becomes active, it affects the sensory nerves of that region, causing an aching pain that follows the nerve paths. Groups of small vesicles develop in the skin that overlies the nerve paths. It is now felt that the same virus that causes the vesicles of shingles also is responsible for the skin vesicles of chickenpox.

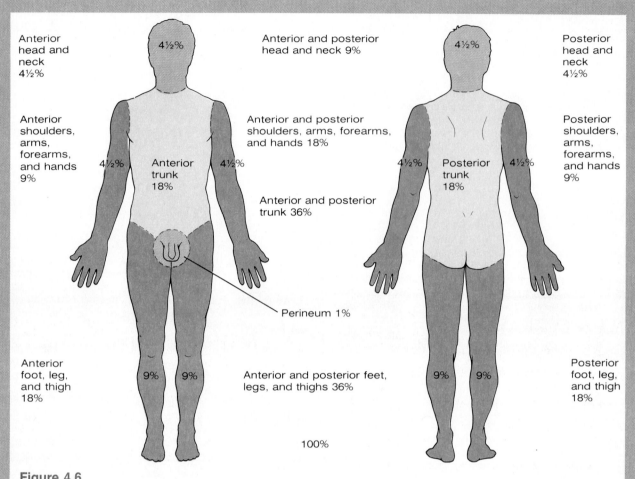

Anterior head and neck 4½%

Anterior and posterior head and neck 9%

Posterior head and neck 4½%

Anterior shoulders, arms, forearms, and hands 9%

Anterior and posterior shoulders, arms, forearms, and hands 18%

Posterior shoulders, arms, forearms, and hands 9%

Anterior trunk 18%

Anterior and posterior trunk 36%

Posterior trunk 18%

Perineum 1%

Anterior foot, leg, and thigh 18%

Anterior and posterior feet, legs, and thighs 36%

Posterior foot, leg, and thigh 18%

100%

Figure 4.6
Estimating the extent of burns on the body surface area by using the rule of nines.

Cancers

Numerous types of tumors arise within the skin. Some originate in the various layers of the epidermis, some in the dermis, and others in the sweat glands and the sebaceous glands. Most of these tumors are benign and do not spread to other parts of the body. Warts are an example of benign tumors. Other tumors are malignant and have the capability of spreading (*metastasizing*) to other regions of the body. It is these tumors that are generally called *cancers*.

The three most common forms of skin cancer are basal cell carcinoma, squamous cell carcinoma, and malignant melanoma. *Basal cell carcinoma* accounts for over 75 percent of all skin cancers, but is the least likely to be malignant. In this condition cells of the stratum basale proliferate, producing growths that may reach the surface of the epidermis, where they often form a crust. If not removed surgically, the growths may invade and destroy the underlying dermis and hypodermis. *Squamous cell carcinoma* begins its development in the stratum spinosum. The tumor is often in the form of a reddish nodule—somewhat similar in appearance to a wart. Squamous cell carcinoma may well become malignant. *Malignant melanoma* is a relatively rare condition but, as its name implies, is highly malignant. It develops in me-

lanocytes, often within a preexisting epidermal mole. However, it can develop anywhere there are melanocytes, and any brown or black patch that develops on the skin should be examined by a physician.

The cause of most skin tumors is not known. However, prolonged overexposure to the ultraviolet rays of sunlight appears to be related directly to the development of many of them. There is a greater incidence of skin tumors in farmers and others whose occupations require that they work outdoors over a period of years. A higher incidence is also noted in the southern United States as compared to the northern regions. Skin cancers are seldom found in dark races, where the skin is heavily pigmented.

Burns

While burns cannot be regarded as pathological conditions of the skin, they disrupt the homeostasis of the body so drastically that we consider them here.

The seriousness of burns results from the destruction of skin, and clearly demonstrates the skin's importance to the rest of the body systems. When the skin is destroyed, there is a large loss of body water (**tissue fluid**) and blood plasma. Lost along with these fluids are plasma proteins, which upsets the osmotic equilibrium of the body, and mineral salts, which results in an electrolyte imbalance. These losses cause dehydration, kidney malfunction, and shock. In addition, with the protection of the skin gone, it is very easy for infectious agents to invade the body.

Burns are classified according to their severity. In *first-degree burns*, only the epidermal layers of the skin are damaged. Their symptoms include localized pain, redness, and swelling. Sunburn is usually a first-degree burn. In *second-degree burns*, there is damage to both the epidermis and the dermis, but the damage is not severe enough to prevent the skin from quickly regenerating. In *third-degree burns*, both the epidermis and the dermis are so severely damaged that they can only regenerate from the edges of the wound. If the burned area is extensive, this regeneration can be a slow process during which body fluids are constantly being lost from the damaged area, and the possibility of infection is high. In addition, such wounds can result in extensive scar tissue formation, which is not only disfiguring but can also restrict movement of the damaged part. To hasten healing (and thus reduce the loss of body fluids) and to minimize scar formation, large burned areas are often covered with *skin grafts* taken from other regions of the body.

Because the treatment of burns depends to some degree on the amount of body surface area that has been damaged, it is useful to be able to estimate quickly the extent of a burn. There are methods by which rather precise estimates can be obtained, but a less exact method is commonly used because it is so easy to apply. This method is called the *rule of nines* (Figure 4.6). In this estimation, the body surface is divided in the following way: each upper limb is considered to have 9% of the body surface area; each lower limb has 18%; the anterior and posterior trunk regions each have 18%; the head and neck have 9%; and the perineum has the remaining 1%.

Effects of Aging on the Integumentary System

With aging, the skin tends to become thin, somewhat wrinkled, dry, and occasionally scaly. The thinning of the epidermis is due in part to an increased scaling off of its cells and a declining rate of cellular division. Because the epidermis becomes thinner, it may become more permeable and allow substances to pass through it more readily. Moreover, with aging the collagen fibers in the dermis become thicker and the elastic fibers less elastic, and there is a gradual decrease in the underlying subcutaneous fat. These changes contribute to the formation of wrinkles and sags that are common in older people. There is also a decrease in the number and activity of the hair follicles, sweat glands, and sebaceous glands. Consequently, aging is often accompanied by a loss of hair, reduced sweating, and decreased oil (sebum) production. Melanocytes tend to atrophy with age, so there is often a graying of the hair. However, the melanocytes that remain in the skin tend to be larger and group together, forming dark pigment plaques called *aging spots* that are typical of older persons.

Skin that has been exposed to sunlight over a lifetime will show changes that are more severe than those that are due to aging alone. Such skin shows more marked wrinkling and furrowing, and may develop nodules of an abnormal type of collagen. Moreover, aging skin that has been exposed to large amounts of sunlight tends to develop more cutaneous cancers than skin that has had less exposure.

STUDY OUTLINE

EPIDERMIS outermost layer of skin; develops from embryonic ectoderm; lacks blood vessels; is generally thin, but can thicken as calluses. Skin color primarily determined by a dark pigment called melanin, also influenced by the yellow pigment carotene and dermal blood vessels.   pp. 77–80

EPIDERMAL LAYERS generally five in number.

STRATUM BASALE deepest layer, where mitosis occurs and supplies epidermis with new cells.

STRATUM SPINOSUM composed of flattened cells with cytoplasmic extensions.

STRATUM GRANULOSUM composed of cells that contain granules of keratohyalin in cytoplasm; as granules expand, the cell nucleus dies, so outermost cells of this layer are dead.

STRATUM LUCIDUM clear band superficial to stratum granulosum; cells of this layer continuously become part of stratum corneum through presence of eleidin, which is transformed into keratin.

STRATUM CORNEUM outermost layer composed of closely packed dead cells filled with fibrous protein (keratin).

NOURISHMENT OF THE EPIDERMIS epidermis obtains nourishment by diffusion from the capillary beds of the dermis.

SKIN COLOR determined primarily by melanin, but also by carotene and blood vessels.

DERMIS lies deep to stratum basale; is the second main layer of skin; is well supplied with blood vessels, lymph vessels, nerves, glands, sense organs; has two indistinctly separated layers.   pp. 80–82

PAPILLARY LAYER next to basal layer of epidermis; contains specialized sensory receptors and capillary loops.

RETICULAR LAYER deep layer consisting of dense bundles of collagenous fibers, continuous with the deeper hypodermis layer.

HYPODERMIS not part of the skin, but important because it attaches skin to underlying structures; composed of loose connective tissue.   p. 82

GLANDS OF THE SKIN   pp. 82–83

SWEAT GLANDS also called sudoriferous glands; distributed over most of the body surface.

ECCRINE SWEAT GLANDS coiled tubules within dermis; they secrete watery solution of salt, with traces of urea, sulfates, and phosphates.

APOCRINE SWEAT GLANDS secrete part of their cell contents, so secretion is more complex than true sweat.

CERUMINOUS GLANDS produce "wax" in ears; are modified sweat glands.

SEBACEOUS GLANDS empty their secretion (sebum) into hair follicles; serve to oil the skin and hair. Especially active in adolescence. In regions of skin lacking hair, glands empty secretions onto epidermal surface.

HAIR covers most of the body; growth due to mitotic activity of epidermal cells at bottom of hair follicles.
pp. 83–84

HAIR FOLLICLES extend from epidermis into dermis; composed of two layers: (1) inner layer that gives rise to hair; (2) outer layer of connective tissue that develops from dermis.

PAPILLAE at bottom of hair follicles; covered by mitotically active matrix; contain blood capillaries for nourishment.

ARRECTOR PILI MUSCLE pulls on follicle, causing hair to "stand up."

SINGLE HAIR consists of root (in follicle) and shaft (above skin surface); shaft has central core (medulla) of loose horny cells, cortex of tightly compressed keratinized cells that surround medulla, and outside cuticle of hard keratinized cells. Hair color primarily due to melanin.

NAILS heavily cornified layers of strata corneum and lucidum. Each nail rests on nail bed of stratum basale. Mitosis that produces nail growth occurs in thickened matrix under proximal end of nail. pp. 84–85

FUNCTIONS OF THE INTEGUMENTARY SYSTEM pp. 85–86

PROTECTION skin forms physical barrier against body's invasion by foreign substances; reduces water loss; melanin protects against ultraviolet radiation; acid fluid film acts as antiseptic layer; macrophages and Langerhans' cells activate immune system.

BODY TEMPERATURE REGULATION

OVERHEATING OF BODY prevented as capillaries in dermis dilate and bring greater volume of blood to body surface to lose heat by radiation; body surface also becomes wet, providing additional cooling by evaporation.

HEAT CONSERVATION accomplished during cold by constriction of dermal capillaries.

EXCRETION some nitrogenous wastes and salt leave the body via sweat.

SENSATION nerve endings and specialized receptors in skin provide body with much information, such as temperature change and increased pressure.

VITAMIN D PRODUCTION occurs in skin in presence of sunlight, ultraviolet radiation; helps maintain optimum levels of calcium and phosphate.

CONDITIONS OF CLINICAL SIGNIFICANCE: THE INTEGUMENTARY SYSTEM pp. 87–89

FUNGI, YEASTS, BACTERIA live on body skin, yet cause no harm unless epidermis is damaged, allowing them to enter the body.

ACNE inflammatory disease caused by bacteria that provoke excessive secretion by sebaceous glands and result in pimples, blackheads, dandruff; most frequent during puberty.

WARTS caused by viral invasion of skin; most common in adolescents and young adults.

DERMATITIS AND ECZEMA general terms for many inflammatory skin conditions.

PSORIASIS common condition of small reddish-brown elevations and patches covered by layers of silvery scales; accompanied by bleeding and tiny abscesses; cause unknown.

IMPETIGO highly contagious infection common in children; pustules form beneath stratum corneum, causing inflammation and swelling; caused by bacteria.

MOLES common pigmented elevations of skin; considered to be congenital; may become malignant, but most grow, stabilize, and finally atrophy.

HERPES SIMPLEX fever blister or cold sore caused by viral activity.

HERPES ZOSTER (SHINGLES) caused by viral activity in spinal nerves; most often affects sensory nerves of thoracic region; causes vesicle formation and pain.

CANCERS numerous types of tumors in epidermis, dermis, and skin glands; most are nonspreading (benign), but some do spread (malignant); cause of most skin tumors unknown, but some may be caused by prolonged overexposure to ultraviolet radiation.

BURNS seriousness results from loss of skin, loss of body water, invasion by microorganisms; extent of a burn is determined by using the rule of nines; burns classified according to severity:

FIRST-DEGREE only epidermal layers of skin are damaged.

SECOND-DEGREE both epidermis and dermis are damaged, but skin quickly regenerates.

THIRD-DEGREE both epidermis and dermis are damaged so extensively that skin can regenerate only from edges of wound.

EFFECTS OF AGING ON THE INTEGUMENTARY SYSTEM with aging, skin tends to become thin, wrinkled, dry, and sometimes scaly. Collagen and elastic fibers change and subcutaneous fat decreases. Number and activity of hair follicles, sweat glands, and sudoriferous glands decreases. Melanocytes tend to form aging spots. Ultraviolet radiation increases aging changes in skin.

SELF-QUIZ

1. The subcutaneous connective tissue is called the: (a) epidermis; (b) hypodermis; (c) stratum corneum.

2. In regions where the epidermis is thin, the stratum lucidum is often absent. True or false?

3. Match the following terms associated with the epidermis to the appropriate description.

   Stratum basale
   Keratohyalin
   Melanin
   Stratum corneum
   Eleidin
   Stratum granulosum
   Carotene
   Stratum lucidum
   Keratin

   (a) The layer that helps restrict the loss of body water.
   (b) The substance that is transformed into keratin.
   (c) A yellow pigment.
   (d) A clear band of several layers of flattened, closely packed cells.
   (e) Where mitosis takes place.
   (f) The deepest layer of the epidermis.
   (g) A fibrous protein.
   (h) Granules associated with the disintegration of the nucleus.
   (i) The stratum deriving its name from the presence of keratohyalin.
   (j) A dark pigment.

4. Skin color is primarily determined by the presence and distribution of: (a) carotene; (b) melanin; (c) hemoglobin.

5. The dermis develops from the mesoderm of the embryo, as do the muscles and the skeleton of the body. True or false?

6. Those glands that empty their secretions into hair follicles are termed: (a) sebaceous; (b) ceruminous; (c) sudoriferous.

7. Hair growth is due to the mitotic activity of epidermal cells at the bottom of the arrector pili. True or false?

8. Nail growth occurs at the: (a) nail tips; (b) nail bed; (c) matrix.

9. The skin tends to be slightly basic, with a pH of 6.8. True or false?

10. Body heat is *not* conserved when the dermal capillaries: (a) dilate; (b) constrict.

11. Vitamin D_3 assists in maintaining the calcium and phosphate levels of the body at optimum levels. True or false?

12. The most abundant organisms on the skin are: (a) yeasts; (b) bacteria; (c) fungi.

13. Acne is an inflammatory disease caused by: (a) bacteria; (b) virus; (c) fungi.

14. The common wart is the result of a viral invasion of the skin and is most common in adolescents and young adults. True or false?

15. The cause of which of these disorders is unknown? (a) warts; (b) eczema; (c) psoriasis.

16. Which of these conditions is congenital? (a) herpes simplex; (b) warts; (c) moles.

17. Which of these is a viral infection of spinal nerves? (a) warts; (b) shingles; (c) psoriasis.

18. Match the common pathologies of the integumentary system with the appropriate description.

   Acne
   Warts
   Psoriasis
   Impetigo
   Moles
   Herpes simplex
   Herpes zoster
   Cancers
   Eczema

   (a) Caused when a dormant virus becomes activated on the skin and mucous membranes.
   (b) The name for any number of inflammatory skin conditions.
   (c) These pigmented elevations of the skin are congenital.
   (d) The formation of pimples, blackheads, and dandruff due to bacteria.
   (e) Metastasizing tumors.
   (f) Small reddish-brown patches covered by layers of silvery scales.
   (g) Caused by a viral invasion of the skin.
   (h) A highly contagious skin infection common to children.
   (i) Aching pain and formation of vesicles along paths of thoracic spinal nerves.

19. A burn victim has experienced damage to both the epidermis and dermis, but the lost tissue will quickly regenerate. What degree is the burn? (a) first; (b) second; (c) third.

20. A burn of the anterior trunk region involves about the same amount of body surface area as a burn of the anterior surfaces of both lower limbs. True or false?

LEARNING OBJECTIVES

After completing this chapter, you should be able to:

- List the functions of the skeleton.

- Describe the two methods by which bone develops in the embryo.

- Distinguish between the axial skeleton and the appendicular skeleton, and name the components of each.

- Name the bones that form the various regions of the skull.

- Name the components of the skeleton of the thorax.

- Distinguish between the functions of the pectoral and pelvic girdles, and name the major components of each.

- Describe the differences between the various types of vertebrae.

- Describe the differences between the pelvises of males and females.

CHAPTER CONTENTS

The Skeletal System

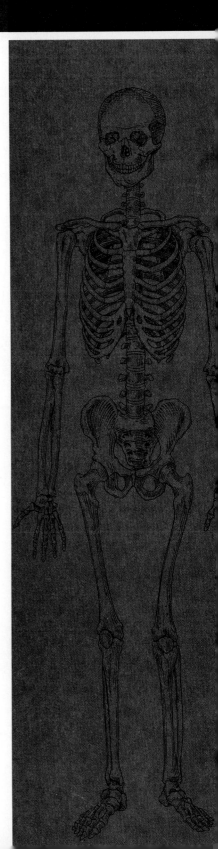

The human skeleton is an *endoskeleton*—that is, it lies within the soft tissues of the body. It is a living structure capable of growth, adaptation, and repair. An endoskeleton differs greatly from the exoskeleton of arthropods such as beetles and crayfish. Because an exoskeleton is a nonliving structure located on the outside of the body, animals that have exoskeletons must shed their skeletal structure and form a new, larger one if they are to continue to grow. As you know from your own development, your skeleton has grown along with the rest of your body structures.

FUNCTIONS OF THE SKELETON

The skeleton performs several important functions: support, movement, protection, storage of minerals, and formation of blood cells (hemopoiesis).

Support
The skeleton acts as the framework of the body, giving support to the soft tissues and providing points of attachment for most of the body muscles.

Movement
Because many of the muscles attach to the skeleton, and many of the bones meet *(articulate)* in movable joints, the skeleton plays an important role in determining the kind and extent of movement of which the body is capable.

Protection
The skeleton protects many of the vital internal organs from injury. The brain is encased within the cranial cavity of the skull, the spinal cord is within the canal formed by the vertebrae, the thoracic organs are protected by the rib cage, and the urinary bladder and internal reproductive organs are protected by the bony pelvis.

Mineral Reservoir
Calcium, phosphorus, sodium, potassium, and other minerals are stored within the bones of the skeleton. These minerals can be mobilized and distributed by the blood vascular system to other regions as they are required by the body. For instance, during pregnancy, calcium is removed from the mother's skeleton and used in the development of the baby's bones if the mother's diet does not include enough calcium. Because of their large mineral content, bones can remain intact for many years after death.

Hemopoiesis (Blood-Cell Formation)
Following birth, the red marrow in certain bones produces the blood cells found in the circulatory system.

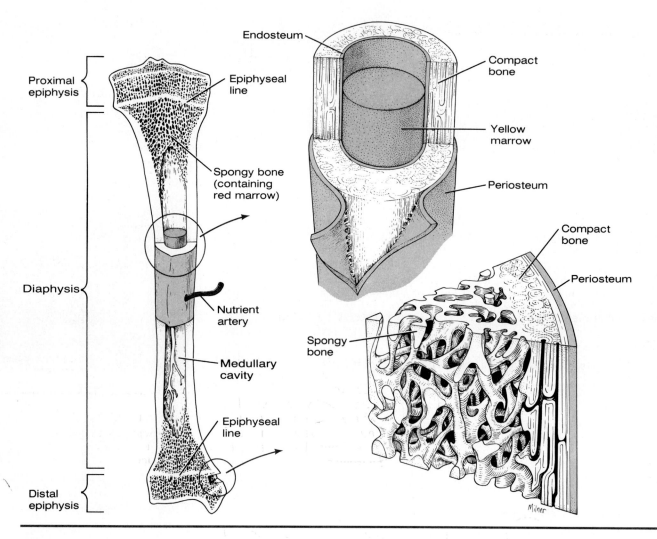

Figure 5.1
Structure of a long bone—longitudinal section. (Inserts are at higher magnification.)

CLASSIFICATION OF BONES

Bones can be classified according to their shape: long, short, flat, or irregular.

Long Bones

Most bones of the upper and lower limbs have a long axis; that is, they are longer than they are wide. These are classified as long bones (humerus, radius, ulna, femur, tibia, fibula, phalanges).

Short Bones

Bones that do not have a long axis, such as those of the wrist (carpals) and ankle (tarsals), are called short bones.

Flat Bones

The rather thin bones that form the roof of the cranial cavity, the ribs, and the sternum (the breastbone) are flat bones.

Irregular Bones

Bones of various shapes that do not fit any of the other categories are classified as irregular bones. Certain skull bones, the vertebrae, and the bones of the pectoral and pelvic girdles are examples of irregular bones.

STRUCTURE OF BONE

It is instructive to study the structure of bone at three different levels: at the gross level—where no microscope is used; at the microscopic level; and at the chemical level.

Gross Anatomy

A typical long bone (Figure 5.1) has a shaft, called a **diaphysis,** and two ends, **F5.1** called **proximal** and **distal epiphyses.** The diaphysis is formed of a hollow cylinder of **compact bone** that surrounds a **medullary cavity.** The medullary cavity, which is used as a fat storage site, is also called the **yellow bone marrow cavity.** It is lined by a thin connective-tissue layer called the **endosteum.** The outer surfaces of the epiphyses are also formed of compact bone, but their central regions are filled with interconnecting plates of **spongy (cancellous) bone.** The cavities between the bony plates of spongy bone are lined with endosteum. The spongy bone in the epiphyses of certain bones contains **red bone marrow.** In children and young adults, the diaphysis and epiphysis are separated by an **epiphyseal cartilage** or **plate** that provides the means for the bone to increase in length. In the adult, when skeletal growth has been completed, the epiphyseal cartilage is replaced by bone, firmly uniting the epiphysis with the rest of the bone. This bony junction is called the **epiphyseal line.**

There is no medullary cavity in a flat bone. This kind of bone is formed of spongy bone called **diploe** that is sandwiched between two surface layers of compact bone (Figure 5.2). The spongy bone contains red marrow. **F5.2**

Bones are covered with a double layer of dense connective tissue called the **periosteum.** There is no periosteum in joints where the bone is covered with an articular cartilage. The outer layer of the periosteum is well supplied with blood vessels and nerves, some of which enter the bone. The inner layer is anchored to the bone by collagenous bundles **(Sharpey's fibers)** that penetrate the bone.

Microscopic Anatomy

When examined under the microscope, compact bone is seen to be composed of many organized systems of interconnecting canals (Figure 5.3, Figure 5.4). **F5.3, F5.4** The unit of structure of adult compact bone is the **haversian system,** or **osteon.** Each haversian system has a central **haversian canal** that is surrounded by concentrically arranged **lamellae** (layers) of bone. Because the haversian systems generally run parallel to the long axis of bone, in longitudinal section the canals appear as long tubes. This orientation of haversian systems contributes to the capacity of bone to resist compressive forces. Located between adjacent lamellae in a haversian system are small cavities called **lacunae.** Each lacuna contains a cell called an **osteocyte.** All the lacunae in a haversian system are interconnected by tiny canals called **canaliculi.** The osteocytes have cytoplasmic processes that extend into the canaliculi and make contact via gap junctions with the cytoplasmic processes of neighboring osteocytes. The gap junctions make possible the rapid passage of nutrients and wastes from one osteocyte to another (Figure 5.5). **F5.5**

Each haversian canal contains at least one blood capillary, which provides a source of nutrients and a means of waste removal for osteocytes that are embedded in the lacunae. The nutrients and wastes need to diffuse only a short distance through the tissue fluid within the lacunae and canaliculi in order to enter or leave the haversian canal. After entering an osteocyte, the nutrients from the blood vessels are distributed to adjacent osteocytes by means of the gap junctions connecting the cytoplasmic processes within the canaliculi. The blood vessels reach the canals from larger vessels that are located either on the surface of the bone (in the vascular layer of the periosteum) or within the marrow cavity. Blood vessels, as well as lymph vessels

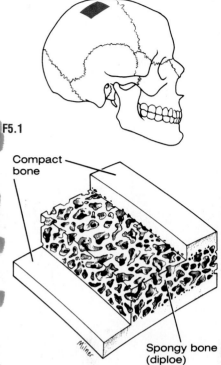

Compact bone

Spongy bone (diploe)

Figure 5.2

Cross section showing the structure of a flat bone.

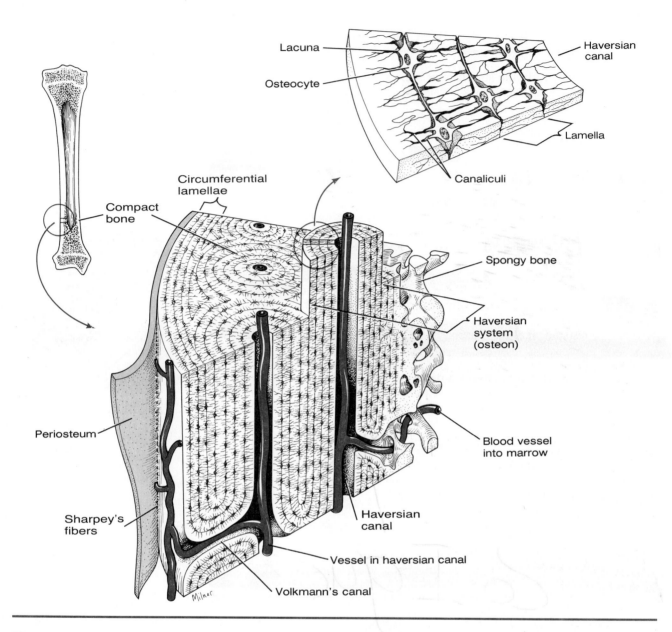

Figure 5.3
Diagram of magnified haversian systems as seen in compact bone tissue. The periosteum has been pulled back to show a blood vessel entering the haversian systems through a Volkmann's canal. The upper inset is a highly magnified sketch showing osteocytes within lacunae. Notice that the lacunae are interconnected by canaliculi.

and nerves, enter and leave the marrow cavity by means of **nutrient canals** that penetrate the bone from the surface and communicate with the marrow cavity. Blood vessels from either of these sources reach the haversian canals through **Volkmann's canals,** which run at right angles to the haversian canals. At the external surface of a bone, just beneath the periosteum, there may be several **circumferential lamellae,** which follow the circumference of the shaft rather than surrounding a haversian canal.

Spongy bone does not show the organization that is characteristic of compact bone. While the osteocytes are embedded in lacunae and the lacunae intercommunicate via canaliculi as in compact bone, the lamellae are not arranged in concentric layers. Rather, they are arranged in various directions that correspond with the lines of maximum pressure or tension. Blood capillaries reach the vicinity of the osteocytes by passing within the bone marrow spaces between the plates of bone formed by the lamellae.

From this consideration of the microscopic anatomy of bone, it is apparent that the skeletal system is a living system well supplied with blood vessels

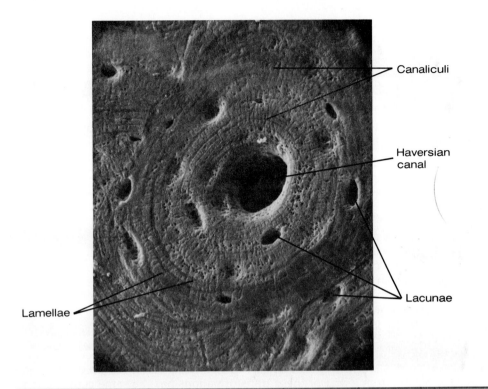

Canaliculi

Haversian
canal

Lacunae

Lamellae

Figure 5.4
Scanning electron
micrograph of a
haversian system (×587).
(From *Tissues and Organs:
A Text-Atlas of Scanning
Electron Microscopy* by
Richard G. Kessel and
Randy H. Kardon. W.H.
Freeman and Company.
Copyright © 1979.)

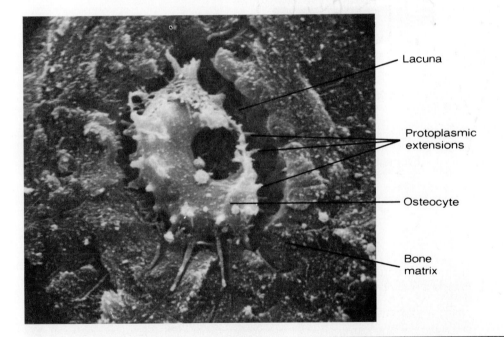

Lacuna

Protoplasmic
extensions

Osteocyte

Bone
matrix

Figure 5.5
Scanning electron
micrograph of an
osteocyte within a lacuna
(×5326). Notice the
protoplasmic extensions
from the surface of the
cell entering the
canaliculi. (From *Tissues
and Organs: A Text-Atlas
of Scanning Electron
Microscopy* by Richard G.
Kessel and Randy H.
Kardon. W.H. Freeman
and Company. Copyright
© 1979.)

and nerves. As such, it is capable of performing the dynamic functions of
forming blood cells and storing minerals, as well as the static functions of
support, movement, and protection.

Composition

The intercellular substance (*matrix*) of bone is composed of two main components: an **organic framework** and **inorganic salts.** The organic framework, in
which the inorganic salts become embedded, is formed of *collagenous fibers*

similar to those found in other connective tissues. Surrounding these fibers is a homogeneous ground substance. The inorganic salts of bone are composed principally of *calcium* and *phosphate*.

Collagen fibers provide bone with great tensile strength—that is, they can resist stretching and twisting. The salts allow bone to withstand compression. This combination of fibers and salts makes bone exceptionally strong without being brittle. The same principle is used in reinforced concrete, where steel rods provide tensile strength, and cement, sand, and gravel give compressional strength.

DEVELOPMENT OF BONE

Early Development of Bone

Box 5.1

The skeleton develops by the transformation of embryonic connective tissue into bone (see Box 5.1). The connective tissues that give rise to most bones are derived from cells of the mesodermal layer of the embryo. When the embryonic tissues are undifferentiated mesoderm (mesenchyme), a relatively simple process called **intramembranous ossification** occurs. If the mesodermal cells transform into cartilage-producing cells before bone formation begins, the process is more complicated and is called **endochondral (intracartilaginous) ossification.** In this process the skeletal structure starts out as cartilage, which is then replaced by bone. The only difference between intramembranous and endochondral bone is the tissue that is replaced. The bone that results from either of these transformations has the same composition.

Intramembranous Ossification

The flat bones of the cranial roof and certain facial bones are formed by the process of intramembranous ossification. The first indication that intramem-

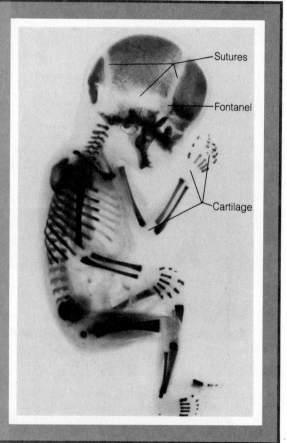

Box 5.1
Fetal Skeleton

Developing bones of a human fetus at about 12 weeks. The darker areas represent ossification (mineral deposition) of the bones. Bones first begin to appear in the fifth week of embryonic development. The clavicle is usually the first bone to ossify. The light areas indicate cartilage that has not yet been replaced by bone. Notice the suture lines separating the individual bones of the skull. Where the sutures meet with one another, a soft spot called a *fontanel* is formed. Some fontanels do not ossify until 18 months after birth. Some bones in the human body are not fully ossified until puberty.

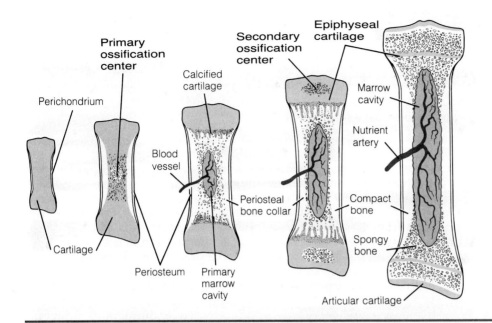

Figure 5.6
Stages of endochondral ossification as seen in a long bone. Cartilage is shown in color. Notice that cartilage remains in the epiphyseal plates and on the articular surfaces. Eventually, the cartilage of the epiphyseal plates will be completely replaced by bone.

branous ossification is occurring is the formation (by fibroblasts) of an organic matrix that has collagenous fibrils extending throughout the tissue. Some of the cells in the matrix increase in size and begin forming bony spicules by calcifying the interstitial substance between the fibrils. These bone-forming cells are **osteoblasts.** Osteoblasts arise from undifferentiated cells called **osteoprogenitor cells,** which develop from embryonic mesenchymal cells. As more layers of bone are deposited, the spicules become heavier and trap the osteoblasts within lacunae. After the osteoblasts are enclosed within lacunae, their activity slows, and they become mature bone cells called **osteocytes.**

Well-developed spicules of bone are known as **trabeculae** ("little beams"). The trabeculae radiate in all directions, uniting with one another to form a network of spongy bone. In areas that will eventually form compact bone, the trabeculae continue to thicken as additional bone is deposited. Gradually, the spaces between the trabeculae are narrowed and bone replaces the spaces, forming compact bone.

During development, most of the external surfaces of bones become covered with a periosteum. This membrane contains osteoprogenitor cells that can give rise to osteoblasts that begin forming bone. Increased activity of these peripheral osteoblasts is responsible for the increase in thickness of the bone.

Even after being calcified, these bones must of necessity be capable of undergoing extensive remodeling in order to meet the changing dimensions of a growing body. This remodeling is accomplished through the reabsorption of previously laid-down bone by large cells called **osteoclasts** and the deposition of new bone (by osteoblasts) in patterns that conform to the growth requirements. Osteoclasts, like osteoblasts, arise from osteoprogenitor cells.

Endochondral Ossification

Most bones form by ossification of hyaline cartilage models that are formed in the early development of the embryo. The cartilage models resemble the shape of the future bone. Each model is surrounded by a fibrous connective tissue membrane called the **perichondrium.** During endochondral ossification, the cartilage of the model degenerates, chondrocytes are replaced by osteoblasts, and bone develops.

Endochondral bone development begins with the transformation of the perichondrium into a bone-producing **periosteum** (Figure 5.6). This change is **F5.6** brought about by the gathering of osteoblasts on the inner surface of the perichondrium. As a result of this process, the cartilage of the diaphysis becomes encased by a collar of compact bone that is laid down by the cells of the periosteum.

While bone is forming on the periphery of the diaphysis, the cartilage cells within the diaphysis enlarge, as do the lacunae that surround the cells. As a result of these enlargements, the cartilage matrix that lies deep to the site of the peripheral ossification becomes reduced to thin partitions and spicules. This matrix begins to calcify, forming a type of tissue called *calcified cartilage*, which makes the diffusion of nutrients to the cartilage cells impossible. The cartilage cells therefore die and degenerate, leaving hollow spaces within the calcified cartilage matrix. This occurs first in the diaphysis. Blood vessels from the periosteum enter the spaces within the matrix and form capillary loops throughout the matrix. The blood vessels transport osteoprogenitor cells and osteoblasts from the periosteum to the spaces within the calcified matrix. The osteoblasts form trabeculae of spongy bone on what is left of the calcified matrix. The region within the diaphysis where this occurs is called the **primary ossification center.** In most bones, the primary ossification center appears in the third month of embryonic life.

The only difference between the formation of trabeculae in endochondral ossification and their formation in intramembranous ossification is that in intramembranous ossification the trabeculae form around collagenous fibers, whereas in endochondral bones they develop around calcified cartilage spicules. As the primary ossification center enlarges in the diaphysis, some of the newly formed spongy bone is broken down by osteoclasts, forming the marrow cavity.

At birth, most bones are in this primary stage of development—that is, they have a diaphysis of compact bone that surrounds remnants of spongy bone and the marrow cavity, while the epiphyses remain as hyaline cartilage. Shortly after birth, the cartilage cells in the epiphyses enlarge, the matrix calcifies, and **secondary ossification centers** appear in both epiphyses. The secondary ossification centers form in the same manner as the primary center, except that there is no marrow cavity formed. These new ossification centers increase in size, and eventually each center occupies an entire epiphysis. The only cartilage that remains is a thin surface layer (which will become the articular cartilage) and a thicker layer that separates each epiphysis from the bony diaphysis. This latter cartilage is called the **epiphyseal cartilage** or **plate.**

Increase in Bone Length and Diameter

As long as the epiphyseal plate remains, the bone can increase in length. It does so in the following manner. Cartilage cells undergo mitosis and therefore tend to increase the size of the epiphyseal plate. At the same time, the side of the plate toward the diaphysis is being replaced by bone tissue. Under normal conditions of growth, these two processes balance one another so that the diaphysis increases in length while the epiphyseal plate remains a constant thickness. As the bone increases in length, it also undergoes a remodeling process by means of selective bone resorption and formation, which maintains the epiphyses at a relatively constant size. These conditions prevail until the late teen years, when the rate of cartilage growth slows and is gradually overtaken by the continuing ossification on the diaphyseal side of the plate. By age 25 in males and several years earlier in females, the cartilage in the epiphyseal plate is completely replaced by bone, leaving only an **epiphyseal line** to mark its previous location. When this has occurred, the bone is no longer able to increase in length. It does, however, retain its ability to increase in diameter.

Bone increases in diameter as a result of the presence of *osteogenic cells*—that is, cells that are capable of transforming into osteoblasts and form bone. These cells, which are located in the inner layer of the periosteum, deposit concentric layers of new bone around the diaphysis, just beneath the periosteum. At the same time as the diameter of the diaphysis is being increased by the addition of these bony layers on the surface, bone is being resorbed from within, beneath the endosteum. Since this resorption increases the volume of the medullary cavity, the thickness of compact bone in the diaphysis may not increase much while the total diameter of the bone does increase.

Theories of Bone Formation

Many of the chemical events of bone formation are still incompletely understood, and it is beyond the scope of this text to examine in depth the processes involved in bone formation. In general, the osteoblasts secrete the organic framework of bone, which includes collagen fibers and glycoproteins such as chondroitin sulfate. The inorganic salts are deposited in the organic framework. It appears that salts such as calcium phosphate [$Ca_3(PO_4)_2$] are formed initially, and then, through a process of substitution and addition of ions, these salts are converted to the inorganic material of bone, which is composed of highly insoluble crystals of hydroxyapatite [$3Ca_3(PO_4)_2 \cdot Ca(OH)_2$].

Although it is still not known what causes mineralization during bone growth and maintenance, one theory is that bone-forming cells concentrate large quantities of calcium and phosphate and that the cells subsequently release calcium phosphate compounds into the extracellular fluid. The small areas of calcium phosphate salts that result may serve as sites for further salt deposition. An alternative theory suggests that the initial formation of bone mineral takes place in membrane-bounded vesicles called matrix vesicles, which are synthesized and secreted into the extracellular fluid by bone-forming cells.

Factors That Affect Bone Development

A number of factors can greatly influence the development of bone. Among the more important factors are *stress*, the amounts of certain *hormones* present in the blood, and the *nutrition* of the individual in which bone is developing.

Stress

Bone is a living tissue that is capable of adjusting its strength in proportion to the degree of stress to which it is subjected. Increased amounts of collagen fibers and inorganic salts can be deposited in a bone in response to prolonged heavy loads. Conversely, if a bone is not subjected to stress, salts are withdrawn from bone.

As new stress patterns occur, the orientation of the collagen fibers in a bone may change so that the fibers are aligned in such a manner as to provide maximal tensile strength to withstand the new stress patterns.

Bone is normally subjected to two major kinds of stresses: *gravitational forces*, such as those that result from supporting the weight of the body, and *functional forces*, such as those that result from the pull exerted on the bones by contracting muscles. There is no consensus as to which of these forces is more important in affecting the form and structure of bone. However, it has been repeatedly demonstrated that bone does not develop normally in the absence of either of these forces. For instance, when gravitational forces are removed for extended periods, such as under the weightless conditions of space travel, or when functional forces are greatly reduced, such as occurs when a limb is paralyzed or immobilized by a cast, the bones in the affected areas do not grow, and may actually atrophy (that is, degenerate). In addition, there are some indications that bone growth is promoted by intermittent forces such as would occur during muscular exercise.

Because of these effects of gravitational and functional forces, regular exercise can alter the form of the skeleton. As weight lifters strengthen their muscles so they can lift heavier weights, the bony skeleton is also strengthened. If this bone strengthening did not occur, a greatly strengthened muscle could break the bones to which it is attached. In a similar manner, the skeleton of an obese person becomes heavier because of the increased forces to which it is constantly subjected.

Another effect of the stresses of exercise on bone involves the presence of the epiphyseal cartilages. The chondrocytes within these cartilages remain active for approximately 20 years, and are susceptible to injury and unusual stress during that time. Injury to the epiphyseal cartilages could interrupt the normal rate of bone growth as well as the pattern of growth. It has not been proven conclusively that contact sports and repeated specialized stresses actually do adversely influence these centers of growth. However, their very presence is reason enough to warrant caution when recommending strenuous exercises for young people.

Hormones

Hormones of the parathyroid and thyroid glands are particularly influential in bone development. Increased levels of the parathyroid hormone **parathormone** increases the resorption of bone by osteoclasts. The hormone **calcitonin** from the thyroid gland has an effect opposite to that of parathormone. Calcitonin decreases the resorptive activity of osteoclasts, and may stimulate the formation of new bone. Any remodeling of bone that occurs involves the interactions of these two hormones.

Nutrition

For normal bone development to occur, it is necessary to follow a diet that provides the body with a variety of essential substances. Of particular importance is vitamin D, which is necessary for the proper absorption of calcium into the bloodstream from the gastrointestinal tract. As you already know, calcium is an important constituent of the inorganic portion of bone matrix.

CONDITIONS OF CLINICAL SIGNIFICANCE

The Skeletal System

Fractures

Even though the composition of bones is such that they are well suited to withstand twisting and compressional forces, it is possible to break them. A broken bone is referred to as a *fracture.*

Types of Fractures

Fractures are named according to various conditions at the site of the break. The following are the more common types of fractures.

SIMPLE FRACTURE   The broken ends of the bone do not penetrate through the skin.

COMPOUND FRACTURE   The broken ends of the bone protrude through the skin.

COMMINUTED FRACTURE   Rather than being broken in a single plane, the bone is splintered into many fragments at the site of the break.

DEPRESSED FRACTURE   The broken region is pushed inward, as often occurs in fractures of the flat skull bones that form the roof over the brain.

IMPACTED FRACTURE   The broken ends of the bone are driven into each other. Such fractures occur in falls in which the person lands on the ends of the bone.

Healing of Fractures

Fractures undergo three progressive changes during the healing process (Figure 5.7): formation of a procallus, formation of a fibrocartilaginous callus, and formation of a bony callus.

FORMATION OF A HEMATOMA   When a bone is broken, bleeding occurs from the vessels of the haversian systems and the periosteum. This produces a swelling and forms a blood clot called a **hematoma.**

FORMATION OF A FIBROCARTILAGINOUS CALLUS   Fibroblasts invade the hematoma and form fibers. Within a few days, the hematoma is replaced with a **fibrocartilaginous callus,** which develops between the ends of the bone and eventually forms a bridge that reunites the fragments. The part of the callus that extends beyond the usual surface of the bone is the **external callus.** The part of the callus between the broken ends of the bone, which occludes the medullary cavity, is called the **internal callus.**

FORMATION OF A BONY CALLUS   Initially the outer portion of the external callus consists of cartilage formed by chondroblasts and chondrocytes that invade the area. Gradually, however, osteoblasts derived from the inner layer of the periosteum and the endosteum form a **bony callus** that knits the ends of the bone firmly together. At first the bony callus is spongy bone, but it is slowly remodeled to form compact bone. The formation of compact bone is partially under the influence of stress as the repaired bone is again used for body movement and support.

Because of the importance of stress in strengthening the bone, prolonged immobilization of a broken bone may be detrimental to the healing process. For this rea-

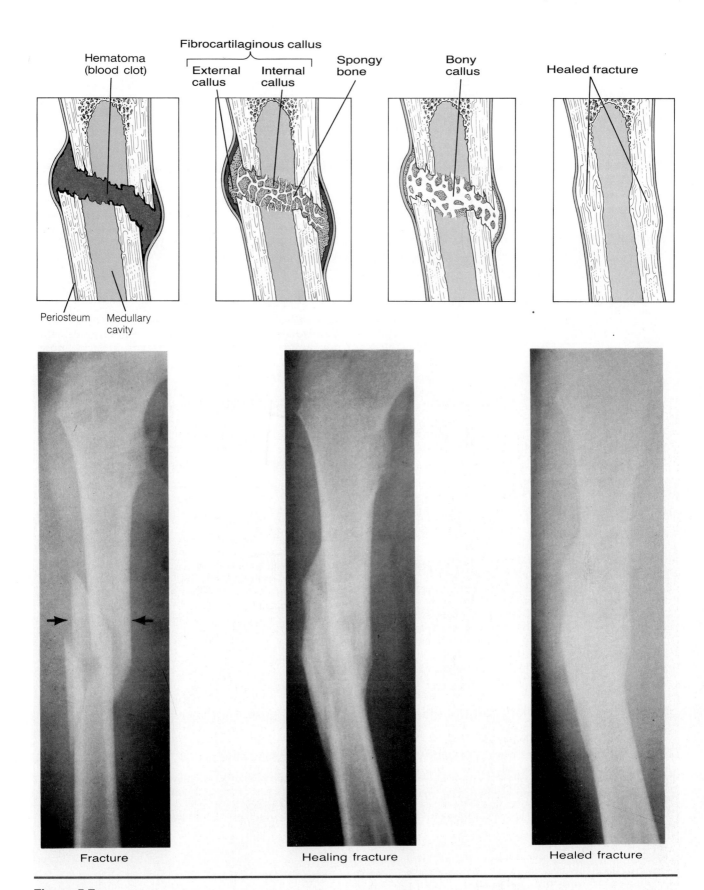

Figure 5.7

Healing of a fracture. The initial repair begins with the formation of a blood clot called a *hematoma*. Connective tissue invades the hematoma, replacing it with a fibrocartilaginous callus. The fibrous callus is eventually replaced by bone that develops from cells of the periosteum.

son, pins are sometimes used to hold the ends of a broken bone together, thus allowing the bone to support weight almost immediately. The stresses of use increase the activity of osteoblasts and facilitate healing.

During the months following the formation of a bony callus, osteoclasts reabsorb the bone of the external callus and the part of the internal callus that blocks the medullary cavity so that eventually only a slight external enlargement is left to mark the location of the break.

Metastatic Calcification

The deposition of calcium in tissues that do not usually become calcified is called *metastatic calcification.* It often results if blood calcium levels become too high, as occurs during some decalcifying diseases of bone, or when the amount of parathyroid hormone is increased. The kidney is the most common site of metastatic calcification (which results in kidney stones), but it occurs in many other tissues as well.

Spina Bifida

Occasionally, the posterior portions of the vertebrae of the spinal column fail to form a complete bony arch around the spinal cord. This condition, which is called *spina bifida,* is most common in the lumbar and sacral regions of the spinal column, but may occur in a vertebra of any region. If the defect is large, the coverings of the spinal cord and the wall of the spinal cord itself may protrude.

Osteoporosis

Osteoporosis, or "porous bone," is a common condition in older people. It also occurs in the bones of paralyzed or immobilized limbs. Osteoporosis is believed to result from a gradual reduction in the rate of bone formation while the rate of bone absorption remains normal. This condition causes the bone to become porous, fragile, and relatively easily broken.

Because some degree of osteoporosis occurs in most women over age 65 but only in relatively few men, hormonal changes associated with menopause are thought to be a major factor in its development. There is evidence that the removal of calcium and other minerals from bone may be slowed by increasing the level of estrogen in the blood. For this reason estrogen therapy has been used in postmenopausal women to treat osteoporosis.

Osteomyelitis

In *osteomyelitis,* the periosteum, the contents of the marrow cavity, and the bone tissue become infected. Since the causative agent is usually *Staphylococcus aureus,* a bacterium that enters the body through a boil or some other break in the skin, osteomyelitis may follow trauma. The initial damage to bone has the appearance of an abscess. The abscess spreads throughout the bone, converting the fatty tissue of the marrow cavity into pus and destroying the bony tissue. The infection can pass from the shaft of the bone, where it generally appears first, to the epiphyses, where it perforates the articular cartilage and enters the joint cavity.

Before the discovery of antibiotics, osteomyelitis had been a very serious condition with a high mortality rate. Because the disease responds well to antibiotic therapy, death from it is now rare.

Tuberculosis of Bone

Tuberculosis of bone is a type of osteomyelitis that is caused by another bacterium, *Mycobacterium tuberculosis.* The bacterium is generally carried by the bloodstream from an infection in the lung or lymph nodes. This disease is characterized by excessive bone destruction.

Rickets and Osteomalacia

Both *rickets* and *osteomalacia* result from the demineralization and subsequent softening of bone. Rickets occurs in children, whereas osteomalacia refers to the softening of adult bones. These conditions are due to deficiencies of calcium, phosphorus, or vitamin D, or to a lack of sunlight. Metabolically transformed vitamin D_3 facilitates the absorption of calcium and phosphate from the intestine into the bloodstream, making them available for bone formation. Ultraviolet rays in sunlight provide the body with vitamin D_3 by converting sterols in the skin into vitamin D_3. In children, the softened bones can develop bends, resulting in bowed legs (femur) or pigeon breast (sternum).

Tumors of Bone

Many types of tumors, both benign and malignant, have been identified in bone. Some malignant bone tumors originate in the lungs, breast, prostate, or other structure and reach the bone via the bloodstream. Tumors weaken the bone by destroying the tissue. Their presence may be detected for the first time when a patient is X-rayed after a fracture.

Abnormal Growth Patterns

The amount of growth hormone secreted by the anterior pituitary gland can have a dramatic effect on bone development. The presence of excessive growth hormone delays the ossification of the epiphyseal cartilage. As a consequence, bone development continues for a longer period than normal, producing a *pituitary giant.* Conversely, a deficiency of growth hormone, below the level needed to maintain active epiphyseal cartilages, results in the early replacement of those cartilages by bone. This closure of the epiphyseal plates causes the development of most bones to halt early in life, producing a midget referred to as a *pituitary dwarf.*

If excessive growth hormone is secreted after the epiphyseal cartilages have been replaced by bone, an abnormal pattern of bone growth occurs that causes enlargement of bones in certain regions, particularly in the hands, face, and feet. This condition is called *acromegaly.*

Sometimes, under the influence of hereditary factors as well as hormonal levels, the epiphyseal cartilages of the bones function for only a short time, while the bones of the rest of the body continue developing fairly normally. This condition, called *achondroplasia,* results in a dwarf with short arms and legs but normal trunk and head. Dachshunds are achondroplastic dogs that have been selectively bred.

Effects of Aging on the Skeletal System

The loss of calcium from the bones, which is the major effect of aging on the skeletal system, is more severe in women than in men. In women, the amount of calcium in the bones steadily decreases after the age of 40, so that by age 70 as much as 28% of the calcium in the skeletal system has been lost. Men tend to have higher bone calcium levels to start with and generally do not begin to lose calcium until after age 60. The loss of calcium results in osteoporosis. The cause of the calcium loss is not known, and there are no certain methods of prevention. However, it has been shown that the loss of calcium from bone is more prevalent in people who are immobilized for prolonged periods. And there is strong evidence that continued weight-bearing activities such as walking can increase the deposition of calcium into bone. Therefore it is very important for elderly persons, particularly women, to exercise regularly.

Normally the matrix of bone is constantly being broken down and replaced by new matrix. In elderly people, however, along with the calcium loss, the rate of protein formation may be so slow that the organic portion of the matrix is not replaced as rapidly as it is broken down. As a consequence, the matrix of bone gradually comes to contain a greater proportion of inorganic salts. This tends to cause the bones of elderly people to become brittle and fracture rather easily.

FRONTIERS IN HEALTH
Prescription for Healthy Bones

As Helen, age 62, gets out of bed and steps onto the floor, she fractures a vertebra. However, she is unaware of the fracture, which is, in fact, her seventh in 2 months. All she notices is a nagging pain in her back—a complaint she ascribes to old age. In a matter of months Helen's spinal column will begin to shrink, and she may develop an unsightly hunchback. Like 17 million other Americans, Helen is suffering from osteoporosis, a progressive loss of bone calcium that leads to increased bone brittleness and a marked increase in bone fractures.

One of every two women in the United States will suffer from postmenopausal osteoporosis, but most will not be aware of their problem until they fracture a hip or vertebra. Each year more than 55,000 Americans—mostly women—die from complications of this disease. Hemorrhage, fat embolisms, and shock are three of the most common causes of death.

New advances in prevention and treatment have begun to offer a ray of hope for millions of potential victims. It has been learned that osteoporosis begins to develop much earlier than previously thought. The disease often begins by the time a woman reaches her mid-20s. By age 30, many women have lost one-third of their bone calcium. The gradual demineralization continues until the bones become so brittle that normal activities, such as getting out of bed or dancing, cause tiny fractures throughout the bones of the body.

Why does calcium loss begin so early? The reasons appear to be mostly dietary. Many women become weight conscious in their mid-20s and avoid fatty foods, such as milk and cheese, that happen to be rich in calcium. Many adults also develop an intolerance to milk and other dairy products. Therefore, even though their bodies require over 1000 milligrams of calcium every day, many women consume as little as 450 milligrams.

It is possible to diet and still not suffer from osteoporosis, however, if dietary supplements are taken or if the intake of calcium is increased. Vitamin D, which increases the intestinal absorption of calcium, can also help prevent osteoporosis. Scientific studies also suggest that weight-bearing exercise, such as aerobics, running, walking, and tennis, can help prevent the disease. Instead of taking the elevator, walk up the stairs in order to keep bones from growing soft. Instead of spending the weekend in front of the television, take a brisk walk, or jog in the park, or play a game of tennis.

But what if the disease has already struck? Is there any way to arrest or reverse it? Studies show that osteoporosis can be halted and reversed, even after it has reached an advanced stage. For example, recent work by Dr. Everitt L. Smith and his colleagues at the University of Wisconsin in Madison demonstrated that 45 minutes of moderate exercise three times a week greatly slows the loss of calcium in older women and, after a year, reverses the demineralization. In this study, an inactive control group of older subjects lost 7.5% of their bone calcium over a 3-year period. A second group, which exercised three times a week, lost 3.8% of their bone calcium the first year, but *gained* calcium the second and third years, giving them a 3-year loss of only 1%. Thus, over the duration of the experiment, the gains nearly offset the first-year losses. These results suggest that continued exercise may yield a long-term increase in bone calcium and, with it, a decrease in bone fractures.

One of the more conventional treatments for osteoporosis in postmenopausal women is the administration of the sex steroid estrogen. Produced by the ovaries until menopause, estrogen promotes bone growth. Fairly low doses of estrogen can stop bone demineralization altogether and may actually promote bone formation. However, women who are given estrogen are at an increased risk of developing endometrial cancer. Physicians therefore often prescribe a mixed dose of estrogen and progesterone. The progesterone lessens the likelihood of cancer.

A better understanding of the prevention and treatment of osteoporosis and the dissemination of this knowledge to the general public will help reduce much of the suffering caused by this disease. For Helen and millions of women in her age group, the goal is to stop the disease before it progresses too far. But for millions of younger women, early detection and sound preventative measures, such as exercise and vitamin D and calcium dietary supplements, can prevent the disease altogether.

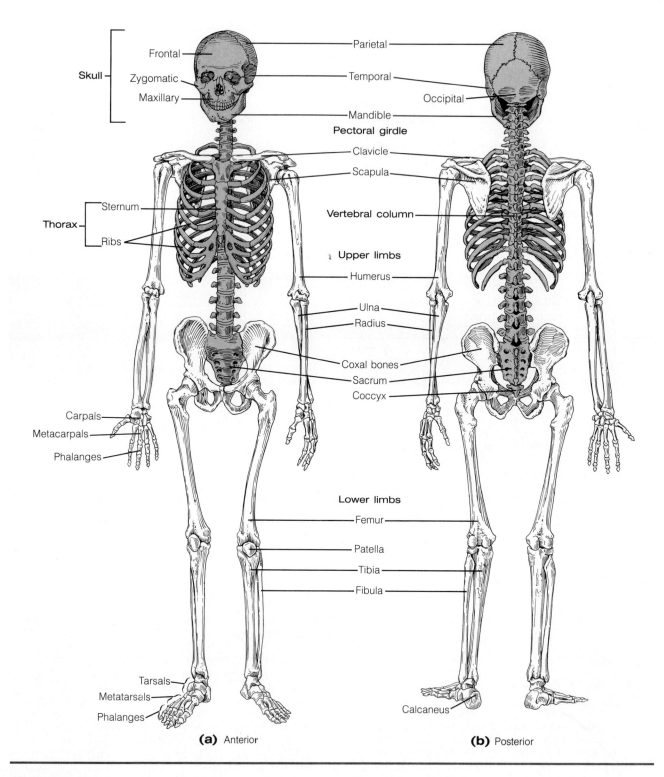

Skull
- Frontal
- Zygomatic
- Maxillary

Parietal
Temporal
Occipital
Mandible

Pectoral girdle
- Clavicle
- Scapula

Vertebral column

Thorax
- Sternum
- Ribs

Upper limbs
- Humerus
- Ulna
- Radius

Coxal bones
Sacrum
Coccyx

Carpals
Metacarpals
Phalanges

Lower limbs
- Femur
- Patella
- Tibia
- Fibula

Tarsals
Metatarsals
Phalanges

Calcaneus

(a) Anterior

(b) Posterior

Figure 5.8
The human skeleton, anterior and posterior views. Axial skeleton in color; appendicular skeleton in white.

INDIVIDUAL BONES OF THE SKELETON

Table 5.1
F5.8

The human skeleton consists of 206 bones. The bones can be grouped into the axial skeleton and the appendicular skeleton (Table 5.1). The two groups are shown in different colors in Figure 5.8. Your study of the skeletal system will be easier if you first familiarize yourself with some of the common terms used

to describe the structural features of bones. These terms are listed in Table 5.2. Table 5.2
Throughout the chapter, each bone is discussed in a general way, and the
reader is usually provided with both a figure that illustrates the bone and a
table that gives more specific information about it.

AXIAL SKELETON

The axial skeleton consists of the bones that form the skull, the vertebral
column, and the thorax. This portion of the skeleton provides the main axial
support for the body and protects the central nervous system and the organs
of the thorax.

Skull

The skull is formed of 29 bones, 11 of which are paired. With the exception of
the mandible (lower jaw) and three small bones (ossicles) within each middle

Table 5.1   Divisions of the Skeleton

Category		Number of Bones
Axial skeleton		80
Skull	29	
Vertebral column	26	
Thorax (ribs and sternum)	25	
Appendicular skeleton		126
Pectoral girdle	4	
Upper limbs	60	
Pelvic girdle	2	
Lower limbs	60	
Total		206

Table 5.2   Common Skeletal Structure Terms

Crest a sharp, prominent bony ridge.

Condyle a rounded prominence that articulates with another bone.

Epicondyle a small projection located on or above a condyle.

Facet a smooth, nearly flat articular surface.

Fissure a narrow cleftlike passage.

Foramen a hole.

Fossa a depression; often used as an articular surface.

Fovea a pit; generally used for attachment rather than for articulation.

Head generally, the larger end of a long bone; often set off from the shaft of the bone by a constricted neck.

Line a slight bony ridge.

Meatus a canal.

Process a prominence or projection.

Ramus a projecting part or elongated process.

Spine a slender pointed projection.

Sulcus a groove.

Trochanter a large, somewhat blunt process.

Tubercle a nodule or small, rounded process.

Tuberosity a broad process, larger than a tubercle.

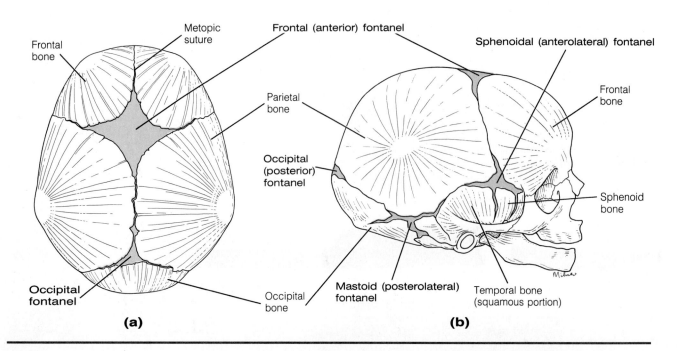

Frontal bone

Metopic suture

Frontal (anterior) fontanel

Sphenoidal (anterolateral) fontanel

Parietal bone

Frontal bone

Occipital (posterior) fontanel

Sphenoid bone

Occipital fontanel

Occipital bone

Mastoid (posterolateral) fontanel

Temporal bone (squamous portion)

(a)   **(b)**

Figure 5.9
Fetal skull, showing the fontanels. **(a)** Superior view. **(b)** Lateral view.

ear cavity, the bones of the adult skull are joined together primarily in immovable joints called **sutures.** At birth and for some years afterward, most of these sutures are held together by fibrous connective tissue rather than bone and are therefore capable of some movement. This flexibility permits the skull to narrow somewhat during birth by allowing the **calvarium** (roofing bones) to override one another when subjected to pressure in the birth canal. The presence of connective tissue also allows for further growth of the skull to accommodate the normal development of the brain. At some points of junction of two or more sutures there are fibrous membrane areas that remain prominent for up to 18 months after birth. These "soft spots" of the skull are called

F5.9 **fontanels** (Figure 5.9).

Calvarium

The calvarium is formed of the *frontal*, *parietal*, and *occipital* bones.

FRONTAL BONE   The **frontal** is a single bone that forms the anterior supe-
F5.10, F5.11 rior region of the skull (Figure 5.10, Figure 5.11). This bone begins its development as two separate bones that meet in a midline **metopic suture,** which is generally not distinguishable in the adult. The frontal bone forms the forehead and the roof of the orbital cavities. Inside the bone, just above its junction with the nasal bones, are the **frontal sinuses.** These, like the other sinuses of the skull, are air spaces lined with mucous membrane. Posteriorly, the frontal bone joins with the two parietal bones, forming the **coronal suture.**
Table 5.3 Table 5.3 lists the features associated with the frontal bone.

PARIETAL BONES   The two **parietal bones** form most of the calvarium (Fig-
F5.10, F5.11, F5.14 ure 5.10, Figure 5.11, Figure 5.14). The parietals meet in the midline, forming the **sagittal suture.** The parietal bones form the **coronal suture** across the top of the skull, where they meet with the frontal bone; the **lambdoidal suture** at the back of the skull, where they meet with the occipital bone; and the **squamosal sutures** along the lower sides of the skull, where they meet with the temporal bones. In a young child's skull, the fibrous fontanels occupy the
F5.9 point of junction of these sutures (Figure 5.9). The **frontal (anterior) fontanel** is located at the junction of the sagittal and coronal sutures. The **occipital (posterior) fontanel** is located at the junction of the sagittal and lambdoidal sutures. There are right and left **sphenoidal (anterolateral) fontanels** at the

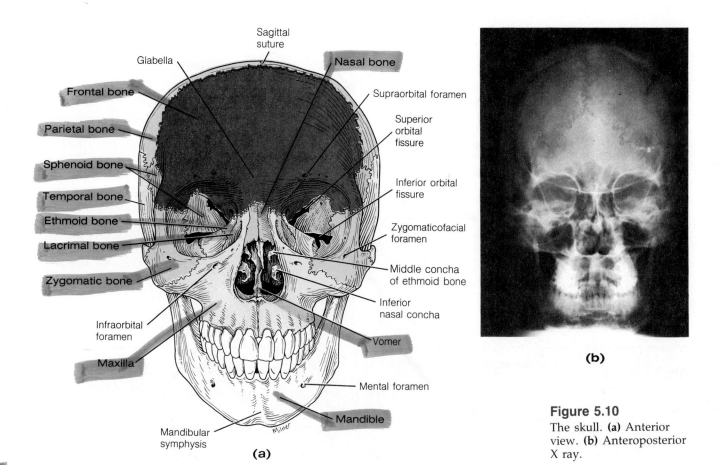

Sagittal suture

Glabella

Nasal bone

Frontal bone

Supraorbital foramen

Superior orbital fissure

Parietal bone

Sphenoid bone

Inferior orbital fissure

Temporal bone

Ethmoid bone

Zygomaticofacial foramen

Lacrimal bone

Middle concha of ethmoid bone

Zygomatic bone

Inferior nasal concha

Infraorbital foramen

Vomer

Maxilla

Mental foramen

Mandibular symphysis

Mandible

Milner

(a)

(b)

Figure 5.10
The skull. **(a)** Anterior view. **(b)** Anteroposterior X ray.

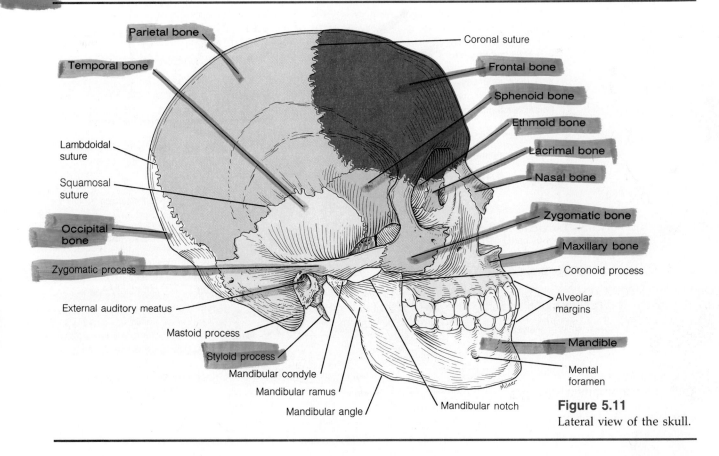

Parietal bone

Coronal suture

Temporal bone

Frontal bone

Sphenoid bone

Ethmoid bone

Lambdoidal suture

Lacrimal bone

Squamosal suture

Nasal bone

Zygomatic bone

Occipital bone

Maxillary bone

Zygomatic process

Coronoid process

External auditory meatus

Alveolar margins

Mastoid process

Mandible

Styloid process

Mental foramen

Mandibular condyle

Mandibular ramus

Mandibular angle

Mandibular notch

Milner

Figure 5.11
Lateral view of the skull.

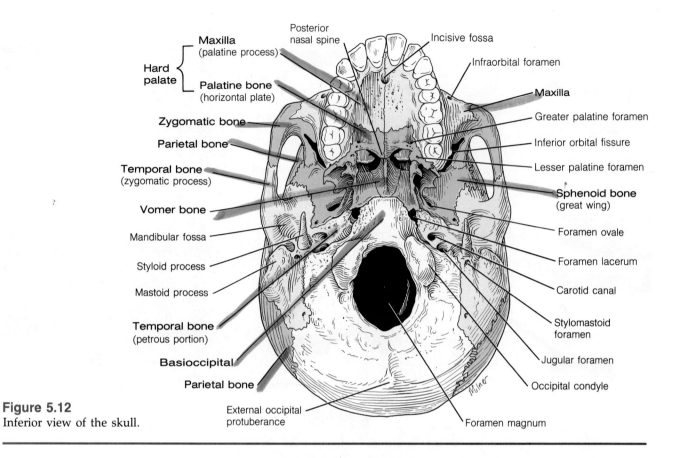

Figure 5.12
Inferior view of the skull.

Labels (clockwise from top): Posterior nasal spine, Incisive fossa, Maxilla (palatine process), Hard palate, Palatine bone (horizontal plate), Zygomatic bone, Parietal bone, Temporal bone (zygomatic process), Vomer bone, Mandibular fossa, Styloid process, Mastoid process, Temporal bone (petrous portion), Basioccipital, Parietal bone, External occipital protuberance, Foramen magnum, Occipital condyle, Jugular foramen, Stylomastoid foramen, Carotid canal, Foramen lacerum, Foramen ovale, Sphenoid bone (great wing), Lesser palatine foramen, Inferior orbital fissure, Greater palatine foramen, Maxilla, Infraorbital foramen

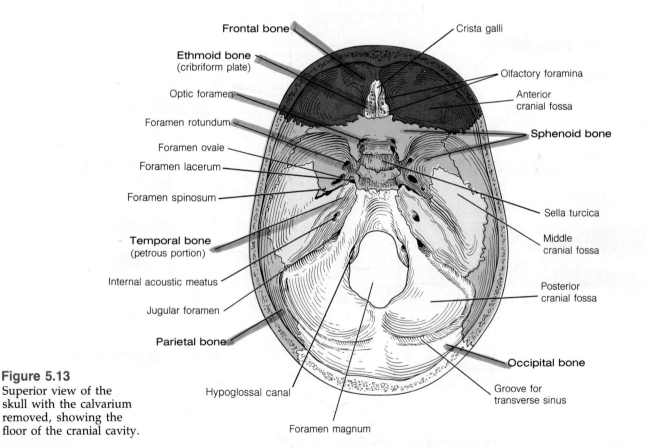

Figure 5.13
Superior view of the skull with the calvarium removed, showing the floor of the cranial cavity.

Labels: Frontal bone, Crista galli, Ethmoid bone (cribriform plate), Olfactory foramina, Optic foramen, Anterior cranial fossa, Foramen rotundum, Sphenoid bone, Foramen ovale, Foramen lacerum, Foramen spinosum, Sella turcica, Middle cranial fossa, Temporal bone (petrous portion), Internal acoustic meatus, Posterior cranial fossa, Jugular foramen, Parietal bone, Occipital bone, Hypoglossal canal, Groove for transverse sinus, Foramen magnum

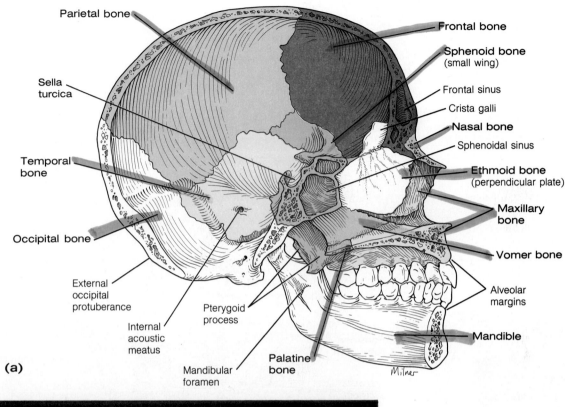

Parietal bone

Frontal bone

Sphenoid bone
(small wing)

Sella
turcica

Frontal sinus

Crista galli

Nasal bone

Sphenoidal sinus

Temporal
bone

Ethmoid bone
(perpendicular plate)

Maxillary
bone

Occipital bone

Vomer bone

External
occipital
protuberance

Alveolar
margins

Internal
acoustic
meatus

Mandible

Pterygoid
process

(a)

Palatine
bone

Mandibular
foramen

Milner

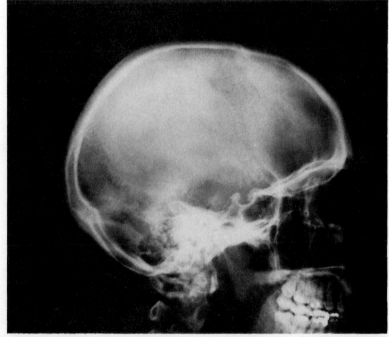

(b)

Figure 5.14
The skull. **(a)** Sagittal
view. **(b)** Lateral X ray.

junction of the coronal and squamosal sutures and **mastoid (posterolateral)
fontanels** where the squamosal suture meets the lambdoidal suture.

OCCIPITAL BONE   The single **occipital bone** forms the lower posterior wall
of the calvarium as well as the posterior portion of the floor of the cranial
cavity (Figure 5.11, Figure 5.12, Figure 5.13, Figure 5.14). Its most obvious
landmark is the large **foramen magnum,** by which the cranial cavity communicates with the vertebral canal. The medulla oblongata of the brain stem
passes through this foramen. Anterior and lateral to the foramen magnum are

F5.11, F5.12, F5.13, F5.14

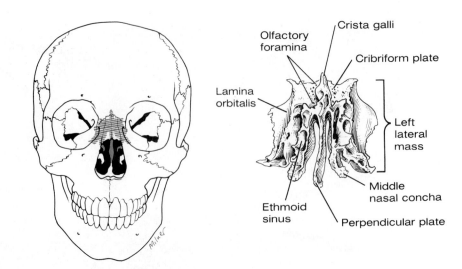

Figure 5.15
Anterior view of the
ethmoid bone.

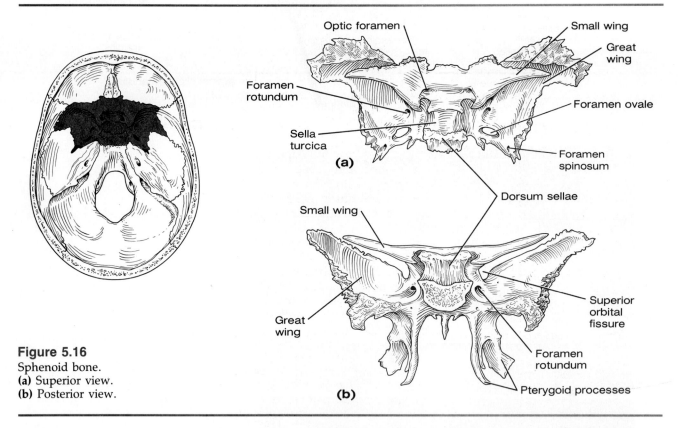

Figure 5.16
Sphenoid bone.
(a) Superior view.
(b) Posterior view.

two **occipital condyles,** one on each side. These condyles articulate with the
superior surface of the first cervical vertebra (the atlas). This is the only connection between the skull and the vertebral column. In addition to articulating with the parietal bones (forming the lambdoidal suture), the occipital
Table 5.3   bone also meets the temporal and sphenoid bones. Table 5.3 lists the features
of the occipital bone.

Bones That Form the Floor of the Cranial Cavity

The floor of the cranial cavity, upon which the brain rests, is formed by six
bones: the midline *frontal, ethmoid, sphenoid, occipital,* and the paired *temporals*
F5.12, F5.13, F5.14   (Figure 5.12, Figure 5.13, Figure 5.14). We have already discussed the frontal
and occipital bones. The frontal bone (together with the ethmoid) forms the
anterior cranial fossa; the occipital bone forms the **posterior cranial fossa.** We
will now consider the ethmoid, the sphenoid, and the temporal bones. The
sphenoid and temporal bones form the **middle cranial fossa.**

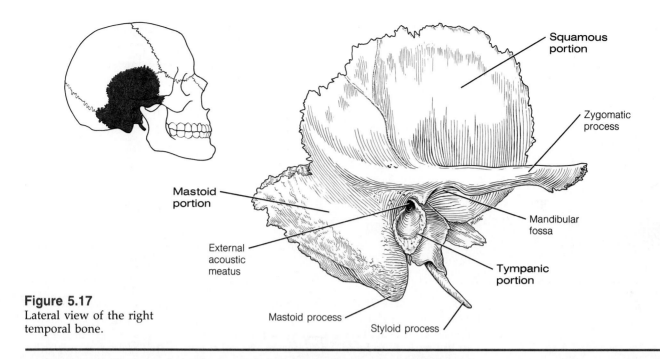

Figure 5.17
Lateral view of the right temporal bone.

ETHMOID BONE The **ethmoid bone** is situated in the middle of the floor of the anterior cranial fossa, where it forms most of the walls of the upper portion of the nasal cavity (Figure 5.13, Figure 5.14, Figure 5.15). The ethmoid bone is lightweight and delicate, containing many air sinuses. It has four parts: the **horizontal (cribriform) plate,** a midline **perpendicular plate,** and two **lateral masses** that project downward from the horizontal plate.

F5.13, F5.14, F5.15

The horizontal plate joins with the frontal bone to form the floor of the anterior cranial fossa. This plate is called the cribriform plate (*cribriform = sievelike*) because it is perforated by many tiny **olfactory foramina.** The olfactory nerves pass through these foramina in traveling between the mucous membranes of the nasal cavity and the olfactory bulbs of the brain. The perpendicular plate forms the major part of the nasal septum, which divides the nose into right and left nasal cavities. On the medial surfaces of the lateral masses there are projections, called **superior** and **middle conchae (turbinates),** that form the sidewalls of the nasal cavities; the smooth lateral surfaces of the lateral masses are referred to as the **lamina orbitalis** because they form part of the medial walls of the orbital cavities. Table 5.3 lists additional information concerning the ethmoid bone.

Table 5.3

SPHENOID BONE The **sphenoid bone** extends completely across the floor of the middle cranial fossa (Figure 5.12, Figure 5.13, Figure 5.14, Figure 5.16). The sphenoid is surrounded on all sides by other bones, articulating posteriorly with the basioccipital portion of the occipital bone, laterally with the temporal and parietal bones, and anteriorly with the frontal and ethmoid bones.

F5.12, F5.13, F5.14, F5.16

The sphenoid has a complex shape, with a central **body** from which pairs of **small (lesser) wings, great wings,** and **pterygoid processes** project. The anterior surfaces of the great wings form most of the posterior walls of the orbital cavities. The **optic foramina,** located in the bases of the small wings, provide for the passage of the optic nerves from the eyes to the base of the brain. The superior surface of the body of the sphenoid contains a deep depression called the **sella turcica** (Turk's saddle). The sella turcica, which houses the pituitary gland, is bounded posteriorly by a ridge of bone called the **dorsum sellae.** Table 5.3 lists additional information concerning the sphenoid bone.

Table 5.3

TEMPORAL BONES The two **temporal bones,** together with the sphenoid bone, form the *middle cranial fossa* (Figure 5.11, Figure 5.12, Figure 5.13, Figure 5.17). Each bone consists of four regions:

F5.11, F5.12, F5.13, F5.17

1. The thin **squamous portion** projects upward, articulating with the parietal bone in the squamous suture.

2. The **tympanic portion** forms the walls of the external auditory meatus and the region of the bone that closely surrounds the meatus.

F5.12, F5.13
3. The **petrous portion** (Figure 5.12, Figure 5.13) projects medially between the sphenoid and occipital bones. It contains the middle and inner ear cavities.

4. The **mastoid portion** is located posterior to the external auditory meatus.

Table 5.3 Table 5.3 lists additional information concerning the temporal bones.

Facial Skeleton

Ten bones form most of the facial skeleton: the unpaired *frontal* and *mandible* bones and the paired *maxillary, zygomatic, lacrimal,* and *nasal* bones (Figure **F5.10** 5.10). We have already discussed the frontal bone.

MAXILLARY BONES   The **maxillary bones (maxillae)** form the central part **F5.18** of the facial skeleton (Figure 5.18). With the exception of the mandible, all the facial bones articulate directly with the maxillae. The two maxillary bones join in the midline to form the upper jaw. In addition, each assists in forming the roof of the mouth, the floor and lateral wall of the nasal cavity, and the floor of the orbit. The large maxillary air sinuses are within the body of the bone. **Table 5.3** Table 5.3 lists additional information concerning the maxillary bones.

ZYGOMATIC BONES   The **zygomatic (malar) bones** articulate with the max- **F5.10,** illary and temporal bones to form the prominences of the cheek (Figure 5.10, **F5.11** Figure 5.11). They also articulate with the frontal and sphenoid (great wing) to form part of the floor and the lateral wall of the orbit. A small **zygomaticofacial foramen** allows for the passage of the zygomaticofacial blood vessels and nerve.

F5.18 NASAL BONES   The **nasal bones** (Figure 5.18) are two small oblong bones that meet at the midline of the face to form the bridge of the nose. In addition, they articulate with the frontal, ethmoid (perpendicular plate), and maxillary bones (frontal process).

F5.18 LACRIMAL BONES   The right and left **lacrimal bones** (Figure 5.18) are small delicate bones that help to form the medial surface of the orbital cavity. They articulate above with the frontal bone, behind with the ethmoid (orbital surfaces of the lateral masses), and in front with the maxillary bones (frontal process). Near their anterior edges, each lacrimal bone has a **lacrimal sulcus** for the lacrimal sac and the nasolacrimal duct. The lacrimal sac and the nasolacrimal duct transport tear fluid from the surface of the eye to the nasal cavity.

F5.11 MANDIBLE   The facial skeleton is completed by the **mandible** (Figure 5.11), which forms the lower jaw. The mandible consists of a horizontal horseshoe-shaped **body** and two perpendicular **rami.** The condyloid processes located on the superior margins of the rami form movable joints with the mandibular **Table 5.3** fossae of the temporal bones. Table 5.3 lists additional information concerning the mandible.

Bones That Form the Nasal Cavity

We have already discussed how most of the nasal septum is formed by the perpendicular plate of the ethmoid. Much of the lateral walls of the nasal cavity are formed by the superior and middle conchae of the ethmoid bone **Table 5.3** (see Table 5.3). Two additional bones also contribute to the formation of the nasal cavity: the *vomer* and the *inferior nasal conchae.*

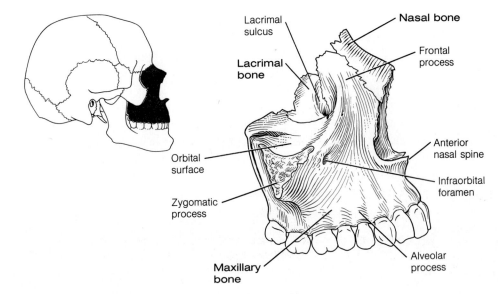

Figure 5.18
Lateral view of the right maxillary, nasal, and lacrimal bones.

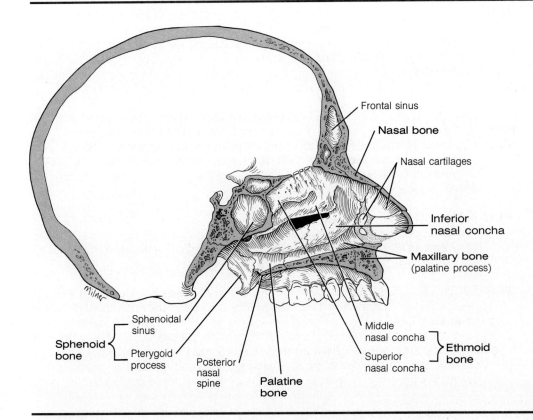

Figure 5.19
Bones that form the left lateral wall of the nasal cavity.

The **vomer** (Figure 5.14) is a thin quadrangular bone that forms the posterior inferior portion of the nasal septum. Its superior border articulates with the sphenoid between the pterygoid processes; its inferior border articulates with the upper surface of the hard palate (maxillae and palatine bones). The upper part of the vomer's anterior border articulates with the perpendicular plate of the ethmoid, and the lower part is in contact with the cartilaginous portion of the nasal septum.   **F5.14**

The paired **inferior nasal conchae** (Figure 5.19) form elongated shelves that protrude medially from the lateral walls of the nasal cavity. They are located just below the middle conchae of the ethmoid bone. The inferior nasal conchae articulate with the maxillae, lacrimals, ethmoid, and palatine bones.   **F5.19**

Table 5.3   Summary of Specific Features of Individual Skull Bones

FRONTAL BONE   *[F5.10; F5.11; F5.13; F5.14; F5.21]*

METOPIC SUTURE The line of junction between the two separate embryonic ossification centers. Generally not present in the adult skull.

FRONTAL SINUSES [F5.14] Mucous-membrane-lined air cavities located within the bone, close to the orbital cavities.

SUPRAORBITAL FORAMINA OR NOTCHES [F5.10] Openings for blood vessels and nerves located just above the orbital cavities. They may appear as holes (foramina) or notches.

GLABELLA [F.10] The smooth area located between the two orbital cavities just above the nose.

OCCIPITAL BONE   *[F5.11; F5.12; F5.13; F5.14]*

FORAMEN MAGNUM [F5.12; F5.13] The opening through which the medulla oblongata of the brain stem leaves the skull to become continuous with the spinal cord.

CONDYLES [F5.12] Smooth, convex external projections on either side of the foramen magnum. They articulate with the first cervical vertebra.

HYPOGLOSSAL CANAL [F5.13] Opening through the base of the condyle for the passage of the hypoglossal nerve and a branch of the ascending pharyngeal artery.

BASIOCCIPITAL [F5.12] A narrow portion that extends anteriorly from the foramen magnum. It articulates with the sphenoid bone.

EXTERNAL OCCIPITAL PROTUBERANCE [F5.12; F5.14] A midline prominence on the outer surface a short distance above the foramen magnum.

NUCHAL LINES* Slight ridges on the external surface. *Medial nuchal line* Runs vertically between the external occipital protuberance and the foramen magnum. *Superior nuchal line* Extends laterally from the external occipital protuberance. *Inferior nuchal line* Extends laterally from the medial nuchal line at about its midpoint.

INTERNAL OCCIPITAL PROTUBERANCE* A prominence on the inner surface of the bone. This marks the confluence of grooves for the sagittal, transverse, and occipital venous blood sinuses of the brain.

ETHMOID BONE   *[F5.13; F5.14; F5.15; F5.19; F5.21]*

HORIZONTAL (CRIBIFORM) PLATE [F5.13; F5.15] The transverse portion that forms the roof of the nasal cavity and floor of the anterior cranial cavity. The plate is perforated by the olfactory foramina to allow passage of the olfactory nerves (first cranial nerve).

CRISTA GALLI [F5.13; F5.14; F5.15] A midline projection from the horizontal plate into the cranial cavity. It serves as the anterior point of attachment for the

falx cerebri, a midline connective tissue septum that anchors the brain within the anterior cranial fossa.

PERPENDICULAR PLATE [F5.14; F5.15] A downward projection from the midline of the undersurface of the horizontal plate. It forms the upper portion of the nasal septum. The remainder of the septum is formed by the vomer bone and hyaline cartilage.

LATERAL MASSES [F5.15] Thin-walled processes that extend downward from the lateral margins of the horizontal plate. They contain the **ethmoid sinuses,** which are mucous-membrane-lined air cavities. The smooth lateral surfaces **(lamina orbitalis)** of the lateral masses form the medial walls of the orbital cavities.

SUPERIOR AND MIDDLE CONCHAE (TURBINATES) [F5.19] Thin plates of bone that form the medial surfaces of the lateral masses. They also form part of the lateral walls of the nasal cavity. Recesses called **superior, middle,** and **inferior meatuses** are located beneath the shelves of the conchae.

SPHENOID BONE   *[F5.10; F5.11; F5.12; F5.13; F5.14; F5.16]*

BODY [F5.16] The central portion of the bone. It contains a large mucous-membrane-lined air sinus.

SELLA TURCICA [F5.13; F5.16] A saddle-shaped depression on the superior surface of the body, bounded posteriorly by the **dorsum sellae.** It serves as the protective cavity for the pituitary gland.

SMALL WINGS [F5.14; F5.16] Sharp lateral projections from the superior portion of the body of the sphenoid. They form part of the posterior walls of the orbital cavities.

OPTIC FORAMINA [F5.13; F5.16] Openings through the bases of each small wing for the passage of the optic nerves (second cranial nerves) and ophthalmic artery into the orbital cavities.

GREAT WINGS [F5.12; F5.16] Large lateral projections from the body of the sphenoid. They form most of the posterior wall of the orbital cavity.

SUPERIOR ORBITAL FISSURES [F5.10; F5.16; F5.21] Slitlike openings between the great and small wings. They allow for the passage of the third, fourth, part of the fifth (ophthalmic division), and sixth cranial nerves from the brain into the orbital cavity.

FORAMEN ROTUNDUM [F5.13; F5.16] The opening through the base of each of the great wings for the passage of the maxillary division of the fifth cranial nerves.

FORAMEN OVALE [F5.12; F5.13; F5.16] The opening through the base of each of the great wings for the passage of the mandibular division of the fifth cranial nerves.

FORAMEN SPINOSUM [F5.13; F5.15] Small opening through posterior angle of the sphenoid for passage of the middle meningeal blood vessels.

PTERYGOID PROCESSES [F5.14; F5.19] Two downward projections from the region where the great wings unite with the body. Each process consists of **medial** and **lateral plates.** The processes articulate anteriorly with the palatine bones.

TEMPORAL BONE   *[F5.10–F5.14; F5.17]*

SQUAMOUS PORTION *[F5.13; F5.17]* The thin vertical projection that forms the anterior and superior portion of the bone. It meets with a parietal bone to form the squamous suture.

ZYGOMATIC PROCESS [F5.17] The anterior projection from the squamous portion. It articulates with the zygomatic (malar) bone to form the cheek (zygomatic arch).

MANDIBULAR FOSSA [F5.17] An oval depression on the inferior surface of the base of the zygomatic process. It articulates with the condyle of the mandible to form the temporomandibular joint.

TYMPANIC PORTION [F5.17] Forms and surrounds the external acoustic meatus.

EXTERNAL ACOUSTIC MEATUS [F5.17] The opening that leads into the middle ear cavity from the exterior of the skull.

PETROUS PORTION [F5.12; F5.13] A medial wedge of bone that forms the floor of the middle cranial fossa between the sphenoid and the occipital bones. It houses the middle and inner ear structures.

INTERNAL ACOUSTIC MEATUS [F5.13; F5.14] The opening on the posterior surface of the petrous portion. It transmits the seventh cranial nerve as it travels to the facial structures and the eighth cranial nerve as it travels to the inner ear.

STYLOID PROCESS [F5.11; F5.12; F5.17] A sharp spine that projects from the inferior lateral surface of the petrous portion. It serves as a point of attachment for the ligament to the hyoid bone and for several ligaments and muscles of the pharynx and tongue.

CAROTID CANAL [F5.12] The passageway for the internal carotid artery as it travels through the petrous portion.

JUGULAR FOSSA [F5.12] The depression for the internal jugular vein on the inferior surface of the petrous portion.

STYLOMASTOID FORAMEN [F5.12] The opening between the styloid process and mastoid process through which the seventh cranial nerve leaves the skull. (The nerve enters through the internal acoustic meatus.)

JUGULAR FORAMEN [F5.13] The large opening that allows for passage of the internal jugular vein and the ninth, tenth, and eleventh cranial nerves. It is located at the junction of the petrous portion with the occipital bone.

MASTOID PROCESS [F5.11; F5.12; F5.17] A prominent downward projection from the mastoid portion, just posterior to the external acoustic meatus.

MASTOID SINUSES * Mucous-membrane lined air spaces within the mastoid process. These sinuses, which communicate with the middle ear cavity, are the only cranial sinuses that do not drain into the nasal cavity.

MAXILLARY BONE (MAXILLA)   *[F5.10; F5.12; F5.14; F5.18]*

MAXILLARY SINUS [F5.22] A large mucous-membrane-lined cavity within the bone.

FRONTAL PROCESS [F5.18] The vertical process that forms part of the bridge of the nose. It articulates above with the frontal bone, anteriorly with the nasal, and posteriorly with the lacrimal.

ZYGOMATIC PROCESS [F5.18] A rough triangular eminence that articulates with the zygomatic (malar) bone.

ALVEOLAR PROCESS [F5.11; F5.18] The inferior border that holds the teeth. When the two maxillae are articulated with each other, their alveolar processes together form the alveolar arch.

PALATINE PROCESS [F5.12; F5.19] The medial horizontal shelf that runs from the inner surface of the alveolar process. It joins with the palatine process of the other maxillary bone to form most of the hard palate.

ANTERIOR NASAL SPINE [F5.18] A pointed process just below the nasal cavity. It joins with the nasal spine of the other maxillary bone to form a point of attachment for the cartilage portion of the nasal septum.

INFRAORBITAL FORAMEN [F5.10; F5.18; F5.21] The opening just below the margin of the orbit. It transmits blood vessels and nerves.

ORBITAL SURFACE [F5.21] The smooth, flat surface that forms the floor of the orbit.

INCISIVE FOSSA [F5.12] Located on the palatine process, just behind the incisor teeth. For the passage of the nasopalatine nerve and descending septal blood vessels.

MANDIBLE   *[F5.10; F5.11; F5.14]*

BODY [F5.11] The curved, horizontal portion that forms the chin.

RAMI [F5.11] Two perpendicular projections that join the posterior lateral margins of the body at approximately right angles.

MANDIBULAR SYMPHYSIS [F5.10] The vertical midline fusion between the two embryonic ossification centers that form the body.

ALVEOLAR BORDER [F5.11] The superior edge of the body that contains the sockets for the teeth.

Table 5.3   Summary of Specific Features of Individual Skull Bones (continued)

MENTAL FORAMINA [F5.10; F5.11] Two openings on the external surface of the body that allow for the passage of mental blood vessels and nerves.

ANGLE [F5.11] A sharp curve on the posterior inferior portion of the ramus.

MANDIBULAR FORAMINA [F5.14] Openings on the inner surfaces of each ramus for the passage of inferior alveolar blood vessels and nerves.

CORONOID PROCESSES [F5.14] Thin upward projections on the anterior surface of each ramus. They provide attachment for the temporalis muscle.

MANDIBULAR CONDYLES [F5.11] Smooth, convex surfaces on the superior borders of each ramus. They articulate with the mandibular fossae of the temporal bones.

MANDIBULAR NOTCHES [F5.11] Deep depression between the coronoid process and the condyle of each ramus.

*Not illustrated.

Bones That Form the Hard Palate

F5.12, F5.14   The hard palate forms the roof of the mouth (Figure 5.12, Figure 5.14). The anterior portion of the hard palate is formed by the **palatine processes** of the **maxillary bones.** The posterior portion is formed by the **horizontal plates** of the two **palatine bones.**

F5.20   Each palatine bone is L-shaped, with a horizontal and a vertical portion (Figure 5.20). On the posterior edge of the horizontal plates, at their midline point of junction, is a sharp projection called the **posterior nasal spine.** This spine serves for the attachment of the uvula, a small fleshy mass that hangs from the soft palate. In the lateral margin of each horizontal plate are **greater** and **lesser palatine foramina** for the passage of greater and lesser palatine nerves and blood vessels. The vertical portion of each palatine bone forms part of the posterior lateral wall of the nasal cavities. A small portion of the vertical plate contributes to the formation of the orbital cavity. Just below the orbital surface a **sphenopalatine foramen** allows for the passage of the pterygopalatine blood vessels and nerves through the vertical portion.

Bones That Form the Orbital Cavity

F5.21,   We have already discussed the contributions of the various bones to the for-
Table 5.4   mation of the orbit (Figure 5.21). These are summarized in Table 5.4.

Passageways Through the Skull

There are numerous openings through the skull for the passage of blood vessels and nerves. Many of these have been mentioned in the discussions of individual skull bones. Table 5.5 summarizes these passageways.

Paranasal Sinuses

Located within the *frontal, ethmoid, maxillary,* and *sphenoid* bones are a series of air spaces called the **paranasal sinuses.** The paranasal sinuses, which drain into the nasal cavity, are lined with ciliated epithelium on a mucous membrane that is continuous with the mucous membrane of the nasal cavity. The **frontal sinuses** (Figure 5.14, Figure 5.19, Figure 5.22) are found above the medial ends of the orbits in the region of the glabella. The **ethmoid sinuses** (Figure 5.15, Figure 5.22) are a series of small spaces located in the lateral masses of the bone. The **maxillary sinuses** (Figure 5.22), the largest of the paranasal sinuses, occupy much of the bone from the orbit to the alveolar processes. The **sphenoidal sinuses** (Figure 5.14, Figure 5.19, Figure 5.22) are contained within the body of the bone.

Auditory Ossicles

F5.23   Three tiny bones called **auditory ossicles** are located in the *middle-ear (tympanic) cavities* (Figure 5.23). The cavities are inside the petrous portion of each temporal bone. The ossicles form a bridge across the tympanic cavity from the

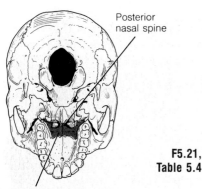

Posterior nasal spine

Horizontal part

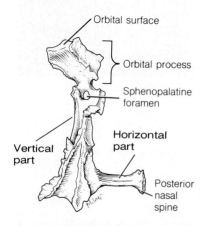

Orbital surface

Orbital process

Sphenopalatine foramen

Vertical part

Horizontal part

Posterior nasal spine

Figure 5.20
Left palatine bone (posterior view).

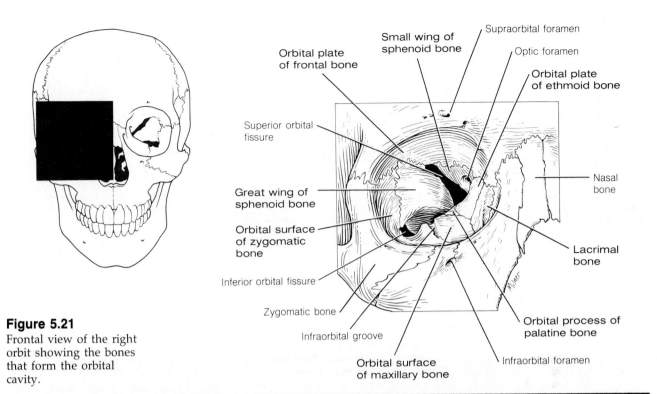

Figure 5.21
Frontal view of the right orbit showing the bones that form the orbital cavity.

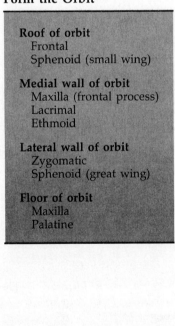

Table 5.4 Bones that Form the Orbit

Roof of orbit
Frontal
Sphenoid (small wing)

Medial wall of orbit
Maxilla (frontal process)
Lacrimal
Ethmoid

Lateral wall of orbit
Zygomatic
Sphenoid (great wing)

Floor of orbit
Maxilla
Palatine

Frontal sinuses

Sphenoidal sinuses

Ethmoid sinuses

Maxillary sinuses

Figure 5.22
X ray of the skull showing several paranasal sinuses.

Table 5.5   Major Passageways Through the Skull

Name	Location	Structures Passing Through
Carotid canal [F5.12]	Temporal (petrous portion)	Internal carotid artery
Hypoglossal canal [F5.13]	Occipital (base of condyles)	Hypoglossal nerve (XII) and branch of ascending pharyngeal artery
Incisive fossa [F5.12]	Maxillary (palatine process, posterior to incisor teeth)	Nasopalatine nerve and descending septal blood vessels
Inferior orbital fissure [F5.10; F5.21]	Between great wing of sphenoid and maxillary	Maxillary division of trigeminal nerve (V) and infraorbital blood vessels
Infraorbital foramen [F5.10; F5.18; F5.21]	Maxillary (below margin of orbit)	Infraorbital nerve and blood vessels
Jugular foramen [F5.12; F5.13]	Temporal (at junction of petrous portion with occipital)	Glossopharyngeal (IX), vagus (X), and accessory (XI) nerves; internal jugular vein
Foramen lacerum [F5.11; F5.12]	Junction of sphenoid anteriorly, petrous portion of temporal posteriorly and laterally, sphenoid and occipital medially	Branch of ascending pharyngeal artery
Foramen magnum [F5.12; F5.13]	Occipital	Medulla oblongata; accessory nerves (XI); vertebral and spinal arteries
Mandibular foramen [F5.14]	Mandible (inner surface of ramus)	Inferior alveolar blood vessels and nerves
Mental foramen [F5.10; F5.11]	Mandible (external surface of body)	Mental blood vessels and nerves
Olfactory foramina [F5.13; F5.15]	Ethmoid (cribriform plate)	Olfactory nerve (I)
Optic canal (foramen) [F5.13; F5.16]	Sphenoid (base of small wing)	Optic nerve (II) and ophthalmic artery
Foramen ovale [F5.12; F5.13; F5.16]	Sphenoid (base of great wing)	Mandibular division of trigeminal nerve (V)
Palatine foramina (greater and lesser) [F5.12]	Palatine (horizontal plate)	Greater and lesser palatine blood vessels and nerves
Foramen rotundum [F5.13; F5.16]	Sphenoid (base of great wing)	Maxillary division of trigeminal nerve (V)
Sphenopalatine foramen [F5.20]	Palatine (below orbital surface)	Pterygopalatine blood vessels and nerves
Foramen spinosum [F5.13; F5.16]	Sphenoid (posterior angle)	Middle meningeal blood vessels
Stylomastoid foramen [F5.12]	Temporal (between styloid and mastoid)	Facial nerve (VII) and stylomastoid artery
Superior orbital fissure [F5.10; F5.16; F5.21]	Sphenoid (between great and small wing)	Oculomotor (III), trochlear (IV), ophthalmic division of trigeminal (V), and abducens (VI) nerves
Supraorbital foramen (or notch) [F5.10]	Frontal (just above margin of orbit)	Supraorbital blood vessels and nerves
Zygomaticofacial foramen [F5.10]	Zygomatic	Zygomaticofacial blood vessels and nerves

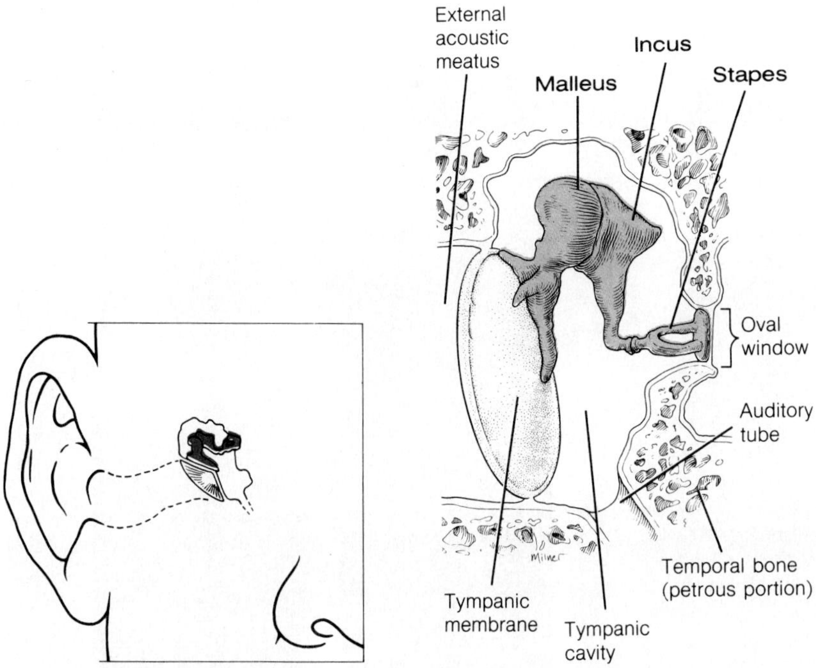

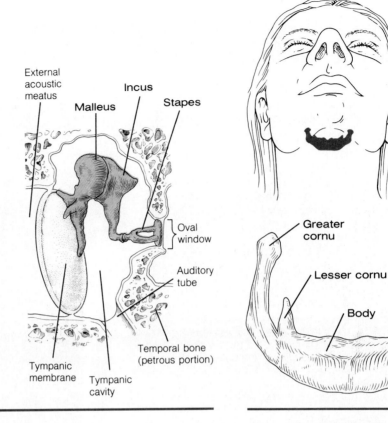

Figure 5.23
The middle ear cavity
and the auditory ossicles.

Figure 5.24
Anterior view of the
hyoid bone.

eardrum (tympanic membrane) to a membrane (oval window) that separates
the middle ear from the internal ear.

The **malleus** (hammer) is attached to the inside surface of the tympanic
membrane. The **stapes** (stirrup) fits against the oval window. The **incus**
(anvil) forms a connection between the malleus and the stapes. The role
played by these bones in transmitting sound is discussed in Chapter 17.

Hyoid Bone

The **hyoid bone** (Figure 5.24) is a U-shaped bone that is suspended by liga- F5.24
ments from the styloid processes of the temporal bones. It is located just
above the larynx. The hyoid consists of a central **body** and pairs of **greater** and
lesser cornua (horns). It serves as points of attachment for muscles of the
tongue and throat.

Vertebral Column

The embryonic vertebral column consists of 33 vertebrae, which are separated
into five different types, depending on the regions of the body in which they
are located. The upper 7 are **cervical** vertebrae, followed by 12 **thoracic,** 5
lumbar, 5 **sacral,** and 4 **coccygeal** vertebrae. In the adult, the sacral vertebrae
fuse into a single **sacrum,** and the coccygeal vertebrae fuse to form a **coccyx.**
Therefore the adult vertebral column has 26 separate bones (Figure 5.25). F5.25

Curvatures of the Vertebral Column

When viewed from the lateral aspect, the vertebral column of a newborn
infant has a single curve, which is convex posteriorly. As the child begins to
raise its head, a **cervical curve** develops that is convex anteriorly. In a similar

FRONTIERS IN HEALTH

Treatment of Scoliosis

Jennifer was a normal child. She ate well and played actively; but at age 10 something troublesome began to occur. From a point midway up her back, Jennifer's spinal column began to curve unnaturally to the right. Her lower spine twisted to the left.

Fortunately for Jennifer, she lived in one of many school districts that have begun to check its students for this disease, which is called scoliosis. Scoliosis, a lateral curvature of the spine, occurs in about 10% of all children. Every year more than 220,000 individuals are diagnosed as having the disease, which appears most frequently in children between the ages of 10 and 15, when the bones are growing rapidly. Mild forms afflict boys and girls in equal numbers, but the more severe cases are more prevalent in girls.

Scoliosis can be crippling if left untreated, bending the spine into an S-shaped or C-shaped form, and its cosmetic effects are demoralizing. Making matters even worse, as the spine curves it also twists, which throws the rib cage out of normal position. The ribs are pushed together on the concave side and separated on the convex side. In severe cases, the distortion of the rib cage can impair breathing.

Treatment for this disease, whose cause is unknown, varies with the severity of the curvature. In mild cases, a body brace is worn 22 hours a day throughout adolescence to hold the spine straight. This can mean wearing a heavy, unsightly brace for 4 or 5 years. The discomfort and appearance of these braces spurred medical researchers to seek a more comfortable alternative.

Orthopedic surgeon Dr. Walter Bobechko of Toronto, Canada, devised a small electrical muscle stimulator that is used by victims of scoliosis only at night. Electrodes are placed on the skin of the child's back and attached to a power pack. Weak electrical impulses travel to the electrodes and cause the underlying muscles to contract several times a minute throughout the night, without interfering with the child's sleep. In the morning, the child can go to school free of any cumbersome brace. Stimulation of the muscles on the convex side of the curvature straightens the back over time.

Transcutaneous muscle stimulation has been very successful. In one clinical study, Dr. Jens Axelgaard and Dr. John Brown of Rancho Los Amigos Hospital in Downey, California, found that the electrical muscle stimulator stopped the curvature in 95 of every 100 patients if used every night as prescribed. Even more important, the device helped return the spine to its correct position, and 2 years after treatment was stopped the spine remained straight. The muscle stimulator halted the disfiguring bending and twisting of the spine regardless of whether the curvature was slight or large, or whether the rate of bending was fast or slow. Furthermore, no side effects were noted. Muscle on the stimulated side appeared normal. The heart seemed totally unaffected, as did the lungs and skin. The muscle stimulator allowed Jennifer to continue normal activities. Only a handful of friends even knew she had the condition.

In severe cases of scoliosis, corrective surgery may be necessary. Surgeons straighten the back and then

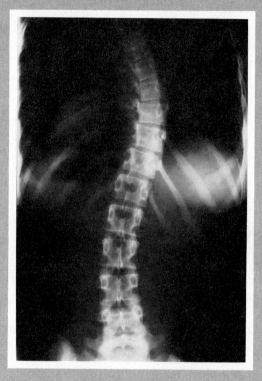

X ray of scoliosis.

place one or two metal rods along the laminae of the vertebrae. The rods are hooked to the laminae and immobilize the straightened spine. To ensure lasting rigidity, surgeons also fuse the vertebrae together by using bone fragments, usually taken from the patient's ilium. After surgery, the patients are encased in a thick cast weighing about 20 pounds and extending from the neck to the thighs. After 6 months, a lighter brace is used to provide support for the next 3 months.

A new technique, devised by a Mexican physician, Dr. Eduardo Luque, promises to reduce the size of the cast. Instead of hooking the metal rods in place, Dr. Luque secures them with wire wrapped tightly around the laminae. This holds the rods more firmly in place. Patients then get all the support they need from a ⅛-inch-thick cast that weighs only 3 pounds and is hardly noticeable beneath their clothing. Dr. Luque's technique cuts the recovery time by a month or more. Recovery is also more comfortable, and the lightweight brace is removable, unlike the heavier body casts that are worn for 6 months and prohibit bathing during that time.

Early detection is the first line of defense in scoliosis. Physicians report that screening programs throughout the country are already proving effective in cutting down on the severity of the disease. It is hoped that screening and electrical muscle stimulation may someday eliminate the need for costly, painful back surgery sometimes necessary to correct scoliosis.

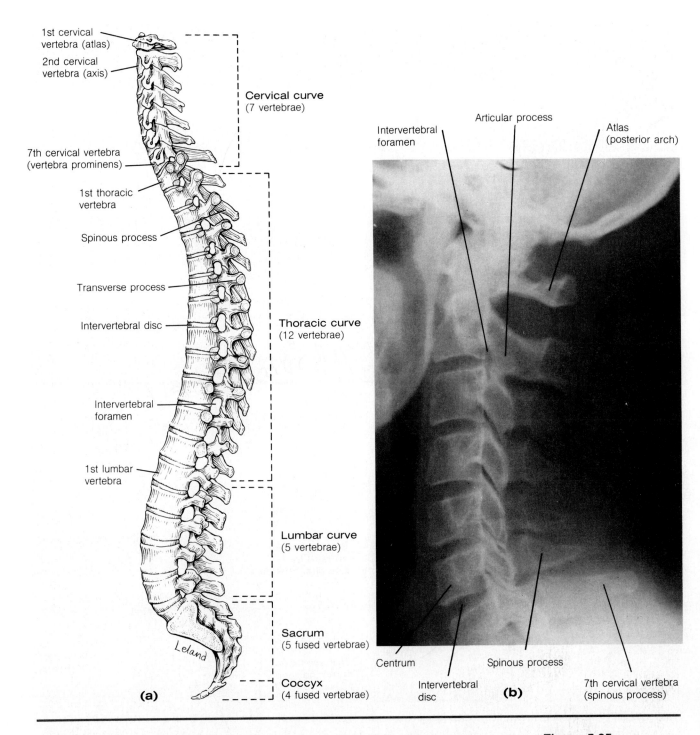

(a)

1st cervical
vertebra (atlas)

2nd cervical
vertebra (axis)

Cervical curve
(7 vertebrae)

7th cervical vertebra
(vertebra prominens)

1st thoracic
vertebra

Spinous process

Transverse process

Intervertebral disc

Thoracic curve
(12 vertebrae)

Intervertebral
foramen

1st lumbar
vertebra

Lumbar curve
(5 vertebrae)

Leland

Sacrum
(5 fused vertebrae)

Coccyx
(4 fused vertebrae)

(b)

Intervertebral
foramen

Articular process

Atlas
(posterior arch)

Centrum

Intervertebral
disc

Spinous process

7th cervical vertebra
(spinous process)

Figure 5.25
(a) Lateral view of the
vertebral column.
(b) Lateral X ray of the
cervical vertebrae.

manner, a secondary **lumbar curve** develops as the child begins to walk. If the anterior lumbar curve is excessive, it is called a *lordosis* (swayback). If the posterior **thoracic curve**—which remains from the primary curve of the newborn—is excessive, it is called a *kyphosis* (hunchback). The vertebral column is normally straight, without any lateral curvatures. If a lateral curve does exist, it is called a *scoliosis*.

Functions of the Vertebral Column

The vertebral column is the main axial support for the body, providing attachments for the skull, the thorax, and the pelvic girdle. Although it is a major support structure, its construction is such that it permits the trunk of the body to have appreciable flexibility. In addition, the vertebral column protects the

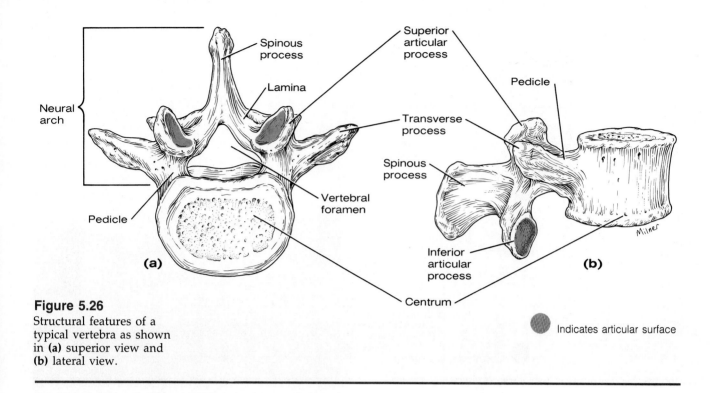

Figure 5.26
Structural features of a typical vertebra as shown in **(a)** superior view and **(b)** lateral view.

Indicates articular surface

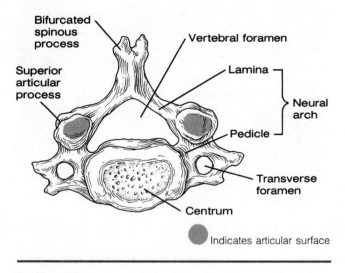

Indicates articular surface

Figure 5.27
Superior view of a typical cervical vertebra.

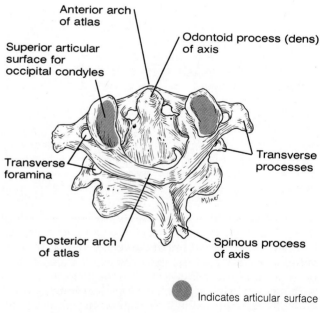

Indicates articular surface

Figure 5.28
Superior lateral view showing the articulated first and second cervical vertebrae (the atlas and the axis).

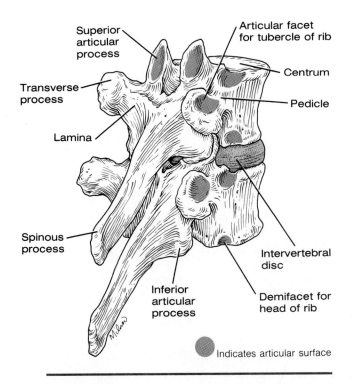

Figure 5.29
Two typical thoracic
vertebrae.

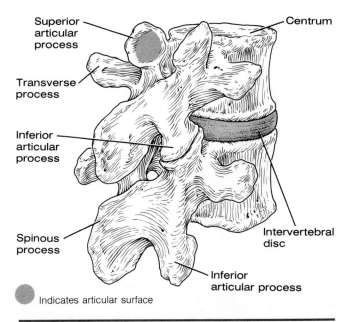

Figure 5.30
Two typical lumbar
vertebrae.

spinal cord while providing openings between adjacent vertebrae for the passage of spinal nerves.

Characteristics of a Typical Vertebra

Although there are differences among the vertebrae of the various regions of the spinal column, enough similarities exist so that it is possible to describe a typical vertebra (Figure 5.26).

A typical vertebra has a thick anterior **centrum** (body), with a **neural (vertebral) arch** that arises from the posterior surface of the centrum. The centra of adjacent vertebrae are separated by fibrocartilaginous **intervertebral discs.** Each neural arch combines with the posterior surface of the centrum and encloses a **vertebral foramen.** The vertebral foramina of adjacent vertebrae are aligned to form a **vertebral canal,** through which the spinal cord passes. **Transverse processes** extend laterally from each neural arch. Projecting from the posterior region of the neural arch is the midline **spinous process.** The spinous and transverse processes allow for the attachment of muscles and ligaments. That portion of the neural arch between the centrum and the transverse process is the **pedicle.** The portion between the transverse process and the spinous process is the **lamina.** Projecting upward from each side of the neural arch is a pair of **superior articulating processes;** their articular surfaces face posteriorly. Projecting downward is a pair of **inferior articulating processes;** their articular surfaces face anteriorly. The smooth articular surface of each process meets with the process of the vertebra above or below it, thereby increasing the rigidity of the vertebral column. The **intervertebral foramina,** through which the spinal nerves pass, are located between the pedicles of adjacent vertebrae.

Regional Differences in Vertebrae

The vertebrae of each region have specific characteristics that vary from the "typical" vertebra and that enable them to be easily identified. These variations are illustrated in Figure 5.27 through Figure 5.31 and discussed in Table 5.6.

F5.26

F5.27–F5.31
Table 5.6

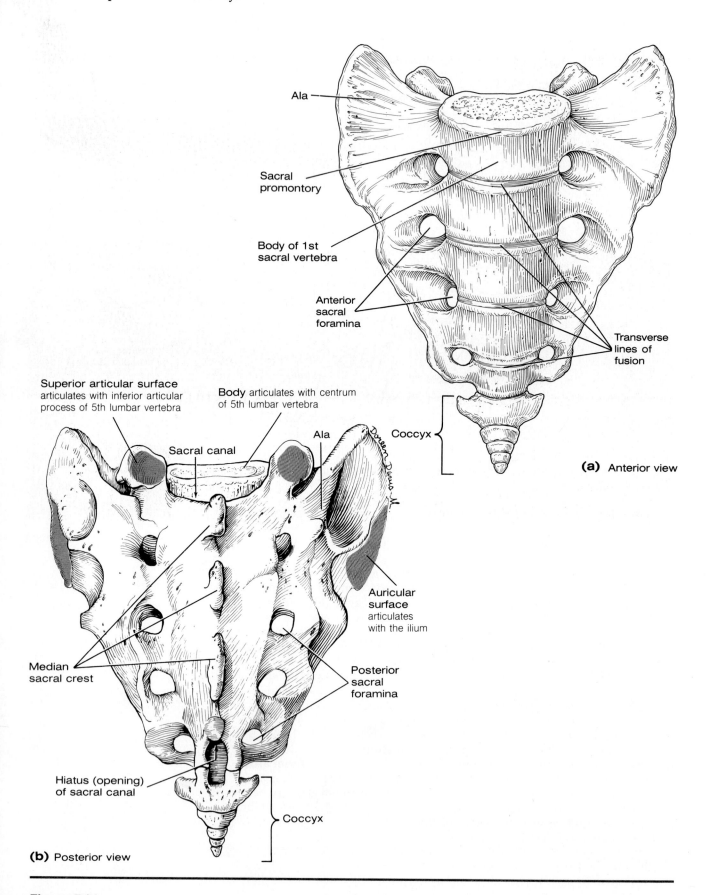

Ala

Sacral
promontory

Body of 1st
sacral vertebra

Anterior
sacral
foramina

Transverse
lines
of
fusion

Coccyx

(a) Anterior view

Superior articular surface
articulates with inferior articular
process of 5th lumbar vertebra

Body articulates with centrum
of 5th lumbar vertebra

Sacral canal

Ala

Coccyx

Auricular
surface
articulates
with the ilium

Median
sacral crest

Posterior
sacral
foramina

Hiatus (opening)
of sacral canal

Coccyx

(b) Posterior view

Figure 5.31
The sacrum and the
coccyx. **(a)** Anterior view.
(b) posterior view.

Table 5.6 Identifying Features of Specific Vertebrae

CERVICAL VERTEBRAE *[F5.25; F5.27; F5.28]*

Transverse foramina [F5.27, F5.28] The openings in the transverse processes of each cervical vertebrae. They allow for the passage of vertebral arteries and veins to and from the brain.

Bifurcated spinous processes [F5.27; F5.28] The spinous processes of the cervical vertebrae have a double tip (with the exception of the first and the seventh).

Vertebra prominens [F5.25] The seventh cervical vertebra, so named because its long, prominent spinous process protrudes beyond those of the other cervical vertebrae, making it useful as a landmark in counting the other spinous processes.

Atlas [F5.25; F5.28] The first cervical vertebra, which articulates with the occipital condyles of the skull, has no centrum or spinous process. It is ringlike, consisting of anterior and posterior arches.

Axis [F5.25; F5.28] The second cervical vertebra has a vertical projection called the **odontoid process,** or **dens,** that arises from the superior surface of its centrum. This process provides a pivot around which the atlas rotates.

THORACIC VERTEBRAE *[F5.25; F5.29]*

Spinous processes [F5.29] Long and slender protuberances that project sharply downward. The downward projection is not as noticeable in the lower thoracic vertebrae.

Facets and demifacets [F5.29] Articular surfaces for the ribs on the transverse processes and the bodies of all thoracic vertebrae. (The eleventh and twelfth vertebrae are exceptions because they do not have articular facets on their transverse processes.)

LUMBAR VERTEBRAE *[F5.25; F5.30]*

Centra [F5.30] Larger and heavier than the centra in other regions.

Spinous processes [F5.30] Short and blunt compared to the spinous processes in other regions.

Articular processes [F5.30] The superior articular processes face inward rather than posteriorly; the inferior articular processes face outward rather than anteriorly. This positioning locks the vertebrae together by preventing rotation.

SACRAL VERTEBRAE *[F5.25; F5.31]* In the adult, the five sacral vertebrae are fused into a single triangular **sacrum.** The transverse lines of fusion are visible on its anterior surface. The spinous processes form the **median sacral crest** on its posterior surface. The fused transverse processes form the **alae** (wings), which articulate with the pelvic bones. The **sacral foramina** represent the intervertebral foramina. The superior edge of the ventral border of the first sacral vertebra forms a projection called the **sacral promontory.**

COCCYX *[F5.25; F5.31]* The fused coccygeal vertebrae. It articulates with the apex of the sacrum.

Thorax

The skeleton of the thorax (Figure 5.32) is formed by the *sternum,* the *ribs,* and the *costal cartilages.* The thoracic vertebrae form its posteriormost portion. **F5.32**

Sternum

The **sternum** is an elongated, flat bone that forms the midline portion of the anterior wall of the thorax (Figure 5.32). It is composed of three parts: the **F5.32** **manubrium,** the **body (gladiolus),** and the **xiphoid process.** The superior portion of the manubrium articulates with the medial end of each clavicle (collarbone). Its lateral margins articulate with the costal cartilages of the first ribs and part of the second ribs. The body of the sternum articulates at its lateral margins with the costal cartilages of the second ribs (which it shares with the manubrium) through the seventh ribs. The small xiphoid process does not articulate with the ribs. It serves as a point of attachment for several ligaments and muscles, including the rectus abdominis muscle. The linea alba, which marks the midline of the abdomen, is also attached to it.

Ribs

There are 12 pairs of **ribs** (Figure 5.32). While all of the ribs articulate posteri- **F5.32** orly with the thoracic vertebrae, only the first seven pairs articulate anteriorly with the sternum, through the costal cartilages. For this reason, the first seven pairs are referred to as the **true** or **vertebrosternal ribs.** The remaining five pairs are called **false ribs.**

The first three pairs of false ribs (that is, the eighth, ninth, and tenth ribs) have their costal cartilages attached to the cartilages of the rib above, rather than directly to the sternum. These are called **vertebrochondral ribs.** The costal cartilages of the eleventh and twelfth ribs are short and have no anterior articulation. For this reason they are called **floating** or **vertebral ribs.**

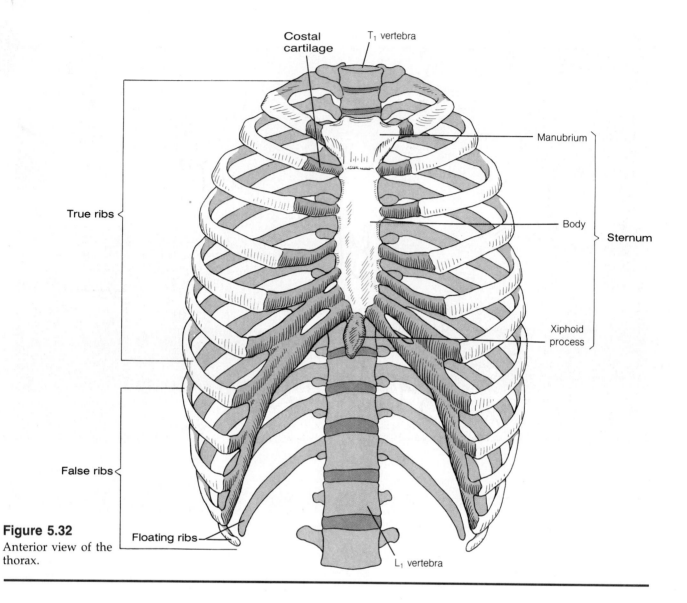

Costal cartilage

T₁ vertebra

Manubrium

True ribs

Body

Sternum

Xiphoid process

False ribs

Figure 5.32
Anterior view of the thorax.

Floating ribs

L₁ vertebra

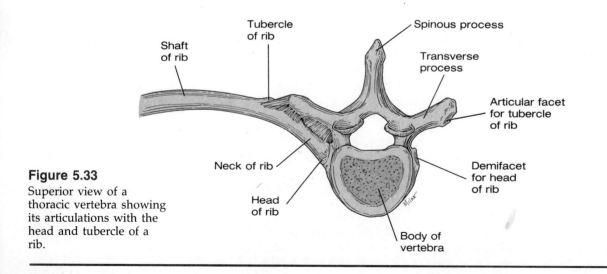

Shaft of rib

Tubercle of rib

Spinous process

Transverse process

Articular facet for tubercle of rib

Neck of rib

Demifacet for head of rib

Head of rib

Figure 5.33
Superior view of a thoracic vertebra showing its articulations with the head and tubercle of a rib.

Body of vertebra

Table 5.7 Summary of Specific Features of the Thoracic Skeleton

STERNUM *[F5.32]*

Manubrium The broad upper segment that articulates with the medial ends of each clavicle, the costal cartilages of the first pair of ribs, and part of the second pair of ribs. It has a small depression called the *jugular notch* on its superior border.

Body (Gladiolus) The elongated middle segment to which the costal cartilages of the second through the seventh ribs attach. It forms the *sternal angle* at its junction with the manubrium.

Xiphoid process A small inferior projection of cartilage that serves for the attachment of several ligaments and muscles.

RIBS *[F5.32; F5.33]*

Head The posterior medial end that articulates with the bodies of the thoracic vertebrae.

Neck The constricted portion just lateral to the head.

Tubercle A small projection just beyond the neck that articulates with the transverse process of a thoracic vertebrae. It is not present in the tenth, eleventh, and twelfth ribs.

TRUE RIBS

Vertebrosternal ribs The first through seventh pairs. They attach directly to the sternum.

FALSE RIBS

Vertebrochondral ribs The eighth, ninth, and tenth pairs. They attach to the costal cartilages of the rib above.

Vertebral (floating) ribs The eleventh and twelfth pairs. They have no anterior attachment.

The head of a typical rib articulates with the demifacets of two adjacent thoracic vertebrae. However, the heads of the first, tenth, eleventh, and twelfth ribs each articulate entirely on the facets of one vertebra (Figure 5.33). **F5.33** A short distance from the head is a **tubercle,** which articulates with the transverse process of a thoracic vertebra. Between the head and the tubercle is a constricted **neck.** Curving anteriorly from the neck is the **shaft** or body of the rib.

Costal Cartilages

The **costal cartilages** are composed of hyaline cartilage. They strengthen the thorax by serving as the anterior anchors for most of the ribs (Figure 5.32). At **F5.32** the same time, because they are cartilage, they provide flexibility that allows the rib cage to expand during respiration.

A summary of the skeletal features of the thorax is listed in Table 5.7. **Table 5.7** Table 5.8 summarizes the bones of the axial skeleton. **Table 5.8**

APPENDICULAR SKELETON

The appendicular skeleton includes the bones of the upper and lower limbs and the bones by which these limbs articulate with the axial skeleton—that is, the pectoral girdle and the pelvic girdle.

The pectoral girdle does not provide very firm support, being attached to the axial skeleton only at the sternum. This support is sufficient, however, since the upper limbs do not bear the body's weight. The pectoral girdle allows for a wide range of movements at the shoulder.

The pelvic girdle, in contrast, does support the body's weight. To accomplish this it not only has more extensive attachments to the axial skeleton through its articulation with the sacrum, but also, the two sides of the girdle

Table 5.8   Summary of Bones That Form the Axial Skeleton

	Number of Bones		Number of Bones
SKULL	29	**VERTEBRAL COLUMN**	26
CRANIUM (calvarium and floor of cranial cavity)	8	Cervical 7	
		Thoracic 12	
Parietal 2		Lumbar 5	
Temporal 2		Sacrum 5 fused to form 1	
Frontal 1		Coccyx 4 fused to form 1	
Occipital 1		**THORAX**	25
Ethmoid 1			
Sphenoid 1		Sternum 1	
		Ribs 24	
FACE AND NASAL CAVITY*	14	**TOTAL AXIAL SKELETON BONES**	80
Maxillary 2			
Zygomatic 2			
Lacrimal 2			
Nasal 2			
Inferior nasal concha 2			
Palatine 2			
Mandible 1			
Vomer 1			
AUDITORY OSSICLES	6		
Malleus 2			
Incus 2			
Stapes 2			
HYOID	1		

*The frontal and ethmoid bones also contribute to the face but are counted under the cranium.

attach to each other at the pubic symphysis. In addition, the pelvic girdle is aligned with the bones of the lower limbs in such a manner that it transfers much of the weight it supports to the skeleton of the lower limbs.

Upper Limbs

Table 5.9   The bones of the upper limbs are listed in Table 5.9.

Pectoral Girdle

F5.34   The **pectoral girdle** (Figure 5.34) straddles the upper part of the thorax. Its only joint with the axial skeleton is where the medial end of each clavicle articulates with the manubrium of the sternum. The lateral end of each clavicle articulates with the acromion process of a scapula. The scapulae are attached to the posterior thorax by muscles, but they do not contact the ribs directly, since they are separated from them by other muscles.

F5.34   CLAVICLE   The **clavicle** (Figure 5.34) is an S-shaped bone that serves as a brace for the scapula. Its **sternal (medial) end,** which articulates with the manubrium of the sternum, is enlarged and blunt. Its **acromial (lateral) end,** which articulates with the acromion process of the scapula, is flattened. Through its articulations with the sternum and the scapula, the clavicle holds the shoulder away from the rib cage, thus allowing the arm to swing freely without first necessitating the lifting of the arm away from the body wall.

Table 5.10   When a clavicle is broken, the entire shoulder collapses. Table 5.10 lists the terms associated with the clavicle.

F5.34   SCAPULA   The **scapula** (Figure 5.34) is a thin, flat, triangular bone that lies over the posterior surfaces of the second to seventh ribs. The **superior border**

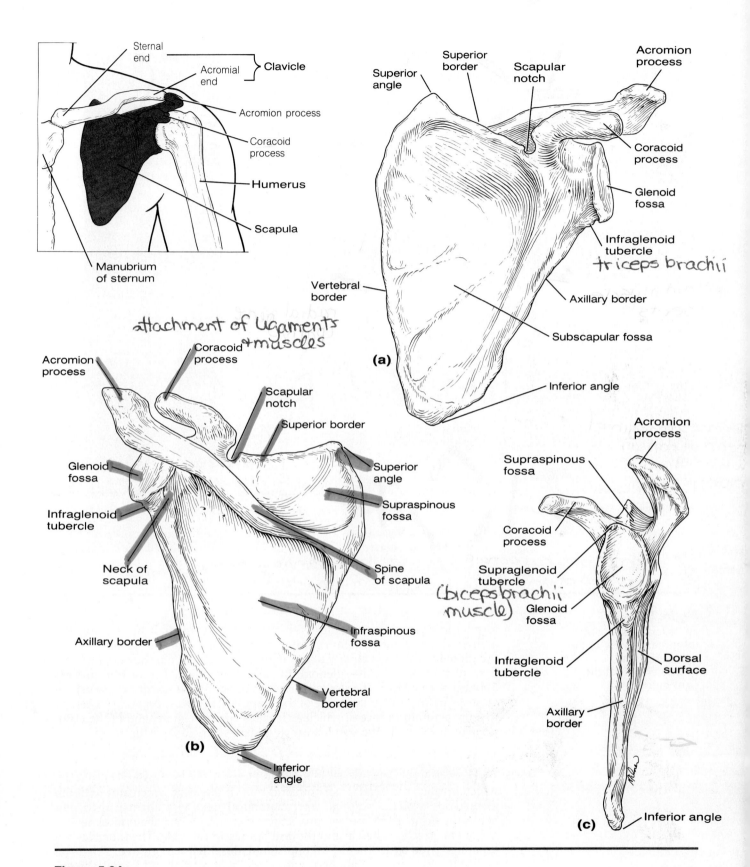

Sternal end
Acromial end
Clavicle
Acromion process
Coracoid process
Humerus
Scapula
Manubrium of sternum

Superior angle
Superior border
Scapular notch
Acromion process
Coracoid process
Glenoid fossa
Infraglenoid tubercle
triceps brachii
Axillary border
Vertebral border
Subscapular fossa
(a)
Inferior angle

attachment of ligaments + muscles
Coracoid process
Scapular notch
Superior border
Acromion process
Superior angle
Glenoid fossa
Supraspinous fossa
Infraglenoid tubercle
Neck of scapula
Spine of scapula
Axillary border
Infraspinous fossa
Vertebral border
(b)
Inferior angle

Acromion process
Supraspinous fossa
Coracoid process
Supraglenoid tubercle
(biceps brachii muscle)
Glenoid fossa
Infraglenoid tubercle
Dorsal surface
Axillary border
(c)
Inferior angle

Figure 5.34
The inset is an anterior view of the pectoral girdle with the rib cage removed. **(a)** Anterior view, **(b)** posterior view, **(c)** lateral view of the left scapula.

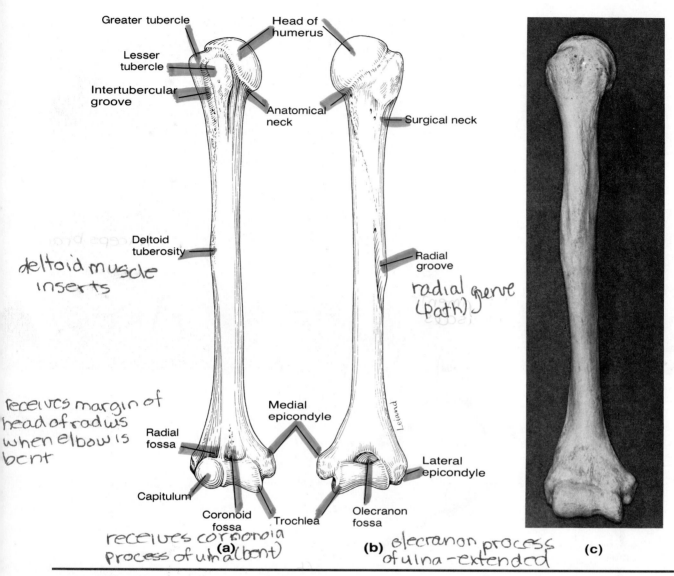

Greater tubercle

Head of humerus

Lesser tubercle

Intertubercular groove

Anatomical neck

Surgical neck

deltoid muscle inserts

Deltoid tuberosity

Radial groove

radial nerve (path)

Leland

Medial epicondyle

recelves margin of head of radius when elbow is bent

Radial fossa

Lateral epicondyle

Capitulum

Coronoid fossa

Trochlea

Olecranon fossa

receives coronoid process of ulna (bent)

(a)

(b) *olecranon process of ulna - extended*

(c)

Figure 5.35
(a) Anterior view and
(b) posterior view of the right humerus.
(c) Photograph of the right humerus.

Table 5.10

forms the base of the triangle, with the **vertebral (medial) border** and the **axillary (lateral) border** joining at the **inferior angle**. The **lateral angle** forms the **glenoid fossa**, which articulates with the humerus. On the superior border, just medial to the glenoid fossa, is the hooked **coracoid process** (*coracoid* = beaklike). The dorsal surface is divided into upper and lower regions by a ridge called the **spine** of the scapula. The spine terminates laterally in a flat **acromion process**, which articulates with the lateral end of the clavicle. The features of the scapula are summarized in Table 5.10.

Arm

F5.35 The **humerus** is the only bone in the arm (Figure 5.35). Its proximal epiphysis, which is called the **head,** is smooth and round. The head articulates with the glenoid fossa of the scapula. The **anatomical neck** is a slight constriction below the head. There are two projections just distal to the anatomical neck—the lateral **greater tubercle** and the anterior **lesser tubercle**. The tubercles are separated from each other by an **intertubercular (bicipital) groove**. The distal end of the humerus is flattened, with prominent **medial** and **lateral epicondyles**. Between the epicondyles are two articular surfaces—the lateral, rounded **capitulum,** which articulates with the radius, and the medial, deeply grooved **trochlea,** which articulates with the ulna. The features of the hu-

Table 5.10 merus are summarized in Table 5.10.

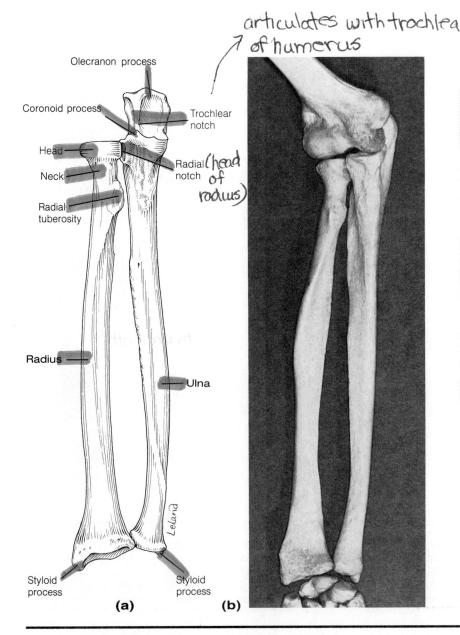

articulates with trochlea of humerus

Olecranon process

Coronoid process

Trochlear notch

Head

Radial notch *(head of radius)*

Neck

Radial tuberosity

Radius

Ulna

Styloid process

Styloid process

(a)   **(b)**

Table 5.9 Bones of the Upper Limbs

	Number of Bones in Each Limb	Number of Bones in Both Limbs
Pectoral girdle		4
Clavicle	1	
Scapula	1	
Arm		2
Humerus	1	
Forearm		4
Ulna	1	
Radius	1	
Hand		54
Carpals	8	
Metacarpals	5	
Phalanges	14	
Total upper limb bones		64

Figure 5.36
(a) Anterior view of the bones of the right forearm. **(b)** Photograph of the right radius and ulna.

Forearm

There are two parallel bones in the forearm. In the anatomical position, the *ulna* is medial and the *radius* is lateral (Figure 5.36). The features of the radius and ulna are summarized in Table 5.10.

F5.36
Table 5.10

ULNA The proximal end of the **ulna** has two prominent processes: the larger posterior **olecranon process** and the smaller anterior **coronoid process** (Figure 5.36). A smooth concave surface, the **trochlear (semilunar) notch,** lies on the anterior surface of the olecranon process and extends onto the superior surface of the coronoid process. The trochlear notch articulates with the trochlea of the humerus. The **radial notch** is a smooth surface on the lateral side of the coronoid process. It articulates with the edge of the head of the radius. Distally, the ulna has a small rounded **head** and a posterior medial **styloid process.**

F5.36

RADIUS In the anatomical position, the **radius** lies lateral to the ulna (Figure 5.36). Its small cylindrical proximal epiphysis is the **head.** The head articulates with the capitulum of the humerus and the radial notch of the ulna. On the medial surface, a short distance below the head, is the **radial tuberosity.** The distal end of the radius is broad and articulates medially with the ulna and

F5.36

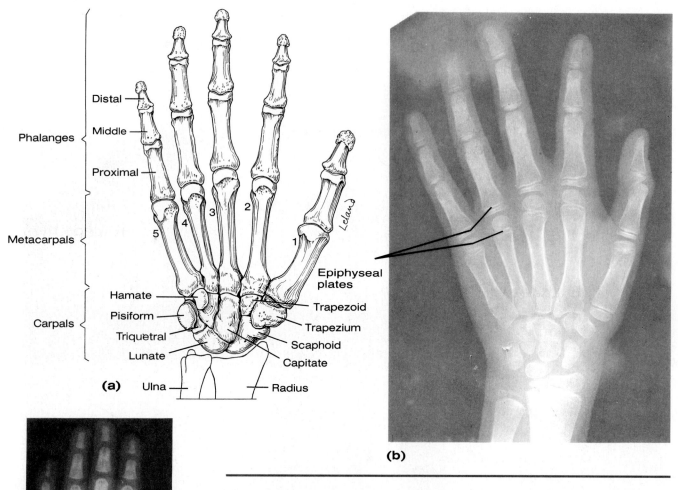

(a)

(b)

(c)

Figure 5.37
(a) Ventral view of the bones of the right hand. X rays of the right hand of **(b)** a 7-year-old boy and **(c)** an 18-month-old child. The spaces between bones in **(c)** are the cartilaginous epiphyses of the various bones. Note that only two carpal bones are ossified. In **(b)** the epiphyses have ossified, leaving thin epiphyseal plates.

distally with the carpal bones of the wrist. It has a conical **styloid process** projecting from its lateral margin. The space between the radius and ulna is occupied by a strong interosseus membrane.

Hand

The skeleton of the hand consists of *carpal bones, metacarpal bones,* and *phalanges.*

F5.37   There are eight **carpal bones** at the wrist, arranged in two transverse rows of four bones each (Figure 5.37). The bones of the proximal row, from lateral to medial, are the **scaphoid, lunate, triquetral,** and **pisiform.** Those of the distal row, from lateral to medial, are the **trapezium, trapezoid, capitate,** and **hamate.*** The scaphoid and lunate articulate with the distal end of the radius to form the wrist joint.

F5.37   Five **metacarpal bones** form the skeleton of the palm of the hand (Figure 5.37). Rather than being named, they are numbered from lateral (thumb) to medial. Proximally, the metacarpals articulate with the distal row of the carpal

---

*When attempting to memorize the names and positions of the carpal bones, it is helpful to associate them with the words of a sentence. For example, the first letter of each word in the sentence "**S**tudents **L**ike **T**he **P**rofessor **T**o **T**each **C**omplex **H**ypotheses" will help you recall the first letter of each carpal bone in order—from lateral to medial, from proximal row to distal row.

bones and with each other. Distally, each articulates with the proximal end of a phalanx.

The skeleton of the fingers is formed by 14 **phalanges.** They are numbered from lateral to medial like the metacarpals. Each finger contains three phalanges: **proximal, middle,** and **distal.** An exception is the first digit (thumb), which has only proximal and distal phalanges.

Table 5.10   **Summary of Specific Features of the Bones of the Upper Limbs**

CLAVICLE *[F5.34]*

Sternal end The blunt medial end that articulates with the manubrium of the sternum.

Acromial end The flattened lateral end that articulates with the acromion process of the scapula.

SCAPULA *[F5.34]*

Superior border The upper horizontal margin.

Vertebral (medial) border The vertical margin just lateral to the vertebral column.

Axillary (lateral) border The thicker oblique lateral margin.

Inferior angle The most inferior point of the bone. It marks the junction of the vertebral and axillary borders.

Superior angle The junction of the vertebral and superior borders.

Lateral angle The junction of the superior and axillary borders. It contains the glenoid fossa.

Glenoid fossa A shallow depression on the lateral angle. It articulates with the humerus.

Supraglenoid tubercle A slight elevation just above the glenoid fossa. This is the point of attachment of the long head of the biceps brachii muscle.

Infraglenoid tubercle The roughened area just below the glenoid fossa. The long head of the triceps brachii muscle originates here.

Coracoid process The projection that hooks anteriorly from the superior border. It provides for the attachment of ligaments and muscles.

Scapular notch A deep notch in the superior border at the base of the coracoid process. It allows for the passage of the suprascapular nerve.

Spine A prominent ridge that runs horizontally across the posterior surface.

Acromion process The flattened lateral end of the spine. It articulates with the clavicle, thus bracing the scapula.

Supraspinous fossa The dorsal surface above the spine.

Infraspinous fossa The dorsal surface below the spine.

Subscapular fossa (costal surface) The slightly concave ventral surface.

HUMERUS *[F.35]*

Head The rounded proximal epiphysis. It articulates with the glenoid fossa to form the shoulder joint.

Anatomical neck A shallow constriction that circles the bone just below the head.

Greater tubercle The rounded projection from the lateral margin of the bone just distal to the anatomical neck.

Lesser tubercle The rounded projection from the anterior surface of the bone just distal to the anatomical neck.

Intertubercular (bicipital) groove A deep groove between the greater and lesser tubercles. The long head of the biceps brachii muscle passes through the groove to reach the supraglenoid tubercle.

Surgical neck A slightly constricted region just inferior to the tubercles. This is frequently the site of fracture.

Deltoid tuberosity A triangular roughened area on the anterior lateral surface near the middle of the shaft. The deltoid muscle inserts here.

Radial groove (sulcus) An oblique groove on the posterior surface just below the deltoid tuberosity. It marks the path of the radial nerve.

Epicondyles (medial and lateral) The projections from the margins of the distal epiphysis.

Capitulum The lateral convex portion of the distal condyles. It articulates with the head of the radius.

Radial fossa A slight depression on the anterior surface above the capitulum. It receives the margin of the head of the radius when the elbow is flexed (bent).

Trochlea The medial concave portion of the distal condyles. It articulates with the semilunar notch of the ulna.

Coronoid fossa A small depression on the anterior surface above the trochlea. It receives the coronoid process of the ulna when the elbow is flexed (bent).

Olecranon fossa A deep depression on the posterior surface above the trochlea. It receives the olecranon process of the ulna when the elbow is extended (straightened).

ULNA *[F5.36]*

Olecranon process The thick posterior projection from the proximal end that forms the point of the elbow. It

Table 5.10 Summary of Specific Features of the Bones of the Upper Limbs (continued)

is received by the olecranon fossa of the humerus when the elbow is extended (straightened).

Coronoid process The anterior projection from the proximal end. It is received by the coronoid fossa of the humerus when the elbow is flexed (bent).

Trochlear notch A curved depression formed by the olecranon and coronoid processes. It articulates with the trochlea of the humerus.

Radial notch A small depression on the lateral side of the coronoid process. It articulates with the margins of the head of the radius, allowing the forearm to rotate, turning the palm down (pronate).

Head The small distal end that articulates with the fibrocartilaginous disc of the wrist joint.

Styloid process The posterior medial projection from the distal end. It serves as the point of attachment for the ulnar collateral ligament of the wrist.

RADIUS *[F5.36]*

Head The proximal, disc-shaped end. The superior surface articulates with the capitulum of the humerus; the edges articulate with the radial notch of the ulna.

Neck The constriction just distal to the head.

Radial tuberosity A flat projection on the medial side, distal to the neck. The biceps brachii muscle inserts here.

Styloid process The lateral downward projection from the distal end. It is the point of attachment for the radial collateral ligament and the brachioradialis muscle.

Ulnar notch A depression on the medial margin of the distal end. It articulates with the ulna.

Table 5.11 Bones of the Lower Limbs

	Number of Bones in Each Limb	Number of Bones in Both Limbs
Pelvic Girdle		2
Coxal bone		
Ilium		
Ischium } fused to form 1		
Pubis		
Thigh		2
Femur 1		
Leg		6
Tibia 1		
Fibula 1		
Patella 1		
Foot		52
Tarsals 7		
Metatarsals 5		
Phalanges 14		
Total lower limb bones		62

Lower Limbs

Table 5.11 The bones of the lower limbs are listed in Table 5.11.

Pelvic Girdle

F5.38 The **pelvic girdle** (Figure 5.38) is formed by a pair of **coxal bones,** or **ossa coxae** (*os* = bone; *ossa* = bones; *coxa* = hip). These bones are also commonly referred to as innominate bones or pelvic bones. The two coxal bones are firmly braced through posterior articulations with the sacrum (forming the sacroiliac joint) and an anterior articulation with each other (forming the symphysis pubis). Each coxal bone is a single bone formed by the fusion of three separate embryonic bones: the *ilium, ischium,* and *pubis.* In the adult bone,

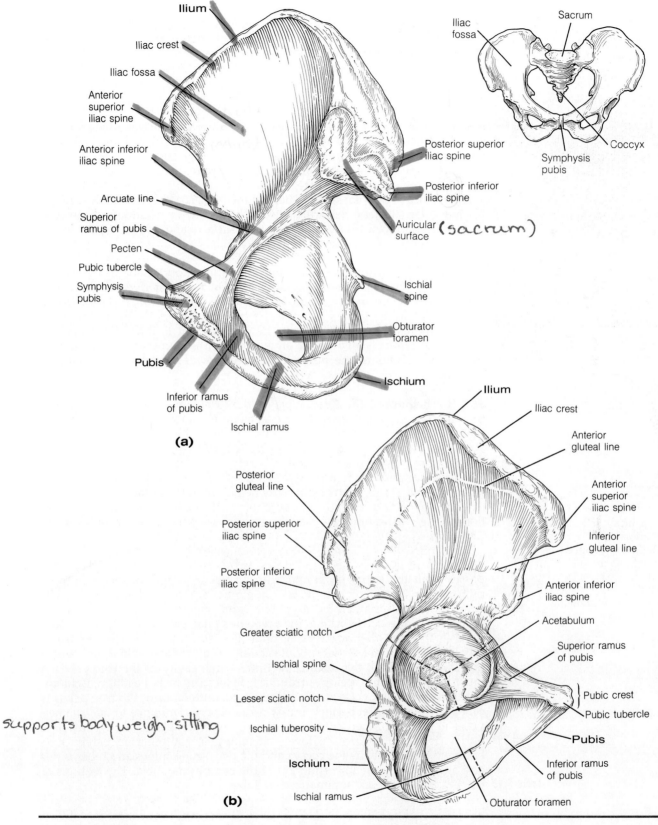

Ilium
Iliac crest
Iliac fossa
Anterior superior iliac spine
Anterior inferior iliac spine
Arcuate line
Superior ramus of pubis
Pecten
Pubic tubercle
Symphysis pubis
Pubis
Inferior ramus of pubis
Ischial ramus

Posterior superior iliac spine
Posterior inferior iliac spine
Auricular surface (sacrum)
Ischial spine
Obturator foramen
Ischium

(a)

Iliac fossa
Sacrum
Coccyx
Symphysis pubis

supports body weigh~ sitting

Posterior gluteal line
Posterior superior iliac spine
Posterior inferior iliac spine
Greater sciatic notch
Ischial spine
Lesser sciatic notch
Ischial tuberosity
Ischium
Ischial ramus

Ilium
Iliac crest
Anterior gluteal line
Anterior superior iliac spine
Inferior gluteal line
Anterior inferior iliac spine
Acetabulum
Superior ramus of pubis
Pubic crest
Pubic tubercle
Pubis
Inferior ramus of pubis
Obturator foramen

(b)

Figure 5-38
Right coxal bone. **(a)** Internal surface. **(b)** External surface.
The dotted lines indicate the approximate junctions
between the ilium, ischium, and pubis, and between the
pubis and ischium.

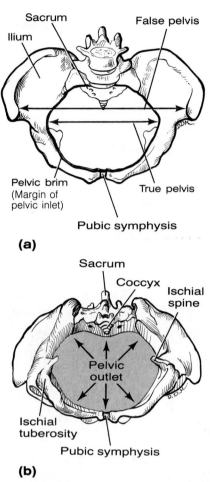

Sacrum

Ilium

False pelvis

Pelvic brim
(Margin of
pelvic inlet)

True pelvis

Pubic symphysis

(a)

Sacrum

Coccyx

Ischial
spine

Pelvic
outlet

Ischial
tuberosity

Pubic symphysis

(b)

Figure 5.39
Pelvic cavities.
(a) Superior view
showing the expanded
false pelvis separated
from the true pelvis by
the pelvic brim.
(b) Inferior view showing
the pelvic outlet.

these individual names are retained for their respective parts. On the lateral surface of the coxal bone, where the ilium, ischium, and pubis bones meet, is a deep cup called the **acetabulum.** The head of the femur articulates with the acetabulum. Below the acetabulum is a large **obturator foramen.** The features of the coxal bones are summarized in Table 5.13.

ILIUM   The **ilium** is a broad, expanded portion of the coxal bone that extends upward from the acetabulum (Figure 5.38). Its superior border is called the **iliac crest.** This crest ends anteriorly at the **anterior superior iliac spine.** A short distance below this spine is the **anterior inferior iliac spine.** The iliac crest ends posteriorly at the **posterior superior iliac spine.** A short distance below this spine is the **posterior inferior iliac spine.** Below the posterior inferior iliac spine is the deep **greater sciatic notch.** The **iliac fossa** is the smooth, slightly concave internal surface. Behind the fossa is a roughened area, called the **auricular surface,** that articulates with the sacrum. Running diagonally downward and forward from this articular surface, and demarking the lower boundary of the iliac fossa, is the **arcuate line.**

ISCHIUM   The **ischium** forms the posterior inferior portion of the coxal bone and part of the acetabulum (Figure 5.38). On the posterior margin of the ischium, below the greater sciatic notch, is the **ischial spine.** The **lesser sciatic notch** is below the spine. The notch is bounded inferiorly by a prominent **ischial tuberosity,** which supports the body weight in the sitting position. The **ischial ramus** is an anterior projection from the tuberosity. It joins with the inferior ramus of the pubis to form the lower border of the obturator foramen.

PUBIS   The **pubis** is the anterior part of the coxal bone (Figure 5.38). It forms the anterior inferior portion of the acetabulum. Its **superior ramus** is supported against the superior ramus of the opposite side, forming the **symphysis pubis.** A projection called the **pubic tubercle** is located on the superior ramus, close to the symphysis pubis. A ridge called the **pecten** extends along the superior ramus, from the pubic tubercle to the arcuate line of the ilium. A short distance lateral to the symphysis pubis, the **inferior ramus** of the pubis extends downward and posteriorly to join with the ischial ramus. The junction of the two inferior rami at the symphysis pubis forms a **pubic arch.**

Pelvic Cavities

The cavity of the pelvis is divided into two parts by a horizontal plane that passes from the sacral promontory to the upper margin of the symphysis pubis, following the arcuate lines on the inner surface of the ilium. The circumference of this plane is the **pelvic brim** (Figure 5.39a). The expanded cavity above the pelvic brim is the **greater** or **false pelvis.** The **lesser** or **true pelvis** is the cavity below the pelvic brim. The false pelvis is bounded posteriorly by the iliac fossa. Anteriorly, it is bounded by the abdominal wall, and is therefore capable of expansion. In contrast, the cavity of the true pelvis is much more restricted, being surrounded on all sides by bone (ilium, ischium, pubis, sacrum, and coccyx). The superior circumference of the true pelvis is called the **pelvic inlet** because it marks the superior entrance to the true pelvis. The margin of the pelvic inlet coincides with the pelvic brim. The **pelvic outlet** (Figure 5.39b) is the lower circumference of the true pelvis. It is bounded posteriorly by the coccyx and the two ischial spines and tuberosities and anteriorly by the lower margin of the symphysis pubis. The features of the pelvic cavities are summarized in Table 5.13.

SEXUAL DIFFERENCES IN THE PELVIS   There are several structural differences between the male and female pelvis, most of which are related to childbearing. Because the fetus must pass from the false pelvis through the true pelvis during birth, the measurements of the pelvic inlet and outlet are of particular importance in the female. Some sexual differences in the pelvis are listed in Table 5.12 and illustrated in Figure 5.40.

F5.39a

F5.39b

Table 5.13

Table 5.12, F5.40

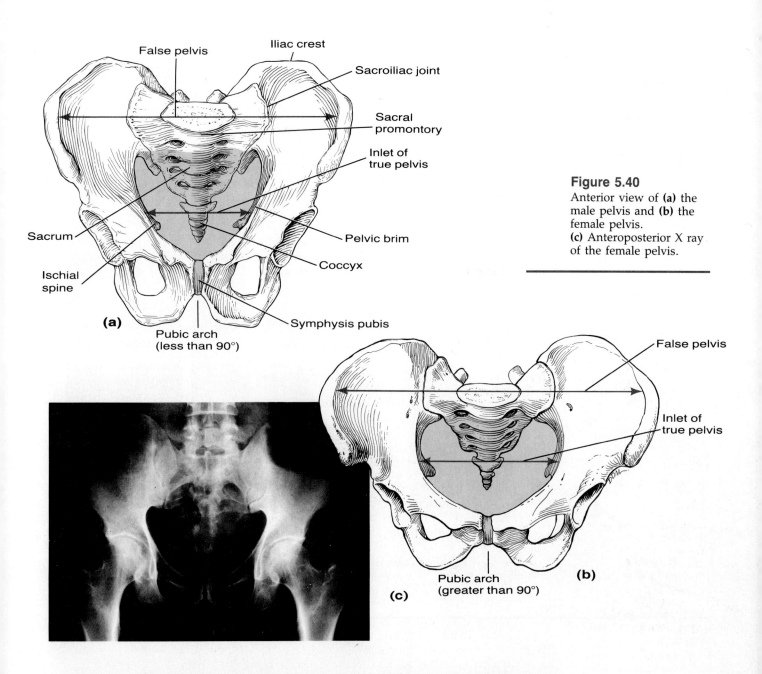

False pelvis

Iliac crest

Sacroiliac joint

Sacral promontory

Inlet of true pelvis

Pelvic brim

Coccyx

Sacrum

Ischial spine

(a)

Pubic arch (less than 90°)

Symphysis pubis

Figure 5.40
Anterior view of **(a)** the male pelvis and **(b)** the female pelvis.
(c) Anteroposterior X ray of the female pelvis.

False pelvis

Inlet of true pelvis

(b)

Pubic arch (greater than 90°)

(c)

Table 5.12   Some Structural Differences Between Female Pelvis and Male Pelvis

Characteristic	Female Pelvis	Male Pelvis
General structure	More delicate	More massive
Anterior iliac spines	More widely separated	Less widely separated
Pelvic inlet	Larger; circular	Heart-shaped
Pelvic outlet	Wider; ischial tuberosities farther apart	Narrower
Pubic arch	Obtuse (greater than 90°)	Acute (less than 90°)
Obturator foramen	Triangular	Oval
Acetabulum	Faces more anteriorly	Faces laterally

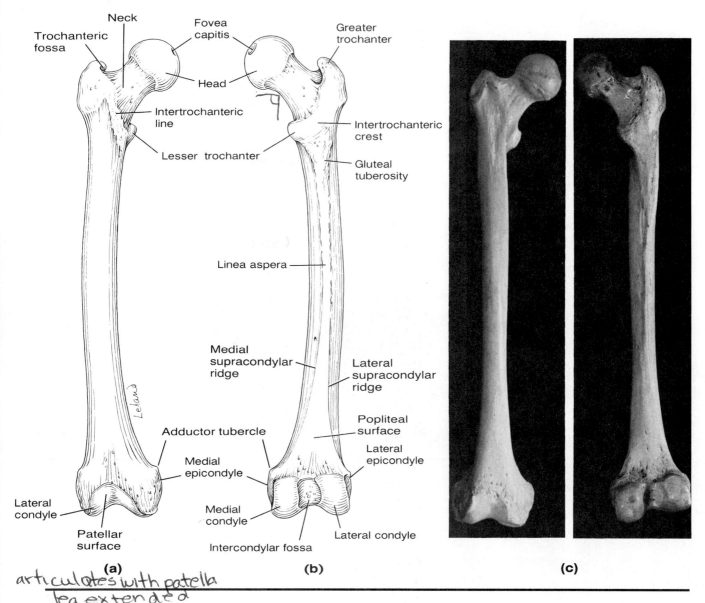

articulates with patella
leg extended

Figure 5.41

Right femur. **(a)** Anterior view. **(b)** Posterior view. **(c)** Photographs of right femur.

Thigh

F5.41 The **femur,** which forms the skeleton of the thigh, is the longest bone of the body (Figure 5.41). The **head** of the femur is the spherical proximal epiphysis, which is directed medially and is received into the acetabulum of the coxal bone. The surface of the head is smooth, except for a central pit called the **fovea capitis.** A strong ligament *(ligamentum teres)* is attached to the fovea and to the acetabulum, helping to maintain the integrity of the hip joint. A constricted **neck** joins the head with the shaft of the bone. The neck is almost at a right angle with the shaft. At the point where the neck joins the shaft are two large processes that serve as sites of muscle attachment. The larger lateral process is the **greater trochanter;** the smaller process, which projects medially and posteriorly, is the **lesser trochanter.** On the posterior surface of the shaft is a distinct longitudinal ridge called the **linea aspera.** At its distal end, the linea aspera divides into **medial** and **lateral supracondylar ridges,** which enclose a flat triangular **popliteal surface** between them.

The distal end of the femur is enlarged into **medial** and **lateral condyles.** The smooth surfaces of the condyles articulate with the tibia. Between the condyles, on the posterior surface, is a deep **intercondylar fossa.** Anteriorly,

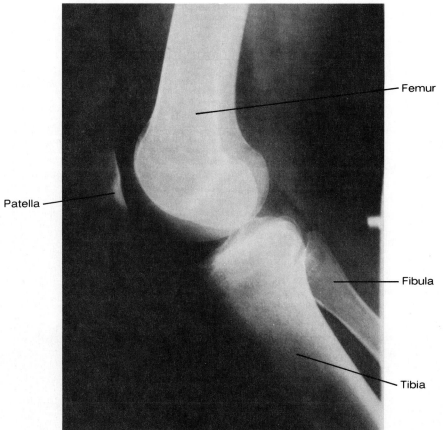

Femur

Patella

Fibula

Tibia

Figure 5.42
Lateral X ray of the knee.

---

there is a smooth **patellar surface** between the condyles. This surface articulates with the patella when the leg is extended. The features of the femur are summarized in Table 5.13.

Table 5.13

Leg

The skeleton of the leg consists of a strong medial *tibia* and a slender lateral *fibula*. Protecting the knee joint, between the thigh and the leg, is the *patella* (kneecap). Table 5.13 summarizes the features of the tibia and fibula.

Table 5.13

PATELLA The **patella** (Figure 5.42) is a *sesamoid bone,* which means that it forms within the tendons of muscles and is not firmly anchored to the skeleton. It is located in front of the knee joint. In addition to protecting the knee joint, the patella improves the leverage of the quadriceps femoris muscle group, in whose tendon it is embedded. The posterior surface of the patella has a smooth articular surface for contact with the patellar surface of the femur.

F5.42

TIBIA The **tibia** (Figure 5.43), which is the medial bone of the leg, supports the body weight that is transmitted to it from the femur. Its proximal end is expanded into **medial** and **lateral condyles.** The superior surfaces of the condyles are smooth and flattened. They articulate with the condyles of the femur. Between these articular surfaces is a prominence called the **intercondylar eminence** or **spine.** The **tibial tuberosity** is a prominent elevation on the anterior surface of the bone, just below the condyles. It provides an anchoring point for the tendons of the muscles of the anterior thigh—that is, those that ensheath the patella and form the *patellar ligament.* A sharp **anterior crest** extends almost the entire length of the shaft. The distal end of the tibia has a

F5.43

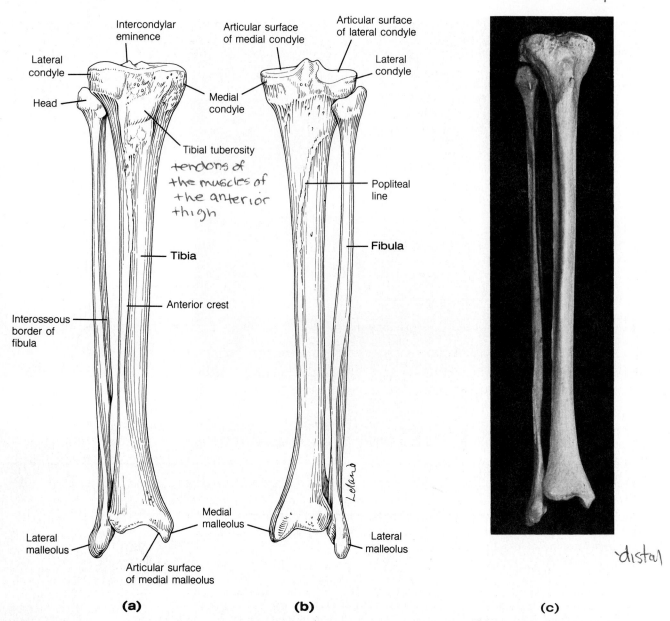

proximal

Intercondylar eminence

Lateral condyle

Head

Articular surface of medial condyle

Medial condyle

Tibial tuberosity

tendons of the muscles of the anterior thigh

Tibia

Anterior crest

Interosseous border of fibula

Lateral malleolus

Articular surface of medial malleolus

Articular surface of lateral condyle

Lateral condyle

Popliteal line

Fibula

Medial malleolus

Lateral malleolus

distal

(a) **(b)** **(c)**

Figure 5.43

(a) Anterior view and **(b)** posterior view of the bones of the right leg. **(c)** Photograph of the right tibia and fibula.

downward projection on its inner side called the **medial malleolus.** This is the prominent lump that can be felt on the medial side of the ankle. The distal surface of the tibia is flattened for articulation with the talus, which is a tarsal bone.

F5.43 FIBULA The **fibula** (Figure 5.43) is a slender bone that is lateral to the tibia. Its proximal end is expanded into a **head** that articulates with the lateral condyle of the tibia. The distal end of the fibula is flattened, forming the **lateral malleolus,** which can be felt on the lateral side of the ankle. The lateral malleolus articulates with the talus and, together with the medial malleolus, forms the ankle joint. Just above the lateral malleolus, the fibula articulates with the distal end of the tibia. The tibia and the fibula are tightly bound together by an interosseous membrane.

Foot

The skeleton of the foot consists of *tarsal bones, metatarsal bones,* and *phalanges.*

FRONTIERS IN HEALTH
Rebuilding Broken Bones

Robert left a local restaurant and drove toward his home in Dumont, New Jersey. He fell asleep while driving, and his car swerved off the road and overturned. When he awoke the next morning surgeons had removed a 4-inch segment of his shattered humerus.

In other times, Robert's arm would have been amputated. Traditional bone grafts would not suffice to rebuild the shattered segment of bone. But today, Robert does yard work and lifts weights with his arm, thanks to the pioneering work of a group of surgeons.

In Robert's case, physicians surgically removed a section of the fibula, the relatively thin bone that lies alongside the tibia. The fibula, which bears very little body weight in humans, has proved useful for replacing segments of other bones. This technique succeeds where conventional bone grafts would have failed because, along with the bone itself, physicians remove the blood vessels that nourish the bone. The graft is then placed in the shattered region of the bone and the blood vessels of the fibula are carefully reconnected with existing arteries and veins by microsurgical techniques, thus ensuring an uninterrupted flow of blood. Ordinary bone grafts often failed because blood flow could not be immediately reestablished.

Supplied with a steady flow of blood, the grafted fibula knits itself into place and gradually thickens, becoming almost indistinguishable in thickness from the bone on either side. Immediately after transplantation, however, the bone must be secured by an external fixator. After 6 to 8 weeks, the fixator is replaced by a cast. Later, a brace is all that is needed to support the bone graft, and eventually even that is no longer necessary.

The success of this technique, which is called the free vascularized fibular graft, has been exciting. Two American teams who have used it on well over 200 patients report outstanding results.

Fibular transplants have been used to repair crushed bones or replace bone segments lost to infection or cancer. The technique was even successfully used to construct a radius in an infant who had been born without one.

Another innovative technique that complements free vascularized fibular grafting is based on the pioneering work of orthopedic surgeon Dr. Marshall Urist, who is developing a method for building new bones from old ones. He obtains bones from human cadavers or even animals and treats them with chemicals to remove the mineral matter. What is left is a rubbery substance, called demineralized bone matter (DBM). Made mostly of collagen, DBM can be surgically implanted in the body to replace lost segments of bone, in much the same way that fibular transplants are done. The implanted DBM is injected with bone cells taken from chips of the patient's own bone. Over time, the rubbery implant is converted into solid bone that is virtually indistinguishable from the normal bone.

Medical scientists are now using DBM to repair congenital birth defects and to fill in cavities formed in bone

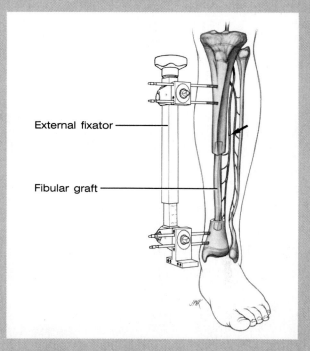

An external fixator stabilizing a fractured tibia, with a fibular graft in place, in a patient with pseudarthrosis (false joints). The arrow points to the anastomosis between the blood vessels of the tibia and the graft.

during surgery to remove cysts, cancer, or serious infections. Oral surgeons are using it to reconstruct jaws damaged in accidents or to rebuild eroded jaw bones in patients with periodontal disease. There is hope that surgeons may be able to use the technique to lengthen arms and legs whose growth was affected by hormonal deficiencies. Perhaps the most widely acclaimed success for DBM was the case of a 6-year-old boy suffering from a rare congenital disease, called cloverleaf deformity, in which bones of the skull fail to grow to accommodate the growing brain. Without treatment the boy would have died. Dr. John Mulliken, a surgeon at Boston's Childrens Hospital, constructed an entirely new skull for the patient out of DBM. Three years later the boy is alive and healthy.

DBM has proved itself in hundreds of cases. Even 5 years after it has been put in place, the new bone grown from DBM appears healthy and strong. The use of DBM has several important advantages. First, it can be produced in large quantities and stored for later use, offering physicians a ready supply of inexpensive, transplantable material. Second, conventional bone grafting and free vascularized fibular grafts require two surgical operations: one to remove the bone and one to put it in its new location. For patients this means more time in surgery, more risk of complications, more pain, higher medical bills, and a longer recovery time. DBM eliminates the need for the extra surgery.

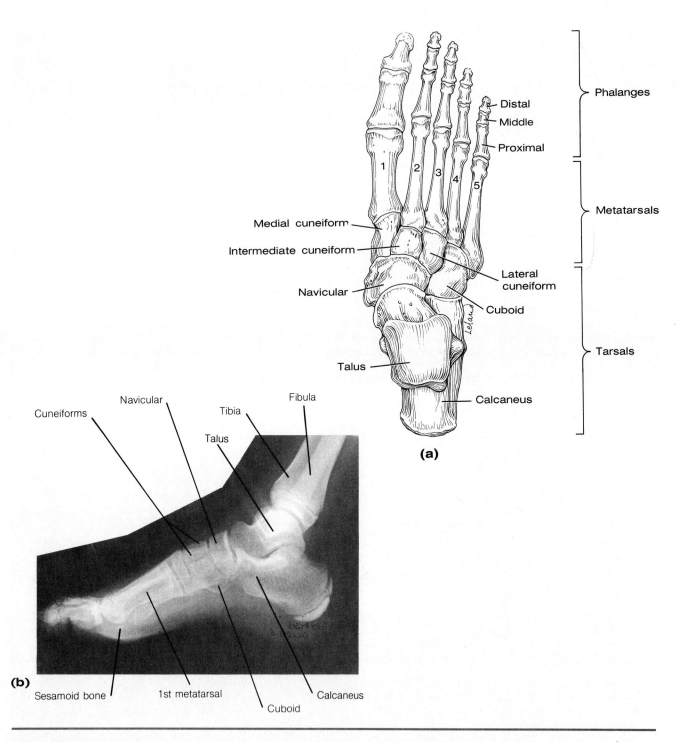

(a)

(b)

Figure 5.44
(a) Superior view of the bones of the right foot.
(b) Lateral X ray of the foot.

F5.44 The proximal portion of the foot, toward the ankle, is composed of seven **tarsal bones** (Figure 5.44): the **talus, calcaneus, navicular, cuboid, medial cuneiform, intermediate cuneiform,** and **lateral cuneiform.***

The calcaneus forms the prominent heel bone, which provides attachment for several muscles of the calf. Resting on the superior surface of the calcaneus is the talus. The talus is located between the malleoli, where it articulates with the tibia and fibula. In this location, the talus receives the

*The first letter of each word in the sentence "Tarsal Common Names Can Cause Constant Confusion" should help you recall the first letter of each tarsal bone. But remember that there are medial, intermediate, and lateral cuneiform bones.

entire weight of the body, which it distributes to the other tarsal bones. The three cuneiforms and the cuboid articulate with the proximal ends of the metatarsal bones.

Five **metatarsal bones** form the skeleton of the intermediate region of the foot (Figure 5.44). They are numbered from medial to lateral. Proximally, the **F5.44** metatarsal bones articulate with the tarsal bones and with each other. Distally, each articulates with the proximal end of a phalanx.

The skeleton of the toes is similar to that of the fingers. It is formed of 14 **phalanges**—three (*proximal, middle,* and *distal*) in each digit, except for the great toe, where there is no middle phalanx. The phalanges of the toes are shorter than those of the fingers.

ARCHES OF THE FOOT   The tarsal and metatarsal bones are joined in such a manner that they form three arches. There are **medial** and **lateral longitudinal arches,** which have the calcaneus as their posterior pillar and the metatarsals and other tarsals as their anterior pillar. The **transverse (metatarsal) arch** is formed mostly by the bases of the metatarsal bones.

The arches are supported primarily by ligaments, though some support is gained from muscles and their tendons. They function to distribute the weight of the body fairly evenly between the heel and the metatarsals. People whose longitudinal arches have collapsed suffer from *flat feet* and often find that their feet tire easily. They may also experience low-back discomfort because the arches normally absorb many of the shocks that occur during walking.

Sesamoid Bones

In certain tendons that are subjected to compression or to unusual tensile stress, small bones called **sesamoid bones** form. The patella is the largest sesamoid bone in the body. Although the locations of sesamoid bones vary, the most common locations include the areas around the metacarpophalangeal and the metatarsophalangeal joints.

Table 5.13   Summary of Specific Features of the Bones of the Lower Limbs

COXAL BONE   [F5.38]

ILIUM [F5.39] The upper, flat, expanded portion of the coxal bone.

ILIAC CREST The superior margin of the ilium.

ILIAC SPINES
  Anterior superior A projection at the anterior end of the iliac crest.
  Anterior inferior A rounded projection just below the anterior superior spine.
  Posterior superior A projection at the posterior end of the iliac crest.
  Posterior inferior A projection just below the posterior superior spine.

GREATER SCIATIC NOTCH A deep notch just below the posterior inferior spine. It provides passage for the sciatic nerve, as well as other nerves, vessels, and the pisiformis muscle.

ARICULAR SURFACE Surface that articulates with the sacrum.

GLUTEAL LINES Three arched lines—posterior, anterior, and inferior—on the outer surface of the ilium. The origins of the three gluteus muscles are located between the lines.

ARCUATE LINE A slight ridge on the internal surface of the ilium. It extends forward and downward from the top of the sacrum to the pecten of the pubis. The line forms the internal margins of the pelvic inlet.

ILIAC FOSSA The smooth, concave internal surface of the ilium, above the arcuate line.

ISCHIUM [F5.38] The posterior inferior portion of the coxal bone.

ISCHIAL SPINE A triangular projection from the posterior border behind the acetabulum.

ISCHIAL TUBEROSITY A roughened enlargement on the posterior inferior margin.

LESSER SCIATIC NOTCH A small indentation that separates the ischial spine and tuberosity. It transmits the tendon of the obturator internus muscle, as well as nerves and blood vessels.

ISCHIAL RAMUS The flattened anterior projection that arises from the tuberosity. It joins with the inferior ramus of the pubis to form the lower border of the obturator foramen.

PUBIS [F5.38] The anterior inferior portion of the coxal bone.

Table 5.13   Summary of Specific Features of the Bones of the Lower Limbs (continued)

SUPERIOR RAMUS Elongated process that extends anteriorly from the acetabulum to form the upper border of the obturator foramen.

SYMPHYSIS PUBIS The midline joint between the superior rami of both pubic bones.

PUBIC TUBERCLE The projection from the superior ramus just lateral to the symphysis pubis.

PECTEN A ridge that extends upward and laterally along the superior ramus. It runs from the pubic tubercle to the arcuate line.

PUBIC CREST A ridge that extends medially from the pubic tubercle to the symphysis pubis.

INFERIOR RAMUS The projection that passes downward and backward from the symphysis pubis. It joins with the ischium to form the lower border of the obturator foramen.

PUBIC ARCH Formed by the convergence of the inferior rami.

ACETABULUM [F5.38] A cup-shaped depression formed by the junction of ilium, ischium, and pubis. It receives the head of the femur to form the hip joint.

OBTURATOR FORAMEN [F5.38] The large opening in the inferior region of coxal bone. It allows for the passage of the obturator nerves and blood vessels.

PELVIC BRIM (INLET) [F5.39] The boundary of the opening that leads into the true pelvis. It is formed by the sacral promontory, arcuate lines, and the superior margin of the symphysis pubis.

FALSE PELVIS [F5.39] The expanded space above the pelvic inlet. It is actually a portion of the abdominopelvic cavity.

TRUE PELVIS [F5.39] The smaller cavity below the pelvic inlet, bounded by the coxal bones, sacrum, and coccyx.

PELVIC OUTLET [F5.39] The lower margin of the true pelvis. It is bounded by the coccyx and the lower parts of the coxal bones.

FEMUR   *[F5.41]*

HEAD The rounded proximal end that fits into the acetabulum.

FOVEA CAPITIS The pit on the head where the ligamentum teres is attached, which in turn is attached to the acetabulum.

NECK The constriction that connects the head with the shaft.

GREATER TROCHANTER The large lateral process just below the neck.

LESSER TROCHANTER The smaller medial process just below the neck.

INTERTROCHANTERIC CREST A ridge on the posterior surface that connects the two trochanters.

INTERTROCHANTERIC LINE A slight line on the anterior surface that connects the two trochanters.

GLUTEAL TUBEROSITY A small projection just below the greater trochanter.

TROCHANTERIC FOSSA A depression on the medial surface of the greater trochanter.

LINEA ASPERA A longitudinal ridge that extends along the middle third of the posterior surface of the shaft.

SUPRACONDYLAR RIDGES The medial and lateral ridges that are formed by the divergence of the linea aspera at its distal end.

POPLITEAL SURFACE A smooth triangular surface on the posterior surface. It is bounded above by the medial and lateral supracondylar ridges.

CONDYLES Rounded medial and lateral enlargements at the distal end of the shaft. They have smooth surfaces for articulation with the tibia.

EPICONDYLES Roughened prominences on the lateral surfaces of the condyles. They provide points of attachment for the medial and lateral collateral ligaments of the knee joint.

ADDUCTOR TUBERCLE A small projection just above the medial condyle at the termination of the medial supracondylar ridge.

INTERCONDYLAR FOSSA The deep notch that separates the condyles posteriorly. It provides points of attachment for the anterior and posterior cruciate ligaments of the knee joint.

PATELLAR SURFACE The smooth anterior surface above the condyles. It articulates with the patella when the leg is extended (straightened).

TIBIA   *[F5.43]*

CONDYLES Flattened enlargements at the proximal end of the shaft. The upper surfaces are smooth for articulation with the condyles of the femur.

INTERCONDYLAR EMINENCE OR SPINE The vertical projection from the superior surface, between the articular surfaces of the condyles.

TIBIAL TUBEROSITY The midline projection from the anterior surface just below the condyles. It serves as the point of attachment for the ligamentum patellae.

ANTERIOR CREST A sharp longitudinal ridge on the anterior surface.

POPLITEAL LINE A ridge on the posterior surface that extends downward and medially from the lateral condyle. It marks the junction between the insertion of the popliteus muscle and the origin of the soleus muscle.

MEDIAL MALLEOLUS A downward projection from the medial side of the distal end of the tibia. It articulates with the talus.

FIBULA *[F5.43]*

HEAD The proximal expanded portion of the bone. It articulates with the lateral condyle of the tibia.

STYLOID PROCESS A rough vertical prominence on the head of the fibula.

LATERAL MALLEOLUS A triangular expansion at the distal end that articulates medially with the distal end of the tibia and with the talus.

STUDY OUTLINE

FUNCTIONS OF THE SKELETON p. 93

SUPPORT body framework.

MOVEMENT muscle attachment to skeleton; movable joints.

PROTECTION of vital internal organs.

MINERAL RESERVOIR storage of calcium, phosphorus, sodium, potassium.

HEMOPOIESIS red marrow of certain bones produces blood cells of adult.

CLASSIFICATION OF BONES p. 94

LONG BONES have a long axis; most bones of upper and lower limbs.

SHORT BONES lack a long axis; carpals and tarsals.

FLAT BONES thin bones; ribs.

IRREGULAR BONES vertebrae.

STRUCTURE OF BONE studied at gross, microscopic, and chemical levels. pp. 95–98

GROSS ANATOMY

DIAPHYSIS bone shaft.

Medullary Cavity fat storage site.

EPIPHYSIS end of bone; contains red bone marrow.

EPIPHYSEAL CARTILAGE OR PLATE separates diaphysis and epiphysis in children; permits increase in bone length.

PERIOSTEUM double-layered connective tissue that covers bone.

MICROSCOPIC ANATOMY OF BONE

COMPACT BONE
1. Unit of structure is haversian system, or osteon.
2. Haversian canals surrounded by concentric lamellae.
3. Osteocytes embedded in lacunae.
4. Lacunae interconnected by canaliculi.

SPONGY BONE osteocytes embedded in lacunae; lamellae not arranged in concentric layers but according to line of tension.

COMPOSITION

ORGANIC FRAMEWORK collagenous fibers in homogeneous ground substance; provides tensile strength.

INORGANIC SALTS calcium and phosphate; allow bone to withstand compression.

DEVELOPMENT OF BONE pp. 98–102

EARLY DEVELOPMENT OF BONE

INTRAMEMBRANOUS OSSIFICATION
1. Flat bones of cranial roof, certain facial bones.
2. Formation of osteoblasts from undifferentiated mesoderm.
3. Osteoblasts form networks of bony spicules (trabeculae).
4. Trabeculae radiate in all directions, uniting with one another to form a network of spongy bone.
5. Osteoclasts reabsorb previously laid-down bone to permit osteoblasts to deposit new bone for growth.

ENDOCHONDRAL OSSIFICATION
1. Most bones of body form this way.
2. Formation of bone from hyaline cartilage models.
3. Perichondrium develops into periosteum that contains osteoblasts.
4. Periosteum produces a collar of bone that covers surface of diaphysis.
5. Cartilage matrix of diaphysis calcifies.
6. Osteoblasts form spicules of bone around calcified cartilage spicules.
7. Primary ossification center in diaphysis.
8. Secondary ossification center in epiphysis.

INCREASE IN BONE LENGTH AND DIAMETER

CARTILAGE CELLS UNDERGO MITOSIS tends to increase size of epiphyseal plate.

DIAPHYSIS SIDE OF EPIPHYSEAL PLATE undergoes calcification, thus increasing length of bone.

INCREASE IN DIAMETER due to bone deposition on outer surface of bone, beneath the periosteum.

THEORIES OF BONE FORMATION
1. Osteoblasts secrete organic framework (collagen fibers and glycoproteins); inorganic salts (calcium phosphate) deposited in organic framework.
2. Calcium and phosphate thought to be concentrated in bone-forming cells and released directly into extracellular fluid or else released in membrane-bounded vesicles.

FACTORS THAT AFFECT BONE DEVELOPMENT

STRESS bone capable of adjusting its strength in

proportion to the degree of stress to which it is subjected. Increased stress results in formation of more collagen fibers and inorganic salts; salts withdrawn in absence of stress. Bone normally subjected to gravitational and functional forces.

HORMONES parathormone from pituitary gland increases resorption of bone by osteoclasts. Calcitonin from thyroid stimulates the formation of new bone.

NUTRITION well-balanced diet required for normal bone development. Vitamin D especially important because of its role in the absorption of calcium into the bloodstream from the gastrointestinal tract.

CONDITIONS OF CLINICAL SIGNIFICANCE: THE SKELETAL SYSTEM pp. 102–105

FRACTURES

TYPES OF FRACTURES

Simple bone ends do not penetrate skin.

Compound broken bone ends penetrate skin.

Comminuted bone splintered at site of break.

Depressed broken region pushed inward.

Impacted broken ends of bone driven into each other.

HEALING OF FRACTURES

Formation of Hematoma (blood clot)

Formation of Fibrocartilaginous Callus

Formation of Bony Callus

METASTATIC CALCIFICATION calcium deposits in tissues that are not normally calcified; kidney is common site.

SPINA BIFIDA failure of posterior portions of vertebrae to form bony arch around spinal cord; most common in lumbosacral area.

OSTEOPOROSIS reduced bone formation rate; normal bone absorption rate.

OSTEOMYELITIS infection of marrow-cavity contents and bone tissue, usually by *Staphylococcus aureus*.

TUBERCULOSIS OF BONE infection of marrow-cavity contents and bone tissue caused by *Mycobacterium tuberculosis*.

RICKETS AND OSTEOMALACIA both due to deficiencies of calcium, phosphorus, vitamin D, or sunlight.

RICKETS demineralization and bone softening in children.

OSTEOMALACIA softening of adult bone.

TUMORS OF BONE benign or malignant tumors.

ABNORMAL GROWTH PATTERNS

PITUITARY GIANT excessive pituitary growth hormone delays ossification of epiphyseal cartilages.

PITUITARY DWARF growth hormone deficiency results in early replacement of epiphyseal cartilages by bone.

ACROMEGALY excess growth hormone secreted after epiphyseal cartilages replaced.

ACHONDROPLASIA epiphyseal cartilages function for short time only, which results in short arms and legs.

EFFECTS OF AGING Gradual loss of calcium from bone (more severe in women).

INDIVIDUAL BONES OF THE SKELETON pp. 106–107

AXIAL SKELETON pp. 107–129

SKULL

CALVARIUM

Frontal Bone anterior superior region of skull.

Two Parietal Bones form most of calvarium.

Occipital Bone posterior floor of cranial cavity.

BONES THAT FORM FLOOR OF CRANIAL CAVITY

Frontal, Occipital, Ethmoid, and *Sphenoid Bones* form midline.

Two Temporal Bones form side of floor.

Sphenoid Bone and *Paired Temporals* form middle cranial cavity.

FACIAL SKELETON

Frontal forehead.

Paired Maxillary upper jaw.

Paired Zygomatic cheek prominences.

Paired Nasal Bones bridge of nose.

Paired Lacrimal medial orbital cavity.

Mandible lower jaw.

BONES THAT FORM THE NASAL CAVITY

Ethmoid perpendicular plate forms part of nasal septum; superior and middle conchae form lateral walls of nasal cavity.

Vomer posterior inferior nasal septum.

Paired Inferior Nasal Conchae form elongated shelves from lateral walls of nasal cavity.

BONES THAT FORM HARD PALATE palatine processes of maxillary bones; horizontal processes of palatine bones.

BONES THAT FORM ORBITAL CAVITY frontal, sphenoid, maxilla, lacrimal, ethmoid, zygomatic, palatine bones.

PASSAGEWAYS THROUGH THE SKULL openings for passage of blood vessels and nerves. (summarized in Table 5.5.)

PARANASAL SINUSES mucous-membrane-lined air spaces in frontal, ethmoid, maxillary, and sphenoid bones.

AUDITORY OSSICLES

Malleus (Hammer)

Stapes (Stirrup)

Incus (Anvil)

HYOID BONE U-shaped; attachment point of tongue and throat muscles.

VERTEBRAL COLUMN

CURVATURES OF THE VERTEBRAL COLUMN

Infant single curve, convex posteriorly

Cervical Curve convex anteriorly; develops with head raising.

Thoracic Curve convex posteriorly; remains from primary curve of newborn.

Lumbar Curve convex anteriorly; develops as child walks.

Abnormalities
1. *Lordosis (Swayback)* excessive anterior lumbar curve.
2. *Kyphosis (Hunchback)* excessive posterior thoracic curve.
3. *Scoliosis* lateral curvature of spine.

FUNCTIONS OF THE VERTEBRAL COLUMN

Support and Flexibility

Spinal Cord Protection intervertebral foramina provide passageway for spinal nerves.

CHARACTERISTICS OF A TYPICAL VERTEBRA
1. Thick anterior centrum with neural arch that arises posteriorly.
2. Spinous process extends from neural arch.
3. Vertebral foramen—passageway for spinal cord.
4. Transverse processes.
5. Superior and inferior articulating processes.

REGIONAL DIFFERENCES IN VERTEBRAE (summarized in Table 5.6)

THORAX

STERNUM elongated, flat bone; forms midline of anterior thorax wall; rib articulation.

RIBS 12 pairs

True Ribs
Vertebrosternal Ribs (pairs 1 through 7); articulate anteriorly with sternum through costal cartilage.

False Ribs
1. *Vertebrochondral Ribs:* pairs 8 through 10; costal cartilages attach to cartilage of rib above.
2. *Vertebral (floating) Ribs:* pairs 11 and 12; short costal cartilages with no anterior articulation.

COSTAL CARTILAGES hyaline cartilage; strengthen thorax and provide flexibility.

APPENDICULAR SKELETON pp. 129–147

UPPER LIMBS

PECTORAL GIRDLE straddles upper thorax.

Clavicle S-shaped bone bracing scapula; holds shoulder away from rib cage.

Scapula thin, flat, triangular bone over posterior surfaces of rib pairs 2 through 7.

ARM *humerus* is only bone.

FOREARM 2 bones

Ulna medial to radius in anatomical position.

Radius lateral to ulna in anatomical position.

HAND 8 *carpal* bones in 2 transverse rows of 4 each at wrist.

Proximal Row lateral to medial; *scaphoid, lunate triquetral, pisiform.*

Distal Row lateral to medial; *trapezium, trapezoid, capitate, hamate.*

PALM OF HAND 5 metacarpal bones.

FINGERS 14 phalanges.

LOWER LIMBS

PELVIC GIRDLE formed by pair of *coxal bones.*

Ilium
Ischium } individual bones during embryonic development; fuse to form adult coxal
Pubis bone.

PELVIC CAVITIES

Greater (False) Pelvis above pelvic brim; expansible.

Lesser (True) Pelvis cavity below pelvic brim; restricted by bone.

Sexual Differences in Pelvis most are related to childbearing.

THIGH *femur* forms thigh skeleton.

LEG

Patella protects knee joint between thigh and leg.

Tibia strong, medial.

Fibula slender, lateral.

FOOT 7 tarsal bones at ankle: *talus, calcaneus, navicular, cuboid, medial cuneiform, intermediate cuneiform, lateral cuneiform;*
5 metatarsal bones.

TOES 14 phalanges.

ARCHES OF FOOT 3 arches formed—2 longitudinal, 1 transverse; supported by ligaments as well as muscles and their tendons; distribute weight evenly between heel and metatarsals.

SESAMOID BONES form in tendons subjected to compression; generally around a joint (for example, patella).

SELF-QUIZ

1. Bones may act as a storehouse for: (a) calcium; (b) phosphorus; (c) both calcium and phosphorus.

2. Flat bones lack: (a) periosteum; (b) a medullary cavity; (c) diploe.

3. The combination of collagen fibers and inorganic salts makes bone exceptionally strong without being brittle. True or false?

4. The connective tissues that give rise to bone are

derived from cells of the ectodermal layer of the embryo. True or false?

5. The reabsorption of previously laid-down bone is accomplished by: (a) osteoblasts; (b) osteoprogenitor cells; (c) osteoclasts.

6. The marrow cavity of a long bone enlarges as the bone grows due to the action of: (a) fibroblasts; (b) osteocytes; (c) osteoclasts.

7. Calcitonin: (a) decreases the resorptive activity of osteoclasts and lowers the blood calcium level; (b) increases the resorptive activity of osteoclasts and raises the blood calcium level; (c) releases calcium ions from the bone that pass to the blood and raise the blood calcium level.

8. A gradual reduction in the rate of bone formation while the rate of bone absorption remains normal results in: (a) spina bifida; (b) osteoporosis; (c) osteomyelitis.

9. The effect of vitamin D is to: (a) soften the matrix of bone; (b) remove calcium from the matrix of bone; (c) increase the absorption of calcium from food by the intestinal wall.

10. Match these skeletal terms with the appropriate description.

Process	(a) A smooth, nearly flat articular surface.
Trochanter	
Tubercle	(b) A pit; generally used for attachment rather than for articulation.
Condyle	
Sulcus	
Crest	(c) A hole.
Facet	(d) A rounded prominence that articulates with another bone.
Fossa	
Fovea	
Foramen	(e) A canal.
Meatus	(f) A large, somewhat blunt process.
	(g) A depression; often used as an articular surface.
	(h) A sharp, prominent bony ridge.
	(i) A prominence or projection.
	(j) A nodule or small rounded process.
	(k) A groove.

11. The single occipital bone forms the lower posterior wall of the calvarium as well as the anterior portion of the floor of the cranial cavity. True or false?

12. An abnormal lateral curve that occurs in the vertebral column is called a: (a) scoliosis; (b) lordosis; (c) kyphosis.

13. Match the structures with the appropriate description.

Fontanels	(a) Form most of the calvarium.
Frontal bone	
Sinuses	(b) Bone of the floor of the cranial cavity.
Parietal bones	
Occipital bone	(c) Form the central part of the facial skeleton.
Ethmoid bone	

Temporal bones	(d) Air spaces in bone lined with mucous membranes.
Maxillary bones	
Zygomatic bones	
	(e) The lower jaw.
Lacrimal bones	(f) "Soft spots" of the skull.
Mandible	
Vomer bone	(g) Forms the posterior portion of the nasal septum.
An auditory ossicle	
Hyoid bone	(h) Its most obvious landmark is the large foramen magnum.
Cervical vertebrae	
	(i) Form the prominences of the cheek.
Thoracic vertebrae	(j) Single bone that forms the anterior superior region of the skull.
Sacral vertebrae	
Kyphosis	(k) Help form the medial surface of the orbital cavity.
	(l) Form part of the middle cranial cavity.
	(m) Form the spinal column in the neck.
	(n) Hunchback condition.
	(o) Serves for attachment of muscles of the tongue and throat.
	(p) Are fused and form a wedge between the coxal bones.
	(q) Malleus.
	(r) There are 12 of these.

14. The xiphoid process is a component of the: (a) sternum; (b) ribs; (c) thoracic vertebrae.

15. Ribs that have their costal cartilages attached to the cartilages of the rib above are called: (a) true ribs; (b) vertebrosternal ribs; (c) false ribs.

16. The first seven pairs of ribs articulate anteriorly with the sternum through the: (a) costal cartilages; (b) tubercles; (c) xiphoid process.

17. Which of the following are part of the appendicular skeleton? (a) pectoral girdle; (b) pelvic girdle; (c) mandible; (d) both a and b.

18. The scaphoid, lunate, triquetral, and pisiform are bones of the: (a) foot; (b) hand; (c) tibia-fibula complex.

19. Which of the following forms the junction of the superior and axillary borders of the scapula and contains the glenoid fossa? (a) inferior angle; (b) superior angle; (c) lateral angle.

20. The coxal bone includes the: (a) ischium; (b) fibula; (c) patella.

21. The lesser or true pelvis is the cavity: (a) above the pelvic brim; (b) anterior to the pelvic brim; (c) below the pelvic brim.

22. Compared with the male pelvis, the female pelvis: (a) is more massive; (b) is narrower at the pelvic outlet; (c) has an obtuse pubic angle.

23. The fibula is a slender bone that is lateral to the radius. True or false?

24. Match the following terms with the appropriate description.

Ilium
Ischium
Pubis
Acetabulum
False pelvis
Pelvic outlet
Femur
Greater
  trochanter
Condyles
Adductor
  tubercle
Tibia
Anterior crest
Fibula

(a) Forms anterior portion of coxal bone.
(b) Expanded space above the pelvic inlet; actually part of abdomino-pelvic cavity.
(c) Rounded medial and lateral enlargements at distal end of femur shaft.
(d) The medial bone of the leg.
(e) Bone that forms the thigh.
(f) Sharp longitudinal ridge on anterior surface of the tibia.
(g) Small projection just above medial condyle of femur.
(h) Upper, flat expanded portion of the coxal bone.
(i) Large lateral process just below neck of the femur.
(j) An opening bounded by coccyx and lower parts of the coxal bone.
(k) Posterior inferior portion of the coxal bone.
(l) A slender bone lateral to tibia.
(m) Cup-shaped depression formed by junction of ilium, ischium, and pubis.

LEARNING OBJECTIVES

After completing this chapter, you should be able to:

- State two especially useful criteria for classifying body joints.

- Distinguish between the two types of fibrous joints and give an example of each.

- Distinguish between the two types of cartilaginous joints and give an example of each.

- Describe the distinguishing features of synovial joints.

- Name the four types of synovial joints and give examples of each.

- Name the ligaments associated with the major joints of the body.

- Describe several common joint disorders.

CHAPTER CONTENTS

FIBROUS JOINTS

CARTILAGINOUS JOINTS

SYNOVIAL JOINTS

CONDITIONS OF CLINICAL SIGNIFICANCE:
  ARTICULATIONS

Articulations

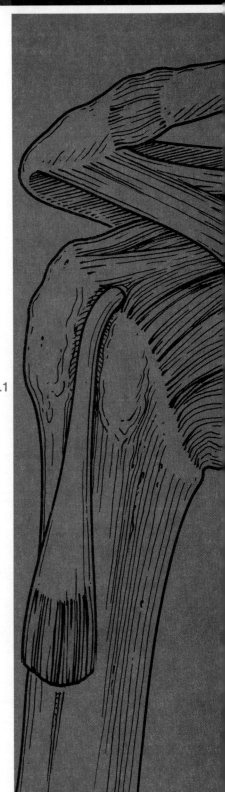

In the previous chapter, we were primarily interested in the support function of the skeleton. Now we will study how the individual bones of the skeleton join, or **articulate.** While some joints are rather rigid and permit little, if any, movement, most joints allow the bones to move in relation to each other. Various criteria can be used to classify the large number of joints in the body. Two of the most useful are (1) according to the *material* that connects the joints and (2) according to the *movement* allowed by the joints.

FIBROUS JOINTS

The **fibrous joints** include all the articulations in which the bones are held tightly together by fibrous connective tissue. Because of its role in strengthening the joint, the connective tissue is referred to as the *sutural ligament.* There is very little material between the ends of the bones, and no appreciable movement is allowed. For this reason fibrous joints are also classified as **synarthroses** (*syn* = together; *arthron* = joint)—that is, a nonmovable joint. However, slight movement is, in fact, permitted in some synarthroses. There are two types of fibrous joints, *sutures* and *syndesmoses*, depending in part on the length of the connective tissue fibers that hold the bones together.

Sutures

In **sutures** (Figure 6.1) the edges of the bones have interdigitations, or grooves, that fit closely and firmly together. Consequently, the connecting fibers are very short, spanning the small gap between the bones. This type of joint is found only between the flat bones of the skull. In early adulthood, the fibers of the suture begin to be replaced by bone. Eventually, if the fibers are replaced completely, the bones on either side of the sutures become firmly fused together. This condition is called *synostosis* (that is, held together by bone). **F6.1**

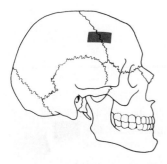
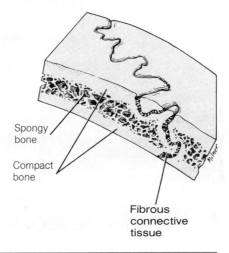

Spongy bone

Compact bone

Fibrous connective tissue

Figure 6.1
Suture. Note the short connective tissue fibers joining the two bones.

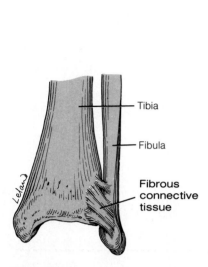

Figure 6.2
Syndesmoses. The distal ends of the tibia and fibula are held together by relatively long connective tissue fibers.

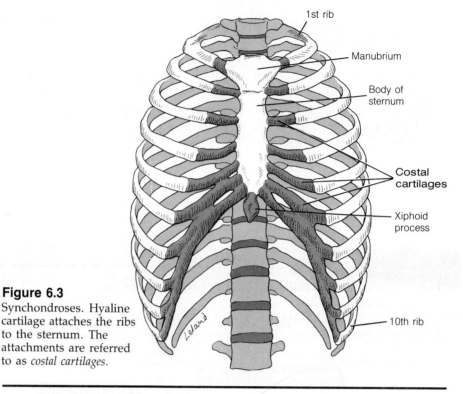

Figure 6.3
Synchondroses. Hyaline cartilage attaches the ribs to the sternum. The attachments are referred to as *costal cartilages*.

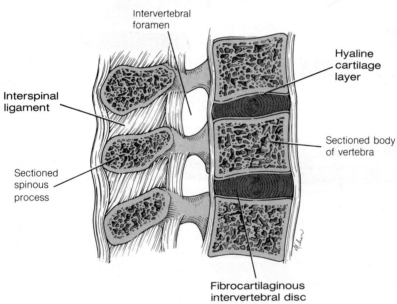

Figure 6.4
Symphyses. A pad of fibrocartilage separates the bodies of the vertebrae.

Syndesmoses

Like a suture, the bones of a syndesmosis joint (*syndesmosis* = held by bands) are joined together by fibrous connective tissue (Figure 6.2). The ends of the bones are farther apart in this type of joint, however. Consequently, the fibers connecting the bones are longer and are therefore generally referred to as ligaments. Bones joined by syndesmoses are not held as firmly as those joined by sutures. Syndesmoses can permit some movement, which is best described as "give," but they do not allow true movement; therefore, these joints are considered to be synarthroses. The joint between the distal ends of the tibia and fibula, and the mid-radius/ulnar and mid-tibia/fibula joints—where the bones are held together by interosseous membranes—are examples of syndesmoses.

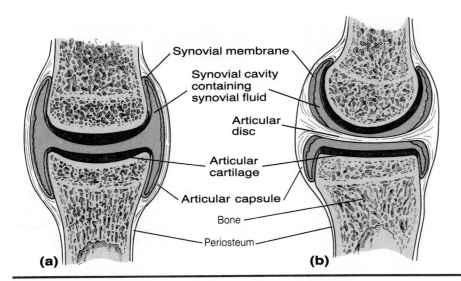

Synovial membrane

Synovial cavity containing synovial fluid

Articular disc

Articular cartilage

Articular capsule

Bone

Periosteum

(a)   **(b)**

Figure 6.5
Structure of synovial joints.

CARTILAGINOUS JOINTS

In **cartilaginous joints,** the bones are united by cartilage. Some cartilaginous joints are immovable and are functionally **synarthroses.** In other cartilaginous joints slight movement is possible and they are classified as **amphiarthroses** (*amphi* = on both sides). There are two types of cartilaginous joints: *synchondroses* and *symphyses.*

Synchondroses

The bones of a **synchondrosis joint** (*synchondrosis* = held by cartilage) are held together by hyaline cartilage (Figure 6.3). Many synchondroses are temporary joints, with the cartilage eventually being replaced by bone. This replacement occurs between the epiphyses and the diaphysis of long bones (where the epiphyseal cartilages are replaced) and between certain skull bones. The joints formed between the first ten ribs and their costal cartilages are permanent synchondroses. **F6.3**

Symphyses

The articular surfaces of bones joined by a **symphysis** are covered with a thin layer of hyaline cartilage (Figure 6.4). Separating the bones within the joint is a fibrocartilaginous pad, which is the distinguishing feature of symphyses. These pads, or discs, are compressible, allowing the symphyses to serve as shock absorbers. The junction of the two pubic bones and the junctions between the bodies of adjacent vertebrae are examples of symphyses. The pads between the vertebrae are called *intervertebral discs.* During development, the two halves of the mandible are joined by a midline symphysis; however, this joint becomes completely ossified by adulthood. **F6.4**

SYNOVIAL JOINTS

Most joints of the body are **synovial joints,** which are characterized by being freely movable. The movement of synovial joints is limited only by ligaments, muscles, tendons, or adjoining bones. Because of this freedom, synovial joints are also referred to as **diarthroses** (*diarthrosis* means "through joint"), indicating that there are only relatively slight limitations to the movement of such joints. Another characteristic of synovial joints is the presence of a fluid-filled joint cavity. Synovial joints have four distinguishing features: an *articular cartilage,* an *articular capsule,* a *synovial membrane,* and *synovial fluid* (Figure 6.5*a*). **F6.5a**

The **articular cartilage** is a thin layer of hyaline cartilage that covers the smooth articular surfaces of the bones. The **articular capsule** is a double-layered membrane that surrounds and encloses the joint. The outer layer of the

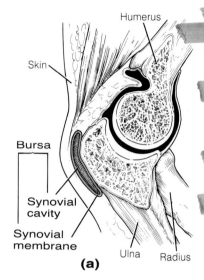

(a)

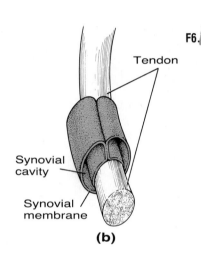

(b)

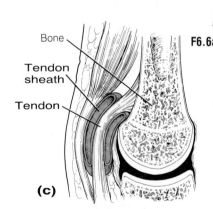

(c)

Figure 6.6
(a) Subcutaneous bursa of the elbow. **(b)** Tendon sheath. **(c)** Longitudinal section showing the position of a tendon sheath.

capsule is composed of dense fibrous connective tissue whose fibers are firmly joined to the periosteum of the bones. Parallel bundles of fibers in the outer layer form ligaments that strengthen the joint. We will name some of these ligaments later in the chapter. The inner layer of the articular capsule is referred to as the **synovial membrane.** The synovial membrane consists of loose connective tissue whose inner surface is well supplied with capillaries. The membrane, which is often thrown into folds that project into the joint cavity, lines the entire joint cavity. However, it does not cover the surfaces of the articular cartilages or the articular disc. The synovial membrane produces a thick fluid called **synovial fluid.** Synovial fluid provides nourishment to the articular cartilages, and lubricates the joint surfaces. In fact, synovial fluid serves as a weight-bearing element in the joint as it keeps the articular cartilages of the bones that form the joint separate, not allowing the cartilages to contact one another. Normally, only enough synovial fluid is secreted to form a thin film over the surfaces within the joint. In a joint that is injured or becomes inflamed, however, fluid production may be stimulated and enough fluid may accumulate to cause swelling and discomfort. The joint capsule is well supplied with nerve fibers, which not only makes the perception of pain possible, but also provides constant information concerning movement and position of the joint.

In addition to these four features, some synovial joints have **articular discs** (Figure 6.5*b*), or in the case of the knee, **menisci,** of fibrocartilage that extend inward from the articular capsule. Articular discs divide the synovial cavity into two separate cavities. In this form of joint, the synovial membrane lining the cavities extends only a short distance onto the surfaces of the disc. The jaw, the sternoclavicular joint, and the distal radial/ulnar joint contain articular discs. The knee joint is only partially subdivided by menisci.

In addition to the strengthening provided by the ligaments formed from the fibrous layer of the articular capsule, various muscles and their tendons, which cross the joints, stabilize the joints while still permitting them to move.

Bursae and Tendon Sheaths

Synovial membranes form two other structures that, while not actually part of the synovial joints, are often associated with them. These are *bursae* and *tendon sheaths.* Both of these structures contain synovial fluid and serve to reduce friction during movement between a structure—such as skin, muscles, tendons, or ligaments—and the bone.

Bursae are small sacs lined with synovial membranes. Because they are filled with synovial fluid they act as cushions between the structures they separate. There are many bursae distributed throughout the body. Some are subcutaneous, lying between the bone and the skin, such as the bursa separating the olecranon process of the elbow from the skin (Figure 6.6*a*). Most bursae are located between the tendons and the bone.

Tendon sheaths (Figure 6.6*b* and *c*) are found where tendons cross joints and where, without the sheaths, the tendons would be subjected to constant friction against the bones, as in the wrist and fingers. The sheaths are cylindrical synovial sacs similar to bursae. They wrap around the tendons, forming fluid-filled, double-walled cushions for the tendons to slide through.

Movements of Synovial Joints

Synovial joints, unlike fibrous and cartilaginous joints, are not classified according to the material that connects the bones. Rather, they are named on the basis of the movements they permit. The shapes of the bony structures that surround a joint, and often the articular surface itself, generally limit its movements. Many joints have *axes of rotation* that allow bones to move in various planes; the plane of movement is generally perpendicular to the axis. For instance, in the movement of the elbow the axis is a horizontal line that passes through the joint from side to side. The bones rotate around this fulcrum (pivot point) in a vertical plane.

Joints that have only one axis of motion and can therefore move in only one plane are called **uniaxial joints;** the elbow is an example (Figure 6.7*a*).

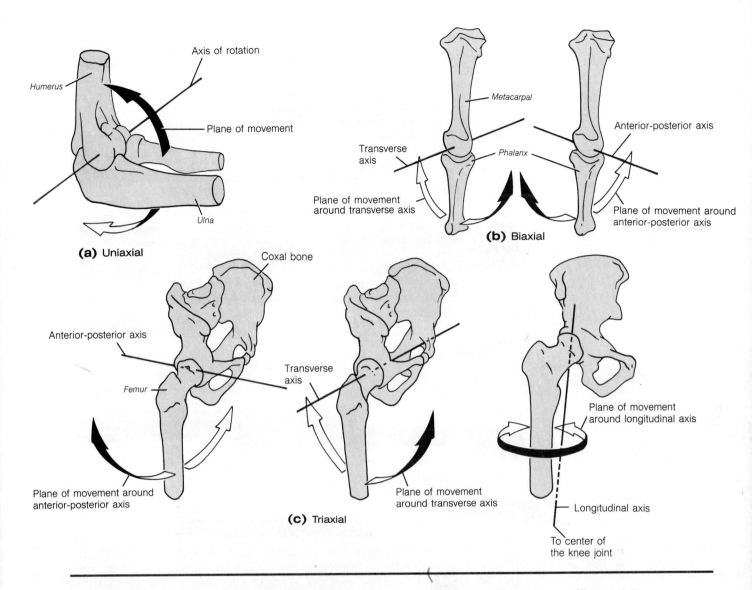

(a) Uniaxial

(b) Biaxial

(c) Triaxial

Note the movement of the knee, a uniaxial joint, in Box 6.1. Some joints have two axes and thus allow movement in two planes that are at right angles to each other. Such joints are called **biaxial joints** (Figure 6.7*b*). Still other joints have three axes and permit movement in three planes. These are called **triaxial joints** (Figure 6.7*c*). In many of the smaller joints, motion is not restricted by the shapes of the articular surfaces and slight movement is possible in any direction. Because their movements do not follow particular axes, these joints are referred to as **nonaxial.**

The general movements allowed in synovial joints can be placed into four groups: *gliding, angular, circumduction,* and *rotation*. In addition, several synovial joints allow movements that are unique to that particular joint. These unique movements are considered in the section ''Special Movements.''

Gliding

The simplest and most common type of motion that can occur in a synovial joint is **gliding**. In this motion, the surfaces of adjoining bones move back and forth upon one another. In many cases the articulating surfaces are flat or slightly concave; but gliding can occur between any two adjoining surfaces regardless of their forms. The joints between the heads of the ribs and the bodies of the vertebrae and between the tubercles of the ribs and the transverse processes of the vertebrae allow gliding movement, as do numerous other joints.

Box 6.1

F6.7b

F6.7c

Figure 6.7
Movements of synovial joints.

Angular Movements

Angular movements increase or decrease the angle between two adjoining bones. There are four angular movements that may occur in various synovial joints: *flexion, extension, abduction,* and *adduction* (Figure 6.8).

F6.8

FLEXION When a bone is moved in an anterior-posterior plane in a way that *decreases* the angle between it and the adjoining bone, **flexion** occurs. Examples include bending the elbow, bringing the thigh up toward the abdomen, and bringing the calf of the leg up toward the back of the thigh. The vertebral column is flexed when it is bent forward. Flexion of the ankle by raising the toe region toward the shin is referred to as **dorsiflexion.**

EXTENSION **Extension,** the opposite of flexion, causes the angle between adjoining bones to *increase.* Extension occurs when a flexed joint is moved back to the anatomical position, as in straightening the arm, thigh, or knee. **Hyperextension** occurs when the part is moved beyond the straight position, as in arching the back or bringing the limbs posteriorly beyond the plane of the body. Extension of the ankle by lowering the toe region is referred to as **plantar flexion.**

ABDUCTION When a part such as a limb is moved away from the midline of the body, **abduction** occurs. In the case of the fingers, abduction involves moving them away from the midline of the hand (third digit). Abduction of the toes is accomplished by moving them away from the longitudinal axis of the second toe.

ADDUCTION **Adduction,** the opposite of abduction, involves the movement of a part toward the midline of the body, back toward the anatomical position. In the case of the fingers, the movement is toward the midline of the hand (third digit). Adduction of the toes is accomplished by moving them toward the longitudinal axis of the second toe.

Circumduction

F6.8 The joint motion known **as circumduction** delineates a cone (Figure 6.8): the base of the cone is outlined by the movement of the distal end of the bone; the apex is in the articular cavity. The movement is actually a sequential combination of flexion, abduction, extension, and adduction. Circumduction is common at the hip and the shoulder joints, but it is possible in other joints also.

Rotation

The motion of a bone around a central axis, without any displacement of that **F6.8** axis, is **rotation** (Figure 6.8). If the anterior surface of a bone such as the humerus or femur moves inward, the movement is called *inward (medial) rotation.* When the anterior surface turns outward, it is *outward (lateral) rotation.*

SUPINATION The term used to describe the outward rotation of the forearm, causing the palms to face anteriorly and the radius and ulna to be parallel, is **supination.** In the anatomical position the forearms are supinated.

PRONATION The term used to describe the inward rotation of the forearm, causing the radius to cross diagonally over the ulna and the palms to face posteriorly, is **pronation.**

Special Movements

Some synovial joints allow special movements that cannot be described by any of the previously mentioned movements. These movements are *elevation, depression, inversion, eversion, protraction,* and *retraction.*

ELEVATION The motion that raises a part is **elevation.** This term is commonly used to refer to the raising of the scapula, as when shrugging the shoulders, or raising the mandible, as when closing the mouth.

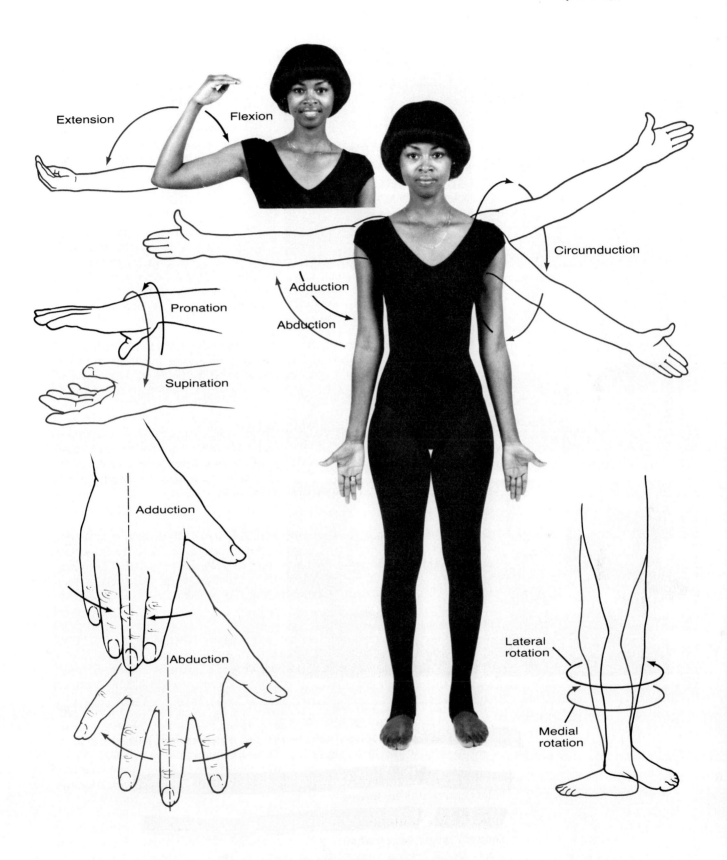

Figure 6.8
Movements typical of synovial joints.

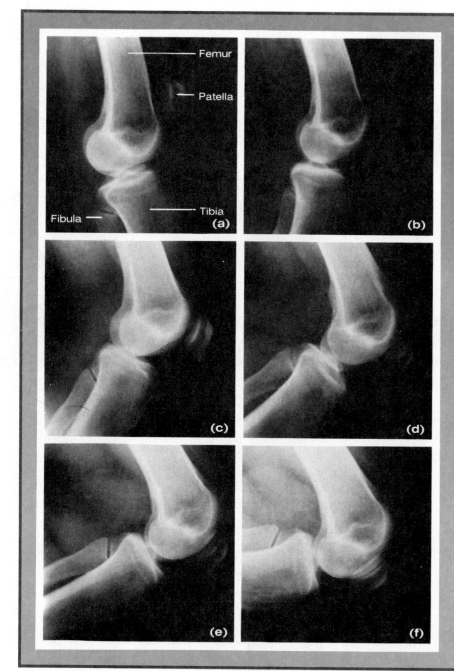

Femur

Patella

Fibula

Tibia

(a) (b) (c) (d) (e) (f)

Box 6.1
Movement of the Knee Joint

These X rays of a normal knee were taken using a method called fluoroscopy. Such X rays are useful in identifying knee injuries— torn cartilage, dislocated kneecaps, and broken bones.

Flexion of the knee, a uniaxial joint, is shown in a series of successive movements. Two different types of movements—hinge and gliding—occur during flexion of the leg. The hinge movement between the femur and the tibia is most easily demonstrated by sitting on a table and dangling the leg over the edge. The knee functions for the leg like a hinge does for a door—allowing it to swing freely. The gliding movement begins as the kneecap (patella) slides over the knee joint.

DEPRESSION   The motion that lowers a part is **depression.** This term is often used to refer to the lowering of the scapula or mandible.

INVERSION   Twisting the foot so that the sole faces inward with its inner margin raised is **inversion.**

EVERSION   Twisting the foot so that the sole faces outward with its outer margin raised is **eversion.**

PROTRACTION   The motion that moves a part (such as the mandible) forward is **protraction.**

RETRACTION   The motion that returns a protracted part to its usual position is **retraction.**

Types of Synovial Joints

On the basis of the movements allowed and the shapes of the articular surfaces involved, it is possible to separate the synovial joints into six types. These types can be grouped according to whether they are *nonaxial, uniaxial, biaxial,* or *triaxial.*

Nonaxial Joints

GLIDING (ARTHRODIAL) JOINTS The **gliding joints** are formed primarily by the apposition of flat, or only slightly curved, articular surfaces. Movement is allowed in any direction, being limited only by ligaments or bony processes that surround the articulation. Gliding joints are found between the articular processes of vertebrae and between most carpal and tarsal bones.

Uniaxial Joints

HINGE (GINGLYMUS) JOINTS In **hinge joints,** the articular surfaces are shaped in such a manner that the only movements possible are flexion and extension. The elbow joint, the knee, and the joints between the phalanges of the fingers and toes (interphalangeal joints) are examples of hinge joints.

PIVOT (TROCHOID) JOINTS The only movement allowed in a **pivot joint** is rotation around the longitudinal axis of the bone. Examples are the rotation of the first cervical vertebra (atlas) around the odontoid process of the second cervical vertebra (axis) and the proximal articulations between the radius and the ulna. In the atlas/axis joint the odontoid process is held against the inside surface of the anterior arch of the atlas by a transverse ligament that passes behind the process while connecting the two sides of the anterior arch. In the radial/ulnar joint, the head of the radius is held firmly against the radial notch of the ulna by a strong annular ligament that encircles its head. The radius rotates within the annular ligament, allowing the forearm to pronate and supinate.

Biaxial Joints

CONDYLOID (ELLIPSOID) JOINTS **Condyloid joints** have one articular surface slightly concave and the other slightly convex; thus movement is allowed in two planes that are at right angles to each other. Flexion, extension, abduction, and adduction can occur in condyloid joints. Circumduction is possible also, but axial rotation is not. The articulations between the radius and the carpals, the occipital condyles of the skull on the first cervical vertebra, and the metacarpophalangeal and metatarsophalangeal joints are condyloid joints.

SADDLE JOINTS **Saddle joints** allow the same movements as condyloid joints: flexion, extension, abduction, adduction, and circumduction. The articular surface of each bone is concave in one direction and convex in the other; therefore, they fit together like two saddles would if their riding surfaces were rotated 90 degrees and placed one on top of the other. The only true saddle joint in the body is the carpometacarpal joint of the thumb.

Triaxial Joints

SPHEROID (BALL AND SOCKET) JOINTS **Spheroid joints** are formed by a spherical head of one bone fitting into a cup-shaped cavity on the other. Such joints allow movement around three axes. In addition to flexion, extension, abduction, adduction, and circumduction, spheroid joints permit medial and lateral rotation. There are only two spheroid joints in the body: the shoulder and the hip.

Figure 6.9 illustrates most of the types of joints; Table 6.1 summarizes all the main joints of the body.   **F6.9, Table 6.1**

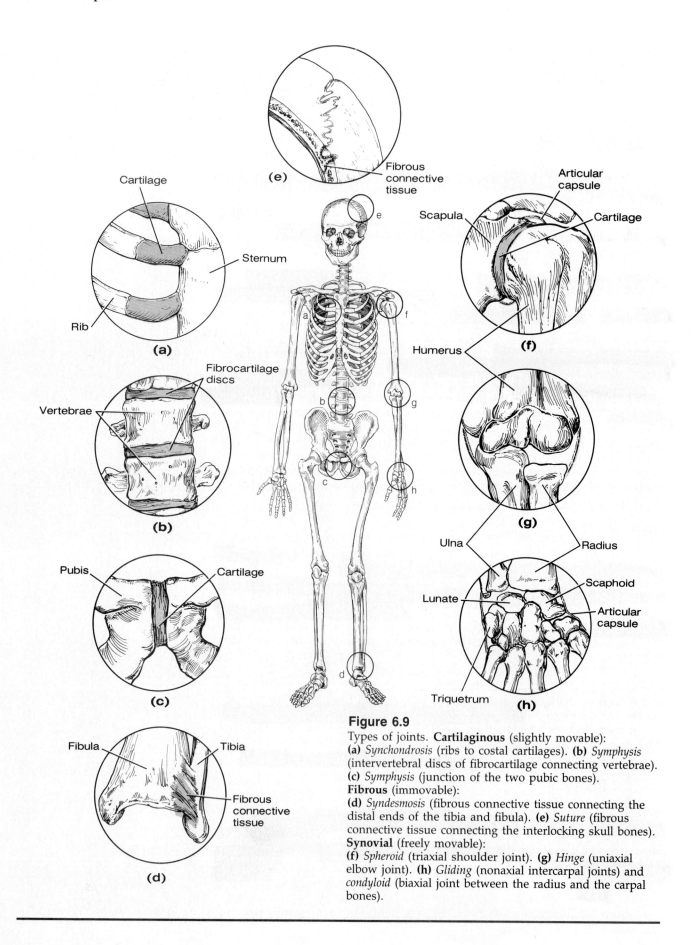

Figure 6.9
Types of joints. **Cartilaginous** (slightly movable):
(a) *Synchondrosis* (ribs to costal cartilages). **(b)** *Symphysis* (intervertebral discs of fibrocartilage connecting vertebrae). **(c)** *Symphysis* (junction of the two pubic bones).
Fibrous (immovable):
(d) *Syndesmosis* (fibrous connective tissue connecting the distal ends of the tibia and fibula). **(e)** *Suture* (fibrous connective tissue connecting the interlocking skull bones).
Synovial (freely movable):
(f) *Spheroid* (triaxial shoulder joint). **(g)** *Hinge* (uniaxial elbow joint). **(h)** *Gliding* (nonaxial intercarpal joints) and *condyloid* (biaxial joint between the radius and the carpal bones).

Table 6.1 Summary of the Main Joints of the Body

Joint	Type	Movement
Between the cranial bones	Fibrous (suture)	No appreciable movement
Between the distal tibia and the fibula	Fibrous (syndesmosis)	Slight movement ("give")
Between the mid-radius and the ulna (interosseous membrane)	Fibrous (syndesmosis)	Slight movement
Between the ribs and the sternum (sternocostal)	Cartilaginous (synchondrosis)	Slight movement
Between the two pubic bones	Cartilaginous (symphysis)	Slight movement
Between the bodies of the vertebrae	Cartilaginous (symphysis)	Slight movement
Between the sacrum and the ilium	Partly cartilaginous (synchondrosis) and partly synovial (gliding)	Generally no movement, but slight gliding movement is possible. In older people, fibers may hold the two bones firmly together.
Between the articular processes of the vertebrae	Synovial (gliding)	Nonaxial; gliding
Between the head of the rib and the body of the vertebra	Synovial (gliding)	Nonaxial; gliding
Between the occipital and the atlas	Synovial (condyloid)	Biaxial; flexion, extension, abduction, adduction, circumduction
Between the atlas and the odontoid process of the axis	Synovial (pivot)	Uniaxial; pivoting around the odontoid process
Between the sternum and the clavicle	Synovial (gliding)	Nonaxial; gliding
Between the acromion of the scapula and the clavicle	Synovial (gliding)	Nonaxial; gliding and rotation of the scapula upon the clavicle
Between the humerus and the scapula	Synovial (spheroid)	Triaxial: flexion, extension, abduction, adduction, circumduction, medial and lateral rotation
Between the ulna and the humerus	Synovial (hinge)	Uniaxial; flexion and extension
Between the head of the radius and the ulna	Synovial (pivot)	Uniaxial; pivoting longitudinal axis, as in pronation and supination
Between the radius and the carpals (scaphoid, lunate)	Synovial (condyloid)	Biaxial; flexion, extension, abduction, adduction, circumduction
Between the carpals	Synovial (gliding)	Nonaxial; gliding
Between the first metacarpal and carpal	Synovial (saddle)	Biaxial; flexion, extension, abduction, adduction, circumduction
Between the second through fifth metacarpals and the carpals	Synovial (gliding)	Nonaxial; gliding
Between the second through fifth metacarpals and the phalanges	Synovial (condyloid)	Biaxial; flexion, extension, abduction, adduction, circumduction
Between the phalanges (hand and foot)	Synovial (hinge)	Uniaxial; flexion, extension

Table 6.1   Summary of the Main Joints of the Body (continued)

Joint	Type	Movement
Between the femur and the coxal bone	Synovial (spheroid)	Triaxial; flexion, extension, abduction, adduction, circumduction, medial and lateral rotation
Between the tibia and the femur	Synovial (hinge)	Uniaxial; flexion, extension (some rotation)
Between the proximal end of the fibula and the tibia	Synovial (gliding)	Nonaxial; gliding
Between the distal ends of the fibula and the tibia and the talus	Synovial (hinge)	Uniaxial; flexion, extension
Between the tarsals	Synovial (gliding)	Nonaxial; gliding
Between the tarsals and the metatarsals	Synovial (gliding)	Nonaxial; gliding
Between the metatarsals and the phalanges	Synovial (condyloid)	Biaxial; flexion, extension, abduction, adduction, circumduction

Ligaments of Selected Joints

Ligaments play an important role in maintaining the proper positioning of bones that articulate in synovial joints while at the same time allowing relatively free movement of the joints. We therefore now turn to the ligaments of several important joints. *These ligaments are summarized in the Study Outline at the end of the chapter.*

Ligaments of the Vertebral Column

A number of ligaments that extend between the atlas, axis, and occipital bones are beyond the scope of this text. Of more importance to us are the ligaments that connect the remainder of the vertebrae into a firm but flexible axial support (Figure 6.10). The joints between adjacent vertebrae consist of two types: a series of cartilaginous symphyses between the vertebral bodies and a series of synovial gliding joints between the articular processes of the vertebrae.

F6.10

The bodies of the vertebrae are held together by fibrous **intervertebral discs.** Each disc is composed of a firm outer portion, called the *anulus fibrosus,* and a softer central portion, the *nucleus pulposus.* Surrounding the synovial joint of each articular process is a fibrous **articular capsule.** Passing along the ventral surface of the bodies of the vertebrae is a strong **anterior longitudinal ligament.** A similar band of fibers, called the **posterior longitudinal ligament,** passes along the dorsal surface of the vertebral bodies. Both the anterior and the posterior longitudinal ligaments extend from the second cervical vertebra to the sacrum. The vertebral arches are interconnected by four groups of ligaments. The **supraspinous ligament** connects the tips of the spinous processes; in the cervical region, from the seventh cervical vertebra to the external occipital protuberance, this ligament continues as the *ligamentum nuchae.* Thin **interspinous ligaments** connect adjoining spinous processes from their roots to the tip of each process. Strong elastic **ligamenta flava** run between the laminae of adjacent vertebrae; **intertransverse ligaments** connect the transverse processes of adjacent vertebrae.

Ligaments of the Clavicular Joints

Each clavicle provides its anchoring function through two synovial gliding joints—medially with the upper lateral portion of the manubrium of the sternum and the costal cartilage of the first rib **(sternoclavicular joint)** and later-

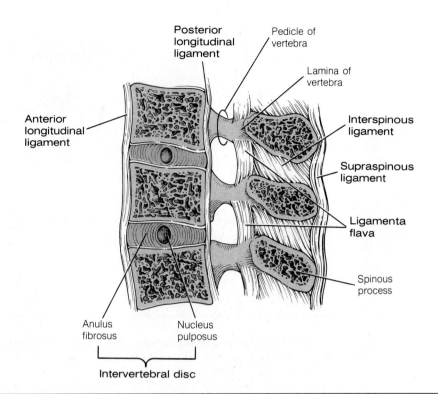

Figure 6.10
Ligaments of the vertebral column as seen in a median sagittal section through two lumbar vertebrae.

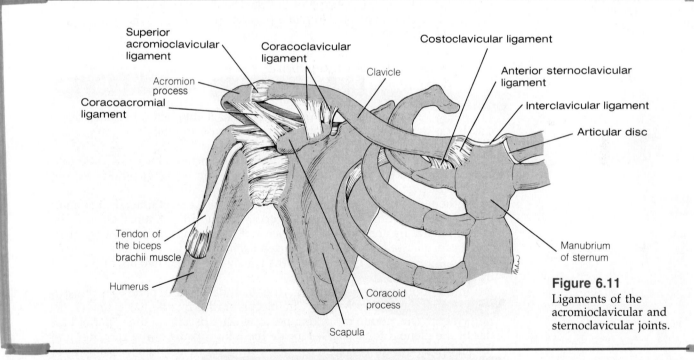

Figure 6.11
Ligaments of the acromioclavicular and sternoclavicular joints.

ally with the acromion process of the scapula **(acromioclavicular joint).** Each of these joints is completely surrounded by an **articular capsule.** An **articular disc** separates the medial end of the clavicle and the sternum, dividing the joint into two portions, each lined with a synovial membrane. An articular disc is sometimes present in the acromioclavicular joint.

The *sternoclavicular joint* is strengthened by four ligaments (Figure 6.11). **F6.11** The **anterior sternoclavicular** and **posterior sternoclavicular ligaments** cover the anterior and posterior surfaces of the articulation. The medial ends of the two clavicles are attached by an **interclavicular ligament,** a band of fibers that passes along the superior margin of the manubrium of the sternum. A **costoclavicular ligament** runs downward from the undersurface of each clavicle near its sternal end to the upper surface of the costal cartilage of the first rib.

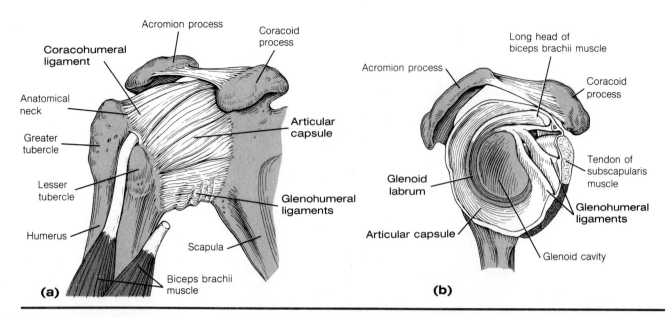

Figure 6.12
(a) Anterior view of ligaments of the right shoulder. **(b)** Lateral view of the right shoulder joint with the humerus removed.

Two ligaments strengthen the *acromioclavicular joint*. The **acromioclavicular ligament** extends between the lateral end of the clavicle and the acromion process of the scapula. For descriptive purposes this ligament is sometimes divided into superior and inferior portions. The **coracoclavicular ligament** passes from the inferior surface of the clavicle to the coracoid process of the scapula. It is often separated into two portions referred to as the *trapezoid* and the *conoid* ligaments.

Ligaments of the Shoulder Joint

In the shoulder joint, the head of the humerus is received by the shallow glenoid fossa of the scapula. The joint is loosely constructed, which permits extremely free movement but also means that it frequently becomes dislocated. It is protected above by the coracoid process and the acromion process of the scapula.

Like all synovial joints, the shoulder joint is enclosed by an **articular capsule** (Figure 6.12). This capsule attaches to the rim of the glenoid fossa and extends outward to the anatomical neck of the humerus. The capsule is strengthened anteriorly by two ligaments: the **coracohumeral ligament,** which extends from the coracoid process to the greater tubercle, and the **glenohumeral ligaments,** which are several thickenings in the lower portion of the capsule itself. The **glenoid labrum** (*labrum* = lip), a rim of fibrocartilage that surrounds the glenoid fossa, adds somewhat to the stability of the joint by deepening the fossa.

In addition to these ligaments, the shoulder joint depends on the surrounding muscles for strength. The biceps brachii muscle has a unique arrangement. The tendon of its long head, which arises from the superior border of the glenoid fossa, passes inside the joint capsule of the shoulder and through the intertubercular groove of the humerus. Thus, in effect, it helps to hold the head of the humerus against the scapula. The tendons of the supraspinatus, infraspinatus, subscapularis, and teres minor muscles also strengthen the shoulder joint by passing along the sides of the joint and fusing with the articular capsule. Because the muscles of these tendons are important in rotation movements of the arm and the tendons almost completely encircle the joint they are collectively referred to as the *rotator cuff.*

Ligaments of the Elbow Joint

The elbow joint is a ginglymus (hinge) joint where the trochlea of the humerus is received into the trochlear notch of the ulna. Closely associated with the elbow joint is a gliding joint between the capitulum of the humerus and the superior surface of the head of the radius. These two articulations share a

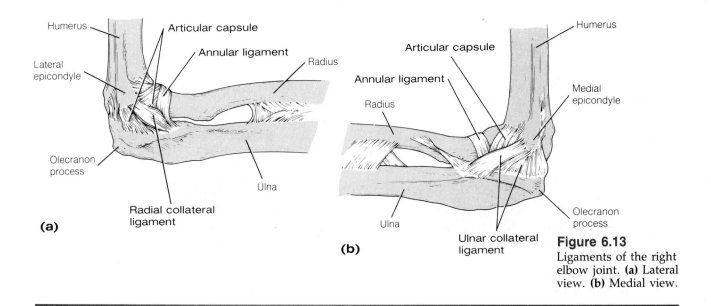

Figure 6.13
Ligaments of the right elbow joint. **(a)** Lateral view. **(b)** Medial view.

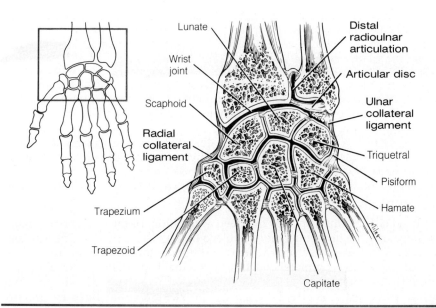

Figure 6.14
Vertical section through the right wrist joint showing the articular disc and collateral ligaments.

common joint cavity and are enclosed by a single fibrous **articular capsule** (Figure 6.13).

Medial and lateral thickenings of the articular capsule serve to stabilize the elbow joint. The medial thickening, the **ulnar collateral ligament,** passes from the medial epicondyle of the humerus to the medial surface of the ulna between the olecranon process and the coronoid process. The lateral thickening, the **radial collateral ligament,** passes from the lateral epicondyle of the humerus to the annular ligament and to the lateral surface of the ulna. The *annular ligament* is a strong band of fibers in the distal margin of the articular capsule of the elbow joint. The annular ligament does not strengthen the elbow joint to any great degree. Instead, it encircles the head and the upper part of the neck of the radius as it passes between the anterior and posterior margins of the trochlear notch of the ulna. In this manner, the radius is held tightly against the ulna and yet is allowed to rotate freely, as occurs during pronation and supination.

Ligaments of the Wrist Joint

The wrist joint (Figure 6.14) is a condyloid joint formed by the distal end of the radius and the articular disc of the distal end of the ulna with the scaphoid, lunate, and triquetral bones.

F6.13

F6.14

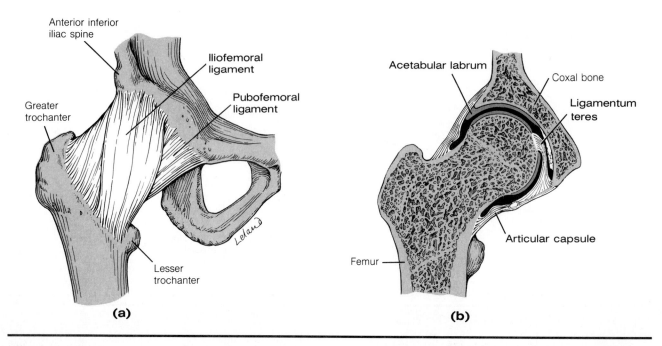

Anterior inferior
iliac spine

Iliofemoral
ligament

Pubofemoral
ligament

Greater
trochanter

Lesser
trochanter

Leland

(a)

Acetabular labrum

Coxal bone

Ligamentum
teres

Articular capsule

Femur

(b)

Figure 6.15
(a) Anterior view of
ligaments of the right hip
joint. **(b)** Frontal section
through the right hip
joint.

The **articular capsule** that envelops the joint is attached to the distal ends of the radius and ulna and to the proximal carpal bones. The wrist joint is strengthened laterally by the **radial collateral ligament,** which extends from the styloid process of the radius to the scaphoid and trapezium bones. The **ulnar collateral ligament** supports the joint medially, extending from the styloid process of the ulna to the triquetral and pisiform bones. The wrist joint is strengthened further by **palmar** and **dorsal radiocarpal ligaments.** These ligaments extend from the distal ends of the radius and ulna to the proximal row of carpal bones on the palmar and dorsal surfaces, respectively.

Ligaments of the Hip Joint

F6.15 The head of the femur fits into the deep acetabulum of the coxal bone (Figure 6.15), making the hip joint a more stable joint than the shoulder. The **articular capsule,** which extends from the margin of the acetabulum to the anatomical neck of the femur, completely encloses the joint. The capsule is strengthened anteriorly by the **iliofemoral** and **pubofemoral ligaments.** On its posterior surface the capsule is strengthened by the **ischiofemoral** ligament (Figure

F6.16 6.16).

The acetabulum is surrounded by a fibrocartilaginous rim called the *acetabular labrum.* The acetabular labrum is incomplete at its inferior margin, which gives it a horseshoe shape. Like the glenoid labrum of the shoulder joint, this labrum deepens the joint cavity. A unique ligament called the **ligamentum teres** extends through the joint cavity from the fovea on the head of the femur to the gap at the lower portion of the acetabular labrum. Because the ligamentum teres is slack during most movements of the hip, it is not believed to contribute significantly to the strength of the joint. It does, however, help prevent the head of the femur from slipping upward.

Ligaments of the Knee Joint

F6.17 The knee joint (Figure 6.17) is a complicated joint that is vulnerable to injury. It is classified as a hinge joint because its movements are restricted, for the most part, to flexion and extension by the surrounding ligaments. It has the structure, however, of a condyloid joint, with the condyles of the femur articulating with the slightly concave condyles of the tibia. The articular surface on the medial condyle of the femur is somewhat longer from front to back and is less curved than the articular surface on the lateral condyle. As a consequence of these structural differences, the final phase of complete extension of the knee joint primarily involves movement of the medial condyle of the femur

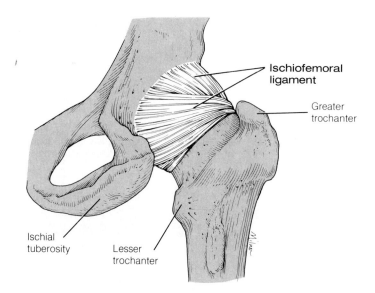

Figure 6.16
Posterior ligaments of the right hip joint.

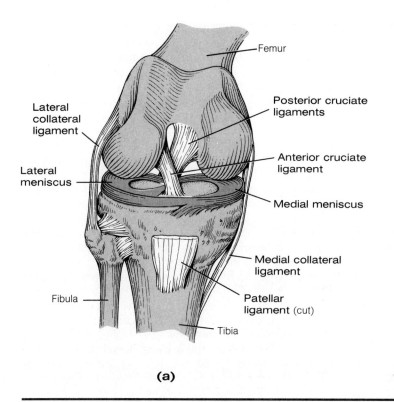

(a)

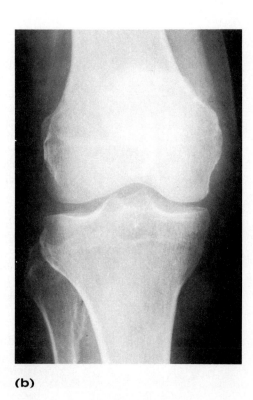

(b)

Figure 6.17
(a) Ligaments of the right knee joint. The femur is flexed slightly to allow the ligaments to be seen. The articular capsule and the patella have been removed.
(b) Anteroposterior X ray of the right knee.

on the tibia. This causes the tibia to undergo some lateral rotation (or the femur to undergo some medial rotation). Similarly, the extended joint must be unlocked by a slight medial rotation of the tibia before flexion of the knee joint can occur. When the knee joint is in a partially flexed position, it is possible for it to undergo even more rotation.

The **articular capsule** of the knee is different from the capsules of most joints in that it does not completely enclose the joint—being absent anteriorly above the patella. The capsule is strengthened posteriorly by the **oblique popliteal ligament** and the **arcuate popliteal ligament** (Figure 6.18). The **F6.18** oblique popliteal ligament is a broad, flat band that is attached proximally to the posterior surface of the femur just above the articular surface of the lateral condyle. From here it extends downward and medially to attach to the posterior surface of the head of the tibia. The arcuate popliteal ligament passes

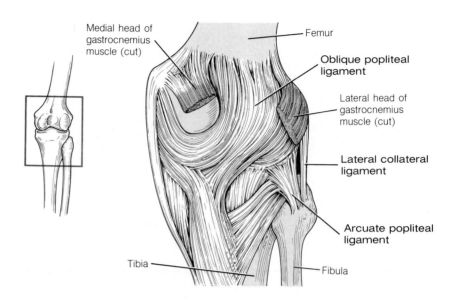

Figure 6.18
Posterior ligaments of the right knee joint.

from the posterior surface of the lateral condyle of the femur to the styloid process of the head of the fibula. The knee joint is stabilized medially and laterally by very strong **medial** and **lateral collateral ligaments,** which extend from the condyles of the femur to the tibia or the fibula. The collateral ligaments limit the amount of rotation that is possible by the knee joint and help prevent hyperextension of the joint. The joint is strengthened anteriorly by the **patellar ligament,** which extends from the patella to the tibial tuberosity. This ligament is a continuation of the central tendon of the quadriceps femoris muscles of the anterior thigh.

The flat superior surfaces of the condyles of the tibia, which are the largest weight-bearing surfaces in the body, are deepened by crescent-shaped cartilages called **medial** and **lateral menisci.** The menisci are attached only at their outer margins and frequently become damaged or are torn loose in athletic injuries.

Additional stability is added to the knee joint by the presence within the joint cavity of **anterior** and **posterior cruciate ligaments** (*cruciate* = cross-shaped). These ligaments extend diagonally from the superior surface of the tibia to the distal end of the femur, between the condyles. They are called cruciate because their paths cross each other. Because of their unique structural arrangements, the cruciate ligaments perform very specialized functions. When the knee is extended, the anterior cruciate ligament is taut, thus guarding against hyperextension of the joint by preventing the anterior **F6.19a** movement of the tibia (Figure 6.19*a*). When the knee is flexed, the posterior cruciate ligament becomes taut, preventing the tibia from slipping posteriorly **F6.19b** (Figure 6.19*b*).

Ligaments of the Ankle Joint

The ankle joint is a hinge joint formed by the distal end of the tibia and its medial malleolus, the lateral malleolus of the fibula, and the upper convex surface of the talus.

An **articular capsule** envelops the joint. On the medial side the capsule is **F6.20a** strengthened by a flat, triangular **deltoid ligament** (Figure 6.20*a*). This ligament extends from the medial malleolus of the tibia to the navicular, talus, and calcaneous bones. The joint is strengthened on the lateral side by three **F6.20b** ligaments (Figure 6.20*b*): the **anterior talofibular ligament,** which passes anteriorly from the malleolus of the fibula to the talus; the **posterior talofibular ligament,** which passes posteriorly from the malleolus of the fibula to the talus; and the **calcaneofibular ligament,** which runs inferiorly from the malleolus of the fibula to the calcaneus.

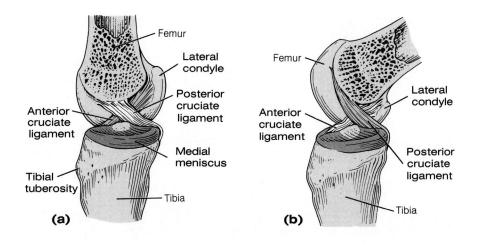

Figure 6.19
Functions of the cruciate ligaments. **(a) Knee extended:** taut anterior cruciate ligament prevents the tibia from moving anteriorly, thereby hindering overextension of the joint. **(b) Knee flexed:** taut posterior cruciate ligament prevents the tibia from slipping posteriorly. (The medial condyle of the femur has been removed to expose the cruciate ligaments.)

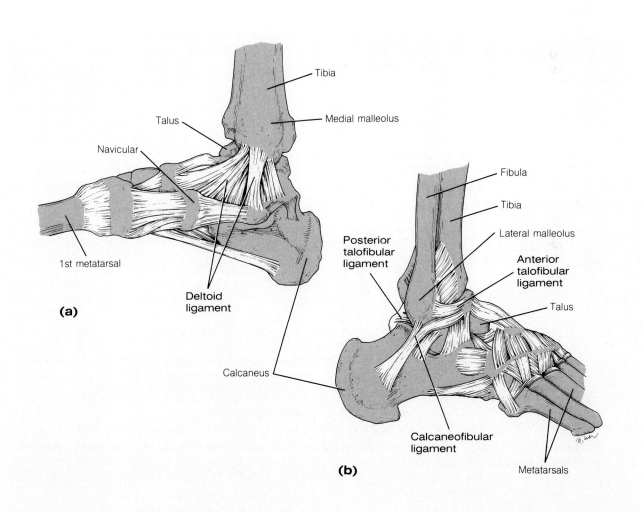

Figure 6.20
Ligaments of the right ankle joint. **(a)** Medial view. **(b)** Lateral view.

CONDITIONS OF CLINICAL SIGNIFICANCE

Articulations

Sprains

Sprains result when twisting or overstretching a joint causes a ligament to tear or to separate from its bony attachment. If the trauma is severe, excessive tissue fluid may accumulate and cause swelling.

Dislocations

When the articular surfaces of the bones are forcibly displaced, *dislocations* occur. When the dislocations are severe, the bones as well as the surrounding tendons and ligaments may be damaged. The most commonly dislocated joints are those of the shoulder, thumb, and fingers.

Bursitis

When one or more of the bursae surrounding a joint become inflamed, the condition is known as *bursitis*. This disorder can result from injury, heavy exercise, or infection. The bursae fill with excessive synovial fluid, causing discomfort and limiting motion in the affected joint.

Tendinitis

Tendinitis is the inflammation of tendon sheaths around a joint. The condition is generally characterized by local tenderness at the point of inflammation and severe pain upon movement of the affected joint. Tendinitis can result from trauma to, or excessive use of, a joint. The wrist, elbow (where it is often referred to as "tennis elbow"), and shoulder joints are most often affected.

Slipped Disc

Among the more common discomforts that people endure are those associated with the back. There are numerous causes of back pain; one cause that involves joints is a *slipped disc.* In this condition, which is most common in the lower back, the semisolid nucleus pulposa within an intervertebral disc is squeezed to one side of the disc. This can result from trauma or the improper distribution of weight along the vertebral column that comes from poor posture or deformity of the vertebrae. The displacement of the nucleus pulposa causes the outer fibrous portion (anulus fibrosus) of the disc to protrude or, in some cases, to rupture. If the disc protrudes into the vertebral canal it can compress spinal nerves or, in the thoracic or cervical regions, the spinal cord itself. The pressure can cause severe pain along the paths of the nerves, as well as numbness of the regions supplied by the nerves. If the pressure upon the spinal nerves continues, actual nerve damage can result. The nerve damage, in turn, can cause weakness and degeneration of muscles supplied by the damaged nerves.

Torn Menisci

The fibrocartilage menisci of the knee joint serve to somewhat adapt the surfaces of the tibia to the shape of the femoral condyles. The menisci are not firmly attached, and during rotation they move slightly on the tibia. Because of this movement, sudden changes of direction while bearing the body weight can cause the menisci to tear loose. Severe pain and swelling of the joint can result.

An innovative surgical technique called *arthroscopic surgery* has greatly increased the success of operations done to repair damaged menisci as well as other joint injuries. In this technique a needlelike viewing instrument called an *arthroscope* is inserted into the joint through a tiny incision (Figure 6.21). The arthroscope contains a fiberoptic light source that makes it possible to examine the interior of the joint visually and to observe the positions of cutting instruments that are inserted through other tiny incisions. This procedure can be done under local anesthesia and the patient can often return home from the hospital the same day.

Arthritis

Many different types of inflammation of the joints fall under the general term *arthritis*. In these conditions there may be pathological changes in the joint membranes, cartilage, and bone that cause swelling and pain. The causes of arthritis are unknown, but trauma to a joint, bacterial infection (staphylococci, streptococci, and gonococci), and metabolic disorders have been implicated. At least one form of arthritis (rheumatoid) is thought to be caused by an immune response by the body to inflammation of the synovial membrane of the affected joint cavity. There is evidence suggesting that arthritis may be genetically inherited, since some families show a predisposition toward the condition.

Osteoarthritis

Osteoarthritis, the most common form of arthritis, is a chronic inflammation that causes the articular cartilage in the affected joint to degenerate gradually. The inflammation is accompanied by pain, swelling, and stiffness. As the articular cartilage degenerates, bony spurs develop from the exposed ends of the bones that form the joint. These spurs tend to restrict the movement of the joint and often cause discomfort.

Rheumatoid Arthritis

Rheumatoid arthritis is a severely damaging form of the disease that tends to affect principally the small joints of the body, such as those in the hands, feet, knees, ankles, elbows, and wrists (see Box 6.2). It affects women more frequently than men. The condition begins with inflammation of the synovial membrane of the joint, causing swelling and pain. The inflamed synovial membrane produces an abnormal tissue known as a *pannus*, which grows over the surface of the articular cartilage. As the disease progresses, the articular cartilage beneath the pannus, and in some cases the bone itself, is gradually destroyed. Eventually the pannus fills the joint space and becomes invaded by fibrous tissue, thereby restricting joint movement. In severe cases, calcification of the pannus may ankylose (fuse) the joint so that it is immovable.

Gouty Arthritis

Gout is a condition characterized by sudden severe pain and swelling of the joints. It affects primarily the toes, insteps, ankles, heels, knees, and wrists. Gout is more common in males. It is due to an inherited genetic defect that causes either an increased production of uric acid or a reduction in the ability of the kidneys to excrete uric acid. In either case the result is an increase in the level of uric acid in the blood *(hyperuricemia).* Uric acid is an end product of purine metabolism. The excessive level of uric acid in the blood causes the body fluids to become supersaturated and eventually results in the formation of sodium urate crystals in the soft tissues of the body as well as in the joints.

When the joints are affected, the condition is called *gouty arthritis.* As crystals accumulate in the joints and the soft tissues surrounding the joints, they cause an inflammation that eventually may erode the articular cartilage and the underlying bone, causing intense pain and immobility of the joint.

Effects of Aging on the Joints

With aging there is a progressive loss of the cartilaginous surface of the joints, which may be accompanied by the appearance of bony, spurlike overgrowths. This development often produces a condition known as *degenerative osteoarthropathy.*

The effects of aging on the joints—particularly that of degenerative osteoarthropathy—vary greatly from individual to individual and are influenced both by genetic factors and by heavy use of certain joints. There is a progressive increase in degenerative osteoarthropathy starting as early as 20 years of age. By the age of 80, virtually everyone has some degenerative osteoarthropathy of the knee and elbow joints, with somewhat lower frequencies for the hip and shoulder joints. Women often develop bony swellings (called Heberden's nodes) on the terminal phalanges. In men, the spine is the most common site of degenerative osteoarthropathy. Here the intervertebral discs degenerate, which results in new bone formation between the bodies of the vertebrae. This degeneration can produce a narrowing of the intervertebral foramina that causes chronic and painful pressure on the nerve roots.

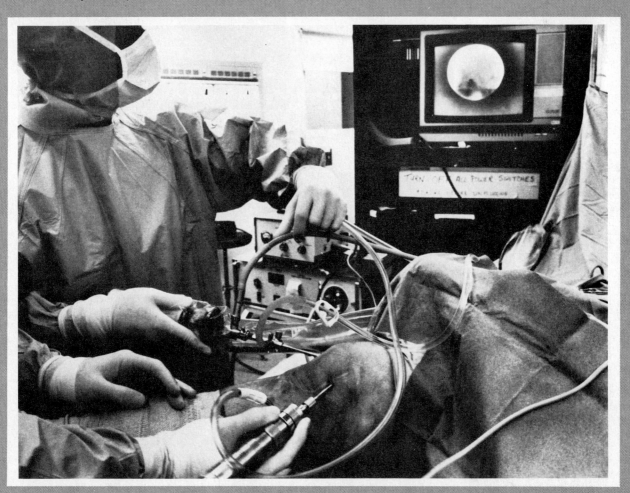

Figure 6.21
Arthroscopic surgery.

FRONTIERS IN HEALTH
New Hope for the Arthritic

Arthritis is this nation's number one crippler, afflicting more than 31 million Americans, or one in seven. Arthritis strikes people of all ages, although most cases occur in people older than 30. The disease claims a million new victims each year and costs the nation more than $14 billion a year in lost wages, medical bills, and disability claims.

Pauline is one of those victims. For her, each day is a long episode of pain and discomfort. Bathing, eating, writing, and even dressing are difficult and painful. She cannot open a screw-top jar or pick up coins. The disfigured, swollen joints of her hand have become practically useless. For Pauline a relaxing walk in the park is out of the question too, as her knee joints have been painfully affected by this incurable disease.

Arthritis literally means inflammation of the joints. According to the Arthritis Foundation, what we call arthritis actually encompasses at least 100 different rheumatic diseases—the painful diseases that affect the joints and are characterized by inflammation and stiffness.

Osteoarthritis and **rheumatoid arthritis** are the most common arthritic diseases. Osteoarthritis, which causes a degeneration of the articular cartilages and the development of bony spurs from the ends of the bones, is believed to result from excessive stress on the joints. It especially affects joints that have been abused by improper exercise or constant demanding work. Obesity contributes to the excessive stress.

Rheumatoid arthritis, whose cause remains obscure, results from an erosion of the articular cartilage in joints and its eventual replacement with fibrous tissue. Rheumatoid arthritis affects mainly the joints of the hands, hips, knees, and feet. Enzymes from phagocytic cells in the synovial fluid and synovial membranes appear to be responsible for the erosion. Inflammation occurs along with the erosion and weakens the joint capsule and supporting ligaments. The affected joint slowly degenerates and becomes stiff, swollen, and sore.

For Pauline and millions of Americans like her, new hope has emerged in recent years with the perfection of artificial joints. One of the most successful of these is the metacarpophalangeal joint. In a recent study carried out by scientists in the United States and the Soviet Union, 100 patients were fitted with artificial metacarpophalangeal joints made of a special type of plastic. The results of this work are very encouraging. All the recipients felt that the newly constructed joints noticeably improved the appearance of their hands. Even more encouraging, the new joints allowed renewed use of the crippled digits: nearly all the patients reported that their ability to grasp objects improved after the operation. Common chores that previously were difficult to perform or required assistance became easier, and many of the tasks made impossible by arthritis were once again accomplishable. The surgery also relieved the constant pain that accompanies arthritis.

Arthroplasty—the surgical replacement of a joint—has improved tremendously in recent years, thanks

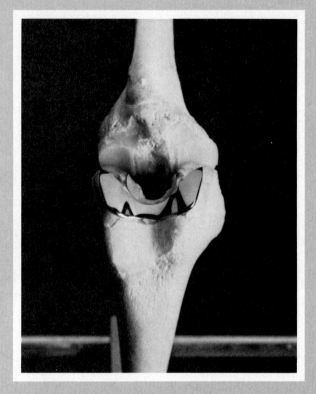

Photo of artificial knee joint.

largely to new glues and new materials that make artificial joints more durable. The computer has also contributed.

Until recently most artificial joints were manufactured by hand. In the first step of this complex process, physicians would X ray a patient's joint. The X rays were used to make blueprints for the replacement joint, and the blueprints were then sent to a laboratory, which constructed the replacement joints by hand. This process has now undergone considerable change, thanks to medical researchers at New York's Hospital for Special Surgery, who are working on a computer-assisted design, computer-assisted manufacturing technique called CAD-CAM.

With CAD-CAM, physicians first make numerous X rays of the joint, so that a complete three-dimensional picture can be composed. This information is fed into a computer, which takes the patient's age and other factors into consideration to construct a workable joint design. The information is then fed into a sophisticated set of lathes and milling machines that build the prescribed joint from a block of titanium.

For the more than 150,000 Americans who receive artificial joints each year, this cooperation of science and technology offers joints that move and a relief from pain.

Box 6.2
Rheumatoid Arthritis

Although arthritis is generally thought to be a disease of the elderly, rheumatoid arthritis commonly occurs between the ages of 20 and 50. The first signs of this crippling disease are usually soreness and swelling of the joints. As the disease progresses, the joints enlarge and the synovial membrane becomes inflamed. The disease is progressive, causing erosion of the articular cartilage. In severe cases, the articular cartilage completely erodes, allowing the bones to fuse together. Such a condition is visible in this X ray in the second finger from the right.

Although the actual cause of arthritis is still unknown, there are a number of theories concerning its origin. Some researchers believe it results from excessive stress caused by sprains or other types of joint injuries. Others believe the body produces a natural antibody that reacts against its own tissues. Other possible causes of arthritis include psychological stress, metabolic disorders, viruses, and allergies. Most likely, a number of these factors can contribute to the onset of arthritis, rather than there being a single causative agent.

While there is no cure for arthritis, relief is possible with the use of aspirin and other pain- and swelling-reducing drugs.

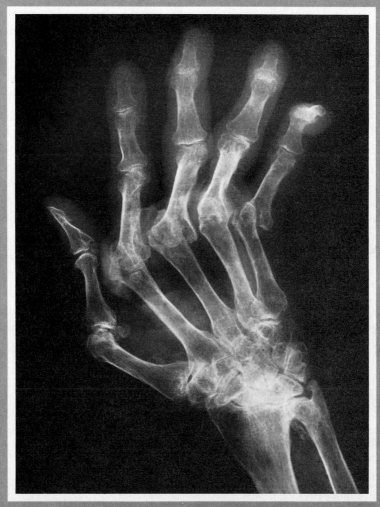

STUDY OUTLINE

FIBROUS JOINTS (Synarthroses) articulations in which bones are held together tightly by fibrous connective tissue; nonmovable joints; two types classed by length of fibers that hold bone together. **pp. 153–154**

  SUTURES grooves on edges of bones fit closely together; short connecting fibers; found only between flat bones of skull.

    SYNOSTOSIS in adulthood, suture fibers replaced by bone; sutures fused together.

  SYNDESMOSES joint fibers longer than in sutures; called ligaments; allow some slight "give" movement; joint between distal ends of tibia and fibula is an example.

CARTILAGINOUS JOINTS (Synarthroses or Amphiarthroses) bones united by cartilage; some immovable; others allow slight movement. **p. 155**

  SYNCHONDROSES joints in which bones are held together by hyaline cartilage.

    TEMPORARY SYNCHONDROSES cartilage replaced by bone (for example, epiphyses of long bones).

PERMANENT SYNCHONDROSES between first ten ribs and their costal cartilages.

SYMPHYSES articular surfaces of bones covered with thin layer of hyaline cartilage. Fibrocartilage pad separates bones within the joint; shock-absorbing action. Junction of pubic bones and junctions between adjacent vertebrae are examples.

SYNOVIAL JOINTS (Diarthroses) freely movable; movement limited only by ligaments, muscles, tendons, and adjoining bones.   pp. 155–171

FLUID-FILLED JOINT CAVITY

ARTICULAR CARTILAGE thin layer of hyaline cartilage that covers smooth articular bone surfaces.

ARTICULAR CAPSULE dense fibrous connective tissue enclosing joint; joined to periosteum.

SYNOVIAL MEMBRANE inner surface of articular capsule; vascular.

SYNOVIAL FLUID clear viscous secretion of synovial membrane; lubricates and nourishes.

ARTICULAR DISCS present in some synovial joints; fibrocartilage that extends inward from capsule, dividing synovial cavity in two; jaw and distal radial/ulnar joints are examples.

BURSAE AND TENDON SHEATHS formed by synovial membranes; reduce friction.

BURSAE sacs lined with synovial membranes, filled with synovial fluid; subcutaneous or between tendon and bone.

TENDON SHEATHS cylindrical synovial sacs around tendons; found where tendons cross joints.

MOVEMENTS OF SYNOVIAL JOINTS

GLIDING bones move back and forth upon one another.

ANGULAR MOVEMENTS increase or decrease angle between two adjoining bones by moving in a single plane.

Flexion bone moved in anterior-posterior plane to decrease angle between it and adjoining bone.

Extension increases angle between adjoining bones.

Abduction body part moved away from midline.

Adduction body part moved toward midline.

CIRCUMDUCTION delineates a cone; base outlined by movement of distal end of bone and apex in articular cavity; hip and shoulder joints are examples.

ROTATION motion of bone around a central axis.

Supination outward rotation of forearm; palms anterior.

Pronation inward rotation of forearm; palms posterior.

SPECIAL MOVEMENTS

Elevation raising a body part.

Depression lowering a body part.

Inversion twisting foot so that sole faces inward.

Eversion twisting food so that sole faces outward.

Protraction moving a part forward.

Retraction returning protracted body part to usual position.

TYPES OF SYNOVIAL JOINTS classed by movements permitted and by shapes of articular surfaces.

NONAXIAL JOINTS

Gliding (Arthrodial) Joints

Articular Surface: formed by apposition of flat or slightly curved surfaces.

Movement: any direction.

Example: found between vertebrae and intercarpal and intertarsal joints.

UNIAXIAL JOINTS

Hinge (Ginglymus) Joints

Movement: flexion and extension.

Example: elbow and interphalangeal joints.

Pivot (Trochoid) Joints

Movement: rotation around longitudinal axis of bone.

Example: proximal articulations of radius and ulna.

BIAXIAL JOINTS

Condyloid (Ellipsoid) Joints

Articular Surface: one slightly concave; other slightly convex.

Movement: in two perpendicular planes (flexion, extension, abduction, adduction, circumduction).

Example: radiocarpal articulations.

Saddle Joints

Articular Surface: each bone is concave in one direction and convex in the other.

Movement: same as condyloid joints.

Example: carpometacarpal joint of thumb.

TRIAXIAL JOINTS

Spheroid (Ball and Socket) Joints

Articular Surface: spherical head of one bone fits in cup-shaped socket of second bone.

Movement: allows medial and lateral rotation plus all condyloid movements.

Example: shoulder and hip joints.

LIGAMENTS OF SELECTED JOINTS

LIGAMENTS OF THE VERTEBRAL COLUMN

Articular Capsule encloses joints between articular processes.

Anterior Longitudinal Ligament along ventral surface of bodies of vertebrae.

Posterior Longitudinal Ligament along dorsal surface of bodies of vertebrae.

Supraspinous Ligament connects tips of adjoining spinous processes.

Ligamentum Nuchae continuation of supraspinous ligament in cervical region.

Interspinous Ligament connects sides of adjoining spinous processes.

Ligamenta Flava connect adjoining laminae.

Intertransverse Ligaments connect adjoining transverse processes.

LIGAMENTS OF THE CLAVICULAR JOINTS

Sternoclavicular Articulation

Articular Capsule encloses joint.

Articular Disc separates clavicle and sternum.

Anterior Sternoclavicular Ligament strengthens anterior surface of joint.

Posterior Sternoclavicular Ligament strengthens posterior surface of joint.

Interclavicular Ligament connects medial ends of both clavicles.

Costoclavicular Ligament connects clavicle to first rib.

Acromioclavicular Articulation

Articular Capsule encloses joint.

Acromioclavicular Ligament attaches clavicle to acromion process.

Coracoclavicular Ligament attaches clavicle to coracoid process.

LIGAMENTS OF THE SHOULDER JOINT

Articular Capsule encloses joint.

Coracohumeral Ligament strengthens joint anteriorly; extends from coracoid process to greater tubercle.

Glenohumeral Ligaments several thickenings of lower portion of capsule.

Glenoid Labrum deepens fossa.

Biceps Brachii Muscle tendon helps strengthen joint.

LIGAMENTS OF THE ELBOW JOINT

Articular Capsule encloses joint.

Ulnar Collateral Ligament from medial epicondyle to ulna.

Radial Collateral Ligament from lateral epicondyle to annular ligament and ulna.

LIGAMENTS OF THE WRIST JOINT

Articular Capsule encloses joint.

Radial Collateral Ligament from styloid process of radius to scaphoid and trapezium.

Ulnar Collateral Ligament from styloid process of ulna to triquetral and pisiform.

Palmar Radiocarpal Ligament from distal end of ulna and radius to ventral surface of carpals.

Dorsal Radiocarpal Ligament from distal end of ulna and radius to dorsal surface of carpals.

LIGAMENTS OF THE HIP JOINT

Articular Capsule encloses joint.

Iliofemoral and *Pubofemoral Ligaments* provide anterior joint strength.

Ischiofemoral Ligament strengthens joint posteriorly.

Acetabular Labrum deepens joint cavity.

Ligamentum Teres extends through joint cavity; no significant contribution to hip-joint strength.

LIGAMENTS OF THE KNEE JOINT

Articular Capsule encloses joint.

Oblique and *Arcuate Popliteal Ligaments* strengthen capsule posteriorly.

Medial Collateral Ligament stabilizes joint medially.

Lateral Collateral Ligament stabilizes joint laterally.

Patellar Ligament strengthens joint anteriorly.

Medial and Lateral Menisci cartilages that deepen condyles of tibia.

Anterior and Posterior Cruciate Ligaments prevent hyperextension and posterior tibial slippage.

LIGAMENTS OF THE ANKLE JOINT

Articular Capsule encloses joint.

Deltoid Ligament from medial malleolus to navicular, talus, and calcaneus.

Anterior Talofibular Ligament from malleolus of fibula to anterior region of talus.

Posterior Talofibular Ligament from malleolus of fibula to posterior region of talus.

Calcaneofibular Ligament from malleolus of fibula to calcaneus.

CONDITIONS OF CLINICAL SIGNIFICANCE: ARTICULATIONS pp. 172–173

SPRAINS joint overstretched or twisted, resulting in ligament tearing or separation.

DISLOCATIONS articular surfaces of bones forcibly displaced.

BURSITIS inflamed bursa resulting from injury, exercise, or infection.

TENDINITIS inflammation of tendon sheath.

SLIPPED DISC nucleus pulposa squeezed to cause protrusion from intervertebral disc; causes pressure on spinal nerves.

TORN MENISCI menisci torn loose; generally occurs during sudden changes of direction.

ARTHRITIS joint inflammation caused by trauma, bacterial infection, metabolic disorders, or other unknown causes; may be inherited.

OSTEOARTHRITIS most common form; gradual degeneration of articular cartilage and development of bony spurs.

RHEUMATOID ARTHRITIS severely damaging form of disease; pannus develops on surface of articular cartilage; articular cartilage and bone beneath it often destroyed; joint may fuse.

GOUTY ARTHRITIS sudden severe pain and swelling; due to excessive production of uric acid or inability to excrete uric acid; sodium urate crystals form in joints and soft tissues; articular cartilage may be eroded.

EFFECTS OF AGING ON THE JOINTS progressive loss of articular cartilage and growth of bony spurs; often results in degenerative osteoarthropathy.

SELF-QUIZ

1. The humerus/ulna joint is an example of a fibrous joint. True or false?

2. The joint found between the flat bones of the skull is classed as: (a) syndesmosis; (b) suture; (c) amphiarthrosis.

3. Most joints of the body are: (a) synchondroses; (b) symphyses; (c) synovial.

4. Which one of the following exemplifies a symphysis? (a) junction of the two pubic bones; (b) junctions between the costal cartilages; (c) the epiphyses of a long bone to the diaphysis.

5. The term *diarthrosis* refers to a synovial joint that is kept apart by a fluid-filled cavity. True or false?

6. Match the terms with the appropriate description.

   Synarthrosis
   Sutures
   Syndesmoses
   Synostosis
   Amphiarthroses
   Synchondroses
   Symphyses
   Synovial

   (a) A joint in which the bones on either side of a suture become firmly fused.
   (b) The fibers in this synarthrosis joint are relatively long.
   (c) These joints usually serve as shock absorbers.
   (d) All fibrous joints.
   (e) Most joints of the body are of this type.
   (f) Articulations that permit slight movement.
   (g) The edges of the bones that form these joints have grooves that fit firmly and closely together.
   (h) Temporary joints in which the original cartilage is eventually replaced by bone.

7. Inability to produce the fluid that keeps most joints moist would likely be due to a disorder in the: (a) bursae; (b) synovial membrane; (c) articular cartilage.

8. Bursae and tendon sheaths are part of the synovial joints. True or false?

9. Synovial joints, like fibrous and cartilaginous joints, are classified according to the material that connects the bones. True or false?

10. The elbow is an example of a: (a) uniaxial joint; (b) biaxial joint; (c) triaxial joint.

11. Match the angular movement terms with the appropriate descriptions. There may be more than one answer per term.

    Flexion
    Extension
    Abduction
    Adduction

    (a) Moving a toe toward the midline of the foot.
    (b) Arching the back.

    Plantar Flexion
    Dorsiflexion

    (c) Moving a limb away from the midline of the body.
    (d) Moving a bone in an anterior-posterior plane, decreasing the angle between it and the adjoining bone.
    (e) Lowering the toe region of the foot.
    (f) The opposite of flexion.
    (g) Raising the toe region of the foot toward the shin.

12. *Supination* is the term used to describe the inward rotation of the forearm, causing the radius to cross diagonally over the ulna. True or false?

13. This movement is characteristic of the hip and shoulder joints: (a) pronation; (b) supination; (c) circumduction.

14. Twisting the foot so that the sole faces outward, with its outer margin raised, is called: (a) inversion; (b) eversion; (c) protraction.

15. The joints between the articular processes of vertebrae and between most carpal and tarsal bones are termed: (a) hinged; (b) gliding; (c) condyloid.

16. Match the following joint types with the appropriate descriptions. There may be more than one answer for each type of joint.

    Nonaxial
    Gliding
    Uniaxial
    Hinge
    Pivot
    Biaxial
    Condyloid
    Saddle
    Triaxial
    Spheroid

    (a) Condyloid (ellipsoid).
    (b) Carpometacarpal joint of the thumb.
    (c) Shoulder and hip joints.
    (d) Gliding (arthrodial).
    (e) The only movement here is rotation around the longitudinal axis of the bone.
    (f) Hinge (ginglymus).
    (g) Found between most carpal bones.
    (h) Articulation between radius and carpals.
    (i) Elbow joint.

17. The first metacarpal/carpal joint is a: (a) saddle; (b) condyloid; (c) suture.

18. Both the shoulder joint and the hip joint contain: (a) an articular capsule; (b) an iliofemoral ligament; (c) ligamentum teres.

19. The ligament that prevents the tibia from slipping posteriorly is the: (a) medial meniscus; (b) posterior cruciate; (c) ligamentum patellae.

20. Match the joint types with the appropriate descriptions. There may be more than one answer for each joint type.

Distal tibia/
    fibula
Pubic/pubic
Ulna/humerus
Sternum/clavicle
Radius/carpals
Tarsal/tarsal
Occipital bone/
    atlas

    (a) Fibrous (syndesmosis).
    (b) Cartilaginous (symphysis).
    (c) Synovial (condyloid).
    (d) Synovial (gliding).
    (e) Nonaxial.
    (f) Uniaxial; flexion; extension.
    (g) This joint is capable of slight movement ("give").

21. Place a "V" by the ligaments associated with the joints of the vertebral column; an "H" by those associated with the hip joint; and a "K" by those associated with the knee joint.

Iliofemoral
Arcuate
    popliteal
Ligamentum
    flava
Posterior
    longitudinal
Pubofemoral
Anterior and
    posterior
    cruciate

Ligamentum
    teres
Supraspinous
Medial and
    lateral
    collateral
Articular
    capsule

22. Arthritis refers to many different types of inflammation of the joints. True or false?

23. A painful condition that is due to an inherited genetic defect that causes an increased level of uric acid in the blood: (a) bursitis; (b) arthritis; (c) gouty arthritis.

LEARNING OBJECTIVES

After completing this chapter, you should be able to:

- List the ways in which muscles are classified, and give an example of each.

- Describe the microscopic structure of skeletal muscle.

- Explain what is meant by prime movers, antagonists, synergists, and fixators.

- Name the major muscles of the head and neck, citing the origin, insertion, action, and innervation of each.

- Name the major muscles of the trunk, citing the origin, insertion, action, and innervation of each.

- Distinguish between synergistic and antagonistic muscles, citing several examples of each.

- Distinguish between flexor and extensor muscles, citing several examples of each.

- Distinguish between intrinsic and extrinsic muscles, citing several examples of each.

- Name the major muscles that move the vertebral column.

- Name the major muscles of the upper limbs, citing the origin, insertion, action, and innervation of each.

- Name the major muscles of the lower limbs, citing the origin, insertion, action, and innervation of each.

CHAPTER CONTENTS

The Muscular System

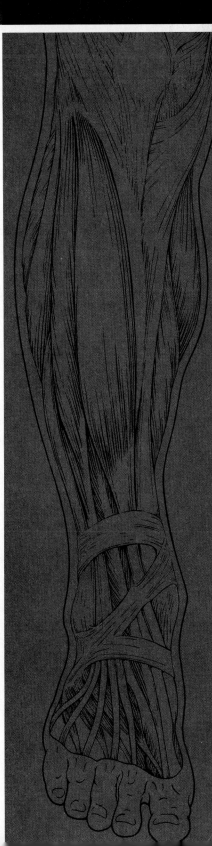

Muscle tissue constitutes almost one-half of the total body weight. Much of the body's form is due to the numerous muscles that attach to the skeleton and underlie the skin. Other muscles are located within the walls of hollow organs and blood vessels.

The functions of the muscles depend on their location. In all cases, however, muscle action is the result of the actions of individual muscle cells. Muscle cells are special in that they are the cells in the body that best exhibit the property of *contractility*—which allows them to shorten and develop tension. As a result, muscle cells are important in such activities as the movement of various body parts, the alteration of the diameters of tubes in the body, the propulsion of materials through the body, and the expulsion of waste substances from the body. In addition, the contraction of skeletal muscles produces significant amounts of heat that can be used to maintain normal body temperature. Because of its many functions, muscle tissue contributes importantly to the maintenance of homeostasis.

MUSCLE TYPES

The body contains three types of muscle—*skeletal muscle, smooth muscle,* and *cardiac muscle.* Muscles can be further classified as *voluntary* or *involuntary*, on the basis of the type of control exercised over their activity.

The contractions of **voluntary muscles** are normally under the conscious control of the individual. However, under many conditions, the contractions of voluntary muscles do not require conscious thought. For example, a person does not usually have to think about contracting the muscles involved in maintaining posture. Voluntary muscles are controlled by the portion of the nervous system known as the *somatic nervous system.*

The contractions of **involuntary muscles** are generally not under the conscious control of the individual. Rather, they are governed by the portion of the nervous system known as the *autonomic nervous system,* as well as by hormones and by factors intrinsic to the muscles themselves.

Skeletal Muscle

As the name implies, most **skeletal muscle** attaches to the bones of the skeleton. The contractions of skeletal muscle exert force on the bones and thus move them. Consequently, skeletal muscle is responsible for activities such as walking and manipulating objects in the external environment.

When viewed microscopically, skeletal muscle cells exhibit alternating transverse light and dark bands that give them a striped appearance. Therefore, skeletal muscles are referred to as *striated muscles.*

Skeletal muscles are the only *voluntary muscles* in the body.

Smooth Muscle

Smooth muscle is so named because its cells lack the striations evident in skeletal muscle cells. It is also called *visceral muscle* because it is found in the walls of hollow organs and tubes such as the stomach, intestines, and blood vessels.

Smooth muscle is *involuntary muscle*, and its contractions govern the movement of materials through the organ systems of the body.

Cardiac Muscle

Cardiac muscle is a specialized type of muscle that forms the wall of the heart. It is *involuntary*, like smooth muscle, and *striated*, like skeletal muscle.

EMBRYONIC DEVELOPMENT OF MUSCLE

Although all muscles are formed from embryonic mesoderm, the three types of muscle—skeletal, smooth, and cardiac—follow somewhat different patterns of development.

Skeletal Muscle

With the exception of the muscles of the limbs and some of the muscles of the head, skeletal muscles develop from **somites,** which are embryonic masses of mesodermal cells. The somites are located dorsally along both sides of the axial skeleton. Only cells from one portion of a somite, called the **myotome,** differentiate into muscle cells. The cells from the myotomes spread downward, between the skin and body cavity, until the right and left sides meet ventrally in the midline. The skeletal muscles of the trunk develop from this sheet of mesoderm. Typical somites do not form in the head of the embryo; consequently, most of the muscles of the head develop from the general mesoderm of that region. The muscles of the limbs begin development from condensations of mesoderm within the embryonic limb buds. Some of the mesodermal cells that form the limb muscles probably migrate to the area from the myotomes.

Smooth Muscle

As the digestive tube and the body organs form in the embryo, mesodermal cells migrate to them and form a thin layer around them. These mesodermal cells develop into the smooth muscles of the body.

Cardiac Muscle

The muscle of the heart forms in a manner similar to that of the smooth muscles. Mesodermal cells migrate to and surround the early heart while it is still in the form of a tubule. Cardiac muscle begins contracting very early in the development of the embryo, even before the peripheral blood vessels have been completely formed.

GROSS ANATOMY OF SKELETAL MUSCLES

When examining skeletal muscles without the aid of a microscope, as is done during gross dissection, it is possible to note various distinguishing features of individual muscles. Among the most obvious features are the *connective-tissue coverings*, the *attachments*, and the *shape* of each muscle.

Connective-Tissue Coverings

Each skeletal muscle is composed of many individual muscle cells, called **muscle fibers.** Muscle fibers are held together by thin sheets of fibrous connective-tissue membranes called *fasciae*. The fascia that envelops an entire F7.1 muscle is called the *epimysium* (Figure 7.1). Fasciae also penetrates muscle, separating the muscle fibers into bundles called **fasciculi.** This fascia is called the **perimysium.** Very thin extensions of the fascia, called the **endomysium,** envelop the cell membrane of each muscle fiber. Blood vessels and nerves pass into the muscle with the fascial sheaths to reach the individual

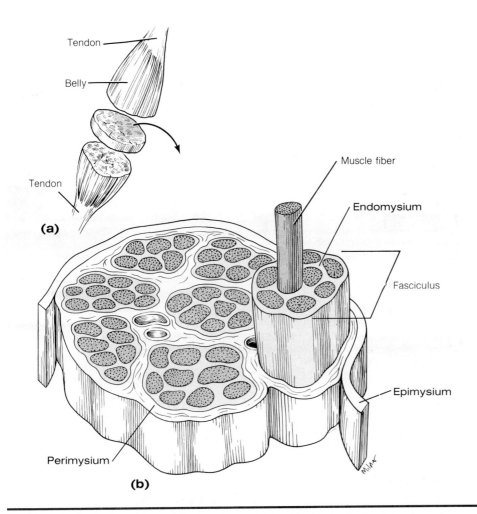

Figure 7.1
Connective tissue membranes of a skeletal muscle. **(a)** Entire muscle with the belly sectioned. **(b)** Enlargement of a cross section of the belly.

muscle cells. Beds of capillaries form between the muscle cells, and each cell is supplied by a branch of a nerve cell.

Skeletal Muscle Attachments

Skeletal muscles are anchored to the skeleton by extensions of the endomysium, perimysium, and epimysium. These connective tissues continue beyond the end of a muscle and either attach directly to the periosteum of a bone, as is often seen at the proximal attachment of a muscle, or they may blend into a strong fibrous connection called a **tendon,** which then becomes continuous with the periosteum of the bone. Some tendons are quite short, whereas others are more than a foot in length. Tendons that take the form of broad, thin sheets are called **aponeuroses.**

The attachments of both ends of a skeletal muscle are given specific names. The **origin** is the less movable end and is generally proximal. The **insertion** is the more movable end and is generally distal. The widest portion of a muscle, between the origin and insertion, is called its **belly.** A muscle is described as *arising* from the origin and *inserting* at the insertion. The origin may be rather broad, arising from several different places on a bone or even from several different bones. The insertion, in contrast, tends to be much more restricted. With the exception of the sphincter muscles that surround body openings, and some of the facial muscles that insert on the skin, joints are located between the origins and insertions of skeletal muscles. When a skeletal muscle contracts it shortens, using the joint as a pivot point to pull the insertion closer to the origin.

The origins and insertions that are described later in this chapter are the most common ones. Keep in mind, however, that for some muscles it is possible to reverse the origin and the insertion functionally—that is, the more fixed end, which is normally called the origin, can be used as the more movable end in some actions. For instance, certain of the superficial muscles of the chest are described as having their origins on the thorax, and their insertions on the humerus. Clearly the humerus is generally more movable than the thorax. However, if you are doing pull-ups, the thorax is moved toward the humerus and is therefore serving as an insertion.

Skeletal Muscle Shapes

The arrangement of the bundles of muscle fibers (fasciculi) varies in the different skeletal muscles (Figure 7.2). In some muscles, the fasciculi run parallel to the long axis of the muscle, forming straplike muscles. The contraction of muscles that possess such **longitudinally** arranged fasciculi produces considerable movement; however, such muscles do not have much power. Less movement, but greater power, is produced by those muscles that have a tendon running the entire length of the muscle, with the fasciculi inserting diagonally into this tendon. In some muscles of this type, all the fasciculi insert onto one side of the tendon; this arrangement is called **unipennate**. **Bipennate** muscles have their fasciculi inserting obliquely on both sides of the tendon. The fasciculi of some muscles have a complex arrangement that involves the convergence of several tendons; these are **multipennate** muscles. In a few muscles, the fasciculi converge from a broad origin into a single narrow tendon; this is a **radiate** (fan-shaped) arrangement.

MICROSCOPIC ANATOMY
OF SKELETAL MUSCLES

When a skeletal muscle is examined with the aid of a microscope, it is apparent that the muscle fibers have a regular subcellular structure. Skeletal muscle fibers are multinucleate cells approximately 10 to 100 microns in diameter that are frequently many centimeters long. Each fiber contains several hundred to several thousand regularly ordered, threadlike **myofibrils** that extend lengthwise throughout the cell (Figure 7.3). When highly magnified, the myofibrils appear to be cross-striated by alternating light and dark bands. The dark bands are called **anisotropic bands,** or **A bands;** the light bands are called **isotropic bands,** or **I bands.** Crossing the center of each I band is a dense **Z line.** The Z lines divide the myofibrils into a series of repeating segments called **sarcomeres.** In the center of a sarcomere, and therefore in the center of

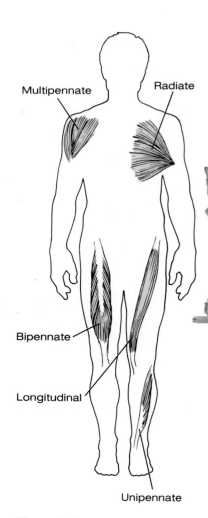

Figure 7.2
Variation in muscle shape.

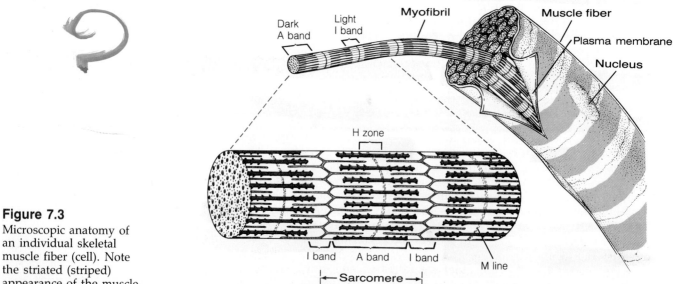

Figure 7.3
Microscopic anatomy of an individual skeletal muscle fiber (cell). Note the striated (striped) appearance of the muscle fiber and the myofibrils.

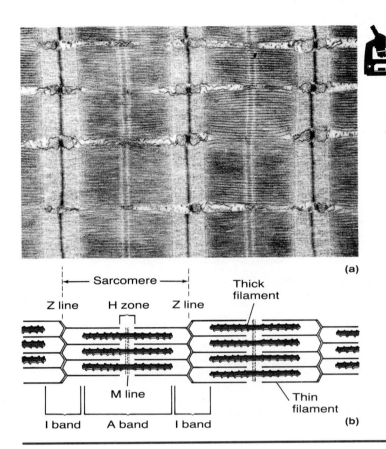

est. × 20,000

(a)

Sarcomere

Z line H zone Z line Thick filament

M line

Thin filament

I band A band I band

(b)

Figure 7.4
Longitudinal view of the structure of sarcomeres.
(a) Electron micrograph of striated muscle showing two sarcomeres.
(b) Schematic representation of the filaments within each sarcomere. The I band of the sarcomere consists only of thin filaments and is divided by the Z line. The A band consists of thick filaments that overlap at either end with the thin filaments. The region where only thick filaments occur is the H zone. A single sarcomere extends from one Z line to the next. Thus half of the I band is associated with one sarcomere and half with the neighboring sarcomere.

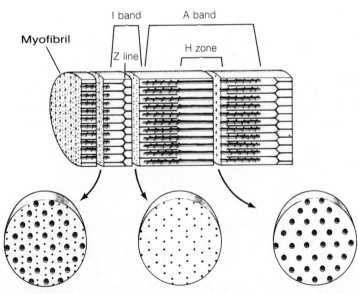

Myofibril

I band A band

Z line H zone

Figure 7.5
Cross sections through different regions of sarcomeres of a myofibril. Note that in the region of the A band where thick and thin filaments overlap, each thick filament is surrounded by six thin filaments, and each thin filament is surrounded by three thick filaments.

the A band, is a somewhat less dense region referred to as the **H zone.** A thin, dark **M line** crosses the center of the H zone.

A sarcomere is composed of two distinct types of longitudinally oriented **myofilaments:** thick filaments and thin filaments (Figure 7.4). **Thick filaments** are found only in the A band. The H zone of the A band contains only thick filaments. The M line is formed by linkages between the thick filaments that hold them in a parallel arrangement. **Thin filaments** occupy the I band and the part of the A band up to the H zone. Thin filaments attach to the Z lines. In the region of the A band where thick and thin filaments overlap, there is a hexagonal arrangement of thin filaments around a thick filament (Figure 7.5).

F7.4

F7.5

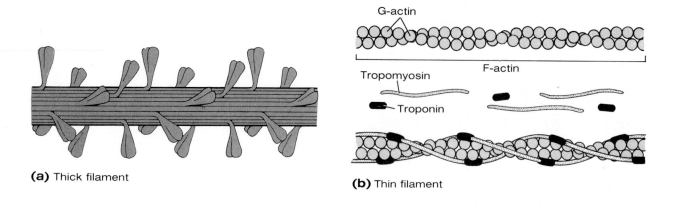

(a) Thick filament

(b) Thin filament

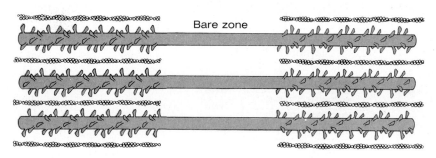

(c) Longitudinal section of filaments

Figure 7.6

Composition of myofilaments. **(a)** The thick filament consists of golf-club-shaped molecules of myosin bundled together so the heads of the molecules project from the shaft of the filaments and spiral around the shaft. **(b)** The thin filament. Individual G-actin subunits link into a double chain of F-actin. In a single thin filament, two chains of G-actin subunits twist around one another. Along the surface of each chain lie threadlike tropomyosin molecules that each cover seven actin subunits. Each tropomyosin molecule is attached to a molecule of troponin. Troponin is also attached to actin. **(c)** Longitudinal view of thick and thin filaments as arranged in a sarcomere. Note that the myosin molecules project in opposite directions on either side of the bare zone. During activity, the heads of the myosin molecules (cross bridges) attach to actin subunits.

Composition of the Myofilaments

Both the thick and the thin filaments are composed of proteins. The thick

F7.6a filaments consist principally of a protein called **myosin** (Figure 7.6*a*). A myosin molecule is made up of two identical subunits, each shaped something like a golf club. The two subunits are tightly wound around each other so that a complete myosin molecule has two rather bulbous heads protruding from one end of a straight shaft. A thick filament contains approximately 200 myosin molecules arranged in such a way that the shafts of the molecules are bundled together with the heads (called **cross bridges**) facing outward (Figure

F7.6a 7.6*a*). The myosin molecules face in opposite directions on either side of the center of the thick filament, with the shafts of the molecules directed toward the center. Because of this arrangement, the central area of the filament con-

F7.6c tains the shaft portions of the molecules but no myosin heads (Figure 7.6*c*).

The thin filaments of the sarcomeres are composed mainly of the proteins

F7.6b **actin, tropomyosin,** and **troponin** (Figure 7.6*b*). The actin portion of the thin filament is composed of globular or spherical subunits of actin called *G-actin* (for globular actin). These G-actin subunits are organized into a double chain called *F-actin* (for fibrous actin). The F-actin structure resembles two strings of pearls that have been twisted around one another in a spiral, with each pearl being equivalent to a G-actin subunit. Although the G-actin subunits are

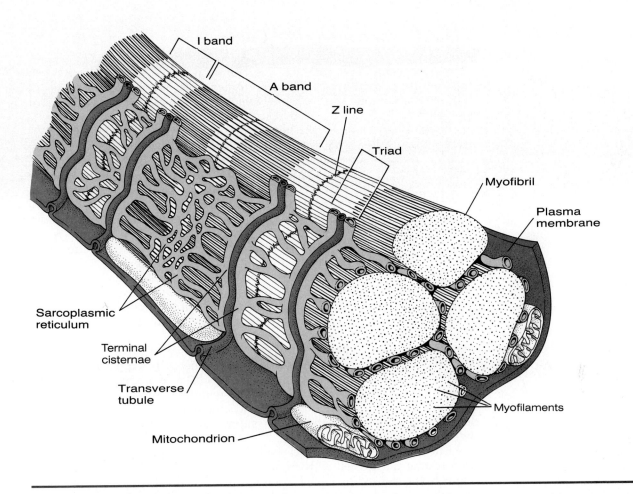

I band

A band

Z line

Triad

Myofibril

Plasma membrane

Sarcoplasmic reticulum

Terminal cisternae

Transverse tubule

Mitochondrion

Myofilaments

Figure 7.7
Schematic representation of transverse tubules and the sarcoplasmic reticulum of skeletal muscle.

spherical, they have a definite polarity and they link to one another from front to back. Associated with each chain of G-actin subunits are threadlike molecules of tropomyosin. The tropomyosin molecules lie end to end along the surfaces of the actin chains, and each tropomyosin molecule covers approximately seven G-actin subunits. Attached to each tropomyosin molecule and also to actin is a smaller molecule of the protein troponin. The arrangement of thick and thin filaments in a sarcomere is illustrated in Figure 7.6c.   **F7.6c**

Transverse Tubules and Sarcoplasmic Reticulum

Tubules known as **transverse tubules (*t* tubules)** pass deep into a skeletal muscle fiber from the plasma membrane (Figure 7.7). In addition, a membranous network, the **sarcoplasmic reticulum,** extends throughout the fiber and surrounds each myofibril. The sarcoplasmic reticulum is the smooth endoplasmic reticulum of muscle cells. Elements of the sarcoplasmic reticulum and *t*-tubule system lie close to one another over the junction of the A and I bands of the sarcomeres. At these locations structures consisting of three tubules (*triads*) are formed.   **F7.7**

CONTRACTION OF SKELETAL MUSCLE

Now that we have discussed the ultrastructure of a skeletal muscle cell, let us briefly consider the events that occur during contraction. In that way the roles of these structures will become clear.

Cellular Events During Contraction

Skeletal muscles are voluntary muscles, requiring stimulation from the nervous system in order to contract. The neurons (a *neuron* is a single nerve cell) that supply skeletal muscle fibers are called *motor neurons*. The endings of these motor neurons approach the membranes of the skeletal muscle cells at specialized points called **neuromuscular (myoneural) junctions.**

When a nerve impulse reaches a neuromuscular junction, a chemical neurotransmitter called *acetylcholine* is released from the terminal ending of the neuron. Acetylcholine causes a change in the permeability of the plasma membrane of the skeletal muscle cell at the neuromuscular junction—a change that results in the generation of a stimulatory impulse that spreads over the plasma membrane. From the plasma membrane, the impulse passes along the *t* tubules into the interior of the cell. By transmitting the impulse into the central areas of the muscle cell so that all areas of the cell receive the impulse at approximately the same time, the *t* tubules help to ensure a uniform, coordinated response by the cell. As the impulse spreads through the cell along the *t* tubules, it can affect the sarcoplasmic reticulum, since, in the triad area, the *t* tubules and sarcoplasmic reticulum lie in close approximation to one another. In fact, the arrival of a stimulatory impulse results in the temporary release of calcium ions from reticular sites called **terminal cisternae** (Figure 7.7). It is the release of calcium ions from the sarcoplasmic reticulum that initiates the events of muscle contraction.

Muscle contraction requires energy. The energy is supplied by a compound called *adenosine triphosphate (ATP)*—a substance produced by cells and widely used by them in a variety of energy-requiring processes. When ATP is split into *adenosine diphosphate (ADP)* and inorganic phosphate, energy is released. It is this energy that is used to support energy-requiring cellular activities such as muscle contraction.

The cellular events of muscle contraction are thought to occur as follows. In a resting muscle cell with an adequate supply of ATP, an ATP molecule binds to the club-shaped head of a myosin molecule. The myosin molecule is capable of enzymatic activity, and it splits the ATP into ADP. The energy released by this action is transferred to the myosin, producing a high-energy form of myosin.

In addition to an ATP binding site, a myosin head contains a binding site that can attach to a complementary site on an actin subunit of a thin filament. The high-energy form of myosin has a strong tendency to bind to actin. In a relaxed, unstimulated muscle fiber, however, this binding is prevented by tropomyosin, which lies along the surface of the actin and physically blocks interaction between high-energy myosins and actin subunits. However, when a muscle fiber is stimulated and calcium ions are released from the sarcoplasmic reticulum, the released calcium ions bind to the troponin molecules that are linked to both tropomyosin and actin. This binding of calcium ions to troponin weakens the linkage between troponin and actin. This allows tropomyosin to move away from its blocking position. With the tropomyosin out of the way, the high-energy myosins can link with G-actin subunits.

When a high-energy myosin combines with an actin subunit, the energy stored within the myosin is discharged. This discharge of energy produces a force that causes the myosin head (cross bridge) that is attached to actin to swivel so that the actin-containing thin filament is pulled toward the center of the sarcomere. Through numerous repetitions of this process, the thin filament is pulled past the thick filament, the Z lines of the sarcomere are drawn closer together, and the muscle cell shortens (Figure 7.8).

The release of calcium ions in response to a stimulus is a temporary event, and the ions are actually free for only a very short time. Following their release, calcium ions are rapidly trapped again by the sarcoplasmic reticulum. With the return of calcium ions to the sarcoplasmic reticulum, the troponin strengthens its connection with actin, pulling the tropomyosin back into its blocking position so that no further interaction between actin and high-energy myosins is possible. This causes the myofilaments to return to their original positions as the muscle relaxes.

F7.7

F7.8

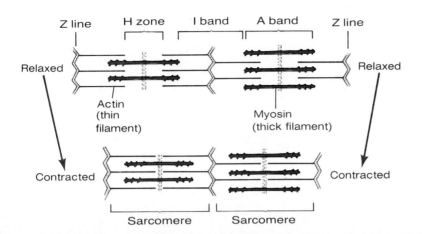

Figure 7.8
A muscle fiber shortens when the thin filaments move past the thick filaments toward the centers of the sarcomeres, and the Z lines are drawn closer together.

Energy Sources for Contraction

The ATP required as a source of energy for muscular contraction is produced by metabolic breakdown of glucose or glycogen in muscle cells. Under some conditions, lipids and even proteins may also be used for energy production. Under conditions of mild to moderate exercise, the production of ATP by muscle cells is the result of metabolic processes that utilize oxygen **(aerobic processes).** However, during periods of intense activity, oxygen cannot be supplied to many muscle fibers fast enough and oxidative metabolism cannot produce all the energy required for contraction. During such periods non-oxygen-requiring metabolic processes **(anaerobic processes)** provide additional ATP. Anaerobic processes break down glucose and stored glycogen to produce a substance called *lactic acid*, which diffuses out of the muscle fibers and enters the blood.

Immediately following the initiation of muscular activity, many actively contracting muscle fibers utilize ATP faster than the metabolic reactions just discussed can supply it. However, muscle fibers contain a substance called *creatine phosphate* that provides an additional source of energy. Creatine phosphate serves essentially as a quickly available energy reserve for muscle, as it contains energy and phosphate that are transferred to ADP to produce ATP:

$$\text{Creatine phosphate} + \text{ADP} \rightleftharpoons \text{ATP} + \text{Creatine}$$

Following the initiation of muscular activity, as ATP is being utilized, this reaction proceeds from left to right and ATP is formed. During inactive periods, when metabolic reactions are producing ATP not immediately required for contraction, the direction of the reaction is from right to left, and creatine phosphate is regenerated.

MUSCLE ACTIONS

It is the contraction of skeletal muscles that causes the various movements at different joints, as described in Chapter 6. Some muscles pass in front of a joint and thus flex the bone to which they are attached; others pass behind a joint and extend the bone to which they are attached. Some muscles move a part away from the midline of the body—or abduct it; others move the part back toward the midline—or adduct it. Some muscles rotate the bones that form certain joints, and so forth.

In order to bring about these movements, muscles usually work in groups rather than individually. Those muscles whose contractions are primarily responsible for a particular movement are called **prime movers.** In any movement there are always some muscles, generally situated on the opposite side of the joint, whose actions oppose the particular movement. These muscles, whose contraction offers resistance to the movement, are called **antagonists.** When prime movers contract and produce a movement, the antagonists are

FRONTIERS IN HEALTH
High Tech Moves into Muscle Building

Milo of Croton was no ordinary Greek. He was both a champion wrestler and an innovator, serving as an inspiration to all the body builders since his time.

Milo utilized a unique method to strengthen his muscles. Beginning with a newborn calf, he lifted it and carried it around with him day after day. As the calf grew, so did Milo's muscles.

The success of Milo's innovative weight-lifting program was obvious, as he went on to win six Olympic championships in wrestling. As effective as carrying a calf might be for developing one's muscles, though, it is not a practical method for everyone. As a result, modern weight lifting came into being.

The theory behind weight training is simple: strengthen the muscles by making them work. As they strengthen, muscle protein builds up, producing bulges that are aesthetically appealing.

Building muscle requires an exercise program in which the rate of muscle formation exceeds the rate of destruction. This requires a regular exercise routine, because muscle tissue turns over rather rapidly. It is estimated that half of the cells in a muscle are replaced every 2 weeks. If you are not faithful to your exercise program, those bulging biceps you have worked so hard to develop will soon begin to disappear.

However, many devotees of weight lifting find their muscle-building efforts boring and time consuming. A new type of machine, known as a progressive resistance machine (PRM), promises to alleviate these problems.

Progressive resistance machines consist of weights connected to levers and chains. The body builders are strapped into the machines and push and pull themselves into physical fitness. Many PRMs are specifically designed to exercise a certain muscle or muscle group. By following various exercise procedures, PRMs make it possible to work just about every muscle in the body.

Most of the health spas in the United States now have PRMs, of which the Nautilus machine is perhaps the most renowned. What makes them so popular is that they are safe. There are no barbells to fall on your chest or feet and no danger of wrenching your back if you make a mistake. Lifting free weights requires technique and training. Working out on the PRMs is simple, even for the uncoordinated.

Another advantage of the PRMs is that they cut the exercise time approximately in half. For example, when working the biceps machine, the biceps muscle is worked while both flexing and extending the arm, if the weight is let down slowly. Furthermore, you can lift heavier weights with the PRMs than with free weights. This is because with free weights you can lift only as heavy a weight as you can handle during the part of the exercise in which your muscles are weakest. Any additional weight will result in a muscle tear or a dropped weight, telling you that you have exceeded the weight-bearing capacity of your muscles. With PRMs, special pulleys avoid this problem by allowing the resistance to vary. The resistance decreases slightly during the weak phase of an exercise to prevent damage, but is full during the rest of the exercises.

Clearly, the development of muscle strength by making muscles work has come a long way since the time of Milo.

A body builder works out at a Nautilus PRM.

stretched. It is important to realize that a particular muscle is not always either a prime mover or an antagonist; rather, its role changes, depending on the movement that is being produced. For instance, when the forearm is flexed, the anterior muscles of the arm are the prime movers and the posterior arm muscles are the antagonists. When the forearm is brought back to the anatomical position—or extended—the posterior muscles of the arm become the prime movers and the anterior arm muscles are the antagonists.

In addition to prime movers and antagonists, most joint movements involve muscles that act as **synergists.** Synergists are muscles that indirectly aid in a movement by steadying a joint—thus preventing unwanted movements and thereby allowing the prime movers to function efficiently—or by otherwise aiding the movement. Muscles that function as prime movers often cause actions other than the desired movement. For example, flexion may be accompanied by rotation in the joint. In this case, the contraction of the synergistic muscles may assist the prime-mover muscles by opposing undesired rotation movements that the prime movers may cause. In a similar manner, if synergistic muscles did not act to immobilize the wrist and thus keep it from moving, the wrist would flex every time a person made a fist, since the muscles that flex the fingers also pass anteriorly across the wrist. Therefore, some synergistic muscles may assist a movement by acting as antagonists.

When a synergist acts to immobilize a joint or an individual bone, it is referred to as a **fixator.** Those muscles described in the previous paragraph as immobilizing the wrist were functioning as fixators. Many of the muscles that attach the scapula to the axial skeleton have important actions as fixators. The scapula is freely movable, and in order for it to serve as a firm origin for the muscles that move the arm, it must be held steady when the muscles contract. The contractions of fixators hold the scapula firmly against the thorax so that contractions of the arm muscles can move only their insertions, which are on the bones of the arm and forearm.

Although it is the prime movers that cause the actual movements, the contractions of antagonistic and synergistic muscles are necessary to produce the smooth coordinated motions that are typical of a normal person. The strength of an antagonist's contraction affects the strength and speed of the prime mover's contraction. For instance, if the extensor muscles of the forearm remain partially contracted—and therefore act as antagonists—while the flexor muscles of the forearm are causing the elbow to bend—and therefore are serving as prime movers—the flexor muscles have to contract harder to overcome the opposition and the joint movement will be slower than it might otherwise be. The actions of antagonists and synergists make very fine and precise movements possible.

RELATIONSHIP BETWEEN LEVERS AND MUSCLE ACTIONS

The movements brought about by the actions of most skeletal muscles involve the use of levers. A **lever** is a rigid structure that is capable of moving around a pivot point, called a **fulcrum,** when a force is applied. In the body, the bones of the skeleton function as levers, the joints serve as fulcrums, and skeletal muscles provide the force to move the bones. Depending on the location of the fulcrum, a lever can make it possible to move heavier loads than could otherwise be moved, or to alter both the rate of movement and the distance over which a load can be moved.

Classes of Levers

There are three *classes of levers:* class I, class II, and class III.

Class I Levers

In **Class I levers,** the fulcrum is located between the point at which the force is applied and the weight that is to be moved (Figure 7.9a). A seesaw is a common example of a Class I lever. In the body, this type of lever is involved in **F7.9a**

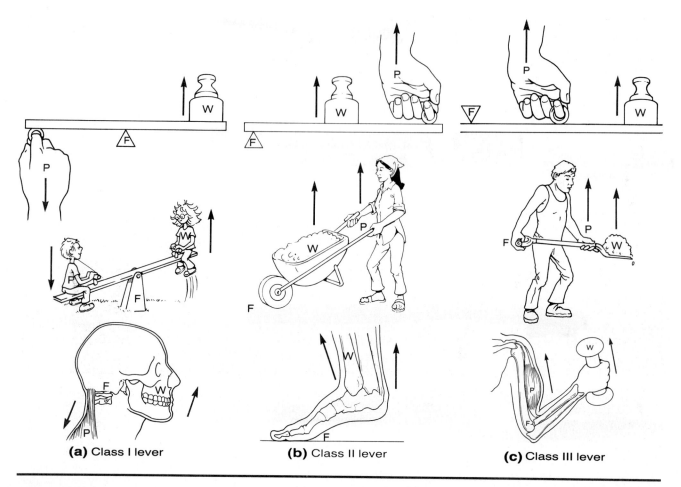

(a) Class I lever   **(b)** Class II lever   **(c)** Class III lever

Figure 7.9
(a) Class I lever. Fulcrum (F) is located between the weight (W) (or resistance) and the force (P) (or pull). Arrows indicate the direction of movement. **(b) Class II lever.** The weight (W) (or resistance) is located between the fulcrum (F) and the point of force (P) (or pull). Arrows indicate the direction of movement. **(c) Class III lever.** The force (P) (or pull) is applied between the fulcrum (F) and the weight (W) (or resistance). Arrows indicate the direction of movement.

raising the face. The occipital condyles on the atlas serve as the fulcrum, the facial portion of the skull is the weight, and the force (pull) is applied to the back of the skull by the posterior muscles of the neck.

Class II Levers

F7.9b In **Class II levers,** the weight to be moved is between the fulcrum and the point of force (Figure 7.9*b*). This type of lever is used in the wheelbarrow. The best example in the body is raising the body on the toes. In this case the base of the toes serves as the fulcrum, the toes support the weight, and the contraction of the posterior muscles of the calf causes a force (pull) to be exerted on the calcaneus bone.

Class III Levers

F7.9c In **Class III levers,** the weight is at one end, the fulcrum is at the other, and the force is applied between them (Figure 7.9*c*). Lifting a shovel utilizes this type of leverage. There are many examples of Class III levers in the body, since it is the most common lever system used. One example is flexion of the forearm, where the weight is at the wrist, the fulcrum is the elbow joint, and the force (pull) is exerted by the contraction of the flexor muscles on the anterior of the arm that insert on the radius or the ulna, between the fulcrum and the weight.

Effects of Levers on Movements

The portion of a lever located between the fulcrum and the point where the force is applied is called the **power arm;** the portion between the fulcrum and the weight is the **weight arm.** When the weight arm is long in relation to the power arm, a weight can be moved rapidly over a considerable distance, but a

CONDITIONS OF CLINICAL SIGNIFICANCE
Skeletal Muscle

Muscular Atrophy

The shrinkage and death of muscle cells—*muscular atrophy*—cause a reduction in the size of the affected muscles. Atrophy can be caused by prolonged disuse or by a number of disorders, most of which reduce the blood supply or interfere with the nerve supply to the muscle. Muscular atrophy can be widespread or localized. Localized atrophy may not reduce the size of the muscle, however, since the unaffected cells can undergo compensatory enlargement (hypertrophy).

Cramps

Cramps are painful, involuntary muscle contractions that are slow to relax. They can occur during exercise or at rest. Their precise cause is not known, but they may be caused by conditions in the muscle itself (such as low oxygen supply) or by stimulation from the nervous system. There is some evidence that cramps that occur during heavy exercise are caused by low levels of sodium and chloride ions in the blood, a result of their loss through sweating. However, it is not clear whether the depletion of sodium chloride acts on the muscle or on the nervous system.

Muscular Dystrophy

The term *muscular dystrophy* refers to a group of diseases characterized by progressive muscular weakness. The weakness is the result of the degeneration of muscle cells, an increase in connective tissue in the muscles, and in some forms, the replacement of muscle cells by fatty tissue. The muscular dystrophies are genetically transmitted. Some forms of muscular dystrophy are fatal, but other forms have a more favorable outlook.

Myasthenia Gravis

Myasthenia gravis is a chronic condition characterized by extreme muscle weakness. It is caused by an abnormal response of the body's immune system that disrupts acetylcholine receptors on the muscle cell membrane at the neuromuscular junctions. This decreases the responsiveness of the muscle fibers to acetylcholine, which normally transmits the motor nerve impulse to the muscle. In a high percentage of people afflicted with myasthenia gravis the thymus gland, which is an important part of the immune system, is enlarged, and its removal often produces dramatic improvement. About 10% of the people who have this disease die from it. However, if an afflicted individual survives the first three years, there is a good chance that the condition will stabilize, with some degree of recovery.

Effects of Aging on the Skeletal Muscles

Starting in the middle twenties, a progressive and continuous loss of skeletal muscle mass begins, and much of the loss is replaced by fat. Since fat weighs less than muscle, the normal body weight at age 50 is less than at age 20. As the muscle mass decreases, so does the maximal strength, which declines by about 50% between 20 and 80 years of age. The amount of muscle mass lost with aging also depends on physical activity, with the rate of loss being slower in those people who maintain a regular regime of physical activity.

strong force is required. Conversely, when the weight arm is short in relation to the power arm, the same weight can be moved with less force, but both the speed of movement and the range of movement are reduced. Thus, depending on the arrangement of the particular muscles and bones, the levers of the body may enable muscles to move loads faster over greater distance than would otherwise be possible or they may enable muscles to move heavier loads than they otherwise could.

SMOOTH MUSCLE

Smooth muscle fibers are uninucleate, spindle-shaped cells that are considerably smaller than skeletal muscle fibers. Smooth muscle fibers have a well-developed capacity for anaerobic metabolism, but their overall metabolic machinery is not as highly developed as that of skeletal muscle fibers.

Smooth muscle fibers possess thick filaments that contain myosin and thin filaments that contain actin and tropomyosin. However, the filaments are not organized into regularly ordered sarcomeres, and smooth muscle fibers are not striated. Nevertheless, smooth muscle uses cross-bridge movements between myosin and actin to generate force, and it is generally believed that smooth muscle contraction occurs by a sliding filament mechanism similar to that in skeletal muscle.

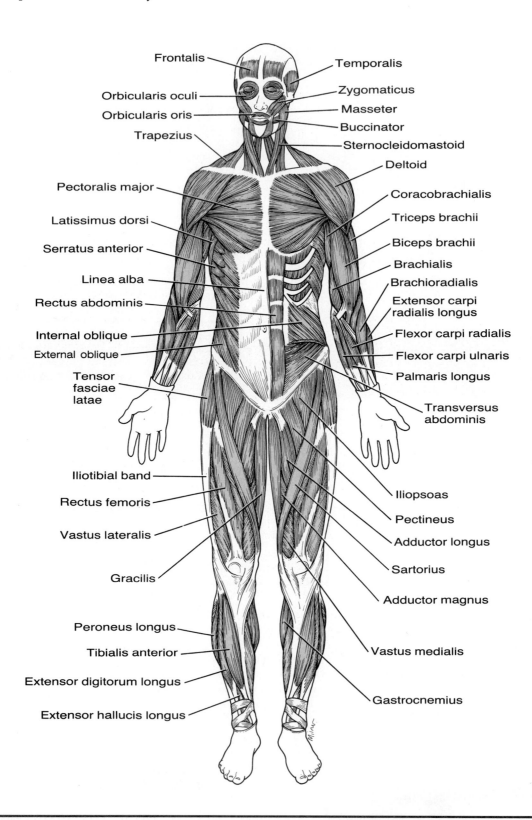

Frontalis

Temporalis

Orbicularis oculi

Zygomaticus

Orbicularis oris

Masseter

Trapezius

Buccinator

Sternocleidomastoid

Deltoid

Pectoralis major

Coracobrachialis

Latissimus dorsi

Triceps brachii

Serratus anterior

Biceps brachii

Linea alba

Brachialis

Rectus abdominis

Brachioradialis

Internal oblique

Extensor carpi radialis longus

External oblique

Flexor carpi radialis

Tensor fasciae latae

Flexor carpi ulnaris

Palmaris longus

Transversus abdominis

Iliotibial band

Rectus femoris

Iliopsoas

Vastus lateralis

Pectineus

Adductor longus

Gracilis

Sartorius

Adductor magnus

Peroneus longus

Tibialis anterior

Vastus medialis

Extensor digitorum longus

Extensor hallucis longus

Gastrocnemius

Figure 7.10
Anterior view of the
muscles of the body.
The left external oblique
muscle has been
removed.

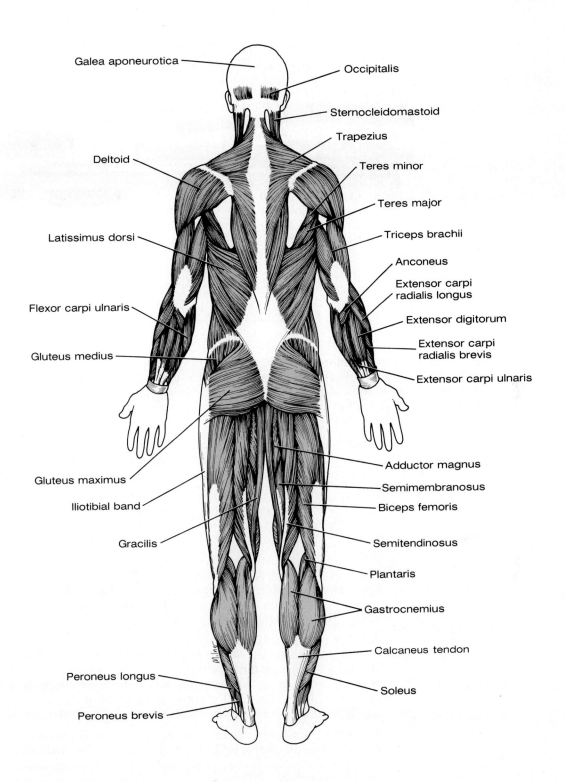

Galea aponeurotica

Occipitalis

Sternocleidomastoid

Trapezius

Deltoid

Teres minor

Teres major

Triceps brachii

Latissimus dorsi

Anconeus

Extensor carpi
radialis longus

Extensor digitorum

Flexor carpi ulnaris

Extensor carpi
radialis brevis

Gluteus medius

Extensor carpi ulnaris

Adductor magnus

Semimembranosus

Biceps femoris

Gluteus maximus

Semitendinosus

Iliotibial band

Plantaris

Gracilis

Gastrocnemius

Calcaneus tendon

Peroneus longus

Soleus

Peroneus brevis

Figure 7.11
Posterior view of the
muscles of the body.

CARDIAC MUSCLE

The individual cells of cardiac muscle interconnect with one another to form branching networks. Where adjoining cells meet end to end, their junctions form structures called *intercalated discs*. Within the intercalated discs are two types of cell-to-cell membrane junctions (see Chapter 3): desmosomes, which attach one cell to another; and gap junctions, which allow electrical impulses to spread from cell to cell. Cardiac muscle cells are cross-striated, and they possess thick, myosin-containing filaments and thin, actin-containing filaments that are arranged into regularly ordered sarcomeres and myofibrils. The basic contractile events that occur in cardiac muscle cells are believed to be similar to those that occur in skeletal muscle cells. However, cardiac muscle cells have larger *t* tubules and a less-developed sarcoplasmic reticulum than skeletal muscle cells.

SKELETAL MUSCLES OF THE HUMAN BODY

F7.10, F7.11 There are over 600 muscles in the human body (Figure 7.10, Figure 7.11). Only the more commonly studied muscles are included in this chapter. Several criteria are used to name muscles; each describes a particular characteristic of the muscle being named, such as its shape, action, or location. You will find it quite useful in your study of the muscles to familiarize yourself with these criteria, which follow.

Shape   The names of some muscles include references to their shape. For example, the trapezius muscles are shaped like trapezoids and the rhomboideus muscles resemble rhomboids.

Action   Various muscle names include references to the actions of the muscle by using the terms flexor, extensor, adductor, or pronator. For example, the flexor carpi radialis muscles flex the hands, and the extensor digitorum longus muscles extend the toes.

Location   It is possible to locate certain muscles by their name. For example, the intercostal muscles (*inter* = between; *costal* = rib) are located between the ribs, and the tibialis anterior muscles lie alongside the anterior margin of each tibia.

Attachments   The attachments of a muscle to the skeleton are included in some names. For example, the sternocleidomastoid muscles have origins on the sternum and clavicles and insert on the mastoid processes of the temporal bones; the coracobrachialis muscles have their origins on the coracoid processes of the scapulae and insert on each brachium—which refers to the arm.

Number of divisions   Some muscles are separated into two, three, or four divisions, and this is indicated in their names. For example, the biceps brachii muscles have two divisions, the triceps brachii have three, and the quadriceps femoris muscles have four.

Size relationships   Terms referring to size are often included in muscle names—for example, the gluteus maximus and gluteus minimus muscles of the buttocks are large and small, respectively, and the peroneus longus and peroneus brevis muscles of the leg are long and short, respectively.

In many cases muscle names include more than one of these criteria. For example, the name of the flexor digitorum longus muscle indicates the muscle's action (flexion), its insertion (digits), and its size relationship (long—in comparison to the flexor digitorum brevis muscle).

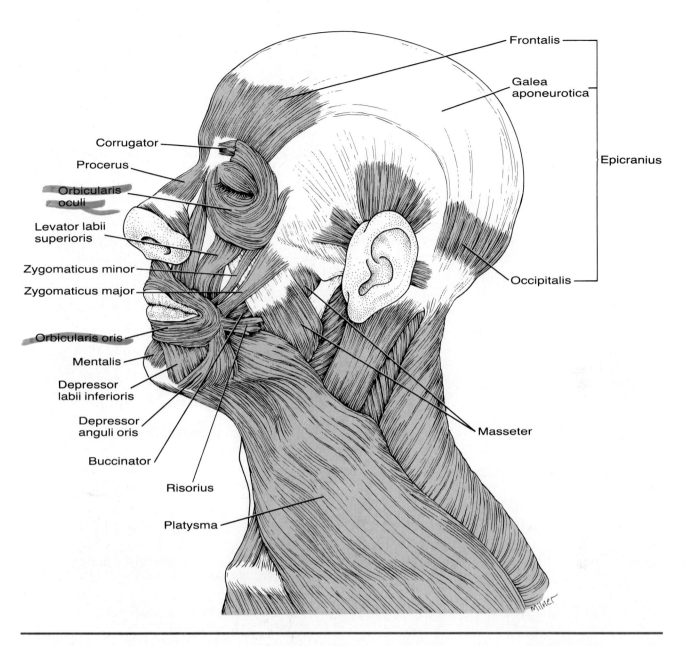

Figure 7.12
Muscles of the face and neck.

To organize their study, we will consider the muscles in various groups (but note that some muscles belong to more than one group). Each group of muscles is discussed in a general way, and the reader is provided with figures that illustrate the muscles as well as tables that give a more detailed description of each muscle, including information on each muscle's origin, insertion, principal actions, and nerve innervation.

MUSCLES OF THE HEAD AND NECK

Muscles of the Face

Because of their actions, the muscles of the face (Figure 7.12, Table 7.1) are also referred to as the muscles of facial expression. Whereas some facial muscles have their origins on the bones of the skull, others arise from the superficial fascia of the face. Most of them insert into the skin of the region and

F7.12, Table 7.1

Table 7.1 Muscles of the Face [F7.12]

Muscle	Origin	Insertion	Action	Innervation
Buccinator	Alveolar process of the mandible and the maxillary bone	Orbicularis oris and skin at the angle of the mouth	Compresses cheek; pulls corner of the mouth laterally	Facial (cranial nerve VII)
Corrugator	Frontal bone, lateral to the glabella	Skin of the eyebrows	Draws the eyebrows together, as in frowning	Facial
Depressor anguli oris	Body of the mandible, below the mental foramen	Skin and muscles at the angle of the mouth	Pulls the angle of the mouth downward	Facial
Depressor labii inferioris	Body of the mandible between the symphysis and the mental foramen	Skin and muscles of the lower lip	Pulls the lower lip downward	Facial
Epicranius *Frontalis*	Galea aponeurotica	Skin and muscles of the forehead	Raises the eyebrows; wrinkles the skin of the forehead	Facial
Occipitalis	Occipital bone (superior nuchal line)	Galea aponeurotica	Draws the scalp posteriorly	Facial
Levator labii superioris	Lower margin of orbit (maxillary and zygomatic bones)	Skin and muscles of the upper lip, and wing of the nose	Raises the upper lip; dilates the nares (nostrils)	Facial
Mentalis	Mandible, near the symphysis	Skin of the chin	Raises and protrudes the lower lip	Facial
Orbicularis oculi	Frontal and maxillary bones; medial palpebral ligament	Circles the orbit and extends within the eyelids	Closes the eyelids; tightens the skin of the forehead	Facial
Orbicularis oris	Muscles surrounding the mouth	Skin surrounding the mouth	Closes and protrudes the lips	Facial
Platysma	Fascia over the pectoralis major and the deltoid muscles	Lower border of the mandible, and the skin of the chin and cheek	Depresses the mandible; draws the angle of the mouth downward; tightens and wrinkles the skin of the neck	Facial
Procerus	Lower portion of the nasal bone; upper part of the lateral nasal cartilage	Skin between the eyebrows	Wrinkles the skin between the eyebrows	Facial
Risorius	Fascia of the masseter muscle	Skin at the angle of the mouth	Pulls the angle of the mouth backward	Facial
Zygomaticus major and zygomaticus minor	Zygomatic bone	Skin and muscles above the angle of the mouth	Raise the angle of the mouth	Facial

therefore move the skin rather than a joint. Among the unusual muscles in this group are the sphincters, which surround the eyes and the mouth. One sphincter, the **orbicularis oculi,** is used in closing the eye, winking, and squinting. Contraction of the other sphincter, the **orbicularis oris,** closes the mouth and purses the lips. Several of the other facial muscles insert onto the fascia that covers the orbicularis oris.

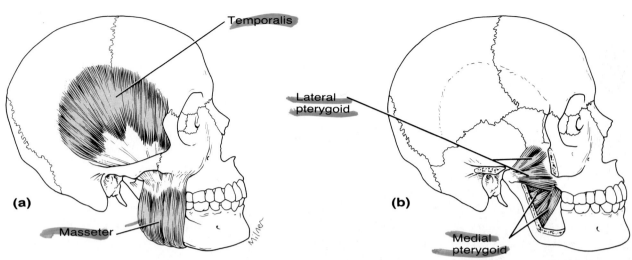

Figure 7.13
Muscles of mastication. **(a)** The temporalis and masseter muscles are the strongest masticatory muscles. **(b)** The temporalis and masseter muscles have been removed and the zygomatic arch and mandible have been sectioned to reveal the pterygoid muscles.

Table 7.2   Muscles of Mastication [*F7.13*]

Muscle	Origin	Insertion	Action	Innervation
Temporalis	Temporal fossa	Coronoid process and ramus of the mandible	Raises the mandible, closing the jaws; retracts the mandible	Trigeminal (cranial nerve V)
Masseter	Zygomatic arch	Angle and ramus of the mandible	Raises the mandible, closing the jaws	Trigeminal
Medial pterygoid	Medial surface of the lateral pterygoid plate of the sphenoid bone, and the tuberosity of the maxillary bone	Inner surface of the mandible, at the angle	Closes the jaws; together with the lateral pterygoid, it aids in sideways movement of the jaws	Trigeminal
Lateral pterygoid	Lateral surface of the lateral pterygoid plate and the great wing of the sphenoid bone	Mandible just below the condyle	Opens and protrudes the mandible; moves the mandible from side to side	Trigeminal

Another unusual facial muscle is the **epicranius.** This muscle has two parts: the anterior **frontalis** and the posterior **occipitalis.** These two muscular portions are connected by a broad, flat tendon, the **galea aponeurotica,** which lies tight against the top of the skull. Contraction of one or the other muscular portion pulls the scalp forward or backward.

While the **platysma** is not actually a facial muscle, we will consider it here because its main actions are on the mandible and the skin around the mouth. It is a superficial, sheet-like muscle that covers the ventral surface of the upper thorax and the neck, and extends over the chin to the region of the mouth. Contraction of the platysma lowers the mandible, the lower lip, and the corners of the mouth, as well as tightening the skin of the neck.

Muscles of Mastication

Four pairs of muscles are involved in biting and chewing. These include the large fan-shaped **temporalis,** which passes deep to the zygomatic arch of the cheek, and the quadrilateral **masseter,** which arises from the zygomatic arch. Both muscles serve to close the mandible (Figure 7.13, Table 7.2). These muscles can be felt when the teeth are forcibly clenched. The other two pairs of   F7.13, Table 7.2

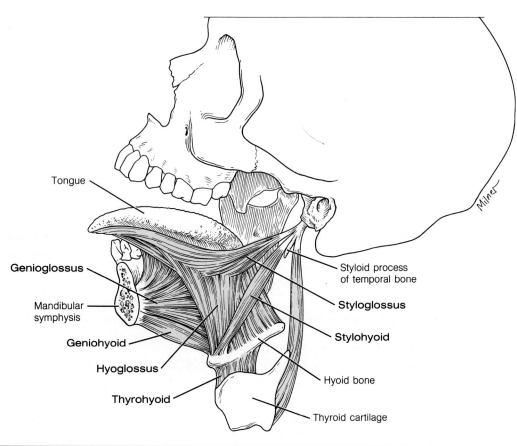

Figure 7.14
Extrinsic muscles of the tongue and the suprahyoid muscles of the neck.

Table 7.3 **Extrinsic Muscles of the Tongue [*F7.14*]**

Muscle	Origin	Insertion	Action	Innervation
Genioglossus	Internal surface of the mandible, near the symphysis	Undersurface of the tongue; body of the hyoid	Protracts, retracts, and depresses the tongue	Hypoglossal (cranial nerve XII)
Hyoglossus	Body and greater cornu of the hyoid bone	Side of the tongue	Depresses the tongue; draws its sides down	Hypoglossal
Styloglossus	Styloid process of the temporal bone	Side of the tongue	Retracts and elevates the tongue	Hypoglossal

muscles involved in mastication are the **medial** and **lateral pterygoid** muscles, which move the mandible sideways in grinding movements, as well as assisting in opening and closing the mouth.

Muscles of the Tongue

The tongue is a muscular organ that is covered with mucous membrane. Some of the muscles, called **intrinsic muscles,** lie entirely within the tongue. The fibers of the intrinsic muscles are arranged in longitudinal, vertical, and horizontal planes; consequently, they squeeze, fold, and curl the tongue when they contract. These actions are particularly useful in speaking and manipulating food in the mouth. The **extrinsic muscles** (Figure 7.14, Table 7.3) anchor the tongue to the skeleton (hyoid, mandible, and temporal bones) and control the extension, retraction, and sideways movement of the tongue.

F7.14, Table 7.3

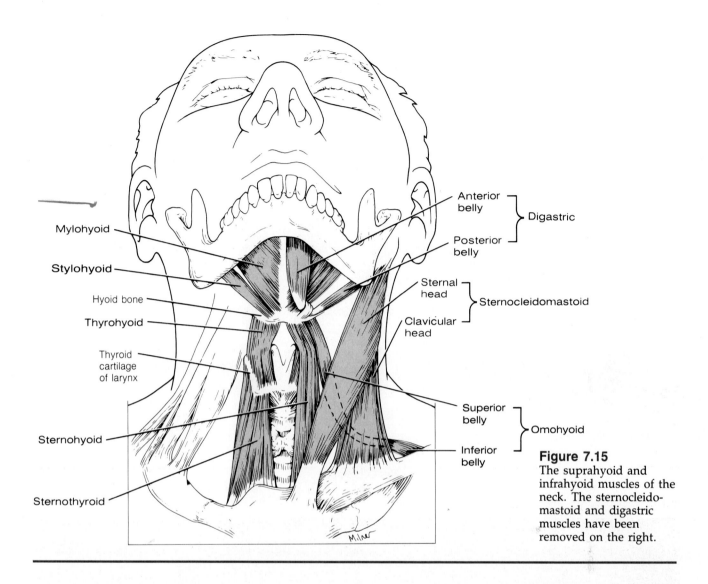

Anterior belly ⎱
Posterior belly ⎰ Digastric

Mylohyoid

Stylohyoid

Hyoid bone

Thyrohyoid

Thyroid cartilage of larynx

Sternohyoid

Sternothyroid

Sternal head ⎱
Clavicular head ⎰ Sternocleidomastoid

Superior belly ⎱
Inferior belly ⎰ Omohyoid

Figure 7.15
The suprahyoid and infrahyoid muscles of the neck. The sternocleido-mastoid and digastric muscles have been removed on the right.

Muscles of the Neck

The muscles of the neck (Figure 7.15, Table 7.4) are often described as being located within one of two triangles. Those in the anterior triangle are separated from those in the posterior triangle by the **sternocleidomastoid** muscle. This muscle runs diagonally across the lateral margins of the neck from the mastoid process of the temporal bone to the sternum and the clavicle. It is beyond the scope of this book to describe all the muscles in these two triangles. Rather, we will consider here only certain muscles of the anterior triangle—namely, the muscles of the throat. The muscles of the posterior triangle are discussed later, along with the muscles that move the vertebral column and the head (Table 7.5).

F7.15, Table 7.4

Table 7.5

Muscles of the Throat

The muscles of the throat are the deep muscles of the anterior triangle (Figure 7.14, Figure 7.15, Table 7.4). They help form the floor of the oral cavity and are attached to the hyoid bone. Because the tongue is also attached to the hyoid bone, these muscles are involved with movements of the tongue. Moreover, some of the throat muscles are attached to the larynx and therefore aid in swallowing. These muscles are often divided into two groups:

F7.14, F7.15, Table 7.4

1. The *suprahyoid muscles:* **digastric, stylohyoid, mylohyoid,** and

Table 7.4   Muscles of the Anterior Triangle of the Neck [*F7.14, F7.15*]

Muscle	Origin	Insertion	Action	Innervation
Sternocleidomastoid	By two heads: the manubrium of the sternum, and the medial portion of the clavicle	Mastoid process of the temporal bone	Both muscles acting together flex the cervical vertebral column; acting singly, each rotates head to the opposite side	Accessory (cranial nerve XI) and upper cervical spinal nerves
SUPRAHYOID MUSCLES				
Digastric	*Anterior belly:* inner surface of the mandibular symphysis *Posterior belly:* mastoid process of the temporal bone	Hyoid bone, via the intermediate tendon	Raises the hyoid and assists in lowering the jaw	Trigeminal (anterior belly); facial (posterior belly)
Stylohyoid	Styloid process of the temporal bone	Hyoid bone	Raises the hyoid and pulls it backward	Facial
Mylohyoid	Inner surface of the mandible, from the symphysis to the angle	Hyoid bone	Raises the hyoid and the floor of the mouth	Trigeminal
Geniohyoid	Inner surface of the mandibular symphysis	Hyoid bone	Pulls the hyoid anteriorly	C_1 (through hypoglossal)
INFRAHYOID MUSCLES				
Sternohyoid	Manubrium and the medial end of the clavicle	Hyoid bone	Pulls the hyoid inferiorly	C_1–C_3 (through ansa cervicalis—see p. 384)
Sternothyroid	Manubrium	Thyroid cartilage of the larynx	Pulls the larynx inferiorly	C_1–C_3 (through ansa cervicalis)
Thyrohyoid	Thyroid cartilage of the larynx	Hyoid bone	Pulls the hyoid inferiorly and raises the larynx	C_1 (through hypoglossal)
Omohyoid	Superior border of the scapula	Hyoid bone	Pulls the hyoid inferiorly	C_1–C_3 (through ansa cervicalis)

geniohyoid. As a group, these muscles raise the hyoid bone during swallowing and lower the jaw when the hyoid bone is fixed.

2. The *infrahyoid muscles:* **sternohyoid, sternothyroid, thyrohyoid,** and **omohyoid.** These muscles pull down on the larynx and hyoid, returning them to their normal position after swallowing.

MUSCLES OF THE TRUNK

The muscles of the trunk include those that are associated with the vertebral column, the back, the thorax, the floor of the pelvic cavity, and the wall of the abdomen. Trunk muscles have various actions, depending on their locations.

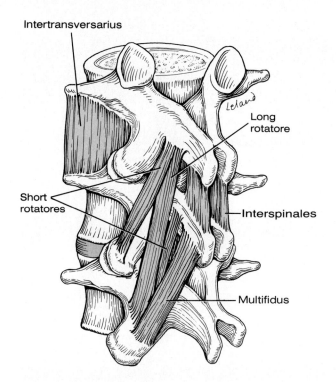

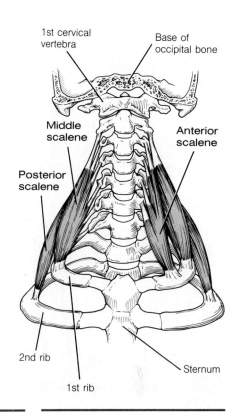

Figure 7.16
Muscles of the vertebral
column.

Figure 7.17
The scalene muscles
viewed from the front.
The right anterior scalene
muscle has been
removed.

Some move the vertebral column, others move the head; some are involved in respiratory movements, others move the upper limbs, and so forth. We will study them in groups according to their actions.

Muscles of the Vertebral Column

Most of the muscles that move the vertebral column are located on the posterior surface of the spine (Table 7.5). A few, such as the **splenius,** have fibers that insert onto the skull and therefore move the head as well as the vertebral column. The deepest of these muscles are located medially and travel only a few segments superiorly before inserting onto the transverse processes or spinous processes of the vertebrae (Figure 7.16). These muscles include the **multifidus, rotatores, interspinales,** and **intertransversarii** and the **semispinalis thoracis, cervicis,** and **capitis.** The **scalenes** (anterior, middle, and posterior) (Figure 7.17) pass from the transverse processes of the cervical vertebrae to the upper two ribs.

Lateral to these muscles, located in the depression between the spinous processes and the transverse processes and the ribs, is a longitudinal muscle mass that extends from the sacrum to the skull. This is the **erector spinae (sacrospinalis)** muscle (Figure 7.18). The erector spinae has three subdivisions in the form of columns. The **iliocostalis,** which is the most lateral column, inserts on the ribs; the medial column is the **spinalis,** the fibers of which insert on the vertebrae; the **longissimus** subdivision is located between the other two columns. Each of these columns is further separated into **lumborum, thoracis, cervicis,** and **capitis** parts, which are named according to their points of insertion. In Table 7.5 the origins, insertions, and actions of these parts have been combined for each subdivision of the erector spinae.

All the muscles that insert on the vertebral column act to extend the spine. When acting on one side only, they bend the vertebral column to that side and may assist in its rotation. These muscles are not the only ones that

Table 7.5

F7.16

F7.17

F7.18

Table 7.5

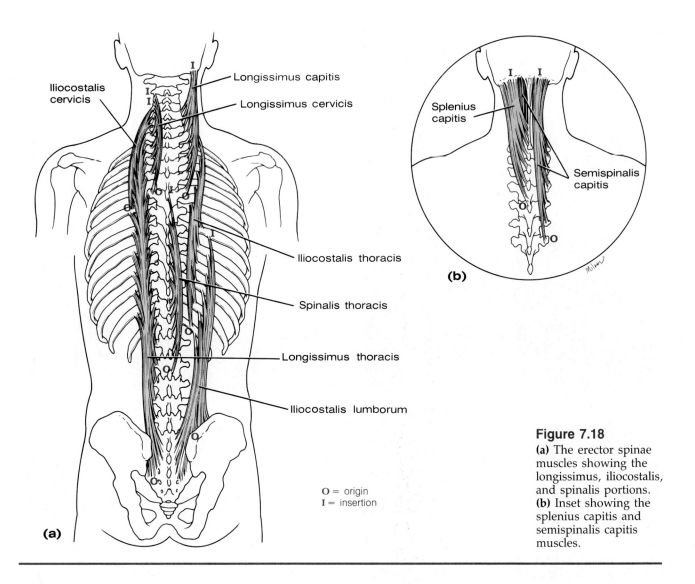

Iliocostalis cervicis

Longissimus capitis

Longissimus cervicis

Iliocostalis thoracis

Spinalis thoracis

Longissimus thoracis

Iliocostalis lumborum

O = origin
I = insertion

(a)

Splenius capitis

Semispinalis capitis

(b)

Figure 7.18
(a) The erector spinae muscles showing the longissimus, iliocostalis, and spinalis portions. **(b)** Inset showing the splenius capitis and semispinalis capitis muscles.

move the vertebral column, however. Certain muscles of the abdominal wall, **Table 7.7** such as the **rectus abdominis** and the **quadratus lumborum** (Table 7.7) also **Table 7.17** act on the vertebral column. In addition, the **psoas major** (Table 7.17), which acts on the hip joint, can cause the vertebral column to flex if the thighs are fixed.

Deep Muscles of the Thorax

Most of the deep muscles of the thorax insert on the ribs and assist in breath-**Table 7.6** ing by drawing the ribs together or elevating or depressing the rib cage (Table 7.6). Since the ribs slope downward as they pass forward, any muscle that elevates them increases the volume of the thoracic cavity, causing inspiration. Conversely, muscles that depress the ribs back to their usual positions decrease the volume of the thoracic cavity, forcing air from the lungs in expiration. The muscles described in Table 7.6 are those involved in normal quiet respirations. In forced breathing, when overexpansion of the thoracic cage is **Table 7.5** beneficial, additional muscles may be involved, such as the **scalenes** (Table **Table 7.4** 7.5), the **sternocleidomastoid** (Table 7.4), and the **quadratus lumborum** **Table 7.7** (Table 7.7). Because these muscles have other actions that are more commonly performed, they are listed with other groups, as their table references indicate.

The spaces between adjacent ribs are reinforced primarily by **external** and **F7.19** **internal intercostal** muscles (Figure 7.19). In addition, small **innermost intercostal** muscles lie deep to the internal intercostals. The fibers of the external

Table 7.5   Muscles That Move the Vertebral Column

Muscle	Origin	Insertion	Action	Innervation
Semispinalis [F7.18] thoracis cervicis capitis	Transverse processes of the thoracic and the seventh cervical vertebrae	Spinous processes of the second cervical through the fourth thoracic vertebrae, and the occipital bone	Extend the vertebral column and the head (capitis); rotate them to the opposite side	Branches of the spinal nerves
Multifidus [F7.16]	Posterior surface of the sacrum and the ilium, and the transverse processes of the lumbar, thoracic, and lower cervical vertebrae	Spinous processes of the lumbar, thoracic, and cervical vertebrae	Extends the vertebral column; rotates it towards the opposite side	Branches of the spinal nerves
Rotatores (long and short) [F7.16]	Transverse processes of all the vertebrae	Base of the spinous process of the vertebra above the vertebra of origin (short) or the second vertebra above (long)	Extend the vertebral column; rotate it towards the opposite side	Branches of the spinal nerves
Interspinales [F7.16]	Superior surface of all the spinous processes	Inferior surface of the spinous process of the vertebra above the vertebra of origin	Extend the vertebral column	Branches of the spinal nerves
Scalenes [F7.17]	Transverse process of cervical vertebrae	Upper two ribs	Flex and rotate the neck; assist in inspiration	Branches of the lower cervical nerves
Intertransversarii [F7.16]	Transverse processes of all the vertebrae	Transverse processes of the vertebra above the vertebra of origin	Bend the vertebral column laterally	Branches of the spinal nerves
Splenius [F7.18] capitis cervicis	Spinous processes of the upper thoracic and the seventh cervical vertebrae, and from the ligamentum nuchae	Occipital bone, mastoid process of the temporal bone, and the transverse processes of the upper three cervical vertebrae	Acting together, they extend the head and the neck; acting singly, they abduct and rotate the head towards the same side	Branches of the spinal nerves

ERECTOR SPINAE (SACROSPINALIS) [F7.18]

Muscle	Origin	Insertion	Action	Innervation
Iliocostalis lumborum thoracis cervicis	Crest of the sacrum; spinous processes of the lumbar and lower thoracic vertebrae; iliac crests; angles of the ribs	Angles of the ribs; transverse processes of the cervical vertebrae	Extend the vertebral column and bend it laterally	Branches of the spinal nerves
Longissimus thoracis cervicis capitis	Transverse processes of the lumbar, thoracic, and lower cervical vertebrae	Transverse processes of the vertebra above the vertebra of origin, and the mastoid process of the temporal bone (capitis)	Extend the vertebral column and head; rotate the head towards the same side	Branches of the spinal nerves
Spinalis thoracis cervicis	Spinous process of the upper lumbar, lower thoracic, and seventh cervical vertebrae	Spinous processes of the upper thoracic and the cervical vertebrae	Extend the vertebral column	Branches of the spinal nerves

Table 7.6 Deep Muscles of the Thorax (Respiratory Muscles)

Muscle	Origin	Insertion	Action	Innervation
Diaphragm [F7.20]	The xiphoid process; inner surfaces of lower six ribs; and the lumbar vertebrae	Central tendon of the diaphragm	Pulls central tendon downward, increasing the size of the thoracic cavity and therefore causing inspiration	Phrenic
External intercostals [F7.19]	Inferior border of the ribs and the costal cartilages	Superior border of the rib below the rib of origin	Draw ribs together, aiding respiration	Intercostal
Internal intercostals [F7.19]	Inner surface of the ribs and the costal cartilages	Superior border of the rib below the rib of origin	Draw ribs together, aiding respiration	Intercostal
Subcostales*	Inner surface of the ribs, near their angles	Inner surface of the second or third rib below the rib of origin	Draw ribs together, aiding expiration	Intercostal
Transversus thoracis*	Inner surface of the sternum and the xiphoid process	Inner surface of the costal cartilages	Draws anterior portion of the rib cage downward, aiding expiration	Intercostal

*Not illustrated.

O = origin
I = insertion

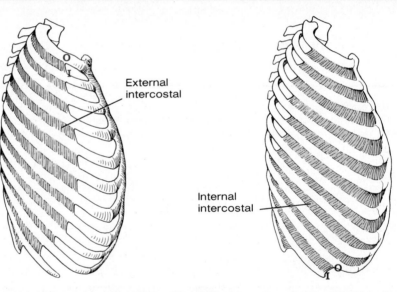

External intercostal

Internal intercostal

Figure 7.19
The external and internal intercostal muscles.

intercostal muscles run at right angles to the fibers of the internal and innermost muscles, as in a bias-ply automobile tire, thus providing a strong muscular wall between the ribs without requiring heavy musculature.

The **diaphragm** is the muscle most responsible for quiet breathing. It is dome-shaped and separates the throacic cavity from the abdominopelvic cavity. The upper surface of the diaphragm is in contact with the heart and lungs; the lower surface contacts the liver, the stomach, the spleen, and the pad of fat that surrounds the suprarenal glands and kidneys.

F7.20   The muscle fibers of the diaphragm are grouped into sternal, costal, and lumbar portions (Figure 7.20). The small *sternal* portion arises from the inner surface of the xiphoid process; the *costal* fibers arise from the inner surfaces of the seventh, eighth, and ninth ribs and the distal ends of the last three ribs; the *lumbar* portion arises from the front of the lumbar vertebrae by two tendinous bands called crura. The muscle fibers from these three portions insert on a common *central tendon*, which they surround. When the diaphragm con-

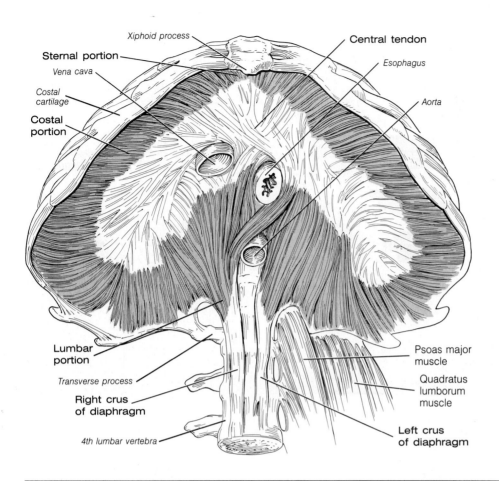

Figure 7.20
The abdominal surface of the diaphragm.

tracts, its dome is pulled downward, flattening the muscle and increasing the volume of the thoracic cavity.

The diaphragm is pierced by a number of openings that permit the passage of structures between the thorax and abdomen. The largest openings are for the aorta, the vena cava, and the esophagus.

Muscles of the Abdominal Wall

Because there are no skeletal supports within the wall of the abdominal cavity, it derives its strength entirely from muscles. There are three layers of muscles in the wall (Figure 7.21, Table 7.7). The fibers of each of the muscles run in different directions, thus providing additional strength. The outermost layer is the **external abdominal oblique** muscle, whose fibers pass medially and downward as a continuation of the external intercostal muscles. The **internal abdominal oblique** muscle lies just deep to the external oblique. Its fibers run upward and medially, becoming continuous over the ribs with the internal intercostal muscles. Deep to both the oblique muscles is a thin muscle whose fibers run horizontally, encircling the abdominal cavity. This is the **transversus abdominis** muscle. The tendons of these three muscles pass medially in the form of broad aponeuroses that insert on a midline **linea alba** (white line). The linea alba is a fibrous band that extends from the xiphoid process to the symphysis pubis.

At the lower margin of the muscle, the fascia of the external abdominal oblique muscle forms a tendinous border called the **inguinal ligament.** This ligament, which runs between the pubic tubercle and the anterior superior iliac spine, marks the separation between the body wall and the thigh. At one point there is an opening between the muscle fascia and the inguinal ligament. This opening, the *superficial inguinal ring,* is the entrance to the **inguinal canal.** The canal passes laterally, above and parallel with the inguinal ligament. About midway between the anterior superior iliac spine and the pubic

F7.21, Table 7.7

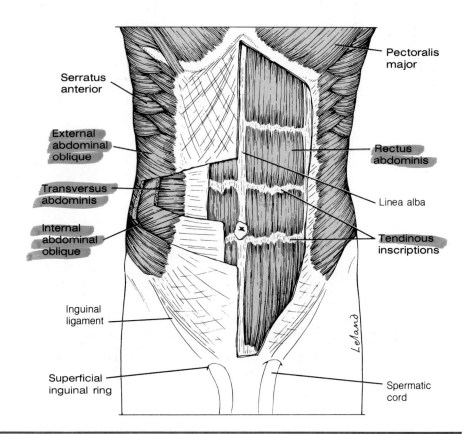

Figure 7.21
Muscles of the abdominal wall.

Table 7.7 Muscles of the Abdominal Wall [F7.21]

Muscle	Origin	Insertion	Action	Innervation
External abdominal oblique	External surface of the lower eight ribs	Linea alba and the anterior half of the iliac crest	Compresses the abdominopelvic cavity; assists in flexing and rotating the vertebral column	Intercostal, iliohypogastric, and ilioinguinal
Internal abdominal oblique	Inguinal ligament, the iliac crest, and the lumbodorsal fascia	Linea alba, the pubic crest, and the lower four ribs	Compresses the abdominopelvic cavity; assisting in flexing and rotating the vertebral column	Intercostals, iliohypogastric, and ilioinguinal
Transversus abdominis	Inguinal ligament, the iliac crest, the lumbodorsal fascia, and the costal cartilages of the last six ribs	Linea alba, and the pubic crest	Compresses the abdominopelvic cavity	Intercostals, iliohypogastric, and ilioinguinal
Rectus abdominis	Pubic crest	Xiphoid process and the costal cartilages of the fifth through the seventh ribs	Compresses the abdominopelvic cavity; flexes the vertebral column	Intercostals
Quadratus lumborum	Iliac crest, and the iliolumbar ligament	Lower border of the twelfth rib; the transverse processes of the upper lumbar vertebrae	Pulls the thoracic cage toward the pelvis; bends the vertebral column laterally toward the side that is being contracted	Twelfth thoracic and first lumbar

Cadaver Photos

Muscles

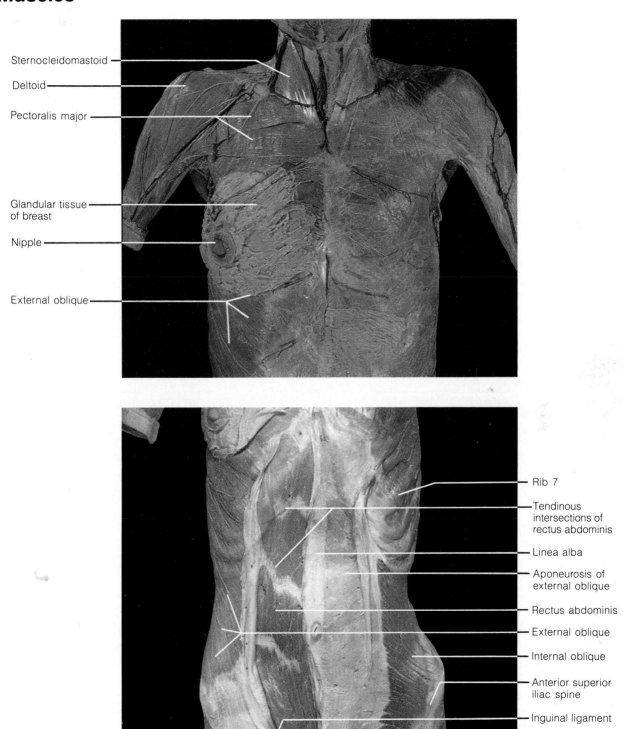

Sternocleidomastoid

Deltoid

Pectoralis major

Glandular tissue
of breast

Nipple

External oblique

Rib 7

Tendinous
intersections of
rectus abdominis

Linea alba

Aponeurosis of
external oblique

Rectus abdominis

External oblique

Internal oblique

Anterior superior
iliac spine

Inguinal ligament

Muscles

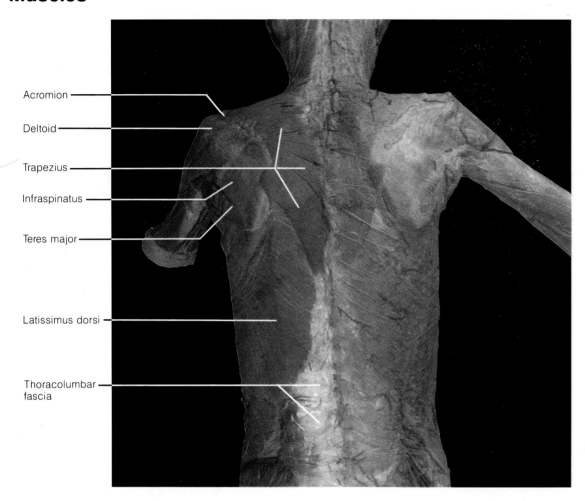

Acromion

Deltoid

Trapezius

Infraspinatus

Teres major

Latissimus dorsi

Thoracolumbar fascia

Cardiovascular

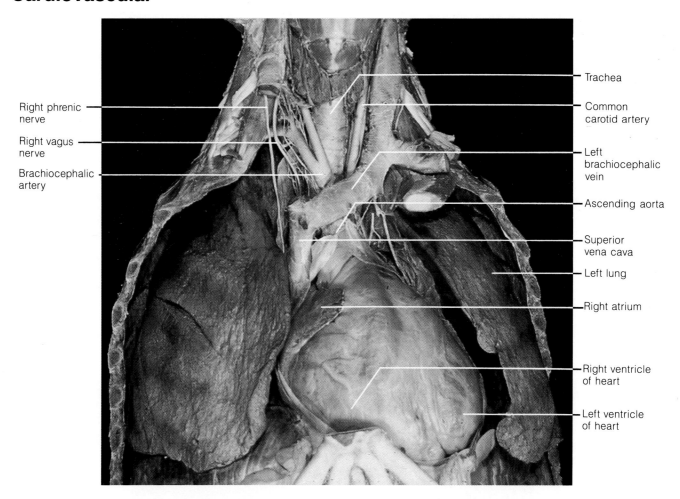

Right phrenic nerve

Right vagus nerve

Brachiocephalic artery

Trachea

Common carotid artery

Left brachiocephalic vein

Ascending aorta

Superior vena cava

Left lung

Right atrium

Right ventricle of heart

Left ventricle of heart

Respiratory

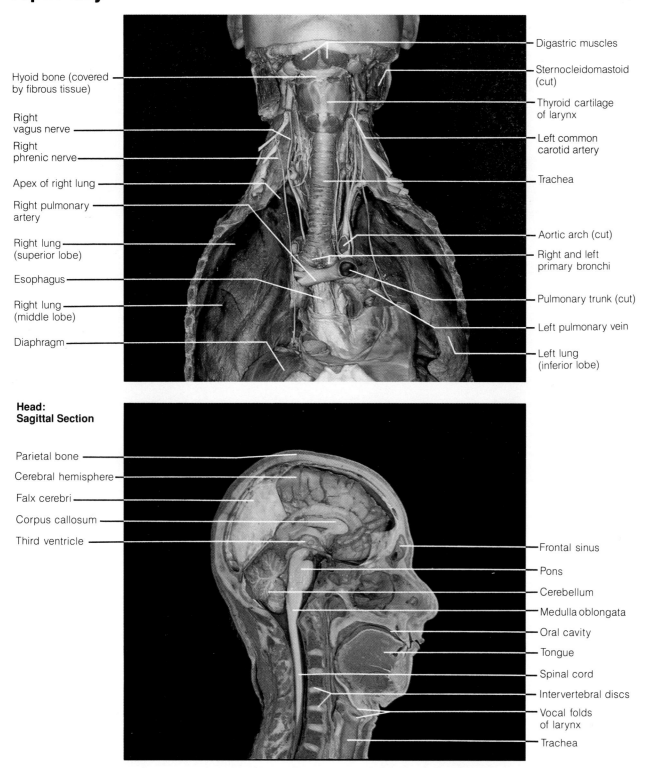

Digastric muscles

Hyoid bone (covered by fibrous tissue)

Sternocleidomastoid (cut)

Right vagus nerve

Thyroid cartilage of larynx

Right phrenic nerve

Left common carotid artery

Apex of right lung

Trachea

Right pulmonary artery

Right lung (superior lobe)

Aortic arch (cut)

Right and left primary bronchi

Esophagus

Right lung (middle lobe)

Pulmonary trunk (cut)

Left pulmonary vein

Diaphragm

Left lung (inferior lobe)

Head:
Sagittal Section

Parietal bone

Cerebral hemisphere

Falx cerebri

Corpus callosum

Third ventricle

Frontal sinus

Pons

Cerebellum

Medulla oblongata

Oral cavity

Tongue

Spinal cord

Intervertebral discs

Vocal folds of larynx

Trachea

Digestive

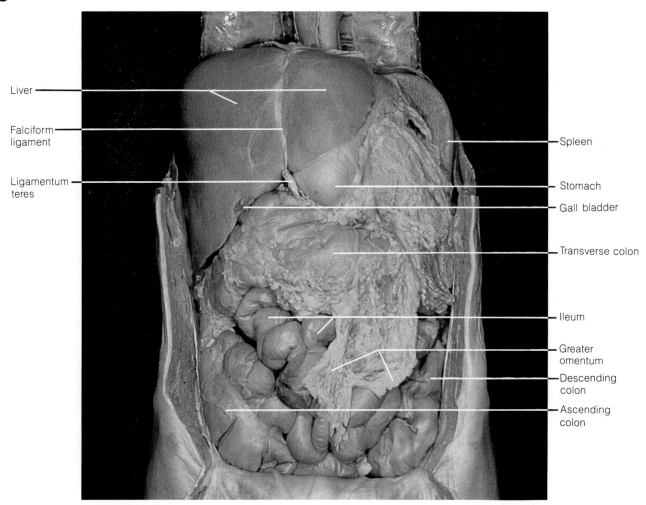

Liver

Falciform ligament

Ligamentum teres

Spleen

Stomach

Gall bladder

Transverse colon

Ileum

Greater omentum

Descending colon

Ascending colon

Urinary

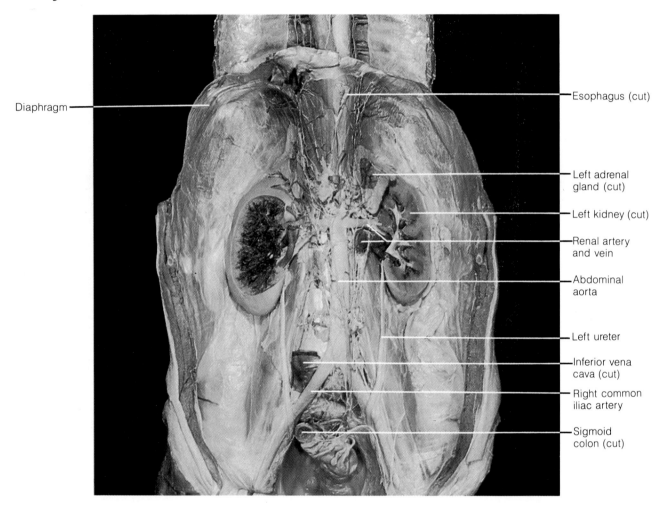

Diaphragm

Esophagus (cut)

Left adrenal
gland (cut)

Left kidney (cut)

Renal artery
and vein

Abdominal
aorta

Left ureter

Inferior vena
cava (cut)

Right common
iliac artery

Sigmoid
colon (cut)

Male Reproductive

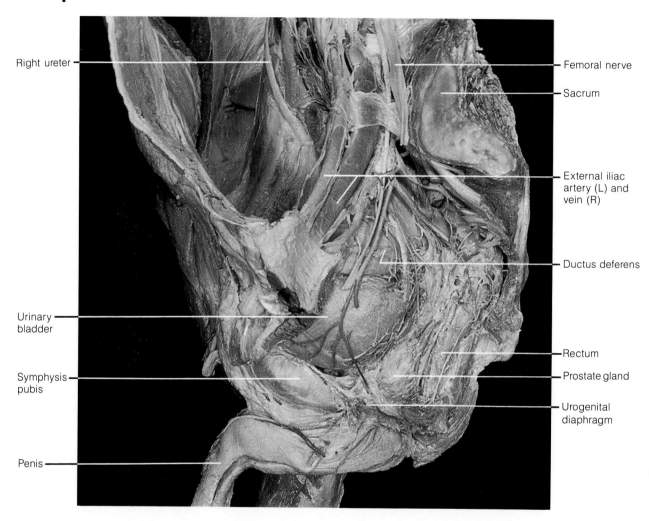

Right ureter

Femoral nerve

Sacrum

External iliac
artery (L) and
vein (R)

Ductus deferens

Urinary
bladder

Rectum

Symphysis
pubis

Prostate gland

Urogenital
diaphragm

Penis

Female Reproductive

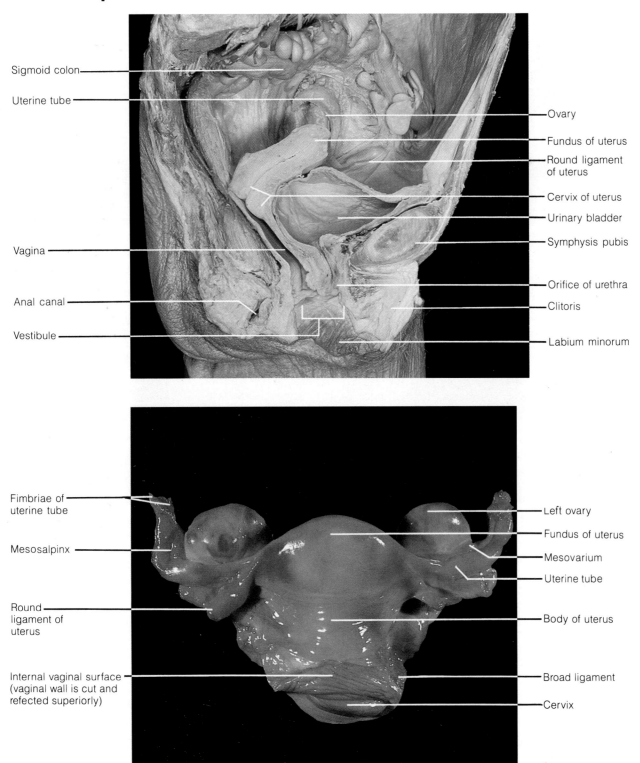

Sigmoid colon

Uterine tube

Vagina

Anal canal

Vestibule

Ovary

Fundus of uterus

Round ligament
of uterus

Cervix of uterus

Urinary bladder

Symphysis pubis

Orifice of urethra

Clitoris

Labium minorum

Fimbriae of
uterine tube

Mesosalpinx

Round
ligament of
uterus

Internal vaginal surface
(vaginal wall is cut and
refected superiorly)

Left ovary

Fundus of uterus

Mesovarium

Uterine tube

Body of uterus

Broad ligament

Cervix

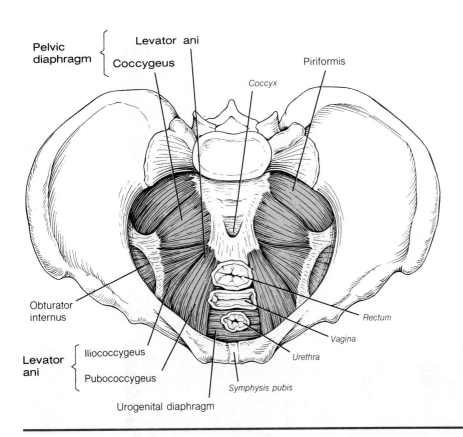

Pelvic diaphragm { Levator ani
                   Coccygeus

Piriformis

Coccyx

Obturator internus

Levator ani { Iliococcygeus
             Pubococcygeus

Urogenital diaphragm

Symphysis pubis

Urethra

Vagina

Rectum

Figure 7.22
Pelvic diaphragm of the female viewed from inside the pelvic cavity.

tubercle the canal opens into the abdominal cavity through an aperture in the fascia of the transversus abdominis muscle. This opening is called the *deep inguinal ring*. In the male, the spermatic cord passes through the inguinal canal; in the female, the canal provides for passage of the round ligament of the uterus. If the deep inguinal ring is weak, increased abdominal pressure can force some of the abdominal contents into the inguinal canal. This condition is called an inguinal *hernia*, or *rupture*. Inguinal hernias are more common in males because during embryonic development their inguinal canals are expanded and weakened as a result of the passage of the testes through them to reach the scrotal sacs.

The **rectus abdominis** is a narrow, flat muscle on the ventral aspect of the abdominal wall. Its fibers run vertically from the pubis to the rib cage on either side of the linea alba. Each rectus abdominis is completely ensheathed by the fasciae of the oblique and transversus abdominal muscles. The fasciae separate in various combinations to pass superficially and deep to the rectus abdominis muscle. Each rectus abdominis is crossed by three transverse fibrous bands called the **tendinous inscriptions.** In a person with rectus abdominis muscles developed through exercise, portions of the muscles between these inscriptions enlarge, causing the inscriptions to be visible through the skin as horizontal depressions.

The three oblique muscles and the rectus abdominis compress the abdominal cavity, assisting in forced expiration, defecation, and urination.

Most of the posterior portion of the abdominal wall is formed by the **quadratus lumborum** muscle (Figure 7.37). This is a broad quadrilateral muscle that runs from the posterior region of the iliac crest to the twelfth rib. The psoas major muscle also forms part of the posterior wall, but it acts primarily on the femur and so is described with that group of muscles (Table 7.17). A small psoas minor muscle, which acts on the lumbar vertebral column, is sometimes present ventral to the psoas major. **F7.37**

Table 7.17

Muscles That Form the Floor of the Abdominopelvic Cavity

The viscera of the abdominopelvic cavity are supported by a muscular floor called the **pelvic diaphragm** (Figure 7.22, Table 7.8). Two muscles, the **levator** **F7.22, Table 7.8**

Table 7.8 Muscles That Form the Floor of the Abdominopelvic Cavity [*F7.22*]

Muscle	Origin	Insertion	Action	Innervation
PELVIC DIAPHRAGM				
Levator ani	Inner surface of the superior ramus of the pubic bone, the lateral pelvic wall, and the spine of the ischium	Inner surface of the coccyx	Supports the pelvic viscera	Third through fifth sacral
Coccygeus	Spine of the ischium and the sacrospinous ligament	Sides of the coccyx and the sacrum	Supports the pelvic viscera	Fourth and fifth sacral

Table 7.9 Muscles of the Perineum [*F7.23*]

Muscle	Origin	Insertion	Action	Innervation
Ischiocavernosus	Tuberosity and rami of ischium	Crus of penis or clitoris	Retards return of blood through veins, thereby maintaining erection of penis or clitoris	Pudendal (second through fifth sacral)
Bulbospongiosus (Bulbocavernosus)	*Male:* from ventral median raphe on base of penis	Encircles base of penis and joins with fibers from opposite side on dorsum of penis	Empties urethral canal; assists in erection of penis	Pudendal
	Female: central tendinous point of perineum near anus	Pass on either side of vagina to insert on base of clitoris	Assists in erection of clitoris	Pudendal
Superficial transverse perineus	Tuberosity of ischium	Central tendinous point of perineum	Fixes (tightens) central tendinous point	Pudendal
Urogenital diaphragm *deep transverse perineus*	Inferior rami of ischium	Median raphe where it joins with corresponding muscle from opposite side ⎱ Both muscles of urogenital diaphragm act as constrictors of urethra		Pudendal
sphincter urethrae	Encircle membranous portion of urethra			

ani and the **coccygeus,** form the pelvic diaphragm. The levator ani is composed of several parts, of which the **pubococcygeus** muscle and the **iliococcygeus** muscle are the most prominent. There are openings through these muscles for the rectum, the urethra, and in the female, the vagina.

Muscles of the Perineum

F7.23
Table 7.9 The perineum is the lower end of the trunk between the thighs (Figure 7.23, Table 7.9). It is bounded anteriorly by the pubic arch, posteriorly by the coccyx, and laterally by the ischiopubic rami and the sacrotuberous ligaments, which run between the ischial tuberosities and the lateral margins of the sacrum and coccyx.

The perineal muscles are located inferior to the pelvic diaphragm. They consist of superficial muscles associated with the external genital organs and deeper muscles that form the **urogenital diaphragm.** The urogenital dia-

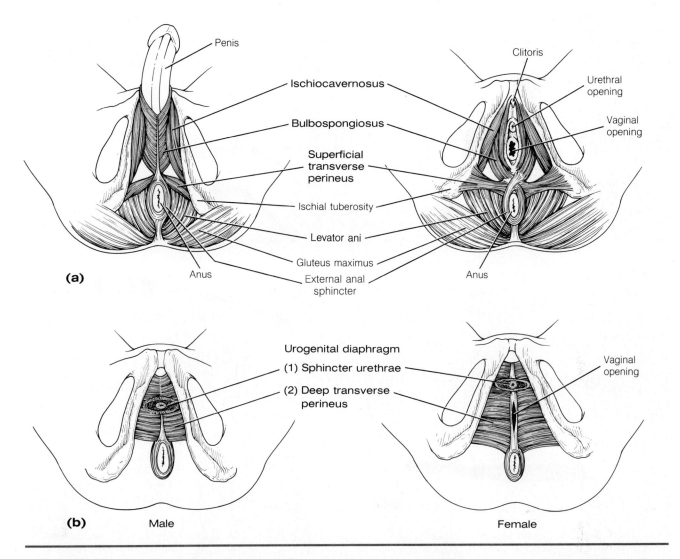

(a)

Penis

Ischiocavernosus

Bulbospongiosus

Superficial
transverse
perineus

Ischial tuberosity

Levator ani

Gluteus maximus

Anus

External anal
sphincter

Clitoris

Urethral
opening

Vaginal
opening

Anus

(b)   Male

Urogenital diaphragm
(1) Sphincter urethrae
(2) Deep transverse
perineus

Vaginal
opening

Female

Figure 7.23
Muscles of the perineum.
(a) Superficial muscles.
(b) Deep muscles forming
the urogenital
diaphragm.

phragm is located just below the anterior portion of the pelvic diaphragm, to which it adds support. The urogenital diaphragm is composed primarily of the **deep transverse perinei** and the **sphincter urethrae** muscles and fascial sheets. The superficial perineal muscles include the **ischiocavernosus, bulbospongiosus (bulbocavernosus),** and **superficial transverse perinei** muscles. After passing through the pelvic diaphragm, the intestine is surrounded by the **external anal sphincter** muscle.

MUSCLES OF THE UPPER LIMBS

Muscles That Act on the Scapula

Included in the muscles of the upper limbs are those muscles of the pectoral girdle that anchor the scapula (and, to a lesser extent, those that anchor the clavicle) to the axial skeleton (Table 7.10). Although it is possible for the scapula to move over the ribs, the main action of these muscles is to act as fixators of the scapula. When it is immobilized by these muscles, the scapula is able to serve as a stable point of origin for most of the muscles that move the arm.

Table 7.10

There are four posterior muscles that anchor the scapula (Figure 7.24): the **trapezius,** the **rhomboideus major** and **minor,** and the **levator scapulae.** The trapezius and rhomboids are named for their shapes, the levator for its action.

F7.24

Because of the shape of the *trapezius,* its fibers pull in various directions and its actions depend on which portion of the muscle contracts. If the upper

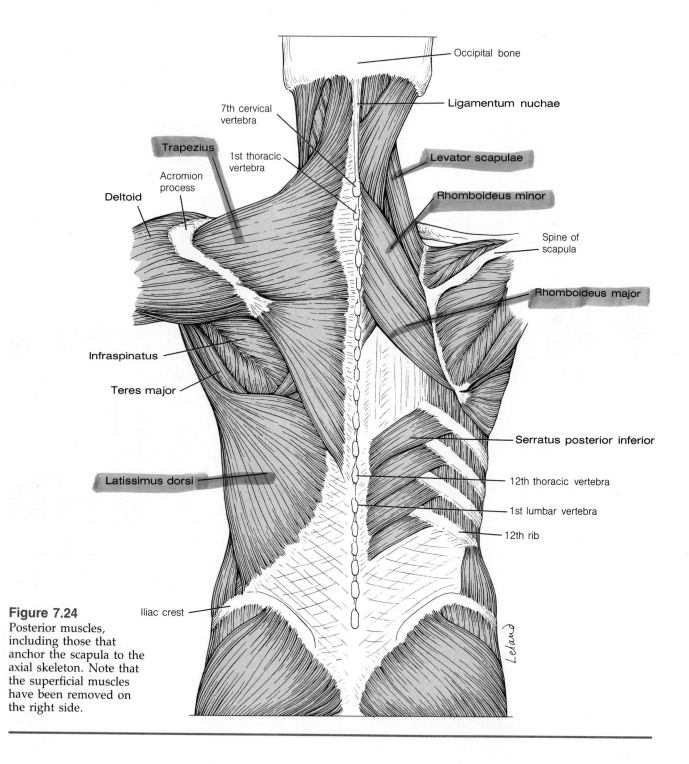

Occipital bone

Ligamentum nuchae

7th cervical vertebra

Trapezius

1st thoracic vertebra

Levator scapulae

Acromion process

Rhomboideus minor

Deltoid

Spine of scapula

Rhomboideus major

Infraspinatus

Teres major

Serratus posterior inferior

Latissimus dorsi

12th thoracic vertebra

1st lumbar vertebra

12th rib

Iliac crest

Figure 7.24
Posterior muscles, including those that anchor the scapula to the axial skeleton. Note that the superficial muscles have been removed on the right side.

portion contracts, the scapula is elevated, as in shrugging the shoulders. If the lower portion contracts, the scapula is depressed. If the entire muscle contracts, the scapula is pulled toward the vertebral column—that is, it is adducted. If the scapula is fixed, the trapezius muscle assists in moving the head posteriorly.

The *rhomboids* insert on the vertebral border of the scapula and pull the scapula medially, as well as downwardly rotating it. The *levator scapulae*, which runs from the upper cervical vertebra to the superior angle of the scapula, elevates the scapula, acting as a synergist to the upper portion of the trapezius muscle.

Table 7.10 Muscles That Act on the Scapula [F7.24, F7.25]

Muscle	Origin	Insertion	Action	Innervation
Trapezius	Occipital bone, the ligamentum nuchae, and the spinous processes of the seventh cervical and all of the thoracic vertebrae	Lateral third of the clavicle, the acromion process, and the spine of the scapula	Elevates (upper portion) or depresses (lower portion), rotates, adducts, and stabilizes the scapula	Accessory (cranial nerve XI)
Rhomboideus major	Spinous processes of the second through the fifth thoracic vertebrae	Vertebral border of the scapula, below the spine of the scapula	Adduct, stabilize, and rotate the scapula, lowering its lateral angle	Dorsal scapular (fifth cervical)
Rhomboideus minor	Spinous processes of the seventh cervical and first thoracic vertebrae	Vertebral border of the scapula, at the base of the spine of the scapula		
Levator scapulae	Transverse processes of the upper four cervical vertebrae	Vertebral border of the scapula, above the spine of the scapula	Elevates scapula and bends the neck laterally when the scapula is fixed	Dorsal scapular
Pectoralis minor	Anterior surface of the third through the fifth ribs	Coracoid process of the scapula	Depresses the scapula and pulls it anteriorly	Medial pectoral (eighth cervical and first thoracic)
Serratus anterior	Outer surface of the first nine ribs	Entire length of the ventral surface of the vertebral border of the scapula	Stabilizes, abducts, and rotates the scapula upward	Long thoracic (fifth through seventh cervical)
Subclavius	Outer surface of the first rib	Inferior surface of the lateral portion of the clavicle	Stabilizes and depresses the pectoral girdle	Fifth and sixth cervical

Two anterior muscles anchor the scapula to the thorax (Figure 7.25). The **pectoralis minor** pulls the scapula anteriorly and downward, lowering the lateral angle. In this manner, it acts antagonistically to the trapezius, rhomboid, and levator scapulae muscles. The **serratus anterior** muscle derives its name from its notched origin from the anterior surfaces of the ribs. From the origin it passes posteriorly between the dorsal surface of the ribs and the subscapular fossa of the scapula to insert on the vertebral border of the scapula. The serratus anterior pulls the scapula laterally; therefore, it acts antagonistically to the rhomboid muscles.

In addition to these muscles, the **subclavius** muscle anchors the pectoral girdle to the thoracic cage through its insertion onto the clavicle. For this reason it is included in Table 7.10 even though it does not act directly on the scapula.

Muscles That Act on the Arm

Nine muscles cross the shoulder joint and insert on the humerus (Table 7.11). Seven of these muscles arise from the scapula, indicating how important it is for the muscles discussed in Table 7.10 to fix the scapula. The remaining two muscles arise from the axial skeleton and have no attachments to the scapula. These are the *pectoralis major* and *latissimus dorsi* muscles. The actions of all these muscles are summarized in Table 7.12.

F7.25

Table 7.10

Table 7.11

Table 7.12

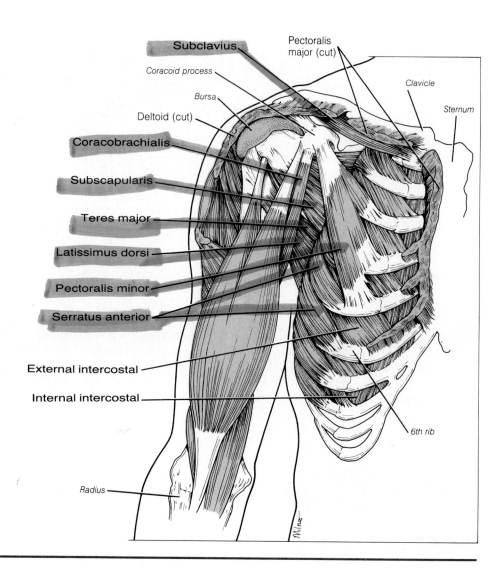

Subclavius

Pectoralis major (cut)

Coracoid process

Clavicle

Bursa

Sternum

Deltoid (cut)

Coracobrachialis

Subscapularis

Teres major

Latissimus dorsi

Pectoralis minor

Serratus anterior

External intercostal

Internal intercostal

6th rib

Radius

Figure 7.25
Deep anterior muscles,
including those that
anchor the scapula and
clavicle to the thorax.

F7.26   The **pectoralis major** (Figure 7.26) is a large, fan-shaped chest muscle that completely covers the smaller pectoralis minor muscle. The pectoralis major passes from the thoracic cage to the humerus, forming the anterior border of the axilla. It acts to adduct, to flex, and, because it passes anterior to the shoulder joint, to rotate the humerus medially.

F7.24   The **latissimus dorsi** (Figure 7.24) arises from the vertebrae of the lower back and from the pelvis. It twists upon itself as it passes between the humerus and the scapula to insert on the anterior surface of the humerus. In doing so, it forms the posterior border of the axilla. The latissimus dorsi extends the arm, pulling it downward or backward, thereby acting antagonistically to the pectoralis major muscle. At the same time, it acts synergistically to the pectoralis major in adducting and, because of the manner in which it wraps around the humerus to insert on the anterior surface, in rotating the arm medially.

F7.24,   The **deltoid** (Figure 7.24, Figure 7.26) is a large muscle that forms a cap
F7.26 over the shoulder. It arises anteriorly from the clavicle and posteriorly from the scapula. Some of its fibers therefore pass in front of the shoulder joint; some pass behind; and some pass directly over the lateral surface of the joint. Because of the different positions of their fibers, the anterior and posterior parts of the muscle have actions that are antagonistic to each other. If the entire muscle contracts, it abducts the arm (antagonistic to the pectoralis major and the latissimus dorsi). The anterior fibers flex and medially rotate the arm (synergistic to the pectoralis major). The posterior fibers extend and laterally rotate the arm (antagonistic to the action of the anterior fibers).

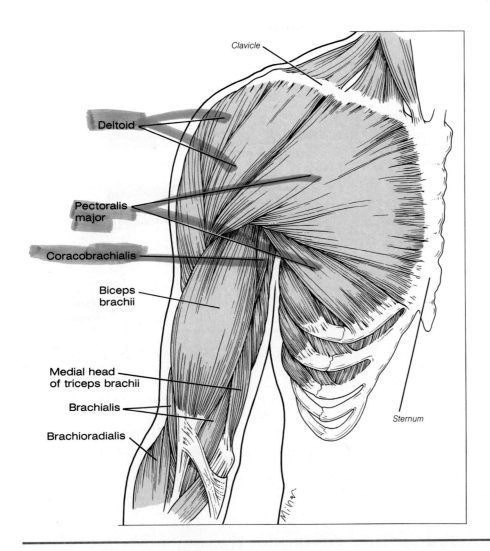

Clavicle

Deltoid

Pectoralis
major

Coracobrachialis

Biceps
brachii

Medial head
of triceps brachii

Brachialis

Brachioradialis

Sternum

Figure 7.26
Superficial muscles of the
chest, shoulder, and
anterior arm.

The origins of the **supraspinatus, infraspinatus,** and **subscapularis** muscles cover most of the ventral and dorsal surfaces of the scapula (Figure 7.25, Figure 7.27). They derive their names from the scapular fossae from which they arise. The *infraspinatus* passes posteriorly to the shoulder joint, acting as a lateral rotator of the arm. The *supraspinatus* crosses the upper part of the shoulder joint, allowing it to serve as an abductor of the arm. The *subscapularis,* whose origin is on the ventral surface of the scapula, separates the scapula from the serratus anterior muscle. It inserts on the anterior surface of the humerus; consequently, it acts as a medial rotator of the arm. In addition to their prime actions, all these deep muscles that cross the shoulder joint assist in strengthening and stabilizing the joint.

F7.25, F7.27

The **teres major** and **teres minor** muscles (Figure 7.27) both originate from the axillary border of the scapula. The teres major is the longer, its origin being from the inferior angle. The two muscles bracket the humerus, the teres major inserting on the anterior surface, the teres minor inserting on the posterior surface. Because of these insertions, the major is a medial rotator and the minor is a lateral rotator of the arm.

F7.27

The **coracobrachialis** (Figure 7.25, Figure 7.29) is the remaining muscle of scapular origin. As the name indicates, it arises from the coracoid process of the scapula and inserts onto the humerus. It is an anterior muscle that assists in flexion and adduction of the arm.

F7.25, F7.29

Two additional muscles, the *biceps brachii* and the *triceps brachii,* have origins on the scapula and pass into the arm. Their major actions are on the elbow joint and are discussed with that group of muscles (Table 7.13).

Table 7.13

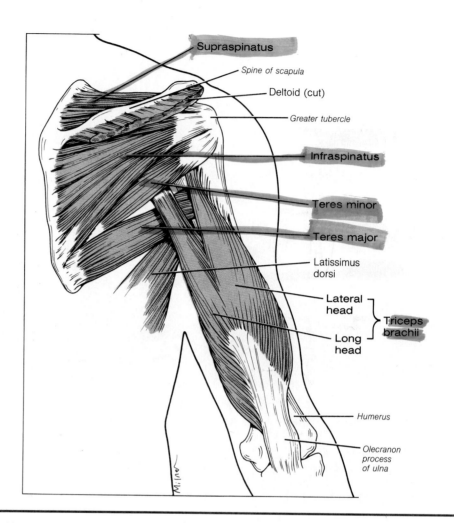

Supraspinatus

Spine of scapula

Deltoid (cut)

Greater tubercle

Infraspinatus

Teres minor

Teres major

Latissimus dorsi

Lateral head ⎤
          ⎬ Triceps brachii
Long head ⎦

Humerus

Olecranon process of ulna

Figure 7.27
Deep posterior muscles that attach the arm to the scapula. Note that the deltoid muscle has been removed.

Muscles That Act on the Forearm

The more powerful of the muscles that move the elbow and/or the proximal radial/ulnar joint (producing supination and pronation) are located in the arm **Table 7.13** (Table 7.13). They are assisted, however, by several muscles whose bellies lie in the forearm.

The two anterior muscles of the arm, the *biceps brachii* and the *brachialis*, flex the forearm. A third anterior arm muscle, the *coracobrachialis*, has its prime action on the shoulder joint and was described earlier with that group **Table 7.11** of muscles (Table 7.11).

F7.26, As the name indicates, the **biceps brachii** (Figure 7.26, Figure 7.28) has **F7.28** two heads, both of which have their origins on the scapula. The tendon of the long head passes over the top of the shoulder joint and travels through the intertubercular groove on the humerus. The two heads blend into the thick belly of the muscle. The main insertion of the biceps brachii is on the tuberosity of the radius. When the forearm is supinated, the biceps flexes it at the elbow. However, when the forearm is pronated with the tuberosity of the radius rotated toward the ulna, the biceps acts to supinate the forearm. In addition, it strengthens the shoulder joint and assists in flexion of the arm.

F7.26, The **brachialis** (Figure 7.26, Figure 7.29), which is deep to the biceps, is **F7.29** also a strong flexor of the forearm. Because it arises on the humerus and inserts on the ulna, it acts only on the elbow joint.

There are two muscles that act to extend the forearm. The **triceps brachii** **F7.27** (Figure 7.27) is the only muscle in the posterior compartment of the arm. One of its heads—the *long head*—arises from the scapula; the other two—the *lateral* and *medial heads*—arise from the humerus. All three heads insert by a

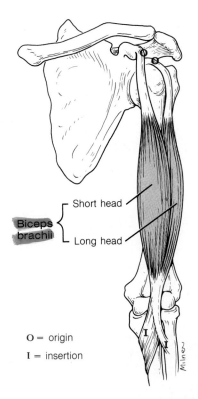

Biceps brachii
 ┌ Short head
 └ Long head

O = origin
I = insertion

Figure 7.28
The biceps brachii muscle.

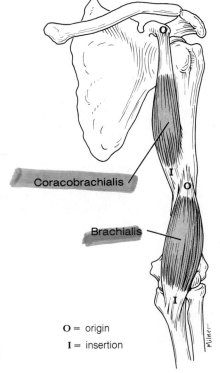

Coracobrachialis

Brachialis

O = origin
I = insertion

Figure 7.29
The brachialis and coracobrachialis muscles. The more superficial biceps brachii muscle has been removed.

common tendon onto the olecranon process of the ulna. Because the long head, which passes between the teres major and minor muscles, crosses the shoulder joint, it acts as a weak synergist to the latissimus dorsi muscle to extend and adduct the arm. Its main action, however, is extension of the forearm.

The second extensor of the forearm is the **anconeus** (Figure 7.33, Figure 7.34), a small muscle that appears to be a lateral continuation of the triceps brachii. It runs from the lateral epicondyle of the humerus to the lateral side of the olecranon process. Although it is considered to be a muscle of the forearm, the anconeus does not act upon the wrist; its only action is to assist the triceps in extending the forearm.

There are two additional muscles that are located in the forearm but act to flex the forearm (Figure 7.30). The **brachioradialis** muscle, which runs from the distal end of the humerus to the distal end of the radius, and the **pronator teres** muscle which has its origin on the distal end of the humerus and the proximal end of the ulna and inserts on the shaft of the radius. The pronator teres, along with the **pronator quadratus,** also acts to pronate the forearm.

F7.33, F7.34

F7.30

Muscles That Act on the Hand and Fingers

Most of the muscles that form the bulge of the proximal end of the forearm have at least a part of their origins on the distal end of the humerus. As a result, they cross the elbow joint as well as the wrist. However, their actions on the elbow joint are very slight. Their prime actions are on the hand and fingers (Table 7.14).

T7.14

Although this is a complex group of muscles with formidable-sounding names, most of the names indicate the muscle's action, origin, or insertion. These muscles can be divided into two groups on the basis of location and function. The muscles of the *anterior group* serve as flexors. Most of these anterior muscles have their origins on the medial epicondyle of the humerus and insert on the carpals, metacarpals, or phalanges. The *posterior group* of forearm muscles serve as extensors. Most of these muscles have their origins on the lateral epicondyle of the humerus and insert on the metacarpals or

Table 7.11 Muscles That Act on the Arm

Muscle	Origin	Insertion	Action	Innervation
ORIGIN ON AXIAL SKELETON				
Pectoralis major [F7.26]	Medial half of the clavicle, the sternum, the costal cartilages of the upper six ribs, and the aponeurosis of the external oblique muscle	Greater tubercle of the humerus	Flexes, adducts, and medially rotates the arm	Medial and lateral pectoral
Latissimus dorsi [F7.24]	Spinous processes of the lower six thoracic and the lumbar vertebrae, the sacrum, the posterior iliac crest—all via the lumbodorsal fascia	Medial margin of the intertubercular groove of the humerus	Extends, adducts, and medially rotates the arm	Thoracodorsal
ORIGIN ON SCAPULA				
Deltoid [F7.24, F7.26]	Lateral third of the clavicle, the acromion process, and the spine of the scapula	Deltoid tuberosity of the humerus	Abducts arm; anterior fibers flex and medially rotate the arm; posterior fibers extend and laterally rotate the arm	Axillary
Supraspinatus [F7.27]	Supraspinous fossa of the scapula	Greater tubercle of the humerus	Abducts the arm; slight lateral rotation	Suprascapular
Infraspinatus [F7.27]	Infraspinous fossa of the scapula	Greater tubercle of the humerus (posterior to the supraspinatus)	Rotates the arm laterally; slight adduction	Suprascapular
Subscapularis [F7.25]	Subscapular fossa of the scapula	Lesser tubercle of the humerus	Rotates the arm medially	Subscapular
Teres major [F7.27]	Dorsal surface of the inferior angle of the scapula	Lesser tubercle of the humerus	Adducts, extends, and medially rotates the arm	Subscapular
Teres minor [F7.27]	Axillary border of the scapula	Greater tubercle of the humerus (posterior to the infraspinatus)	Rotates the arm laterally; weakly adducts and extends the arm	Axillary
Coracobrachialis [F7.29]	Coracoid process of the scapula	Middle of the humerus, medial surface	Flexes and adducts the arm	Musculocutaneous

Table 7.12 Summary of Muscle Actions on the Arm

Flexion	Extension	Adduction	Abduction	Medial rotation	Lateral rotation
Deltoid	Deltoid	Pectoralis major	Deltoid	Pectoralis major	Deltoid
Pectoralis major	Latissimus dorsi	Latissimus dorsi	Supraspinatus	Latissimus dorsi	Supraspinatus
Coracobrachialis	Teres major	Teres major Coracobrachialis Infraspinatus		Deltoid Subscapularis Teres major	Infraspinatus Teres minor

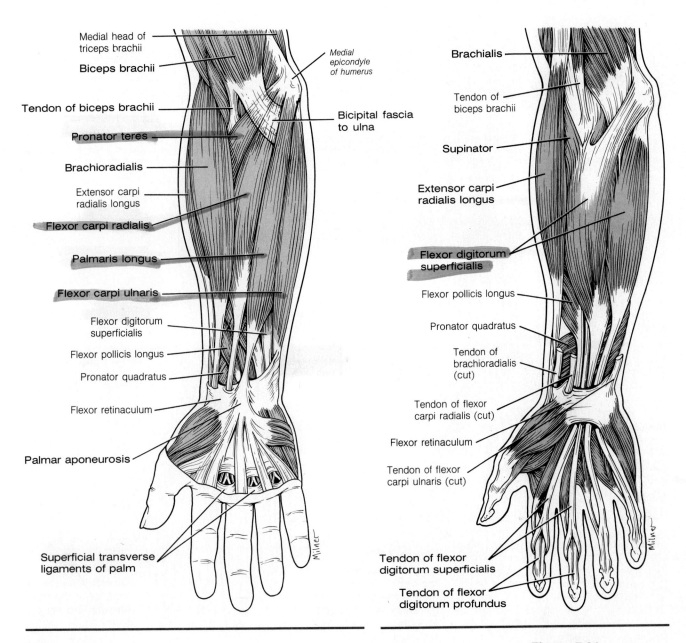

Figure 7.30
Superficial anterior muscles of the right forearm and hand.

Labels (left figure):
Medial head of triceps brachii
Biceps brachii
Tendon of biceps brachii
Pronator teres
Brachioradialis
Extensor carpi radialis longus
Flexor carpi radialis
Palmaris longus
Flexor carpi ulnaris
Flexor digitorum superficialis
Flexor pollicis longus
Pronator quadratus
Flexor retinaculum
Palmar aponeurosis
Superficial transverse ligaments of palm
Medial epicondyle of humerus
Bicipital fascia to ulna

Figure 7.31
Second layer of anterior muscles of the right forearm. The superficial muscles have been removed.

Labels (right figure):
Brachialis
Tendon of biceps brachii
Supinator
Extensor carpi radialis longus
Flexor digitorum superficialis
Flexor pollicis longus
Pronator quadratus
Tendon of brachioradialis (cut)
Tendon of flexor carpi radialis (cut)
Flexor retinaculum
Tendon of flexor carpi ulnaris (cut)
Tendon of flexor digitorum superficialis
Tendon of flexor digitorum profundus

phalanges. Both the anterior and posterior groups can be divided further into superficial and deep muscles. The tendons of these forearm muscles are held down at the wrist by heavy thickenings of the fascia called **flexor** and **extensor retinacula** (*retinaculum* = halter). If these transverse bands of fascia were not present, the tendons would protrude when the hand is flexed or extended.

Although the muscles that act on the hand and fingers are not described here individually, they are illustrated in Figure 7.30 through Figure 7.34 and their precise locations and actions are listed in Table 7.14. In addition, Table 7.15 summarizes their actions and makes it possible to easily identify synergists and antagonists. Notice that while the three muscles that extend the wrist are antagonistic to the flexors of the forearm, the **extensor carpi radialis longus** acts synergistically with the **flexor carpi radialis** to abduct the hand. In a similar manner, the **extensor carpi ulnaris** acts synergistically with the **flexor carpi ulnaris** to adduct the hand.

F7.30–F7.34
Table 7.14
Table 7.15

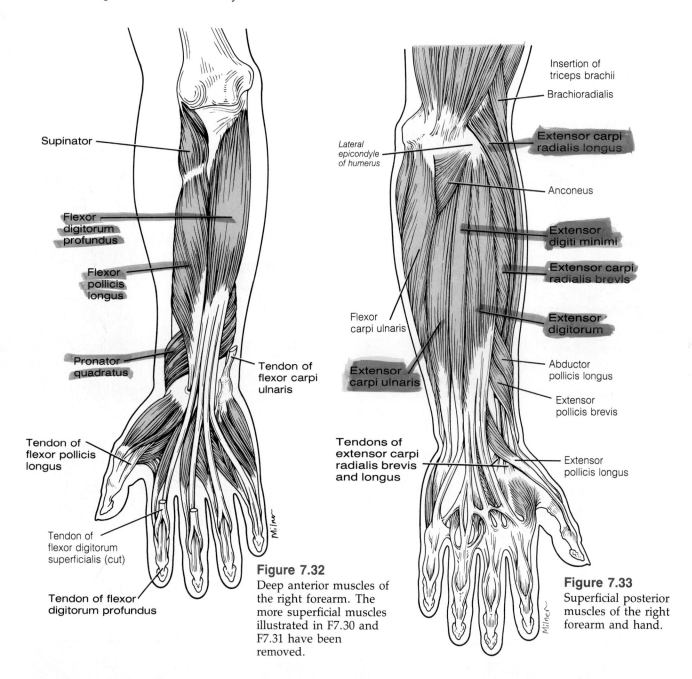

Supinator

Flexor digitorum profundus

Flexor pollicis longus

Pronator quadratus

Tendon of flexor pollicis longus

Tendon of flexor digitorum superficialis (cut)

Tendon of flexor digitorum profundus

Tendon of flexor carpi ulnaris

Figure 7.32
Deep anterior muscles of the right forearm. The more superficial muscles illustrated in F7.30 and F7.31 have been removed.

Insertion of triceps brachii

Brachioradialis

Extensor carpi radialis longus

Lateral epicondyle of humerus

Anconeus

Extensor digiti minimi

Extensor carpi radialis brevis

Extensor digitorum

Flexor carpi ulnaris

Extensor carpi ulnaris

Abductor pollicis longus

Extensor pollicis brevis

Tendons of extensor carpi radialis brevis and longus

Extensor pollicis longus

Figure 7.33
Superficial posterior muscles of the right forearm and hand.

Table 7.13   Muscles That Act on the Forearm

Muscle	Origin	Insertion	Action	Innervation
Biceps brachii [F7.28]	*Long head:* supraglenoid tubercle of the scapula *Short head:* coracoid process of the scapula	Tuberosity of the radius	Flexes the forearm and the arm; supinates the forearm	Musculocutaneous
Brachialis [F7.29]	Anterior surface of the distal half of the humerus	Coronoid process of the ulna	Flexes the forearm	Musculocutaneous
Triceps brachii [F7.27]	*Long head:* infraglenoid tubercle of the scapula	Olecranon process of the ulna	Extends the forearm; long head also extends the arm	Radial

Table 7.13   Muscles That Act on the Forearm (continued)

Muscle	Origin	Insertion	Action	Innervation
	Lateral head: posterior surface of the humerus above the radial groove			
	Medial head: posterior surface of the humerus below the radial groove			
MUSCLES OF FOREARM				
Anconeus [F7.33; F7.34]	Lateral epicondyle of the humerus	Lateral surface of the olecranon process of the ulna	Extends the forearm	Radial
Brachioradialis [F7.30]	Lateral supracondylar ridge of the humerus	Styloid process of the radius	Flexes the forearm	Radial
Pronator teres [F7.30]	Medial epicondyle of the humerus and the coronoid process of the ulna	Middle of the lateral surface of the shaft of the radius	Pronates and weakly flexes the forearm	Median
Pronator quadratus [F7.30, F7.31, F7.32]	Distal ventral surface of the ulna	Distal ventral surface of the radius	Pronates the forearm	Median
Supinator [F7.31, F7.32, F7.34]	Lateral epicondyle of the humerus	Proximal end of the lateral surface of the shaft of the radius	Supinates the forearm	Radial

Table 7.14   Muscles That Act on the Hand and Fingers

Muscle	Origin	Insertion	Action	Innervation
ANTERIOR GROUP *SUPERFICIAL MUSCLES (LISTED FROM LATERAL TO MEDIAL)*				
Flexor carpi radialis [F7.30]	Medial epicondyle of the humerus	Ventral surface of the second and third metacarpals	Flexes and abducts the hand; aids in flexion and pronation of the forearm	Median
Palmaris longus [F7.30]	Medial epicondyle of the humerus	Palmar aponeurosis	Flexes the hand	Median
Flexor carpi ulnaris [F7.30, F7.33]	Medial epicondyle of the humerus, olecranon process, and the proximal two-thirds of the posterior surface of the ulna	Pisiform, hamate, and fifth metacarpal	Flexes and adducts the hand	Ulnar
Flexor digitorum superficialis (beneath the other superficial muscles) [F7.30, F7.31]	Medial epicondyle of the humerus, coronoid process of the ulna, and the anterior surface of the radius	Ventral surface of the middle phalanges of the second through the fifth fingers	Flexes the phalanges and the hand	Median

Table 7.14   Muscles That Act on the Hand and Fingers (continued)

Muscle	Origin	Insertion	Action	Innervation
DEEP MUSCLES				
Flexor digitorum profundus (*profundus* = deep) [F7.32]	Upper one-half of anterior and medial surfaces of the ulna, the coronoid process, and the interosseous membrane	Ventral surface base of the distal phalanges of the second through the fifth fingers	Flexes the phalanges and the hand	Median and ulnar
Flexor pollicis longus (*pollex* = thumb) [F7.30, F7.31, F7.32]	Ventral surface of the radius and the interosseus membrane	Ventral surface base of the distal phalanx of the thumb	Flexes the thumb; aids in flexing the hand	Median
POSTERIOR GROUP SUPERFICIAL* MUSCLES (LISTED FROM LATERAL TO MEDIAL)				
Extensor carpi radialis longus [F7.30, F7.31, F7.33]	Lateral supracondylar ridge of the humerus	Dorsal surface of the base of the second metacarpal	Extends and abducts the hand	Radial
Extensor carpi radialis brevis (*brevis* = short) [F7.33]	Lateral epicondyle of the humerus	Dorsal surface of the base of the third metacarpal	Extends the hand	Radial
Extensor digitorum [F7.33]	Lateral epicondyle of the humerus	Dorsal surface of the phalanges of the second through the fifth fingers	Extends the fingers and the hand	Radial
Extensor digiti minimi (little finger) [F7.33]	Tendon of the extensor digitorum	Tendon of the extensor digitorum on the dorsum of the little finger	Extends the little finger	Radial
Extensor carpi ulnaris [F7.33]	Lateral epicondyle of the humerus	Base of the fifth metacarpal	Extends and adducts the hand	Radial
POSTERIOR GROUP DEEP MUSCLES (LISTED LATERAL TO MEDIAL)				
Abductor pollicis longus [F7.33, F7.34]	Posterior surface of the middle of the radius and ulna and the interosseus membrane	Base of the first metacarpal	Extends the thumb and abducts the hand	Radial
Extensor pollicis brevis [F7.33, F7.34]	Posterior surface of the middle of the radius, and the interosseus membrane	Base of the first phalanx of the thumb	Extends the thumb and abducts the hand	Radial
Extensor pollicis longus [F7.33, F7.34]	Posterior surface of the middle of the ulna, and the interosseus membrane	Base of the last phalanx of the thumb	Extends the thumb and abducts the hand	Radial
Extensor indicis [F7.34]	Posterior surface of the distal end of the ulna, and the interosseus membrane	Tendon of the extensor digitorum to the index finger	Extends the index finger	Radial

*The brachioradialis is located with the posterior superficial muscles but is described in Table 7.13 because it is a flexor of the forearm.

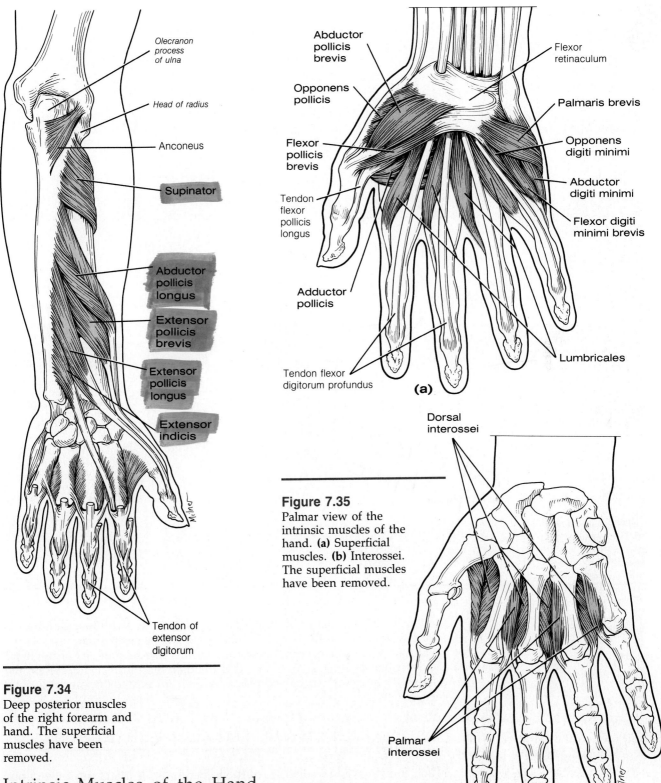

Olecranon process of ulna

Head of radius

Anconeus

Supinator

Abductor pollicis longus

Extensor pollicis brevis

Extensor pollicis longus

Extensor indicis

Tendon of extensor digitorum

Abductor pollicis brevis

Opponens pollicis

Flexor pollicis brevis

Tendon flexor pollicis longus

Adductor pollicis

Tendon flexor digitorum profundus

Flexor retinaculum

Palmaris brevis

Opponens digiti minimi

Abductor digiti minimi

Flexor digiti minimi brevis

Lumbricales

(a)

Dorsal interossei

Palmar interossei

(b)

Figure 7.34
Deep posterior muscles of the right forearm and hand. The superficial muscles have been removed.

Figure 7.35
Palmar view of the intrinsic muscles of the hand. **(a)** Superficial muscles. **(b)** Interossei. The superficial muscles have been removed.

Intrinsic Muscles of the Hand

We have seen that several muscles of the forearm have long tendons that reach the phalanges and serve to move the fingers. Apart from these forearm muscles, there are several groups of small muscles whose origin and insertion are both in the hand. As we learned earlier, such muscles are called *intrinsic* muscles. The intrinsic muscles make possible the fine and precise movements that are typical of the fingers.

The intrinsic muscles of the hand (Figure 7.35) are divided into three groups. Those that act on the thumb form the **thenar eminence** at the base of

F7.35

Table 7.15   Summary of Muscle Actions on the Hand

Flexion	Extension	Adduction	Abduction
Flexor carpi radialis	Extensor carpi radialis longus	Flexor carpi ulnaris	Flexor carpi radialis
Palmaris longus		Extensor carpi ulnaris	Extensor carpi radialis longus
Flexor carpi ulnaris	Extensor carpi radialis brevis		Abductor pollicis longus
Flexor digitorum superficialis	Extensor carpi ulnaris		Extensor pollicis brevis
Flexor digitorum profundus	Extensor digitorum		Extensor pollicis longus
Flexor pollicis longus			

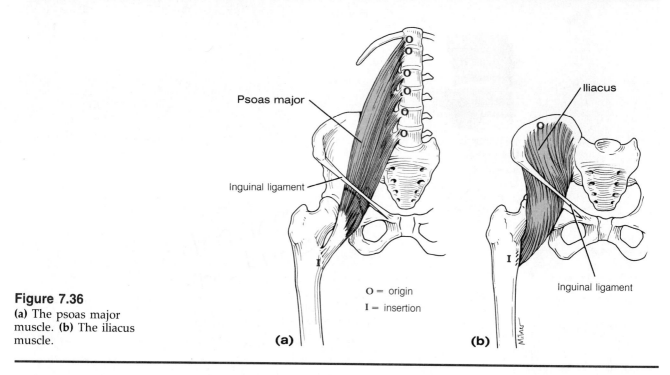

Figure 7.36
(a) The psoas major muscle. **(b)** The iliacus muscle.

the thumb. Those that act on the little finger form the **hypothenar eminence** on the medial side of the hand. The intermediate, or **midpalmar,** muscles act on all the phalanges except the thumb. These intrinsic muscles of the hand are described in Table 7.16. Notice that there are no intrinsic muscles on the dorsum of the hand, since the dorsal interossei are located between the metacarpal bones.

Table 7.16

MUSCLES OF THE LOWER LIMBS

When compared to the muscles of the upper limbs, those of the lower limbs tend to be bulkier and more powerful. The versatile movements characteristic of the upper limbs are somewhat sacrificed in the lower limbs in favor of strength, stability, and locomotion. Many of the muscles of the lower limbs are used in maintaining upright posture and therefore must constantly resist the pull of gravity. Unlike the pectoral girdle, the pelvis does not rely entirely on muscles to stabilize or fix it. The only movement possible in the bony pelvis is a slight gliding between the sacrum and ilium. Many muscles of the lower limbs cross two joints—either the hip and the knee or the knee and the

Table 7.16 Intrinsic Muscles of the Hand

Muscle	Origin	Insertion	Action	Innervation
THENAR MUSCLES				
Abductor pollicis brevis [F7.35a]	Flexor retinaculum, scaphoid, and trapezium	Proximal phalanx of the thumb	Abducts the thumb	Median
Opponents pollicis (opponens = one that opposes) [F7.35a]	Flexor retinaculum and trapezium	Lateral border of the metacarpal of the thumb	Pulls the thumb in front of the palm to meet the little finger	Median
Flexor pollicis brevis [F7.35a]	Flexor retinaculum, trapezium, and first metacarpal	Base of the proximal phalanx of the thumb	Flexes and adducts the thumb	Median and ulnar
Adductor pollicis [F7.35a]	Capitate, and second and third metacarpals	Proximal phalanx of the thumb	Adducts the thumb	Ulnar
HYPOTHENAR MUSCLES				
Palmaris brevis [F7.35a]	Flexor retinaculum	Skin on the ulnar border of the hand	Pulls the skin toward the middle of the palm	Ulnar
Abductor digiti minimi [F7.35a]	Pisiform, and the tendon of the flexor carpi ulnaris	Base of the proximal phalanx of the little finger	Abducts the little finger	Ulnar
Flexor digiti minimi brevis [F7.35a]	Flexor retinaculum and hamate	Base of the proximal phalanx of the little finger	Flexes the little finger	Ulnar
Opponens digiti minimi [F7.35a]	Flexor retinaculum and hamate	Metacarpal of the little finger	Brings the little finger out to meet the thumb	Ulnar
MIDPALMAR MUSCLES				
Lumbricales [F7.35a]	Tendons of the flexor digitorum profundus	Tendons of the extensor digitorum	Flex the proximal phalanx and extend the middle and distal phalanges of the second through fifth fingers	Median and ulnar
Dorsal interossei (4) [F7.35b]	Adjacent sides of all of the metacarpals	Proximal phalanx of second, third, and fourth fingers	Abduct the fingers from the middle finger; flex the proximal phalanx	Ulnar
Palmar interossei (3) [F7.35b]	Medial side of the second metacarpal, and lateral side of the fourth and fifth metacarpals	Proximal phalanx of the same finger	Adduct the fingers toward the middle finger; flex the proximal phalanx	Ulnar

ankle—and act equally strongly on both joints. These double actions are listed in Tables 7.17, 7.18, and 7.19.

Tables 7.17–7.19

Muscles That Act on the Thigh

Most of the muscles that act on the femur arise from the pelvis. One muscle, the **psoas major** (Figure 7.36a, Figure 7.37), arises from the lumbar vertebrae. Because the psoas major muscle has a common insertion with, and acts synergistically with, the **iliacus** muscle (Figure 7.36b, Figure 7.37), the two muscles

F7.36a,
F7.37
F7.36b,
F7.37

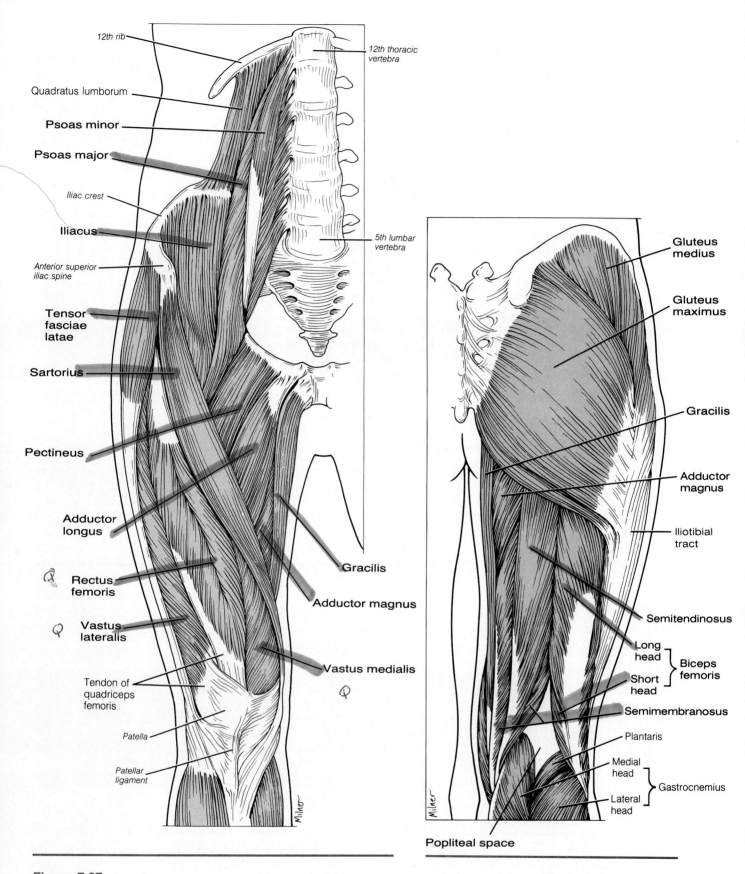

12th rib

12th thoracic vertebra

Quadratus lumborum

Psoas minor

Psoas major

Iliac crest

5th lumbar vertebra

Iliacus

Anterior superior iliac spine

Tensor fasciae latae

Sartorius

Pectineus

Adductor longus

Gracilis

Adductor magnus

Rectus femoris

Vastus lateralis

Vastus medialis

Tendon of quadriceps femoris

Patella

Patellar ligament

Gluteus medius

Gluteus maximus

Gracilis

Adductor magnus

Iliotibial tract

Semitendinosus

Long head

Short head

Biceps femoris

Semimembranosus

Plantaris

Medial head

Lateral head

Gastrocnemius

Popliteal space

Figure 7.37
Anterior view of the muscles that attach the femur to the pelvis and the lumbar vertebrae.

Figure 7.38
Superficial muscles of the posterior hip and thigh.

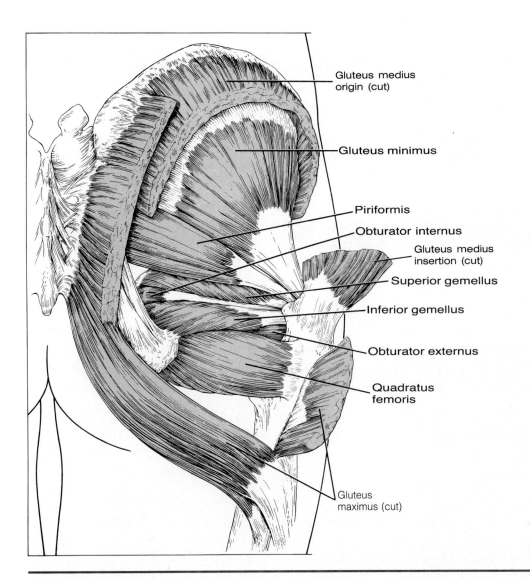

Gluteus medius
origin (cut)

Gluteus minimus

Piriformis

Obturator internus

Gluteus medius
insertion (cut)

Superior gemellus

Inferior gemellus

Obturator externus

Quadratus
femoris

Gluteus
maximus (cut)

Figure 7.39
Deep muscles of the
posterior hip. The
gluteus maximus and
gluteus medius have
been cut to expose the
deep muscles.

are often referred to as the **iliopsoas** muscle. The tendon of the iliopsoas passes beneath the inguinal ligament to reach the femur, which it flexes. When the lower limbs are fixed, the psoas muscles flex the vertebral column, as when bending over.

Three large gluteal muscles (Figure 7.38, Figure 7.39) give shape to the buttocks and serve as powerful mobilizers of the hip joint. The largest and most superficial is the **gluteus maximus.** The gluteus maximus covers the posterior third of the smaller **gluteus medius;** deep to the gluteus medius is the still smaller **gluteus minimus.** The broad tendon of the gluteus maximus passes behind the hip joint, causing it to extend and laterally rotate the femur. In contrast, the tendon of the gluteus medius passes above the hip joint and that of the minimus passes in front of it, causing these muscles to abduct and medially rotate the femur. Therefore, in rotation of the thigh, the gluteus maximus is an antagonist to the two smaller gluteal muscles. Acting synergistically with the gluteus maximus in rotating the thigh laterally are six deep muscles that extend from the sacrum and the coxal bone to the posterior surface of the proximal end of the femur. These lateral rotators of the hip are included in Tables 7.17 and 7.19.

The **tensor fasciae latae** (Figure 7.37) is a lateral hip muscle that inserts on a strong band of connective tissue called the **iliotibial tract** of the **fascia lata** (broad fascia). The fascia lata invests all of the muscles of the thigh, but it is

**F7.38,
F7.39**

**Tables 7.17,
7.19**

F7.37

Table 7.17 Muscles That Act on the Thigh

Muscle	Origin	Insertion	Action	Innervation
Iliopsoas				
Psoas major [F7.36, F7.37]	Transverse processes and bodies of the last thoracic and all of the lumbar vertebrae	Lesser trochanter of the femur	Flex the thigh; flex the trunk on the femur	Femoral and first lumbar
Iliacus [F7.36, F7.37]	Iliac crest and fossa			
Gluteus maximus [F7.38]	Posterior gluteal line of the ilium, and the posterior surface of the sacrum and the coccyx	Gluteal tuberosity of the femur; iliotibial tract	Extends and laterally rotates the thigh	Inferior gluteal
Gluteus medius [F7.38, F7.39]	Outer surface of the ilium, between the posterior and the anterior gluteal lines	Lateral surface of the greater trochanter of the femur	Abducts and medially rotates the thigh	Superior gluteal
Gluteus minimus [F7.39]	Outer surface of the ilium, between the anterior and the inferior gluteal lines	Anterior surface of the greater trochanter of the femur	Abducts and medially rotates the thigh	Superior gluteal
Tensor fasciae latae [F7.39]	Anterior portion of the iliac crest, and the anterior superior iliac spine	Iliotibial tract of the fascia latae	Tenses the fascia lata; assists in flexion, abduction, and medial rotation of the thigh	Superior gluteal
Piriformis [F7.39]	Anterior surface of the sacrum	Superior border of the greater trochanter of the femur	Rotates the thigh laterally; assists in extending and abducting the thigh	Second sacral
Obturator internus [F7.39]	Inner surface of the obturator membrane, and the bony margins of obturator foramen	Greater trochanter of the femur	Rotates the thigh laterally	Fifth lumbar, and first and second sacral
Obturator externus [F7.39, F7.42]	Outer surface of the obturator membrane, and the bony margins of the obturator foramen	Trochanteric fossa of the femur	Rotates the thigh laterally	Obturator
Superior gemellus [F7.39]	Ischial spine	Greater trochanter of the femur	Rotates the thigh laterally	Fifth lumbar, and first and second sacral
Inferior gemellus [F7.39]	Ischial tuberosity	Greater trochanter of the femur	Rotates the thigh laterally	Fourth and fifth lumbar, and first sacral
Quadratus femoris [F7.39]	Ischial tuberosity	Shaft of the femur just below the greater trochanter	Rotates the thigh laterally	Fourth and fifth lumbar
Adductor **magnus*** [F7.40]	**Inferior ramus of the pubis and the ischium, and the ischial tuberosity**	Most of the length of the linea aspera, and the adductor tubercle of the femur	Adducts and laterally rotates the thigh; assists in extending the thigh	Obturator and sciatic

Table 7.17   Muscles That Act on the Thigh (continued)

Muscle	Origin	Insertion	Action	Innervation
Adductor longus [F7.40]	Crest and the symphysis of the pubis	Middle third of the linea aspera of the femur	Adducts and laterally rotates the thigh; assists in flexion of the thigh	Obturator
Adductor brevis [F7.40]	Inferior ramus of the pubis	Upper part of the linea aspera of the femur	Adducts and laterally rotates the thigh	Obturator
Pectineus [F7.37]	Superior ramus of the pubis	Posterior surface of the femur just below the lesser trochanter	Adducts, flexes, and laterally rotates the thigh	Obturator and femoral

*The three adductor muscles and the pectineus are in the medial compartment [Figure 7.41] of the thigh musculature, but they are included in this table because they act on the femur.

especially thick laterally—thus forming the iliotibial tract. The tensor serves primarily to stabilize the knee by tightening the fascia lata, but it also assists in flexing the thigh.

Three other large muscles that act on the femur are the medial muscles of the thigh. All of these muscles originate from the pubis and pull the femur toward the midline. They are therefore named *adductors* (Figure 7.40). Because they are inserted onto the posterior surface of the femur, the adductor muscles also rotate the femur laterally, acting synergistically with the gluteus maximus and the six small, deep muscles mentioned above. The *pectineus* (Figure 7.37), which is included with the adductors in the medial group of thigh muscles, acts synergistically with the adductor muscles. **F7.40** **F7.37**

Muscles That Act on the Leg

The muscles that act on the leg are located in the thigh, with their tendons crossing the knee joint. The thigh muscles are grouped by connective tissue sheets into *anterior, posterior,* and *medial compartments* (Figure 7.41, Table 7.18). **F7.41, Table 7.18**

The *medial compartment* (Figure 7.40) is also referred to as the *adductor compartment* because the muscles in this compartment function to adduct the femur. The **adductors magnus, longus,** and **brevis** and the **pectineus** muscles (Figure 7.37) are included in Table 7.17 because they act entirely on the femur. The one remaining muscle of the medial compartment of the thigh is the **gracilis** (Figure 7.42). In addition to adducting the thigh, the gracilis assists in flexing the leg and is therefore listed in Table 7.18. **F7.40** **F7.37, Table 7.17** **F7.42** **Table 7.18**

The *anterior compartment* of the thigh is called the *extensor compartment* because the muscles that it contains function primarily to extend the leg. There are five muscles in the anterior compartment. Four of them, although individually named—**rectus femoris, vastus intermedius, vastus lateralis,** and **vastus medialis**—are often grouped as the **quadriceps femoris** muscle (Figure 7.43). The four heads of the quadriceps muscle have a common insertion on the patella, or kneecap, which is a large sesamoid bone that lies within the tendon of the quadriceps as it crosses in front of the knee joint. A strong band of connective tissue called the **patellar ligament** (ligamentum patellae) extends from the patella to the tibial tuberosity. Therefore, functionally the four heads of the quadriceps insert on the tibial tuberosity. The **sartorius** (Figure 7.44) is the remaining muscle of the anterior compartment. It is a long, narrow muscle that forms a band diagonally across the thigh, from the ilium to the medial side of the tibia. It crosses both the hip and the knee, producing flexion at both joints. It is unusual for an anterior thigh muscle to flex the leg; it does so because it passes posterior to the medial condyle of the femur. **F7.43** **F7.44**

The muscles in the *posterior,* or *flexor, compartment* of the thigh function to flex the leg. They also extend the thigh. The three muscles in this compartment (Figure 7.38)—**biceps femoris, semimembranosus,** and **semitendinosus**—are known as the **hamstring muscles.** They have a common origin from the ischial tuberosity and insert on the tibia and fibula. Consequently, **F7.38**

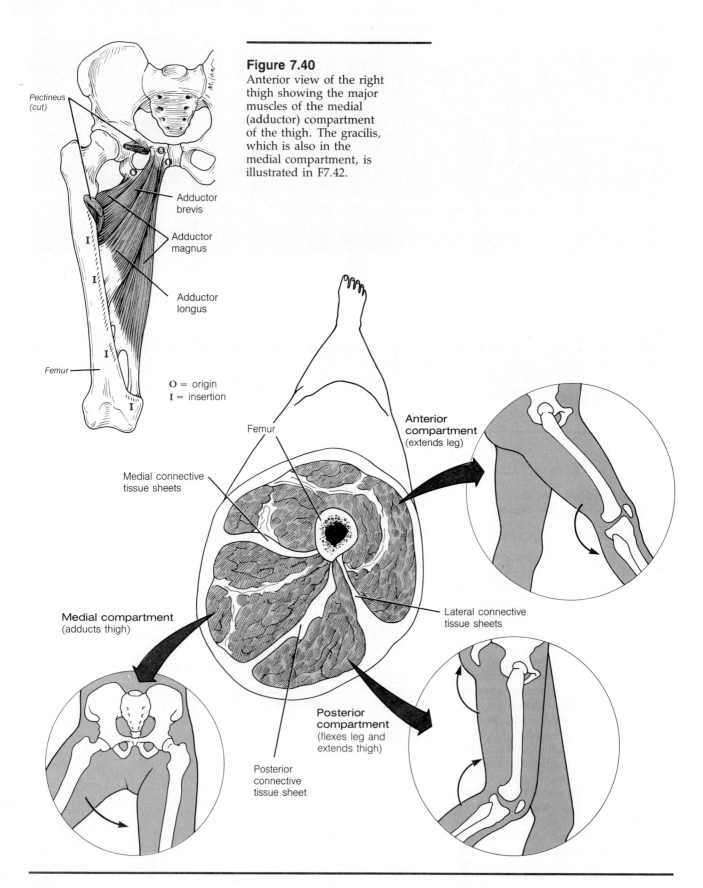

Pectineus
(cut)

Adductor
brevis

Adductor
magnus

Adductor
longus

Femur

O = origin
I = insertion

Figure 7.40
Anterior view of the right thigh showing the major muscles of the medial (adductor) compartment of the thigh. The gracilis, which is also in the medial compartment, is illustrated in F7.42.

Femur

Medial connective
tissue sheets

Anterior
compartment
(extends leg)

Medial compartment
(adducts thigh)

Lateral connective
tissue sheets

Posterior
compartment
(flexes leg and
extends thigh)

Posterior
connective
tissue sheet

Figure 7.41
Cross section of the right thigh showing the three compartments and general actions of the muscles in them.

Table 7.18   Muscles That Act on the Leg (Muscles of the Thigh)

Muscle	Origin	Insertion	Action	Innervation
MEDIAL COMPARTMENT				
Adductor magnus Adductor longus Adductor brevis Pectineus	These muscles act only on the femur. They are illustrated in Figure 7.40 and described in Table 7.17.		Adduct and laterally rotate the thigh	Obturator
Gracilis [F7.42]	Symphysis pubis and the pubic arch	Medial surface of the tibia just below the condyle	Adduct the thigh; flex the leg	Obturator
ANTERIOR COMPARTMENT				
Sartorius [F7.44]	Anterior superior iliac spine	Proximal medial surface of the tibia, below the tuberosity	Flex the thigh and the leg; laterally rotate the thigh	Femoral
Quadriceps femoris				
Rectus femoris [F7.43]	Anterior inferior iliac spine and just above the acetabulum of the coxal bone			
Vastus lateralis [F7.43]	Greater trochanter and lateral lip of the linea aspera of the femur	Tibial tuberosity, via the patella and the patellar ligament	Extend the leg; the rectus femoris also flexes the thigh	Femoral
Vastus medialis [F7.43]	Medial lip of the linea aspera of the femur			
Vastus intermedius [F7.43]	Anterior surface of the shaft of the femur			
POSTERIOR COMPARTMENT				
Hamstrings				
Biceps femoris [F7.38]	*Long head:* ischial tuberosity *Short head:* lateral lip of the linea aspera	Lateral surface of the head of the fibula, and the lateral condyle of the tibia	Flexes the leg; long head extends the thigh	Sciatic
Semitendinosus [F7.38]	Ischial tuberosity	Medial surface of the proximal end of the tibia	Flexes the leg; extends the thigh	Tibial
Semimembranosus [F7.38]	Ischial tuberosity	Medial surface of the proximal end of the tibia	Flexes the leg; extends the thigh	Tibial

they act on the hip as well as on the knee joint. The tendons of insertion of the semimembranosus and the semitendinosus pass medially behind the knee; that of the biceps femoris passes laterally. Between these tendons, on the posterior surface of the knee, is a triangular *popliteal space*.

Notice in Table 7.18 that the muscles in a compartment of the thigh are all innervated by a common nerve: those in the anterior compartment by the femoral nerve; those in the medial compartment by the obturator nerve; and those in the posterior compartment by the sciatic nerve.

Table 7.19 summarizes the actions of various muscles on the thigh.

Table 7.18

Table 7.19

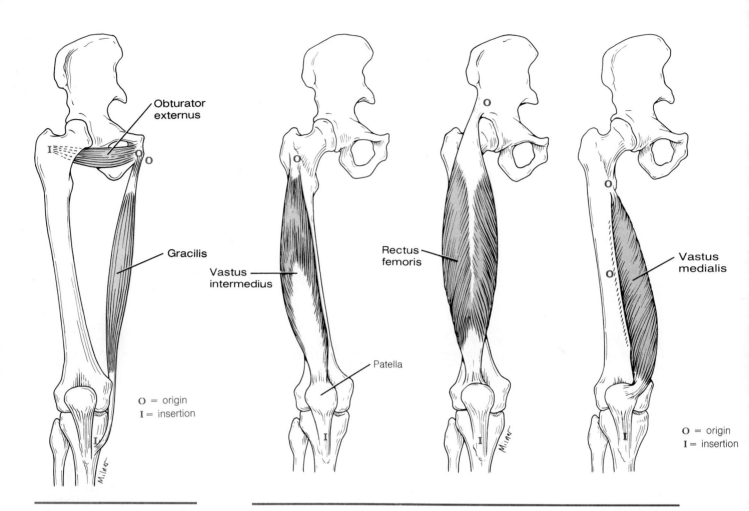

Figure 7.42
The gracilis and obturator externus muscles.

Figure 7.43
The individual muscles that form the quadriceps femoris muscle.

Table 7.19   Summary of Muscle Actions on the Thigh (Including Muscles of the Thigh)

Flexion	Extension	Adduction	Abduction	Medial rotation	Lateral rotation
Illiopsoas	Gluteus maximus	Adductor magnus	Gluteus medius	Gluteus medius	Gluteus maximus
Sartorius	Biceps femoris	Adductor longus	Gluteus minimus	Gluteus minimus	Piriformis
Rectus femoris	Semitendinosus	Adductor brevis	Tensor fasciae latae	Tensor fasciae latae	Obturator internus
Pectineus	Semimembranosus	Pectineus	Piriformis		Obturator externus
Adductor longus	Piriformis	Gracilis			Superior and inferior gemelli
Tensor fasciae latae	Adductor magnus				Quadratus femoris Adductor magnus Adductor longus Adductor brevis Pectineus

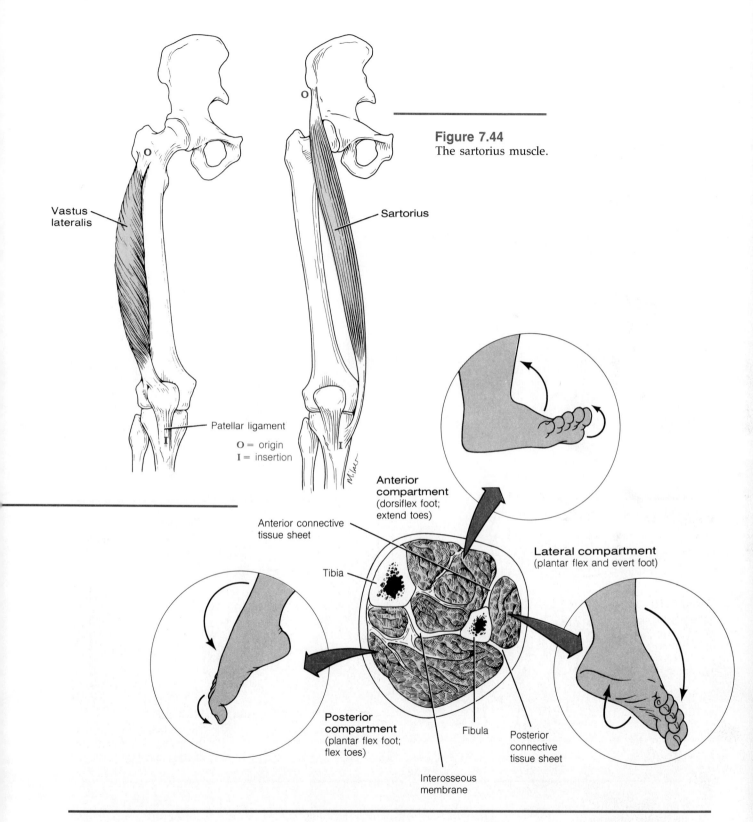

Figure 7.44
The sartorius muscle.

Vastus
lateralis

Sartorius

Patellar ligament

O = origin
I = insertion

Anterior
compartment
(dorsiflex foot;
extend toes)

Anterior connective
tissue sheet

Lateral compartment
(plantar flex and evert foot)

Tibia

Posterior
compartment
(plantar flex foot;
flex toes)

Fibula

Posterior
connective
tissue sheet

Interosseous
membrane

Muscles That Act on the Foot and Toes

The muscles of the leg, like those of the thigh, are grouped into three compartments (Figure 7.45, Table 7.20). The interosseous membrane between the tibia and fibula, and connective tissue sheets that extend anteriorly and posteriorly from the fibula, separate the muscles into *anterior*, *posterior*, and *lateral compartments*.

F7.45,
Table 7.20

Figure 7.45
Cross section of the right leg showing the three compartments and general actions of the muscles in them.

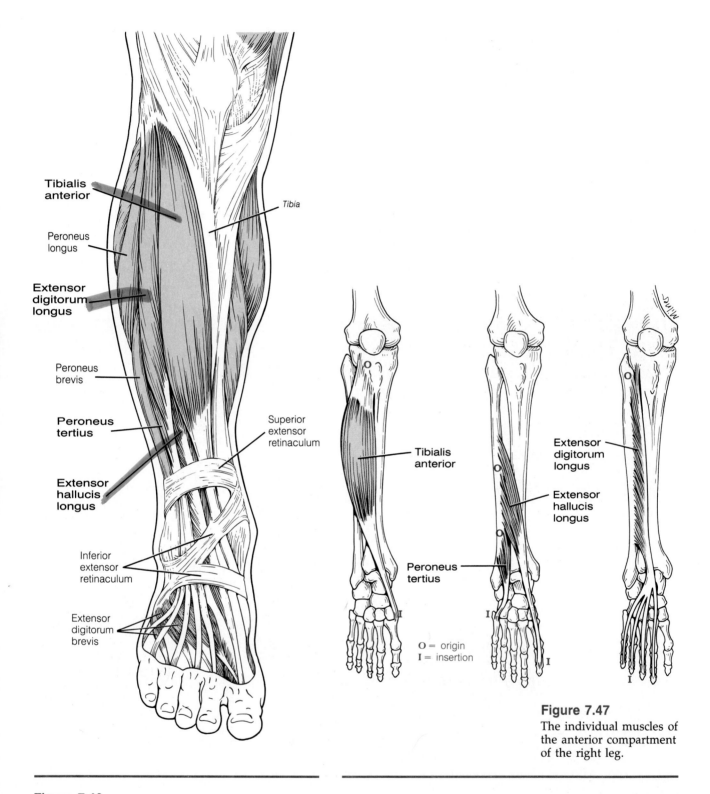

Tibialis
anterior

Tibia

Peroneus
longus

Extensor
digitorum
longus

Peroneus
brevis

Superior
extensor
retinaculum

Peroneus
tertius

Extensor
hallucis
longus

Inferior
extensor
retinaculum

Extensor
digitorum
brevis

Tibialis
anterior

Extensor
digitorum
longus

Extensor
hallucis
longus

Peroneus
tertius

O = origin
I = insertion

Figure 7.47
The individual muscles of
the anterior compartment
of the right leg.

Figure 7.46
Muscles of the anterior
compartment of the right
leg.

F7.46,
F7.47   The muscles of the *anterior compartment* (Figure 7.46, Figure 7.47) extend
the toes and/or dorsiflex the foot. The tendons of these anterior muscles are
held firmly to the ankle by **superior** and **inferior extensor retinaculae** in a
manner similar to the tendons at the wrist.

F7.48,
F7.49   The *lateral*, or *peroneal*, *compartment* (Figure 7.48, Figure 7.49) contains two
of the three peroneal muscles. The **peroneus tertius** is fused to the extensor
digitorum longus muscle and is included in the anterior compartment. The
tendons of the **peroneus longus** and **brevis** pass behind the lateral malleolus
of the fibula to insert on the plantar and lateral surfaces of the metatarsals.

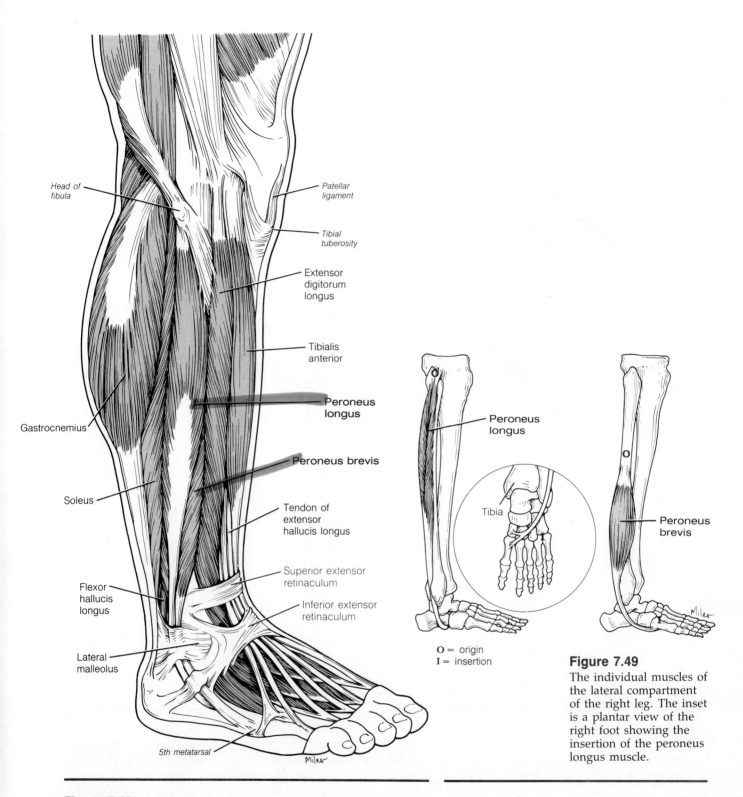

Figure 7.48
Muscles of the lateral
compartment of the right
leg.

Figure 7.49
The individual muscles of
the lateral compartment
of the right leg. The inset
is a plantar view of the
right foot showing the
insertion of the peroneus
longus muscle.

The peroneus longus tendon crosses the sole of the foot obliquely from the
lateral edge to the medial tarsal and metatarsal bones. By using the lateral
malleolus as a pulley, the peroneal muscles of the lateral compartment act to
plantar flex and evert the foot.

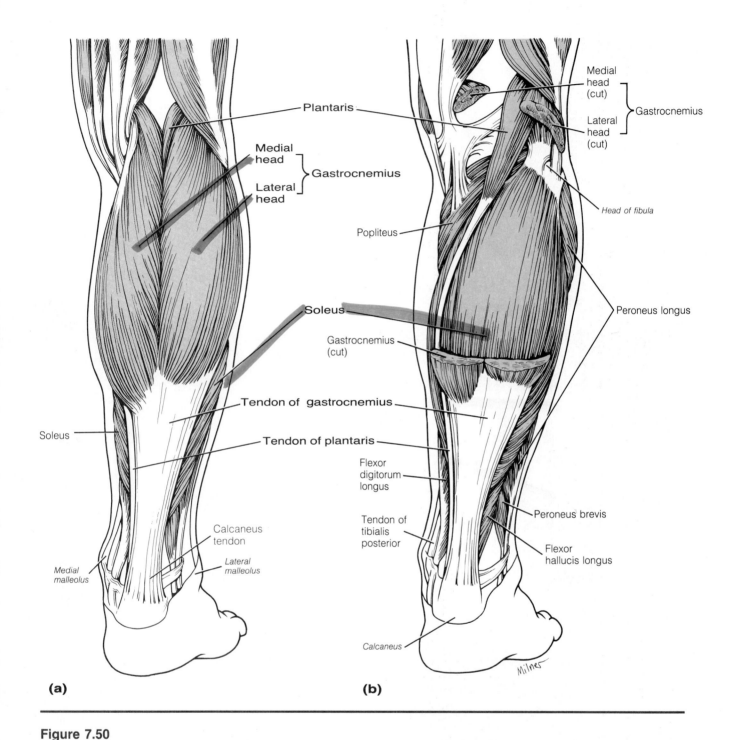

(a)

Plantaris

Medial head ⎫
Lateral head ⎭ Gastrocnemius

Soleus

Soleus

Calcaneus tendon

Medial malleolus

Lateral malleolus

(b)

Medial head (cut) ⎫
Lateral head (cut) ⎭ Gastrocnemius

Head of fibula

Popliteus

Peroneus longus

Soleus

Gastrocnemius (cut)

Tendon of gastrocnemius

Tendon of plantaris

Flexor digitorum longus

Peroneus brevis

Tendon of tibialis posterior

Flexor hallucis longus

Calcaneus

Milner

Figure 7.50
Superficial muscles of the posterior compartment of the right leg.
(a) Superficial layer.
(b) Note that the origin of the gastrocnemius has been removed.

F7.50
F7.51,
F7.52

Table 7.20

The muscle of the *posterior compartment* are grouped into three superficial posterior muscles (Figure 7.50)—the **gastrocnemius, soleus,** and **plantaris**—and four deep posterior muscles (Figure 7.51, Figure 7.52)—the **popliteus, flexor hallucis longus, flexor digitorum longus,** and **tibialis posterior.** The tendons of the latter three muscles pass behind the medial malleolus of the tibia and use it as a pulley. The gastrocnemius and the underlying soleus form the characteristic bulge of the calf. They share a common tendon, the **calcaneus (Achilles) tendon,** which inserts onto the calcaneus. With the exception of the popliteus, which flexes and medially rotates the leg, the muscles of the posterior compartment plantar flex the foot, and two of them flex the toes.

As is the case in the thigh, the muscles in each compartment of the leg are innervated by a common nerve (Table 7.20). The muscles in the anterior compartment are innervated by the deep peroneal nerve; those in the lateral

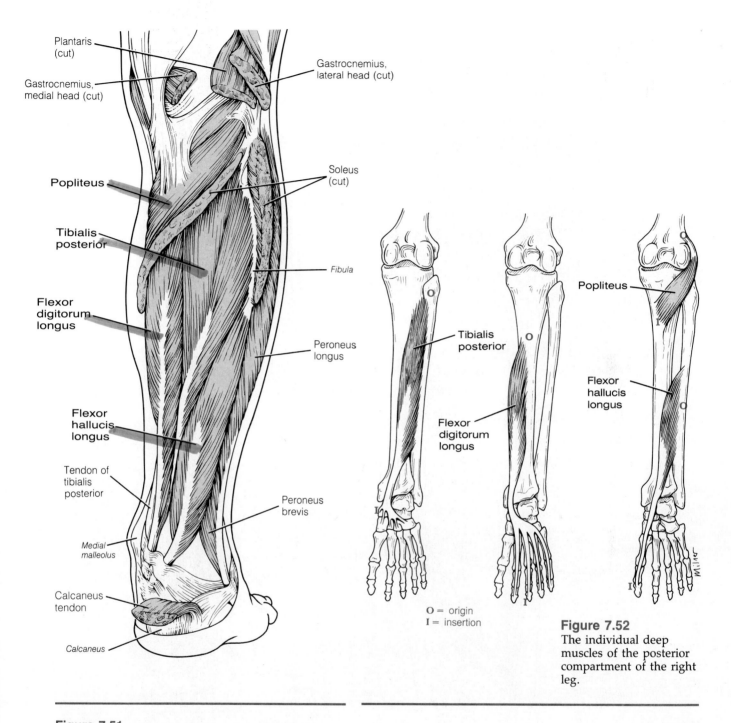

Plantaris (cut)

Gastrocnemius, medial head (cut)

Gastrocnemius, lateral head (cut)

Popliteus

Soleus (cut)

Tibialis posterior

Fibula

Flexor digitorum longus

Peroneus longus

Flexor hallucis longus

Tendon of tibialis posterior

Peroneus brevis

Medial malleolus

Calcaneus tendon

Calcaneus

Tibialis posterior

Flexor digitorum longus

Popliteus

Flexor hallucis longus

O = origin
I = insertion

Figure 7.52
The individual deep muscles of the posterior compartment of the right leg.

Figure 7.51
Deep muscles of the posterior compartment of the right leg. The gastrocnemius and soleus muscles have been removed.

compartment by the superficial peroneal nerve; and those in the posterior compartment by the tibial nerve.

Table 7.21 summarizes the muscle actions on the foot.   Table 7.21

Intrinsic Muscles of the Foot

The skeletal structure of the foot is very similar to that of the hand; however, it is required to perform quite different functions. Whereas the hand has great versatility, being capable of grasping and other fine movements, the foot is adapted for support and locomotion. Consequently, the muscles of the foot

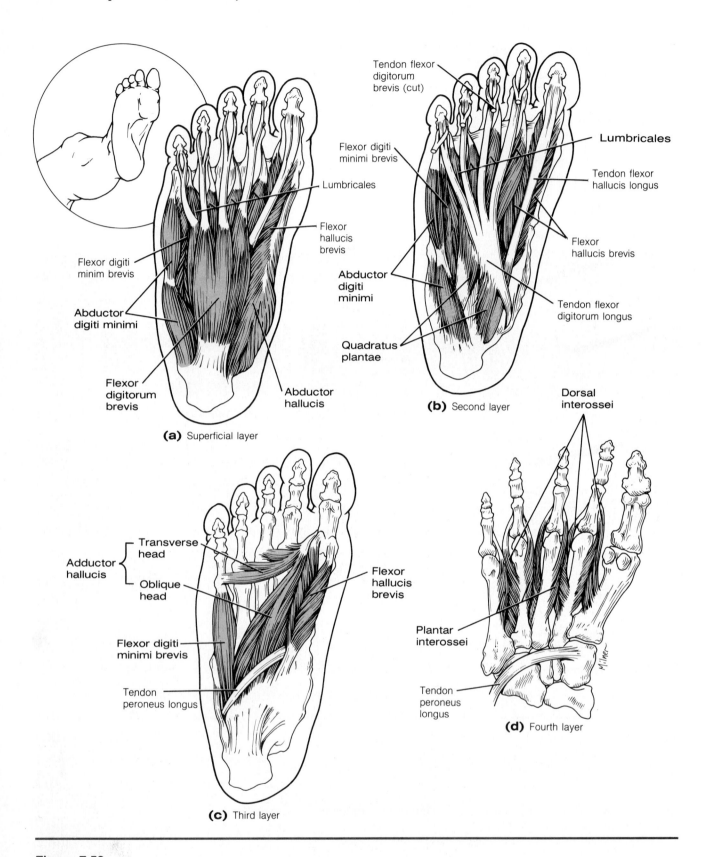

Tendon flexor
digitorum
brevis (cut)

Flexor digiti
minim brevis

Lumbricales

Flexor
hallucis
brevis

Flexor digiti
minim brevis

Abductor
digiti minimi

Flexor
digitorum
brevis

Abductor
hallucis

(a) Superficial layer

Flexor digiti
minimi brevis

Lumbricales

Tendon flexor
hallucis longus

Flexor
hallucis brevis

Abductor
digiti
minimi

Tendon flexor
digitorum longus

Quadratus
plantae

(b) Second layer

Adductor
hallucis

Transverse
head

Oblique
head

Flexor
hallucis
brevis

Flexor digiti
minimi brevis

Tendon
peroneus longus

(c) Third layer

Dorsal
interossei

Plantar
interossei

Tendon
peroneus
longus

(d) Fourth layer

Figure 7.53
Plantar view of the
intrinsic muscles of the
right foot showing
successively deeper
layers of muscles.

tend to be heavier than those of the hand, thus making it possible for them to support the arches of the foot. The intrinsic (and extrinsic) muscles are aided in this support by a tough, fibrous **plantar aponeurosis** that extends from the calcaneus to the phalanges. Another difference in the musculature of the foot compared to that of the hand is the presence of an intrinsic muscle, the **extensor digitorum brevis,** on the dorsum of the foot. Recall that there are no intrinsic muscles on the dorsum of the hand.

The remaining intrinsic muscles are on the plantar surface. They are illustrated in Figure 7.53 and described in Table 7.22. These muscles are separated into four layers (as illustrated), and can be divided functionally into those that act on the great toe, those that move only the small toe, and those that move all the toes except the great toe.

F7.53, Table 7.22

Table 7.20   Muscles That Act on the Foot and Toes (Muscles of the Leg)

Muscle	Origin	Insertion	Action	Innervation
ANTERIOR COMPARTMENT				
Tibialis anterior [F7.47]	Lateral condyle and proximal two-thirds of the shaft of the tibia, and the interosseous membrane	Medial surface of first cuneiform and first metatarsal	Dorsiflexes and inverts foot	Deep peroneal
Extensor hallucis longus (*hallux* = great toe) [F7.47]	Anterior surface of the middle of the fibula, and the interosseous membrane	Dorsal surface of the distal phalanx of the great toe	Dorsiflexes and inverts foot; extends the great toe	Deep peroneal
Extensor digitorum longus [F7.47]	Lateral condyle of the tibia, proximal three-fourths of the anterior surface of the fibula, and the interosseous membrane	Dorsal surface of the phalanges of the second through fifth toes	Dorsiflexes and everts the foot; extends the toes	Deep peroneal
Peroneus tertius (*tertius* = third) [F7.47]	Distal third of the anterior surface of the fibula, and the interosseous membrane	Dorsal surface of the fifth metatarsal	Dorsiflexes and everts the foot	Deep peroneal
LATERAL COMPARTMENT				
Peroneus longus [F7.48]	Proximal two-thirds of the lateral surface of the fibula	Ventral surface of the first metatarsal and the medial cuneiform	Plantar flexes and everts the foot	Superficial peroneal
Peroneus brevis [F7.48]	Distal two-thirds of the fibula	Lateral side of the fifth metatarsal	Plantar flexes and everts the foot	Superficial peroneal
POSTERIOR COMPARTMENT				
SUPERFICIAL MUSCLES				
Gastrocnemius [F7.50]	Medial and lateral condyles of the femur	Calcaneus, via the calcaneus tendon	Flexes the leg; plantar flexes the foot	Tibial
Soleus [F7.50]	Posterior surface of the proximal third of the fibula, and the middle third of the tibia	Calcaneus, via the calcaneus tendon	Plantar flexes the foot	Tibial
Plantaris [F7.50]	Posterior surface of the femur above the lateral condyle	Calcaneus	Flexes the leg; plantar flexes the foot	Tibial

Table 7.20   Muscles That Act on the Foot and Toes (Muscles of the Leg) (continued)

Muscle	Origin	Insertion	Action	Innervation
DEEP MUSCLES				
Popliteus [F7.51]	Lateral condyle of the femur	Proximal portion of the tibia	Flexes the leg and rotates it medially	Tibial
Flexor hallucis longus [F7.51]	Lower two-thirds of the fibula	Distal phalanx of the great toe	Flexes the great toe; plantar flexes and inverts the foot	Tibial
Flexor digitorum longus [F7.51]	Posterior surface of the tibia	Distal phalanx of the second through fifth toes	Flexes the toes; plantar flexes and inverts the foot	Tibial
Tibialis posterior [F7.51]	Posterior surface of the interosseous membrane, the tibia, and the fibula	Navicular, cuneiforms, cuboid; second through fourth metatarsals	Plantar flexes and inverts the foot	Tibial

Table 7.21   Summary of Muscle Actions on the Foot

Dorsiflexion	Plantar flexion	Adduction plus inversion	Abduction plus eversion
Tibialis anterior	Peroneus longus	Tibialis anterior	Peroneus tertius
Extensor hallucis longus	Peroneus brevis	Tibialis posterior	Peroneus longus
Extensor digitorum longus	Gastrocnemius	Extensor hallucis longus	Peroneus brevis
Peroneus tertius	Soleus	Flexor hallucis longus	Extensor digitorum longus
	Plantaris	Flexor digitorum longus	
	Tibialis posterior		
	Flexor hallucis longus		
	Flexor digitorum longus		

Table 7.22   Intrinsic Muscles of the Foot

Muscle	Origin	Insertion	Action	Innervation
DORSAL MUSCLE				
Extensor digitorum brevis [F7.46]	Lateral surface of the calcaneus	Proximal phalanx of the great toe and the tendons of the extensor digitorum longus	Extends the first through fourth toes	Deep peroneal
PLANTAR MUSCLES				
Superficial layer				
Abductor hallucis [F7.53a]	Calcaneus	Proximal phalanx of the great toe (with the tendon of the flexor hallucis brevis)	Abducts the great toe	Medial plantar

Table 7.22 Intrinsic Muscles of the Foot (continued)

Muscle	Origin	Insertion	Action	Innervation
Flexor digitorum brevis [F7.53a]	Calcaneus and plantar aponeurosis	Middle phalanx of the second through fifth toes	Flexes the second through fifth toes	Medial plantar
Abductor digiti minimi [F7.53a]	Calcaneus and plantar aponeurosis	Proximal phalanx of the small toe	Abducts the small toe	Lateral plantar
Second layer				
Quadratus plantae [F7.53b]	Calcaneus	Into tendons of the flexor digitorum longus	Aids in flexing the second through fifth toes by straightening the pull of the flexor digitorum longus	Lateral plantar
Lumbricales (4) [F7.53b]	From tendons of the flexor digitorum longus	Into tendons of the extensor digitorum longus	Flex the proximal phalanx, extend the middle and distal phalanges of the second through fifth toes	Medial and lateral plantar
Third layer				
Flexor hallucis brevis [F7.53c]	Cuboid and lateral cuneiform	Proximal phalanx of the great toe	Flexes the great toe	Medial plantar
Adductor hallucis [F7.53c]	*Oblique head:* second, third, and fourth metatarsals *Transverse head:* ligaments of the metatarsophalangeal joints	Proximal phalanx of the great toe	Adducts the great toe	Lateral plantar
Flexor digiti minimi brevis [F7.53c]	Fifth metatarsal	Proximal phalanx of the small toe	Flexes the small toe	Lateral plantar
Fourth layer				
Plantar interossei (3) [F7.53d]	Third, fourth, and fifth metatarsals	Proximal phalanx of the same toe	Adduct the toes toward the second toe	Lateral plantar
Dorsal interossei (4) [F7.53d]	Bases of the adjacent metatarsals	Proximal phalanges; both sides of the second toe; lateral side of the third and fourth toes	Abduct the toes from the second toe; move the second toe medially and laterally	Lateral plantar

STUDY OUTLINE

MUSCLE TYPES pp. 181–182

VOLUNTARY MUSCLES controlled by somatic nervous system; skeletal muscles.

INVOLUNTARY MUSCLES controlled by autonomic nervous system, hormones, and intrinsic factors; cardiac and smooth muscles.

SKELETAL MUSCLE most attaches to skeleton; striated cells; voluntary control.

SMOOTH MUSCLE in the walls of hollow organs and tubes; the cells lack striations; also called visceral muscle; involuntary control.

CARDIAC MUSCLE forms the wall of the heart; striated cells; involuntary control.

EMBRYONIC DEVELOPMENT OF MUSCLE p. 182

SKELETAL MUSCLE except for the muscles of the

head and limbs, skeletal muscles develop from embryonic masses of mesoderm called somites—located dorsally along axial skeleton. Myotome is the portion of somite that differentiates into muscle cells. Head musculature develops from general mesoderm of that region. Limb musculature develops from mesodermal condensations within embryonic limb buds.

SMOOTH MUSCLE (visceral muscle) mesodermal cells migrate to embryonic digestive tube and body organs and surround them in thin layer.

CARDIAC MUSCLE mesodermal cells migrate to and surround early tubular heart.

GROSS ANATOMY OF SKELETAL MUSCLES pp. 182–184

CONNECTIVE-TISSUE COVERINGS skeletal muscle is composed of muscle cells—called *muscle fibers*—that are held together by thin sheets of fibrous connective tissue membranes called *fascia*.

EPIMYSIUM fascia that envelops entire muscle.

PERIMYSIUM fascia that penetrates muscle; separates fibers into bundles called *fasciculi*.

ENDOMYSIUM thin extensions of fascia; envelop cell membrane of each muscle fiber.

SKELETAL MUSCLE ATTACHMENTS extensions of endomysium, perimysium, and epimysium may directly attach to bone or may blend into a strong fibrous connection called a tendon.

TENDONS extensions of connective tissue beyond end of muscle; length varies; continuous with periosteum.

APONEUROSES broad thin tendon sheets.

ORIGIN less movable end of muscle; usually proximal.

INSERTION more movable end of muscle; usually distal.

BELLY widest portion of a muscle, between its origin and insertion.

SKELETAL MUSCLE SHAPES fasciculi may run parallel to the long axis of muscle, producing considerable movement but little strength; or fasciculi may insert diagonally into a tendon running the length of muscle, producing less movement but greater power.

UNIPENNATE all fasciculi insert on one side of tendon.

BIPENNATE fasciculi insert on both sides of tendon.

MULTIPENNATE convergence of several tendons.

RADIATE fasciculi converge from broad origin to a single narrow tendon.

MICROSCOPIC ANATOMY OF SKELETAL MUSCLE longitudinal threadlike arrays of proteins (called *myofibrils*) in sarcoplasm; appear cross-striated because of alternating light and dark bands.
pp. 184–187

ANISOTROPIC (A, DARK) BANDS have less-dense H zone in center.

ISOTROPIC (I, LIGHT) BANDS have dense Z lines crossing center and dividing myofibrils into sarcomeres.

SARCOMERE repeating units of myofibrils; contain filamentous structures called *myofilaments*.

THICK FILAMENTS only in A band; H zone has only thick filaments.

THIN FILAMENTS I band; part of A band; attach to Z lines.

COMPOSITION OF THE MYOFILAMENTS

THICK FILAMENTS composed of the protein myosin.

THIN FILAMENTS composed of the proteins actin, tropomyosin, and troponin.

TRANSVERSE TUBULES AND SARCOPLASMIC RETICULUM

TRANSVERSE TUBULES (T TUBULES) run deep into muscle cell from sarcolemma at junction of A and I bands.

SARCOPLASMIC RETICULUM membranous network throughout cell.

CONTRACTION OF SKELETAL MUSCLE pp. 187–189

CELLULAR EVENTS DURING CONTRACTION
1. Nerve impulse reaches neuromuscular junction; acetylcholine released, changes permeability of plasma membrane at neuromuscular junction.
2. Impulse transmitted along *t* tubules to cell interior.
3. Calcium ions released from reticular sites called *lateral sacs*; muscle contraction is initiated.
4. Calcium ions bind to troponin molecules attached to tropomyosin and actin; tropomyosin molecules move aside from blocking position.
5. High-energy myosins link with G-actin subunits.
6. ATP splits into ADP and inorganic phosphate; myosin acts as enzyme; myosin head (cross bridge) swivels, and pulls actin-containing thin filament toward center of the sarcomere.
7. With the linkage of another ATP to myosin, the process repeats, Z lines draw close together, muscle contracts.
8. Calcium ions return to sarcoplasmic reticulum; tropomyosin returns to blocking position; muscle relaxes.

ENERGY SOURCES FOR CONTRACTION

OXIDATIVE METABOLISM provides ATP (aerobic process).

ATP AND LACTIC ACID produced during intense activity by nonoxidative (anaerobic) metabolism.

CREATINE PHOSPHATE alternative energy source; rapidly produces ATP.

MUSCLE ACTIONS pp. 189–191

PRIME MOVERS those muscles whose contractions are primarily responsible for a movement.

ANTAGONISTS those muscles whose contraction offers resistance to a movement; generally on opposite side of joint.

SYNERGISTS those muscles that indirectly aid a movement by preventing unwanted movement (steadying a joint, for example). Actions of synergists and antagonists allow for smooth, coordinated, and precise movements.

FIXATOR a synergist acting to immobilize joint.

RELATIONSHIP BETWEEN LEVERS AND MUSCLE ACTIONS pp. 191–193

LEVER rigid structure capable of moving around a pivot point (fulcrum) when force is applied.

CLASSES OF LEVERS

CLASS I LEVER fulcrum is between point of force and the weight to be moved (examples: a seesaw, raising the face).

CLASS II LEVER weight to be moved is between the fulcrum and the point of force (examples: a wheelbarrow, raising body on toes).

CLASS III LEVER weight is at one end, fulcrum at other, and the point of force is between them (examples: lifting shovel, flexing forearm).

EFFECTS OF LEVERS ON MOVEMENTS

POWER ARM portion of lever between fulcrum and point of force.

Longer Power Arm great strength, but slower movement and less range of movement.

WEIGHT ARM portion of lever between fulcrum and weight.

Longer Weight Arm weight moved faster and over greater range, but stronger contraction required.

CONDITIONS OF CLINICAL SIGNIFICANCE: SKELETAL MUSCLE p. 193

MUSCULAR ATROPHY reduction in size of muscles due to shrinkage and death of muscle cells; may be generalized or local.

CRAMPS involuntary painful muscle contractions; slow to relax; may be caused by low oxygen supply in muscles, by nervous system stimulation, or by heavy exercise (due to low levels of sodium and chloride ions in blood).

MUSCULAR DYSTROPHY genetically transmitted progressive muscular weakness due to muscle cell degeneration, increase in connective tissue, or replacement of muscle cells by fatty tissue.

MYASTHENIA GRAVIS chronic condition of extreme muscle weakness; due to abnormal reaction of immune system causing interference of nerve impulse transmission at neuromuscular junction.

EFFECTS OF AGING ON THE SKELETAL MUSCLES progressive loss of skeletal muscle mass; replacement of muscle by fat.

SMOOTH MUSCLE uninucleate; spindle-shaped; has thick filaments of myosin and thin filaments of actin and tropomyosin, but not organized into sarcomeres. pp. 193–195

CARDIAC MUSCLE cells form branching networks that are connected by intercalated discs; striated, with thick and thin filaments arranged in sarcomeres and myofibrils. p. 196

SKELETAL MUSCLES OF THE HUMAN BODY pp. 196–197

Over 600 muscles; criteria used to name muscles include:
1. *Muscle shape* for example, rhomboid, trapezius.

2. *Muscle action* for example, flexor, extensor, adductor, pronator.
3. *Muscle location* for example, intercostal, tibialis anterior.
4. *Muscle attachments* for example, sternocleidomastoid, coracobrachialis.
5. *Muscle divisions* for example, biceps brachii, triceps brachii, quadriceps femoris.
6. *Muscle size relationships* for example, gluteus maximus, gluteus minimus, peroneus longus, peroneus brevis.

MUSCLES OF THE HEAD AND NECK pp. 197–202

MUSCLES OF THE FACE (Table 7.1) also called muscles of facial expression. Some arise from skull bones, others from superficial fascia of face. Most insert into skin of the region and thus move the skin rather than joints. Facial muscles control mouth and lips; compress cheeks; control eyebrow, eyelid, and scalp movement; dilate nares; tighten and wrinkle neck and forehead skin; depress mandible.

MUSCLES OF MASTICATION (Table 7.2) four pairs of muscles involved in biting and chewing: two pairs close mandible; two pairs move mandible sideways and assist in opening and closing mouth.

MUSCLES OF THE TONGUE intrinsic muscles of tongue squeeze, fold, and curl tongue; extrinsic muscles of tongue (Table 7.3) anchor it to skeleton and control its protrusion, retraction, and sideways movement.

MUSCLES OF THE NECK include muscles grouped within one of two triangles—anterior (Table 7.4) or posterior—that are separated by sternocleidomastoid muscle.

MUSCLES OF THE THROAT consist of deep muscles of anterior triangle; help form floor of oral cavity and are attached to hyoid bone; aid in movements of tongue and swallowing.

MUSCLES OF THE TRUNK include muscles of vertebral column, the back, thorax, abdominal wall, and muscles that form floor of abdominopelvic cavity. pp. 202–211

MUSCLES OF THE VERTEBRAL COLUMN (Table 7.5) most are located on posterior surface of spine. Some have fibers that insert onto skull and thus also move head. Others (some that comprise the abdominal wall and those that act on the hip) also act on vertebral column.

DEEP MUSCLES OF THE THORAX (Table 7.6) most insert on the ribs and assist in breathing by elevating or depressing rib cage; the spaces between ribs are reinforced by a double layer of intercostal muscles; diaphragm is muscle most responsible for quiet breathing.

MUSCLES OF THE ABDOMINAL WALL (Table 7.7) wall of abdominal cavity lacks skeletal support and thus derives strength entirely from muscles, which are composed of three layers:
1. *External abdominal oblique:* outermost layer, with fibers that pass medially and downward.
2. *Internal abdominal oblique:* layer immediately beneath external oblique; has fibers that run upward and medially.

3. *Transverse abdominis muscle:* layer deep to both oblique muscles; a thin muscle whose fibers run horizontally and encircle abdominal cavity.

RECTUS ABDOMINIS MUSCLE ensheathed by the fasciae of the oblique and transversus muscles; passes vertically from pubic symphysis to rib cage.

QUADRATUS LUMBORUM MUSCLE forms most of posterior abdominal wall.

MUSCLES THAT FORM THE FLOOR OF THE ABDOMINOPELVIC CAVITY (Table 7.8) viscera of abdominopelvic cavity are supported by muscular floor called pelvic diaphragm, which is composed of two principal muscles: levator ani and coccygeus.

MUSCLES OF THE PERINEUM (Table 7.9) five muscles in two layers; located below pelvic diaphragm; urogenital diaphragm formed by two of the muscles.

MUSCLES OF THE UPPER LIMBS include muscles that act on scapula, clavicle, arm and forearm, hand, and fingers. **pp. 211–224**

MUSCLES THAT ACT ON THE SCAPULA (Table 7.10) six muscles—four posterior and two anterior—act directly on scapula and anchor it to thorax.

MUSCLES THAT ACT ON THE ARM (Tables 7.11 and 7.12) nine muscles cross shoulder joint and insert on humerus; seven of these arise from scapula.

MUSCLES THAT ACT ON THE FOREARM (Table 7.13) elbow joint and/or proximal radial/ulnar joints are moved by powerful muscles in arm, but are assisted by muscles in forearm.

MUSCLES THAT ACT ON THE HAND AND FINGERS (Tables 7.14 and 7.15) moved by muscles that form bulge of proximal end of forearm; anterior group of muscles serves as flexors and/or pronators; posterior group serves as extensors and/or supinators.

INTRINSIC MUSCLES OF THE HAND (Table 7.16) several muscles of forearm move fingers; intrinsic finger muscles produce fine, precise finger movement.

MUSCLES OF THE LOWER LIMBS include muscles that act on thigh, leg, foot, and toes, and intrinsic muscles of foot. Lower limb muscles tend to be bulkier and more powerful than those of the upper limb because the former are concerned with support and locomotion. Many muscles of lower limbs cross and act on two joints: either hip and knee or knee and ankle. **pp. 224–241**

MUSCLES THAT ACT ON THE THIGH (Tables 7.17 and 7.19) most arise from pelvis.

MUSCLES THAT ACT ON THE LEG (Table 7.18) are in thigh; their tendons cross knee; muscles acting on leg grouped into anterior (extensor), posterior (flexor), and medial (adductor) compartments.

MUSCLES THAT ACT ON THE FOOT AND TOES (Tables 7.20 and 7.21) most located in leg; grouped into anterior (dorsiflex ankle and/or extend toes), posterior (plantar flex foot and/or flex toes), and lateral (plantar flex and evert foot) compartments.

INTRINSIC MUSCLES OF THE FOOT (Table 7.22) four layers of intrinsic foot muscles on plantar surface, and one intrinsic muscle on dorsum of foot; foot muscles tend to be heavier than those of hand, since foot is adapted for support and locomotion.

SELF-QUIZ

1. Match the following terms with the appropriate lettered description.

    Epimysium
    Perimysium
    Endomysium
    Aponeuroses
    Origin
    Insertion
    Unipennate
    Bipennate
    Multipennate

    (a) The less movable end of a skeletal muscle.
    (b) When all the fasciculi insert onto one side of the tendon.
    (c) The fascia that envelops an entire muscle.
    (d) Tendons in the form of long, thin sheets.
    (e) When the fasciculi of certain muscles have a complex arrangement involving the convergence of several tendons.
    (f) Thin extensions of the fascia that envelop the plasma membrane of each muscle fiber.
    (g) The more movable end of a skeletal muscle.
    (h) Fascia that penetrates a muscle, separating the fibers into bundles.
    (i) These muscles have fasciculi inserting obliquely on both sides of the tendon.

2. The threadlike molecule that lies along the surface of actin is: (a) myosin; (b) tropomyosin; (c) troponin.

3. The thick filament of a sarcomere of a skeletal muscle cell contains: (a) actin; (b) troponin; (c) myosin; (d) both actin and troponin.

4. The tubular structures that run from the plasma membrane into the interior of a muscle cell form the: (a) *t*-tubule network; (b) sarcoplasmic reticulum; (c) myofibers.

5. Skeletal muscle cells: (a) exhibit cross striations; (b) are principally innervated by the autonomic nervous system; (c) are typically found in the walls of hollow organs.

6. Match each facial muscle with the correct statement about its action.

    Depressor labii inferioris
    Epicranius occipitalis
    Corrugator
    Orbicularis oculi
    Orbicularis oris
    Procerus
    Risorius

    (a) Pulls angle of mouth backward.
    (b) Draws eyebrows together in frowning.
    (c) Draws scalp posteriorly.
    (d) Pulls lower lip down.
    (e) Wrinkles skin between eyebrows.
    (f) Closes eyelids; tightens skin on forehead.
    (g) Closes and protrudes lips.

7. Raising the hyoid, pulling it backward and anteriorly, and raising the floor of the mouth are actions performed by the: (a) infrahyoid muscles; (b) suprahyoid muscles; (c) sternothyroid muscle.

8. Match each vertebral column muscle with the correct statement.

    Multifidus
    Interspinales
    Splenius capitis
      and cervicis
    Spinalis thoracis
      and cervicis

    (a) Origin: spinous processes of upper lumbar, lower thoracic, and seventh cervical vertebrae.
    (b) Insertion: occipital bone, mastoid process of temporal bone, and transverse processes of upper three cervical vertebrae.
    (c) Action: extends vertebral column; rotates it toward opposite side.
    (d) Origin: superior surface of all spinous processes.

9. Select the correct statement about the temporalis muscle: (a) it originates in the zygomatic arch; (b) it opens and protrudes the mandible; (c) it is innervated by the trigeminal nerve.

10. Inguinal hernias are more common in females than in males. True or false?

11. Which muscle of the posterior abdominal wall pulls the thoracic cage toward the pelvis and bends the vertebral column toward the side that is being contracted? (a) quadratus lumborum; (b) internal abdominal oblique; (c) rectus abdominis.

12. A muscle that forms the floor of the abdominopelvic cavity is the: (a) transversus abdominis; (b) levator ani; (c) internal abdominal oblique.

13. The urogenital diaphragm is composed principally of which muscle? (a) deep transversus perineus; (b) ischiocavernosus; (c) levator ani.

14. Match the following muscles that act on the scapula with the correct statement about their action.

    Subclavius
    Rhomboideus
      major
    Pectoralis minor

    (a) Depresses the scapula, and pulls it anteriorly.
    (b) Stabilizes and depresses the pectoral girdle.
    (c) Adducts, stabilizes, and rotates the scapula, lowering its lateral angle.

15. Match the muscle action with the appropriate muscles. (*Note:* a muscle may have more than one action associated with it.)

    Deltoid
    Latissimus dorsi
    Teres minor
    Teres major
    Coracobrachialis
    Infraspinatus

    (a) Flexion
    (b) Extension
    (c) Adduction
    (d) Abduction
    (e) Medial rotation
    (f) Lateral rotation

16. Which of the following muscles flex the forearm? (a) biceps brachii; (b) brachialis; (c) triceps brachii; (d) anconeus; (e) brachioradialis.

17. Match the muscle action with the appropriate wrist, hand, or finger muscle.

    Flexor carpi
      radialis
    Extensor indicis
    Extensor carpi
      ulnaris
    Extensor pollicis
      longus
    Extensor carpi
      radialis
      longus
    Extensor digiti
      minimi
    Flexor carpi
      ulnaris
    Flexor digitorum
      superficialis

    (a) Flexes the hand and phalanges.
    (b) Extends and abducts the hand.
    (c) Extends the little finger.
    (d) Extends the index finger.
    (e) Extends and adducts the hand.
    (f) Extends the thumb and abducts the hand.
    (g) Flexes and adducts the hand.
    (h) Flexes and abducts the hand.

18. Match the muscle action with the appropriate muscle.

    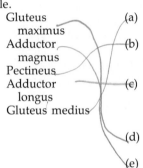

    Gluteus
      maximus
    Adductor
      magnus
    Pectineus
    Adductor
      longus
    Gluteus medius

    (a) Abducts and medially rotates the thigh.
    (b) Adducts, flexes, and laterally rotates the thigh.
    (c) Adducts and laterally rotates the thigh; assists in flexion of the thigh.
    (d) Extends and laterally rotates the thigh.
    (e) Adducts and laterally rotates the thigh; assists in extension of the thigh.

19. The muscles of the thigh that flex the leg and extend the thigh are located in the: (a) medial compartment; (b) anterior compartment; (c) posterior compartment.

20. What is the action of the muscles located in the anterior compartment of the leg? (a) flex the leg; (b) dorsiflex the foot and/or extend the toes; (c) plantar flex the foot and flex the toes.

21. Match the muscle action with the appropriate leg muscle that acts on the foot and/or toes.

    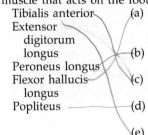

    Tibialis anterior
    Extensor
      digitorum
      longus
    Peroneus longus
    Flexor hallucis
      longus
    Popliteus

    (a) Flexes the great toe; plantar flexes and inverts the foot.
    (b) Plantar flexes and everts the foot.
    (c) Dorsiflexes and inverts the foot.
    (d) Flexes the leg and rotates it medially.
    (e) Dorsiflexes and everts the foot; extends the toes.

LEARNING OBJECTIVES

After completing this chapter, you should be able to:

- Identify visually or by palpation various structures of the head, neck, chest, abdomen, back, and limbs.

- Visualize the actions of various muscles by considering their location.

CHAPTER CONTENTS

Surface Anatomy

Now that you have studied the skeletal and muscular systems, it will help you to remember some skeletal structures and the locations of various muscles if we identify the more prominent features visible on the body's surface. The study of the structures that give form to the body and provide landmarks on its surface is called **surface anatomy.**

Bony structures can be located by **palpation**—feeling with the hand. Muscles can best be identified by causing them to contract and perform their actions. Therefore, surface anatomy also provides a means by which muscle actions, as listed in the tables in Chapter 7, can be studied. We will study the body's surface anatomy by regions, identifying only the more prominent features. The descriptions that follow can be used to help you locate the structures on your own body. The illustrations show the more obvious landmarks.

THE HEAD

Since the muscles and fasciae that cover the cranial part of the head are rather flat, the conformation of the head (Figure 8.1, Figure 8.2) is largely determined by the underlying frontal, parietal, and occipital bones. These bones are smooth and have few palpable landmarks. On the back of the head, near the base of the skull, the *external occipital protuberance* is easily palpable. Extending laterally from the external occipital protuberance are the *superior nuchal lines,* which mark the uppermost margin of the muscles of the neck. On each side of the forehead are small rounded elevations called the *frontal eminences.* The lateral margins of the cranium are formed by ridges called the *parietal eminences.* The *mastoid process* of the temporal bone can be palpated just behind the lower portion of the ear.

In the region of the face it is possible to palpate the *orbital margins* formed by the frontal, zygomatic, and maxillary bones. The *superciliary ridges* of the frontal bone run parallel with the upper portion of the orbital margins deep to the eyebrows. *Supraorbital foramina* and *notches* can be palpated on the ridges. The smooth area between the superciliary ridges is the *glabella.* Just inferior to the glabella the nasal bones form the *bridge* of the nose. The *zygomatic arch* (cheek) is formed by the zygomatic bone and processes from the maxillary and temporal bones. By pressing upward against the base of the nasal septum it is possible to locate the *anterior nasal spine* of the maxillary bone.

The entire mandible can be recognized by palpation, particularly the *symphysis* and the *angle.* The *temporomandibular joint* can be located by palpating just anterior to the *external acoustic meatus* while opening the jaws.

Most muscles of the head are too small to be readily identified by palpation, but a few can be located. If you frown, you can feel the *corrugator* muscles toward the medial ends of the eyebrows. The *orbicularis oculi* is obvious while squinting, and the *orbicularis oris* while pursing the lips. The *mentalis* muscle can be palpated just above the mandibular symphysis while protruding the lower lip. By palpating superior to the zygomatic arch while opening and closing the mouth it is possible to locate the *temporalis* muscle. The *masseter* muscle becomes obvious by palpating over the ramus of the mandible while clenching the teeth.

247

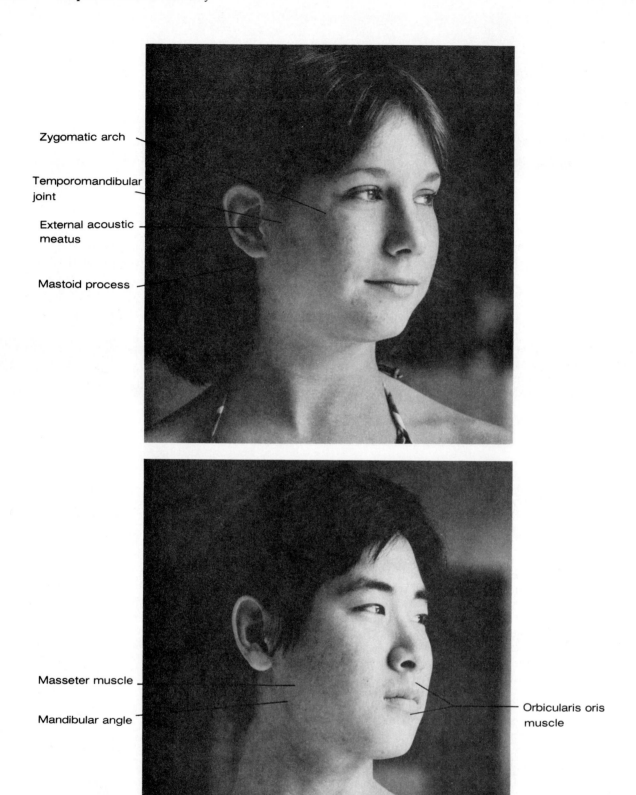

Zygomatic arch

Temporomandibular joint

External acoustic meatus

Mastoid process

Masseter muscle

Mandibular angle

Orbicularis oris muscle

Figure 8.1
Surface anatomy of the lateral side of the head.

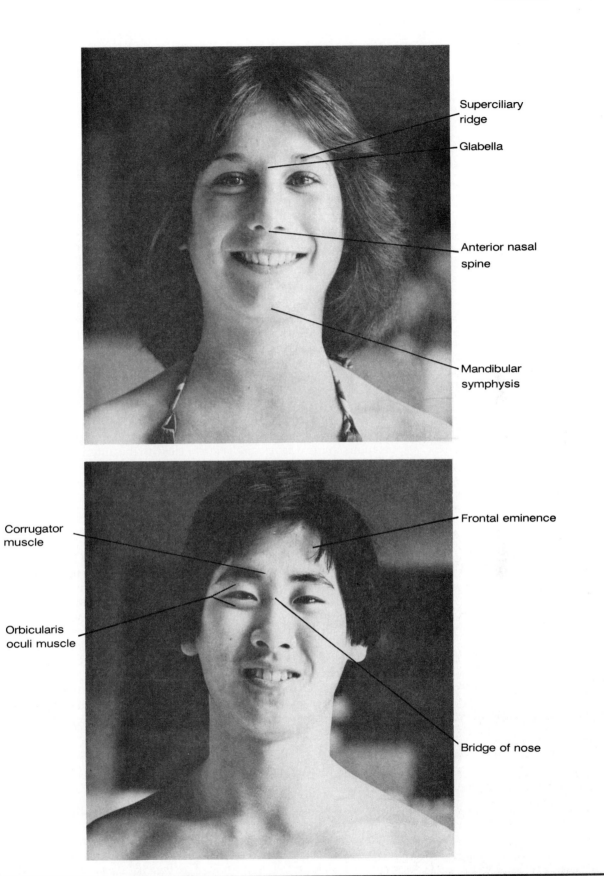

Superciliary ridge

Glabella

Anterior nasal spine

Mandibular symphysis

Corrugator muscle

Frontal eminence

Orbicularis oculi muscle

Bridge of nose

Figure 8.2
Surface anatomy of the face.

The most prominent blood vessel of the head is the *superficial temporal artery*. Its pulsation can be felt just above the zygomatic arch anterior to the ear.

THE NECK

The most prominent structure visible on the anterior surface of the neck, especially in males, is the *laryngeal prominence*, which is formed by the *thyroid cartilage* of the larynx (voice box) (Figure 8.3). This structure is discussed in Chapter 19. Above the thyroid cartilage may be felt the *hyoid bone*, which is located in the base of the tongue. On the posterior surface of the neck the *spinous processes* of the cervical vertebrae may be palpated through the ligamentum nuchae, which covers them. The spinous process of the seventh cervical vertebra is exceptionally long and is clearly visible when the neck is flexed. For this reason, it is referred to as the *vertebra prominens*.

F8.3

The *sternocleidomastoid* is the most prominent muscle of the neck. This muscle passes diagonally in the neck, dividing the region into an anterior triangle in front of it and a posterior triangle behind it. When the sternocleidomastoid muscle is strongly contracted, it is possible to identify its sternal and clavicular attachments. The margin of the *trapezius* muscle forms the upper sloping border of the shoulder and the posterior boundary of the posterior triangle of the neck. When the neck muscles are tensed, the *platysma* muscle can be observed crossing obliquely upward and medially across the side of the neck.

The *external jugular vein*, passing down the neck, is sometimes visible in the vicinity of the angle of the jaw. By palpation it is possible to locate the pulsations in the *external carotid artery* close to the anterior margin of the sternocleidomastoid muscle.

THE CHEST

F8.4

The *clavicle, sternum,* and *ribs* (Figure 8.4) can be palpated and may be visible, depending on the development of the chest muscles and the person's weight. The *jugular notch* is a depression located on the upper border of the manubrium of the sternum between the medial ends of the clavicles. A ridge called the *sternal angle* can be palpated along the line of junction between the manubrium and the body of the sternum. The sternal angle serves as a landmark for locating the junction of the second rib with the sternum, from which point it is possible to identify all the other ribs. The *costal margin,* marking the costal cartilages of the seventh through the tenth ribs, can be palpated running diagonally downward and laterally along the lower edge of the chest.

The *nipples* of the breasts are generally located in the fourth intercostal space. This location can vary in females, depending on the size of the breasts.

The most prominent muscles of the chest are the *pectoralis major* and the *serratus anterior*. The depression of the *axilla* is located between *anterior* and *posterior axillary folds*. The anterior fold is formed by the pectoralis major muscle, and the posterior fold by the latissimus dorsi muscle.

THE ABDOMEN

F8.4

The abdomen (Figure 8.4) extends from the costal margins of the thorax to the iliac crests. The *anterior superior iliac spine* can be palpated on the anterior margin of each iliac crest (Figure 8.5). The *inguinal ligament* extends downward and medially from the anterior superior iliac spine, forming a groove along the groin. The *pubic tubercle* and the *pubic symphysis* can be palpated at the medial end of the ligament. Overlying the pubic symphysis in the female is a pad of fat called the *mons pubis*.

F8.5

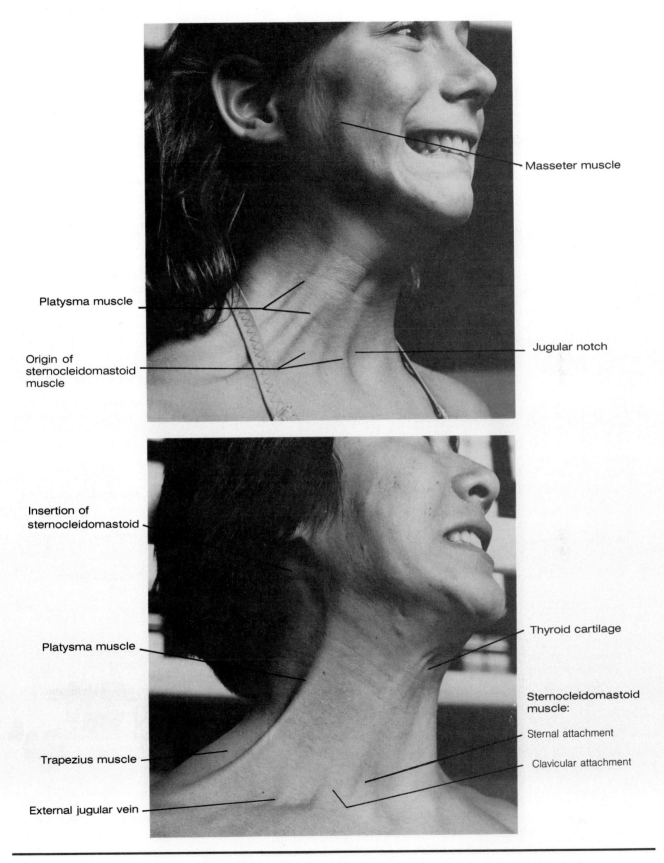

Masseter muscle

Platysma muscle

Origin of
sternocleidomastoid
muscle

Jugular notch

Insertion of
sternocleidomastoid

Thyroid cartilage

Platysma muscle

Sternocleidomastoid
muscle:

Sternal attachment

Trapezius muscle

Clavicular attachment

External jugular vein

Figure 8.3
Surface anatomy of the
anterior neck region.

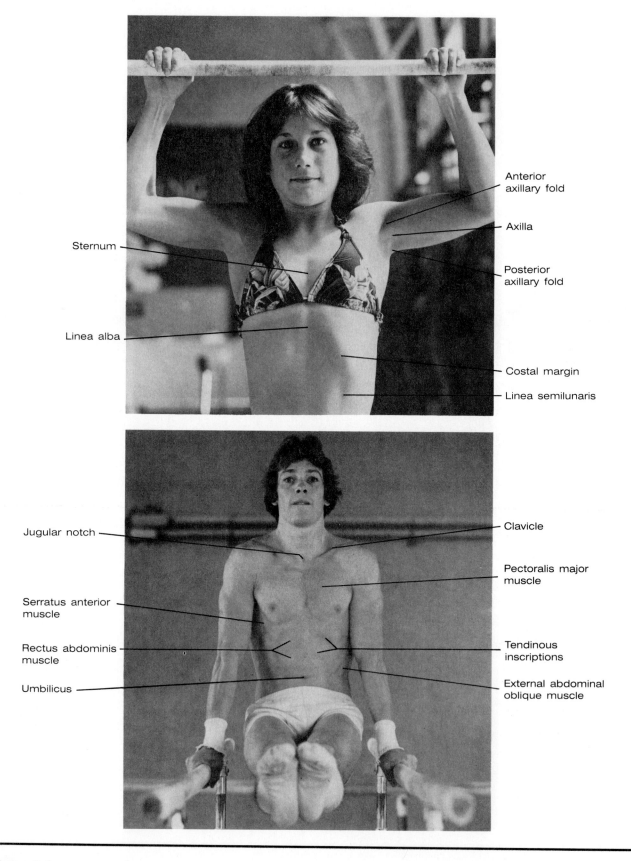

Figure 8.4
Surface anatomy of the
chest and abdomen.

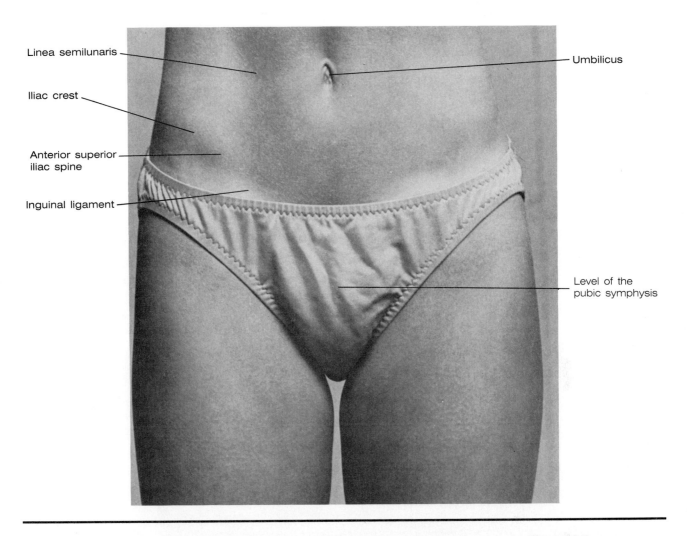

Linea semilunaris

Iliac crest

Anterior superior iliac spine

Inguinal ligament

Umbilicus

Level of the pubic symphysis

Figure 8.5
Surface anatomy of the abdomen and pelvis.

The *umbilicus* is the most prominent landmark of the abdomen. It is generally located at about the middle of the abdomen on a level with the third or fourth lumbar vertebra. The *rectus abdominis* muscles pass vertically on either side of the umbilicus. In a person who has developed these muscles especially well, several transverse furrows, called *tendinous inscriptions*, can be observed. A vertical groove indicating the location of the *linea alba* separates the medial margins of the two recti muscles. The lateral margins of the recti muscles are marked by a groove called the *linea semilunaris*. The *external abdominal oblique* muscles are lateral to the recti muscles.

THE BACK

The *spinous processes* of the vertebrae are easily palpable down the middle of the back (Figure 8.6). Using the vertebra prominens (seventh cervical) as a **F8.6** landmark, the spinous processes are helpful in locating other bony structures of the back. For instance, the *spine of the scapula*, at its medial border, is generally level with the spinous process of the third thoracic vertebra; the *inferior angle* of the scapula is opposite the spinous process of the seventh thoracic vertebra; the highest point of the *iliac crest* is level with the fourth lumbar spinous process; and the *posterior superior iliac spine* forms a depression that is opposite the spinous process of the second sacral vertebra.

Several muscles are visible on the back, including the *sacrospinalis, trapezius, infraspinatus, teres major,* and *latissimus dorsi*. The latter three muscles have their major actions on the arm.

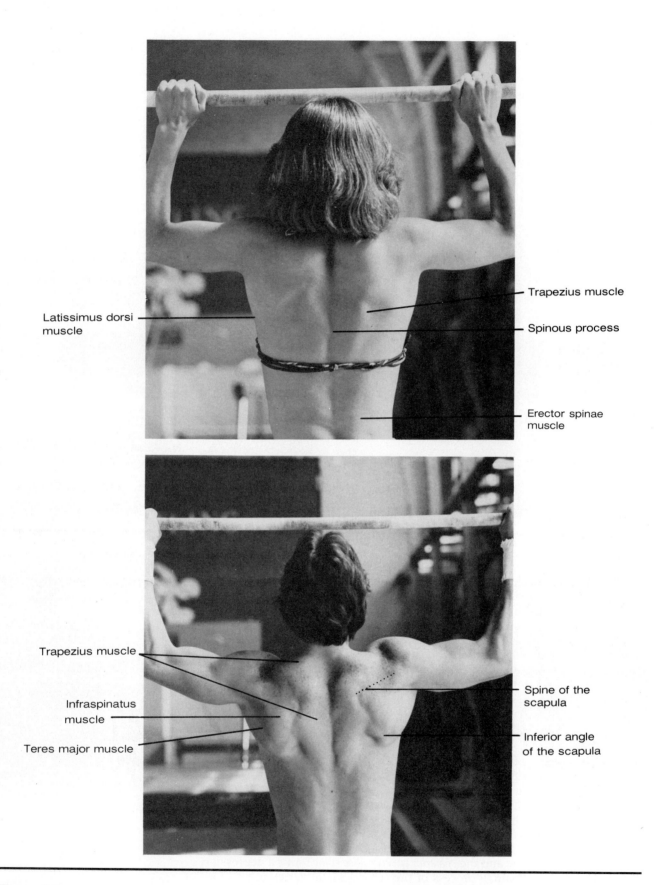

Trapezius muscle

Latissimus dorsi
muscle

Spinous process

Erector spinae
muscle

Trapezius muscle

Infraspinatus
muscle

Teres major muscle

Spine of the
scapula

Inferior angle
of the scapula

Figure 8.6
Surface anatomy of the
back.

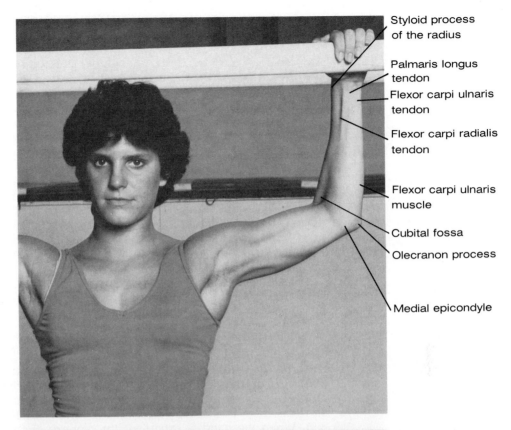

Styloid process
of the radius

Palmaris longus
tendon

Flexor carpi ulnaris
tendon

Flexor carpi radialis
tendon

Flexor carpi ulnaris
muscle

Cubital fossa

Olecranon process

Medial epicondyle

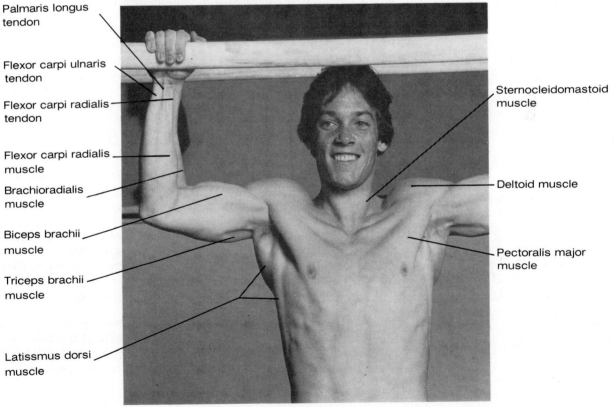

Palmaris longus
tendon

Flexor carpi ulnaris
tendon

Flexor carpi radialis
tendon

Flexor carpi radialis
muscle

Brachioradialis
muscle

Biceps brachii
muscle

Triceps brachii
muscle

Latissmus dorsi
muscle

Sternocleidomastoid
muscle

Deltoid muscle

Pectoralis major
muscle

Figure 8.7
Surface anatomy of the
anterior surface of the
upper limb.

THE UPPER LIMBS

F8.7, F8.8 A number of bony structures of the upper limb are visible or can be palpated and thus serve as landmarks (Figure 8.7, Figure 8.8). The *acromion process* of the scapula articulates with the lateral end of the clavicle to form the point of the shoulder. The *coracoid process* of the scapula is located a short distance anterior to and below the acromion process. The *greater tubercle* of the humerus can be palpated just lateral to and below the acromion process. On the margins of the arm, just above the elbow joint, are the prominent *medial* and *lateral epicondyles* of the humerus. The *olecranon process* of the ulna forms the point of the elbow and is easily identified. The *head* of the radius can be palpated just below the lateral epicondyle of the humerus, especially while the forearm is undergoing supination and pronation. When the hand is abducted, both the *styloid process* of the ulna and, just distal to it, the *pisiform* bone form projections from the medial margin of the forearm. The *styloid process* of the radius can be palpated on the lateral margin of the forearm when the hand is adducted.

The *deltoid* muscle forms a heavy cap over the shoulder and is visible on the anterior as well as the posterior surfaces. The *biceps brachii* is the most prominent muscle of the anterior surface of the arm. It flexes the forearm. The triangular space on the anterior surface of the elbow joint, at the lower end of the biceps muscle, is the *cubital fossa*. The *triceps brachii* muscle, which extends the forearm, occupies the posterior surface of the arm. The *brachioradialis* muscle forms the lateral margin of the forearm. The medial margin of the forearm is formed by the *flexor carpi ulnaris* muscle. When the hand is flexed, the tendons of the *flexor carpi radialis, palmaris longus, flexor digitorum superficialis,* and *flexor carpi ulnaris* are visible on the ventral surface of the wrist (Figure F8.9 8.9). When the hand is extended, the tendons of the *extensor carpi radialis longus* and *brevis* and the *extensor carpi ulnaris* can be palpated on the posterior surface of the wrist. The tendons of the *extensor digitorum* are prominent on the back of the hand when the fingers, as well as the hand, are extended F8.10 (Figure 8.10). When the thumb is extended, two tendons become prominent on the dorsal surface of the base of the thumb. The anteriormost tendon (in the anatomical position) is that of the *extensor pollicis brevis;* the other is the tendon of the *extensor pollicis longus.*

THE LOWER LIMBS

Several of the bony landmarks of the pelvic girdle have already been identified in the discussion of the surface anatomy of the abdomen and back. These include the *iliac crests,* the *anterior* and *posterior superior iliac spines,* the *public tubercle,* and the *symphysis pubis.* In addition, it is possible to locate the *ischial tuberosities* by deep palpation of the lower region of the buttocks while the hip is flexed. The *greater trochanter,* the *medial* and *lateral epicondyles,* and the *adductor tubercle*—all located on the femur—are visible or can be palpated (Figure F8.11–F8.13 8.11, Figure 8.12, Figure 8.13). The *patella,* with its associated *patellar ligament,* is prominent on the anterior surface of the knee. On the margins of the limb, just below the knee, the *medial* and *lateral condyles* of the tibia can be palpated, and the *tibial tuberosity* forms a prominent bulge on the anterior surface. Extending downward from the tuberosity, the *anterior crest* of the tibia can be easily palpated for its entire length. At the ankle the *medial malleolus* of the tibia and the *lateral malleolus* of the fibula form prominent eminences. The *calcaneus* bone forms the entire heel.

Most of the anterior surface of the thigh is formed by the *tensor fasciae latae, sartorius, rectus femoris, vastus lateralis,* and *vastus medialis* muscles. These muscles flex the thigh and/or extend the leg. The three *adductor* muscles and the *gracilis* form the medial region of the thigh. The prominence of the buttock is formed by the gluteal muscles, particularly the *gluteus maximus.* The *ham-string* muscles are visible on the posterior surface of the thigh. When the leg is

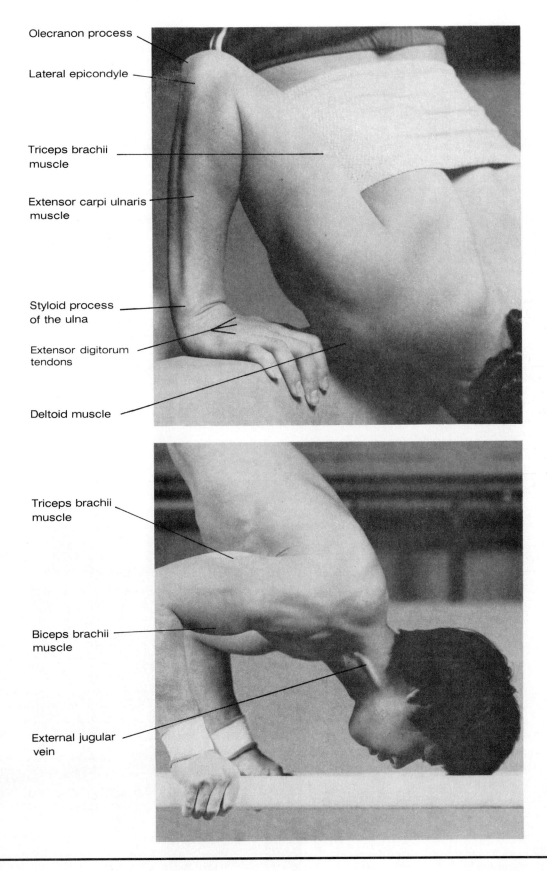

Olecranon process

Lateral epicondyle

Triceps brachii
muscle

Extensor carpi ulnaris
muscle

Styloid process
of the ulna

Extensor digitorum
tendons

Deltoid muscle

Triceps brachii
muscle

Biceps brachii
muscle

External jugular
vein

Figure 8.8
Surface anatomy of the
posterior surface of the
upper limb.

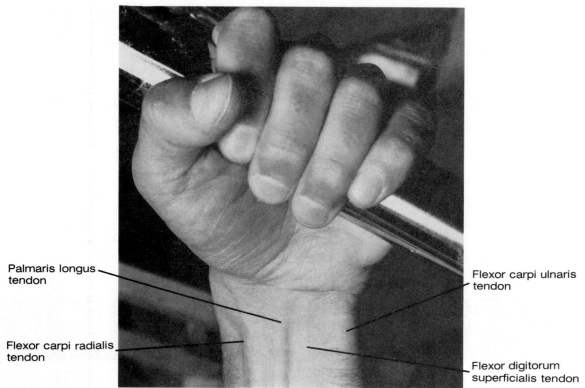

Palmaris longus
tendon

Flexor carpi radialis
tendon

Flexor carpi ulnaris
tendon

Flexor digitorum
superficialis tendon

Figure 8.9
Surface anatomy of the
ventral surface of the
wrist.

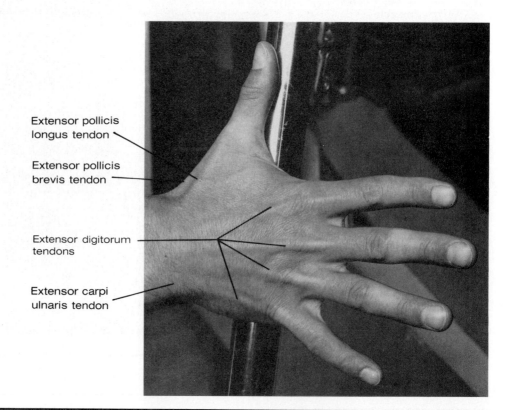

Extensor pollicis
longus tendon

Extensor pollicis
brevis tendon

Extensor digitorum
tendons

Extensor carpi
ulnaris tendon

Figure 8.10
Surface anatomy of the
dorsal surface of the
wrist and hand.

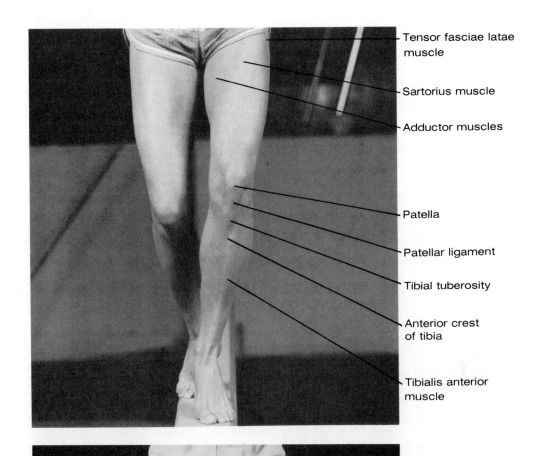

Tensor fasciae latae muscle

Sartorius muscle

Adductor muscles

Patella

Patellar ligament

Tibial tuberosity

Anterior crest of tibia

Tibialis anterior muscle

Rectus femoris muscle

Vastus lateralis muscle

Vastus medialis muscle

Medial malleolus

Lateral malleolus

Tendon of tibialis anterior muscle

Figure 8.11
Surface anatomy of the anterior surface of the lower limbs.

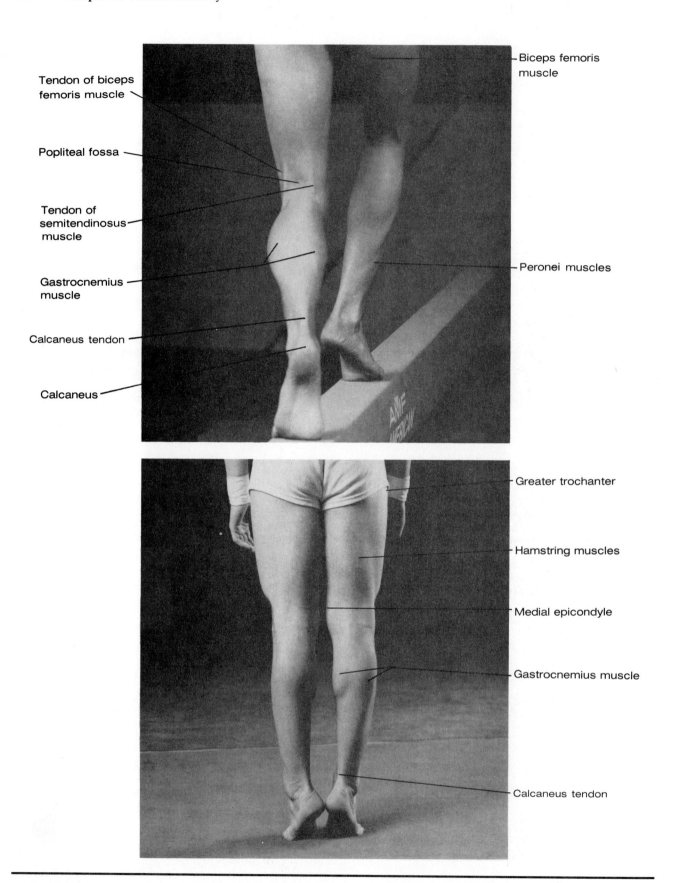

Tendon of biceps femoris muscle

Popliteal fossa

Tendon of semitendinosus muscle

Gastrocnemius muscle

Calcaneus tendon

Calcaneus

Biceps femoris muscle

Peronei muscles

Greater trochanter

Hamstring muscles

Medial epicondyle

Gastrocnemius muscle

Calcaneus tendon

Figure 8.12
Surface anatomy of the posterior surface of the lower limbs.

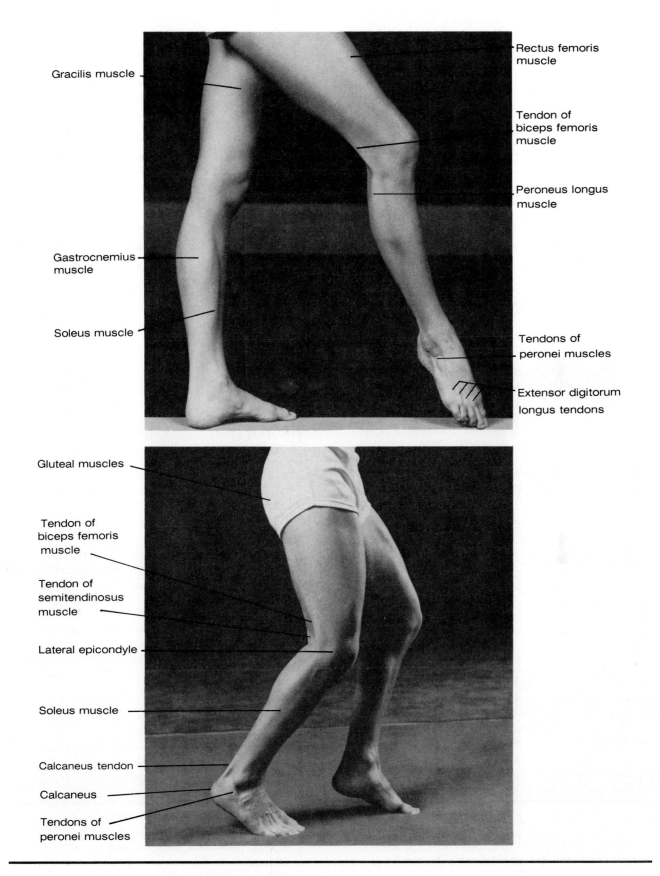

Gracilis muscle

Gastrocnemius muscle

Soleus muscle

Rectus femoris muscle

Tendon of biceps femoris muscle

Peroneus longus muscle

Tendons of peronei muscles

Extensor digitorum longus tendons

Gluteal muscles

Tendon of biceps femoris muscle

Tendon of semitendinosus muscle

Lateral epicondyle

Soleus muscle

Calcaneus tendon

Calcaneus

Tendons of peronei muscles

Figure 8.13
Surface anatomy of the lateral surface of the lower limbs.

flexed, the tendon of the *biceps femoris* can be seen passing laterally behind the knee and the tendons of the *semimembranosus* and *semitendinosus* can be identified passing medially behind the knee. Between these two sets of ligaments is a depression called the *popliteal fossa*. In the leg the superficial muscles that can be observed are the *tibialis anterior*, just lateral to the anterior crest of the tibia; the *peroneus longus* and *peroneus brevis*, on the lateral margin; and the *gastrocnemius* and *soleus*, forming the calf of the leg. The common tendon of the latter two muscles, called the *calcaneus (Achilles) tendon*, forms a prominent ridge as it passes to the calcaneus bone. The tendons of the peroneus longus and peroneus brevis muscles can be seen to pass behind the lateral malleolus of the fibula.

STUDY OUTLINE

SURFACE ANATOMY study of structures that give form to the body and provide landmarks on its surface. Bony structures can be located by palpation—that is, feeling with the hand. Muscles are easiest to identify when they are contracted and thus performing their actions. **p. 247**

THE HEAD structures that can be identified include: external occipital protuberance; superior nuchal lines; frontal eminences; parietal eminences; mastoid process; orbital margins; superciliary ridges; supraorbital foramina or notches; glabella; bridge of nose; zygomatic arch; anterior nasal spine; mandibular symphysis; mandibular angle; temporomandibular joint; external acoustic meatus; corrugator; orbicularis oculi; orbicularis oris; mentalis; temporalis; masseter; superficial temporal artery. **pp. 247–250**

THE NECK structures that can be identified include: thyroid cartilage; spinous processes; hyoid bone; sternocleidomastoid; trapezius; external jugular vein; external carotid artery. **p. 250**

THE CHEST structures that can be identified include: clavicle; sternum; ribs; jugular notch; sternal angle; costal margin; nipples; pectoralis major; serratus anterior; axilla; anterior and posterior axillary folds. **p. 250**

THE ABDOMEN structures that can be identified include: iliac crest; anterior superior iliac spine; inguinal ligament; pubic tubercle; pubic symphysis; mons pubis; umbilicus; rectus abdominis; tendinous inscriptions; linea alba; linea semilunaris; external abdominal oblique. **pp. 250–253**

THE BACK structures that can be identified include: spinous processes; spine and inferior angle of scapula; iliac crest; posterior superior iliac spine; sacrospinalis; trapezius; infraspinatus; teres major; latissimus dorsi. **p. 253**

THE UPPER LIMBS structures that can be identified include: acromion process and coracoid process of scapula; greater tubercle and medial and lateral epicondyles of humerus; olecranon process and styloid process of ulna; head and styloid process of radius; pisiform; deltoid; biceps brachii; cubital fossa; triceps brachii; brachioradialis; flexor carpi ulnaris; flexor carpi radialis; palmaris longus; flexor digitorum superficialis; extensor carpi radialis longus and brevis; extensor carpi ulnaris; extensor digitorum; extensor pollicis brevis; extensor pollicis longus. **p. 256**

THE LOWER LIMBS structures that can be identified include: iliac crests; anterior and posterior superior iliac spines; pubic tubercle; symphysis pubis; ischial tuberosity; greater trochanter, medial and lateral condyles, tibial tuberosity, anterior crest, and medial malleolus of tibia; lateral malleolus of fibula; calcaneus; tensor fasciae latae; sartorius; rectus femoris; vastus lateralis; vastus medialis; gracilis; adductor magnus, brevis, and longus; gluteus maximus; biceps femoris; semimembranosus; semitendinosus; tibialis anterior; peroneus longus; peroneus brevis; gastrocnemius; soleus; calcaneus tendon. **pp. 256–262**

SELF-QUIZ

1. Which muscle separates the anterior triangle of the neck from the posterior triangle? (a) trapezius; (b) sternocleidomastoid; (c) pectoralis major.

2. The depression on the upper border of the manubrium of the sternum is called the: (a) jugular notch; (b) sternal angle; (c) cubital fossa.

3. The axilla is a depression between which two muscles? (a) teres major and pectoralis major; (b) serratus anterior and latissimus dorsi; (c) pectoralis major and latissimus dorsi.

4. Tendinous inscriptions may be present in a well-developed: (a) external abdominal oblique; (b) serratus anterior; (c) rectus abdominis.

5. The inferior angle of the scapula is usually level with which vertebra? (a) third thoracic; (b) first lumbar; (c) ninth thoracic.

6. When the hand is adducted, what structure can be palpated on the lateral margin of the forearm? (a) olecranon process; (b) styloid process of the radius; (c) pisiform bone.

7. Which muscle caps the shoulder and is visible from the front as well as the back? (a) deltoid; (b) trapezius; (c) biceps brachii.

8. Which muscle forms the lateral margin of the forearm? (a) flexor carpi ulnaris; (b) brachioradialis; (c) palmaris longus.

9. What structure on the humerus can be palpated on the lateral margin of the arm a short distance below the point of the shoulder? (a) lateral epicondyle; (b) coracoid process; (c) greater tubercle.

10. When the fingers are extended, the tendons of which muscles are prominent on the back of the hand? (a) extensor digitorum; (b) extensor carpi radialis longus; (c) extensor carpi ulnaris.

11. Which structure would you expect to be able to palpate on the medial side of the knee? (a) tibial tuberosity; (b) medial epicondyle of the femur; (c) greater trochanter of the femur.

12. What is the name of the space on the posterior surface of the knee? (a) cubital fossa; (b) posterior triangle; (c) popliteal fossa.

13. What is the name of the prominent bulge on the inside of the ankle? (a) medial malleolus; (b) calcaneus; (c) adductor tubercle.

14. Which of the following muscles is/are not located on the anterior surface of the thigh? (a) semimembranosus; (b) rectus femoris; (c) vastus lateralis.

15. Which of the following muscles can be observed on the lateral margin of the leg? (a) peroneus longus; (b) biceps femoris; (c) peroneus brevis; (d) both a and c.

LEARNING OBJECTIVES

After completing this chapter, you should be able to:

- List the principal functions of the blood.
- List the important proteins and formed elements of the blood.
- Distinguish between plasma, erythrocytes, platelets, and leukocytes, and cite the functions of each.
- Describe the sources of the formed elements of the blood.
- Describe the hemoglobin molecule.
- Name and describe the various types of leukocytes, and give the relative abundance of each.
- Explain the basis by which blood is typed in the ABO and Rh groups.

CHAPTER CONTENTS

The Circulatory System: The Blood

The circulatory system and its fluid component, the **blood**, link the internal environment of the body to the external environment. The blood transports materials between these two environments and between the different cells and tissues of the body.

The volume of blood in both lean males and lean females varies almost directly with body weight and averages approximately 79 milliliters (ml) of blood per kilogram of body weight. However, fat tissue has little vascular volume, and the volume of blood per unit of body weight declines as the proportion of adipose (fat) tissue in the body increases. Since the average female has a greater fat/lean-tissue ratio than the average male, women tend to have a lower blood volume per kilogram of body weight than men. Because of this difference, as well as differences in body size, the general range of total blood volume is 4 to 5 liters in females and 5 to 6 liters in males.

FUNCTIONS OF BLOOD

Since the blood travels throughout the body and supplies just about all the various tissues, it is not surprising that several of its functions depend on its role of providing *transportation*. The main functions of blood include:

1. **Transporting respiratory gases.** The blood carries oxygen from the lungs to the cells of the body and carbon dioxide from the cells to the lungs.

2. **Transporting food materials** from the digestive organs to the cells of the body.

3. **Transporting waste products** from the cells of the body to the kidneys.

4. **Transporting cellular products** such as hormones to the cells of the body.

5. **Maintaining homeostasis** by regulating the pH of body tissues. It does this by means of buffers transported in the blood.

6. **Aiding the regulation of body temperature** by providing a means for the dissipation of heat.

7. **Protecting tissues against toxic foreign materials and organisms** by means of phagocytic cells and antibodies in the blood.

8. **Preventing excessive loss of body fluids,** through the blood-clotting mechanism.

9. **Aiding the regulation of tissue fluid volume and content.**

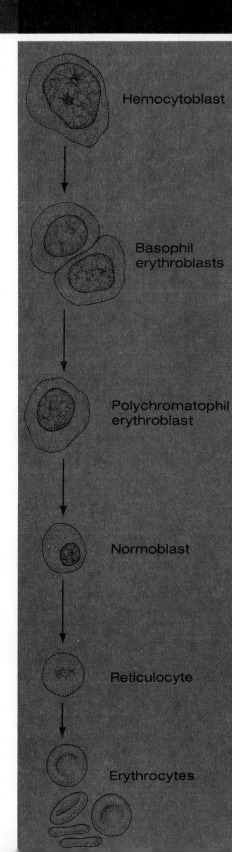

Hemocytoblast

Basophil erythroblasts

Polychromatophil erythroblast

Normoblast

Reticulocyte

Erythrocytes

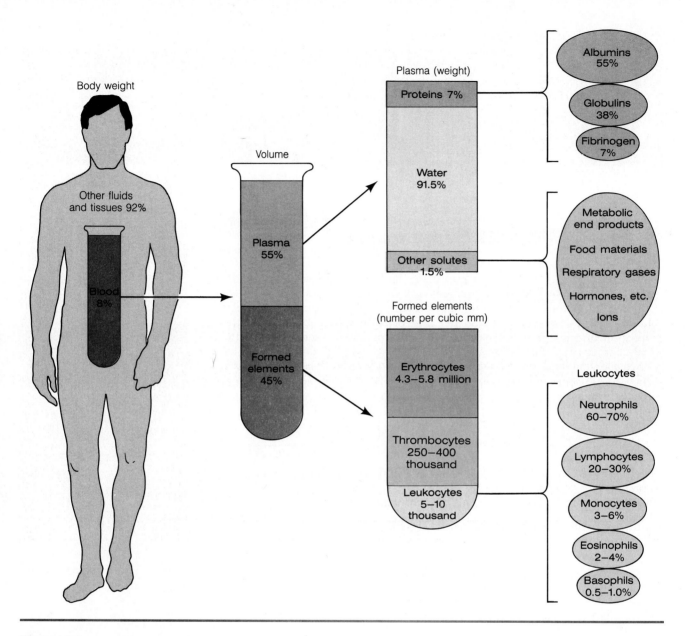

Figure 9.1
Components of blood.

COMPOSITION OF BLOOD

The blood consists of both a liquid component, called *plasma*, and structures in the plasma that are collectively called *formed elements*. The major compo-
F9.1 nents of blood are outlined in Figure 9.1.

Plasma

Plasma, which is approximately 90% water, accounts for about 55% of total blood volume. The portion of plasma that is not water consists of various dissolved or colloidal materials, which give plasma a yellowish color. Hormones and other cellular products are transported by the plasma, as are metabolic end products such as urea. The plasma also contains proteins. Among the most important plasma proteins are various *albumins, fibrinogen* (which is involved in blood-clotting), and *globulins* (some of which act as antibodies in immune responses, others of which serve as transport molecules). The plasma proteins contribute importantly to the osmotic pressure of the plasma

Table 9.1 Formed Elements of the Blood

Formed Element	Approximate Diameter (millimicrons)	Approximate Abundance (per mm³)	Function
ERYTHROCYTES	7.4	Males: 5.1 to 5.8 million Females: 4.3 to 5.2 million	Transport oxygen and aid in the transport of carbon dioxide
PLATELETS	2.5	250,000 to 400,000	Involved in the processes of hemostasis and blood coagulation
LEUKOCYTES			
Granulocytes			
Neutrophils	12 to 14	3000 to 7000 (60 to 70% of total leukocytes)	Phagocytic cells that are capable of ameboid movements
Eosinophils	12	50 to 400 (1 to 4% of total leukocytes)	Phagocytic cells that are believed to destroy antigen–antibody complex
Basophils	9	0 to 50 (0 to 1% of total leukocytes)	Release chemicals such as histamine and heparin, and are involved in inflammatory and allergic responses
Agranulocytes			
Monocytes	15 to 20	100 to 600 (2 to 6% of total leukocytes)	Develop into phagocytic cells called macrophages within the tissue spaces
Lymphocytes	9 (small) 12 to 14 (large)	1000 to 3000 (20 to 30% of total leukocytes)	Involved in specific immune responses, including antibody production

and also to its viscosity. Ions such as sodium (Na^+), chloride (Cl^-), and bicarbonate (HCO_3^-) are also present in the plasma and contribute to its osmotic pressure. Plasma contains food materials such as carbohydrates (for example, glucose), amino acids, and lipids, as well as gases such as oxygen, nitrogen, and carbon dioxide.

Formed Elements

The portion of the blood that is not plasma consists of **formed elements** (Table 9.1). The formed elements include *erythrocytes (red blood cells)*, various types of *leukocytes (white blood cells)*, and *platelets (thrombocytes)*.

Table 9.1

The process by which blood cells are formed is called **hemopoiesis.** Prior to birth, blood cells are produced in a number of tissues, including the liver, spleen, bone marrow, thymus gland, and lymph nodes. The exact source of formed elements of the blood following birth has been a cause of disagreement over the years. It is now generally believed that all types of blood cells are derived from primitive stem cells, called **hemocytoblasts,** that are present in the **red bone marrow** (*myeloid tissue*).

The bone marrow's cellular composition varies in different regions of the skeleton. It also varies with the age of the individual. Nearly all the bones of the fetus contain red marrow; the color indicates that the marrow is capable of producing blood cells. Following birth, the number of active hemocytoblasts decreases in most areas of bone marrow, and they are replaced with fat cells. The abundance of fat cells causes the color of the marrow to change from red to yellow. Therefore, the marrow in which this replacement has occurred is

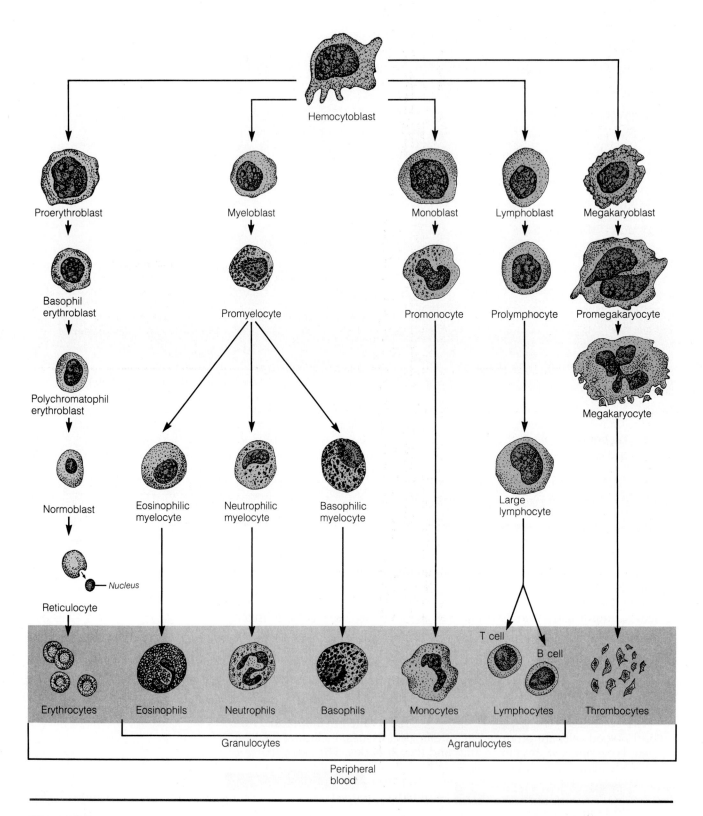

Figure 9.2
Origin and development
of the formed elements
of the blood.

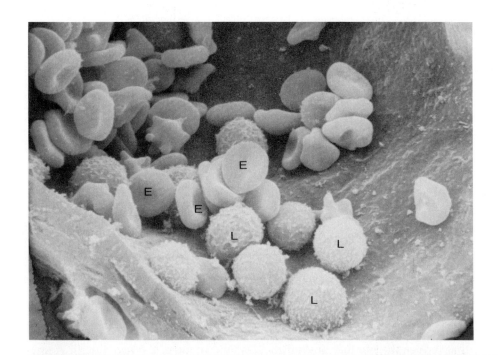

Figure 9.3
Scanning electron micrograph of blood cells within a blood vessel. E: erythrocytes or red blood cells. L: leukocytes or white blood cells (×2480).

referred to as *yellow bone marrow*. Yellow marrow is used primarily for fat storage, and generally is not actively forming blood cells. In the adult most bones contain yellow marrow, with red marrow present only in the ends of certain long bones, the ribs, sternum, vertebrae, and pelvis. The formation of the formed elements of blood therefore normally occurs only in those regions.

Bone marrow consists primarily of immature blood cells and fat cells packed within a fibrous meshwork of reticular connective tissue. It is well supplied with blood from the nutrient artery of the bone. Capillaries from this artery form a system of thin-walled sinusoids, which allow for the slow flow of blood throughout the marrow. The lining of the sinusoids permits blood cells that are produced in the marrow to pass through the walls of the sinusoids and enter the blood.

The hemocytoblasts in red bone marrow are immature cells capable of developing into five types of cells that, in turn, give rise to the mature blood cells (Figure 9.2):   **F9.2**

1. **Proerythroblasts** form red blood cells.

2. **Myeloblasts** form three types of white blood cells called granulocytes.

3. **Lymphoblasts** form a type of white blood cell called lymphocytes.

4. **Monoblasts** form a type of white blood cell called monocytes.

5. **Megakaryoblasts** form cell fragments called platelets.

Erythrocytes

Erythrocytes, or **red blood cells,** are small, circular, biconcave discs approximately 7.5 microns in diameter; they have no nuclei (Figure 9.2, Figure 9.3).   **F9.2, F9.3**
They are the most numerous cell type in the blood. Although their numbers are quite variable, a cubic millimeter of blood may contain about 5.1 to 5.8 million of these cells in males and about 4.3 to 5.2 million in females. The percentage of erythrocytes, by volume, in whole blood is called the *hematocrit.* The hematocrit is determined by centrifuging a sample of blood in a hematocrit tube in order to pack the erythrocytes at the bottom of the tube. The hematocrit is then measured as the ratio or percentage of the packed erythrocyte volume to the total sample volume.

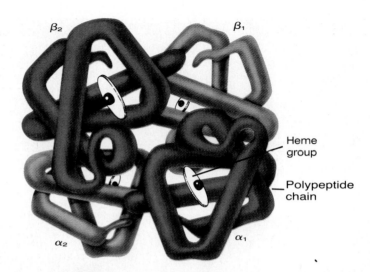

β_2 β_1

Heme group

Polypeptide chain

α_2 α_1

Figure 9.4
A hemoglobin molecule showing its four polypeptide chains, designated α_1, α_2, β_1, and β_2. The structure in the center of each chain represents an iron-containing heme group.

Normal hematocrits average about 45% for males and 42% for females, but considerable variability is noted. Conditions that reduce the fluid portion of the blood, such as dehydration, tend to raise the hematocrit. Conditions that reduce the number of erythrocytes tend to lower the hematocrit. A lowered hematocrit may be observed following excessive blood loss because the lost erythrocytes are replaced much more slowly than the lost fluid. Erythrocytes contribute importantly to total blood viscosity (which is normally 3.5 to 5.5 times that of water). Thus the higher the hematocrit, the greater the viscosity of the blood. As the viscosity of the blood increases, it becomes more difficult to move the blood through the vessels and the heart must work harder to maintain circulation.

FORMATION OF ERYTHROCYTES   During erythrocyte formation, some of the primitive stem cells (hemocytoblasts) in the red bone marrow give rise to
F9.2 cells called **proerythroblasts** (Figure 9.2). The proerythroblasts, in turn, form cells called *basophil erythroblasts*, which synthesize **hemoglobin.** As hemoglobin synthesis continues, the basophil erythroblasts differentiate into cells called *polychromatophil erythroblasts*; these, in turn, give rise to cells called *normoblasts.* When the normoblast cytoplasm has attained a sufficient hemoglobin concentration, the nucleus is pinched off from the rest of the cell. The nonnucleated cells that result are called *reticulocytes.* Reticulocytes are essentially young erythrocytes. It is these cells that are usually released from the bone marrow into the blood. Reticulocytes generally become mature erythrocytes within one or two days after their release from the bone marrow.

The process of erythrocyte formation is called **erythropoiesis.** During erythropoiesis the various cells continue to divide through the normoblast stage of development so that greater and greater numbers of cells are formed. Occasionally, when erythrocytes are being manufactured very rapidly, immature nucleated cells appear in the circulation, but this is not usual.

FUNCTION OF ERYTHROCYTES   Erythrocytes function in the transport of gases by the blood, particularly oxygen and carbon dioxide. It is the hemoglobin in the erythrocytes that is responsible for this activity. Hemoglobin consists of the protein *globin* combined with four nonprotein groups called *hemes*
F9.4 (Figure 9.4). The globin protein consists of four polypeptide chains each of which has a heme group bound to it. Each heme group contains an iron atom that can combine reversibly with one molecule of oxygen. Therefore, each hemoglobin molecule can potentially associate with four oxygen molecules. When hemoglobin is combined with oxygen, it is called *oxyhemoglobin;* when it is not carrying oxygen, it is called *reduced hemoglobin.* In the body, virtually all the hemoglobin that travels to the tissues is oxygenated, whereas about 25% of the hemoglobin that returns from the tissues is in the reduced form. When we consider that a single red blood cell may contain up to 300 million

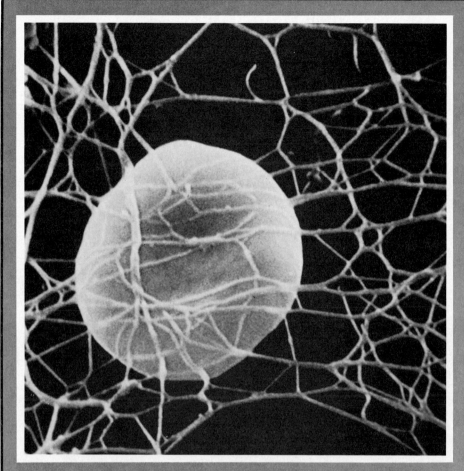

Box 9.1
Erythrocyte in a Fibrin Mesh

An erythrocyte caught in a fibrin mesh of a blood clot. Fibrinogen, a protein produced in the liver, is indispensable to the clotting process. Fibrinogen molecules, through the action of platelets and a number of enzymes, form an insoluble mesh of fibrin. The fibrin captures blood cells and blocks the flow of blood, thus preventing excessive bleeding whenever a blood vessel is damaged.

Blood clotting is a complicated process. To date, 13 different factors have been identified that must be present in the blood for clotting to occur.

 est. × 12,000

hemoglobin molecules and that there may be 5 million erythrocytes per cubic millimeter of blood, the substantial oxygen-carrying capacity of these cells becomes readily apparent. Approximately 198 ml of oxygen can be transported to the tissues by every liter of blood, and about 195 ml of this is associated with hemoglobin molecules within erythrocytes. The remaining 3 ml is carried in physical solution dissolved in the plasma.

In addition to transporting oxygen, hemoglobin can also carry carbon dioxide. However, in contrast to oxygen, which is carried in association with the iron in the heme groups of hemoglobin, carbon dioxide is carried in reversible association with the protein portion of the hemoglobin molecule. Carbon dioxide–protein complexes that involve the globin of hemoglobin are called *carbamino compounds.*

FATE OF ERYTHROCYTES   The average life of an erythrocyte is approximately 120 days in males and 109 days in females. Aged, abnormal, or damaged erythrocytes are disposed of by phagocytic cells called *macrophages.* These cells are found in such tissues as the spleen, liver, and bone marrow. The macrophages degrade hemoglobin to heme and globin. The globin may be further broken down into its component amino acids, which may be resynthesized into new proteins. The iron is liberated from the heme and may be reused in the formation of new hemoglobin or it may be stored, principally in the liver. The remnants of the heme groups are converted to a pigmented compound called *bilirubin,* which enters the blood and is eventually secreted by the liver in bile.

CONTROL OF ERYTHROCYTE PRODUCTION   In order to maintain the proper level of erythrocytes in the blood, erythrocyte production must balance the destruction and removal of erythrocytes from the system. Erythrocyte production is stimulated by a hormone called **erythropoietin,** which is

FRONTIERS IN HEALTH

A Bloodless Battle

In 1966, Dr. Leland C. Clark of the University of Cincinnati's College of Medicine stunned the medical world by immersing a mouse in a bubbling solution of fluorocarbon. The mouse's lungs filled with the oxygen-rich liquid, but nothing traumatic happened. His heart kept beating, and he showed no signs of anoxia (lack of oxygen). After a while, the mouse was pulled from the solution, apparently none the worse for his experience.

What Dr. Clark and his colleague, Dr. Frank Gollan, had done was develop the first prototype of an artificial blood. But was there a need for artificial blood?

Consider the fate of an accident victim in rural America. Miles from a hospital and the closest blood bank, a victim of a severe accident often has little chance of surviving long enough to reach the emergency room. Drs. Clark and Gollan began to work on an artificial blood—or more correctly, a blood substitute—that could be readily available, even in rural areas, and would have the potential of saving thousands of lives each year.

The result of their work was a fluorocarbon emulsion called fluosol. Fluosol has now been approved for use in Canada, Holland, and Italy. The milky white solution contains two fluorocarbons, a number of salts, and water. It carries twice as much oxygen as blood does and contains fine particles one-seventieth the size of red blood corpuscles. Because the particles are so small, fluosol can pass through the occluded arteries of victims of stroke and heart attack.

The Japanese have performed much of the pioneering work on human subjects. The first volunteers to receive the blood substitute were Japanese, and the first clinical trial also took place in Japan. Reported in 1980, this test was performed by Dr. Kenji Honda and his colleagues at the Fukushima Medical College. A 65-year-old patient suffering from a massive bleeding ulcer was rushed to surgery to have his stomach removed, but the blood bank could not supply the matched blood on time. The patient's blood pressure fell dangerously low. To prevent his death, Dr. Honda administered some fluosol; within minutes the blood pressure rose. The patient's life had been saved. Two hours later, when whole blood became available, surgeons completed the removal of the damaged stomach.

Since that historic event, fluosol has been used in dozens of patients requiring emergency surgery when no whole blood was available, or in anemic patients who refused conventional blood transplants before more routine surgery. There have been very few side effects, and it is possible that the Food and Drug Administration will approve fluosol for widespread use in the United States.

Fluosol's applications may extend beyond emergency surgery or surgery on anemic patients. Consider the benefits of using it to sustain the tissues of a brain-dead person, whose organs can then be made available for transplant in various distant hospitals. Consider the benefits of using artificial blood during a war. Finally, consider the case of a patient who had fallen into a deep coma caused by infectious hepatitis. The diseased liver poured toxins into the blood, and the toxins poisoned the liver cells, thus creating a vicious cycle of liver destruction. Dr. Gerald Klebanhoff of Lackland Air Force Base Medical Center used artificial blood to break this cycle. He drained the patient's blood entirely and replaced it with artificial blood. This removed the toxins and allowed the beleaguered liver to begin to recover, while life was sustained by the blood substitute. After a short time, the artificial blood was drained and replaced with whole blood. The comatose patient awoke in the recovery room a few hours after the procedure, attesting to the success of the total blood replacement.

Dr. Anthony Hunt of the University of California, San Francisco, and Dr. Ronald Burnette of the University of Wisconsin, Madison, have recently perfected a new method of producing artificial blood. They have devised a way to use discarded hemoglobin to make miniature red blood corpuscles (RBC), which they call neohemocytes. Neohemocytes, which are microscopic spheres of hemoglobin coated with lipids, are capable of carrying oxygen and are proving to be another substitute for blood.

One of the exciting findings of this research was that neohemocytes are more stable than RBCs. They can be stored in refrigerators for 2 months, which is 5 weeks longer than the body's RBCs can be stored. Because of their smaller size, neohemocytes can slip through constricted blood vessels clogged by cholesterol and blood clots. Victims of heart attack and stroke may therefore have faster recoveries and less tissue damage from transfusions with these promising RBC substitutes.

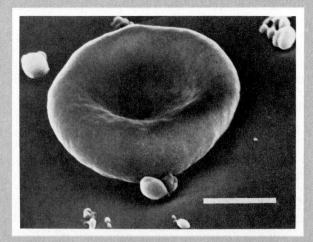

A scanning electron micrograph of a single human erythrocyte surrounded by several neohemocytes. Each neohemocyte encapsulates purified hemoglobin. (The bar is 2.0 microns.) Courtesy of Dr. C. A. Hunt, University of California, San Francisco.

Anemia

Anemia is a condition characterized by either a decreased number of erythrocytes in the blood or by a decreased concentration of hemoglobin. Whatever the cause, anemia decreases the blood's ability to transport oxygen to the tissues. Because tissues cannot function at optimum levels without adequate supplies of oxygen, anemia is frequently associated with listlessness, fatigue, and a lack of energy. Anemia can result from a number of conditions.

Polycythemia

When there is a large increase (6 to 10 million cells per cubic millimeter) in the number of erythrocytes in the blood, the condition is called **polycythemia.** In many cases the cause of the polycythemia is unknown. In other cases, however, it is a response to a deficiency in the supply of oxygen to the tissues (as in chronic lung disease) or to increased levels of erythropoietin (as in some kidney diseases and certain tumors). While polycythemia does increase the amount of oxygen that can be carried in the blood, it also has serious detrimental effects. The presence of large numbers of erythrocytes increases the blood's total volume and its viscosity. These changes, in turn, may elevate the blood pressure. Polycythemia also leads to an increased tendency for clotting to occur in the blood vessels.

produced by cells in the kidneys and possibly elsewhere. Erythropoietin acts mainly on the stem cells in the red bone marrow to increase the rate of formation of hemocytoblasts and to cause proliferation of the hemocytoblasts themselves.

Normally, a relatively small amount of erythropoietin circulates in the blood. This low level of erythropoietin causes a basal level of erythrocyte production. Additional production of erythropoietin, and thus increased erythrocyte production, is triggered by a decreased oxygen supply to the tissues. Conversely, an increased oxygen supply to the tissues causes a decrease in erythropoietin levels and, as a result, a decrease in erythropoiesis. Since a major function of the erythrocytes is to deliver oxygen to the cells of the body, the control of erythrocyte production by tissue oxygen levels through the action of erythropoietin is a logical mechanism.

Platelets

Another formed element found in the blood is the **platelet,** or **thrombocyte** (Figure 9.2). Platelets are small cytoplasmic fragments about 2.5 microns in diameter that have no nuclei and contain many granules. They are formed in the red bone marrow as pinched-off portions of large cells called **megakaryocytes.** There are about 250,000 to 400,000 platelets per cubic millimeter of blood. Platelets are involved in blood-clotting (see Box 9.1), and they have also been implicated in other processes that stop the flow of blood.

F9.2

Box 9.1

Leukocytes

Leukocytes, or **white blood cells** (Figure 9.2, Figure 9.3), are formed elements in the blood that are involved in the body's defense against disease and infection. By means of phagocytosis, they guard the tissues against invasion by foreign organisms or chemicals, and some remove the debris that results from dead or injured cells. Some leukocytes are able to pass through the intact walls of blood vessels and enter the tissue spaces by a process called *diapedesis.* Consequently, they act primarily in the loose connective tissue rather than in the blood. Leukocytes are present within the blood in much smaller numbers than erythrocytes, with between 5,000 and 10,000 leukocytes in a cubic millimeter of blood. Those within the blood are mainly being transported throughout the body from the bone marrow, where they are formed. Many leukocytes are found in lymphoid tissues such as the tonsils, thymus, lymph nodes, spleen, and lymphatic nodules in the linings of the gastrointestinal tract. There are two major classes of leukocytes: *granulocytes* and *agranulocytes.*

F9.2, F9.3

CONDITIONS OF CLINICAL SIGNIFICANCE
Leukocytes

Leukemia

Leukemia is a cancerous condition in which an uncontrolled proliferation of leukocytes leads to a diffuse and almost total replacement of the red bone marrow with leukemic cells. These cells often replace the cells that form erythrocytes, and anemia results. In addition, there is frequently a decrease in the number of blood platelets (which are also formed in the red bone marrow). Since platelets are involved in blood clotting, leukemia may be accompanied by bleeding and hemorrhage. In fact, one of the causes of death in leukemia is internal hemorrhage, especially cerebral hemorrhage. Many of the leukocytes formed in leukemic red bone marrow are immature or abnormal. These abnormal leukocytes are unable to adequately defend the body against invasion by foreign organisms, so leukemia victims can also die from infection.

Infectious Mononucleosis

Infectious mononucleosis, which occurs mostly in children and young adults, is caused by a virus called the Epstein-Barr virus. The disease is characterized by an increase in the relative and absolute numbers of lymphocytes in the blood, and many of the lymphocytes are atypical. The symptoms of infectious mononucleosis include fatigue, sore throat, and slight fever. Recovery is usually complete and without any ill effects.

GRANULOCYTES **Granulocytes** have clearly evident granules in their cytoplasm. They are formed in the red bone marrow from stem cells called **myeloblasts.** Three types of granulocytes are distinguished on the basis of their reaction to certain stains. They are *neutrophils, eosinophils,* and *basophils* **F9.2** (Figure 9.2).

Neutrophils possess very small cytoplasmic granules that appear light reddish-purple when stained with Wright's blood stain. Because they typically have a nucleus that varies in shape and consists of two or more lobes connected by narrow strands, neutrophils are also referred to as *polymorphonuclear leukocytes.* (However, this term is also sometimes used to refer to all three types of granulocytes, since the shape of the nucleus varies somewhat in all of them.) Neutrophils are the most abundant type of leukocyte, comprising approximately 60% of the total white blood cell count. They are phagocytic cells capable of ameboid movement. Neutrophils are able to leave the blood vessels and enter the tissues, where they protect the body by ingesting bacteria and other foreign substances.

Eosinophils possess coarse cytoplasmic granules that appear reddish-orange when stained with Wright's stain. Their nucleus generally has two lobes connected by a narrow strand. Eosinophils are capable of both phagocytosis and ameboid movement, and they are believed to ingest and destroy antigen–antibody complexes. Less than 4% of leukocytes are eosinophils; however, their number increases during certain parasitic infections and in conditions involving allergic hypersensitivity (for example, asthma and hay fever).

Basophils possess relatively large cytoplasmic granules that appear purplish-blue when stained with Wright's stain. The nucleus of basophils is elongated and often bent in a U or S shape. Basophils comprise less than 1% of all white blood cells. They are not phagocytic; rather, it is believed, they are able to release the chemical substances histamine and heparin. Histamine causes vascular dilation and increased blood vessel permeability in inflammation and contributes to allergic responses. Heparin can prevent blood coagulation. Basophils function similarly to *mast cells,* which are found in connective tissue.

AGRANULOCYTES Some leukocytes, called **agranulocytes** or *nongranular leukocytes,* do not have prominent granules in their cytoplasm. There are two **F9.2** types of agranulocytes: *monocytes* and *lymphocytes* (Figure 9.2).

Monocytes have a single, large nucleus that is oval or indented. They are the largest of the leukocytes. They comprise up to 6% of the white blood cells. Monocytes are derived from **monoblasts.** They are capable of ameboid move-

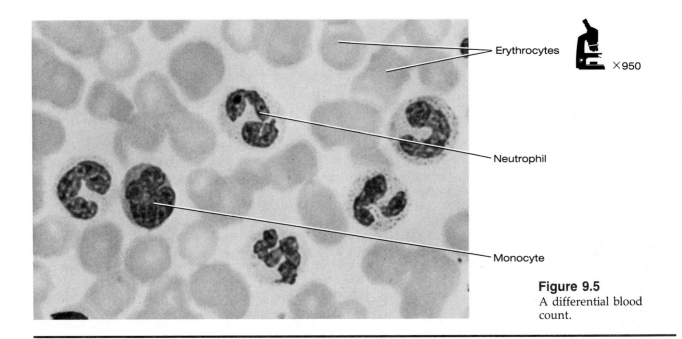

Erythrocytes

×950

Neutrophil

Monocyte

Figure 9.5
A differential blood
count.

ment and leave the blood vessels and enter the loose connective tissue, where they develop into large phagocytic cells called *macrophages,* which can ingest bacteria and other foreign substances. However, monocytes are not the body's only source of macrophages. Other macrophages are capable of undergoing mitosis to increase their numbers.

Lymphocytes are the second most abundant type of leukocyte (after neutrophils), comprising up to 30% of the circulating white blood cells. They are formed from **lymphoblasts.** Lymphocytes are small, being only slightly larger than erythrocytes, and each has a nucleus that is either round or slightly indented on one side. Only a thin rim of cytoplasm surrounds the nucleus. On the basis of their size, lymphocytes are classified as *small* or *large.* Comparatively few of the total number of lymphocytes are found in the circulation; most are lodged in the lymphoid tissues of the body. Lymphocytes are important in the body's specific immune responses, including antibody production.

It is important for diagnostic purposes to be able to estimate the relative abundance of each type of leukocyte in the blood. This can be done by a procedure called a **differential count** (Figure 9.5), which involves staining a **F9.5** blood smear, then identifying and counting the leukocytes under a microscope. A differential blood count can also be obtained by placing a blood sample in a specialized machine that automatically identifies and tabulates the various types of leukocytes. The count provides a percentage figure for each type of leukocyte. The percentages of the different types of leukocytes present in the blood change in certain diseases and can, in some cases, greatly aid in their diagnosis.

BLOOD GROUPS

Antibodies are proteins produced in the body in response to the presence of substances called *antigens.* The antibodies neutralize or otherwise react with the specific antigens that stimulate their production. The surfaces of erythrocytes contain antigens that may react with the appropriate antibodies present in the plasma of the blood. The reactions between antigens and antibodies form the basis of the various blood classifications. The surface antigens most often considered are those of the ABO system.

ABO System

The antigens of the **ABO system,** which are inherited, are designated A and B. The absence of A and B antigens on the erythrocytes is designated O. A person has either antigen A, antigen B, both antigen A and antigen B (AB), or neither antigen A nor antigen B (O). In contrast to the normal immune response, which requires prior exposure to an antigen for the production of antibodies, antibodies against A and B are normally found in relatively high concentrations in the plasma. A person has antibodies to those antigens that are not on his or her erythrocytes but does not have antibodies to those antigens that are present on his or her erythrocytes. Thus, people with antigen A have anti-B antibodies in their plasma; people with antigen B have anti-A antibodies; people who have neither antigen A nor antigen B (O) have both anti-A and anti-B antibodies; and people who possess both antigen A and antigen B on their erythrocytes have neither anti-A nor anti-B antibodies in their plasma. The individual's blood type indicates the *antigens* he or she possesses, not the antibodies (Table 9.2).

Table 9.2

In blood transfusions, difficulties can arise from the mixing of incompatible blood types. If a person with type A blood (antigen A on erythrocytes, anti-B antibodies in plasma), for example, were transfused with type B blood (antigen B on erythrocytes, anti-A antibodies in plasma), the incoming type B erythrocytes would be attacked and clumped by the recipient's anti-B antibodies, thus causing hemoglobin to be released into the plasma. At the same time, the incoming anti-A antibodies of the type B blood could attack the type A erythrocytes of the recipient, although this problem is not usually as serious, since the incoming antibodies are diluted in the recipient's plasma. Since type AB individuals possess no antibodies to attack incoming erythrocytes, they are often called *universal recipients.* On the other hand, since the erythrocytes of type O individuals have no antigens and therefore will not be attacked by either anti-A or anti-B antibodies, these people are called *universal donors.* These terms are misleading, however, because there are many other erythrocyte antigens and plasma antibodies that can cause transfusion problems. Therefore, blood for transfusion should be closely matched to the blood of the recipient.

Rh System

Another erythrocyte antigen–antibody system is the **Rh system** (so named because of its initial study in rhesus monkeys). The Rh system consists of a group of surface antigens on erythrocytes, and the antibody component of the system—the anti-Rh antibodies—is normally *not* present in the plasma. In order to have anti-Rh antibodies produced, sensitization to the Rh antigens must take place. As with the ABO system, a person with Rh antigens does not produce anti-Rh antibodies against his or her own antigens. However, a person without the Rh antigens may produce anti-Rh antibodies upon sensitization to the Rh antigens. Such a sensitization can occur when a mother without the Rh antigens on her erythrocytes (designated *Rh negative*) carries a fetus who possesses the antigens (designated *Rh positive*) inherited from the father. In this case, it may be possible for some of the fetal Rh antigens to enter the maternal circulation through tears in the placenta and thereby sensitize the mother by stimulating the production of anti-Rh antibodies in her plasma. If, after having been sensitized, a mother carries an Rh-positive fetus, the maternal antibodies may enter the fetal circulation, where they will attack and rupture fetal erythrocytes. This can result in *erythroblastosis fetalis (hemolytic disease of the newborn),* a severe anemic condition in the fetus.

Only about 5% of Rh-negative mothers actually produce anti-Rh antibodies while carrying an Rh-positive fetus, and the first baby is almost always safe; later pregnancies, however, are more risky because the immune system has, in effect, a memory. It is now possible immediately following delivery to inject Rh-negative mothers with agents that prevent or limit sensitization to Rh antigens. Thus the Rh-incompatibility problem between mother and fetus has greatly lessened in recent years.

Table 9.2   Summary of the ABO System

Blood Type	Antigens on Erythrocytes	Antibodies in Plasma
A	A	Anti-B
B	B	Anti-A
AB	Both A and B	Neither anti-A nor anti-B
O	Neither A nor B	Both anti-A and anti-B

STUDY OUTLINE

FUNCTIONS OF BLOOD include: transportation of respiratory gases, food materials, waste products, antibodies, and cellular products; regulation of pH, body temperature, and body fluids; protection against foreign materials. **p. 265**

COMPOSITION OF BLOOD **pp. 266–275**

PLASMA 90% water; hormones and metabolic end products; proteins (albumins, fibrinogen, globulins); ions (sodium, chloride, bicarbonate); food materials; gases.

FORMED ELEMENTS all develop from stem cells called *hemocytoblasts* that are present in the red bone marrow of certain long bones and such bones as the ribs, sternum, vertebrae, and pelvis. Hemocytoblasts are capable of differentiating into five cell types: (1) proerythroblasts, (2) myeloblasts, (3) lymphoblasts, (4) monoblasts, (5) megakaryoblasts.

ERYTHROCYTES (RED BLOOD CELLS) small, circular, biconcave discs with no nuclei.

Formation of Erythrocytes (Erythropoiesis)
Hemocytoblasts → proerythroblasts → basophil erythroblasts → polychromatophil erythroblasts → normoblasts → reticulocytes → erythrocytes.

Function of Erythrocytes gas transport:
1. Oxygen carried by iron of hemoglobin (oxyhemoglobin).
2. Carbon dioxide carried by protein of hemoglobin (carbamino compounds).

Fate of Erythrocytes 120-day life.
1. Macrophages of spleen, liver, and bone marrow degrade hemoglobin to heme and globin.
2. Iron liberated from heme is stored or used in the formation of new hemoglobin.

Control of Erythrocyte Production
1. Erythropoietin, produced by cells in kidneys, stimulates stem cells in red bone marrow to increase hemocytoblast formation and proliferation.
2. Tissue oxygen levels influence production of erythropoietin.

CONDITIONS OF CLINICAL SIGNIFICANCE: ERYTHROCYTES **p. 273**

ANEMIA decreased number of erythrocytes or decreased concentration of hemoglobin in the blood.

POLYCYTHEMIA increased number of erythrocytes in blood.

PLATELETS (THROMBOCYTES) granular cytoplasmic fragments with no nuclei; formed from megakaryocytes in red bone marrow; involved in blood-clotting.

LEUKOCYTES (WHITE BLOOD CELLS) in blood and lymphoid tissues; some are phagocytic.

Granulocytes granules in cytoplasm; formed from myeloblasts; protect outside circulatory system. There are three types:
1. *Neutrophils:* most abundant type; granules stain light reddish-purple; phagocytic; nuclei have several lobes.
2. *Eosinophils:* granules stain reddish-orange; phagocytize antigen–antibody complexes.
3. *Basophils:* granules stain purplish-blue; release histamine and heparin; functionally similar to mast cells.

Agranulocytes no granules in cytoplasm. There are two types:
1. *Monocytes:* single nucleus; large cell; formed from monoblasts; develop into macrophages in loose connective tissue.
2. *Lymphocytes:* most are located in lymphoid tissue; formed from lymphoblasts; important in body's specific immune responses, including antibody production.

CONDITIONS OF CLINICAL SIGNIFICANCE: LEUKOCYTES **p. 274**

LEUKEMIA a cancerous condition in which there is excessive, uncontrolled proliferation of leukocytes, leading to diffuse and almost total replacement of red bone marrow with leukemic cells.

INFECTIOUS MONONUCLEOSIS viral disease characterized by an increase in relative and absolute numbers of lymphocytes in the blood.

BLOOD GROUPS   pp. 275–277

ABO SYSTEM

1. Based on inherited antigens (A, B) on erythrocyte surfaces.
2. Person does not have antibodies to antigens present on erythrocytes.
3. Person has antibodies to those antigens not present on erythrocytes.

4. Incompatible blood types cause clumping and rupture of erythrocytes during transfusions.

RH SYSTEM

1. Surface antigens present on erythrocytes; anti-Rh antibodies not normally present in plasma.
2. Rh incompatibility between Rh-negative mother and Rh-positive fetus.

SELF-QUIZ

1. As the unit of adipose (fat) tissue increases, the volume of blood per unit of body weight: (a) increases; (b) declines; (c) remains constant.

2. Among the substances included in the blood's formed elements are: (a) albumins; (b) leukocytes; (c) platelets; (d) globulins; (e) b and c; (f) a and d; (g) all of these.

3. All the formed elements of the blood develop from stem cells called hemocytoblasts that are located in the red bone marrow. True or false?

4. Match the terms associated with erythrocytes with the appropriate description.

   Hemoglobin
   Heme
   Erythropoietin
   Reticulocytes
   Globin
   Macrophages

   (a) Nonnucleated cells that develop into erythrocytes.
   (b) These cells dispose of hemoglobin released from old, damaged, or fragmented erythrocytes.
   (c) An oxygen-carrying compound in erythrocytes.
   (d) The component of hemoglobin that contains iron.
   (e) A compound that controls the rate of red blood cell production.
   (f) The protein component of hemoglobin.

5. Anemia is a condition that can be characterized by a decreased: (a) number of erythrocytes; (b) percentage of hemoglobin; (c) number of granulocytes; (d) number of platelets; (e) a and b; (f) a, b, c; (g) all of these.

6. Platelets are cytoplasmic fragments that are formed from megakaryocytes. True or false?

7. Some leukocytes leave the blood vessels and destroy pathogens outside the circulatory system. This involves: (a) diapedesis; (b) elevated blood pressure; (c) phagocytosis; (d) decrease in hemoglobin content; (e) both a and c.

8. Which is the most common type of leukocyte? (a) basophil; (b) monocyte; (c) neutrophil; (d) eosinophil; (e) platelet.

9. All agranulocytes develop from stem cells located in the lymphoid tissue. True or false?

10. The most strongly phagocytic granulocytes are the: (a) neutrophils; (b) lymphocytes; (c) basophils.

11. Match the following formed elements of the blood with the appropriate description.

    Erythrocytes
    Platelets
    Neutrophils
    Eosinophils
    Basophils
    Monocytes
    Lymphocytes

    (a) The most abundant type of leukocyte.
    (b) These cells are involved in specific immune responses.
    (c) These cells are converted to macrophages in the tissue.
    (d) These cells phagocytize antigen–antibody complexes.
    (e) These cells are involved in the transport of CO_2 and O_2.
    (f) These cells release histamine and are involved in allergic responses.
    (g) These cells are involved in blood coagulation.

12. Blood types are named according to the reactions between antigens on the erythrocyte and antibodies in the plasma. True or false?

13. A person who has antigen B also has: (a) anti-B antibodies; (b) anti-A antibodies; (c) both types of antibodies.

14. People with type O blood are considered to be universal recipients because they lack antibodies in their serum. True or false?

15. Like the ABO blood classification, the Rh system also consists of surface antigens on the erythrocyte and the presence or absence of antibodies in the plasma. True or false?

LEARNING OBJECTIVES

After completing this chapter, you should be able to:

- Locate the borders and valves of the heart on the surface of the thorax.

- Describe the coverings of the heart.

- Describe the path of bloodflow through the heart.

- Describe the structure of the heart, including its "skeleton."

- Explain the mechanical actions that produce the normal heart sounds ("lub-dup").

- Describe the conducting system of the heart.

- Distinguish between bradycardia and tachycardia, and describe the conditions called flutter and fibrillation.

- Describe some common cardiac abnormalities.

CHAPTER CONTENTS

EMBRYONIC DEVELOPMENT OF THE HEART

POSITION OF THE HEART

COVERINGS OF THE HEART

ANATOMY OF THE HEART

CIRCULATION THROUGH THE HEART

CONDUCTING SYSTEM OF THE HEART

CONDITIONS OF CLINICAL SIGNIFICANCE:
  THE HEART

The Circulatory System: The Heart

The circulatory system can be separated into two divisions: the **cardiovascular system** and the **lymphatic system.** The cardiovascular system includes the *heart*, which serves as a pump for the blood, and the *blood vessels*, which transport the blood throughout the body. The lymphatic system consists of *organs* that play a role in specific immune responses (tonsils, thymus, spleen, lymph nodes, and lymphatic nodules) and *vessels* that collect tissue fluid from between the cells of the body and transport it to the cardiovascular system. In this chapter we will study the heart. The blood vessels are discussed in Chapter 11 and the lymphatic system in Chapter 12.

The cardiovascular system is a closed circular system. Confined to the heart and the numerous vessels, blood continuously travels a circular route through the heart, into arteries, then to capillaries, into veins, and back to the heart. Normally, blood does not leave this system, although some of the fluid part of the blood does pass through the walls of the capillaries to join the tissue fluid between the cells. However, even this fluid is returned to the cardiovascular system directly or by way of the lymphatic system. The heart is the pump that provides the force necessary to keep the blood flowing through the system of vessels.

EMBRYONIC DEVELOPMENT OF THE HEART

Early in embryonic development the future heart is a simple pulsating tube that receives blood from the veins at its posterior end and pumps it into the arterial system through its anterior end (Figure 10.1*a*). From this simple beginning the heart must not only respond to the changing needs of the developing embryo but must also be capable of functioning under the vastly different conditions immediately following birth. To accomplish this, the tubular heart must develop into a four-chambered organ complete with valves and a midline partition. And the heart must undergo these alterations without interrupting its delivery of blood to the developing embryo.

By the fifth week of development it becomes apparent that the heart is beginning to undergo changes as it grows rapidly (Figure 10.1*b*) and evolves into an S-shaped structure (Figure 10.1*c*). With continued development, the anterior vessel, which carries blood away from the heart, divides into two vessels: the future *pulmonary trunk*, which supplies blood to the lungs, and the *aorta*, which carries blood to the vessels that supply the rest of the body. At the same time a midline septum (wall) is developing within the heart. When completed, this wall will separate the bloodflow through the heart into two channels: the blood on one side of the wall passes through the pulmonary trunk to the lungs; that on the other side is directed into the aorta and thus to the rest of the body (Figure 10.1*d*).

By the seventh week the embryonic heart has further divided into the four chambers it will retain in the adult—two atria and two ventricles. Through a series of changes, including the development of new segments in some veins and the degeneration of portions of other veins, the vessels that

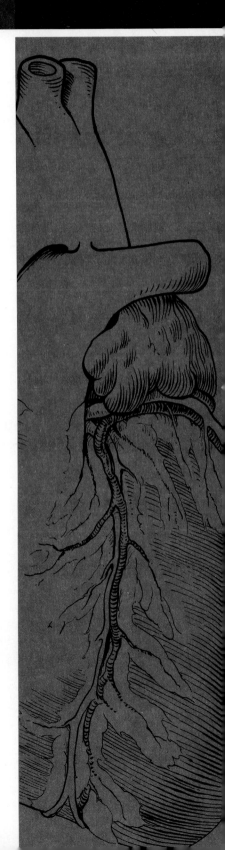

281

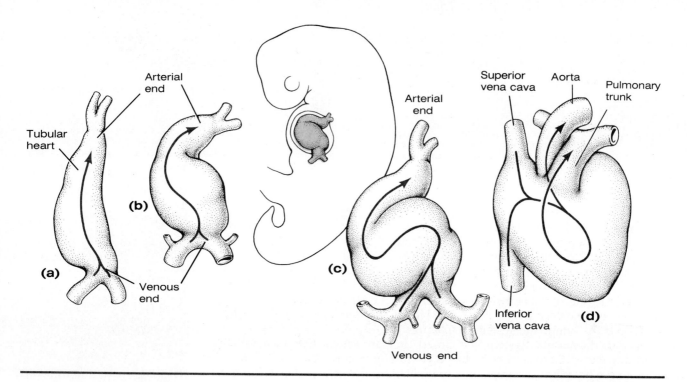

Figure 10.1
Successive stages in the
embryonic development
of the heart. The arrows
indicate the direction of
bloodflow.

enter the posterior (venous) region of the heart develop into *superior* and *inferior vena cavae,* which return all the venous blood from the body to the heart. No other major changes occur in the heart until birth.

POSITION OF THE HEART

F10.2 The adult heart is a cone-shaped organ about the size of a closed fist. It is located between the lungs in the space called the **mediastinum** (Figure 10.2). The heart lies obliquely in the mediastinum, and is described as having a base and an apex, diaphragmatic and sternocostal surfaces, and four borders.

The *base* of the heart faces posteriorly and is located superiorly and to the right, behind the sternum, at about the level of the second and third ribs. It consists mainly of the left atrium, part of the right atrium, and the proximal portions of the large veins that enter the posterior wall of the heart. From the base the heart projects downward, anteriorly, and to the left, ending in a blunt *apex.* The apex reaches the fifth intercostal space, about 8 cm to the left of the midsternal line. The *diaphragmatic surface* of the heart is that part between the base and the apex that rests upon the diaphragm. It involves the left and right ventricles. The anterior surface of the heart, which is formed mainly by the right ventricle and the right atrium, is referred to as the *sternocostal surface.*

The *superior border* of the heart is formed by both atria, and is the region where the great vessels enter and leave the heart. It lies at about the level of the second intercostal space. The *inferior border* extends from behind the lower portion of the sternum to the left fifth intercostal space, where it ends at the apex. It is formed mostly by the right ventricle, plus a small portion of the left ventricle at the apex. The *right border* of the heart is formed by the right atrium, and is located about 2.5 cm to the right of the sternum. The *left border* is formed mainly by the left ventricle, with the left atrium forming the upper portion. The left border extends to the apex of the heart from the level of the junction of the left second rib with its costal cartilage.

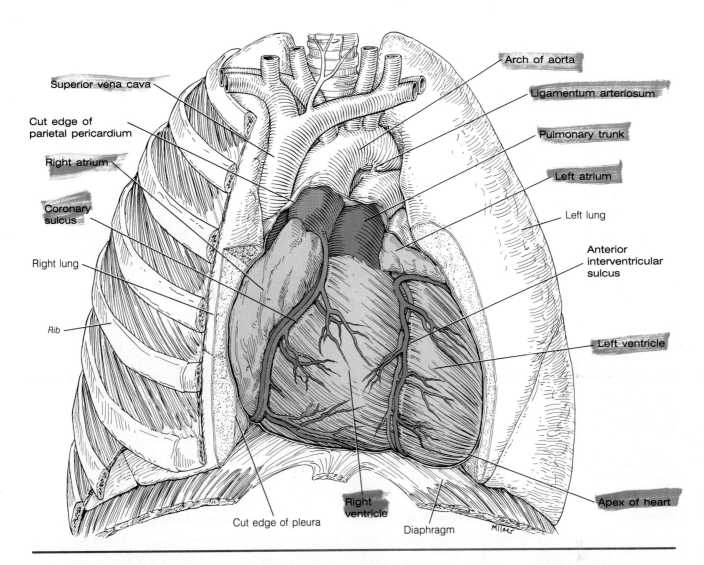

Superior vena cava

Cut edge of
parietal pericardium

Right atrium

Coronary
sulcus

Right lung

Rib

Arch of aorta

Ligamentum arteriosum

Pulmonary trunk

Left atrium

Left lung

Anterior
interventricular
sulcus

Left ventricle

Apex of heart

Cut edge of pleura

Right
ventricle

Diaphragm

Figure 10.2
Frontal view of the
thorax showing the
position of the heart in
the mediastinum.

COVERINGS OF THE HEART

The heart is enclosed in a double-walled membranous sac called the **pericardium.** The inner layer of the pericardium, called the **epicardium** or the **visceral pericardium,** is a serous membrane with a surface layer of mesothelium overlying a thin layer of loose connective tissue that adheres to the outer surface of the heart (Figure 10.3). At the point where the large vessels enter and leave the heart, the serous layer of the visceral pericardium folds back and is continuous with the outer layer of the pericardium, the **parietal pericardium.** The parietal pericardium is composed of two layers: an outer *fibrous* layer, which strengthens it and anchors it within the mediastinum, and an inner *serous* layer, which lines the inside of the fibrous layer and is continuous with the serous layer of the visceral pericardium. Between the serous membranes of the visceral and parietal layers is a small space called the **pericardial cavity.** This cavity contains **pericardial fluid,** which is secreted by the cells of the serous membranes of the pericardium. The fluid lubricates the membranes, permitting them to slide over one another with a minimum of friction as the heart beats.

Inflammation of the pericardium, which is referred to as *pericarditis*, can result from a variety of causes. The amount and character of the pericardial fluid varies in the different forms of pericarditis. In some cases, the pericardial fluid is scanty; in others, it is abundant; and some infections produce fibrin,

F10.3

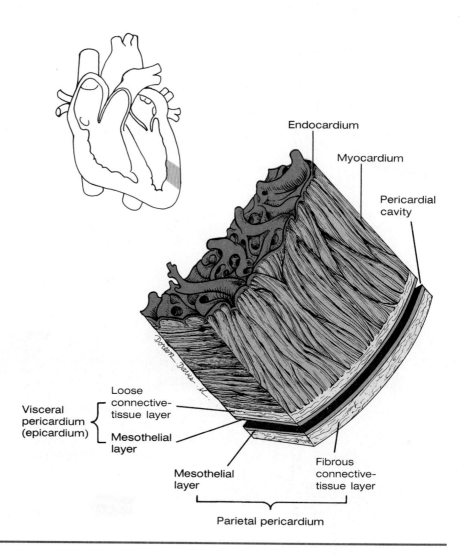

Figure 10.3
Section through the wall of the heart showing the parietal and visceral layers of the pericardium (heart sac), the myocardium (heart muscle), and the endocardium (inner lining of the heart chamber).

others produce pus, in the pericardial cavity. In pericarditis, the serous layers of the pericardium become roughened, which causes pain as they move over one another and also interferes with the normal filling of the heart chambers.

ANATOMY OF THE HEART

In order to function as a pump, the heart must have both receiving and delivery chambers, valves to direct the flow of blood through the heart, a wall that is strongly compressible and thus provides the force to propel blood, and vessels to deliver blood to and from the heart.

Chambers of the Heart

F10.4–F10.6

The heart consists of four chambers: **right** and **left atria** and **right** and **left ventricles** (Figure 10.4, Figure 10.5, Figure 10.6). The atria are small and are located toward the superior region of the heart. The ventricles are larger and compose the bulk of the heart. Located inferiorly, they form the apex of the heart. The right ventricle forms most of the heart's anterior surface; the left ventricle forms most of the inferior surface and left margin of the heart. The atria are separated from one another by an **interatrial septum.** The ventricles are separated by an **interventricular septum.**

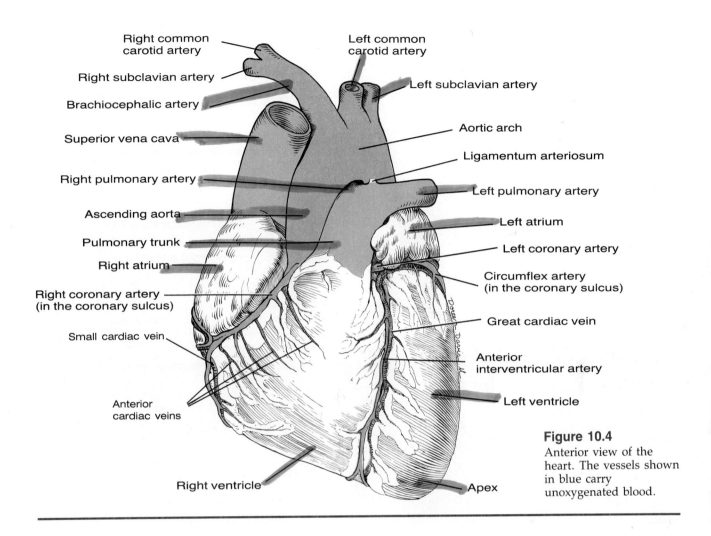

Right common carotid artery

Left common carotid artery

Right subclavian artery

Left subclavian artery

Brachiocephalic artery

Aortic arch

Superior vena cava

Ligamentum arteriosum

Right pulmonary artery

Left pulmonary artery

Ascending aorta

Left atrium

Pulmonary trunk

Left coronary artery

Right atrium

Circumflex artery (in the coronary sulcus)

Right coronary artery (in the coronary sulcus)

Great cardiac vein

Small cardiac vein

Anterior interventricular artery

Left ventricle

Anterior cardiac veins

Right ventricle

Apex

Figure 10.4
Anterior view of the heart. The vessels shown in blue carry unoxygenated blood.

Vessels Associated with the Heart

Several large vessels enter or leave the base and the superior border of the heart (Figure 10.6):

F10.6

1. the **superior** and **inferior venae cavae,** which return venous blood from the vessels of the body to the right atrium.

2. the **pulmonary trunk,** which divides into **right** and **left pulmonary arteries** and carries blood from the right ventricle to the lungs.

3. the four **pulmonary veins,** which carry blood from the lungs to the left atrium.

4. the **aorta,** which carries blood from the left ventricle into the vessels that supply the body.

Wall of the Heart

The heart is composed primarily of cardiac muscle anchored to a fibrous skeleton.

Epicardium, Myocardium, and Endocardium

The wall of the heart is formed of three layers: the epicardium, the myocardium, and the endocardium (Figure 10.3). The **epicardium (visceral pericardium)** is a thin serous membrane that adheres to the outer surface of the

F10.3

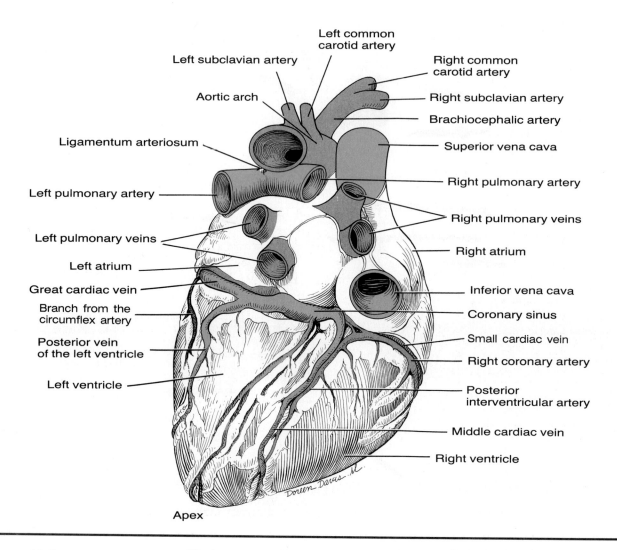

Left common carotid artery

Left subclavian artery

Aortic arch

Right common carotid artery

Right subclavian artery

Brachiocephalic artery

Ligamentum arteriosum

Superior vena cava

Left pulmonary artery

Right pulmonary artery

Right pulmonary veins

Left pulmonary veins

Right atrium

Left atrium

Great cardiac vein

Inferior vena cava

Branch from the circumflex artery

Coronary sinus

Posterior vein of the left ventricle

Small cardiac vein

Right coronary artery

Left ventricle

Posterior interventricular artery

Middle cardiac vein

Right ventricle

Apex

Figure 10.5
Posterior-inferior view of the heart. The vessels shown in blue carry unoxygenated blood.

heart. The thickest layer of the wall of the heart, the **myocardium,** is composed of cardiac muscle. The myocardium is lined on the inside by the **endocardium,** which is composed of connective tissue and a surface layer of squamous cells. Foldings of the endocardium form the valves that separate the atria from the ventricles—the *atrioventricular valves*—and the ventricles from the aorta and the pulmonary trunk—the *semilunar valves*. The endocardial lining of the heart is continuous with the endothelium that lines all the arteries, veins, and capillaries of the body.

The myocardium varies considerably in thickness from one heart chamber to another. Its thickness is related to the resistance encountered in pumping the blood from the different chambers. Since the muscles of the atria meet little resistance in pushing the blood into the ventricles, the walls of the atria **F10.6** are the thinnest part of the myocardium (Figure 10.6). In contrast, the ventricles must move the blood through the blood vessels of either the lungs or the rest of the body and back into a receiving chamber. Consequently, the myocardium of the ventricles is thicker than that of the atria. Moreover, the left ventricle, which propels blood through all parts of the body (other than the lungs) and back into the right atrium, has thicker walls than the right ventricle, which only moves the blood through the blood vessels of the lungs and back into the left atrium. However, even though the left side of the heart is capable of pumping with greater force than the right side, equal amounts of blood are normally moved by each side. If this were not the case, the relative amounts of blood in the blood vessels of the lungs and those in the rest of the body would fluctuate. Such fluctuation can be a problem in some diseases of the heart.

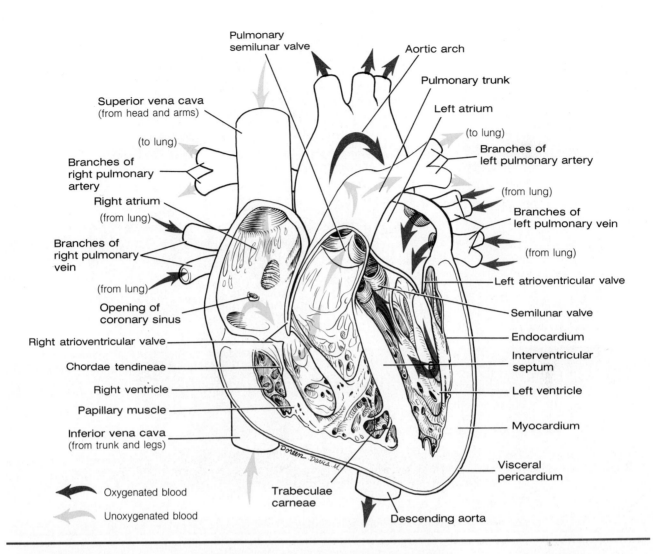

Pulmonary
semilunar valve

Aortic arch

Pulmonary trunk

Superior vena cava
(from head and arms)

Left atrium

(to lung)

(to lung)

Branches of
left pulmonary artery

Branches of
right pulmonary
artery

(from lung)

Right atrium

Branches of
left pulmonary vein

(from lung)

(from lung)

Branches of
right pulmonary
vein

Left atrioventricular valve

(from lung)

Semilunar valve

Opening of
coronary sinus

Endocardium

Right atrioventricular valve

Interventricular
septum

Chordae tendineae

Right ventricle

Left ventricle

Papillary muscle

Myocardium

Inferior vena cava
(from trunk and legs)

Visceral
pericardium

Oxygenated blood

Trabeculae
carneae

Descending aorta

Unoxygenated blood

The inner surface of the myocardium of the ventricles is irregular, having folds and bridges called **trabeculae carneae** and cone-shaped **papillary muscles** that project into the lumen (Figure 10.6). Strong fibrous strands called **chordae tendineae** run from the papillary muscles to the cusps (flaps) of the atrioventricular valves.

F10.6

Figure 10.6
Frontal section of the heart. The arrows indicate the path of bloodflow through the chambers, the valves, and the major vessels. Note that the branches of the right pulmonary vein pass behind the heart and enter the left atrium.

Skeleton of the Heart

Horizontal **fibrous rings** surround the atrioventricular openings and the openings of the aorta and pulmonary trunk. The rings are fused together by additional fibrous tissue called **fibrous trigones** (Figure 10.7). Collectively, these fibrous supports are referred to as the **skeleton of the heart.** This fibrous skeleton not only serves as the attachment for the heart muscle and valves but also helps to form the septa that separate the atria from the ventricles.

F10.7

Vessels of the Myocardium

The myocardium receives an abundant blood supply through the **right** and **left coronary arteries** (Figure 10.4, Figure 10.5, Box 10.1). These arteries arise from the aorta just as it leaves the superior border of the heart (Figure 10.7). The coronary arteries obtain their blood from sinuses behind the cusps of the aortic semilunar valve (Figure 10.8).

F10.4, F10.5,
Box 10.1
F10.7

F10.8

The **right coronary artery** arises from the right anterior surface of the aorta and passes to the right margin of the heart in a groove called the **coronary sulcus** (Figure 10.4). The coronary sulcus separates the atria from the ventricles. The right coronary artery extends around the margin of the heart to the posterior surface, sending branches to the right atrium and the right ventricle.

F10.4

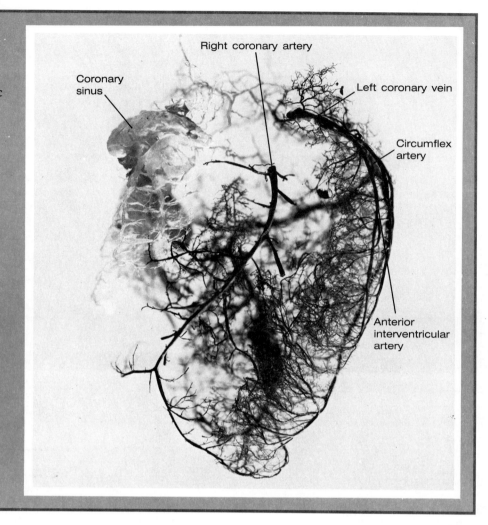

On the posterior surface of the heart, the main branch of the artery turns downward toward the apex of the heart, running in the groove between the right and left ventricles. This branch, the **posterior interventricular artery,** F10.5 supplies smaller branches to both ventricles (Figure 10.5).

The **left coronary artery** arises from the left anterior surface of the aorta, behind the pulmonary trunk; after passing a short distance toward the left margin of the heart, it divides into anterior interventricular and circumflex branches. The **anterior interventricular artery** travels down the anterior surface of the interventricular septum toward the apex of the heart. It supplies branches to both ventricles. The **circumflex artery** travels in the coronary sulcus, between the left atrium and the left ventricle, to the left margin of the heart. After passing around the left margin, the circumflex artery supplies branches to the posterior surface of the left atrium and the diaphragmatic surface of the left ventricle.

Any narrowing or blockage of the coronary arteries interferes with the oxygen supply to the myocardium, which causes regions of the cardiac muscle to die. This condition, which can be disabling or fatal, is commonly called a heart attack. If the blockage of the coronary arteries is only temporary, resulting in an inadequate oxygen supply to the myocardium for only a few seconds, a sharp pain in the chest, which often radiates down the left arm, may result. This pain is called *angina pectoris.*

After passing through an extensive capillary network, the blood from the coronary arteries enters the **cardiac veins** that travel alongside the arteries F10.4, (Figure 10.4, Figure 10.5). The cardiac veins join together to form an enlarged F10.5 vessel, the **coronary sinus,** which is located on the posterior surface of the heart between the atria and ventricles in the coronary sulcus.

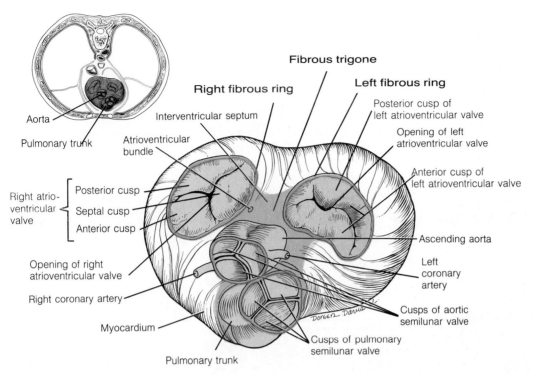

Figure 10.7

Superior view of the heart with the atria removed showing the fibrous skeleton (in color) that surrounds the valve openings.

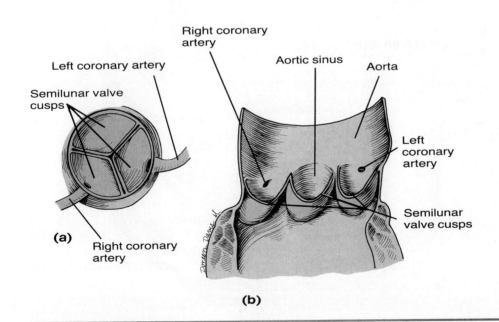

Figure 10.8

The origin of the coronary arteries. **(a)** A closed aortic valve viewed from above. **(b)** An aortic orifice cut and opened to show the semilunar valves.

The anterior surface of the heart is drained primarily by the **great cardiac vein,** which passes upward alongside the anterior interventricular artery. The great cardiac vein begins at the apex of the heart and ascends to the base of the ventricles, where it becomes continuous with the coronary sinus. The largest veins on the posterior-inferior surface of the heart are the **posterior vein of the left ventricle,** which accompanies a branch of the circumflex artery, and the **middle cardiac vein,** which travels alongside the posterior interventricular artery. These veins, like the great cardiac vein, empty into the

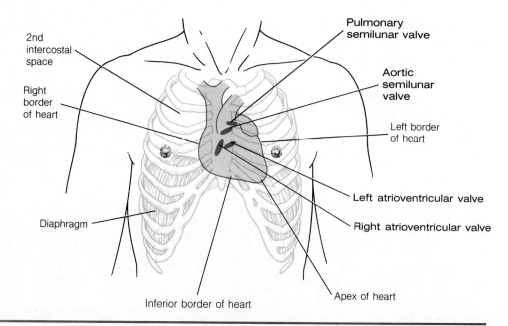

Figure 10.9
Anterior view of the thorax showing the position of the heart and the heart valves in relationship to the ribs, sternum, and diaphragm.

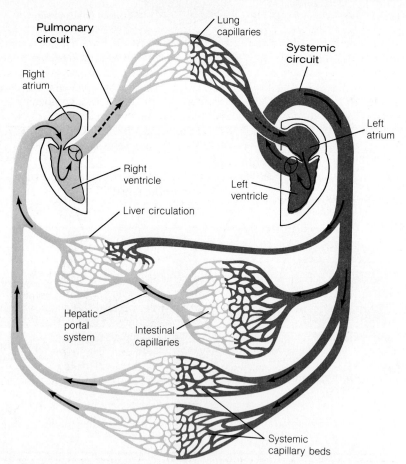

Figure 10.10
Schematic diagram of circulation. Note that the right heart chambers propel blood into the pulmonary circuit (dotted arrows) and the left heart chambers propel blood into the systemic circuit (solid arrows).

coronary sinus. The coronary sinus, in turn, empties into the right atrium. Also draining into the coronary sinus is a **small cardiac vein,** which travels along the right margin of the heart and enters the coronary sulcus on the posterior aspect of the heart. In addition to these veins there are several small **anterior cardiac veins,** which drain the anterior surface of the right ventricle directly into the right atrium.

Valves of the Heart

There are four sets of valves that keep the blood flowing in the proper direction through the chambers of the heart—two sets of *atrioventricular valves* and two sets of *semilunar valves.*

Atrioventricular Valves

Located between the atria and the ventricles, the two **atrioventricular (AV) valves** are flaps of endocardium with an inner framework of fibrous connective tissue (Figure 10.6, Figure 10.7). The flaps are anchored to the papillary muscles of the ventricles by the chordae tendineae. The papillary muscles, which are extensions of the myocardium, exert tension on the valve cusps, thus preventing them from being forced into the atria when the ventricles are contracting. The right atrioventricular valve, which separates the right atrium from the right ventricle, has three flaps, or **cusps,** and is therefore sometimes referred to as the *tricuspid valve.* The left atrioventricular valve is also called the *bicuspid* or *mitral valve* because it has only two cusps. Both of the atrioventricular valves are forced shut as the pressure in the ventricles increases, thus preventing the flow of blood back into the atria when the ventricles are contracting. **F10.6, F10.7**

Semilunar Valves

Blood is prevented by **semilunar valves** from returning to the ventricles after they have completed their contractions (Figure 10.6, Figure 10.7). The **pulmonary semilunar valve** is located in the proximal end of the pulmonary trunk; the **aortic semilunar valve** is in the proximal end of the aorta. Both sets of semilunar valves have three cusps. Each cusp resembles a shallow cup that has been cut in half vertically, with the cut edges attached to the walls of the vessel (Figure 10.8). When the ventricles contract, the force of the blood pushes the cusps against the vessel wall. When the ventricles relax, the blood starts to flow back into them. As this occurs, blood fills the cusps and causes their free margins to meet in the middle of the vessel, thus preventing any further backflow. **F10.6, F10.7** **F10.8**

The openings of the superior and inferior venae cavae, which empty into the right atrium, and those of the pulmonary veins, which empty into the left atrium, are not guarded by functional valves.

Surface Locations of the Valves

It is helpful to be able to locate the valves of the heart on the surface of the thorax (Figure 10.9). The semilunar valves are located toward the base of the heart. The aortic semilunar valve is behind the left side of the sternum, opposite the third intercostal space. The pulmonary semilunar valve lies slightly above and to the left of the aortic semilunar valve, behind the costal cartilage of the third rib. The atrioventricular valves are situated more centrally in the heart than are the semilunar valves. The right atrioventricular valve lies almost directly behind the sternum, extending between the levels where the fourth and fifth costal cartilages join with the sternum. The left atrioventricular valve is located at the level of the fourth costal cartilage, behind the left side of the sternum. **F10.9**

CIRCULATION THROUGH THE HEART

Because the chambers on the right side of the heart are separated from those on the left by septa (the interatrial and interventricular septa), the heart functions as a double pump. Each pump has a receiving chamber (the atrium) and a propulsion chamber (the ventricle).

The right pump receives blood that has passed through the vessels of the body, and sends it to the lungs—that is, through the *pulmonary circuit* (Figure 10.10). Venous blood from the body enters the right atrium through: **F10.10**

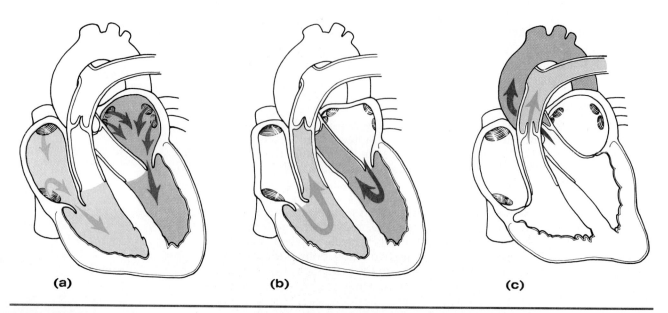

(a)   **(b)**   **(c)**

Figure 10.11
Bloodflow through the heart during a single cardiac cycle. **(a)** Blood fills both atria and enters both ventricles. **(b)** The atria contract, which squeezes more blood into the ventricles. **(c)** The ventricles contract, which forces blood into the aorta and the pulmonary trunk.

1. the *superior vena cava*, which brings blood from the head, thorax, and upper limbs.

2. the *inferior vena cava*, which returns blood from the trunk, lower limbs, and abdominal viscera.

3. the *coronary sinus* and *anterior cardiac veins* that drain the myocardium.

From the right atrium, blood enters the right ventricle, which pumps it through the pulmonary trunk and pulmonary arteries to the capillary network of the lungs. In the lungs, the blood gives up carbon dioxide and receives oxygen.

The left pump receives newly oxygenated blood from the lungs and sends it to the body—that is, through the *systemic circuit* (Figure 10.10). Blood from the lungs is returned to the left atrium by pulmonary veins. From the left atrium the blood enters the left ventricle, which pumps it through the aorta to the body.

The right and left pumps of the heart work in unison. When the heart beats, both atria contract simultaneously, and then both ventricles do the same.

The period from the end of one heartbeat to the end of the next is called the *cardiac cycle*. During this cycle, blood from the superior and inferior venae cavae (as well as from the coronary sinus and anterior cardiac veins) moves through the right atrium, past the open right atrioventricular valve, and into the right ventricle (Figure 10.11). At the same time, blood from the pulmonary veins moves through the left atrium, past the open left atrioventricular valve, and into the left ventricle. The simultaneous contraction of both atria then squeezes more blood into the ventricles. Subsequently, the simultaneous contraction of both ventricles closes the atrioventricular valves and forces blood past the semilunar valves and into the pulmonary trunk (from the right ventricle) and the aorta (from the left ventricle).

Note from this discussion that the contraction of the atria is not essential for the movement of blood into the ventricles. In fact, even if the atria fail to function, the ventricles can still pump considerable quantities of blood.

F10.10

F10.11

CONDUCTING SYSTEM OF THE HEART

The heart contracts approximately 72 times each minute. Within the heart are some specialized cardiac muscle cells that initiate the impulses that cause it to

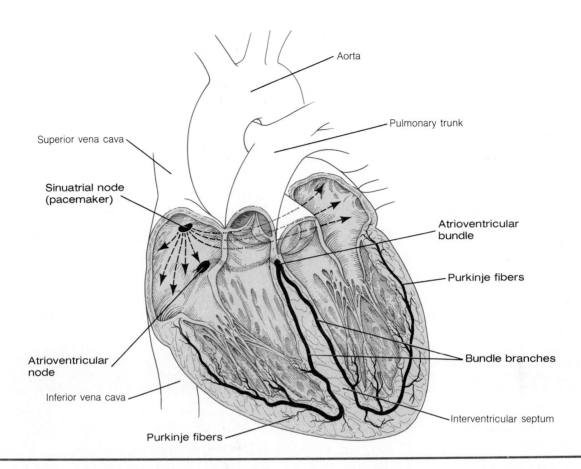

Aorta

Pulmonary trunk

Superior vena cava

Sinuatrial node
(pacemaker)

Atrioventricular
bundle

Purkinje fibers

Atrioventricular
node

Bundle branches

Inferior vena cava

Interventricular septum

Purkinje fibers

Figure 10.12

The conducting system of
the heart. The dotted
lines indicate the action
potential from the
sinuatrial node as it
travels through the
myocardium of both
atria.

contract. There are other specialized cardiac muscle cells that conduct the
impulses throughout the myocardium. This *conducting system* coordinates the
heartbeat, producing an efficient pumping action.

Like other cells, cardiac muscle cells maintain an unequal distribution of
ions on either side of their cell membranes, and they are electrically polarized.
If the cell membrane of a cardiac muscle cell becomes depolarized, allowing
the rapid movement of certain ions across it, a stimulation, or action potential, results, which causes the cell to contract. Because adjacent cardiac muscle
cells are tightly bound by gap junctions that form the intercalated discs that
separate the cells, the action potential can be transmitted from cell to cell, thus
causing each cell to contract. This is a relatively slow process, however, and
the presence of a conducting system in the heart allows for the rapid transmission of the action potential throughout the heart and provides for the
coordination of its beat—thereby making it more efficient.

Sinuatrial Node

In the wall of the right atrium, near the entrance of the superior vena cava, is
a small mass of specialized cardiac muscle cells called the **sinuatrial node (SA
node or sinoatrial node)** (Figure 10.12). (The term *sinu* is used because during
embryonic development this region is called the *sinus venosus*). Under resting
conditions, the cells of this node spontaneously depolarize, without any external stimulus, and generate an action potential approximately 70 to 80 times
each minute (that is, every 0.8 second).

Other areas of the myocardium can also undergo spontaneous depolarization and generate action potentials. However, they do so at slower rates
than the sinuatrial node. As a result, impulses from the sinuatrial node
spread to these areas and stimulate them so frequently that they do not generate action potentials at their own inherent rates. Thus, the rate of discharge of
the sinuatrial node sets the rhythm for the entire heart, and it is for this
reason that the sinuatrial node is called the **pacemaker** of the heart. The contractions initiated by the sinuatrial node begin at the upper regions of the atria

F10.12

FRONTIERS IN HEALTH

New Weapons in the Fight Against Heart Attacks

Jerry collapsed while playing tennis. When he arrived at the hospital his heart was beating, but only weakly, and only because of the heroic efforts of the emergency medical personnel who had transported him to the hospital.

Still dazed and only barely aware of what was going on, the 52-year-old patient was ushered into the cardiology ward. After a series of tests, his physicians made an incision in Jerry's groin and inserted a small plastic catheter into the femoral artery. With the aid of special equipment, they snaked the tube up through the aorta and into the coronary artery until it was stopped by a blood clot. The physician then injected streptokinase, an experimental drug, directly into the clot. This enzyme began to dissolve the clot and within 30 minutes had opened up the blood vessel, restoring blood flow to the oxygen-starved heart muscle.

Serious heart attacks strike thousands of Americans every year, and many of the victims never recover. The most common cause of a heart attack is atherosclerosis, a thickening of the arterial walls that narrows the lumen of the vessel. What is responsible for this thickening? The complete answer is not known, but too many cigarettes, too much alcohol, too much animal (saturated) fat, too much cholesterol, and too little physical activity have all been implicated.

Atherosclerosis restricts the flow of blood, which may then be insufficient to meet the needs of the body when it becomes necessary for the heart to work harder. Lacking sufficient oxygen, muscle cells go into spasms and can die if the occlusion is severe. Making matters worse, and for reasons unknown, blood clots often accompany heart attacks. The clots complicate matters by further restricting blood flow and thus increasing the extent of damage.

Medical scientists are now seeking new ways to reduce the severity of heart attacks. Streptokinase, the enzyme given to Jerry, is one promising answer. It has been shown that, when given within 4 hours of the onset of an attack, streptokinase opens up clogged arteries and restores blood flow. This treatment reduces damage to heart muscle and hastens a patient's recovery. In fact, in several clinical studies streptokinase reduced the number of deaths by 75%. Recent work shows that streptokinase is effective even when given intravenously. This means that even hospitals without expensive cardiac catheterization facilities can administer the drug and improve the prognosis for hundreds of victims of heart attacks.

Streptokinase is a bacterial enzyme and unfortunately evokes an allergic reaction in many patients. As a result, it cannot be readministered to a patient for at least 6 months without possible serious consequences. Some patients may react violently to the foreign protein and die of anaphylactic shock. Streptokinase may also be hazardous if surgery is required, since small breaks in arteries may cause uncontrollable hemorrhaging in patients who have received the enzyme.

Because of the allergic response to streptokinase, scientists have been working with new chemicals that

A cardiopulmonary catheter in use at the Beth Israel Hospital.

they hope will not cause such a response. Urokinase, an enzyme extracted from human tissue, is currently under study. Urokinase dissolves blood clots and, because it is chemically the same in all humans, will not trigger an immune reaction. In time this substance may prove to be a suitable replacement for streptokinase. Mass production by genetic engineering could help lower its cost, which now is prohibitively high.

Medical researchers are also experimenting with another naturally occurring clot dissolver, called plasminogen activator, or TPA. TPA has already been produced successfully using genetic engineering, and medical researchers have established a cell line that produces large quantities of TPA when maintained in culture. TPA has proved remarkably effective when given to dogs. It works fast, breaking up the clot and reestablishing blood flow in as little as 7 minutes. This quick reestablishment of blood flow reduces damage to cardiac muscle by 30 to 70%. TPA has been tested in humans and appears free of the side effects of streptokinase.

and travel toward the atrioventricular valves. This action helps move the blood in the atria into the ventricles.

Atrioventricular Node

An impulse generated in the sinuatrial node spreads from cell to cell throughout the myocardium of the atria, stimulating the atria to contract. However, the fibrous skeleton of the heart that surrounds the openings between the atria and the ventricles as well as the openings of the aorta and the pulmonary trunk (Figure 10.7) cannot depolarize. Therefore, a stimulatory impulse trans- **F10.7** mitted from the sinuatrial node throughout the atria cannot pass directly to the myocardium of the ventricles. Instead, the impulse reaches the ventricles by way of a specialized conducting system.

   A group of specialized cardiac muscle cells called the **atrioventricular node (AV node)** is located within the interatrial septum just above the junction of the atria and the ventricles (Figure 10.12). From the atrioventricular **F10.12** node, a bundle of specialized muscle tissue called the **atrioventricular bundle** (*bundle of His*) passes to the ventricles. The atrioventricular bundle enters the interventricular septum, where it divides into **right** and **left bundle branches** (or *crural*) that travel down the septum. Small groups of terminal conducting fibers, called **Purkinje fibers** (or *crural rami*), exit from the bundle branches and end on the true cardiac muscle cells of the ventricles. At the apex of the heart, the Purkinje fibers pass to the outer wall of the ventricles and travel back toward the base of the heart.

Pathway of Conduction Through the Heart

As a stimulatory impulse from the sinuatrial node spreads throughout the atrial myocardium, it reaches the atrioventricular node. After a delay of approximately 0.10 second, the atrioventricular node depolarizes and the action potential travels from it through the atrioventricular bundle and Purkinje fibers to reach the cells of the ventricular walls. The rate of transmission of the action potential is much faster through these specialized conductive muscle fibers than it is through the muscle cells themselves, as occurs in the atria. Therefore, the entire ventricular myocardium contracts almost immediately after the depolarization of the atrioventricular node. The contraction begins at the apex, where the bundle branches leave the interventricular septum, and proceeds toward the superior border of the heart, where the openings that lead to the pulmonary trunk and the aorta are located.

   The importance of the sinuatrial node as the regulator of heart action is recognized in heart transplant operations. When possible, the region of the diseased heart that contains the node is left intact in the recipient. An attempt is also made to leave intact the nerves that supply the node, although the left vagus and the sympathetic nerves must generally be severed. The transplanted heart is therefore subjected to some of the same control mechanisms as was the recipient's original heart.

CONDITIONS OF CLINICAL SIGNIFICANCE

The Heart

Normal Heart Sounds

There are two principal sounds that normally occur as blood moves through the heart during a cardiac cycle. These sounds are best described as "lub-dup." The first heart sound (the "lub") is associated with the closure of the atrioventricular valves at the beginning of ventricular contraction (*systole*). It is largely due to vibrations of the taut atrioventricular valves immediately after closure and to the vibration of the walls of the heart and major vessels around the heart. The second sound (the "dup") is associated with the closure of the semilunar valves as the ventricles begin to relax (*diastole*) following their contraction. This sound is due largely to vibrations of the taut, closed semilunar valves and to the vibration of the

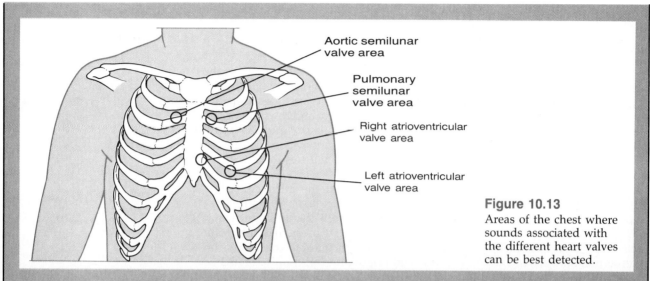

Aortic semilunar valve area

Pulmonary semilunar valve area

Right atrioventricular valve area

Left atrioventricular valve area

Figure 10.13
Areas of the chest where sounds associated with the different heart valves can be best detected.

walls of the pulmonary artery, the aorta, and to some extent the ventricles. The areas of the chest where a stethoscope can be placed to detect most effectively the sounds associated with the different valves are indicated in Figure 10.13.

Abnormal Heart Sounds

By listening to the heart, a trained person can obtain considerable information concerning its condition. Abnormal sounds, which are referred to as *heart murmurs*, can be indicative of particular heart problems. These sounds, which are described as blowing or vibrating sounds, are caused by a turbulent flow of blood as it passes through the heart.

There are two basic conditions of the valves that cause heart murmurs: valvular regurgitation and valvular stenosis (see *Valvular Malfunctions*). In valvular regurgitation, a secure seal is not formed when the valve closes and blood leaks back into the chamber from which it had been pumped. The regurgitated blood interferes with the incoming stream of blood and causes a detectable turbulence. In valvular stenosis, growths narrow the opening and interfere with the blood flowing through it. Increased pressure is then necessary to force blood from the chamber, and the walls around a narrowed valve are often roughened, both of which contribute to the turbulence and murmur.

There are murmurs, called *functional murmurs*, that are not pathological and are considered normal. The turbulence that produces functional murmurs can be caused during heavy exercise, when the blood is being moved rapidly through the heart. Functional murmurs are particularly common in young people.

Abnormal Heart Rates

In a resting adult, the heart normally contracts about 72 times per minute. When the rate drops below 60 beats per minute, the condition is referred to as *bradycardia*. Bradycardia is generally not considered to be pathological. Much more serious, and generally associated with cardiovascular pathology, is *tachycardia*, which is a heart rate over 100 beats per minute. When the heart rate is very fast—for example, over 200 beats per minute—the

chambers of the heart do not have time to fill properly between beats, and the movement of blood through the heart is therefore inefficient. If the rhythm of contractions remains normally coordinated during tachycardia, the condition is called *flutter;* when the contraction is uncoordinated, it is referred to as *fibrillation.*

Valvular Malfunctions

Valvular malfunctions can interfere with the normal movement of blood through the heart. They can decrease the amount of blood pumped out of a ventricle with each contraction, and thereby make it necessary for the heart to work harder to maintain a given cardiac output. Valvular regurgitation and valvular stenosis are two of the most common valvular malfunctions.

Valvular Regurgitation

Valvular regurgitation occurs when the cusps of a valve do not form a tight seal when the valve is closed. As a result, blood leaks back, or regurgitates, into the chamber from which it came. If an atrioventricular valve does not close completely, blood flows back into the atrium when the ventricle contracts, and less than the normal amount of blood may be moved into the aorta or pulmonary trunk. Similarly, if a semilunar valve does not close completely, blood that moves into the aorta or pulmonary trunk during ventricular contraction flows back into the ventricle when it relaxes. Growths or scar tissue that form on a valve as a result of diseases such as rheumatic fever can prevent the valve from closing securely and cause valvular regurgitation.

Valvular Stenosis

Valvular stenosis is a condition in which the opening of a valve becomes so narrow that it interferes with the flow of blood through it. If the opening of an atrioventricular valve is too narrow, the ventricle may not fill completely with blood, and a lower than normal amount of blood may be pumped out when the ventricle contracts. If the opening of a semilunar valve is too narrow, the ventricle may not eject a normal amount of blood into the aorta or pulmonary trunk when it contracts. Growths or scar tissue on a valve can cause valvular stenosis as well as valvular regurgitation. In many cases valvular regurgitation and valvular stenosis occur in the same valve.

Heart Block

Damage to the atrioventricular node or a weak impulse from the sinuatrial node can result in the failure to transmit some or all of the impulses from the sinuatrial node to the ventricles. In some cases, every second (or third, or fourth, and so on) atrial contraction is followed by a ventricular contraction. In this condition, known as a *partial heart block*, the heart retains a definite, though altered, rhythm. When the condition of the atrioventricular node is so severe that it does not allow any impulses from the sinuatrial node to be conducted to the ventricles, it is called *complete heart block*. In complete heart block, the ventricles contract at their own inherent rate, which is slower than the rate of the atria and completely independent of it.

Bundle Branch Block

If there is interference in one of the branches of the atrioventricular bundle, the impulses reach one ventricle slightly later than the other. This condition, where the contraction of one ventricle is delayed, is known as *bundle branch block*. The heart's efficiency is diminished in such cases because of the asynchronous contractions of the ventricles.

Arteriosclerotic Heart Disease

This is the most common type of heart disease. Because of a hardening and narrowing of the coronary arteries, the blood supply to the myocardium is diminished. Usually the heart is smaller than normal. Arteriosclerotic heart disease is a slow disease that progresses gradually and is often associated with older people. If a coronary artery should become completely blocked, a myocardial infarction may occur.

Myocardial Infarction

If the blood flow through the coronary arteries to the myocardium is blocked, the muscle cells supplied by the blocked vessel die. The region of dead tissue is called an *infarct*. If the area of the infarct is massive the individual may die. If the infarct is smaller the individual may recover after a period of complete bed rest. Muscle cells are not capable of regenerating, however, and dead tissue of a myocardial infarct is replaced by scar tissue. The scar tissue can interrupt the conducting system of the heart or affect its efficiency in other ways.

Endocarditis

Endocarditis is an infection of the lining of the heart—usually the valves—by bacteria, fungi, or perhaps viruses. The infection generally becomes established at a site in the heart that had been damaged by previous heart disease, such as rheumatic heart disease or congenital malformations of the heart.

The most common result of endocardial infection is the development of bulky masses of the causative organisms on the leaflets of the valves of the heart. These masses not only interfere with proper valve functioning, often causing them to leak, but also serve as sites from which the infection can be spread to other organs and tissues of the body.

Myocarditis

When the myocardium of the heart becomes infected, the condition is called *myocarditis*. Myocarditis can be secondary, resulting from the spread of some systemic disease, or primary, affecting the heart directly. Primary myocarditis is caused by bacteria, viruses, or parasites. Inflammation and swelling of the heart occurs, and there may be degeneration of muscle cells. A variety of symptoms can accompany myocarditis, including conduction defects that cause abnormal heartbeat rhythms. Most cases of myocarditis subside within a month or two.

Effects of Aging on the Heart

In the absence of heart disease the size of the heart does not change much with aging. However, because diseases of the heart and the blood vessels are so common, the heart is, in fact, often enlarged in older persons.

Because of reduced demands placed on the heart by lower levels of physical activity in older persons, and because cardiac muscle cells are not capable of mitosis and therefore are not replaced as they die, there is a progressive decline in the strength of heart contraction and in the volume of blood the heart is capable of pumping with each contraction (*cardiac output*). There also is thought to be a tendency for a progressive reduction in the capillaries supplying the myocardium, which further affects the ability of the heart to contract.

With aging the endocardium tends to become thicker due to the deposition of connective tissue and, accompanying this, the heart valves tend to thicken and become more rigid. It is not uncommon for some calcification of the valves to occur after age 60.

There is an increase in fibrous connective tissue within the conducting system of the heart with aging. This may interfere with the initiation and transmission of the nerve impulses that regulate heart contraction, leading to irregular beats or heart block.

Numerous studies have shown that older persons who follow a continuous exercise program can increase their cardiac output significantly. However, even if a person does not follow a regular program of exercise the aging changes that occur in his or her heart generally do not hinder it from adequately maintaining the reduced level of activity characteristic of most older persons. Only a disease condition reduces the capacity of the heart to maintain normal activities.

STUDY OUTLINE

EMBRYONIC DEVELOPMENT OF THE HEART   pp. 281–282
1. Begins as pulsating tubule.
2. Tubule becomes S-shaped.
3. Anterior vessel becomes the pulmonary trunk and aorta.
4. Posterior vessel becomes the superior and inferior vena cavae.
5. Four chambers develop—two atria and two ventricles.

POSITION OF THE HEART   cone-shaped organ in mediastinum; size of closed fist. Base behind the sternum at level of second and third ribs. Apex to the left of midsternal line at level of fifth intercostal space. Has diaphragmatic and sternocostal surfaces.   p. 282

COVERINGS OF THE HEART   pp. 283–284
1. Pericardium is a double-walled membranous sac (visceral and parietal layers).
2. Parietal pericardium has fibrous and serous layers.
3. Pericardial cavity contains pericardial fluid.

ANATOMY OF THE HEART   pp. 284–291

CHAMBERS OF THE HEART

ATRIA (RIGHT AND LEFT) smaller, located toward superior region of heart; separated by interatrial septum.

VENTRICLES (RIGHT AND LEFT) larger, located at apex of heart; separated by interventricular septum.

VESSELS ASSOCIATED WITH THE HEART

SUPERIOR AND INFERIOR VENAE CAVAE return venous blood from body to right atrium.

PULMONARY TRUNK from right ventricle to lungs.

PULMONARY VEINS from lungs to left atrium.

AORTA from left ventricle to body.

WALL OF THE HEART

EPICARDIUM serous membrane that adheres to outer surface of heart.

MYOCARDIUM cardiac muscle.

ENDOCARDIUM connective tissue and squamous cells; folds to form heart valves.

SKELETON OF THE HEART fibrous rings separating atria from ventricles.

VESSELS OF THE MYOCARDIUM supplied by coronary arteries; cardiac veins drain into coronary sinus.

VALVES OF THE HEART

ATRIOVENTRICULAR VALVES
1. Right and left AV valves; tricuspid valve on right; bicuspid (mitral) valve on left.
2. Prevent blood backflow into atria during ventricular contraction.

SEMILUNAR VALVES
1. Pulmonary and aortic semilunar valves.
2. Prevent return of blood to ventricles from aorta and pulmonary trunk after contraction.

SURFACE LOCATION OF THE VALVES
1. Aortic semilunar valve—opposite left third intercostal space.
2. Pulmonary semilunar valve—behind left third costal cartilage.
3. Right AV valve—behind sternum, at level of fourth and fifth costal cartilages.
4. Left AV valve—at level of left fourth costal cartilage.

CIRCULATION THROUGH THE HEART   pp. 291–292
1. Heart functions as a double pump: Right pump receives blood from systemic circuit and pumps it into pulmonary circuit; left pump receives blood from pulmonary circuit and pumps it into systemic circuit.
2. Superior and inferior venae cavae, coronary sinus, and anterior cardiac veins return blood from body to right atrium. Blood then moves into right ventricle, which pumps it through pulmonary trunk and pulmonary veins. Blood then moves into left ventricle, which pumps it through aorta to body.
3. Both atria contract simultaneously, followed by simultaneous contraction of both ventricles.

CONDUCTING SYSTEM OF THE HEART   pp. 292–295

SINUATRIAL (SA) NODE called the pacemaker; located on the right atrium wall; depolarizes spontaneously 70 to 80 times each minute to initiate heart contractions.

ATRIOVENTRICULAR (AV) NODE located in the interatrial septum; atrioventricular bundle and Purkinje fibers pass from AV node to ventricles.

PATHWAY OF CONDUCTION THROUGH THE HEART SA node generates action potential that travels through muscles of atria; AV node depolarizes; action potential carried through the atrioventricular bundle and Purkinje fibers; ventricles contract.

CONDITIONS OF CLINICAL SIGNIFICANCE: THE HEART   pp. 295–297

NORMAL HEART SOUNDS first heart sound is associated with closure of atrioventricular valves at beginning of systole. The second heart sound is associated with closure of semilunar valves as ventricles begin to relax (diastole) following their contraction.

ABNORMAL HEART SOUNDS are often due to turbulent flow and are called *murmurs;* may be caused by valvular regurgitation or valvular stenosis.

ABNORMAL HEART RATES bradycardia, tachycardia, flutter, fibrillation.

VALVULAR MALFUNCTIONS can make it necessary for heart to work harder to maintain a given cardiac output.

VALVULAR REGURGITATION occurs if cusps of a valve do not form a tight seal when valve is closed. As a result, blood leaks into chamber from which it came.

VALVULAR STENOSIS a condition in which opening of valve becomes so narrow it interferes with flow of blood through it.

HEART BLOCK when impulses from SA node are not transmitted to ventricles; may be partial or complete.

BUNDLE BRANCH BLOCK delayed contraction of one ventricle due to interference in conduction within a branch of AV bundle.

ARTERIOSCLEROTIC HEART DISEASE gradual hardening and narrowing of coronary arteries, diminishing blood supply to myocardium; may cause myocardial infarction.

MYOCARDIAL INFARCTION myocardial cells die because of interrupted blood supply; dead cells replaced by scar tissue.

ENDOCARDITIS infection of lining of heart by bacteria, fungi, or viruses; often forms masses on valves, thus interfering with proper valve functioning.

MYOCARDITIS infection of myocardium by bacteria, viruses, or parasites.

EFFECTS OF AGING While there is a progressive decline in cardiac output and strength of contraction with age, a healthy heart is quite capable of meeting the reduced physical demands of older persons.

SELF-QUIZ

1. The heart is enclosed in a double-walled membranous sac called the: (a) mediastinum; (b) pericardium; (c) epicardium.

2. Match the various terms associated with the heart with the appropriate description.

Pericardial cavity
Atria
Ventricles
Interatrial septum
Inferior vena cava
Aorta
Epicardium
Myocardium
Endocardium
Trabeculae carneae

   (a) Separates the two atria.
   (b) Carries blood from the left ventricle into the systemic circuit.
   (c) The muscular layer of the wall of the heart.
   (d) Small chambers located toward the superior region of the heart.
   (e) Folds and bridges of the inner surface of the myocardium.
   (f) Conducts blood to the right atrium.
   (g) space located between the visceral and parietal layers.
   (h) Foldings of this structure form the valves that separate the atria from the ventricles.
   (i) Large chambers that compose the bulk of the heart.
   (j) A thin serous membrane that adheres to the outer surface of the heart.

3. Blood is prevented from returning to the ventricles after they have completed their contractions by the: (a) atrioventricular valves; (b) tricuspid valve; (c) semilunar valve.

4. The openings of the superior and inferior venae cavae are not guarded by functional valves. True or false?

5. During the cardiac cycle, both atria contract simultaneously, followed by the simultaneous contraction of the two ventricles. True or false?

6. Especially in young people, turbulence caused by strenuous exercise may produce: (a) functional murmurs; (b) valvular stenosis; (c) valvular regurgitation.

7. The heart sounds are caused by: (a) closure of the AV valves; (b) closure of the semilunar valves; (c) both a and b.

8. When locating the heart sounds on the surface of the thorax, those of the cuspid valves are more centrally located in the heart than those of the semilunar valves. True or false?

9. Heart murmurs can be caused by: (a) valvular regurgitation; (b) valvular stenosis; (c) both a and b.

10. The skeleton of the heart is composed of cartilage and serves as the attachment for the heart muscles and valves. True or false?

11. The AV valves are prevented from being forced into the atria while the ventricles are contracting by: (a) trabeculae carneae; (b) papillary muscles; (c) fibrous trigones.

12. Which is located in the coronary sulcus? (a) coronary arteries; (b) coronary sinus; (c) both a and b.

13. The frequency of heart contractions is regulated by the: (a) AV node; (b) SA node; (c) Purkinje fibers.

14. The atrioventricular bundle transmits impulses from the: (a) AV node; (b) SA node; (c) pacemaker.

15. Since there are no specialized conducting fibers in the myocardium of the atria, the action potential spreads from muscle cell to muscle cell. True or false?

16. The condition in which some impulses from the SA node do not reach the AV node is called: (a) myocardial infarction; (b) heart block; (c) bundle branch block.

17. Bacteria and viruses can cause an infection of the heart known as: (a) endocarditis; (b) myocarditis; (c) both a and b.

18. If the blood supply to an area of the heart is blocked, the muscle cells in that area may die. This region of dead tissue is called an embolism. True or false?

LEARNING OBJECTIVES

After completing this chapter, you should be able to:

- Describe the structure of arteries, veins, and capillaries.

- Describe the common diseases of the blood vessels and cite causes of each.

- Distinguish between the pulmonary and the systemic circuits.

- Identify the principal pulmonary arteries and veins.

- Identify the principal systemic arteries and veins.

CHAPTER CONTENTS

The Circulatory System: Blood Vessels

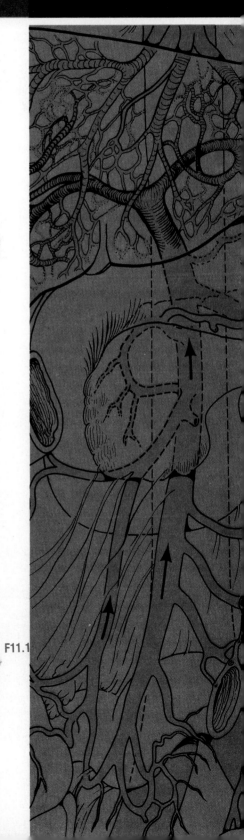

Upon leaving the heart, the blood enters the vascular system, which is composed of numerous **blood vessels.** The vessels transport the blood to all parts of the body, permit the exchange of nutrients, metabolic end products, hormones, and other substances between the blood and the interstitial fluid, and ultimately return the blood to the heart. Both the size of the vessels and the thickness of their walls vary, as does the blood pressure within them.

TYPES OF VESSELS

Vessels called **arteries** carry blood *away* from the heart. Compared to the other types of blood vessels, arteries must be able to withstand the greatest internal pressure. The major arteries divide into smaller arteries, then into still smaller **arterioles,** and finally into tiny **capillaries.** With the progressive change from arteries to capillaries, there is a decrease in the diameter of the vessels, in the thickness of the vessel walls, in the pressure within the vessels, and in the velocity at which blood travels through the vessels. Although the size of the internal cavity *(lumen)* of the individual vessels decreases from arteries to arterioles to capillaries, there is such a large number of arterioles and capillaries that the *sum* of the diameters of the lumina of parallel branches of these vessels actually increases. In fact, the smallest individual vessels, the capillaries, are so numerous that they have the greatest total cross-sectional diameter (and area) of the entire circulatory system.

The capillaries converge into very small vessels called **venules,** which, in turn, join to form larger vessels called **veins.** The major veins *return* blood to the atria of the heart. After the blood leaves the capillaries, its pressure continues to drop; it is lowest near the right atrium of the heart in the superior and inferior venae cavae. With the progressive change from capillaries to venules to veins, the diameters of individual vessels and the thickness of their walls steadily increase, whereas the total cross-sectional area of parallel vessels decreases. Venous pressure is always lower than arterial pressure, and the walls of the veins are never as thick as the walls of the corresponding arteries.

GENERAL STRUCTURE OF BLOOD VESSEL WALLS

The variation in thickness of the blood vessel walls is due to the presence or absence of one or more of three layers of tissues and to the differences in their thicknesses (Figure 11.1). The **tunica intima** *(tunic* = coat) is the innermost layer. It is formed of a layer of *simple squamous epithelium* called the **endothelium,** a layer of *connective tissue,* and a *basement membrane.* The endothelium of the tunica intima is the only layer present in vessels of all sizes. Moreover, it is continuous with the endocardium of the heart. The middle layer, the **tunica media,** is often quite thick and is composed of *smooth muscle fibers*—mostly circularly arranged—mixed with *elastic fibers.* The outermost layer is the **tunica externa** (or **adventitia**). This relatively thin layer of *connective tissue* con-

F11.1

303

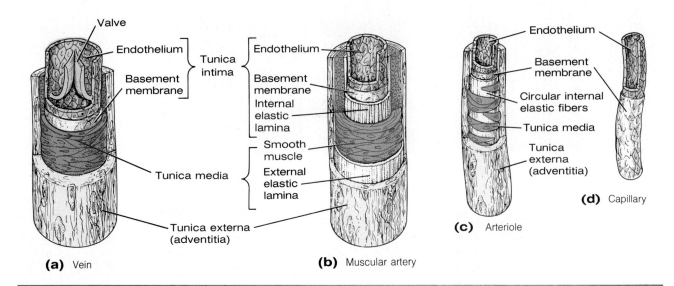

(a) Vein

(b) Muscular artery

(c) Arteriole

(d) Capillary

Figure 11.1
Comparison of the structure of blood vessels.

tains elastic and collagenous fibers that run parallel to the long axis of the vessel. Separating the tunica intima and the tunica media is a thin layer of elastic fibers called the **internal elastic lamina.** There is often a similar layer of elastic fibers—the **external elastic lamina**—between the tunica media and the tunica externa. The walls of the larger vessels are too thick to be nourished by diffusion from the blood in the vessel. Instead, they are supplied by their own small nutrient vessels—the **vasa vasorum** (= vessels of the vessels)—which are located in the tunica externa and arise either from the blood vessel itself or from other vessels located close by.

STRUCTURE OF ARTERIES

The composition of the walls of the arteries differs, depending upon the size of the vessels.

Elastic Arteries

The large arteries, such as the aorta and its major branches and the pulmonary trunk, are called **elastic arteries.** The walls of these arteries are composed of the three tunics just described. The tunica media of the large arteries is quite thick, and it contains, in addition to smooth muscle fibers, many elastic fibers. During ventricular systole, elastic arteries are passively stretched as blood is ejected from the heart. During ventricular diastole, the recoil of the walls of the elastic arteries helps maintain pressure within the vessels.

Muscular Arteries

The tunica media of the walls of most smaller arteries consists almost entirely of smooth muscle cells, with relatively few elastic fibers. Such arteries are called **muscular arteries** or, because they carry blood throughout the body, **distributing arteries.** Muscular arteries have well-defined internal and external elastic laminae (Figure 11.2).

STRUCTURE OF ARTERIOLES

When the diameter of an arterial vessel is less than 0.5 mm, it is referred to as an **arteriole.** Arterioles have a small lumen and a relatively thick tunica media that is composed almost entirely of smooth muscle, with very little elastic tissue. In the smallest arterioles—that is, those closest to the capillaries—the external elastic membrane is lost and the tunica media is gradually reduced until it is composed of only a few scattered smooth muscle cells.

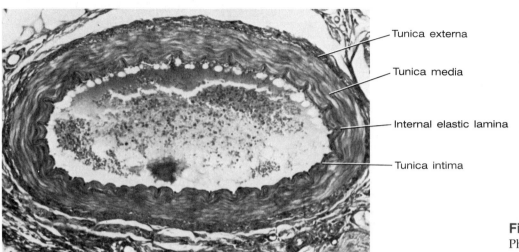

- Tunica externa
- Tunica media
- Internal elastic lamina
- Tunica intima

Figure 11.2
Photomicrograph of a muscular artery.

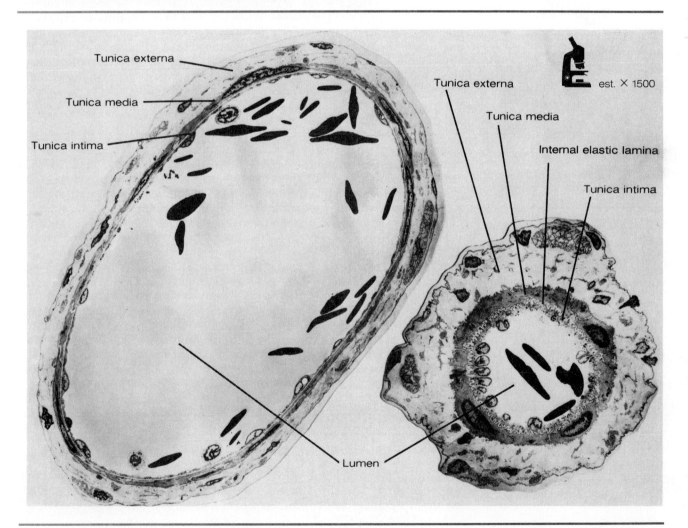

- Tunica externa
- Tunica media
- Tunica intima

- Tunica externa
- Tunica media
- Internal elastic lamina
- Tunica intima

est. × 1500

Lumen

Figure 11.3
Photomicrograph of cross sections of dilated arteriole (left) and constricted arteriole (right).

The arterioles play a major role in regulating the flow of blood into the capillaries. When the smooth muscle of the tunica media contracts, the internal cavities, or **lumens,** of the vessels are narrowed—that is, the vessels undergo *vasoconstriction*, which restricts the flow of blood into the capillaries (Figure 11.3). When the muscles relax, the lumens of the arterioles enlarge—that is, the vessels undergo *vasodilation*, which allows the blood to enter the capillaries freely.

F11.3

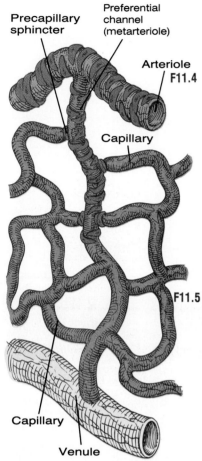

Precapillary sphincter

Preferential channel (metarteriole)

Arteriole
F11.4

Capillary

F11.5

Capillary

Venule

Figure 11.4
A capillary network. Thoroughfare channels directly connect arterioles and venules; true capillaries branch from and join with the thoroughfare channels.

STRUCTURE OF CAPILLARIES

In most tissues, a capillary network contains two types of vessels: *thoroughfare channels,* which directly connect arterioles and venules, and *true capillaries,* which branch from and join with the thoroughfare channels (Figure 11.4). A ring of smooth muscle called a *precapillary sphincter* usually surrounds each true capillary at the point where it arises from a thoroughfare channel. The contraction and relaxation of the sphincters helps regulate the flow of blood through the capillaries.

Capillaries have extremely thin walls. As a consequence, they are sites at which the exchange of materials between the blood and the interstitial fluid takes place. Capillary structure varies from one part of the body to another, but in general, a capillary consists of a single layer of endothelial cells surrounded by a thin basal lamina of the tunica intima. There is no tunica media or tunica externa present. Endothelial cells are held to one another by tight junctions. Water-filled clefts occur between adjacent endothelial cells.

In the capillaries of muscle, connective tissue, and nervous tissue the endothelium forms a single uninterrupted layer around the entire circumference of the capillary. These vessels are referred to as **continuous capillaries** (Figure 11.5). In the capillaries of endocrine glands, intestines, and kidneys the membrane of the endothelial cells are pierced by numerous tiny pores called *fenestrations.* Such vessels are referred to as **fenestrated capillaries.** In most vessels the fenestrae are closed by a very thin diaphragm; however in the capillaries of the kidneys they lack a diaphragm and appear to be open. As a consequence, fluid moves across the walls of the capillaries of the kidneys much faster than it does in the capillaries of muscles.

Although a single capillary is only about 0.5 to 1 mm long and 0.01 mm in diameter, capillaries are so numerous that their total surface area within the body has been estimated to be more than 600 square meters. This provides a large surface across which the exchange of materials can occur. Substances may enter or leave capillaries by four possible routes: (1) through the tight junctions that anchor the endothelial cells together; (2) directly through the cell membrane; (3) within small membrane-bounded vesicles; or, (4) in the case of capillaries that contain fenestrations, through pores in the membrane of the endothelial cells.

The arterioles in certain structures, rather than connecting with capillaries, empty into relatively large vascular channels that have very thin walls. These channels are **sinusoids.** Sinusoids are so thin-walled that they generally conform in shape to the space in which they are located rather than being cylindrical. In some sinusoids the endothelium that lines them is discontinuous, with gaps present between the cells; in others the endothelium has small pores that are closed by thin diaphragms, as in fenestrated capillaries. Sinusoids are characteristic of the liver, spleen, bone marrow, and some endocrine glands.

STRUCTURE OF VENULES

In the venules closest to the capillaries, the walls have an inner lining composed of the endothelium of the tunica intima, surrounded by a very thin tunica externa. The larger venules that are further from the capillaries are encircled by a few smooth muscle fibers that form a thin tunica media.

STRUCTURE OF VEINS

The veins, which receive blood from venules, have the same three coats as the arteries. In general however, the tunica media of the veins is quite thin and has few muscle fibers. The tunica externa forms the greatest part of the wall, often being several times thicker than the media, although there is very little

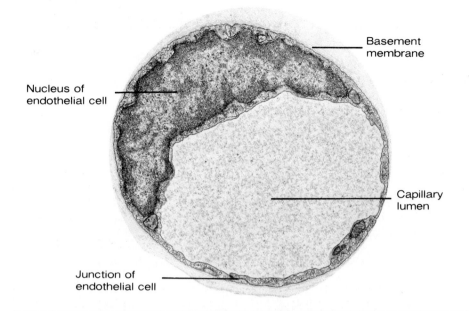

Basement membrane

Nucleus of endothelial cell

Capillary lumen

Junction of endothelial cell

Figure 11.5
Electron micrograph of cross section of a continuous capillary. Note that the entire circumference of the capillary is made up of a single endothelial cell.

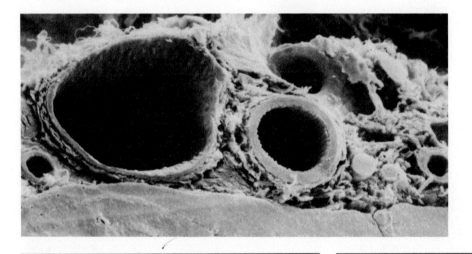

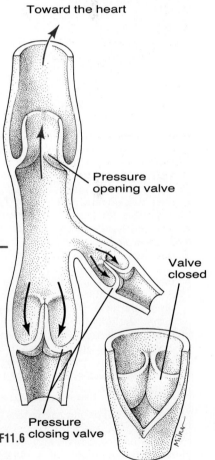

Toward the heart

Pressure opening valve

Valve closed

Pressure closing valve

Figure 11.6
Scanning electron micrograph of the lumens of a medium-sized vein (left) and its accompanying muscular artery (right). (×229). (From *Tissues and Organs: A Text-Atlas of Scanning Electron Microscopy* by Richard G. Kessel and Randy H. Kardon. W.H. Freeman and Company, Copyright © 1979.)

Figure 11.7
Valves of a vein. The arrows indicate that the valves are forced open by pressure from below and shut by pressure from above. This allows blood to move in only one direction—toward the heart.

F11.6
F11.7

elastic tissue present. Veins have no internal or external elastic laminae. They tend to have a larger lumen and thinner walls than the arteries they accompany (Figure 11.6). The walls of the veins, like the walls of the arteries, receive nourishment through tiny vasa vasorum. Some veins contain valves that allow the one-way flow of blood toward the heart (Figure 11.7). These valves are folds of the tunica intima and have a form similar to the semilunar valves of the heart. When blood in the veins attempts to flow backward, away from the heart, the flaps of the valves fill with blood and block the vessel. Valves are most common in the veins of the lower limbs, where the blood is carried against the force of gravity and its movement depends largely on the contraction of the surrounding skeletal muscles.

FRONTIERS IN HEALTH

Wired Arteries

Your heart and arteries get little rest. From the day you are born to the day you die, they pump and transport blood to the oxygen-hungry cells of your body. Beating more than 10,000 times each day, your heart is unmatched in its durability and unfailing performance.

The stress and strain of pumping blood, along with problems resulting from various ways in which we abuse our bodies, however, do take their toll in some people. Arteries clog and hearts deteriorate. Clogged arteries can cut off blood flow to the body's cells, causing them to die. The areas of cell death are called infarctions. Clogged arteries that are hardened by atherosclerosis also have a tendency to weaken and can eventually rupture. Degeneration of the muscle layer of an artery (the tunica media) can result in a ballooning of the inner and outer layers of the vessel wall. This is known as an aneurysm. An aneurysm in an arterial wall resembles a worn spot on a tire, in that when pressure builds up inside or when the wall thins too much, it can rupture. In a tire, this is called a blowout. A similar violent event can occur in a blood vessel.

Blood spilling into the chest cavity (in the case of a ruptured aortic aneurysm) or the brain (in the case of a ruptured cerebral aneurysm) usually leads to death. In the United States alone, an estimated 30,000 people die each year from ruptured cerebral aneurysms and 2500 die from ruptured aortic aneurysms.

The first line of defense against ruptured aneurysms is prevention. The best means of prevention is to reduce the two main causes, atherosclerosis and high blood pressure. The second line of defense lies in detecting aneurysms and treating them before they burst. Pain in the affected area generally alerts the physician to their possible presence, and they are fairly easy to detect by X ray. Once a defective wall is found, surgeons must act fast. How they correct the aneurysm depends on the location and severity of the damage. For example, in cases of popliteal aneurysm, surgeons often remove the aneurysm and graft a vein in its place. To repair a cerebral aneurysm, which most often occurs in the cerebral arterial circle at the base of the brain, surgeons clamp or tie off the neck of the bulge on the arterial wall. This lowers the pressure in the bulge and prevents rupture. In larger arteries, however, surgeons use grafts of dacron or other synthetic materials to replace the damaged section of the artery. Regardless of which technique is used, the repair of aneurysms involves major surgery that is costly and risky.

Now, however, thanks to an innovative procedure, the surgical repair of aneurysms may become much less frequent. Dr. Andrew Cragg, a radiologist from the University of Minnesota, has been working with an alloy of nickel and titanium, called nitinol, that may someday be widely used to repair aneurysms quickly and safely, without the need for major surgery.

Nitinol is a "metal with a memory." When a fine nitinol wire is coiled around a cylinder and heated, it forms a tight spring. Cooling the spring causes it to revert to the form of a straight wire. When the wire is heated again, the coil reforms.

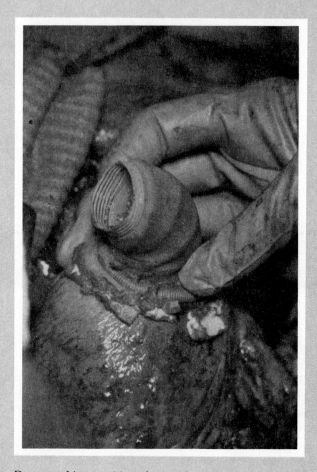

Dacron tubing positioned to replace an aneurysm.

Dr. Cragg has developed a method for repairing aneurysms by capitalizing on nitinol's "memory." He and his colleagues first make a nitinol coil the same size as the internal diameter of a damaged artery. They then push the wire through a specially cooled catheter, which is inserted into the damaged region of the artery. As the wire comes out the end of the catheter, body heat causes it to coil back into the shape it was "taught." The nitinol coil forms inside the weakened portion of the arterial wall, strengthening the wall and preventing its rupture.

Nitinol has yet to be tested on humans, but experiments with dogs have been very encouraging. Dr. Cragg reports that 4 hours after emplacement, a fine layer of fibrin (a protein derived from the blood) covers the coil; after 8 weeks, endothelial cells have grown over the coil, completely incorporating it into the arterial wall.

Because the insertion of wire coils can quickly strengthen the walls of arteries damaged by atherosclerosis and high blood pressure without the need for major surgery, the procedure holds much promise for the thousands of people who suffer from aneurysms.

CONDITIONS OF CLINICAL SIGNIFICANCE
Blood Vessels

High Blood Pressure (Hypertension)

There are two pressures that can be measured in the blood vessels. *Systolic pressure* results when the ventricles of the heart contract and force blood into the vessels. *Diastolic pressure* exists in the vessels while the ventricles are not contracting, and is the result of the force of the elastic recoil of the large elastic arteries. Blood pressure is reported as a fraction such as 120/80—where the numerator is systolic pressure and the denominator is diastolic pressure. These pressures are measured in millimeters of mercury (mm Hg). Both diastolic pressure and systolic pressure are important in hypertension, since both affect the mean pressure. In general, pressures above 150/90 mm Hg are considered to be dangerously elevated. Approximately 20% of the population can expect to have high blood pressure at some time during their lives.

In most cases of high blood pressure the cause is unknown. Since cardiac output is usually only slightly elevated, however, increased peripheral resistance in the blood vessels is suspected of being responsible for elevating the arterial pressure. Constriction of the arterioles may be the cause of the high peripheral resistance.

There are several suspected causes of the general vasoconstriction that can accompany high blood pressure. *Renin,* which is produced by a special group of cells in the kidneys, is involved in the formation in the bloodstream of a vasoconstrictor substance called *angiotensin II.* In addition to causing vasoconstriction of the systemic arterioles, angiotensin II increases the production of *aldosterone* by the adrenal cortex, which in turn, causes the kidneys to retain sodium. The retention of sodium may lead to the retention of water and an increase in blood volume, thereby elevating the arterial pressure. Another suspected cause of high blood pressure is a high dietary intake of sodium, as in table salt. The mechanisms by which sodium affects blood pressure are unknown.

Arteriosclerosis and Atherosclerosis

As they grow older, a great number of people experience degenerative changes in the walls of their arteries that are known as *arteriosclerosis,* or "hardening of the arteries." Arteriosclerosis can greatly reduce the elasticity of the vessel walls. As the walls become less elastic, the vessels are unable to properly expand and recoil in response to the pressure changes produced by a beating heart. Consequently, in severe cases, the pressure in the vessels rises quite high during systole and falls unusually low during diastole.

One form of arteriosclerosis that can cause serious disease is *atherosclerosis.* Atherosclerosis is characterized by nodular deposits of lipid materials, primarily in the tunica intima of the arteries. Some vessels may have heavy deposits; others may have none. There is evidence that in the early stages of their development the deposits (plaques) that form on the inside of the vessel walls are composed primarily of smooth muscle cells. These cells are similar to those normally found in the walls of blood vessels except that in many instances they contain accumulations of lipid droplets. It is not known definitely what causes these smooth muscle cells to multiply at certain places in some vessels.

The atherosclerotic deposits can gradually narrow the lumen of blood vessels, as well as promote the formation of blood clots within the vessels. In either case the flow of blood through the vessels is restricted or blocked. The first indication of atherosclerosis is often the dysfunction of the organ supplied by the affected vessel. If the vessel supplies the brain, the lack of blood to a particular region may cause the brain tissue in that region to die. This is what occurs in a "stroke." If the coronary arteries become occluded, the heart muscle is affected, which causes a "heart attack."

Aneurysms

An *aneurysm* is a local dilation of an artery due to a weakness of the artery wall. The most frequent site of aneurysms is the aortic arch. The greatest danger from an aneurysm is that the affected vessel will rupture; however, aortic aneurysms can also exert pressure on the trachea, which can cause death from choking, or exert pressure on nerves, producing pain and possible paralysis.

Phlebitis

The inflammation of a vein is called *phlebitis.* Phlebitis can result from a number of conditions. One type of phlebitis is caused by bacteria that invade a vein, perhaps from an abscess or where the vein passes through an area of inflammation. In phlebitis, the inflammation of a vein may lead to the formation of a clot (a *thrombus*) in the blood vessel, with all its accompanying dangers. This condition is called *thrombophlebitis.*

Varicose Veins

Varicose veins are veins that are dilated, lengthened, and tortuous. Varicose veins are often brought about by a congenital and inherited weakness of the walls and valves of the veins. Such veins are unable to efficiently carry blood toward the heart, thus allowing the blood to accumulate in the lower limbs when the body is in an upright position. Other varicosities may be caused by pressure on veins, such as might occur during pregnancy, in obesity, or from an abdominal tumor. In severe cases, varicose veins can interfere with the return of blood to the extent that muscle cramps and edema (swelling) occur. Hemorrhage, phlebitis, and thrombosis are also possible complications.

Effects of Aging on the Vascular System

One of the most common age-related structural changes in blood vessels, especially in the larger arteries, is a reduction in the elasticity of their walls. The diminished elasticity affects the capacity of arteries to resist stretching and to rebound after being stretched, therefore it may affect blood pressure. The reduced elasticity is due primarily to a progressive replacement of elastic fibers in the walls of arteries with collagenous fibers. In addition, there is a tendency for calcium to bind to elastic fibers in older persons, and this also reduces the elasticity of arterial walls. In a similar manner, the walls of veins may thicken with age because of an increase in connective tissue and calcium deposits. However, because the blood pressure is so low in veins this is not thought to affect cardiovascular functioning significantly.

Diseases of the vascular system cause more than half the deaths of the population over age 65 in the United States. Not all of these vascular changes are necessarily due to aging alone, however. Evidence suggests that in nonindustrialized societies there is little vascular disease. By comparison, approximately 20% of the population in industrialized societies suffer from hypertension by the age of 65. Although age and genetic factors probably contribute to the changes in the walls of blood vessels that lead to myocardial infarction or stroke, many investigators believe that environmental factors, especially diet and exercise, play a more significant role than age or genetic inheritance.

ANATOMY OF THE VASCULAR SYSTEM

The blood of the circulatory system is carried throughout the body in a complex organization of vessels. Only the major blood vessels are named in this chapter, but you should bear in mind that there are many other smaller vessels that are not mentioned. The arteries branch into many small arterioles, which, in turn, branch into numerous microscopic capillaries, where the exchange between tissues and blood occurs. From the capillaries, the blood enters tiny venules that join together to form the larger veins.

The vessels of the circulatory system can be divided into two separate circuits, each of which leaves and returns to the heart. The **pulmonary circuit** carries blood from the right side of the heart to the lungs and back to the left side of the heart. The **systemic circuit** carries blood that has just left the pulmonary circuit to the rest of the body and back to the right side of the heart.

Pulmonary Circuit

F11.8 Blood enters the pulmonary circuit from the right ventricle of the heart through the **pulmonary trunk** (Figure 11.8). The short pulmonary trunk divides into **left** and **right pulmonary arteries,** which pass directly to the left and right lungs. As they enter the lungs, the pulmonary arteries divide into several **lobar branches,** one to each lobe of the lungs. The right pulmonary artery divides into three lobar arteries; the left divides into two. (The right lung has three lobes and the left lung has two.) The lobar arteries divide into numerous orders of smaller arteries and arterioles and, eventually, into capillary plexuses located in the walls of the tiny air sacs (alveoli) of the lungs. It is through these capillary plexuses that gases are exchanged between the blood and the air. In the alveoli of the lungs, the blood (in capillaries) and the air (in alveoli) are separated only by the thin alveolar epithelium and the endothelium of the capillary. The total thickness of these membranes is less than half a micron (one micron equals 0.001 mm).

From the capillaries, the blood is collected into venules and then into progressively larger veins that lead into left and right **pulmonary veins,** generally two from each lung. The pulmonary veins empty into the left atrium. The pulmonary arteries carry blood that has a high carbon dioxide content and is low in oxygen. The blood in the pulmonary veins, however, has a high oxygen content and is low in carbon dioxide. This pattern is the exact opposite of that of the systemic arteries and veins. In other words, the vessels are named according to the direction of bloodflow within them, not the condition of the blood they carry. Thus any vessel that carries blood away from the heart is an *artery* and any vessel that returns blood to the heart is a *vein*.

The vessels of the pulmonary circuit, unlike those of the systemic circuit,

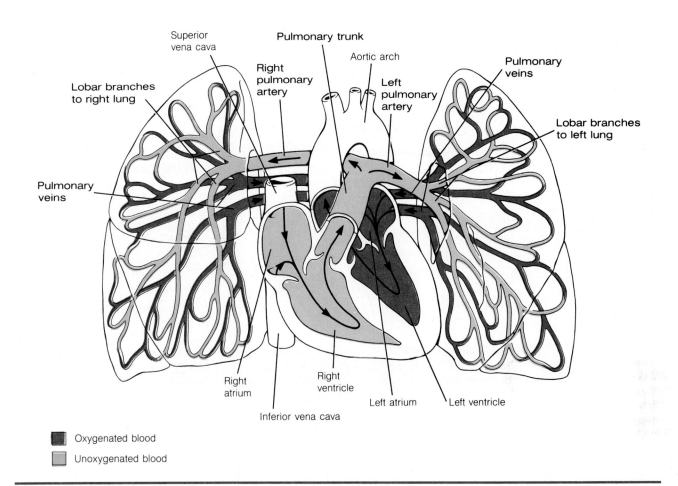

Oxygenated blood

Unoxygenated blood

do not supply oxygen and nutrients to the lung tissues. The metabolic needs of the lungs are provided for by the small *bronchial vessels* of the systemic circuit.

Systemic Circuit

Blood that is not in the pulmonary circuit is in the systemic circuit. The vessels of the systemic circuit transport blood to all tissues and organs of the body except the alveoli of the lungs.

Systemic Arteries

Blood from the left ventricle enters the systemic circuit through the **aorta,** from which all the arteries of the systemic circuit branch (Figure 11.9). For descriptive purposes, it is convenient to divide the aorta into the **ascending aorta,** the **aortic arch,** and the **descending aorta,** which has **thoracic** and **abdominal** portions. The branches of these various portions are summarized in Table 11.1.

F11.9

Table 11.1

BRANCHES OF THE ASCENDING AORTA The ascending aorta is a short vessel that passes in front of the right pulmonary artery. Its only branches are the **right** and **left coronary arteries,** which arise from sinuses behind the cusps of the aortic valve and supply the heart muscle (Figure 10.7, Figure 10.8).

F10.7, F10.8

BRANCHES OF THE AORTIC ARCH As the aorta leaves the pericardial sac it arches dorsally and to the left, forming the aortic arch. Three branches arise from the aortic arch: the **brachiocephalic artery,** the **left common carotid artery,** and the **left subclavian artery** (Figure 11.9). These arteries furnish all the blood to the head, neck, and upper limbs.

F11.9

Figure 11.8
Pulmonary circulation showing the pulmonary trunk dividing into right and left pulmonary arteries, which, in turn, divide into lobar branches. Blood from the lungs returns to the left atrium through the pulmonary veins. The arrows indicate the direction of bloodflow.

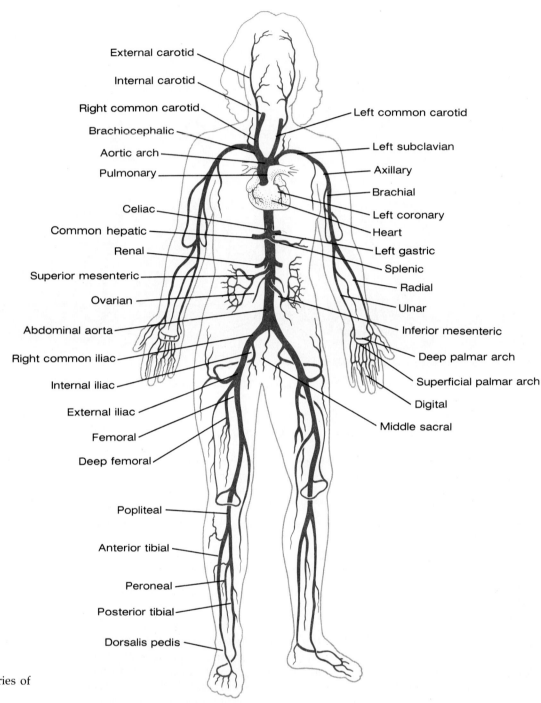

External carotid

Internal carotid

Right common carotid

Brachiocephalic

Aortic arch

Pulmonary

Celiac

Common hepatic

Renal

Superior mesenteric

Ovarian

Abdominal aorta

Right common iliac

Internal iliac

External iliac

Femoral

Deep femoral

Popliteal

Anterior tibial

Peroneal

Posterior tibial

Dorsalis pedis

Left common carotid

Left subclavian

Axillary

Brachial

Left coronary

Heart

Left gastric

Splenic

Radial

Ulnar

Inferior mesenteric

Deep palmar arch

Superficial palmar arch

Digital

Middle sacral

Figure 11.9
Major systemic arteries of
the body.

ARTERIES OF THE HEAD AND NECK Shortly after arising from the arch of the aorta, the brachiocephalic artery divides into a **right subclavian artery** and a **right common carotid artery** (Figure 11.10). The right common carotid artery and the **left common carotid artery** (which arises directly from the arch of the aorta) supply most of the blood to the head and neck. The common carotid arteries course upward alongside the trachea a short distance before each bifurcates (divides) into an **internal carotid artery** and an **external carotid artery.** At this point of bifurcation, the vessels enlarge slightly, forming the **carotid sinus.** The sinus contains pressoreceptors that assist in regulating blood pressure. Near this same region is a small, oval **carotid body** that con-

F11.10

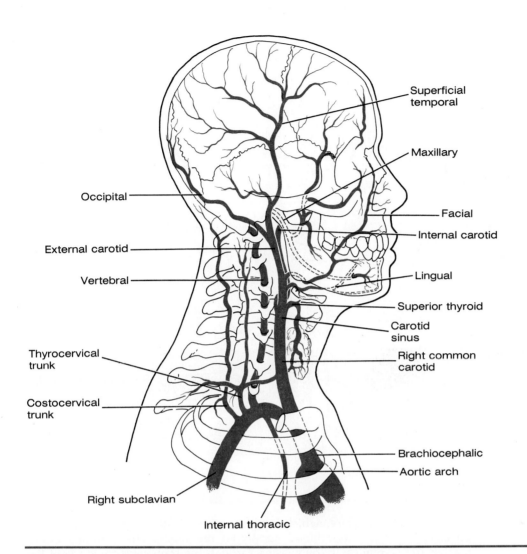

Superficial temporal

Maxillary

Occipital

Facial

Internal carotid

External carotid

Vertebral

Lingual

Superior thyroid

Carotid sinus

Right common carotid

Thyrocervical trunk

Costocervical trunk

Brachiocephalic

Aortic arch

Right subclavian

Internal thoracic

Figure 11.10
Arteries of the right side of the head and neck.

tains chemoreceptors sensitive to changing levels of oxygen, carbon dioxide, and pH in the blood that travels toward the brain.

The **external carotid artery,** which supplies most of the head and neck except for the brain, lies more superficially than the internal carotid artery. It passes through the parotid salivary gland and terminates as the **maxillary** and **superficial temporal arteries** (Figure 11.10). Just before terminating, the external carotid artery gives off the **occipital artery** to the posterior region of the scalp. Through these branches the external carotid artery supplies the muscles of mastication, the mucous membranes of the nose, the pharynx, and the palate, the teeth of the upper and lower jaws, some neck muscles, and the entire scalp region. While still in the neck, it sends branches to the thyroid gland **(superior thyroid artery),** tongue **(lingual artery),** and the skin and the muscles of the face **(facial artery).**

The **internal carotid artery** enters the cranial cavity through the carotid ⟶ *foramen lacerum* canal in the lower surface of the petrous portion of the temporal bone. It travels forward in the canal, and soon leaves the bone to pass forward and upward along the side of the sella turcica. The internal carotid artery gives off an **ophthalmic artery,** which follows the lower surface of the optic nerve and enters the orbit. The internal carotid artery supplies the brain through its terminal branches: the **anterior cerebral artery** and the **middle cerebral artery.** A small **anterior communicating artery** connects the right and left anterior cerebral arteries. These vessels also help form the **cerebral arterial circle (circle of Willis)** (Figure 11.11), which surrounds the infundibulum of the pituitary gland.

F11.10

F11.11

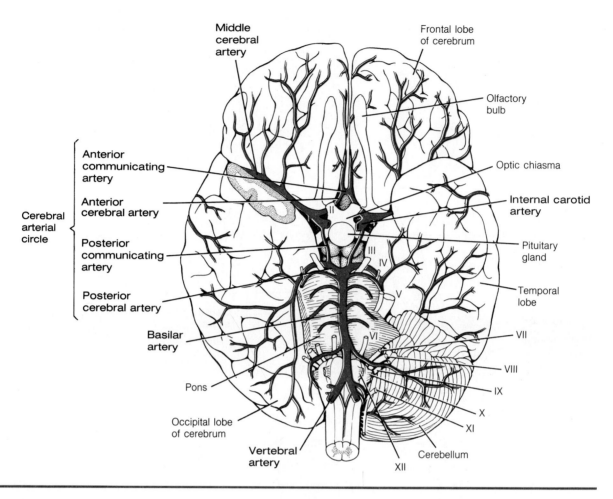

Middle cerebral artery

Frontal lobe of cerebrum

Olfactory bulb

Anterior communicating artery

Anterior cerebral artery

Cerebral arterial circle

Optic chiasma

Internal carotid artery

Pituitary gland

Posterior communicating artery

Posterior cerebral artery

Temporal lobe

II

III

IV

V

VI

VII

VIII

IX

X

XI

Basilar artery

Pons

Occipital lobe of cerebrum

Vertebral artery

Cerebellum

XII

Figure 11.11

Arteries of the base of the brain, forming the cerebral arterial circle around the pituitary gland. To provide an unobstructed view, part of the right temporal lobe and the right cerebellar hemisphere have been removed. Roman numerals indicate cranial nerves.

F11.11

The **vertebral artery,** which is the first branch of the subclavian artery, provides another major blood supply to the brain. The vertebral artery reaches the cranial cavity by passing through the transverse foramina of the cervical vertebrae and the foramen magnum. The right and left vertebral arteries join on the ventral surface of the pons of the brain, forming the **basilar artery** (Figure 11.11). The basilar artery continues forward and terminates as **right** and **left posterior cerebral arteries,** which supply the posterior regions of the cerebral hemispheres of the brain. The basilar artery also gives off branches that supply the pons and the cerebellum. The **posterior communicating arteries** from the internal carotid arteries join with the posterior cerebral arteries to complete the cerebral arterial circle around the infundibulum.

ARTERIES OF THE UPPER LIMBS The upper limbs are supplied by the **subclavian arteries.** As noted earlier, the right subclavian artery is a branch of the brachiocephalic artery, whereas the left subclavian arises directly from the aortic arch. The first part of each subclavian artery gives off branches that do

F11.12 not enter the upper limbs (Figure 11.12). Just lateral to the vertebral artery, the **thyrocervical trunk** and **costocervical trunk** send branches to the lower neck, the dorsal scapular, and the intercostal regions. The **internal thoracic (internal mammary) artery** arises from the subclavian artery and passes downward beneath the costal cartilage. It supplies branches to the thoracic and intercostal muscles, the mediastinum, and the diaphragm.

After passing beneath the clavicle, the subclavian artery becomes the **axillary artery.** The axillary artery gives off branches to the lateral wall of the thorax (**thoracoacromial trunk, lateral thoracic artery, subscapular artery,** and **thoracodorsal artery**) and to the region around the proximal end of the humerus (**anterior** and **posterior humeral circumflex arteries**). It then continues into the arm as the **brachial artery,** which travels down the medial side of the humerus, supplying the muscles of the arm.

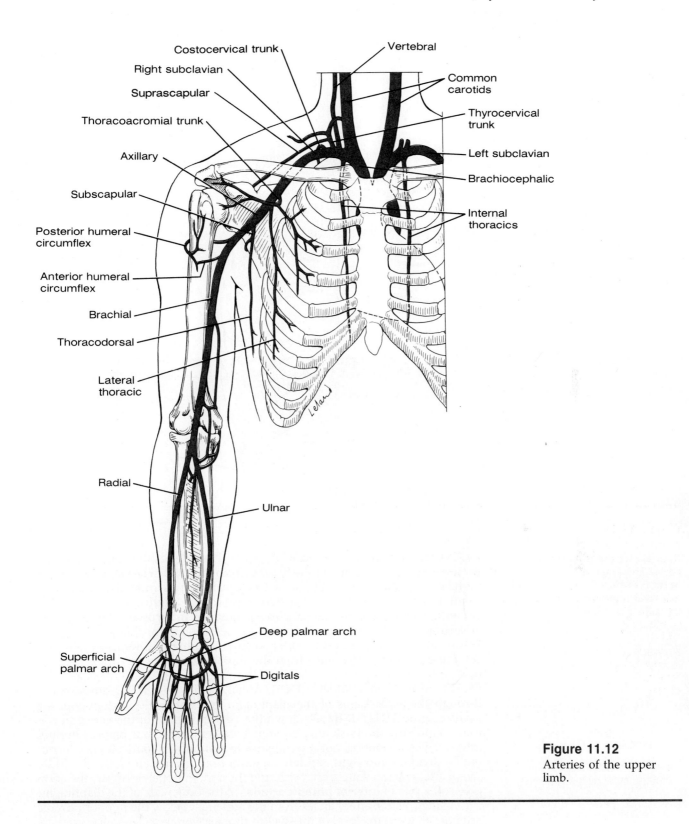

Costocervical trunk

Right subclavian

Suprascapular

Thoracoacromial trunk

Axillary

Subscapular

Posterior humeral circumflex

Anterior humeral circumflex

Brachial

Thoracodorsal

Lateral thoracic

Radial

Superficial palmar arch

Vertebral

Common carotids

Thyrocervical trunk

Left subclavian

Brachiocephalic

Internal thoracics

Ulnar

Deep palmar arch

Digitals

Figure 11.12
Arteries of the upper limb.

The brachial artery crosses the anterior surface of the elbow and divides into **radial** and **ulnar arteries**. Both vessels travel down the anterior surface of the forearm—the radial artery alongside the radius, the ulnar artery along the ulna. The radial artery is easily palpated on the lateral side of the wrist, where it is used to take the pulse. The two arteries join in the hand through interconnecting branches of **superficial** and **deep palmar arches. Digital arteries** from the palmar arches supply the fingers.

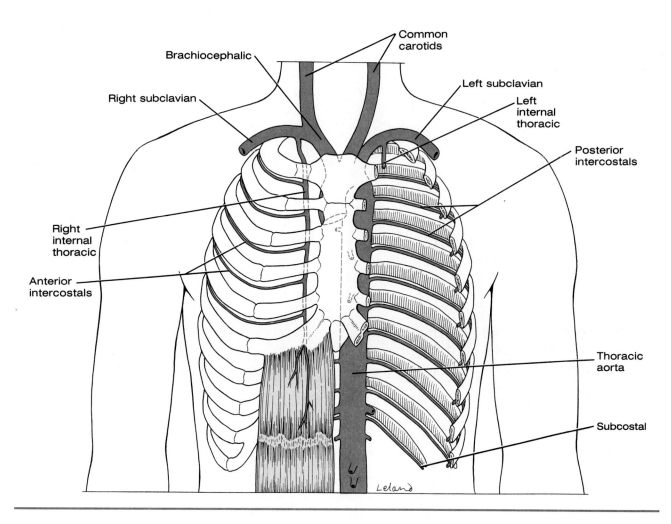

Figure 11.13
Branches of the thoracic aorta. The anterior portion of the rib cage has been removed on the left side.

BRANCHES OF THE THORACIC AORTA The arteries that supply the thorax arise from the first part of the descending aorta, which lies to the left of the vertebral column (Figure 11.13). All of the branches from this region are small. Most prominent are the **posterior intercostal arteries,** which travel between the ribs and anastomose with the **anterior intercostal arteries,** which arise from the internal thoracic artery. In addition, **bronchial, esophageal, subcostal** (beneath the twelfth rib), and **superior phrenic arteries** (which supply the diaphragm) also arise from the thoracic aorta.

BRANCHES OF THE ABDOMINAL AORTA The descending aorta passes through the *aortic hiatus* of the diaphragm and enters the abdominopelvic cavity (Figure 11.14). It travels down the ventral surface of the vertebral column, supplying the posterior abdominal walls through four pairs of **lumbar arteries.** The abdominal aorta terminates in front of the fourth lumbar vertebra by dividing into **right** and **left common iliac arteries** and a small **middle sacral artery.** Immediately after entering the abdominopelvic cavity, the aorta gives off a pair of **inferior phrenic arteries** to the underside of the diaphragm.

A single **celiac artery** arises from the aorta just below the inferior phrenic arteries, at about the level of the twelfth thoracic vertebra. The celiac artery is very short and almost immediately divides into *left gastric, splenic,* and *common hepatic arteries* (Figure 11.15). The **left gastric artery** follows the lesser curvature of the stomach, supplying the lower part of the esophagus and part of the stomach. The **splenic artery,** which supplies the spleen, also gives off small branches to the pancreas and stomach, as well as the large **left gastroepiploic artery,** which supplies the greater curvature of the stomach. As it travels to the liver, the **common hepatic artery** sends branches to the stomach,

F11.13

F11.14

F11.15

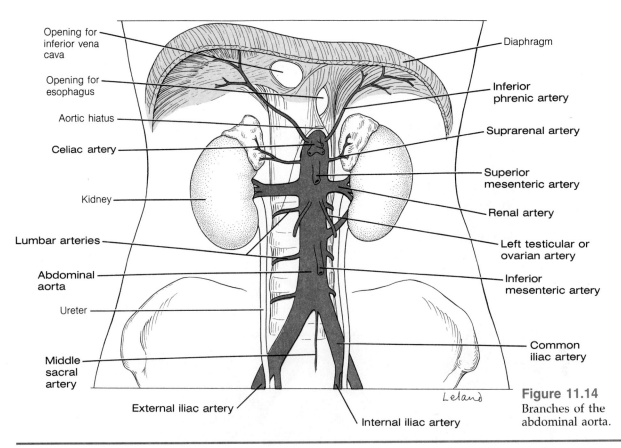

Opening for inferior vena cava

Opening for esophagus

Aortic hiatus

Celiac artery

Kidney

Lumbar arteries

Abdominal aorta

Ureter

Middle sacral artery

External iliac artery

Internal iliac artery

Diaphragm

Inferior phrenic artery

Suprarenal artery

Superior mesenteric artery

Renal artery

Left testicular or ovarian artery

Inferior mesenteric artery

Common iliac artery

Leland

Figure 11.14
Branches of the abdominal aorta.

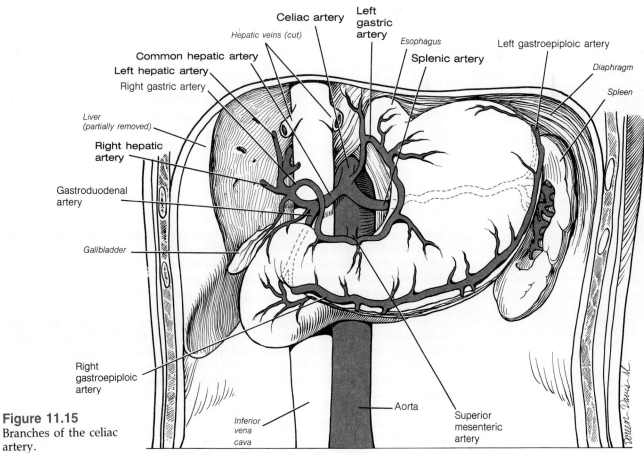

Celiac artery

Left gastric artery

Hepatic veins (cut)

Common hepatic artery

Left hepatic artery

Right gastric artery

Liver (partially removed)

Right hepatic artery

Gastroduodenal artery

Gallbladder

Right gastroepiploic artery

Esophagus

Splenic artery

Left gastroepiploic artery

Diaphragm

Spleen

Inferior vena cava

Aorta

Superior mesenteric artery

Figure 11.15
Branches of the celiac artery.

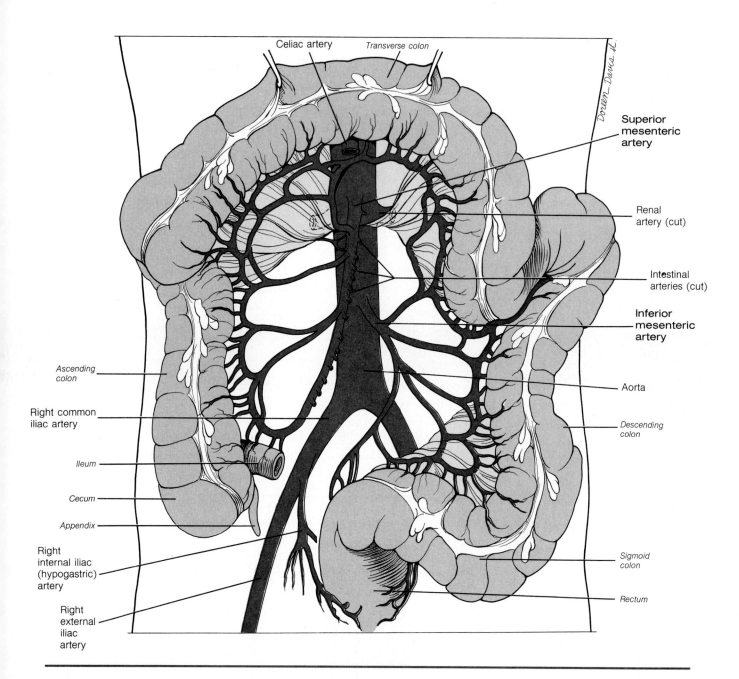

Figure 11.16
The superior and inferior mesenteric arteries. The transverse colon has been lifted up from its normal position.

duodenum, and pancreas through **gastroduodenal** and **right gastric** branches. The right gastric artery supplies the lesser curvature of the stomach, and the **right gastroepiploic artery,** a branch of the gastroduodenal artery, supplies the greater curvature. After giving off these branches the common hepatic artery divides into **right** and **left hepatic arteries,** which supply the liver and gall bladder.

The single **superior mesenteric artery** arises from the aorta just below the celiac artery. It travels within the mesentery of the intestines, where it gives **F11.16** off several branches that form anastomosing loops (Figure 11.16). Branches of the superior mesenteric artery supply all of the small intestine, as well as the cecum, the ascending colon, and most of the transverse colon of the large intestine.

Branching laterally from the aorta at the level of the superior mesenteric arteries are two **suprarenal arteries,** which supply the adrenal glands **F11.14** (Figure 11.14). These glands are also supplied by branches from the inferior **F11.14** phrenic and renal arteries. The paired **renal arteries** (Figure 11.14), which supply the kidneys, arise from the lateral margins of the aorta just inferior to the superior mesenteric artery.

The arteries that supply the gonads (ovaries and testes) (Figure 11.14) arise from the ventral surface of the aorta a short distance below the renal arteries. In the female, the **ovarian arteries** pass downward and laterally into the pelvic cavity to supply the ovaries; in addition, they give off branches to the ureters and uterine tubes. The **testicular (internal spermatic) arteries** of the male are longer than the ovarian arteries, since they pass through the inguinal canal and enter the scrotum, where they supply the testes. **F11.14**

The **inferior mesenteric artery** is the final branch of the abdominal aorta (Figure 11.16). It is a single vessel that arises from the ventral surface of the aorta just above its bifurcation into the **common iliac arteries.** Branches of the inferior mesenteric artery supply part of the transverse colon, the descending colon, the sigmoid colon, and most of the rectum. Branches from this artery anastomose with branches of the superior mesenteric artery. **F11.16**

ARTERIES OF THE PELVIC REGION   Each of the common iliac arteries divides in front of the sacroiliac articulation into *internal* and *external iliac* arteries (Figure 11.17). The **internal iliac** *(hypogastric)* **arteries** enter the pelvic cavity and divide into branches that supply the pelvic viscera (urinary bladder, uterus, vagina, and rectum). In addition, branches from the internal iliac arteries supply the muscles of the gluteal and lumbar regions, the walls of the pelvis, the external genitalia, and the medial region of the thigh. **F11.17**

ARTERIES OF THE LOWER LIMBS   The **external iliac artery** is actually a continuation of the common iliac artery. It travels downward and laterally through the iliac fossa and enters the thigh as it passes beneath the inguinal ligament. While in the pelvic cavity, the external iliac artery supplies branches to the muscles and skin of the lower abdominal wall.

Upon entering the thigh, the external iliac artery becomes the **femoral artery** (Figure 11.17). The femoral artery passes along the anterior medial region of the thigh. In the lower thigh, it passes to the posterior surface of the knee through an opening *(adductor hiatus)* in the tendon of the adductor magnus muscle, and is then known as the **popliteal artery.** While in the thigh, the femoral artery sends several branches to the skin and muscles of the region. The **lateral** and **medial femoral circumflex arteries** supply the region around the proximal end of the femur, whereas the **profunda (deep) femoral artery** passes posteriorly to supply muscles of the posterior compartment of the thigh. **F11.17**

The **popliteal artery** is the continuation of the femoral artery. It passes behind the knee (through the popliteal fossa), supplying the muscles and skin of the area, and then divides into an *anterior tibial artery* and a *posterior tibial artery* (Figure 11.18). **F11.18**

The **posterior tibial artery** continues downward behind the tibia, supplying the muscles of the posterior compartment of the leg. Behind the ankle, it divides into **medial** and **lateral plantar arteries,** which supply the sole of the foot and form the **plantar arch. Digital arteries** arise from the plantar arch. Near its origin the posterior tibial artery gives rise to the **peroneal artery,** which supplies the peroneal muscles in the lateral compartment of the leg.

The **anterior tibial artery** passes through the interosseous membrane that connects the fibula to the tibia. It then travels down the ventral surface of the membrane, supplying the muscles of the anterior compartment of the leg. The anterior tibial artery passes in front of the ankle and ends on the dorsum of the foot as the **dorsalis pedis artery.** Branches of the dorsalis pedis artery supply the dorsum of the foot and anastomose with the plantar arch on the sole of the foot.

Systemic Veins

The blood that has been delivered throughout the body by the arteries is collected and returned to the heart through the veins. Since veins tend to be larger and more numerous than arteries, the capacity of the venous system is

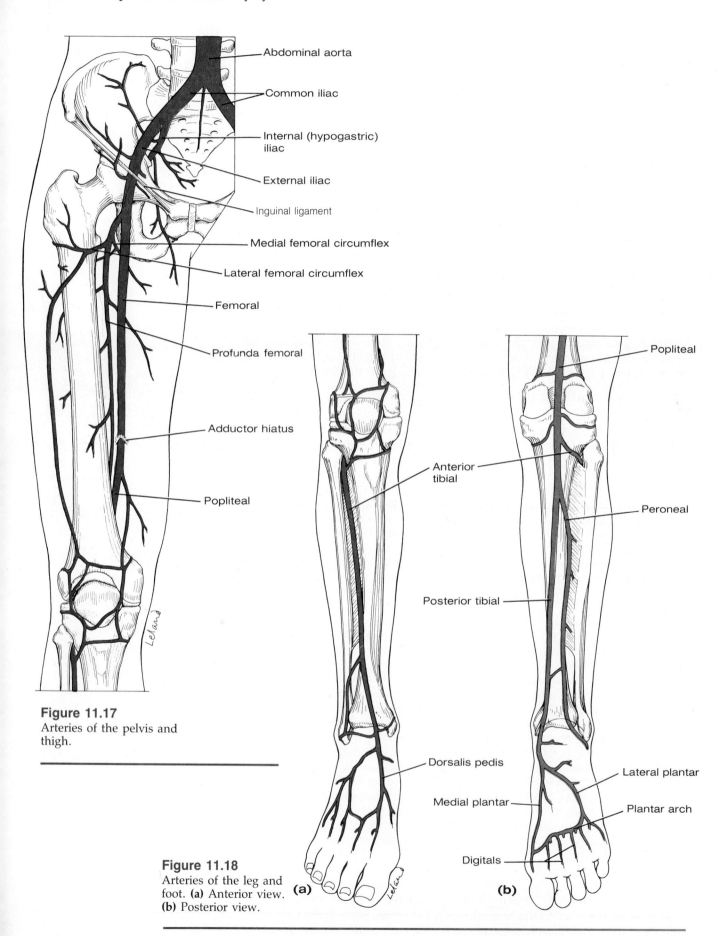

Abdominal aorta

Common iliac

Internal (hypogastric) iliac

External iliac

Inguinal ligament

Medial femoral circumflex

Lateral femoral circumflex

Femoral

Profunda femoral

Adductor hiatus

Popliteal

Figure 11.17
Arteries of the pelvis and thigh.

Anterior tibial

Posterior tibial

Dorsalis pedis

Medial plantar

Figure 11.18
Arteries of the leg and foot. **(a)** Anterior view. **(b)** Posterior view.

(a)

Popliteal

Peroneal

Lateral plantar

Plantar arch

Digitals

(b)

Table 11.1   Schematic Summary of the Systemic Arteries That Arise from Various Regions of the Aorta

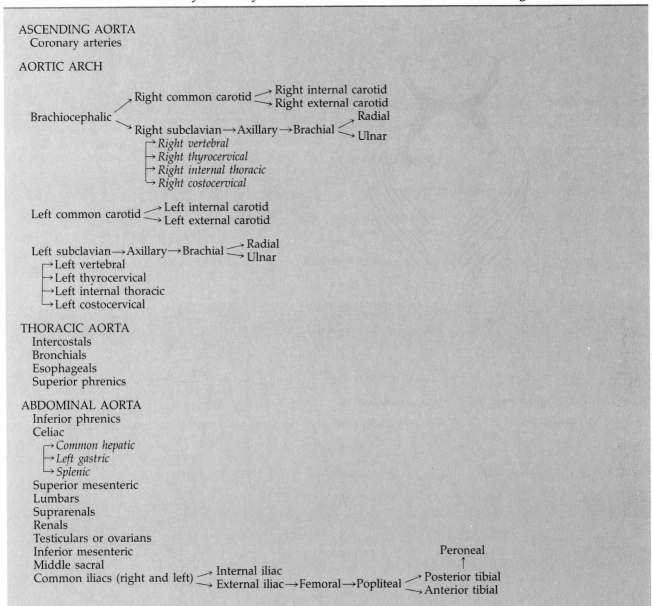

ASCENDING AORTA
  Coronary arteries

AORTIC ARCH

THORACIC AORTA
  Intercostals
  Bronchials
  Esophageals
  Superior phrenics

ABDOMINAL AORTA
  Inferior phrenics
  Celiac
  Superior mesenteric
  Lumbars
  Suprarenals
  Renals
  Testiculars or ovarians
  Inferior mesenteric
  Middle sacral
  Common iliacs (right and left)

greater than that of the arterial system. The major veins, summarized in Table 11.2, tend to travel alongside the arteries. Although the general pattern of veins (Figure 11.19) is quite similar to the pattern of arteries, there are some important differences, which are explained in this chapter. Systemic veins are classified as *deep veins, superficial veins,* or *venous sinuses.*

Table 11.2
F11.19

Most of the **deep veins** travel alongside an artery and have the same name as the artery. Certain deep veins of the head and vertebral column are exceptions to this pattern, however.

The **superficial veins,** which lie just beneath the skin (in the hypodermis), return blood from the skin and the subcutaneous regions to the deep veins. Since there are no large superficial arteries, the names of the superficial veins do not correspond to the names of arteries.

Venous sinuses are not actually vessels, but are spaces that collect blood in certain regions and return it to the veins. The walls of venous sinuses are composed of connective tissue, with no muscle present, and they are lined with endothelium that is continuous with the endothelium of the capillaries

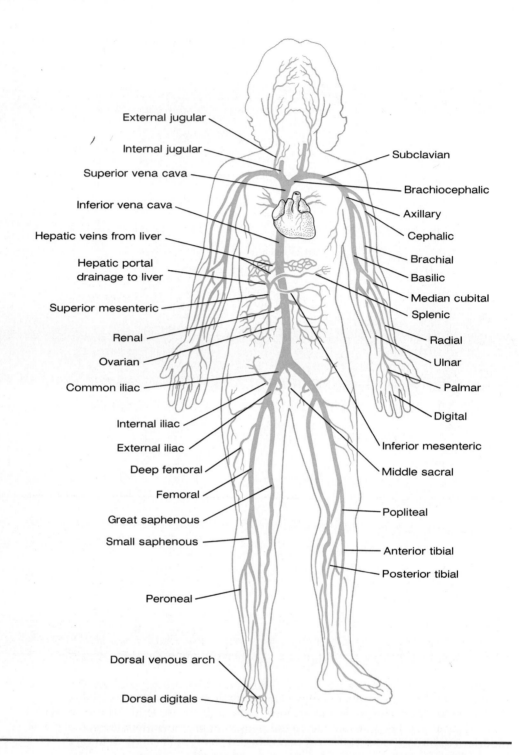

Figure 11.19
Major systemic veins of
the body.

and veins. The coronary sinus is a venous sinus that collects blood from the cardiac veins of the heart and returns it to the right atrium. Still larger venous sinuses are located in the dura mater, the outer meningeal covering of the
F11.20 brain (Figure 11.20). Most of the blood from the brain travels through these **dural sinuses** before entering the veins that return blood to the heart.

VEINS OF THE HEAD AND NECK   Most of the venous blood from the head and neck regions returns to the heart through the *internal jugular veins*, the
F11.21 *external jugular veins*, and the *vertebral veins* (Figure 11.21).
The **internal jugular veins** are larger and deeper than the external jugular veins. Each internal jugular vein drains a **transverse sinus** that receives blood

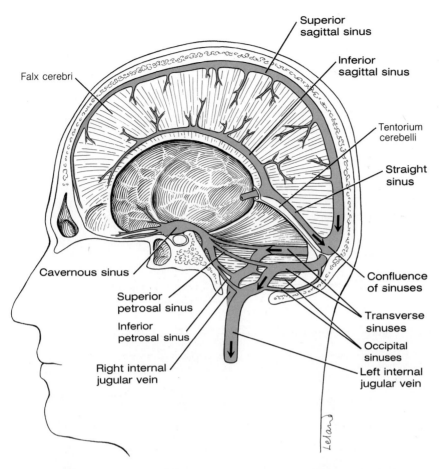

Superior
sagittal sinus

Inferior
sagittal sinus

Falx cerebri

Tentorium
cerebelli

Straight
sinus

Cavernous sinus

Confluence
of sinuses

Superior
petrosal sinus

Transverse
sinuses

Inferior
petrosal sinus

Occipital
sinuses

Right internal
jugular vein

Left internal
jugular vein

Figure 11.20
Venous sinuses of the
brain. The arrows
indicate the direction of
bloodflow. Note that
blood from all of the
dural sinuses enters the
internal jugular vein.

from a **cavernous sinus,** the **superior sagittal sinus,** the **inferior sagittal sinus,** and the **straight sinus** (Figure 11.20). The internal jugular veins there- **F11.20** fore serve as the major venous drainage of the brain. Each internal jugular vein leaves the skull through a jugular foramen, which is located in the junction between the petrous portion of the temporal bone and the occipital bone, and travels through the neck alongside a common carotid artery and a vagus nerve. The internal jugular vein joins with a **subclavian vein** to form a **brachiocephalic vein** (Figure 11.21). **F11.21**

The **vertebral veins** drain the posterior regions of the head (Figure 11.21). **F11.21** Each vertebral vein passes through the transverse foramina of the cervical vertebrae and joins a brachiocephalic vein.

The superficial regions of the head and neck, which are supplied by the external carotid arteries, are drained by the **external jugular veins** (Figure 11.21). Each external jugular vein passes superficially down the neck **F11.21** and empties into a **subclavian vein** lateral to the point at which the subclavian and internal jugular veins join to form a brachiocephalic vein. The right and left brachiocephalic veins join to form the superior vena cava, which empties into the right atrium of the heart.

VEINS OF THE UPPER LIMBS The deep veins of the upper limbs follow the paths of the arteries and are given the same names: **axillary, brachial, radial, and ulnar veins** (Figure 11.19). The superficial veins (Figure 11.22) begin from **F11.19,** venous networks that cover the dorsal and palmar surfaces of the hand. The **F11.22** **cephalic vein** arises from the lateral side of the dorsal veins of the hand, crosses to the ventral-lateral side of the forearm, and continues up the lateral side of the arm. At the shoulder, it goes deep and empties into the axillary vein. The **basilic vein** is the other major superficial vein of the upper limb. It arises from the medial side of the dorsal veins of the hand and ascends along

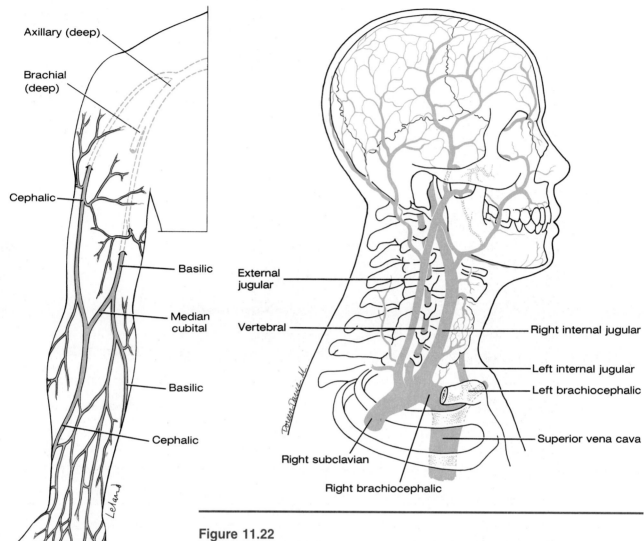

Figure 11.21
Veins of the head and
neck.

Figure 11.22
Superficial veins of the
upper limb.

the medial-posterior side of the forearm. Just below the elbow, the basilic vein travels to the front of the arm and goes deep above the elbow to join with the brachial vein, thus forming the axillary vein. There are several superficial branches between the basilic and cephalic veins. One of these, the **median cubital vein,** which connects the two vessels in front of the elbow, is commonly used to give or receive blood. All the venous blood from the head, neck, and upper limbs is returned to the heart through the superior vena cava.

VEINS OF THE THORAX   The venous blood of the wall of the thorax also empties into the superior vena cava—in this case by way of an *azygos system* of F11.23 veins (Figure 11.23). The **azygos vein** begins in the upper right lumbar region and passes through the diaphragm via the *aortic hiatus* (along with the aorta and the thoracic duct of the lymphatic system) to enter the thorax. It continues up the posterior wall of the thorax, to the right of the vertebral column, and empties into the superior vena cava. The **hemiazygos vein** follows a similar course on the left side, passing in front of the vertebral column to empty into the azygos vein in the middle of the thorax. Most of the upper left region of the thorax is drained by an **accessory hemiazygos vein,** which also crosses the vertebral column to empty into the azygos vein. The upper three left intercostal veins drain directly into the left brachiocephalic vein through

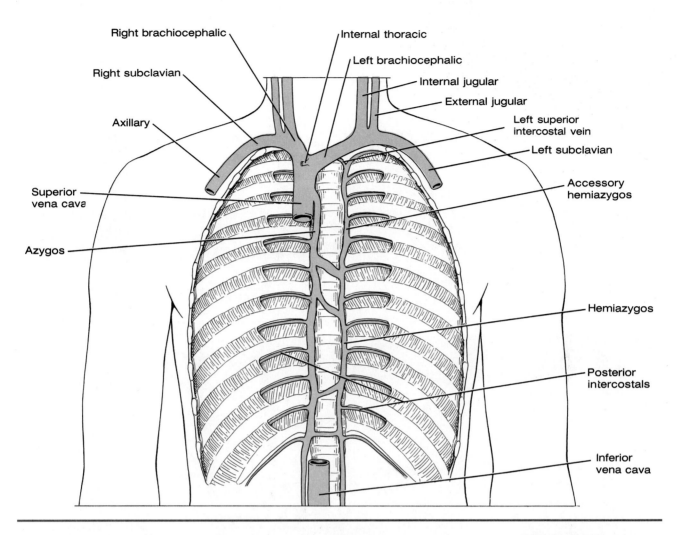

Right brachiocephalic

Right subclavian

Axillary

Superior vena cava

Azygos

Internal thoracic

Left brachiocephalic

Internal jugular

External jugular

Left superior intercostal vein

Left subclavian

Accessory hemiazygos

Hemiazygos

Posterior intercostals

Inferior vena cava

Figure 11.23
Veins of the posterior wall of the thorax. The anterior wall of the thoracic cage has been removed.

the left **superior intercostal vein.** The azygos system of veins also receives the **posterior intercostal, bronchial, esophageal,** and **pericardial veins.**

VEINS OF THE ABDOMEN AND PELVIS Venous blood from the lower regions of the body is returned to the heart through the **inferior vena cava.** This large vessel is formed by the confluence of the **right** and **left common iliac veins** at the level of the fifth lumbar vertebra. It ascends along the posterior body wall to enter the right atrium of the heart. As it passes through the abdomen, the inferior vena cava receives tributaries that correspond to most of the arteries that arise from the abdominal aorta—for example, **lumbar, renal, suprarenal, hepatic,** and **right testicular** or **ovarian veins.** (The **left testicular** or **ovarian veins** empty into the **left renal vein.**) Notice, however, that the inferior vena cava does not receive blood directly from the digestive tract, pancreas, or spleen. The veins from these regions form the **hepatic portal system,** which is discussed in the next section.

The veins of the pelvic region follow the pattern of the arteries and are given the same names. Most of them empty into the **internal iliac veins,** which join with the **external iliac veins** from the lower limbs to form the **right** and **left common iliac veins.**

THE HEPATIC PORTAL SYSTEM Venous blood from the stomach, intestines, spleen, and pancreas is carried by small veins, most of which empty into three large vessels: the *splenic vein,* the *inferior mesenteric vein,* and the *superior mesenteric vein.* The **splenic vein** returns blood from the spleen. As it travels toward the midline of the body it receives tributaries from the stomach

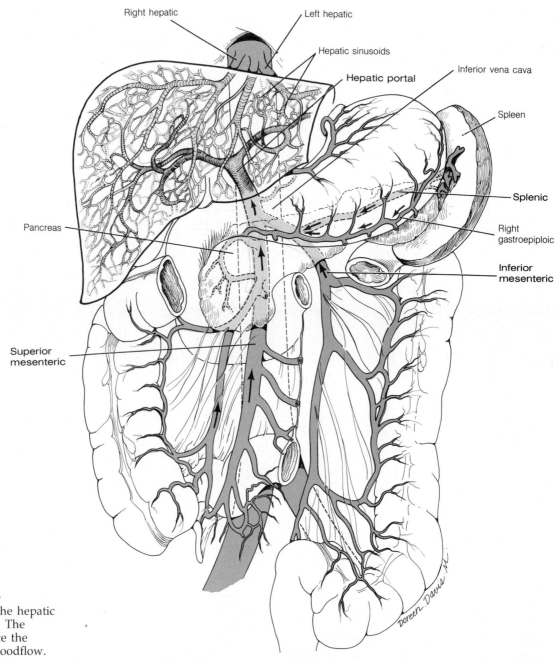

Right hepatic

Left hepatic

Hepatic sinusoids

Hepatic portal

Inferior vena cava

Spleen

Splenic

Right gastroepiploic

Inferior mesenteric

Pancreas

Superior mesenteric

Doreen Davis M.

Figure 11.24
The veins of the hepatic portal system. The arrows indicate the direction of bloodflow.

and pancreas. The **inferior mesenteric vein** returns blood from the rectum and the descending limb of the large intestine. It ascends beneath the parietal peritoneum and joins the splenic vein behind the pancreas. The common vessel formed by the junction of the inferior mesenteric vein and the splenic vein joins the superior mesenteric vein behind the neck of the pancreas. The **superior mesenteric vein** returns blood from the small intestine, the cecum, and the ascending and transverse limbs of the large intestine. The junction of the superior mesenteric vein and the splenic vein forms the **hepatic portal vein** (Figure 11.24). The hepatic portal vein ascends in the right border of the lesser omentum and enters the inferior surface of the liver, through which it follows a unique pathway that is discussed in Chapter 20. After leaving the liver, the blood travels through the hepatic veins to enter the inferior vena cava.

F11.24

FRONTIERS IN HEALTH

Cleaning Out Clogged Arteries

Medical researchers are experimenting with a variety of techniques for cleaning out blood vessels that are clogged by blood clots and the buildup of atherosclerotic plaque. Several chemical agents have been found that dissolve the blood clots that often accompany heart attacks. These substances help restore blood flow after an attack and significantly reduce heart muscle damage. Their use promises to spare thousands of lives every year.

In conjunction with the clot dissolvers, some medical researchers are using a small catheter with a tiny balloon attached to its tip. After the chemical clot dissolvers are administered and the catheter is maneuvered to the location of the clot, the balloon is inflated. The balloon causes the artery to expand and apparently loosens the plaque from the wall, thereby freeing the dangerous accumulation of cholesterol. This technique, called balloon angioplasty, is not yet completely proven, but early results have been very promising.

Scientists are also looking for high-tech weapons to add to their arsenal against clogged arteries. Researchers at Stanford University Medical Center, for example, are experimenting to see if they can use lasers to break up atherosclerotic blockages in arteries. Clinical tests of this technique on human subjects have also been encouraging.

Medical researchers in France have used lasers to open up blood clots in patients undergoing open-heart surgery. A catheter containing fine glass fibers was inserted into the occluded arteries during surgery. Laser beams were than transmitted through the fiber-optic device, quickly burning away the blockage. However, even though the laser opened up the arteries, they all were blocked again within 3 months. In an attempt to determine why the arteries clogged again so quickly,

workers are using a special catheter that not only delivers a laser beam but also uses a beam of light to illuminate the occlusion and a fiber to transmit the image back to the surgeon.

If reocclusion can be prevented, it is hoped that atherosclerotic blockages may be removed without opening a patient's chest. By inserting the laser catheter into the femoral artery and guiding it into the clogged vessel, surgeons may someday be able to remove atherosclerotic plaque with a fraction of the trauma incurred by open-heart surgery.

Medical opinions on the potential of lasers vary, however. Some cardiologists consider the technique ready for widespread use. Others believe cautious optimism is called for because of the potential risk of burning holes in arterial walls or weakening them, or even coagulating blood with the laser's heat.

One of the most exciting developments aimed at minimizing the potential damage is the "cool" laser, which emits ultraviolet light in short bursts, allowing the tissue to cool between pulses. This instrument could greatly reduce the possibility of tissue damage.

Scientists at the University of Florida at Gainesville are working on another technique to remove blood clots and atherosclerotic plaque from arteries. They have devised a special catheter, which is inserted into clogged vessels. Inside the catheter is a small revolving screw that, when turned, creates a mild suction. Tests in dogs indicate that the device is quite successful in vacuuming out blood clots and atherosclerotic plaque.

Cardiovascular disease is often the result of stress and other factors that are by-products of the modern technological society in which we live. It is fitting, then, that technology should offer hope by helping physicians treat the disease.

— Fatty plaque

Coronary artery showing fatty plaque layer.

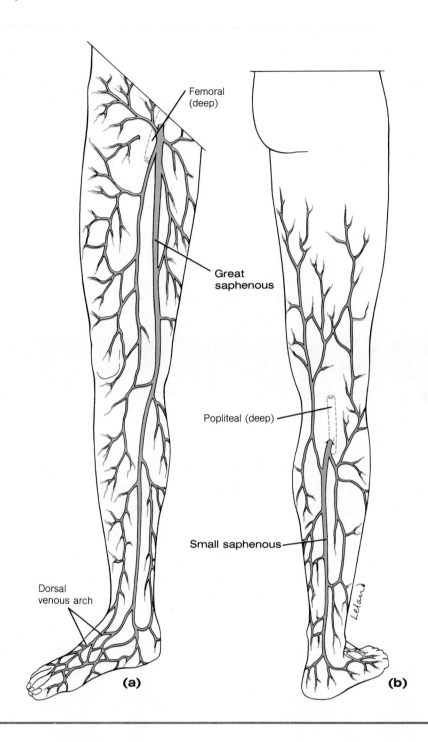

Figure 11.25
Superficial veins of the lower limb. **(a)** Anterior view. **(b)** Posterior view.

This modification of the circulatory system, whereby blood from the digestive tract, pancreas, and spleen passes through the liver before returning to the heart, is very significant. The venous blood in the hepatic portal vein contains nutrients and other substances that have been absorbed from the digestive tract. The hepatic portal system carries these substances directly to the liver, where they can be removed from the blood and stored, metabolized, or, in the case of harmful substances, detoxified.

VEINS OF THE LOWER LIMBS The deep veins of the lower limbs, like those of the upper limbs, travel along with the arteries and are given corresponding names: **external iliac, femoral, popliteal, anterior** and **posterior tibial,** and **peroneal veins** (Figure 11.19).

F11.19

Table 11.2 Schematic Summary of the Major Veins That Empty into the Superior and Inferior Venae Cavae

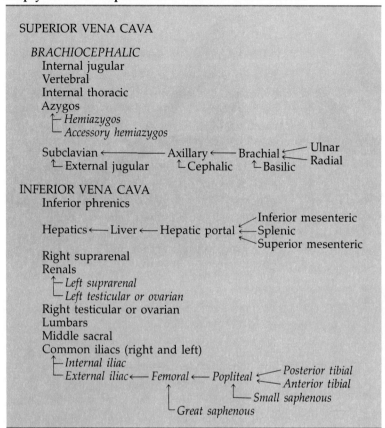

SUPERIOR VENA CAVA

BRACHIOCEPHALIC
Internal jugular
Vertebral
Internal thoracic
Azygos
┌─*Hemiazygos*
└─*Accessory hemiazygos*

Subclavian ◄──────── Axillary ◄──── Brachial ◄─ Ulnar
└─*External jugular*   └─*Cephalic*   └─*Basilic* Radial

INFERIOR VENA CAVA
Inferior phrenics

Hepatics ◄──── Liver ◄──── Hepatic portal ◄─ ┌─Inferior mesenteric
└─Splenic
└─Superior mesenteric

Right suprarenal
Renals
┌─*Left suprarenal*
└─*Left testicular or ovarian*
Right testicular or ovarian
Lumbars
Middle sacral
Common iliacs (right and left)
┌─*Internal iliac*
└─*External iliac* ◄──── *Femoral* ◄──── *Popliteal* ◄─ ┌─*Posterior tibial*
└─*Anterior tibial*
└─*Small saphenous*
└─*Great saphenous*

Two large superficial veins of the lower limbs arise from a **dorsal venous arch** on the top of the foot (Figure 11.25): the *great* and *small saphenous veins*. **F11.25** The **great saphenous vein** is the longest vein in the body. It travels along the medial side of the foot, leg, and thigh, where it joins with the femoral vein just below the inguinal ligament. The **small saphenous vein** travels along the lateral side of the foot, crosses to the posterior surface of the leg, and joins the popliteal vein behind the knee. There are numerous connections between the great and small saphenous veins, as well as between them and the deep veins. Both the deep and the superficial veins of the lower limbs contain valves that assist in returning the blood to the heart against the force of gravity.

STUDY OUTLINE

TYPES OF VESSELS p. 303

ARTERIES large vessels that carry blood away from heart; arterioles and capillaries are progressively smaller vessels.

VENULES, VEINS vessels that return blood to heart.

GENERAL STRUCTURE OF BLOOD VESSEL WALLS pp. 303–304

TUNICA INTIMA composed of endothelium, connective tissue, and basement membrane; this layer is present in all vessels.

TUNICA MEDIA composed of elastic fibers and smooth muscle; often separated from tunica intima by internal elastic lamina, and from tunica externa by external elastic lamina.

TUNICA EXTERNA composed of collagen fibers.

STRUCTURE OF ARTERIES p. 304

ELASTIC ARTERIES large arteries; thick tunica media with many elastic fibers.

MUSCULAR ARTERIES (DISTRIBUTING ARTERIES) tunica media of smooth muscle cells; well-defined internal and external elastic laminae.

STRUCTURE OF ARTERIOLES   diameter less than 0.5 mm; small lumen; thick tunica media of mostly muscle cells.   **pp. 304–305**

STRUCTURE OF CAPILLARIES   0.5 to 1 mm long; 0.01 mm diameter; walls have endothelial layer only. Substances pass through walls through tight junctions, through cell membrane, within vesicles, or through pores in the endothelial cells; sinusoids are large thin-walled channels that receive blood from some arterioles.   **p. 306**

STRUCTURE OF VENULES   inner endothelium and some outer fibrous tissue; large venules encircled by a few smooth muscle fibers.   **p. 306**

STRUCTURE OF VEINS   thin tunica media with few muscle fibers; thick tunica externa with little elastic tissue; larger lumen, thinner walls than arteries; some have valves and vasa vasorum.   **pp. 306–307**

CONDITIONS OF CLINICAL SIGNIFICANCE: BLOOD VESSELS   **pp. 309–310**

HIGH BLOOD PRESSURE (HYPERTENSION) systolic pressure/diastolic pressure above 150/90. Possible causes include:

VASOCONSTRICTION   renin involved in formation of angiotensin II, thus causing vasoconstriction; aldosterone causes sodium retention by kidneys, possibly elevating blood volume and pressure.

HIGH DIETARY SODIUM INTAKE   suspected cause of high blood pressure, but mechanisms unknown.

ARTERIOSCLEROSIS AND ATHEROSCLEROSIS

ARTERIOSCLEROSIS   hardening of arteries; causes large fluctuation between systolic and diastolic pressures.

ATHEROSCLEROSIS   nodular lipid deposits in tunica intima of arteries; may restrict bloodflow through vessels and cause:

Stroke   lack of blood to brain.

Heart Attack   lack of blood to heart muscle.

ANEURYSMS   local dilations of arteries due to wall weakness; danger of vessel rupture.

PHLEBITIS   vein inflammation.

VARICOSE VEINS   dilated, lengthened, tortuous veins.

EFFECTS OF AGING ON THE VASCULAR SYSTEM while there is a reduction in the elasticity of arteries with age, diet and lifestyle seem to affect vascular system more than aging does.

ANATOMY OF THE VASCULAR SYSTEM   **pp. 310–329**

PULMONARY CIRCUIT
1. From right ventricle to pulmonary trunk to pulmonary arteries to lobar arteries to capillary plexuses (for gas exchange) to venules to pulmonary vein to left atrium.
2. Transports oxygenated venous blood and oxygen-poor arterial blood.

SYSTEMIC CIRCUIT

SYSTEMIC ARTERIES   arise from aorta.

Branches of the Ascending Aorta   coronary arteries.

Branches of the Aortic Arch   brachiocephalic, left common carotid, and left subclavian arteries.

Arteries of the Head and Neck   common carotids, internal and external carotids, temporals, cerebral and vertebral arteries, and cerebral arterial circle.

Arteries of the Upper Limbs   subclavian, axillary, brachial, radial, ulnar, and digital arteries.

Branches of the Thoracic Aorta   intercostal, bronchial, esophageal, and superior phrenic arteries.

Branches of the Abdominal Aorta   lumbar, common iliac, middle sacral, inferior phrenic, celiac (gastric, splenic, common hepatic), suprarenal, renal, superior and inferior mesenteric, ovarian or testicular arteries.

Arteries of the Pelvic Region   internal and external iliac arteries.

Arteries of the Lower Limbs   femoral, popliteal, anterior and posterior tibials, peroneal, plantar, digital, and dorsalis pedis arteries.

SYSTEMIC VEINS   deep veins, superficial veins, venous sinuses.

Veins of the Head and Neck   internal and external jugulars, subclavian, brachiocephalic, superior vena cava, and vertebral veins.

Veins of the Upper Limbs   *Deep:* axillary, brachial, radial, and ulnar veins. *Superficial:* cephalic, basilic, and median cubital veins.

Veins of the Thorax   azygos system; intercostal, bronchial, esophageal, and pericardial veins.

Veins of the Abdomen and Pelvis   inferior vena cava; common iliac, lumbar, renal, suprarenal, hepatic, right testicular or ovarian veins.

The Hepatic Portal System   hepatic portal vein formed by joining of inferior mesenteric, superior mesenteric, and splenic veins; carries blood from stomach, intestines, spleen, and pancreas; enters liver for removal of nutrients.

Veins of the Lower Limbs   contain valves. *Deep:* external iliac, femoral, popliteal, anterior and posterior tibials, and peroneal veins. *Superficial:* great and small saphenous veins.

SELF-QUIZ

1. The greatest total cross-sectional diameter of the entire circulatory system occurs in the: (a) arteries; (b) veins; (c) capillaries.

2. The tunica media and tunica externa are not present in: (a) veins; (b) capillaries; (c) arterioles.

3. The endothelium is a part of the: (a) tunica externa; (b) tunica media; (c) tunica intima; (d) adventitia.

4. The walls of the larger arteries and veins are so thick that they are supplied by small nutrient vessels called vasa vasorum. True or false?

5. The walls of distributing arteries are composed mostly of smooth muscle. True or false?

6. The movement of substances into and out of the vessels of the cardiovascular system generally occurs within: (a) veins; (b) venules; (c) capillaries; (d) arteries.

7. Valves are typically found in arteries as well as in the larger veins. True or false?

8. Which of the following blood vessel layers is present in vessels of all sizes? (a) endothelium; (b) tunica media; (c) tunica externa.

9. Compared to corresponding arteries, veins have: (a) a better-developed tunica media; (b) a generally larger lumen; (c) generally thinner walls; (d) b and c; (e) all of these.

10. In a blood pressure reading of 120/80, 120 is: (a) pulse pressure; (b) systolic pressure; (c) diastolic pressure.

11. An inflammation of a vein is called: (a) varicose vein; (b) aneurysm; (c) phlebitis; (d) atherosclerosis.

12. Atherosclerosis is a form of arteriosclerosis that is characterized by the presence of an aneurysm. True or false?

13. Environmental factors such as diet may play a more significant role in vascular disorders than age or genetic inheritance. True or false?

14. The pulmonary arteries carry blood that has a low carbon dioxide content and a high oxygen content. True or false?

15. Match the terms associated with the systemic arteries with the appropriate description.

Aorta	(a) A major blood supplier to the brain; passes through the transverse foramina.
Coronary arteries	
Internal carotid artery	
Vertebral artery	(b) Supply the underside of the diaphragm.
External carotid artery	(c) Supplies all of the small intestine.
Axillary artery	
Radial artery	(d) Formed from bifurcation of abdominal aorta.
Inferior phrenic arteries	

Left gastroepiploic arteries	(e) Terminates as the maxillary and superficial temporal arteries.
Superior mesenteric artery	(f) Vessel from which the pulse is usually taken.
Inferior mesenteric artery	(g) Branches of ascending aorta.
Common iliac arteries	(h) Supplies the muscles of the anterior compartment of the leg.
Popliteal artery	(i) Continuation of the subclavian artery.
Anterior tibial artery	(j) Passes behind the knee and supplies muscles and skin of the area around the knee.
	(k) Blood from the left ventricle enters systemic circuit through this vessel.
	(l) Supplies the greater curvature of the stomach.
	(m) Supplies most of the large intestine.
	(n) Supplies the brain through its terminal branches—the anterior and middle cerebral arteries.

16. Match the terms associated with the systemic veins with the appropriate description.

Sinusoids	(a) At the shoulder, this superficial vein goes deep and empties into the axillary vein.
Brachiocephalic vein	
Superior vena cava	(b) Venous blood from the lower regions of the body is returned to the heart through this vessel.
Cephalic vein	
Median cubital vein	(c) This vein enters the liver and carries blood from the digestive tract.
Azygos vein	
Inferior vena cava	(d) This vein forms part of the hepatic portal system.
Splenic vein	
Great saphenous vein	(e) The internal jugular vein joins with the subclavian vein to form this vein.
Hepatic portal vein	(f) Longest vein in the body.
	(g) Blood-filled spaces found in the liver.
	(h) This vein begins in upper right lumbar

region, passes through the diaphragm, and eventually empties into the superior vena cava.

    (i)   This vein passes in front of the elbow and is commonly used to give or receive blood.

    (j)   The right and left brachiocephalic veins join to form this vein.

17. The vessels of the pulmonary circuit do not supply oxygen and nutrients to the lung tissues. True or false?

18. Blood from the pulmonary circuit enters the right atrium. True or false?

19. The coronary arteries arise from the: (a) inferior vena cava; (b) superior vena cava; (c) right atrium; (d) ascending aorta.

20. Veins that drain the cardiac muscle empty into: (a) left ventricle; (b) right atrium; (c) coronary sinus; (d) both b and c.

21. Which veins return blood to the superior vena cava from the wall of the thorax? (a) hepatic portal; (b) azygos; (c) internal jugular; (d) cephalic.

22. Which artery is the main source for blood to the lower limbs? (a) femoral; (b) internal iliac; (c) common carotid; (d) inferior phrenic.

23. Which vein transports blood from the stomach, intestines, spleen, and pancreas to the hepatic portal vein? (a) splenic; (b) inferior mesenteric; (c) superior mesenteric; (d) all of these.

24. Which vessel is not a branch of the abdominal aorta? (a) lumbar; (b) celiac; (c) subclavian; (d) superior mesenteric.

25. Which vessel is not a branch of the thoracic aorta? (a) azygos; (b) intercostal; (c) bronchial; (d) all of these are branches of the thoracic aorta.

LEARNING OBJECTIVES

After completing this chapter, you should be able to:

- Describe the lymphatic system of vessels.

- Explain how excess interstitial fluid and plasma proteins are returned to the bloodstream by the lymphatic vessels.

- Discuss the functions of the lymphatic system, including its immune functions.

- Describe the structure of the lymph nodes, spleen, thymus gland, and tonsils.

CHAPTER CONTENTS

The Lymphatic System

12

The **lymphatic system** consists of (1) an extensive network of *capillaries* and larger *collecting vessels* which receive fluid from the loose connective tissues throughout the body and transport it to the cardiovascular system; (2) *lymph nodes*, which serve as filters of the fluid within the collecting vessels; and (3) the *lymphoid organs,* including lymphatic nodules, tonsils, the spleen, and the thymus gland.

The lymphatic system is closely related both anatomically and functionally to the cardiovascular system. Fluid that accumulates in the spaces between the cells of loose connective tissue is called **interstitial fluid,** or **tissue fluid.** Under normal conditions, slightly more liquid tends to leave the capillaries of the cardiovascular system than enters them. Plasma proteins do not easily pass through the walls of capillaries; however, as the liquid portion of the blood moves into the intercellular spaces, it carries a small amount of plasma proteins with it. If this fluid (and the plasma proteins) is allowed to accumulate, the tissues swell, producing a condition called *edema.* The plasma proteins are unable to reenter the blood through the capillaries of the blood vascular system; however, they can enter the vessels of the lymphatic system. As a consequence, one role of the lymphatic system is to return the excess interstitial fluid and the plasma proteins back to the bloodstream and thus prevent edema.

LYMPHATIC VESSELS

Interstitial fluid enters the lymphatic system by passing through the extremely thin walls of **lymphatic capillaries.** After entering the vessels of the lymphatic system, interstitial fluid is called **lymph.** Lymph has a composition quite similar to that of blood plasma; it consists mostly of water, electrolytes, and variable amounts of plasma proteins that have escaped from the blood through the capillaries of the cardiovascular system. Lymph differs from blood primarily by the absence of blood corpuscles.

The lymphatic capillaries are dead-end vessels (Figure 12.1). Therefore, the lymphatic system is a one-way system—that is, it only *returns* fluid to the bloodstream. The walls of lymphatic capillaries, like the walls of blood capillaries, are composed of a single layer of endothelium. However, lymphatic capillaries lack the surrounding basement membrane that ensheathes blood capillaries. Another difference between lymphatic and blood capillaries is that the edges of adjacent endothelial cells in lymphatic capillaries are only loosely attached, and they often overlap. This arrangement forms a functional one-way valve. The pressure of the interstitial fluid outside lymphatic capillaries pushes the edges of the endothelial cells inward, allowing the interstitial fluid to enter the capillaries. But once within the lymphatic capillary, this fluid cannot reenter the intercellular spaces because pressure from within the capillaries forces the edges of the endothelial cells together, closing the "valve." Because of this structural arrangement, lymphatic capillaries are more permeable than most blood capillaries, and virtually all the components of the interstitial fluid, including proteins and other large particles (such as disease-causing organisms), can enter these vessels and are transported throughout the

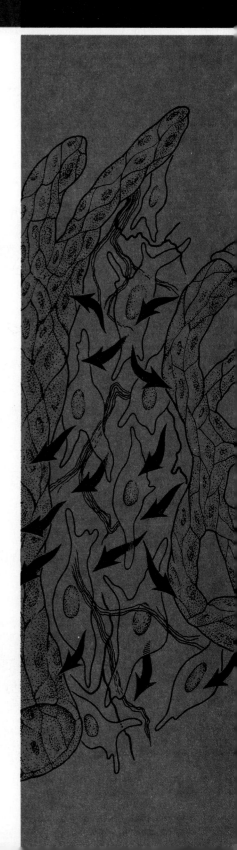

335

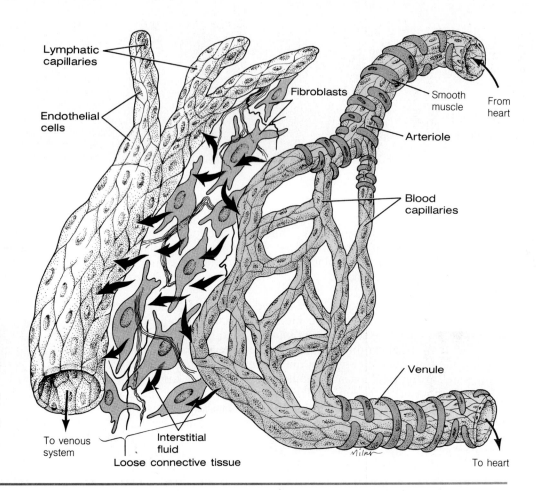

Figure 12.1
Schematic representation of the lymphatic capillaries showing their relationship to the tissue fluid, the blood vascular system, and the tissue cells. The arrows indicate the directions of fluid movement. Note that the lymphatic capillaries begin as dead-end vessels.

body. Most tissues contain plexuses of lymphatic capillaries located among the vascular capillaries. Special lymphatic capillaries called *lacteals* are located in the villi of the intestine. The lacteals aid in the absorption of fat from the digestive tract and carry it as a milky fluid called *chyle* to the bloodstream.

F12.2a   The lymphatic capillaries, which are widely distributed throughout the interstitial spaces of the body, join together to form progressively larger lymphatic vessels (Figure 12.2*a*). Generally, the larger lymphatic vessels, which are called *collecting vessels,* travel alongside arteries and veins of the cardiovascular system and pass through one or more *lymph nodes* before emptying into either the *thoracic duct* or the *right lymphatic duct,* which return the lymph to the bloodstream. The walls of the lymphatic collecting vessels are similar to the walls of veins, although they are thinner and—like veins—they contain valves that occur in pairs, each on opposite sides of the vessel, with their free edges pointing in the direction of lymph flow. The valves aid in the movement of lymph by preventing its backflow.

LYMPH NODES

F12.3   **Lymph nodes** are small, round or bean-shaped organs that are distributed along the course of many of the lymphatic vessels (Figure 12.3). There are groups of lymph nodes in the groin, axilla, and neck, as well as in numerous other deeper locations.

F12.4   Each node is enclosed in a **fibrous capsule** (Figure 12.4). Strands of connective tissue called **trabeculae** extend inward from the capsule and divide the node into several compartments. These compartments are further subdivided by a network of reticular fibers that extend between the trabeculae. The

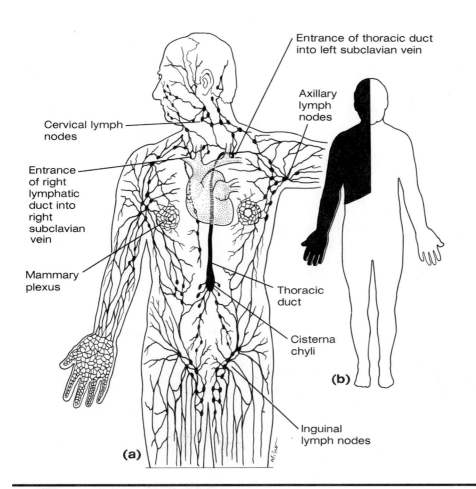

Figure 12.2
The lymphatic system.
(a) Major lymphatic vessels and groups of lymph nodes. **(b)** Lymph from the darkened area returns to the blood vascular system through the right lymphatic duct. Lymph from the remainder of the body travels through the thoracic duct.

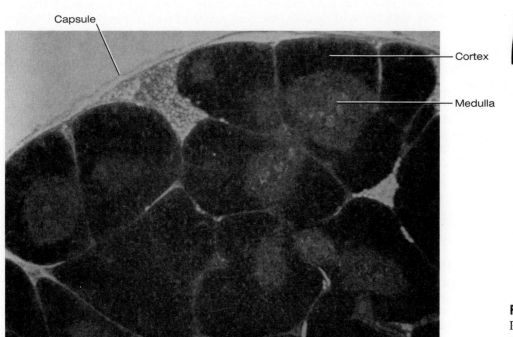

Figure 12.3
Photomicrograph of a portion of a lymph node.

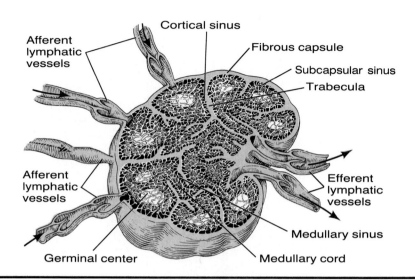

Figure 12.4
Cross section of a lymph node. Note that there are more afferent vessels than efferent vessels, which has the effect of slowing the rate of lymph flow through the node. The arrows indicate the direction of lymph flow.

node consists of an outer *cortical* region and an inner *medullary* region. Within the cortex of each node are separate masses of lymphoid tissue called **germinal centers,** which serve as a source of lymphocytes. *Lymphoid tissue* is a modified type of loose connective tissue, with a prominent network of reticular fibers between which there are many lymphocytes and macrophages. The cells of the medulla are arranged in the form of strands called **medullary cords.**

Lymph enters the convex surface of the node through several **afferent lymphatic vessels,** and filters slowly through irregular channels in the node, called **sinuses.** The sinuses are spanned by networks of reticular fibers. Lymph from the afferent lymphatic vessel enters a *subcapsular sinus* located just beneath the capsule. From here it percolates through *cortical sinuses,* which penetrate the cortex, and enters *medullary sinuses* located between the medullary cords. From the medullary sinuses lymph leaves the node by way of **efferent lymphatic vessels** at a small indentation called the *hilus.* There are fewer efferent vessels than afferent vessels, and this has the effect of slowing the rate of lymph flow through the nodes, which allows them to function more effectively. Typically, all lymph has passed through several lymph nodes before it is returned to the cardiovascular system.

Lymph that flows into the lymph nodes contains many foreign particles and microorganisms, some of which can cause diseases if they are not destroyed. As lymph percolates slowly through the sinuses within a node, large foreign particles such as bacteria become entrapped in the meshes of the reticular fibers that span the sinuses. The entrapped particles are soon attacked and destroyed by phagocytic cells, called *macrophages,* that line the sinuses. In addition, between the sinuses are masses of lymphoid tissue containing *lymphocytes* and *plasma cells* that produce antibodies for destroying certain foreign substances known as antigens. Unfortunately, not all disease-producing cells, such as cancer cells, are destroyed within the lymph nodes. Some cancer cells can survive and multiply within the nodes, thus serving as a site from which they can spread throughout the body by way of the cardiovascular system. For this reason, swollen lymph nodes near cancer sites are often surgically removed.

The efferent collecting vessels from the lymph nodes of most regions of the body converge into larger vessels called *lymph trunks.* There are five major lymph trunks, four of which are paired: (1) the unpaired *intestinal trunk,* which receives lymph from abdominal organs; (2) the *lumbar trunks,* which drain the lower limbs and some pelvic organs; (3) the *subclavian trunks,* which drain the arm and parts of the thorax and the back; (4) the *jugular trunks,* which drain the head and neck regions; and (5) the *bronchomediastinal trunks,* which drain the thorax. The lymph trunks, in turn, empty into either the thoracic duct or the right lymphatic duct.

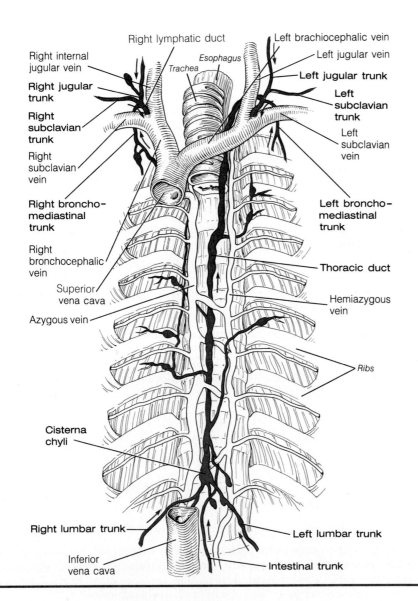

Right lymphatic duct

Right internal jugular vein

Right jugular trunk

Right subclavian trunk

Right subclavian vein

Right broncho-mediastinal trunk

Right bronchocephalic vein

Superior vena cava

Azygous vein

Cisterna chyli

Right lumbar trunk

Inferior vena cava

Left brachiocephalic vein

Left jugular vein

Left jugular trunk

Left subclavian trunk

Left subclavian vein

Left broncho-mediastinal trunk

Thoracic duct

Hemiazygous vein

Ribs

Left lumbar trunk

Intestinal trunk

Esophagus

Trachea

Figure 12.5
Relationship of the thoracic duct and the right lymphatic duct to the blood vascular system.

LYMPHATIC DUCTS

The **thoracic duct** arises from the **cisterna chyli,** which is a saclike enlargement that lies in front of the second lumbar vertebra (Figure 12.5). The cisterna chyli receives lymph from the lumbar and intestinal trunks, which drain the lower limbs and the viscera of the abdominopelvic cavity. As the thoracic duct passes through the diaphragm (along with the aorta) and travels upward through the thoracic cavity in front of the vertebral column, it receives lymphatic vessels that drain the left side of the thorax. Behind the brachiocephalic vein, the thoracic duct curves to the left and usually receives the left subclavian and left jugular trunks from the upper limbs and the left side of the head and neck before emptying into the *left subclavian vein* near its junction with the internal jugular vein.

As we have seen, the thoracic duct returns lymph to the bloodstream from the entire body *except* the upper right limb and the right side of the thorax, neck, and head (Figure 12.2*b*). Lymph from these areas is returned to the *right subclavian vein* through a **right lymphatic duct.** The right lymphatic duct is a small vessel formed by the joining together of the right jugular, right subclavian, and right bronchomediastinal trunks, which drain lymph from these regions. The two lymphatic ducts thus convey all of the lymph that has been collected and filtered from throughout the body back into the cardiovascular system, and the lymph begins the circuit again as blood plasma.

F12.5

F12.2b

MECHANISMS OF LYMPH FLOW

Lymph flows slowly; approximately 3 liters of lymph enter the cardiovascular system every 24 hours. The flow is slow because, unlike the cardiovascular system, the lymphatic system does not have a pump like the heart to keep it moving. Rather, lymph flow depends on more subtle forces, such as contractions of skeletal muscles, which apply pressure on the lymph vessels and compress them. This action forces lymph along the vessels. In a similar manner, the pulsing of nearby arteries can compress lymph vessels and move the lymph within them. In addition, when a section of a lymph vessel is distended by lymph it apparently can contract slightly and thereby propel the lymph along in the vessels. The valves within lymph vessels contribute to the effectiveness of these activities by permitting lymph in the vessels to flow only toward the bloodstream.

FUNCTIONS OF THE LYMPHATIC SYSTEM

The lymphatic system has several important functions. These have been mentioned in the previous sections and are summarized here. The functions of the lymphatic system include the *destruction of bacteria*, the *removal of foreign particles from the lymph*, *specific immune responses*, and the *return of interstitial fluid to the bloodstream*.

Destruction of Bacteria and Removal of Foreign Particles from Lymph

Bacteria and other foreign substances are removed from the lymph by phagocytes—primarily macrophages—that are present in the lymph nodes. During infection, the rate of formation of macrophages within the nodes is so great that the nodes enlarge and become tender.

Specific Immune Responses

In response to the presence of bacteria or other foreign substances, lymphocytes and plasma cells participate in specific immune responses, such as the manufacture of antibodies that destroy the foreign substances. These responses are described in greater detail later in this chapter.

Return of Interstitial Fluid to the Bloodstream

As mentioned earlier, edema would result if excess tissue fluid were allowed to accumulate in the intercellular spaces. Moreover, the volume of blood would be reduced, since it is the major source of interstitial fluid. Perhaps of greater importance is the small amount of protein lost from the blood capillaries into the interstitial fluid but normally returned to the blood by the lymphatic system. If this protein remained in the intercellular spaces, it would increase the osmotic pressure of the interstitial fluid and thereby affect the exchange of materials between the blood, the interstitial fluid, and the lymph.

LYMPHOID ORGANS

In addition to the lymph nodes, several organs are lymphoid in nature. These include the *spleen*, the *thymus gland*, and the *tonsils*. These organs, which have no direct association with the lymphatic system of vessels or with the lymph, are an integral part of the body's immune system.

Spleen

The **spleen** is the largest lymphoid organ. It lies in the left hypochondriac region, between the fundus of the stomach and the diaphragm. The spleen is usually about 12 cm in length; however, its size and weight vary from person

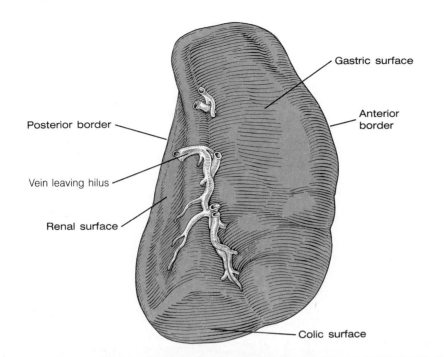

Gastric surface

Anterior border

Posterior border

Vein leaving hilus

Renal surface

Colic surface

Figure 12.6
The visceral surface of the spleen.

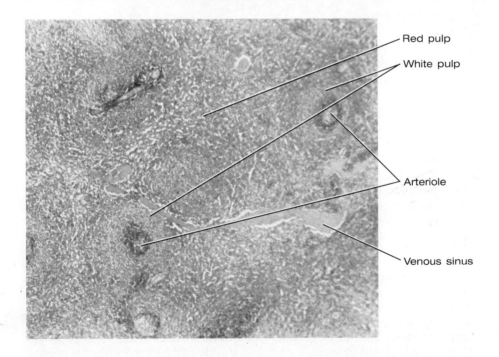

Red pulp

White pulp

Arteriole

Venous sinus

Figure 12.7
Photomicrograph of a portion of the spleen (×50).

to person, as well as in the same individual under different conditions. The *diaphragmatic surface* of the spleen is smooth and convex, conforming to the undersurface of the diaphragm, with which it is in contact. The *visceral surface* is divided into gastric, renal, and colic surfaces that conform to the organs adjacent to it (Figure 12.6). Blood vessels enter and leave the spleen through a **F12.6** region on the visceral surface called the *hilus.*

Like the lymph nodes, the spleen is covered by a strong fibrous capsule with trabeculae that extend into the organ and divide it into compartments. However, in the spleen, smooth muscle cells are also present in the capsule and the trabeculae. Two types of tissue, called *red pulp* and *white pulp,* are distinguishable within the compartments (Figure 12.7). **Red pulp** is the more **F12.7**

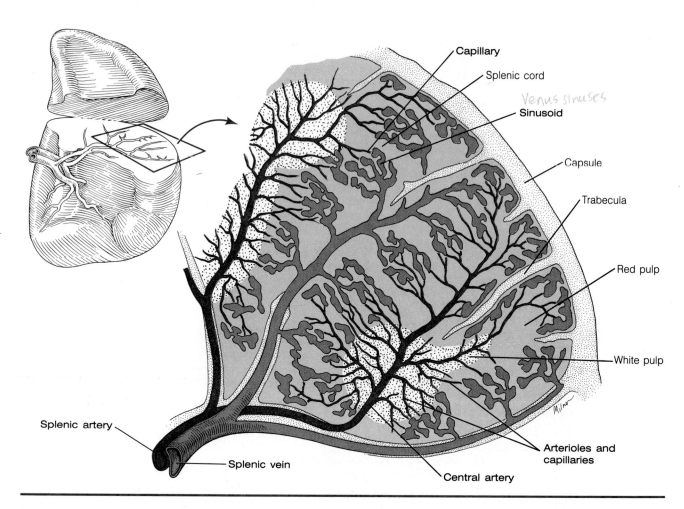

Capillary
Splenic cord
Venus sinuses
Sinusoid
Capsule
Trabecula
Red pulp
White pulp
Splenic artery
Splenic vein
Arterioles and capillaries
Central artery

Figure 12.8
The path of blood through the spleen.

F12.8

abundant. It consists of branching venous sinuses separated from each other by columns of splenic tissue called *splenic cords*. Like other lymphoid tissues, the red pulp contains lymphocytes and macrophages; but it also contains red blood cells, which give the pulp its color. Scattered throughout the red pulp are round masses of **white pulp,** each surrounding an arteriole. White pulp, which does not contain red blood cells, is lymphoid tissue, and large numbers of lymphocytes are present in it. Blood enters the spleen at the hilus through the splenic artery. It may remain in the vessels that are surrounded by white pulp and enter the venous sinuses of the red pulp, or it may pass through the walls of the capillaries and filter between the cells of the spleen before entering the venous sinuses (Figure 12.8). From the venous sinuses, blood leaves the spleen through the splenic vein and travels to the liver within the hepatic portal vein.

The spleen acts as a filter for the bloodstream much as the lymph nodes filter lymph. Like other lymphoid tissues, the spleen produces lymphocytes and plasma cells, which manufacture antibodies against foreign antigens. These activities are carried out primarily in the white pulp. In addition, the macrophages of the red pulp phagocytose old red blood cells, as well as bacteria and other foreign particles. The spleen also serves to a limited extent as a blood reservoir. About 200 ml of the blood contained within the venous sinuses of the red pulp may be forced from the spleen into the general vascular system by contraction of the smooth muscles of the capsule. This can help to offset blood lost through hemorrhage. In the developing embryo, the spleen forms red blood cells; but except for some abnormal conditions that require the replacement of large numbers of red blood cells, this capability is lost following birth.

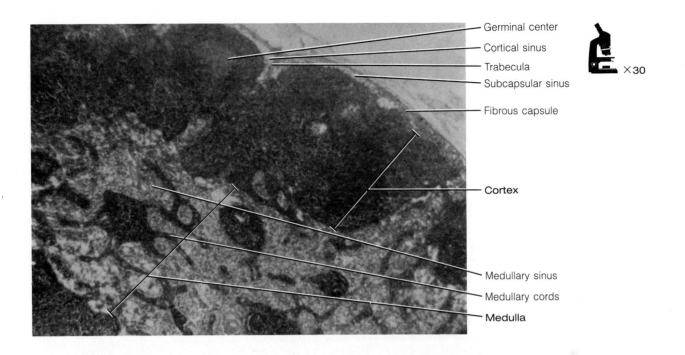

- Germinal center
- Cortical sinus
- Trabecula
- Subcapsular sinus
- Fibrous capsule
- **Cortex**
- Medullary sinus
- Medullary cords
- **Medulla**

×30

Thymus Gland

The **thymus gland** is a bilobed mass of lymphoid tissue located deep to the sternum in the anterior region of the mediastinum. It increases in size during early childhood, then begins to atrophy slowly, and diminishes following puberty. In the adult, it may be entirely replaced by adipose tissue. The thymus gland is covered by a connective-tissue capsule that separates it into smaller lobules (Figure 12.9). Each lobule has an outer *cortex* and an inner *medulla*. The cortex is composed of densely packed lymphocytes. The medulla also contains lymphocytes; but in addition it has organized groups of cells referred to as **thymic corpuscles** (or *Hassall's corpuscles*).

F12.9

The thymus confers on certain lymphocytes the ability to differentiate and mature into cells that can carry out the processes of cell-mediated immunity. There is some evidence that the thymus also releases a hormone that can continue to influence the lymphocytes after they have left the gland.

Tonsils

The **tonsils** are small masses of lymphoid tissue embedded in the mucous membrane lining of the oral and pharyngeal cavities. The **palatine tonsils** are located on the posterior-lateral walls of the throat, one on each side. These are the tonsils that become noticeably enlarged when you suffer from a "sore throat." The **pharyngeal tonsils** are located on the posterior wall of the nasopharynx. They are at their peak of development during childhood, and when enlarged they are referred to as *adenoids*. On the dorsal surface of the tongue near its base is a group of **lingual tonsils.** Being composed of lymphoid tissue and surrounding the junction of the oral and nasal passageways, the tonsils serve as an additional defense against bacterial invasion of the body.

IMMUNE FUNCTIONS OF LYMPHATIC ORGANS

The body is continually exposed to a wide variety of potentially harmful factors, including bacteria, viruses, and hazardous chemicals. Pathogenic bacteria that invade the body often release enzymes that break down cell membranes and organelles or give off toxins that disrupt the functions of organs

Figure 12.9
Photomicrograph of a portion of the thymus gland.

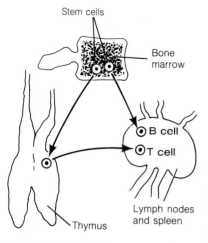

Figure 12.10
Formation and differentiation of B cells and T cells.

and tissues. Viruses can enter cells and utilize cellular facilities to reproduce more virus particles. They can kill cells by depleting them of essential components or by causing them to produce toxic substances.

Fortunately, the body is able to resist many organisms and chemicals that can damage tissues. This ability is called **immunity.** Some of the body's immunity is provided by nonspecific defense mechanisms that do not require previous exposure to a particular foreign substance in order to react to it. These mechanisms, which constitute the body's first line of defense, include the barriers formed by the body surface—specifically, the skin and mucous membranes—and inflammatory responses that occur when tissues are injured. The inflammatory response depends largely on phagocytic activity by macrophages. The body's second line of defense is provided by specific immune responses that depend on prior exposure to a specific foreign material, recognition of it upon subsequent exposure, and reaction to it. These responses are mediated by specialized lymphocytes located within the lymphoid organs.

In order to better appreciate the important role that lymphoid organs have in protecting the body against invasion by foreign particles, we will briefly consider the specific immune responses.

Specific Immune Responses

The specific immune responses involve the reaction of antibodies with antigens; therefore, they require previous exposure to a foreign agent or organism to be most effective. An *antigen* is any substance that causes certain cells within the body to produce specific *antibodies* when it is present in the body. The antibodies, in turn, destroy the foreign antigen that stimulated its production. As we will see, antibodies are produced by cells located in the lymphoid organs.

There are two major aspects of the specific immune responses: *humoral immunity* and *cell-mediated immunity.* Humoral immunity requires the presence of specialized lymphocytes called *B cells.* Cell-mediated immunity requires the presence of specialized lymphocytes called *T cells.*

B cells are lymphocytes that are committed to differentiate into the antibody-producing *plasma cells* that are involved in humoral immunity. Like all lymphocytes, B cells are derived from stem cell precursors in the red bone marrow (Figure 12.10). The stem cells proliferate and develop into B cells within the bone marrow. B cells continually leave the bone marrow and circulate through the blood, the loose connective tissues, and lymph, and at any one time large numbers of B cells are localized in the lymph nodes, spleen and other lymphoid tissues.

T cells are lymphocytes whose stem cells from the red bone marrow leave the marrow and enter the thymus gland, where they proliferate and differentiate (Figure 12.10). T cells continually leave the thymus gland and, like B cells, circulate through the blood, the loose connective tissues, and lymph, and at any one time large numbers of T cells are localized in the lymph nodes, spleen, and other lymphoid tissues.

Humoral Immunity

Humoral immunity involves the production and release into the blood and lymph of antibodies to various antigens that the body recognizes as foreign. Humoral immune responses are important in providing specific immune resistance to most bacteria, bacterial toxins, and the extracellular phase of viral infections. The humoral system responds to an antigen as follows.

When an antigen reaches lymphoid tissues such as lymph nodes or the spleen, a tiny fraction of the B lymphocytes are stimulated to undergo rapid cell division (Figure 12.11). Most of the stimulated cells develop into *plasma cells*, which produce antibodies that are released into the lymph and enter the blood. The antibodies then combine with the specific antigen that stimulated their production, and they destroy the antigen, either directly or by enhancing phagocytosis of the antigen by macrophages.

Some of the stimulated B cells do not differentiate into plasma cells, but instead form *memory cells*, which provide the humoral system with a "mem-

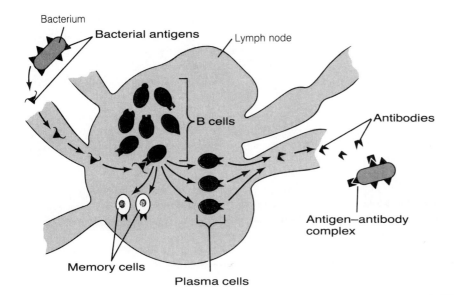

Figure 12.11
Diagrammatic representation of the events of a humoral immune response. In response to the presence of a foreign antigen, antibodies are produced by plasma cells, which develop from B lymphocytes. These antibodies are released into the lymph and blood, where they destroy the specific antigen that stimulated their development. Some B cells remain in the lymph nodes as memory cells.

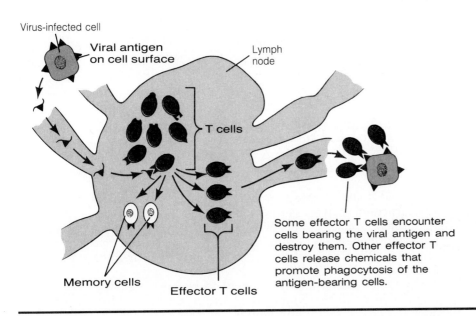

Some effector T cells encounter cells bearing the viral antigen and destroy them. Other effector T cells release chemicals that promote phagocytosis of the antigen-bearing cells.

Figure 12.12
Diagrammatic representation of the events of a cell-mediated immune response. The presence of foreign antigens stimulates some T lymphocytes to form effector T cells, which travel throughout the body and destroy the cells that bear the specific antigen, either directly or through the release of chemicals that promote phagocytosis of the foreign cells. Other T cells remain in the lymph node as memory cells.

ory'' of exposure to the antigen. The system responds rather slowly following an initial antigen exposure, and usually several days are required to build up substantial levels of antibodies. Because of the memory component, however, a subsequent exposure to the same antigen produces a very rapid outpouring of antibodies.

Cell-Mediated Immunity

The cell-mediated immune responses depend on T cells, and particularly on a group of T cells called *effector cells*. These responses are especially effective against cells that possess surface antigens the body recognizes as foreign. For example, most virus-infected cells acquire viral antigens on their surfaces, and T cells produced by the cell-mediated immune system in response to the antigens can attack and destroy the virus-infected cells.

T cells possess specific surface receptors. When an antigen combines with receptors on the surfaces of particular T cells, the T cells become sensitized and undergo division (Figure 12.12). There are many different types of T cells, each sensitized by, and capable of responding to, a different antigen. Some of the new T cells serve as a memory component of the system, and others—the

F12.12

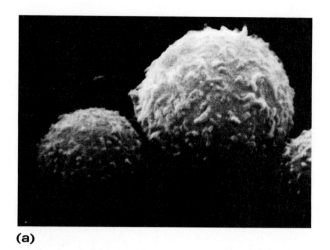

(a)

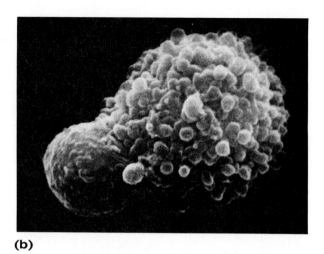

(b)

Figure 12.13
(a) Effector T cell (smaller sphere at lower left) attaches to a cell recognized as foreign.
(b) Destruction of the foreign cell is evidenced by the deep folds in its surface membrane (×7250).

F12.13

effector T cells—leave the lymphoid tissue and travel throughout the body. When these traveling T cells encounter cells bearing the specific antigen that initiated their production, they combine with them and destroy them (Figure 12.13). Some effector T cells directly destroy the cells bearing the antigen on their surfaces. Others release chemicals that enhance the inflammatory reaction, and macrophages—which are attracted to the area by the chemicals—destroy the antigen-bearing cells via phagocytosis.

EFFECTS OF AGING ON THE LYMPHATIC SYSTEM

The lymphoid tissues within the thymus, spleen, lymph nodes, and tonsils undergo structural changes as a person ages, reaching their maximum development at about puberty, after which they slowly regress. The thymus regresses most dramatically, losing up to 90% of its cellular mass by age 50. Accompanying the structural changes, the ability of the thymus to transform stem lymphocytes into T cells declines with age.

A functional change that is thought to be associated with age-related structural changes in the lymphoid tissues is a tendency for autoimmune reactions to become more prevalent as one ages. That is, the antibodies of older persons are less able to recognize their own body tissues, and may attack them in the same manner as they attack foreign antigens.

STUDY OUTLINE

LYMPHATIC VESSELS pp. 335–336

LYMPHATIC CAPILLARIES return interstitial fluid to the bloodstream.

LYMPH NODES lymphatic tissue enclosed in fibrous capsule; afferent lymphatic vessels transport lymph into nodes; efferent lymphatic vessels transport lymph out of nodes; efferent collecting vessels form into five major lymph trunks, which empty into thoracic duct or right lymphatic duct. pp. 336–338

LYMPHATIC DUCTS p. 339

THORACIC DUCT from cisterna chyli; drains most of body; transports lymph to left subclavian vein.

RIGHT LYMPHATIC DUCT drains right upper portion of body; transports lymph to right subclavian vein.

MECHANISMS OF LYMPH FLOW flows very slowly; depends on contractions of skeletal muscles squeezing on lymph vessels, the pulsing of nearby arteries, the presence of valves, etc. **p. 340**

FUNCTIONS OF THE LYMPHATIC SYSTEM
p. 340
1. Destruction of bacteria and removal of foreign particles from lymph via phagocytes in nodes.
2. Specific immune responses, such as the manufacture of antibodies.
3. Return of interstitial fluid and plasma proteins to the bloodstream, thus preventing edema.

LYMPHOID ORGANS have no direct association with lymphatic system; are an integral part of the immune system. **pp. 340–343**

SPLEEN fibrous capsule with trabeculae; contains red and white pulp. Functions include: manufacture of antibodies; phagocytosis of old red blood cells and foreign particles; site of red blood cell formation in the embryo.

THYMUS GLAND gives certain lymphocytes ability to differentiate for cell-mediated immunity.

TONSILS palatine, pharyngeal, and lingual; provide defense against bacteria and other foreign particles.

IMMUNE FUNCTIONS OF LYMPHATIC ORGANS lymphoid organs provide defense against invasion by foreign particles through specific immune responses mediated by specialized lymphocytes. **pp. 343–346**

SPECIFIC IMMUNE RESPONSES involve formation of antibodies by lymphocytes in response to presence of specific foreign antigens. Two types: humoral immunity and cell-mediated immunity.

HUMORAL IMMUNITY involves production and release of antibodies against specific antigens by plasma cells, which develop from B lymphocytes. Some B cells remain as memory cells.

CELL-MEDIATED IMMUNITY T lymphocytes become sensitized by contact with a foreign antigen. Some sensitized T cells remain as memory cells; others leave lymphoid tissues and destroy specific antigen that initiated their production, either directly or by release of chemicals that enhance inflammatory reaction.

EFFECTS OF AGING ON THE LYMPHATIC SYSTEM lymphoid tissues slowly regress after puberty; ability of thymus gland to form T cells declines; autoimmune reactions increase. **p. 346**

SELF-QUIZ

1. Normally, slightly more liquid tends to enter the capillaries of the cardiovascular system than leaves it. True or false?

2. The walls of the lymphatic capillaries are composed of: (a) columnar epithelium; (b) adventitia; (c) a single layer of endothelium.

3. Lymphatic capillaries in the villi of the intestine are called: (a) lacteals; (b) cisterna chyli; (c) efferent lymphatic vessels; (d) trabeculae.

4. Rather large vessels that carry lymph to lymph nodes are called: (a) lymphatic trunks; (b) collecting tubules; (c) thoracic duct; (d) efferent lymphatic vessels.

5. Lymphoid tissue is a modified type of loose connective tissue, with a network of reticular fibers and containing many lymphocytes. True or false?

6. All of the major lymph trunks generally convey lymph into either the thoracic duct or the right lymphatic duct. True or false?

7. Which of the following assist in the flow of lymph within lymph vessels? (a) contraction of skeletal muscles; (b) pulsation of adjacent arteries; (c) valves within lymph vessels; (d) all of these.

8. Which is a function of the lymphatic system? (a) specific immune responses; (b) the return of tissue fluid to the bloodstream; (c) the destruction of bacteria and removal of foreign particles; (d) all of these.

9. The red pulp of the spleen consists of venous sinuses separated from each other by columns of splenic tissue called splenic cords. True or false?

10. The direction of bloodflow through the spleen is from arteries into the venous sinuses, through arteries surrounded by white pulp, and out the hilus in veins. True or false?

11. Both humoral immunity and cell-mediated immunity rely on the production of antigens by specific lymphocytes located in lymphoid organs. True or false?

LEARNING OBJECTIVES

After completing this chapter, you should be able to:

- Distinguish between receptors and effectors.
- Name the two divisions of the nervous system according to structural location, and list the principal components of each division.
- Name the divisions of the nervous system according to function.
- Distinguish between the somatic nervous system and the autonomic nervous system.
- Differentiate between myelinated and unmyelinated nerve fibers.

- Identify the connective-tissue sheaths that cover a nerve.
- Distinguish between a dendrite and an axon.
- Name the types of neuroglia and state the principal function of each.
- Name the types of neurons according to structural and functional classification.
- Cite the various ways receptors may be classified.
- Describe the specialized peripheral neuron endings.

CHAPTER CONTENTS

ORGANIZATION OF THE NERVOUS SYSTEM

EMBRYONIC DEVELOPMENT OF THE NERVOUS SYSTEM

COMPONENTS OF THE NERVOUS SYSTEM

EFFECTS OF AGING ON THE NERVOUS SYSTEM

The Nervous System: Organization and Components

13

The survival of any multicellular organism depends on some means of regulating and coordinating the activities of its cells. If there were no coordination, each cell would function as it is genetically programmed to do—regardless of what functions the other cells were performing. At best, such a haphazard organization would be inefficient; at worst, it would probably be fatal. The human organism, which is composed of billions of cells, has two systems that serve primarily as means of *internal communication* among the cells: the nervous system and the endocrine system.

Because both systems cooperate to provide the body's internal communications, they can be viewed as closely allied networks that operate together to ensure proper body function. By monitoring and regulating how the various body functions interrelate, the nervous and the endocrine systems help to maintain the homeostasis of the body. In Chapters 13 through 17 we consider the nervous system. The endocrine system is discussed in Chapter 18.

The nerve cells of the nervous system carry messages in the form of **nerve impulses.** Nerve impulses often originate within nerve cells as a result of the activity of sensitive structures called **receptors.** Receptors are generally activated by changes in either the internal or external environments of the body. The changes that activate the receptors are called **stimuli.** As a result of receptor activity, nerve impulses are initiated in **sensory nerve cells.** These impulses are carried by the sensory nerve cells to the spinal cord and brain. In the spinal cord and brain, other nerve cells may be activated and conduct nerve impulses to various locations within the spinal cord and brain. Ultimately, nerve impulses carried by **motor nerve cells** leave the brain and spinal cord and bring about responses to environmental changes by selectively activating various **effectors** (Figure 13.1). Effectors capable of responding to **F13.1** nerve impulses include muscle cells and the secretory cells of glands and organs. In addition to responding to environmental stimuli, the human nervous system has the ability to integrate and store the information it receives, thus providing for such capacities as memory, abstract reasoning, and conceptualizing.

ORGANIZATION OF THE NERVOUS SYSTEM

Although there is actually only one nervous system, it can be separated into various divisions based on either structural locations or functional characteristics. It should be kept in mind, however, that these divisions—which are themselves called *nervous systems*—are all integral parts of a single nervous system. Structurally, the nervous system may be divided into two parts: the *central nervous system* and the *peripheral nervous system* (Figure 13.2). **F13.2**

Central Nervous System

The **central nervous system (CNS)** consists of the brain and the spinal cord. It is completely encased in bony structures—the brain within the cranial cavity of the skull and the spinal cord within the vertebral canal of the spinal column. The central nervous system is the *integrative and control center* of the

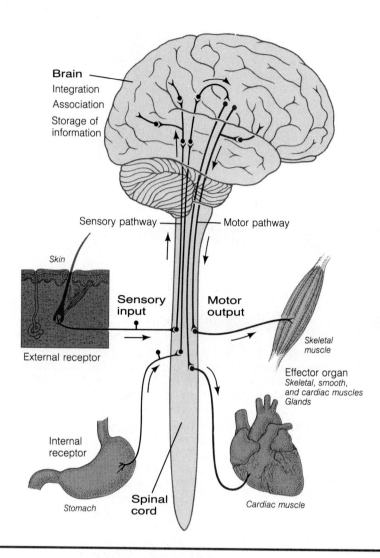

Figure 13.1
Schematic representation of the basic organization of the nervous system. The arrows follow a typical pathway of nerve impulses from receptor to effector.

nervous system. It receives sensory input from the peripheral nervous system and formulates responses to this input.

Peripheral Nervous System

The **peripheral nervous system (PNS)** is composed of all nervous structures located outside of the central nervous system. Specifically:

1. **nerves** that connect the outlying parts of the body and their receptors with the central nervous system

2. **ganglia** (groups of nerve cell bodies) associated with the nerves

The peripheral nervous system includes 12 pairs of **cranial nerves**, which arise from the brain and the brain stem and leave the cranial cavity through foramina in the skull, and 31 pairs of **spinal nerves**, which arise from the spinal cord and leave the vertebral canal through the intervertebral foramina. The paired spinal nerves include 8 cervical, 12 thoracic, 5 lumbar, 5 sacral, and 1 coccygeal nerve (Figure 13.2). The peripheral nervous system can be divided functionally into *afferent (sensory)* and *efferent (motor) divisions*.

F13.2

Afferent Division

The **afferent division** of the peripheral nervous system includes *somatic sensory* nerve cells, which carry impulses to the central nervous system from receptors located in the skin, the fascia, and around the joints, and *visceral*

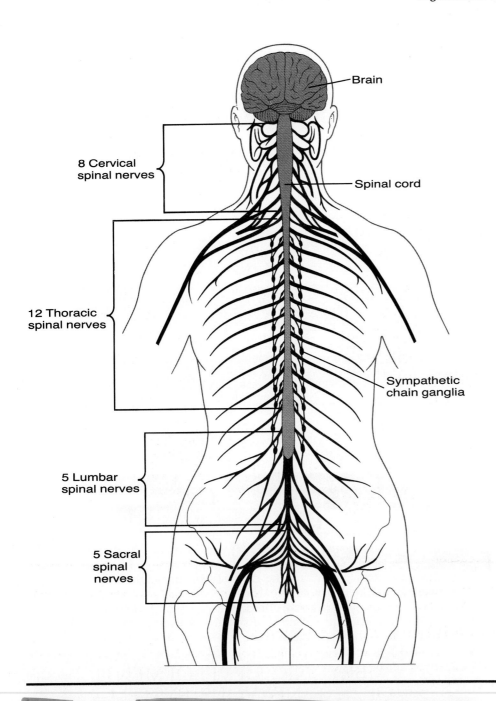

8 Cervical
spinal nerves

12 Thoracic
spinal nerves

5 Lumbar
spinal nerves

5 Sacral
spinal
nerves

Brain

Spinal cord

Sympathetic
chain ganglia

Figure 13.2
The central nervous
system and the proximal
portions of the peripheral
nervous system.

sensory nerve cells, which carry impulses from the viscera of the body to the central nervous system.

Efferent Division

The efferent division of the peripheral nervous system is divided into the *somatic nervous system* and the *autonomic nervous system:*

1. The **somatic nervous system** is also called the **voluntary nervous system** because its motor functions may be consciously controlled. It includes *somatic motor* nerve cells, which carry impulses from the central nervous system to the skeletal muscles. The impulses carried by somatic motor nerves produce contractions of the skeletal muscles. Muscle contractions that are brought about by the somatic nervous system may be under the conscious control of the individual, or, in the case of reflex responses, they may not be consciously controlled.

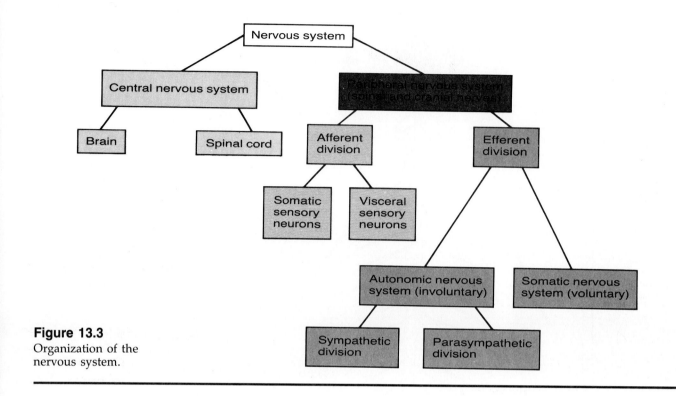

Figure 13.3
Organization of the
nervous system.

2. The **autonomic nervous system**—or **involuntary nervous system**—in contrast to the somatic nervous system, is composed of *visceral motor* nerve cells, which transmit impulses to smooth muscles, cardiac muscle, and glands. Visceral motor impulses generally cannot be consciously controlled.

The autonomic nervous system may be subdivided functionally into **sympathetic** and **parasympathetic divisions**, each of which is studied in more detail in Chapter 16. As noted earlier, the autonomic and somatic nervous "systems" are actually overlapping divisions of the single nervous system. Moreover, although the motor neurons of both the somatic and the autonomic divisions are generally considered to be parts of the peripheral nervous system, they are under the control of centers located in the central nervous
F13.3 system. The organization of the nervous system is summarized in Figure 13.3.

EMBRYONIC DEVELOPMENT OF THE NERVOUS SYSTEM

The nervous system develops from ectodermal cells that, by the second week of gestation, form a flat thickening on the dorsal surface of the embryo. This thickening, called the **neural plate,** gives rise to all the nerve cells of the nervous system. Neural plate cells also give rise to most of the supportive cells **(neuroglial cells)** of the nervous system.

As development continues, the midline of the neural plate sinks inward. At the same time, the proliferation of cells along the margins of the plate produces elevations. The result is a **neural groove** flanked by **neural folds**
F13.4 that extend the entire length of the embryo (Figure 13.4).

The neural groove deepens as the neural folds increase in height. Eventually, the folds meet and fuse in the midline, converting the groove into the
F13.5 **neural tube** (Figure 13.5). The neural tube separates from the surface ectoderm and subsequently develops into the brain and the spinal cord. Moreover, cells in the neural tube send out processes to peripheral structures.

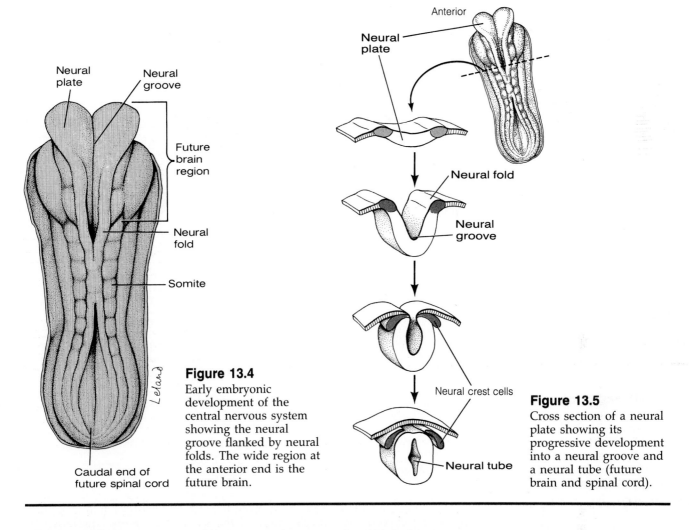

Figure 13.4
Early embryonic development of the central nervous system showing the neural groove flanked by neural folds. The wide region at the anterior end is the future brain.

Figure 13.5
Cross section of a neural plate showing its progressive development into a neural groove and a neural tube (future brain and spinal cord).

These cells and their processes form the motor nerve cells of both the somatic and the autonomic nervous systems.

While the neural folds are fusing, ectodermal cells from the tops of the folds move laterally to form columns of cells on each side of the neural tube. The cells in these columns are called **neural crest cells.** These neural crest cells develop connections with the central nervous system and with peripheral structures, thus forming sensory nerve cells within cranial nerves and spinal nerves. Some neural crest cells migrate to other locations and develop into motor nerve cells of the sympathetic nervous system, into Schwann cells that wrap around nerve cells, or into other structures not directly associated with the nervous system.

The anterior end of the neural tube undergoes the most rapid growth and gives rise to the brain. By the fourth week of gestation, the brain is in the form of three fluid-filled enlargements, or vesicles: the **prosencephalon (forebrain);** the **mesencephalon (midbrain);** and the **rhombencephalon (hindbrain)** (Figure 13.6). During subsequent development, this region undergoes several bendings or flexures and the vesicles subdivide further. As a result, by the fifth week the brain consists of five vesicles that are tightly curved upon themselves at the anterior end of the embryo (Figure 13.7). The prosencephalon subdivides into two vesicles: an anterior **telencephalon** and, just behind it, the **diencephalon.** The mesencephalon does not further subdivide, but the rhombencephalon divides into the **metencephalon** and the most posterior brain vesicle, the **myelencephalon.** The cavities in these brain vesicles will become the ventricles of the adult brain, and the fluid in them is the **cerebrospinal fluid.** Brain development is discussed further in Chapter 14.

F13.6

F13.7

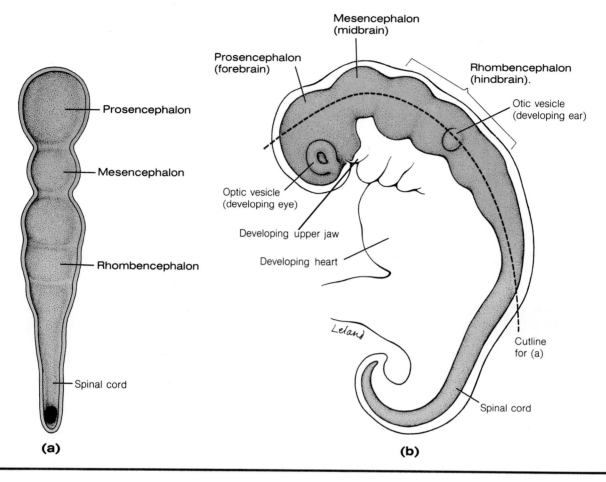

Figure 13.6
Division of the early embryonic brain (three to four weeks) into prosencephalon, mesencephalon, and rhombencephalon.
(a) Frontal section.
(b) Lateral view.

The region of the neural tube posterior to the myelencephalon gives rise to the spinal cord. The development of the spinal cord is also discussed in Chapter 14.

COMPONENTS OF THE NERVOUS SYSTEM

Nervous tissue is composed of two classes of cells that are different structurally and functionally: (1) **neurons (nerve cells),** which are specialized for transmitting and processing nerve impulses; and (2) **supporting cells,** which do not generate nerve impulses, but provide structural and functional support to neurons. The supporting cells are called **neuroglial cells** or **glial cells.** Neuroglial cells can be further subdivided into supporting cells of the central nervous system (astrocytes, oligodendrocytes, microglia, and ependymal cells) and supporting cells of the peripheral nervous system (Schwann cells and satellite cells).

Neurons

The **neuron (nerve cell)** is the basic structural and functional component of the nervous system. It has the ability to respond to stimulation by initiating and conducting electrical signals. Some other cells, such as muscle fibers, are also capable of conducting electrical signals, but the unique shape of the nerve cell makes it especially well suited to serve as a communicator. Neurons have processes that can be quite long. For instance, a single neuron—that is, a single cell—extends from the spinal cord to the tip of a toe!

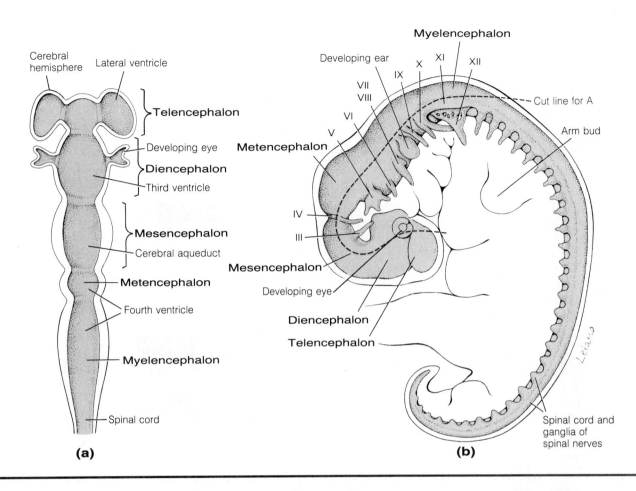

(a)

(b)

Figure 13.7
Later embryonic development of the brain (five weeks) into telencephalon, diencephalon, mesencephalon, metencephalon, and myelencephalon.
(a) Frontal section.
(b) Lateral view. Roman numerals indicate cranial nerves.

With specific exceptions, mature neurons are unable to undergo mitosis. For this reason, once embryonic development has been completed, most nerve cells cannot be replaced when they die or are destroyed. Under the proper conditions, however, peripheral neuronal processes that have been damaged or severed can be repaired or regenerated, thus reestablishing the nerve supply to the affected structures.

Structure of a Neuron

Each neuron is composed of a **cell body (soma,** or **perikaryon)** and one or more processes containing cytoplasm that extend from the cell body (Figure 13.8). Within the cytoplasm of the cell body is a large nucleus containing a **F13.8** prominent nucleolus. As in most cells, the cytoplasm of neurons also contains mitochondria and Golgi apparatus. In addition, the cytoplasm contains dark-staining **chromatophilic substance** (*Nissl bodies*) and slender fibrils called **neurofibrils.** Under the electron microscope, the chromatophilic substance is seen to be composed of free ribosomes and parallel layers of rough endoplasmic reticulum, and the neurofibrils are seen to be composed of microfilaments, intermediate filaments, and microtubules. It is thought that the fibrils provide support to the cell, whereas the microtubules may be involved in the transport of materials within the cell.

Locations of the Cell Bodies of Neurons

The cell bodies of most neurons are located in the central nervous system, although there are large numbers located in the peripheral system. Nerve cell bodies in the CNS tend to be located in clusters called **nuclei.** (Note that this use of the term *nucleus* has no relation to the nucleus of a cell.) A nucleus or group of nuclei whose neurons have related functions is referred to as a **cen-**

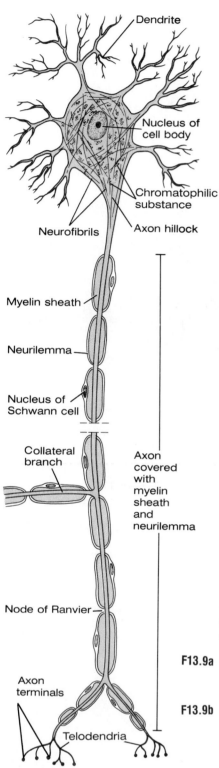

Figure 13.8
Structure of a typical motor neuron.

ter. Nerve cell bodies located outside the CNS, in the peripheral nervous system, are generally found in groups called **ganglia.**

Processes of Neurons

The processes associated with a neuron are very thin extensions of the cell. There are two types of processes: *dendrites* and *axons.* According to classical usage of the terms, dendrites are the neuronal process or processes that conduct electrical signals *toward* the cell body; an axon (of which there is only one per cell) is the neuronal process that conducts electrical signals *away* from the cell body. It is becoming more common, however, to differentiate dendrites and axons in the following way.

DENDRITES   The **dendritic zone** is the *receptive* portion of a neuron, where the electrical signals originate. The dendritic zone may include the cell body as well as branching processes, or **dendrites,** that are either direct extensions of the cell body or more remote branchings separated from the cell body by an axon. The number, length, and extent of branching of the dendrites varies in different types of neurons. Dendrites possess many of the structures found in the cell body, including filaments and microtubules that are oriented parallel to the long axis of the dendrite.

AXONS   The **axon,** or **nerve fiber,** is the *conductive* process of the neuron—that is, the part that transmits the electrical signals. In motor neurons the cell body is located between the axon and the dendrites, but in sensory neurons it is located to one side of the axon. The axon frequently arises from a cone-shaped process on the cell body called the **axon hillock.** The lengths of axons vary considerably. They may be quite short, traveling only a short distance in the central nervous system, or they may be long and travel a considerable distance within the CNS. Some axons extend a meter or more out into the peripheral nervous system. Each neuron has only one axon, but each axon generally has several branches called **collaterals.** Axons contain mitochondria, microtubules, and neurofibrils, but they lack chromatophilic substance. An axon and its collaterals end by separating into many fine branches called **telodendria.** The distal end of each telodendron expands into small bulblike structures called **axon terminals,** or **synaptic knobs.**

Types of Neurons

Neurons can be classified according to their form or structure, and according to their function—that is, the role they perform in the nervous system.

CLASSIFICATION ACCORDING TO STRUCTURE   Structurally, neurons can be classified into three types on the basis of the number of processes that extend from the cell body:

1. **Bipolar neurons** (Figure 13.9*a*) have two processes, one extending from each end of the cell body. There are only a few examples of bipolar neurons in the body.

2. **Unipolar neurons** (Figure 13.9*b*) are formed during embryonic development when the two processes of certain bipolar neurons fuse together so that only a single process arises from the cell body. Beyond this point of fusion the two processes remain separate, with the central branch functioning as an axon and the peripheral branch functioning as a dendrite.

3. **Multipolar neurons** (Figure 13.9*c*), the most common type of neuron, have one long process that arises from the cell body and functions as an axon; numerous other processes that arise from the cell body function as dendrites.

CLASSIFICATION ACCORDING TO FUNCTION   Functionally, there are also three types of neurons:

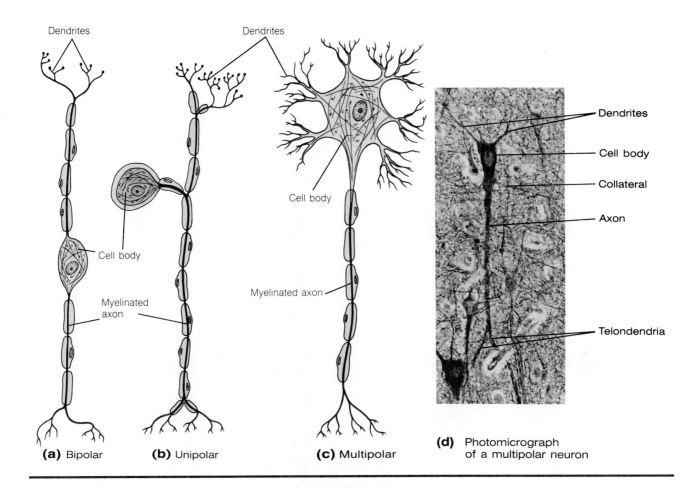

(a) Bipolar   (b) Unipolar   (c) Multipolar   (d) Photomicrograph of a multipolar neuron

Figure 13.9
Types of neurons.

1. **Motor (efferent) neurons,** which transmit impulses away from the CNS to an effector, or from a higher center within the CNS to a lower center.

2. **Sensory (afferent) neurons,** which carry impulses from receptors to the CNS, or from a lower center within the CNS to a higher center.

3. **Interneurons (association neurons),** which, when present, transmit impulses from one neuron to another.

Most interneurons, which are multipolar, are found in the central nervous system. Motor neurons are also multipolar. Most sensory neurons are unipolar, but those that carry impulses from the retina of the eye, inner ear, and the olfactory epithelium are bipolar.

Formation of Myelinated Neurons

Most axons are covered with a fatty substance called **myelin.** Such axons are called **myelinated fibers;** axons not covered with myelin are called **unmyelinated fibers.** Because of the lipid content of myelin, myelinated fibers appear white; when traveling in groups, they form white pathways in the nervous system.

When myelinated axons of the peripheral nervous system are viewed under the light microscope, there appears to be a thin membrane between the myelin and the connective tissue sheath (endoneurium) surrounding the axon. This membrane is called the **neurilemma** or **sheath of Schwann** (Figure 13.8). The neurilemma and the myelin of these myelinated axons are interrupted at regular intervals along the length of the axon. Each of these points of interruption is called a **node of Ranvier** (Figure 13.8).

F13.8

F13.8

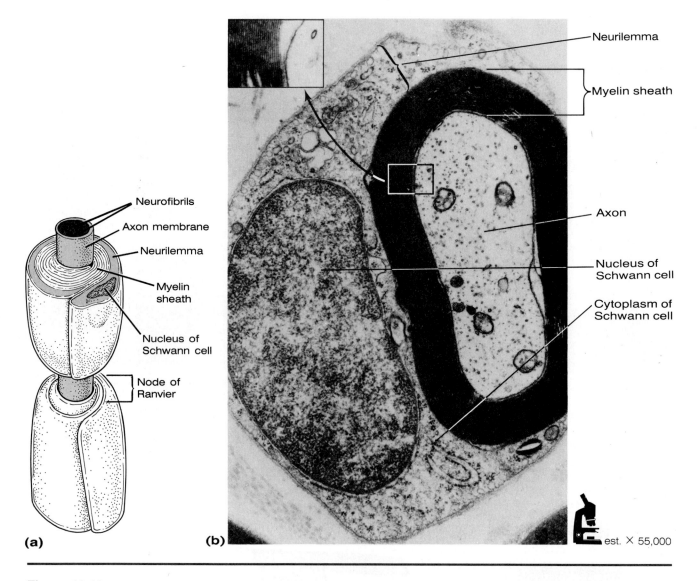

Neurilemma

Myelin sheath

Axon

Nucleus of Schwann cell

Cytoplasm of Schwann cell

est. × 55,000

Neurofibrils

Axon membrane

Neurilemma

Myelin sheath

Nucleus of Schwann cell

Node of Ranvier

(a)

(b)

Figure 13.10

The ultrastructure of myelinated neurons.
(a) Schematic.
(b) Photomicrograph of a cross section of a nerve fiber surrounded by a myelin sheath. The inset is a higher magnification of a portion of the myelin sheath.

Both the myelinated and unmyelinated axons of the peripheral nervous system are surrounded by **Schwann cells** arranged sequentially along the length of the axon. During embryonic development, Schwann cells, which are derived from embryonic neural crest cells, migrate along the axon and envelop it.

We consider the myelinated axons first. Under the electron microscope, it can be seen that the myelin sheath of myelinated axons is formed when each Schwann cell wraps itself around the axon several times, pushing the cytoplasm and nucleus of the Schwann cell peripherally (Figure 13.10). Consequently, the myelin sheath is composed of alternating layers of the proteins and lipids that constitute the plasma membrane of the Schwann cell. The outermost part of the Schwann cell, which contains the cytoplasm and nucleus that were pushed peripherally as the Schwann cell wrapped around the axon, is the neurilemma (sheath of Schwann). The nodes of Ranvier are the junctions between two successive Schwann cells. It is at the nodes of Ranvier that collateral branches of the axon may occur. The nodes are also instrumental in increasing the rate of transmission of the nerve impulse by allowing it to jump directly from one node to the next for the length of the neuron.

F13.11   Not all axons are myelinated, however. As shown in Figure 13.11, it is not unusual for the axons of several neurons to become embedded within the cytoplasm of a single Schwann cell. The Schwann cell may simply envelop some of these axons without forming a myelin sheath around them. Such axons are unmyelinated. The unmyelinated axons of the peripheral nervous system are covered by a neurilemma formed by the Schwann cells. However, they lack the additional spiral wrappings of the Schwann cells that form the myelin covering.

Many neurons in the central nervous system are myelinated, but there are no Schwann cells in the CNS. Instead, the central nervous system contains a type of neuroglial cell—called an **oligodendrocyte**—that sends out processes that spiral around axons, thus forming a myelin covering. Yet, because in this case the myelin sheath is formed by coiling of processes from the cells—rather than by the entire cell—the cytoplasm squeezed out of the coiled regions is forced back toward the cell body. Thus, there is no neurilemma formed in the central nervous system. Myelinated axons form the white matter of the central nervous system.

The Nerve Impulse

An unstimulated neuron that is not conducting a nerve impulse exists in an electrically polarized state: the inside of the cell is negative relative to the outside. This polarity is due primarily to an unequal distribution of ions across the nerve cell membrane. In an unstimulated neuron, potassium ions occur in higher concentration inside the cell than outside, and sodium ions occur in higher concentration outside the cell than inside. The difference in electrical potential produced by the unequal distribution of these ions is referred to as the **resting membrane potential.**

When a neuron is stimulated, changes occur in the cell membrane that tend to upset this characteristic distribution of ions on either side of the membrane. The membrane becomes more permeable to sodium ions in the area of stimulation, and it allows greater numbers of sodium ions to enter the cell here as compared to unstimulated areas of the cell. Since the unstimulated neuron is polarized (the inside of the cell is negative relative to the outside) and since sodium ions are positively charged, the increased movement of sodium ions into the cell in the stimulated area causes the inside of the cell in this area to become relatively less negative. Thus the resting, polarized state of the cell diminishes, and the cell undergoes some degree of depolarization in the area of stimulation.

If a stimulus of sufficient intensity is applied to a neuron, the local depolarization of the cell can reach a critical level known as the **threshold,** and an **action potential** may occur. Once a threshold level is reached, the cell membrane in the depolarized area becomes very permeable to sodium ions and they rapidly enter the cell. As a result, the original polarity of the cell is drastically reduced in the region of depolarization. In fact, the cell's polarity actually becomes reversed: the inside of the cell becomes positive relative to the outside.

The occurrence of an action potential in one area of an axon normally causes an increase in the sodium permeability of the immediately adjacent area of the nerve cell membrane and the generation of an action potential there. In turn, this causes the next adjacent area of the cell membrane to become more permeable to sodium ions, and in this area an action potential is generated. This activity continues along the length of the neuronal membrane, producing a wave of depolarization called a **propagated action potential,** or **nerve impulse,** that moves along the axon.

Synapses

Although a nerve impulse travels the entire length of a single neuron, chains of neurons must be traversed if information is to be transmitted throughout the entire nervous system. This transmission requires a means of passing information (the nerve impulse) from neuron to neuron as well as a method of transmitting information along a single neuron. Most individual neurons do not contact one another directly. Instead, a process of one neuron remains separated from one another by a small space. This type of neuronal junction is called a **synapse,** and the space itself is called the **synaptic cleft** (Figure 13.12). **F13.12**
Many synapses occur between the axon of one neuron and the dendrite of

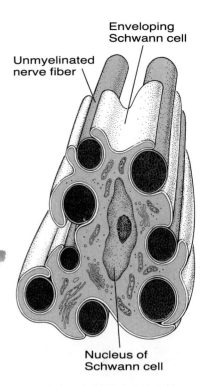

Enveloping Schwann cell

Unmyelinated nerve fiber

Nucleus of Schwann cell

Figure 13.11
A single Schwann cell encompassing eight unmyelinated nerve fibers.

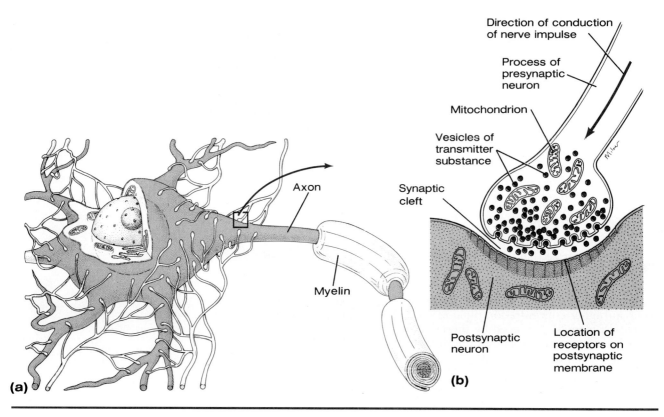

(a)

(b)

Figure 13.12
Synapses between
neurons. **(a)** Axons of
several neurons
synapsing with the
dendrites and cell body
of a neuron. **(b)** A single
synapse at higher
magnification.

another. However, some synapses are between an axon and a cell body.

Because the two neurons at a synapse do not actually contact one another, a nerve impulse that travels along the first neuron, or **presynaptic neuron,** cannot cross to the second neuron, or **postsynaptic neuron.** Instead, the arrival of a nerve impulse at the ending of a presynaptic neuron results in the release of a chemical transmitter substance that diffuses across the synaptic cleft and attaches to receptors located on the membrane of the postsynaptic neuron.

A number of different chemical transmitter substances are known, and each has either a stimulatory or an inhibitory effect on a postsynaptic neuron. If a chemical transmitter is stimulatory, it causes the membrane of the postsynaptic neuron to depolarize. This depolarization is called an **excitatory postsynaptic potential.** If the stimulation of the postsynaptic neuron is of sufficient intensity, the membrane of the postsynaptic neuron depolarizes to threshold, and an electrical signal is then established in the dendrite of the postsynaptic neuron. The electrical signal, in turn, may trigger a nerve impulse when it reaches the axon of the postsynaptic neuron. If a chemical transmitter is inhibitory, it increases the resting polarity of the postsynaptic neuron. This hyperpolarization is called an **inhibitory postsynaptic potential.** As a result of this hyperpolarization, a stronger-than-normal excitatory stimulus is required to elicit a nerve impulse in the postsynaptic neuron.

Once the transmitter substances are released, they remain active for only a short time. Following their release, they are rapidly removed, either by chemical inactivation (via enzymes), by diffusion away from the vicinity of the receptors on the postsynaptic membrane, or by entering the presynaptic neuron.

Nerves

F13.13 A **nerve** is composed of the processes of many neurons held together by connective-tissue sheaths (Figure 13.13). Each nerve fiber is individually wrapped in a thin connective-tissue sheath called the **endoneurium.** The processes of the neurons are separated into groups within the nerve by another

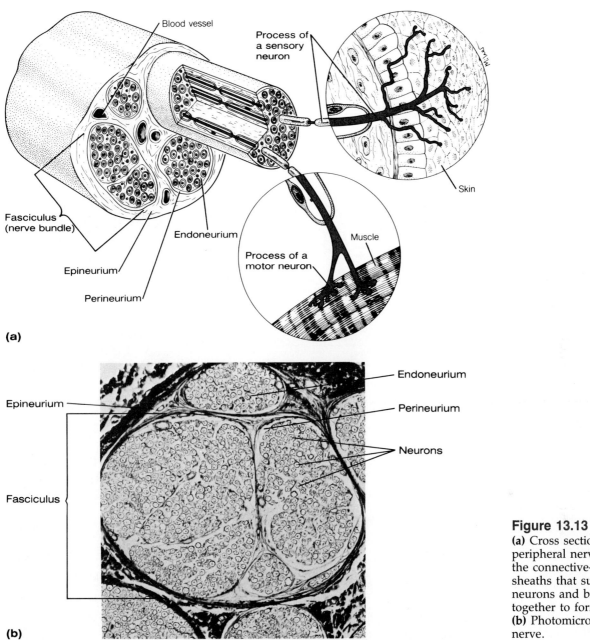

(a)

(b)

Figure 13.13
(a) Cross section of a peripheral nerve showing the connective-tissue sheaths that surround the neurons and bind them together to form a nerve. **(b)** Photomicrograph of a nerve.

connective-tissue wrapping, the **perineurium.** Each bundle of nerve fibers surrounded by perineurium is called a **fasciculus.** Several fasciculi surrounded by a connective-tissue sheath called the **epineurium** constitute a single nerve. Blood vessels and lymphatic vessels travel within the connective tissue sheaths to supply the neurons.

Nerves are found only in the peripheral nervous system and they vary in size and composition. Nerves that contain only the processes of sensory neurons and carry nerve impulses to the CNS are **sensory nerves.** A few of the cranial nerves are sensory nerves. Some cranial nerves contain only the processes of motor neurons and carry nerve impulses from the CNS to the peripheral nervous system. These nerves are **motor nerves.** Other cranial nerves and all spinal nerves are **mixed nerves**—that is, they contain the processes of both sensory and motor neurons. As a result of their composition, nerve impulses within mixed nerves travel both to and from the central nervous system.

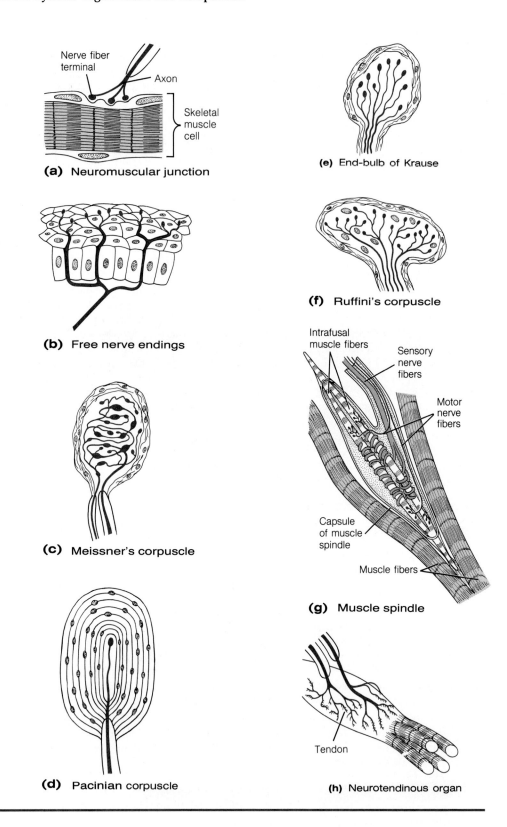

Figure 13.14
Specialized nerve
endings.

(a) Neuromuscular junction
(b) Free nerve endings
(c) Meissner's corpuscle
(d) Pacinian corpuscle
(e) End-bulb of Krause
(f) Ruffini's corpuscle
(g) Muscle spindle
(h) Neurotendinous organ

Table 13.1   To help you understand the terminology used for the main components of the nervous system, Table 13.1 provides a reference for distinguishing between a neuron, a nerve fiber, and a nerve.

Specialized Peripheral Neuron Endings

Neuronal endings are important because they provide the means for communication between two neurons, between neurons and sensory receptors, and

Table 13.1 Comparison of Neurons, Nerve Fibers, and Nerves

Neuron	A nerve cell
Nerve Fiber	Any long process of a neuron. The term usually refers to axons, but also includes the peripheral processes of sensory neurons.
Nerve	A collection of nerve fibers in the peripheral nervous system.

between neurons and their effectors (muscles or glands). The endings of neuronal processes in the peripheral regions of the body are generally specialized structures. They range from simple free nerve endings to complex encapsulated structures (Figure 13.14).

F13.14

Motor Neuron Endings

Motor neurons **(somatic motor neurons, or efferent neurons)** form a **neuromuscular (myoneural) junction** with the skeletal muscles they supply (Figure 13.14*a*). The axon of a somatic motor neuron divides into many terminal branches that end in neuromuscular junctions at skeletal muscle cells. The myelin sheath of the axon does not extend to the terminal end of the process; thus, the bare axon—which is covered on its outer surface with a neurilemma—comes very close to the membrane of the muscle cell. The membranes of the nerve and the muscle do not actually touch, however, but remain separated by a very small gap (300–400 Å). In the terminal branches of the axon are accumulations of mitochondria and small vesicles that contain a chemical substance called *acetylcholine*. Acetylcholine transmits stimulating impulses from the neuron across the gap of the neuromuscular junction to the muscle cell in a way similar to the transmission of a nerve impulse across a synapse, as described earlier in this chapter.

F13.14a

Sensory Neuron Endings

The terminal portions of the peripheral processes of sensory neurons are dendrites. The terminal ends of the dendrites of many sensory neurons are sensitive to changes in their environment. For this reason, the dendritic endings of many sensory neurons function as **receptors.** For example, in the skin there are modified dendrites that serve as receptors associated with the various cutaneous senses.

The cutaneous senses include those of pain, touch, pressure, heat, and cold. For many years it has been believed that our ability to distinguish these cutaneous senses is due to the presence of structurally distinct and functionally specialized peripheral receptors, a different type for the reception of each sensation. There is now some doubt about the accuracy of this one-to-one structural-functional specificity. It may be that each anatomically distinct type of receptor in the skin is, in fact, sensitive to a variety of stimuli rather than to only one basic type of stimulus. According to this new view, one type of receptor may have a somewhat different sensitivity to a variety of stimuli than another type. Although the following discussion indicates that each type of receptor may be responsible for sensing a specific kind of stimulus, bear in mind that this notion may require modification as additional information becomes available. Regardless of their precise function, however, the presence of these structurally distinct types of receptors in the skin has been clearly established.

FREE NERVE ENDINGS **Free nerve endings** are the least modified receptors. They consist of bare dendrites (Figure 13.14*b*). The free nerve endings branch between epithelial cells, connective tissue cells, muscle cells, cells of mucous membranes, and so forth. They are thought to serve primarily as *pain* receptors of the body, although free nerve endings that surround hair follicles are believed to be important *touch* receptors, and some free nerve endings serve as *heat* and *cold* receptors.

F13.14b

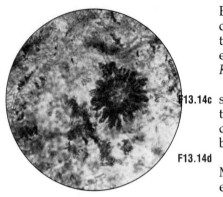

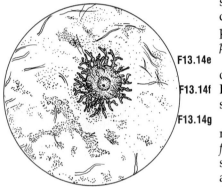

(a) Astrocytes

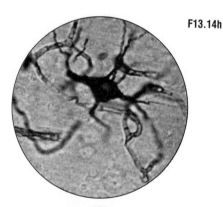

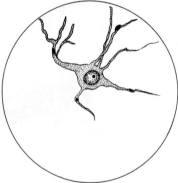

(b) Oligodendrocytes

Figure 13.15
Types of neuroglial cells.

ENCAPSULATED SENSORY ENDINGS   The rest of the general sensory receptors are surrounded by connective tissue capsules that contribute, in ways that are not completely understood, to the activity of the receptors. These encapsulated receptors are *Meissner's corpuscles, Pacinian corpuscles, end bulbs of Krause, Ruffini's corpuscles, muscle spindles,* and *neurotendinous organs.*

F13.14c   **Meissner's corpuscles** are small, elliptical connective-tissue capsules that surround a spiraled ending of a dendrite (Figure 13.14c). These receptors are thought to be particularly sensitive to *light touch* and are abundant in the dermal papillae of the skin (especially the finger tips), in the mucous membranes of the tongue, and in other sensitive regions of the body.

F13.14d   **Pacinian corpuscles** (Figure 13.14d), which are located deeper than the Meissner's corpuscles, are found in the deeper layers of the skin, in the mesenteries, and in loose connective tissue. They are composed of a dendrite and several concentric layers of connective tissue surrounding the dendrite. Because they are surrounded by a comparatively heavy capsule, Pacinian corpuscles are thought to be sensitive not to light touch but rather to heavier *pressure.*

F13.14e   **End-bulbs of Krause** (Figure 13.14e), which are quite common throughout the body, are thought to serve as *cold* receptors. Oval capsules called
F13.14f   **Ruffini's corpuscles** (Figure 13.14f) are found primarily in subcutaneous tissue and are believed to function as *heat* receptors.

F13.14g   **Muscle spindles** (Figure 13.14g) are complex capsules found in skeletal muscles. Within the capsules are thin skeletal muscle fibers called *intrafusal fibers.* The intrafusal fibers are supplied by sensory neurons, some of which spiral around the fibers. When the muscle is stretched, the intrafusal fibers are also stretched. This increases the frequency of nerve impulses carried back to the spinal cord by the neurons that surround the intrafusal muscle fibers. In response to the sensory nerve impulses produced by stretching the skeletal muscle, there is an increase in the frequency of motor nerve impulses to the same muscle. In this way, the stretch is reflexly resisted.

F13.14h   **Neurotendinous organs** (*Golgi tendon organs*) (Figure 13.14h) function in close association with the muscle spindles. They are composed of dendrites enclosed within thin connective-tissue capsules. The dendrites divide into many small branches in a tendon, near its junction with a muscle. The contraction of the muscle causes varying amounts of tension to be exerted on the tendon. The development of excessive tension by the muscle activates the neurotendinous organs. Sensory impulses from the organs are then carried back to the spinal cord, where motor neurons to the same muscle are inhibited, thus relaxing the muscle.

Types of Receptors

Receptors are structures that are generally activated by changes (stimuli) in either the internal or external environment of the body. As a result of the activity of the receptors, nerve impulses are initiated in sensory nerve cells. Receptors may be the endings of peripheral sensory neurons such as those described above, or they may be specialized cells associated with these peripheral endings. Receptors can be classified in two ways: according to the *location* of the stimulus or according to the *type* of stimulus.

Classification According to Location of Stimulus

One means of classification is according to the location of the stimuli to which the receptors respond. Thus, **exteroceptors** respond to stimuli from the body surface, including touch, pressure, pain, temperature, light, and sound. **Interoceptors (visceroceptors)** are sensitive to pressure, pain, and chemical changes that occur in the body. Much of the information transmitted over interoceptors does not reach the level of consciousness. Therefore, a person normally is not aware of small changes in blood pressure, the amount of gases carried in the blood, or contractions of the smooth muscles of the organs. A type of interoceptor called **proprioceptors** provides information concerning the position of body parts without the necessity of visually observing the parts. Thus, for example, it is possible to know the position of the fingers even when the eyes are closed. The muscle spindles and the neurotendinous organs are proprioceptors. There are also sensory receptors that monitor the degree of stretch in joint capsules. Since these stretch receptors provide information concerning the position of joints, they, too, serve as proprioceptors.

Classification According to Type of Stimulus

Receptors can also be classified according to the type of stimulus that causes them to generate a nerve impulse. **Mechanoreceptors** are sensitive to physical deformation, such as pressure or stretch. These receptors include Pacinian corpuscles, muscle spindles, neurotendinous organs, and perhaps Meissner's corpuscles. Receptors that respond to changes in temperature are known as **thermoreceptors.** Ruffini's corpuscles and the end-bulbs of Krause are generally classified as thermoreceptors, but some researchers consider them to be mechanoreceptors sensitive to touch and closely related to Meissner's corpuscles. **Chemoreceptors** are stimulated by various chemicals in food, in the air, or in the blood. Thus, the senses of taste and smell rely on chemical receptors. The cells in the aortic and carotid bodies that monitor the pH and levels of gases in the blood are also chemoreceptors. Specialized receptor cells that can convert light energy into nerve impulses are **photoreceptors.** The only photoreceptors in the body are located in the retina of the eyes. These receptors, along with the chemoreceptors associated with the senses of taste and smell, are considered to be organs of special senses and are discussed in Chapter 17.

Neuroglia

There are billions of neurons in the central nervous system, but there are even more supportive cells distributed among the neurons. These cells are called **neuroglia (glial cells).** Some neuroglia provide structural support for the neurons. Others are involved in the transfer of nutrients from the blood vessels to the neurons and assist in the removal of waste products from the neurons. Still others serve as active phagocytes within the central nervous system. One type of neuroglial cell prevents the electrical activity of a neuron in the central nervous system from interfering with that of adjacent neurons. In contrast to neurons, neuroglia are capable of undergoing mitotic division. Included in the neuroglia of the central nervous system are *astrocytes, oligodendrocytes, microglia,* and *ependymal cells.* With the exception of the microglia, all neuroglial cells are of embryonic ectodermal origin. Microglia originate from the embryonic mesoderm that forms connective tissues.

There are also two types of cells present in the peripheral nervous system that are often considered to be neuroglia. These are *satellite cells,* which are found in the capsule that surrounds the body of sensory neurons, and *Schwann cells,* which are located around axons of peripheral nerves.

Astrocytes

Astrocytes are rather large cells with star-shaped bodies from which numerous processes radiate outward (Figure 13.15*a*). They are the most numerous of the neuroglia and provide most of the structural support for the CNS. Moreover, many of the processes of astrocytes contact blood vessels in the CNS. Because of this contact, astrocytes are thought to be involved in the formation of the *blood-brain barrier,* which surrounds the capillaries of the CNS. In this role, astrocytes are thought to regulate the transport of substances between capillaries and neurons.

The proposed existence of a blood-brain barrier is based on observations that only water, oxygen, and carbon dioxide can readily enter or leave the capillaries of the CNS. All other substances that move across the capillary walls of the central nervous system do so at a slower rate. This impediment to exchange in the brain is in contrast to all other regions of the body, where many substances move rather freely and quickly across capillary walls. Apparently the role of the blood-brain barrier is to prevent sudden and extreme fluctuations in the composition of the tissue fluid of the CNS, thus protecting the neurons, which are irreplaceable.

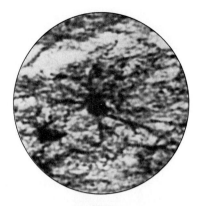

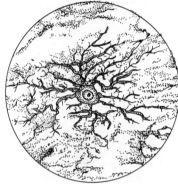

(c) Microglia

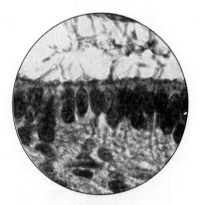

F13.15a

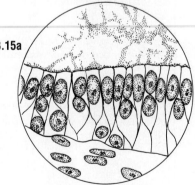

(d) Ependymal cells

Figure 13.15 *continued*
Types of neuroglial cells.

FRONTIERS IN HEALTH

Electric Pain Relief

People who experience pain can tell you that it exists in dozens of forms, from intense to subtle, from nagging to stabbing, from chronic to acute. Pain is a leading affliction in the United States, with nearly one in every three Americans suffering from chronic pain. It has been estimated that pain costs $70 billion a year in medical costs and lost working days. The personal cost of chronic pain, of course, is immeasurable.

Ironically, pain is one of the least understood of all medical conditions. The study of pain falls outside the realm of internal medicine, surgery, or any of the other traditional subdivisions of medicine. But pain is now being studied more than it has been, and several recent techniques offer sufferers of chronic pain new hope.

Pain has traditionally been treated with painkillers, with acupuncture, biofeedback, and hypnosis to complement these medications. For severe pain, surgeons have resorted to cutting nerves or destroying areas of the brain that perceive pain. However, a large percentage of patients who have undergone such treatment report the recurrence of pain, usually after about a year, and often with greater intensity.

One technique that offers great promise is deep brain stimulation. Surgeons find that electrodes implanted in certain regions of the brainstem and midbrain can block out pain impulses that are transmitted to the brain. After the electrodes are inserted through a tiny hole in the skull of pain victims, they are pushed into specific regions of the brain and connected to a small battery worn on the belt, or sometimes implanted under the skin. When pain begins, the patient turns on the current, giving himself or herself a small electric shock that blocks out the pain. Deep brain stimulation successfully blocks out a wide spectrum of painful stimuli without noticeably affecting other brain functions.

Scientists at the Johns Hopkins University Laboratory of Applied Physics, working with Pacesetter Systems in Sylmar, California, have developed an implantable battery pack that can be used in deep brain stimulation. No bigger than a deck of cards, the stimulator is implanted beneath the skin on the lower part of the rib cage. Once they become accustomed to its presence, patients hardly know the device is there. Fine wires run from the unit to the brain or to nerve bundles. According to the developers, the stimulator can last for 10 years with only monthly recharging, which can be done transcutaneously by simply holding an alternating magnetic field over the device.

Deep brain stimulation requires surgery, and with that come potential complications. But with a success rate of 75% most patients are willing to accept the risks in order to rid themselves of the excruciating and often disabling pain that racks their bodies day and night.

For many people, relief may come in a less costly and simpler form of treatment, called transcutaneous electrical nerve stimulation (TENS). In this method of pain control, patients wear a small battery pack, which is connected by wires to small electrodes. Rather than being implanted, however, the electrodes are located on

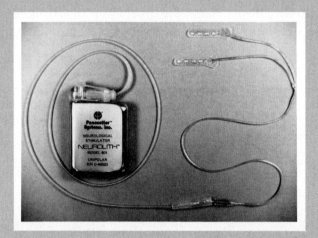

This cerebellar stimulator was designed by Pacesetter Systems for patients with cerebral palsy. Dr. Joseph H. Schulman of the Neurodyne Corporation is monitoring clinical testing, under the authorization of the U.S. F.D.A.

the body's surface, over nerve fibers that transmit pain. When a patient feels pain, he or she simply turns on the electricity. The small current generated by the battery stimulates the skin and blocks the pain fibers.

Medical researchers are not entirely certain how TENS works, but many think that it blocks pain by overloading the neuronal circuitry. Two types of nerve fibers are thought to carry sensory information to the spinal cord. Small-diameter fibers carry pain of many varieties, while larger-diameter fibers carry other forms of sensory information, such as pressure and light touch, from receptors in the skin. Both types of sensory fibers converge in the spinal cord. Nerve impulses traveling from the pain receptors are thought to be blocked by simultaneous stimulation of the larger fibers, either with acupuncture or TENS.

TENS has been used successfully to reduce the pain experienced after major surgery. In one study, throughout the postoperative period, patients using TENS needed only one-third as much narcotic pain killer as patients not using the method. Moreover, patients using TENS were able to leave the hospital several days sooner than those who were treated conventionally.

Alleviation of postoperative pain is only one use for this relatively inexpensive and safe treatment. Dentists have found that TENS can virtually eliminate the need for local anesthesia in routine dental work. It could also find an important niche in obstetrics by reducing the pain of childbirth. Athletic injuries, arthritic pain, low-back pain, neck pain, and painful joints are additional applications for this promising form of electric pain relief.

Oligodendrocytes

Oligodendrocytes (oligodendroglia) are smaller than astrocytes and have fewer processes (Figure 13.15*b*). Some of these processes wrap around axons **F13.15b** in the central nervous system, forming a myelin sheath, much like the Schwann cells do in the peripheral nervous system.

Microglia

Microglia are the smallest types of neuroglial cells (Figure 13.15*c*). They are **F13.15c** phagocytic and therefore function in the removal of dead tissue or foreign materials from the CNS. For this reason, microglia are considered to be a part of the macrophage system (Chapter 3).

Ependymal Cells

Ependymal cells are neuroglial cells that line the ventricles (cavities) of the brain and the central canal of the spinal cord (Figure 13.15*d*). Some of these **F13.15d** cuboidal epithelial cells are ciliated. The ependymal cells in the ventricles of the brain are further modified, having microvilli extending from their free surfaces. These modified ependymal cells are closely associated with invaginations from a covering of the brain called the pia mater and with networks of capillaries, forming structures called **choroid plexuses.** The choroid plexuses produce cerebrospinal fluid, which fills the ventricles.

EFFECTS OF AGING ON THE NERVOUS SYSTEM

Because most mature neurons are unable to undergo mitosis, any nerve cells that die or are destroyed are not replaced. In nervous tissue, as in all tissues, many cells die each day in the normal course of aging. Consequently, with aging there are fewer neurons within the nervous system. How these reductions in neurons affect the functioning of the nervous system varies among individuals, depending on the regions of the nervous system in which the reductions occur. The loss of neurons is thought to be the reason that a brain may lose as much as 10% of its weight by the age of 90 years.

With aging there seems to be some loss of myelin from the neurilemma, as the rate of conduction of impulses by neurons decreases in older persons. There is also an increase in the time required to transmit impulses from one neuron to another—that is, to cross a synapse. This increase is thought to be due to an age-related reduction in the transmitter substances released by the presynaptic neurons and a reduction in the numbers or sensitivity of the receptors on the postsynaptic neurons.

Age-related changes within the neurons themselves include a reduction in the amount of Golgi apparatus and the chromatophilic substance. The chromatophilic substance contains RNA, therefore its reduction could indicate a decline in metabolism within the neurons.

STUDY OUTLINE

ORGANIZATION OF THE NERVOUS SYSTEM pp. 349–352

CENTRAL NERVOUS SYSTEM (CNS) brain and spinal cord; integrative and control centers.

PERIPHERAL NERVOUS SYSTEM (PNS) spinal nerves, cranial nerves, ganglia, receptors.

AFFERENT DIVISION includes somatic and visceral sensory nerve cells.

EFFERENT DIVISION includes somatic and visceral motor nerve cells.

Somatic Nervous System (Voluntary) carries somatic motor impulses to skeletal muscles.

Autonomic Nervous System (Involuntary) transmits visceral motor impulses to cardiac and smooth muscles and glands; sympathetic and parasympathetic divisions.

EMBRYONIC DEVELOPMENT OF THE NERVOUS SYSTEM pp. 352–354

ECTODERMAL CELLS thicken and form neural plate.

NEURAL GROOVE becomes neural tube, develops into brain and spinal cord.

NEURAL CREST CELLS form sensory nerve cells in cranial and spinal nerves, sympathetic motor nerves, and Schwann cells.

ANTERIOR END OF NEURAL TUBE forms brain.

PROSENCEPHALON (FOREBRAIN)

Telencephalon most anterior

Diencephalon

MESENCEPHALON (MIDBRAIN)

RHOMBENCEPHALON (HINDBRAIN)

Metencephalon

Myelencephalon most posterior

CEREBROSPINAL FLUID within cavities of brain.

COMPONENTS OF THE NERVOUS SYSTEM pp. 354–367

NEURONS

STRUCTURE OF A NEURON

Cell Body (Soma, Perikaryon)

One or More Processes contain cytoplasm.

Large Nucleus, Prominent Nucleolus

Chromatophilic Substance parallel layers of rough endoplasmic reticulum.

Neurofibrils fibrils for cell support and microtubules for transport within cell.

LOCATIONS OF THE CELL BODIES OF NEURONS

Nuclei clusters of cell bodies in CNS.

Center group of nuclei with related functions.

Ganglia groups of nerve cell bodies in PNS.

PROCESSES OF NEURONS

Dendrites (Dendritic Zone) receptive portion of neuron; where electrical signal originates.

Axon conductive process of neuron along which transmission of electrical signal occurs; one axon per neuron; axon branches called collaterals; an axon plus its covering is called a nerve fiber; lack chromatophilic substance.

TYPES OF NEURONS

Classification According to Structure
1. Bipolar Neurons two processes; found mostly in the embryo.
2. Unipolar Neurons single process that arises from cell body.
3. Multipolar Neurons one axon; several dendrites.

Classification According to Function
1. Motor Neurons (Efferent) transmit from CNS to effector, or to a lower center in CNS; multipolar.
2. Sensory Neurons (Afferent) transmit from receptor to CNS, or to a higher center in CNS; most are unipolar; a few are bipolar.
3. Interneurons (Association Neurons) transmit from neuron to neuron; multipolar; in CNS only.

FORMATION OF MYELINATED NEURONS

Unmyelinated Fibers axons not covered with myelin; enveloped by Schwann cells.

Myelinated Fibers axons covered with myelin, a fatty substance that causes them to appear white; spiral wrappings of Schwann cells form myelin sheath.
1. Neurilemma (Sheath of Schwann) outermost part of Schwann cell.
2. Node of Ranvier regular points of interruption of neurilemma and myelin, where Schwann cells meet end to end.
3. Myelin Sheath composed of alternate layers of protein and lipid from plasma membrane of Schwann cells.
4. Myelinated CNS Neurons have no Schwann cells; oligodendrocytes spiral around axons, forming myelin; no neurilemma in CNS.

THE NERVE IMPULSE

Unstimulated Neuron electrically polarized: inside of cell negative relative to outside. Polarity due to unequal distribution of potassium and sodium ions across cell membrane; called resting membrane potential.

Stimulated Neuron sodium ions rapidly enter cell in greater numbers. If stimulus is strong enough, membrane depolarizes, with inside of cell becoming positive relative to outside; produces an action potential.

Action Potential continues along length of neuronal membrane, producing a nerve impulse.

SYNAPSES

Neurons in a Chain are joined via synapses; at synapses neurons separated by small space called synaptic cleft.

Presynaptic Neuron releases chemical transmitter substance into synaptic cleft.

If Transmitter Substance Is Stimulatory postsynaptic neuron depolarizes; called excitatory postsynaptic potential.

If Transmitter Substance Is Inhibitory postsynaptic neuron hyperpolarizes; called inhibitory postsynaptic potential. Transmitter substances rapidly inactivated or removed.

NERVES
1. Individual neuron (nerve cell) wrapped in connective tissue endoneurium.
2. Neuron processes in bundles wrapped in perineurium; bundle termed a fasciculus.
3. Nerve is a group of fasciculi surrounded by connective tissue epineurium.
4. Sensory nerves contain only sensory neurons; transmit to CNS.
5. Motor nerves contain only motor neurons; transmit to PNS.

6. Mixed nerves contain sensory and motor neuron processes; transmit to and from CNS.

SPECIALIZED PERIPHERAL NEURON ENDINGS

MOTOR NEURON ENDINGS

Neuromuscular or Myoneural Junctions terminal branches of axon meet but do not touch skeletal muscles.

Myelin not present at end of axon.

Small Vesicles contain acetylcholine; present in axon endings.

SENSORY NEURON ENDINGS
dendrites function as receptors, of which there are structurally distinct types. Each type of receptor may or may not be responsible for sending just one specific kind of stimulus.

Free Nerve Endings bare dendrites; primarily pain receptors; also touch and temperature.

Encapsulated Sensory Endings surrounded by connective tissue capsules.
1. *Meissner's corpuscles:* elliptical capsules; sensitive to light touch.
2. *Pacinian corpuscles:* concentric layers of connective tissue surround dendrite; sensitive to heavy pressure.
3. *End-bulbs of Krause:* cold receptors.
4. *Ruffini's corpuscles:* oval capsules; heat receptors.
5. *Muscle spindles:* complex capsules; contain thin skeletal muscle fibers called *intrafusal fibers;* respond to stretching.
6. *Neurotendinous organs:* dendrites that ramify in a tendon near its junction with a muscle; regulate tension.

TYPES OF RECEPTORS

CLASSIFICATION ACCORDING TO LOCATION OF STIMULUS

Exteroceptors respond to body surface stimuli, such as touch and light.

Interoceptors (Visceroceptors) respond to pressure, pain, and chemical changes in internal body environment. *Proprioceptors* respond to position of body parts (muscle spindles, neurotendinous organs).

CLASSIFICATION ACCORDING TO TYPE OF STIMULUS

Mechanoreceptors sensitive to physical deformations, such as pressure or stretch (Pacinian corpuscles, muscle spindles, neurotendinous organs).

Thermoreceptors sensitive to temperature changes (Ruffini's corpuscles, end-bulbs of Krause).

Chemoreceptors sensitive to chemicals, taste, smell, pH.

Photoreceptors sensitive to light (retina of eyes).

NEUROGLIA (GLIAL CELLS)
structural support for neurons; nutrient transfer; phagocytosis; insulation of electrical activity. Capable of mitotic division; all are ectodermal in origin except microglia, which are mesodermal. Satellite cells and Schwann cells considered to be peripheral neuroglia.

ASTROCYTES
large star-shaped bodies with many processes.

CNS Structural Support

Blood-Brain Barrier decreases passage of substances other than water, oxygen, and carbon dioxide between capillaries and neurons.

OLIGODENDROCYTES
smaller; form myelin sheath in CNS.

MICROGLIA
smallest; considered part of macrophage system; remove dead tissue and foreign materials from CNS (phagocytic).

EPENDYMAL CELLS
cuboidal, some ciliated; line ventricles of brain and central canal of spinal cord; some in ventricles have microvilli—associated with choroid plexuses.

EFFECTS OF AGING ON THE NERVOUS SYSTEM
Reduction in number of neurons, as they are not replaced when they die; reduction in myelin slows impulse conduction rate; reduction in Golgi apparatus and chromatophilic substance. **p. 367**

SELF-QUIZ

1. There are two systems that serve primarily as means of internal communication—the nervous system and the endocrine system. True or false?

2. The central nervous system includes: (a) ganglia; (b) an autonomic division; (c) the spinal cord.

3. The peripheral nervous system includes: (a) the somatic nervous system; (b) the brain; (c) sensory input centers.

4. Impulses that produce contractions of skeletal muscles are carried by: (a) somatic motor nerves; (b) visceral motor nerves; (c) visceral sensory nerves.

5. Both the somatic and the autonomic divisions are under the control of centers located in the peripheral nervous system. True or false?

6. The brain develops from embryonic: (a) mesoderm; (b) endoderm; (c) ectoderm.

7. Nerve impulses in mixed nerves travel: (a) both to and from the central nervous system; (b) only from the CNS; (c) only to the CNS.

8. Mature neurons are generally able to undergo mitosis. True or false?

9. Clusters of nerve cell bodies in the central nervous system are generally called: (a) ganglia; (b) soma; (c) nuclei.

10. Match the following terms with the appropriate lettered item.

Nuclei
Soma
Center
Ganglia
Dendrite
Axon
Nerve fiber
Sheath of
  Schwann
Myelin

(a) Groups of nerve cell bodies in the peripheral nervous system.
(b) The portion of the neuron in which the nerve impulse is generated.
(c) An axon together with certain sheaths.
(d) Clusters of cell bodies of neurons in the CNS.
(e) A thin membrane between the myelin and the endoneurium of myelinated axons in the PNS.
(f) The name given the cell body of each neuron.
(g) A fatty substance that covers most axons; formed by spiraling of Schwann cell.
(h) A group of nuclei whose neurons all have a specific function.
(i) A neuronal process that conducts a nerve impulse.

11. Axon processes in the peripheral nervous system may have the following associated with them: (a) neurilemma; (b) nodes of Ranvier; (c) both.

12. Schwann cells are not found in the central nervous system. True or false?

13. A threshold stimulus applied to a neuron produces a local depolarization called an action potential. True or false?

14. A nerve impulse is a propagated action potential. True or false?

15. If the chemical transmitter substance released into a synapse is stimulatory, it causes: (a) an inhibitory postsynaptic potential; (b) an excitatory postsynaptic potential; (c) a resting membrane potential.

16. Neuroglia that act as phagocytes in the central nervous system are: (a) oligodendrocytes; (b) microglia; (c) ependymal cells.

17. Which of these neuroglial cell types is *not* derived from embryonic ectoderm? (a) microglia; (b) astrocytes; (c) ependymal cells.

18. The most common type of neurons are those called: (a) unipolar; (b) bipolar; (c) multipolar.

19. These efferent neurons transmit impulses away from the central nervous system to an effector: (a) motor; (b) sensory; (c) interneurons.

20. The terminal portions of the peripheral processes of sensory neurons are dendrites. True or false?

21. Match the following terms with the appropriate lettered item.

Acetylcholine
Motor neuron
  endings
Sensory neuron
  endings
Pacinian
  corpuscles
End-bulbs of
  Krause
Ruffini's
  corpuscles
Muscle spindles
Neurotendinous
  organs
Free nerve
  endings
Meissner's
  corpuscles

(a) Modified dendrites that serve as receptors associated with cutaneous senses.
(b) These receptors are stimulated by heavy pressure.
(c) These are the least modified receptors, consisting of bare dendrites.
(d) Heat receptors.
(e) Form a neuromuscular junction with the muscles.
(f) Cold receptors.
(g) Transmits impulses across the gap of the neuromuscular junction.
(h) These receptors are particularly sensitive to light touch.
(i) Complex capsules in skeletal muscles.
(j) Function in close association with muscle spindles.

22. These receptors provide information concerning the position of body parts without the necessity of observing the parts: (a) exteroceptors; (b) proprioceptors; (c) mechanoreceptors.

23. Ruffini's corpuscles and the end-bulbs of Krause are thought to be: (a) thermoreceptors; (b) chemoreceptors; (c) photoreceptors.

24. Receptors that are sensitive to physical deformation, such as pressure or stretch, are called: (a) mechanoreceptors; (b) chemoreceptors; (c) photoreceptors.

25. The receptors located in the retina of the eyes that convert light energy into nerve impulses are called: (a) mechanoreceptors; (b) chemoreceptors; (c) photoreceptors.

26. Receptors that respond to various stimuli on the body surface are called: (a) interoceptors; (b) proprioceptors; (c) exteroceptors.

27. In contrast to neurons, neuroglia are capable of undergoing mitotic division. True or false?

28. Which type of neuroglia cells are thought to be involved in forming the blood-brain barrier? (a) microglia; (b) astrocytes; (c) oligodendrocytes.

29. Some neurons in the central nervous system have a myelin sheath around them that is formed by: (a) Schwann cells; (b) microglia; (c) oligodendrocytes.

30. The ventricles of the brian and the central canal of the spinal cord are lined with: (a) ependymal cells; (b) microglia; (c) astrocytes.

LEARNING OBJECTIVES

After completing this chapter, you should be able to:

- Distinguish between the mesencephalon, metencephalon, and myelencephalon, and list several activities of each.

- List the principal components of the diencephalon, and give at least one function for each.

- Describe the telencephalon, listing its parts and functional areas.

- Name the types of tracts in the white matter of the brain, and state the function of each.

- Describe the ventricles of the brain and the function and flow of cerebrospinal fluid.

- Distinguish between the white and gray matter of the spinal cord in terms of structure and function.

- Name and describe the ascending and descending tracts of the spinal cord.

- Distinguish between a stretch reflex and a tendon reflex.

- Name and describe the meninges.

- Correlate the symptoms of several dysfunctions of the CNS with the region affected.

CHAPTER CONTENTS

THE BRAIN

THE SPINAL CORD

NEURON POOLS

The Central Nervous System

14

The **central nervous system** consists of the brain and spinal cord, both of which develop from the embryonic neural tube (see Chapter 13). The spinal cord is essentially an extension of the nerve tracts of the brain. Neurons in the spinal cord carry messages to and from the brain, and the spinal cord serves as a center for reflexes. For convenience, the brain and spinal cord are discussed as if they were separate entities, but at the same time their interrelationship is emphasized.

THE BRAIN

The brain is formed by the extensive development of the anterior end of the embryonic neural tube. In Chapter 13 it was described how three enlargements, called the *prosencephalon, mesencephalon,* and *rhombencephalon,* form in the region of the future brain (Figure 13.6). We also saw how these three **F13.6** enlargements divide into five chambers: the prosencephalon divides into the *telencephalon* and *diencephalon;* the *mesencephalon* remains as a single chamber; and the rhombencephalon divides into the *metencephalon* and *myelencephalon.* This chapter discusses the adult structures that develop from these five divisions. Table 14.1 lists some of these structures and names the portion of the **Table** neural tube (canal) that is within each subdivision of the brain. The blood **14.1** supply to the brain is discussed in Chapter 11.

Telencephalon

In the adult, the telencephalon consists of right and left **cerebral hemispheres,** which together are referred to as the **cerebrum** (see Box 14.1). Be- **Box** cause the cerebrum grows so extensively, it completely envelops the dien- **14.1** cephalon and obscures much of the rhombencephalon. The cerebrum has an outer surface of **gray matter,** which is composed primarily of nerve cell bodies and unmyelinated nerve fibers. This surface layer is called the **cerebral cortex.** Deep inside each cerebral hemisphere are several additional structures of gray matter called the **basal ganglia** (or **basal nuclei**) (Figure 14.1). The gray matter **F14.1** of the cortex is separated from the basal ganglia by **white matter,** which is composed of tracts of myelinated nerve fibers. A fluid-filled cavity called a **lateral ventricle** is located within each cerebral hemisphere.

Myelinated Nerve Fiber Tracts

In the central nervous system, bundles of nerve fibers are called **tracts.** There are three types of tracts in the white matter of the cerebrum:

1. **Projection tracts** are pathways formed by **projection fibers.** These fibers carry either descending (motor) nerve impulses from the cerebral cortex to other regions of the brain and spinal cord or ascending (sensory) impulses from the spinal cord and lower regions of the brain (such as the thalamus) to the cerebral cortex.

2. **Association tracts** are pathways formed by **association fibers,** which connect various areas of the cerebral cortex within the

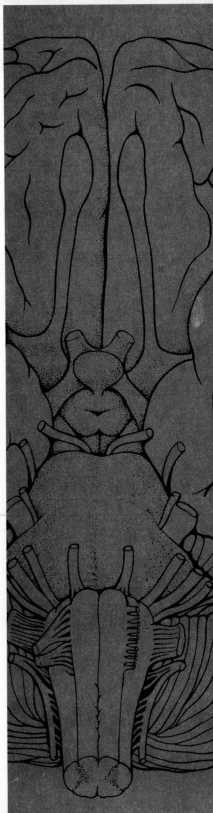

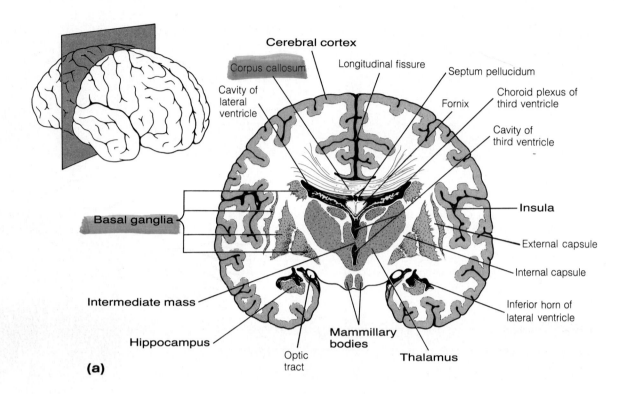

(a)

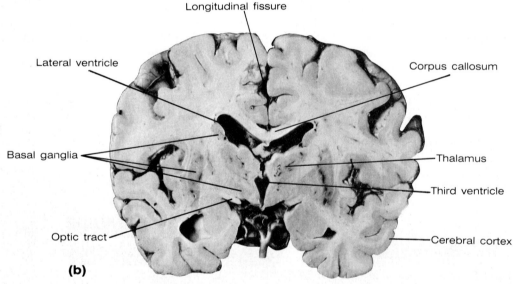

(b)

Figure 14.1
(a) Frontal section of the cerebrum and the diencephalon (color) showing the cerebral cortex (gray matter) surrounding the white matter and, deep within the white matter, the basal ganglia.
(b) Photograph of a frontal section of the brain.

same hemisphere. Association fibers vary in length: some are quite short, whereas others extend the entire length of the hemisphere.

3. **Commissural tracts are pathways formed by commissural fibers.** These fibers connect the left and right cerebral hemispheres. There are two main commissural tracts that connect the cerebral hemispheres: the **anterior commissure** and the large **corpus callosum.**

Rather surprising results are obtained if the corpus callosum is severed, as is sometimes done to provide relief from epileptic convulsions. After recovery, the person appears to function entirely normally. With special testing procedures, however, it is possible to show that following the cutting of the

Table 14.1   Subdivisions of the Neural Tube and the Major Adult Structures Derived from Each

Primary Division	Subdivision	Adult Brain Structures	Neural Canal Region
Prosencephalon (forebrain)	Telencephalon	Cerebral hemispheres (cerebrum)	Lateral ventricles and upper portion of the third ventricle
		Cerebral cortex	
		Basal ganglia	
		Olfactory bulbs and tracts	
	Diencephalon	Epithalamus	Most of the third ventricle
		Thalamus	
		Hypothalamus	
Mesencephalon (midbrain)	Mesencephalon	Corpora quadrigemina	Cerebral aqueduct
		Cerebral peduncles	
Rhombencephalon (hindbrain)	Metencephalon	Cerebellum	Fourth ventricle
		Pons	
	Myelencephalon	Medulla oblongata	Part of the fourth ventricle
Spinal cord	Spinal cord	Spinal cord	Central canal

Box 14.1
CT Scan of the Human Brain

This is a CT scan of the central portion of the human brain (see Box 5.2 for fuller explanation). The thick white area indicates the skull, and the outer thin white area shows the scalp. Gray represents the brain. Symmetric darkened areas represent the ventricular system. The whitened areas within the brain indicate that there is some calcification of the blood vessels, which may result in blood vessel constriction. This is a scan of an older person who may have atherosclerosis, a vascular disease. If these whitish areas of calcified vessels appear in the brain, it is likely they are present throughout the entire body.

CT scans, which are primarily used to determine the presence of tumors, lesions, and vascular disease, reduce the need for exploratory surgery in certain situations. For instance, in the case of a head injury where there may be some concern regarding the possible presence of internal bleeding, a CT scan is an efficient, rapid, and low-risk way of ascertaining whether or not bleeding is present.

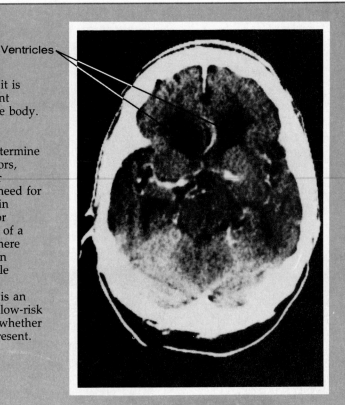

Ventricles

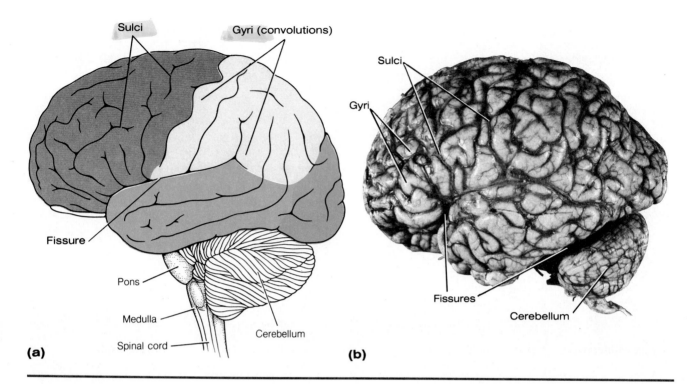

(a)

Sulci

Gyri (convolutions)

Fissure

Pons

Medulla

Spinal cord

Cerebellum

(b)

Sulci

Gyri

Fissures

Cerebellum

Figure 14.2
(a) Lateral view of the surface of the brain.
(b) Photograph. The surface of the cerebral hemisphere has numerous convolutions separated by either sulci or fissures.

commissure, a task learned with one hand cannot be performed by the other hand unless the task is relearned using this hand. A person whose corpus callosum is intact can generally perform a task with either hand, although perhaps not with equal dexterity. Therefore, it appears that the corpus callosum makes possible the transfer of information between cerebral hemispheres. If the commissure is severed, information learned by one cerebral hemisphere is not available to the other.

Gyri, Fissures, Sulci, and Lobes of the Cerebrum

F14.2 The surface of the cerebrum has many rounded ridges called **convolutions,** or **gyri** (singular: *gyrus*) (Figure 14.2). Separating the gyri are furrows. The deeper furrows are called **fissures;** the shallower ones are **sulci** (singular: *sulcus*). The folding of the cortex that produces the gyri and sulci makes the surface area of the cerebral cortex much greater than it would be if the brain's surface were smooth. As it is, a significant percentage of the cerebral cortex is located in the fissures and sulci and is not visible from the surface.

The patterns of the gyri and fissures or sulci vary somewhat from one brain to another. Nevertheless, the locations of certain fissures and sulci are constant enough to serve as surface landmarks by which each hemisphere can
F14.3 be divided into *frontal, parietal, temporal,* and *occipital lobes* (Figure 14.3). Each lobe is located in the same general region as the correspondingly named skull bones.

F14.1 The **longitudinal fissure** (Figure 14.1) is a deep furrow that extends down to the corpus callosum in the central region of the cerebrum. It runs anteriorly and posteriorly, dividing the cerebrum into right and left hemispheres. Each hemisphere is further divided into a **frontal lobe** and a **parietal lobe** by the
F14.3 **central sulcus** *(fissure of Rolando)* (Figure 14.3), which runs at right angles to the longitudinal fissure. Two gyri run parallel to the central sulcus: the one anterior to the sulcus is the **precentral gyrus,** whereas the one that runs posterior to the sulcus is the **postcentral gyrus.** The functional significance of these gyri is explained later.

The parietal lobe is separated posteriorly from the **occipital lobe** by an indistinct **parieto-occipital sulcus.** The **temporal lobes** extend forward along the lateral sides of the cerebral hemispheres. Each temporal lobe is separated from the lower portions of the frontal and parietal lobes by a deep **lateral**
F14.3 **fissure** *(fissure of Sylvius)* (Figure 14.3). The **insula,** a portion of the cerebral

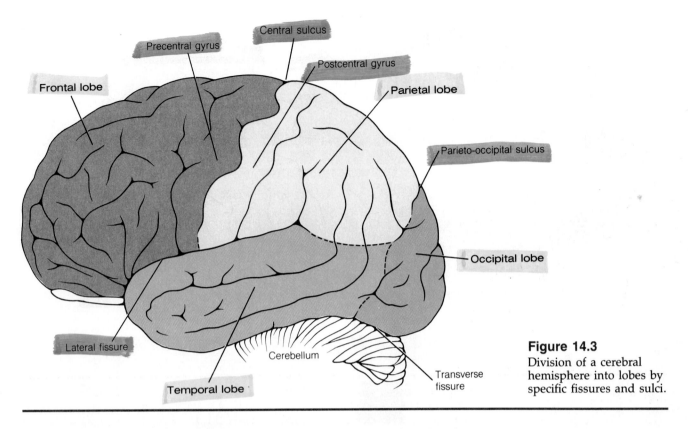

Figure 14.3
Division of a cerebral hemisphere into lobes by specific fissures and sulci.

cortex considered to be a fifth lobe of the cerebrum, is located deep within the lateral fissure (Figure 14.1). The insula is covered by portions of the frontal, parietal, and temporal lobes. The cerebrum is completely separated posteriorly from the cerebellum by a deep **transverse fissure** (Figure 14.3).

F14.1

F14.3

Functional Areas of the Cerebral Cortex

On the basis of the effects of electrical stimulation of specific areas of the cerebral cortex in humans, from observing the clinical manifestations of brain disease or damage in humans, and from the results obtained from detailed experiments on other mammals, it has been determined that certain areas of the cortex are related to specific functions. Some of these areas have been precisely mapped and numbered in a system called the *Brodmann classification*, but for our purposes it is sufficient to consider only the general locations of the major functional areas (Figure 14.4). Keep in mind, however, that Figure 14.4 represents an oversimplification of a very complex organ: no area of the brain functions alone. Because of extensive interconnections between various cortical areas by commissural and association fibers, any function attributed to a specific cortical area actually probably involves several cortical areas.

F14.4

PRIMARY MOTOR AREA   The **primary motor area** is located in the precentral gyrus of the frontal lobe just anterior to the central sulcus. Since the neurons in this gyrus control the conscious and precise voluntary contractions of skeletal muscles, this area is also referred to as the **primary somatic motor area**. The neurons of the primary motor area are distributed in an organized manner: those controlling toe movements are located medially, deep in the longitudinal fissure; those controlling all other body parts are located in a regular but disproportionate sequence laterally along the gyrus (Figure 14.5). Originating in the precentral gyrus are descending motor nerve fiber tracts called **pyramidal tracts** because they form pyramid-shaped structures that are visible on the ventral surface of the medulla.

F14.5

PREMOTOR AREA   Located just anterior to the primary motor area is a region referred to as the **premotor area.** The neurons in the premotor area

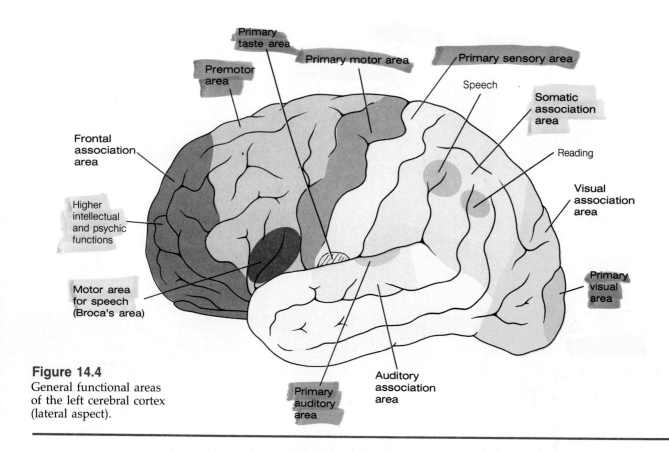

Figure 14.4
General functional areas of the left cerebral cortex (lateral aspect).

cause groups of muscles to contract in a specific sequence, thereby producing stereotyped movements. These repetitive movements are involved in learned activities such as playing a musical instrument and typing. The motor neurons that have their cell bodies located in the premotor area travel within the **extrapyramidal tracts,** which include all motor neurons from the cerebral cortex other than those within the pyramidal tracts.

At the lower margin of the premotor area is a motor area associated with the ability to speak. This region, called **Broca's area,** seems to be located in the left cerebral hemisphere in most individuals.

PRIMARY SENSORY AREA   Located just posterior to the central sulcus in the postcentral gyrus of the parietal lobe is the **primary sensory area.** Within this area are the terminations of the sensory pathways that carry general sensory information concerning temperature, touch, pressure, pain, and proprioception from the body to the cortex of the brain. These general sensations reach the conscious level in the cortex of the primary sensory area—an area also called the **primary somatic sensory area** (or *somesthetic area*). The sensory neurons are organized within the postcentral gyrus in a manner similar to the motor neurons in the precentral gyrus (Figure 14.5).

F14.5

SPECIAL SENSES AREAS   The **primary visual area** is in the posterior portion of the occipital lobe. Located along the upper margin of the temporal lobe is the **primary auditory area,** which receives neural impulses associated with hearing. The area concerned with the sense of smell, the **primary olfactory area,** is located on the medial surface of the temporal lobe. The **primary taste area** is located in the parietal lobe, near the bottom of the postcentral gyrus.

ASSOCIATION AREAS   Surrounding these primary motor and sensory cortical areas are several general association areas containing neurons that interconnect the various motor and sensory areas. The association areas therefore correlate the many activities that occur in the brain.

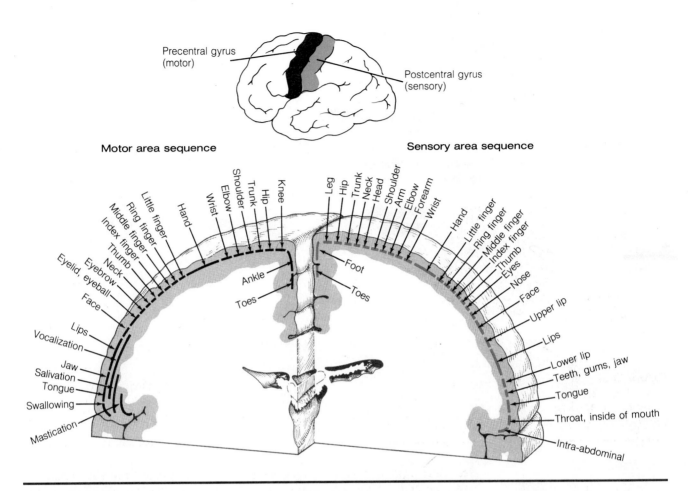

The **frontal association area,** located anterior to the premotor area, is considered to be the site of origin of the higher intellectual activities characteristic of humans. These activities include foresight, the ability to make judgments, and the capacity to select appropriate behavior for a variety of circumstances.

The **somatic association area** is located on the parietal lobe, posterior to the primary sensory area. This center of integration and interpretation makes it possible to determine an object's shape, texture, and orientation without viewing it and provides information concerning the relationships of body parts to one another.

Located posterior to the somatic association area is a **visual association area,** and in the temporal lobe is an **auditory association area.** These areas contribute to the interpretation of visual and auditory experiences.

Basal Ganglia

Located deep within each cerebral hemisphere are several masses of gray matter known collectively as the **basal ganglia** (basal nuclei) (Figure 14.6). The **F14.6** ganglia, which are surrounded by white matter, are composed of groups of nerve cell bodies. Included in the basal ganglia are: the long, arching **caudate nucleus;** the **amygdaloid nucleus,** which is located at the tip of the tail of the caudate nucleus; the **lentiform nucleus,** which is subdivided into the **putamen** and the **globus pallidus;** and the **claustrum,** a thin layer of gray matter just deep to the cortex of the insula (Figure 14.7). The band of white matter **F14.7** located between the basal ganglia and the thalamus is called the **internal capsule.** The internal capsule is composed of projection fibers of the major motor and sensory tracts as they pass to and from the cerebral cortex. Because of their appearance, the caudate nucleus, the internal capsule, and the lentiform nucleus are sometimes referred to as the **corpus striatum** ("striped body").

Figure 14.5
Frontal section of the cerebrum. *Left half:* through the precentral gyrus, showing the locations of neurons within the cerebral cortex that control voluntary motor movement of specific structures. *Right half:* through the postcentral gyrus, showing the locations of regions of the cerebral cortex that receive sensory nerve impulses from specific body structures.

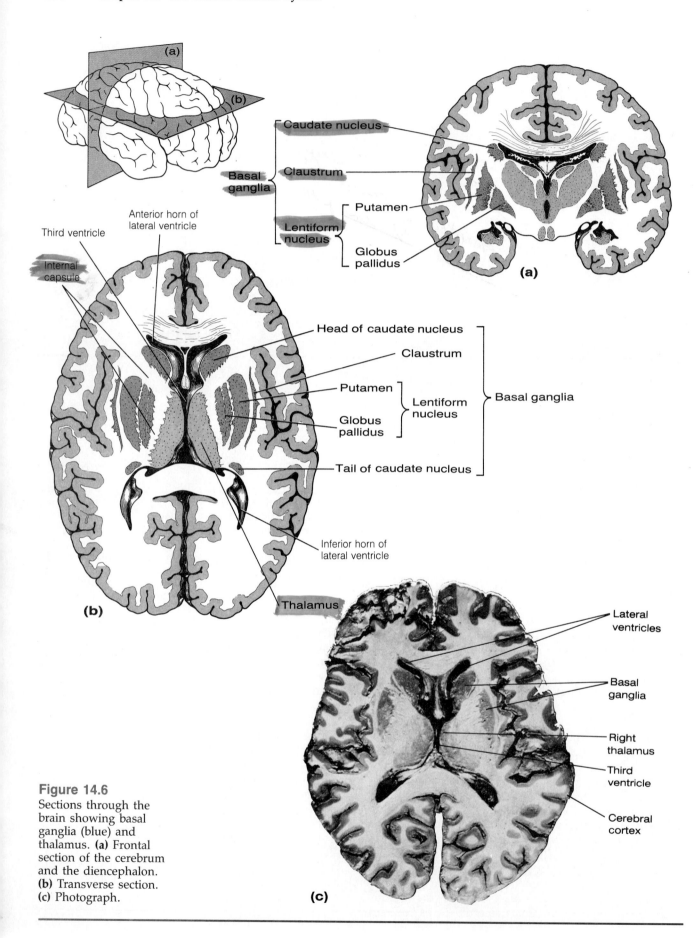

Figure 14.6
Sections through the brain showing basal ganglia (blue) and thalamus. **(a)** Frontal section of the cerebrum and the diencephalon. **(b)** Transverse section. **(c)** Photograph.

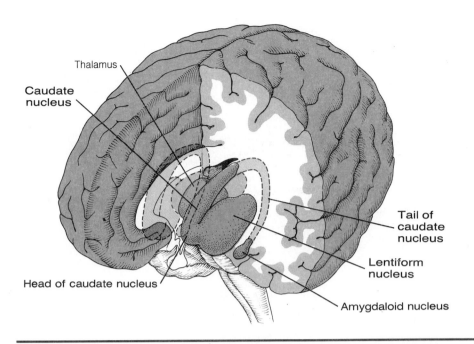

Thalamus

Caudate nucleus

Head of caudate nucleus

Tail of caudate nucleus

Lentiform nucleus

Amygdaloid nucleus

Figure 14.7
Three-dimensional relationship among the structures that comprise the basal ganglia.

The basal ganglia, like the neurons of the precentral gyrus, are involved in somatic motor functions. Because they are located outside the precentral gyrus, the basal ganglia are part of the *extrapyramidal system*. In other words, the somatic motor activities of the body are controlled both by the pyramidal tracts that originate in the cerebral cortex and by motor neurons located elsewhere in the brain (extrapyramidal system), including the basal ganglia and the premotor area. In contrast to the pyramidal tracts, however, the neurons of the basal ganglia seem to *inhibit* motor function. This inhibition, together with the stimulatory effects of the pyramidal system, provides a means by which muscular movements can be precisely controlled. Disorders of the basal ganglia result in involuntary contractions of skeletal muscles such as the muscular rigidity and persistent tremors of the limbs associated with Parkinson's disease.

Olfactory Bulbs

On the ventral surface of each cerebral hemisphere is a small **olfactory bulb** and its associated **olfactory tract** (Figure 14.8). These structures, which are **F14.8** associated with the sense of smell, are located in the portion of the brain called the *rhinencephalon*. The neurons of the **olfactory nerve (cranial nerve I)** pass from the nasal mucosa, through the cribriform plate of the ethmoid bone, and into the olfactory bulb, where they synapse with neurons of the olfactory tract. The neurons of the tracts pass to the olfactory area of the cortex on the medial surface of the temporal lobe. In addition to its olfactory function, the rhinencephalon is thought to be involved in some obscure manner with certain emotional and behavioral responses. Those portions of the rhinencephalon that are not involved with olfaction are considered to be part of the *limbic system*, which we will consider shortly.

Diencephalon

The second subdivision of the forebrain is the **diencephalon.** Since the cerebral hemispheres extend downward and almost completely surround the diencephalon, it is not visible from the exterior of the brain, except for a portion that can be seen when the brain is viewed ventrally. The **third ventricle** forms a midplane cavity within the diencephalon (Figure 14.9). The most important **F14.9** parts of the diencephalon are the *thalamus, hypothalamus,* and *epithalamus.*

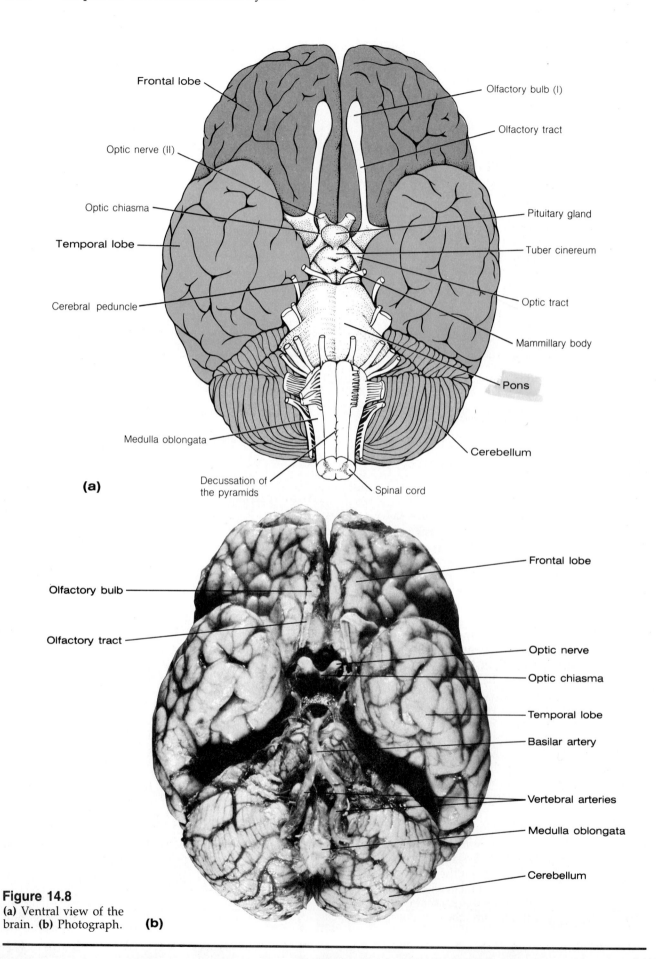

Frontal lobe

Olfactory bulb (I)

Olfactory tract

Optic nerve (II)

Optic chiasma

Pituitary gland

Temporal lobe

Tuber cinereum

Optic tract

Cerebral peduncle

Mammillary body

Pons

Medulla oblongata

Cerebellum

Decussation of the pyramids

Spinal cord

(a)

Olfactory bulb

Frontal lobe

Olfactory tract

Optic nerve

Optic chiasma

Temporal lobe

Basilar artery

Vertebral arteries

Medulla oblongata

Cerebellum

Figure 14.8
(a) Ventral view of the brain. (b) Photograph.   **(b)**

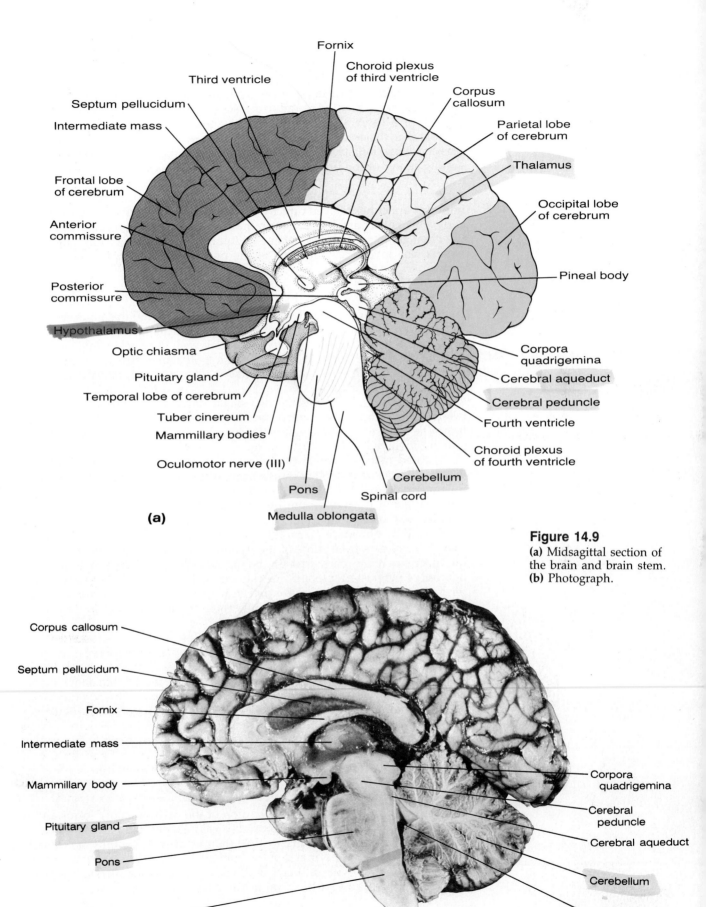

Fornix

Third ventricle

Choroid plexus
of third ventricle

Septum pellucidum

Corpus
callosum

Intermediate mass

Parietal lobe
of cerebrum

Thalamus

Frontal lobe
of cerebrum

Occipital lobe
of cerebrum

Anterior
commissure

Pineal body

Posterior
commissure

Hypothalamus

Corpora
quadrigemina

Optic chiasma

Cerebral aqueduct

Pituitary gland

Cerebral peduncle

Temporal lobe of cerebrum

Fourth ventricle

Tuber cinereum

Mammillary bodies

Choroid plexus
of fourth ventricle

Oculomotor nerve (III)

Cerebellum

Pons

Spinal cord

Medulla oblongata

(a)

Figure 14.9
(a) Midsagittal section of
the brain and brain stem.
(b) Photograph.

Corpus callosum

Septum pellucidum

Fornix

Intermediate mass

Corpora
quadrigemina

Mammillary body

Cerebral
peduncle

Pituitary gland

Cerebral aqueduct

Pons

Cerebellum

Medulla oblongata

(b)

Fourth ventricle

Thalamus

F14.6

The **thalamus** consists of two oval masses of nerve-cell bodies (gray matter) that form the lateral walls of the third ventricle (Figure 14.6). A small bridge called the **intermediate mass** (*massa intermedia*) passes across the third ventricle and connects the two thalamic masses. Each thalamic mass is deeply embedded in a cerebral hemisphere and is bounded laterally by the internal capsule. The thalamus contains over 20 functionally separate nuclei (cell-body masses). Functionally, the thalamus acts as a major sensory relay and integrating center of the brain. Except for the tracts associated with olfaction, all sensory fiber tracts that travel to the conscious perceptive areas in the cerebral cortex synapse within one of the thalamic nuclei. From the thalamus, the impulses are either relayed directly to a specific sensory region of the cerebral cortex or to other areas of the brain that serve as association centers (such as the basal ganglia and hypothalamus). Apart from its sensory role, the thalamus is also involved with some of the motor tracts that leave the cerebral cortex.

Hypothalamus

F14.10

As the name indicates, the **hypothalamus** lies below the thalamus, where it forms part of the walls and floor of the third ventricle. Like the thalamus, the hypothalamus is composed of several nuclei, each of which is involved with specific functions (Figure 14.10). Several hypothalamic structures are visible externally, including the *mammillary bodies, tuber cinereum, infundibulum,* and the *optic chiasma* (or *chiasm*).

The **mammillary bodies** are two small, round nuclear masses that form external bulges from the undersurface of the brain posterior to the infundibulum. The mammillary bodies function as relay stations for olfactory neurons and are involved in olfactory reflexes.

Located just anterior to the mammillary bodies is the **tuber cinereum,** which contains neurons that transport regulatory hormones (or factors) from the hypothalamus, through the infundibulum, to the adenohypophysis of the pituitary gland by way of the hypophyseal portal veins (Chapter 18). The neurons of the tuber cinereum form the *tuberohypophyseal tract*.

Extending downward from the tuber cinereum is the stalklike **infundibulum.** Nerve fibers from some of the hypothalamic nuclei, as well as those of the tuber cinereum, pass through the infundibulum on their way to the pars nervosa of the posterior lobe of the pituitary gland. These nerve fibers, which form the *hypothalamic-hypophyseal tract*, transport various hormones (oxytocin and ADH) synthesized in hypothalamic nuclei to the posterior pituitary, from which they are released.

Anterior to the infundibulum is the **optic chiasma,** which is formed by the decussation (crossing) of some of the neurons in the optic nerves.

The hypothalamus controls many vital processes, most of them associated with the autonomic nervous system. Some of the hypothalamic nuclei have been shown experimentally to regulate sympathetic activity; others control parasympathetic functions. The hypothalamus is involved in regulating body temperature, water balance, appetite, gastrointestinal activity, sexual activity, and even emotions such as fear and rage. The hypothalamus also regulates the release of the hormones of the pituitary gland and thus, to a large extent, it controls the endocrine system (see Chapter 18).

Epithalamus

The **epithalamus,** the most dorsal portion of the diencephalon, forms a thin roof over the third ventricle. The roof has a vascular choroid plexus located on its internal surface. A small mass called the **pineal body (epiphysis)** extends outward from the posterior end of the epithalamus. The possible neuroendocrine function of the pineal body is discussed in Chapter 18. The **posterior commissure** is located just ventral to the pineal body.

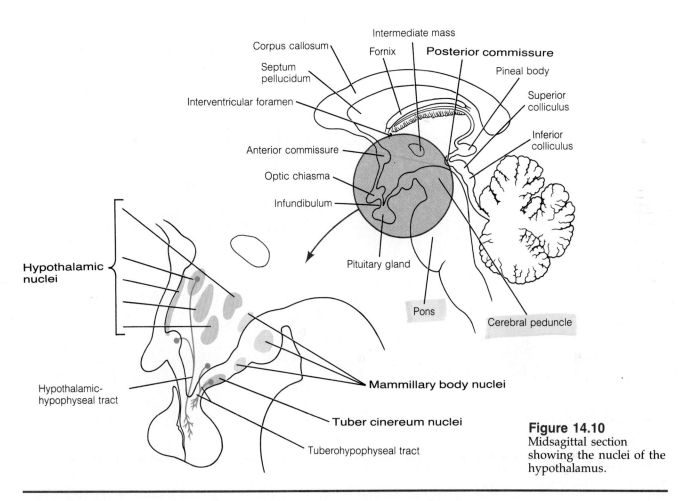

Figure 14.10
Midsagittal section showing the nuclei of the hypothalamus.

The Limbic System

Although the control of emotions is influenced by the hypothalamus, such reactions involve a complex interaction of structures in several different regions of the brain, including the cerebrum and the diencephalon. A group of structures collectively referred to as the **limbic system** play a major role in emotional responses (Figure 14.11). The limbic system includes the **olfactory bulbs;** a band of fibers called the **fornix,** which passes from beneath the corpus callosum to the mammillary bodies of the hypothalamus; a cerebral gyrus called the **cingulate gyrus,** which is located just above the corpus callosum; the **amygdaloid nucleus** of the basal ganglia; the **hippocampus,** which is a part of the cerebrum located in the floor of the lateral ventricle close to the amygdaloid nucleus; the **mammillary bodies;** and various **thalamic** and **hypothalamic nuclei.** F14.11

In animal experiments, it is possible to locate centers within the limbic system that respond to electrical stimulation as if it were pleasant ("pleasure centers") and other centers that respond to the stimulation as if it were unpleasant ("punishment centers"). Under experimental conditions, a wide variety of emotional behavior patterns have been produced by stimulating or removing specific regions of the limbic system. It is assumed therefore that interactions among the structures of the limbic system regulate emotional behavior.

Mesencephalon

The **mesencephalon (midbrain)** is a short, constricted region between the forebrain and the hindbrain. Within the mesencephalon is a small tunnel

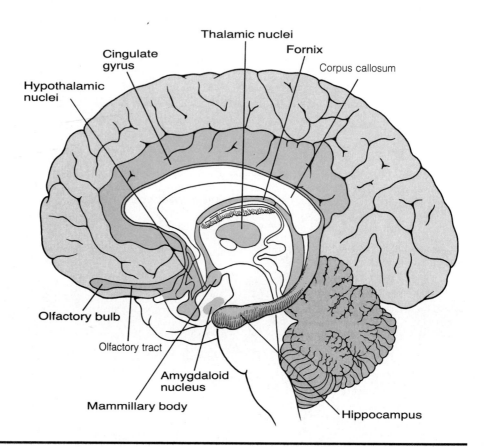

Figure 14.11
The structures that constitute the limbic system.

F14.14 called the **cerebral aqueduct** *(aqueduct of Sylvius)* (Figure 14.14), which connects the third ventricle of the diencephalon with the fourth ventricle of the metencephalon.

Cerebral Peduncles

On the ventral surface of the mesencephalon are two cylindrical bulges called
F14.12 the **cerebral peduncles** (Figure 14.12). The peduncles are composed of motor nerve fibers that travel from the primary motor area of the cerebral cortex to the pons and spinal cord and sensory nerve fibers that travel from the spinal cord to the thalamus. The **oculomotor nerves (cranial nerve III)** emerge between the peduncles. Deeper within the mesencephalon, between the peduncles and cerebral aqueduct, is a small island of gray matter called the **red nucleus.** The cell bodies of neurons that compose the *rubrospinal tract* (discussed later in the chapter) are located in the red nucleus. The red nucleus serves as a relay station by coordinating impulses between the cerebellum and the cerebral hemispheres, thereby contributing to the coordination of movements and to the sense of balance.

Corpora Quadrigemina

The dorsal surface of the mesencephalon, which forms the roof of the cerebral aqueduct, consists of four rounded prominences called the **corpora**
F14.12 **quadrigemina** (four twin bodies) (Figure 14.12). The upper pair of prominences are called the **superior colliculi.** Some neurons of the optic tracts from the retina of the eyes travel to the superior colliculi, where they participate in activities concerned with certain reflex responses to visual stimuli. The lower pair of prominences, the **inferior colliculi,** serve as relay stations and reflex centers for auditory stimuli. The **trochlear nerves (cranial nerve IV)** emerge from the roof of the mesencephalon just below the inferior colliculi.

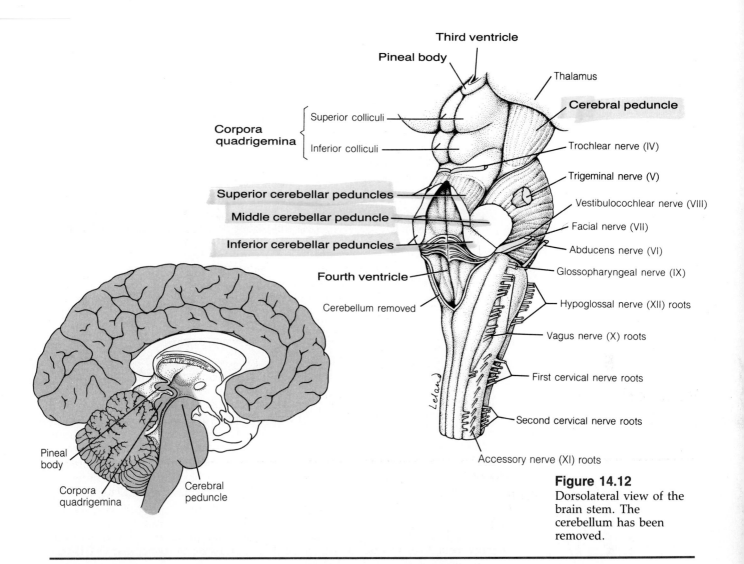

Figure 14.12
Dorsolateral view of the brain stem. The cerebellum has been removed.

Metencephalon

The major structures of the **metencephalon,** which is the most superior portion of the hindbrain, are the *cerebellum* and the *pons*. The cerebral aqueduct of the mesencephalon expands into the **fourth ventricle** in the metencephalon. The inferior portion of the fourth ventricle extends into the myelencephalon. As is true of all the ventricles of the brain, there is a vascular choroid plexus in the thin membrane that forms the roof of the fourth ventricle.

Cerebellum

Projecting from the dorsal surface of the metencephalon, the **cerebellum** (Figure 14.8, Figure 14.9) is separated from the cerebral hemispheres by a strong **F14.8,** membrane called the *tentorium cerebelli*. The tentorium lies within the trans- **F14.9** verse fissure of the brain and supports the occipital lobes of the cerebrum, thus minimizing the pressure that the lobes exert on the cerebellum.

The cerebellum is composed of two lateral **cerebellar hemispheres** connected in the midline by a structure called the **vermis.** The surface of the cerebellum consists of a thin cortex of gray matter. The cortex dips deeply below the apparent surface of the cerebellum in a manner similar to the fissures and sulci of the cerebrum, although the cerebellar fissures are more parallel, giving the appearance of a series of flattened plates. The ridges between the fissures are called **folia.**

The cerebellum is connected to the mesencephalon by a pair of nerve-fiber tracts called the **superior cerebellar peduncles;** to the pons by a pair of **middle cerebellar peduncles;** and to the medulla oblongata by a pair of **inferior cerebellar peduncles** (Figure 14.12). The superior cerebellar peduncles are composed principally of efferent nerve fibers from the cerebellum; the middle and inferior cerebellar peduncles are composed mostly of afferent nerve fibers that transmit impulses from the pons and the medulla oblongata to the cerebellum. These extensive interconnections with other regions of the central nervous system provide the cerebellum with widespread input and output capabilities.

The cerebellum coordinates the activities of the skeletal muscles through sensory information carried to it from receptors for proprioception, equilibrium, and balance. Moreover, the cerebellum receives some sensory information concerning touch, vision, and sound. Further coordination occurs by way of motor impulses sent from the cerebellum to higher brain centers. In particular, nerve impulses from the cerebellum may dictate specific movement sequences to the primary motor area of the cerebral cortex, which then carries them out. A person whose cerebellum has been damaged experiences muscular weakness, a loss of muscular tone, and uncoordinated movements. All the functions with which the cerebellum is concerned remain below the level of consciousness. Thus, the cerebellum is able to mediate certain responses without having them reach the conscious level.

Pons

The **pons** (bridge), which is located on the ventral surface of the metencephalon, consists of bands of nerve-fiber tracts and several nuclei (Figure 14.8, Figure 14.9). The tracts in the pons are both transverse and longitudinal. The transverse tracts consist of neurons that enter the cerebellar hemispheres through the middle cerebellar peduncles. The longitudinal tracts are composed of neurons that travel between the brain stem (see the next section) and the cerebrum. The pons, therefore, functions primarily to connect the cerebellum with the cerebrum and the brain stem, thus providing connections between upper and lower levels of the central nervous system. The cerebral cortex, in particular, achieves most of its connections to the cerebellum by way of nerve-fiber tracts that pass through the pons. In addition, the stimulation of nuclei within the pons affects the rate of respiration. The nuclei of the **trigeminal (V), abducens (VI), facial (VII),** and **vestibulocochlear (VIII)** cranial nerves are located in the pons.

Myelencephalon

The **myelencephalon,** the most inferior division of the brain, is also known as the **medulla oblongata.** The medulla, the pons, and the mesencephalon together form the **brain stem.** Caudally, the medulla is continuous with the spinal cord. The cavity in the medulla forms the lower portion of the fourth ventricle and continues into the spinal cord as the central canal of the cord.

On the ventral surface of the medulla are two large columns of nerve-fiber tracts called the **pyramids.** The pyramids contain the same motor tracts that are found in the cerebral peduncles. Therefore, the tracts in the pyramids carry the voluntary motor output from the primary motor area of the cerebral cortex. The tracts of the pyramids continue into the spinal cord as the *corticospinal tracts* (discussed later in this chapter). Some of the tracts in the pyramids cross from one pyramid to another. This crossing, called the *decussation of the pyramids,* is visible on the ventral surface of the medulla in the groove that separates the pyramids (Figure 14.8). As a consequence of the decussation of these nerve tracts, motor areas located on one side of the cerebral cortex can control muscular movement on the opposite side of the body.

The medulla oblongata also contains nuclei that give rise to the last four cranial nerves: the **glossopharyngeal (IX),** the **vagus (X),** the cranial portion of the **accessory (XI),** and the **hypoglossal (XII).**

Reticular Formation

Inside the medulla is a region of gray matter containing a network of interlacing nerve fibers called the **reticular formation.** The reticular formation extends throughout the brain stem and up into the diencephalon. In addition to receiving nerve impulses from the cerebellum, from the basal ganglia, and from various other nuclei in the brain, the reticular formation also receives input from all the sensory tracts as they ascend through the medulla. Selected impulses that pass through the reticular formation are relayed to the cerebral cortex and activate it (Figure 14.13). Because it exerts this control over the cerebral cortex, the reticular formation is considered to be an activating or arousal system that is essential in maintaining wakefulness and alertness. For this reason, it is also referred to as the **reticular activating system.** Injury or diseases that affect the reticular activating system often produce coma.

Located in the reticular formation of the medulla are the *medullary centers,* which are groups of neurons involved in the control of a variety of vital functions, such as heart rate, respiration, dilation and constriction of blood vessels, coughing, swallowing, and vomiting.

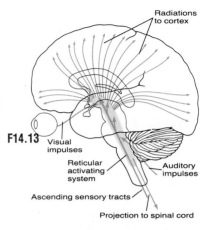

F14.13

Figure 14.13
Schematic representation of the reticular formation. The arrows indicate input to and output from the reticular activating system.

Ventricles of the Brain

The ventricles of the brain (Figure 14.14) develop as expansions of the lumen of the anterior region of the embryonic neural tube and form a continuous fluid-filled system in the brain. Their early development is discussed in Chapter 13 and illustrated in Figure 13.7*a.* The roof of each ventricle is thin and contains no neurons. Each ventricle has a network of capillaries—called a **choroid plexus**—associated with it. These plexuses, together with the ependymal cells that cover them, are the sites of production of **cerebrospinal fluid.** The fluid fills the ventricles of the brain, the central canal of the spinal cord, and the subarachnoid space (Figure 14.15).   **F14.14**

If air is injected into the ventricles, they become distinguishable on an X ray. This procedure is used to detect the presence of tumors or brain damage, both of which distort the normal outlines of the ventricles.   **F14.15**

Lateral Ventricles

Within each cerebral hemisphere is a **lateral ventricle** that has its major portion located in the parietal lobe. Extensions from this portion protrude into the frontal lobe *(anterior horn),* the occipital lobe *(posterior horn),* and the temporal lobe *(inferior horn).* The lateral ventricles are separated from each other medially by a thin vertical partition called the **septum pellucidum** (Figure 14.9). Each lateral ventricle communicates with the third ventricle by a small opening called the **interventricular foramen** *(foramen of Monro).*   **F14.9**

Third Ventricle

The **third ventricle** is a narrow midline chamber in the diencephalon. The right and left masses of the thalamus form most of its lateral walls. A commissure called the *intermediate mass* passes through the ventricle. The third ventricle opens into the fourth ventricle by means of the **cerebral aqueduct** of the mesencephalon.

Fourth Ventricle

The **fourth ventricle** is a pyramidal cavity located in the hindbrain just ventral to the cerebellum. There are two openings in the lateral walls of the fourth ventricle that are called the **lateral apertures** *(foramina of Luschka)* (Figure 14.15). In the roof is a single opening, the **median aperture** *(foramen of Magendie).* The ventricles communicate through these three openings with a *subarachnoid space* that surrounds the brain and spinal cord. Inferiorly, the fourth ventricle is continuous with the narrow **central canal** that extends the length of the spinal cord.   **F14.15**

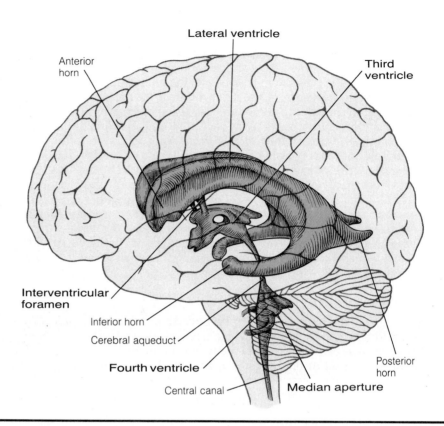

Figure 14.14
Ventricles of the brain viewed as if they could be seen from the surface of the brain.

The Meninges

The entire central nervous system is covered by three layers of connective tissue called the **meninges** (singular: *meninx*). The meninges are composed of the *dura mater*, the *arachnoid*, and the *pia mater*.

Dura Mater

F14.15, F14.16 The **dura mater** is the outermost meninx. It is a strong membrane composed of fibrous connective tissue. Around the brain, the dura mater is a double-layered structure (Figure 14.15, Figure 14.16). The outer layer of the dura mater adheres closely to the bones of the skull, serving as the periosteum of the cranial bones. The inner layer of the dura mater is continuous with the dura mater of the spinal cord. The two layers of the dura that surround the brain are fused together over most of the brain. In certain regions, however, the layers are separated, forming venous sinuses (*dural sinuses*) that carry blood to the internal jugular veins of the neck. The inner layer of the dura dips into the longitudinal fissure that separates the cerebral hemispheres, forming a strong septum called the **falx cerebri,** which is anchored anteriorly to the crista galli of the ethmoid bone. The inner layer of the dura also forms the **falx cerebelli,** which is a strong septum located between the cerebellar hemispheres. Another extension of the dura mater passes transversely in the fissure that separates the cerebrum and the cerebellum, where it forms a septum called the **tentorium cerebelli.** All of these dural extensions anchor the brain to the inside of the cranial cavity.

Arachnoid

F14.16 The middle of the three meninges is the **arachnoid** meninx (Figure 14.16), which is located deep to the dura mater. The arachnoid, a delicate membrane, is closely adherent to the inner surface of the dura mater, with only a very narrow **subdural space** separating the two membranes. Between the arachnoid and the deepest meninx, the pia mater, is the **subarachnoid space.** The subarachnoid and subdural spaces contain cerebrospinal fluid. The subarachnoid space is bridged by weblike strands of the arachnoid.

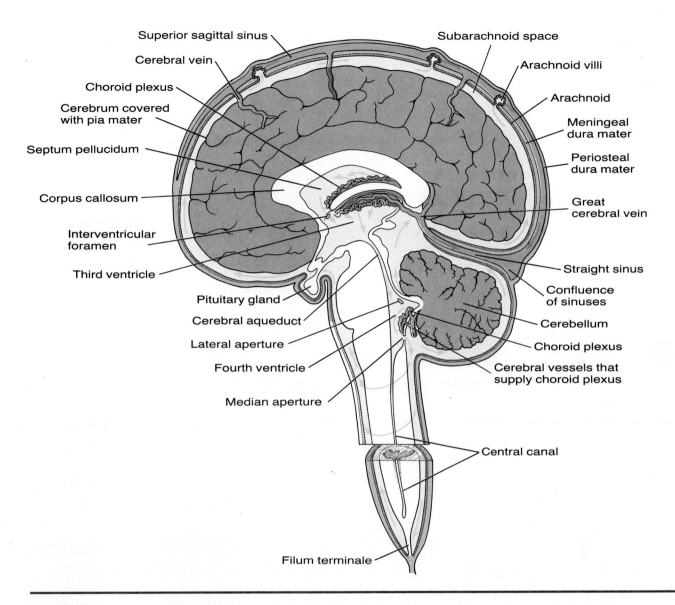

Superior sagittal sinus

Cerebral vein

Choroid plexus

Cerebrum covered with pia mater

Septum pellucidum

Corpus callosum

Interventricular foramen

Third ventricle

Pituitary gland

Cerebral aqueduct

Lateral aperture

Fourth ventricle

Median aperture

Filum terminale

Subarachnoid space

Arachnoid villi

Arachnoid

Meningeal dura mater

Periosteal dura mater

Great cerebral vein

Straight sinus

Confluence of sinuses

Cerebellum

Choroid plexus

Cerebral vessels that supply choroid plexus

Central canal

Figure 14.15
Location of the cerebrospinal fluid (blue) that surrounds the brain and spinal cord. The arrows indicate the direction of the fluid's flow. Blood is shown in orange.

Pia Mater

The innermost meninx is the **pia mater.** A delicate vascular membrane of loose connective tissue, the pia mater adheres closely to the brain and spinal cord, dipping deeply into the fissures and the sulci. In the ventricles, the pia mater plus associated ependymal cells become modified and contribute to the formation of cerebrospinal fluid by the choroid plexuses.

Cerebrospinal Fluid

Cerebrospinal fluid is a watery fluid with a composition similar to that of blood plasma and interstitial fluid. It serves as a cushion for the entire central nervous system, protecting the soft tissue from jolts and blows. Besides filling the ventricles of the brain, the fluid also surrounds the brain and spinal cord, so the central nervous system actually floats in the fluid and is effectively lightened by it. Cerebrospinal fluid is secreted into the ventricles by the choroid plexuses.

There is normally a slight pressure in the ventricles, and the cerebrospinal fluid circulates slowly from the lateral ventricles into the third and then the fourth ventricle. From the fourth ventricle, some cerebrospinal fluid flows into the central canal of the spinal cord, but most of it passes through the

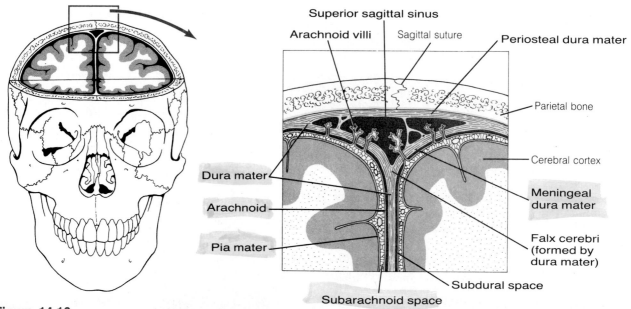

Figure 14.16
Frontal section showing the relationships of the dural venous sinuses, meninges, and the subarachnoid space.

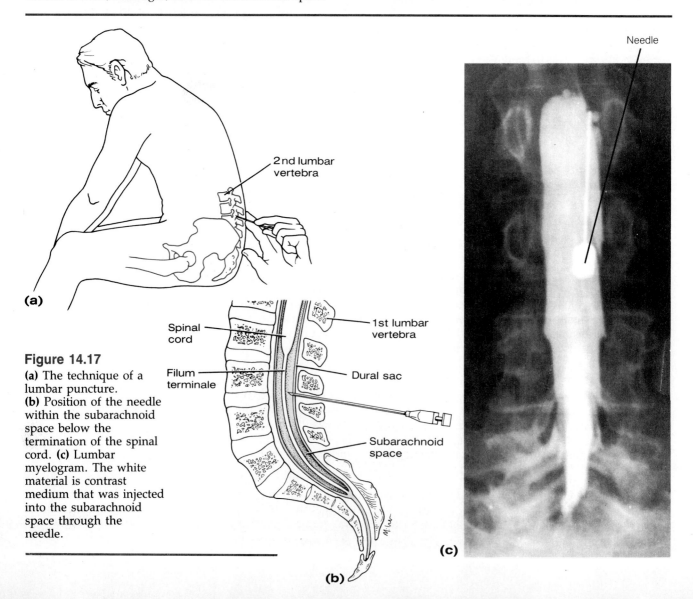

Figure 14.17
(a) The technique of a lumbar puncture. **(b)** Position of the needle within the subarachnoid space below the termination of the spinal cord. **(c)** Lumbar myelogram. The white material is contrast medium that was injected into the subarachnoid space through the needle.

apertures (one median; two lateral) in the roof of the fourth ventricle and enters the subarachnoid space of the meninges. In the subarachnoid space, the cerebrospinal fluid circulates slowly down the posterior surface of the spinal cord, around the cord, and ascends in the anterior portion of the space to reach the brain (Figure 14.15).   **F14.15**

If cerebrospinal fluid were allowed to accumulate, it would exert enough pressure to compress and damage the brain. If the circulation of cerebrospinal fluid is blocked during infancy, before the skull bones have united firmly, the head enlarges as the pressure within the brain increases. This condition is called *hydrocephalus*. Normally, however, cerebrospinal fluid is reabsorbed into the blood at the same rate that it is formed. This reabsorption is accomplished through thin projections of the arachnoid meninx called **arachnoid villi,** which project into the largest dural venous sinuses of the skull (Figure 14.16). Cerebrospinal fluid is therefore formed from the blood and, after circu-   **F14.16** lating through and around the central nervous system, it returns to the blood.

Because of the cerebrospinal fluid's intimate relationship with the central nervous system, examination of cerebrospinal fluid provides a means of determining the presence of infectious agents in the CNS. Samples of cerebrospinal fluid are withdrawn for diagnostic purposes by inserting a needle between the third and fourth lumbar vertebrae into the subarachnoid space—a procedure called a **lumbar puncture** (Figure 14.17). By inserting the needle at   **F14.17** this level, there is little danger of damaging the spinal cord, which ends at the level of the first or second lumbar vertebra. Spinal anesthesias *(spinal blocks)* are sometimes administered in a similar manner. To identify damaged intervertebral discs or the presence of other structures (such as tumors) that might cause pressure on the spinal cord, contrast medium that appears opaque on an X ray is sometimes injected into the subarachnoid space. The X ray taken after the injection of the contrast medium is referred to as a *lumbar myelogram* (Figure 14.17c). Any obstruction of the contrast medium may be due to con-   **F14.17c** striction of the subarachnoid space by a ruptured intervertebral disc or by a tumor.

Brain Functioning

The brain is an extremely complex organ, and even after years of intensive study only enticing bits of understanding of brain activity have been achieved. The thought processes involved in concept formation, abstract reasoning, learning, memory, and so forth have largely proved elusive to investigators. Since so little of brain activity is really understood, it is important to stress that no area or structure of the brain seems to act entirely on its own. The removal of portions of the brain or the severing of tracts in the brain— both of which have been done experimentally in other animals or as the result of trauma in humans—reveal that the most complex higher functions of the brain are generally whole-brain functions. In many cases, the results show that more than one area can control a specific function. This means that each area of the brain is probably involved in many functions. Even those functions that occur automatically—that is, below the conscious level—are generally the result of input from several different sources. Thus, the functions we have attributed to particular regions or structures of the brain are meant to serve only as general references; no attempt has been made to describe the areas of overlap.

THE SPINAL CORD

Below the level of the medulla, the central nervous system continues as the **spinal cord.** The spinal cord performs two main functions: (1) it *conducts* nerve impulses to and from the brain; (2) it *processes* sensory information in a limited manner, making it possible for the cord to initiate stereotyped reflex actions (called *spinal reflexes*) without input from higher centers in the brain.

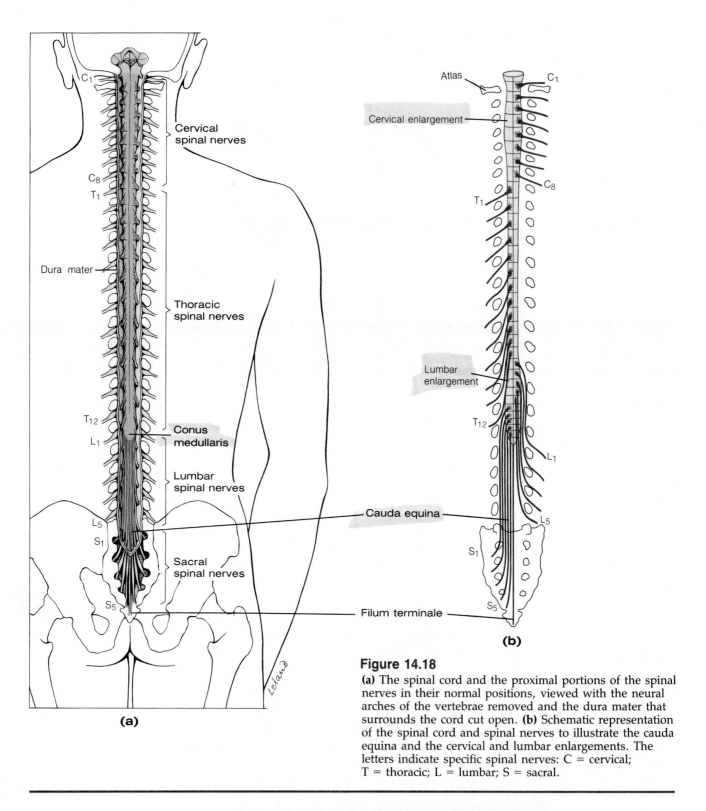

(a)

(b)

Figure 14.18

(a) The spinal cord and the proximal portions of the spinal nerves in their normal positions, viewed with the neural arches of the vertebrae removed and the dura mater that surrounds the cord cut open. (b) Schematic representation of the spinal cord and spinal nerves to illustrate the cauda equina and the cervical and lumbar enlargements. The letters indicate specific spinal nerves: C = cervical; T = thoracic; L = lumbar; S = sacral.

General Structure of the Spinal Cord

F14.18a The spinal cord passes through the vertebral canal of the vertebrae. It extends from the foramen magnum of the skull to the level of the first or second lumbar vertebra (Figure 14.18a). Until the third month of fetal development, the spinal cord is as long as the vertebral column. As the embryo continues to develop, the vertebral column grows at a faster rate than the spinal cord. As a result, the spinal cord does not extend the entire length of the vertebral col-

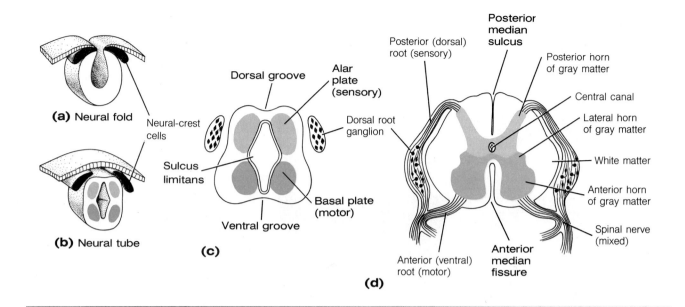

(a) Neural fold

Neural-crest cells

(b) Neural tube

Dorsal groove

Alar plate (sensory)

Dorsal root ganglion

Sulcus limitans

Basal plate (motor)

Ventral groove

(c)

Posterior median sulcus

Posterior (dorsal) root (sensory)

Posterior horn of gray matter

Central canal

Lateral horn of gray matter

White matter

Anterior horn of gray matter

Spinal nerve (mixed)

Anterior (ventral) root (motor)

Anterior median fissure

(d)

Figure 14.19
Development of alar and basal plates within the spinal cord. **(a)** Neural fold stage. **(b)** Neural tube stage. **(c)** Separation of alar and basal plates by the sulcus limitans. **(d)** Fully developed spinal cord with a pair of spinal nerves.

umn in the adult. A thin fibrous filament of the spinal meninges called the **filum terminale** extends from the tip of the spinal cord (the **conus medullaris**) to the coccyx.

Thirty-one pairs of spinal nerves arise from the spinal cord and pass through the intervertebral foramina between adjacent vertebrae of the vertebral column. The spinal nerves that leave the vertebral column between the cervical vertebrae are called *cervical nerves;* those that leave the vertebral column between the thoracic vertebrae are called *thoracic nerves.* Similarly, *lumbar, sacral,* and *coccygeal* nerves leave the vertebral column between the lumbar, sacral, and coccygeal vertebrae, respectively. Each portion of the cord that gives rise to a pair of spinal nerves is called a *spinal segment.* Because the vertebral column grows at a faster rate than the spinal cord, the spinal nerves are pulled downward as the column lengthens. As a result, the roots of the lower spinal nerves pass some distance inferiorly before reaching the appropriate intervertebral foramina (Figure 14.18*b*). At the end of the spinal cord, **F14.18b** the mass of descending lumbar and sacral nerve roots has the appearance of a horse's tail and is therefore called the **cauda equina.**

The spinal cord displays two prominent enlargements (Figure 14.18*b*). **F14.18b** The *cervical enlargement* is located in the portion of the cord that gives rise to the spinal nerves supplying the upper limbs. These nerves form the brachial plexus (see Chapter 15). The *lumbar enlargement* is in the region of the cord that gives rise to the nerves innervating the lower limbs. These nerves form the lumbosacral plexus.

In Chapter 13 we discussed how the spinal cord and brain develop from a neural groove into a neural tube. With further development the lateral walls of the neural tube (and brain stem) show greater development than the roof or floor of the tube (Figure 14.19). These lateral thickenings become separated **F14.19** into dorsal and ventral portions by a groove **(sulcus limitans)** along each wall of the central canal. The two dorsal thickenings are called **alar plates.** The neurons in the alar plates are *interneurons* that receive *sensory* and *coordinative* information from afferent neurons whose cell bodies are located in an outlying structure called the dorsal root ganglion. The two ventral thickenings of the developing neural tube are called **basal plates.** The neurons in the basal plates develop into *motor* neurons. As the lateral regions of the neural tube develop, grooves form along the roof and floor of the tube. A fairly deep longitudinal fissure (the **anterior median fissure**) forms on the ventral surface of the spinal cord (Figure 14.20). A shallow longitudinal groove (the **posterior F14.20 median sulcus**) is located on the posterior surface of the spinal cord.

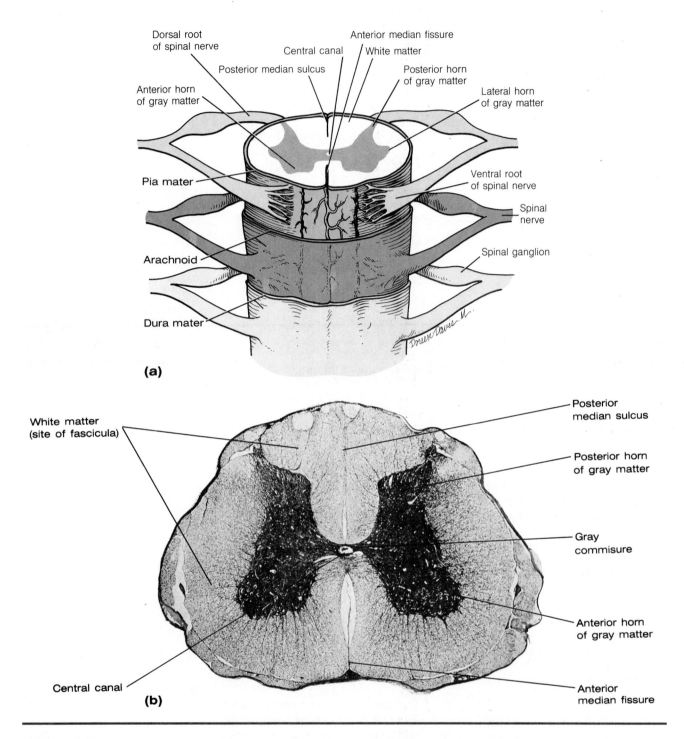

(a)

(b)

Figure 14.20

(a) General internal
structure of the spinal
cord and the meninges
surrounding it.
(b) Photograph of cross
section of a spinal cord.

F14.20

F14.18a

Meninges of the Spinal Cord

The spinal cord is covered with the same three meninges that cover the brain:
the dura mater, the arachnoid, and the pia mater (Figure 14.20).

The spinal **dura mater** is continuous at the foramen magnum with the
inner layer of the dura mater of the brain. Unlike in the skull, the spinal dura
does not fuse to the bone of the surrounding vertebrae. Therefore, there is a
small epidural space between the spinal dura and the vertebral column. The
spinal dura extends beyond the lower end of the cord, enclosing the cauda
equina (Figure 14.18a). In the sacral region, it forms a covering around the
filum terminale. The dura extends laterally to blend with the connective tissue
that covers each spinal nerve.

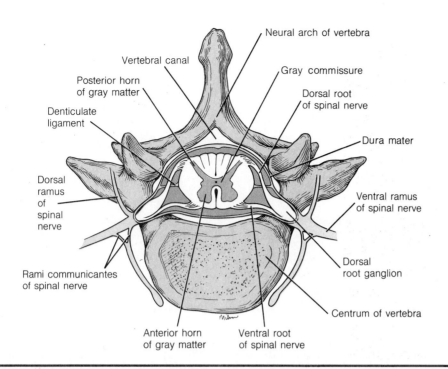

Neural arch of vertebra

Vertebral canal

Posterior horn
of gray matter

Denticulate
ligament

Dorsal
ramus
of
spinal
nerve

Rami communicantes
of spinal nerve

Anterior horn
of gray matter

Ventral root
of spinal nerve

Gray commissure

Dorsal root
of spinal nerve

Dura mater

Ventral ramus
of spinal nerve

Dorsal
root ganglion

Centrum of vertebra

Figure 14.21
Cross section showing
the relationships of the
spinal cord and spinal
nerve roots to a vertebra.

As is the case in the coverings of the brain, the spinal **arachnoid** forms a close lining of the dura mater. The subarachnoid space, which contains cerebrospinal fluid, is largest in the region of the cauda equina.

The innermost spinal meninx is the **pia mater.** It adheres closely to the surface of the cord and the roots of the spinal nerves. The pia mater contains a rich network of blood vessels.

The spinal cord is held in a somewhat fixed position within the meninges by fibrous bridges that cross the subarachnoid space, joining the pia mater with the arachnoid and dura mater. The heaviest of the bridges are the **denticulate ligaments** (Figure 14.21), located along the lateral margins of the spinal cord.   **F14.21**

Composition of the Spinal Cord

The spinal cord, like the brain, consists of areas of white matter and gray matter. As in the brain, the white matter is composed primarily of the myelinated processes of neurons, whereas the gray areas are composed primarily of nerve-cell bodies and unmyelinated internuncial nerve fibers. Neuroglia are present in both the white and gray matter. As we have seen, the gray matter in the cerebrum and cerebellum of the brain forms the surface layer (cortex). By contrast, the gray matter of the spinal cord is centrally located and is surrounded by the white matter (Figures 14.20, 14.21, 14.22).   **F14.20–**
F14.22

Gray Matter of the Spinal Cord

The gray matter of the spinal cord is roughly in the form of a letter H. The transverse bar of gray matter that connects the two lateral gray areas is the **gray commissure** (Figure 14.21). Within the gray commissure is the narrow, **F14.21** fluid-filled **central canal,** which is continuous with the fourth ventricle. The vertical bars of the gray H, on either side of the gray commissure, are separated into a pair of **posterior (dorsal) horns,** or **columns,** and a pair of **anterior (ventral) horns,** or **columns.** In the thoracic and upper lumbar regions, the spinal cord also has a pair of **lateral horns,** or **columns,** of gray matter located between the other two gray horns.

The posterior horns of gray matter develop from the alar plates (sensory) of the embryonic neural tube and are composed of axons of sensory neurons from the spinal nerves and interneurons that transmit sensory information

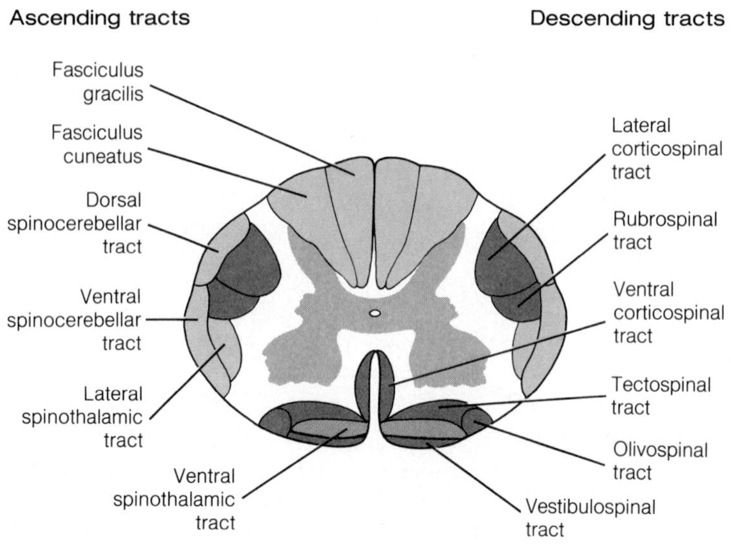

Ascending tracts   Descending tracts

Figure 14.22
Main fasciculi of the
spinal cord. The
ascending (sensory) tracts
are shown in gray and
are labeled only on the
left side. The descending
(motor) tracts are shown
in color and are labeled
only on the right side.

within the central nervous system. Some of the axons of the sensory neurons
enter the posterior area (white matter) of the spinal cord. These axons then
travel to higher levels of the cord or the brain. Other sensory axons enter the
gray substance and synapse either directly with neurons in the anterior horns
(thus forming a spinal reflex arc) or with interneurons. The interneurons, in
turn, may synapse with anterior horn motor neurons at the same level, pass
to higher or lower levels within the spinal cord, or travel all the way up to
various regions in the brain.

The anterior and lateral horns of gray matter develop from the basal
plates (motor) of the embryonic neural tube. The anterior horns contain the
cell bodies of somatic motor (voluntary) neurons whose axons leave the cord
and enter a spinal nerve. The cell bodies of visceral motor (involuntary) neurons are found in the lateral horns (see Chapter 16).

Dorsal and Ventral Roots of Spinal Nerves

Groups of nerve fibers called **dorsal roots** enter the spinal cord where the tips
of the posterior horns of gray matter come close to the surface of the cord.
Similarly, groups of nerve fibers called **ventral roots** leave the spinal cord
where the tips of the anterior horns of gray matter come close to the surface of
the cord. The dorsal and ventral root on each side of each spinal segment
F14.20, unite to form a **spinal nerve** (Figure 14.20, Figure 14.21).
F14.21   The dorsal roots contain only axons of sensory neurons (somatic and
visceral) that pass from the spinal nerve into the posterior horn of gray matter
in the spinal cord. The cell bodies of these sensory neurons are located outside the spinal cord within enlargements of the dorsal roots. These enlargements, which lie in the intervertebral foramina, are called **dorsal root ganglia,**
F14.21 or **spinal ganglia** (Figure 14.21).

The ventral roots are formed by the axons of neurons located in the anterior and lateral horns of gray matter. The ventral roots therefore contribute
processes of both somatic motor and visceral motor (autonomic) neurons to
the spinal nerves.

White Matter of the Spinal Cord

The white matter of the spinal cord completely surrounds the gray matter. It
is composed primarily of myelinated axons. These axons travel in three directions: (1) up the spinal cord to higher levels in the cord or brain; (2) down the
spinal cord from the brain or higher levels of the cord; or (3) across the cord,
transmitting impulses from one side to the other.

In each half of the spinal cord the white matter is divided by the gray matter into three areas: the **posterior funiculus** (cord), the **lateral funiculus,** and the **anterior funiculus.** Within the funiculi are smaller bundles of nerve fibers called **tracts,** or **fasciculi** (bundles) (Figure 14.22). The tracts are com- **F14.22** posed of the processes of neurons that carry similar types of impulses to a specific destination. Some tracts are *ascending (sensory)* tracts—carrying impulses that reach the spinal cord through afferent neurons of a spinal nerve up to the brain. Other tracts are *descending (motor)* tracts—carrying impulses from the brain down to the motor neurons in the anterior or lateral gray horns of the spinal cord. The tracts are not visibly discernible, but their locations have been determined by experimental methods. Most of the spinal tracts have descriptive names that indicate where they begin and where they terminate (Figure 14.22).

F14.22

ASCENDING (SENSORY) SPINAL TRACTS The ascending tracts of the spinal cord carry afferent (sensory) impulses from peripheral sensory receptors to various centers in the brain. These tracts generally contain three successive neurons called first-order, second-order, and third-order neurons. A **first-order neuron** is a unipolar neuron whose cell body is located in a dorsal root ganglion. Its peripheral branch, which functions as a dendrite, travels in a spinal nerve and serves to transmit information from various sensory receptor endings and sensory structures in the periphery to the central nervous system. The central branch of a first-order neuron, which functions as an axon, passes through the dorsal root of a spinal nerve and enters the spinal cord, where it synapses with a second-order neuron. A **second-order neuron** is a neuron whose cell body is located in the spinal cord or the medulla. A second-order neuron transmits the nerve impulse from the first-order neuron to a **third-order neuron,** whose cell body is located in the thalamus. The third-order neuron transmits the sensory information to the cerebral cortex, where it reaches the conscious level. All of the ascending spinal tracts cross to the other side of the central nervous system—either at the level of entry into the spinal cord, or a few segments above the entry level, or within the medulla. As a result, sensory information received by receptors on the right side of the body is interpreted in the left cerebral cortex and sensory information from the left side of the body is interpreted in the right cerebral cortex. The major ascending (sensory) spinal tracts are:

1. the fasciculus gracilis
2. the fasciculus cuneatus
3. the spinothalamic tracts
4. the spinocerebellar tracts

The **fasciculus gracilis** and **fasciculus cuneatus** are two tracts in the posterior funiculus that carry similar types of sensory information concerning *proprioceptive* information from muscles and joints, and *fine-touch localization* from different parts of the body (Figure 14.23). Information carried by these **F14.23** tracts makes it possible to know the position of a body part without having to see it. This information also enables a person to locate where an object is touching the body and assists in identifying the object by shape, texture, weight, and so on. The fasciculi gracilis and cuneatus receive impulses generated by proprioceptors in the muscles and joints and mechanoreceptors in the skin. The fasciculus cuneatus, which is located lateral to the fasciculus gracilis, transmits these impulses from the upper limb, the trunk, and the neck. The fasciculus gracilis transmits impulses that arise from receptors in the lower limbs and lower trunk.

Nerve fibers of the fasciculus gracilis synapse in a center in the medulla called the **nucleus gracilis.** Nerve fibers of the fasciculus cuneatus synapse in a medullary center called the **nucleus cuneatus.** The processes of the second-order neurons that leave the nucleus gracilis or nucleus cuneatus cross to the

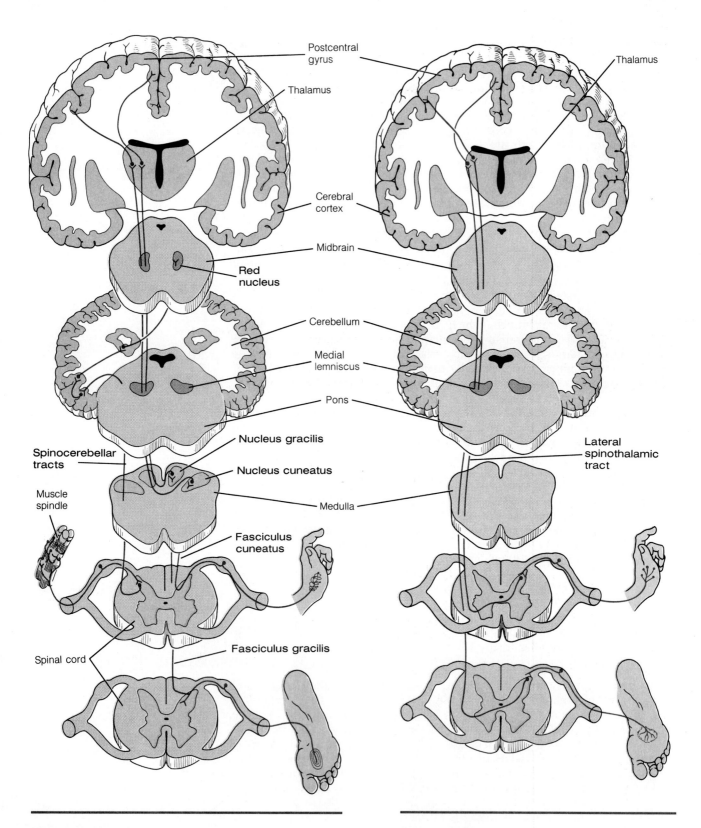

Figure 14.23

Sensory pathways for touch, pressure, and proprioception (conscious and unconscious) within the fasciculi gracilis, the fasciculi cuneatus, and the spinocerebellar tracts.

Figure 14.24

Sensory pathways for pain and temperature within the lateral spinothalamic tracts.

other side of the medulla and give rise to a tract called the *medial lemniscus.* The nerve fibers in the lemniscus synapse in the thalamus with third-order neurons that ascend to the cortex of the postcentral gyrus.

In the spinothalamic tracts most of the second-order neurons cross within the cord after synapsing with a sensory neuron from a spinal nerve and ascend in the white matter of the opposite side as the lateral and ventral spinothalamic tracts. The nerve fibers of the spinothalamic tracts synapse in the thalamus. From the thalamus, third-order neurons ascend to the postcentral gyrus of the cerebral cortex. The **lateral spinothalamic tracts** (Figure 14.24) **F14.24** convey impulses concerned with *pain* and *temperature.* The **ventral spinothalamic tracts** carry impulses from receptors sensitive to *touch* and *pressure.*

There are four spinocerebellar tracts: a pair of **dorsal** and a pair of **ventral spinocerebellar tracts.** The spinocerebellar tracts carry information concerning unconscious *proprioception* from neuromuscular receptors. As the name of the tracts indicates, their nerve fibers end in the cerebellum. The nerve fibers of these tracts do not synapse with higher-order neurons that pass to the cerebral cortex, or even to the thalamus. Therefore, the impulses they carry never reach a conscious level (Figure 14.25). Nerve fibers of the spinocerebel- **F14.25** lar tracts do, however, synapse with neurons in the cerebellum that can cause contraction of skeletal muscles (by way of the red nucleus of the midbrain). Some of the nerve fibers in the spinocerebellar tracts cross to the opposite side of the cord, whereas others remain on the side of the cord that they enter and project directly to the cerebellum. The nerve fibers of the tracts enter the cerebellum by way of the inferior cerebellar peduncles.

DESCENDING (MOTOR) SPINAL TRACTS   The descending spinal tracts carry impulses from the brain to lower motor neurons that regulate the activity of skeletal muscles. All of these tracts cross from one side of the central nervous system to the other, and they all contain two or three consecutive neurons. There are two types of descending (motor) spinal tracts:

1.   pyramidal tracts

2.   extrapyramidal tracts

Pyramidal tracts are motor tracts that originate primarily from large cells (pyramidal cells) in the cortex of the precentral gyrus and travel through the cerebral peduncles of the midbrain and the pyramids of the medulla. The pyramidal tracts are also called **corticospinal tracts,** a name that indicates their origin (cerebral cortex) and termination (spinal cord). From the cerebral cortex these tracts descend through the internal capsule, midbrain, pons, and pyramids of the medulla. The nerve fibers of these tracts (which are called **upper motor neurons**) synapse primarily with motor neurons in the anterior horns of the gray matter of the spinal cord (Figure 14.26), although some of **F14.26** them synapse with motor neurons associated with certain cranial nerves. Most of the upper motor neurons in the corticospinal tracts cross to the opposite side of the cord in the medulla and form the **lateral corticospinal tracts.** The remaining upper motor neurons descend uncrossed as the **ventral corticospinal tracts.** Some of the neurons in the ventral tracts cross at the levels at which they synapse with **lower motor neurons.** Both corticospinal tracts carry *motor stimuli* to *skeletal muscles.* The lateral corticospinal tracts, which extend the entire length of the spinal cord, are the major tracts involved in the voluntary control of skeletal muscles. The ventral corticospinal tracts generally do not extend below the thoracic level of the cord.

The remaining motor tracts, which originate from various regions of the cerebral cortex and subcortical areas, are referred to as **extrapyramidal tracts.** These tracts descend from various nuclei in the brain stem and influence muscular actions, coordination, balance, visual and auditory stimuli, and other functions. They include the **rubrospinal tracts, vestibulospinal tracts, tectospinal tracts,** and **olivospinal tracts** (Figure 14.22). Because there seems **F14.22** to be considerable overlap between the actions of these tracts and the corticospinal tracts, it is not possible to separate clearly the effects of each. In gen-

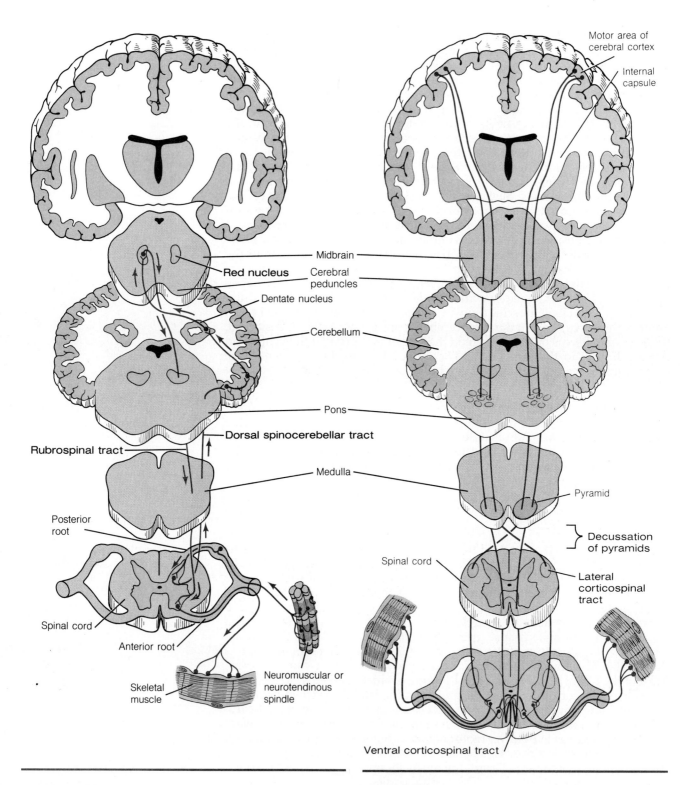

Motor area of
cerebral cortex

Internal
capsule

Midbrain

Red nucleus

Cerebral
peduncles

Dentate nucleus

Cerebellum

Pons

Dorsal spinocerebellar tract

Rubrospinal tract

Medulla

Pyramid

Posterior
root

Decussation
of pyramids

Spinal cord

Lateral
corticospinal
tract

Spinal cord

Anterior root

Skeletal
muscle

Neuromuscular or
neurotendinous
spindle

Ventral corticospinal tract

Figure 14.25
Sensory and motor
pathways for
unconscious muscle
movement within the
dorsal spinocerebellar
and rubrospinal tracts.

Figure 14.26
Pathways of the
pyramidal tracts (lateral
and ventral corticospinal
tracts) carrying motor
impulses to skeletal
muscles.

eral, the pyramidal tracts control the muscles involved in fine movements of the body, whereas the extrapyramidal tracts tend to modify muscular contractions related to *posture* and *balance.* Some extrapyramidal tracts have been found to be inhibitory of movements rather than excitatory.

The rubrospinal tracts, which originate in the red nucleus of the midbrain, are illustrated in Figure 14.25. Notice that the proprioceptive informa- **F14.25** tion that is carried to the cerebellum over the spinocerebellar tracts is ultimately conveyed to the rubrospinal tracts. In this manner, the rubrospinal tracts help coordinate the reflexes that are concerned with postural adjustment by affecting muscle tone. The fibers of the vestibulospinal tracts also affect skeletal muscle tone, and thus maintain posture and equilibrium. However, the impulses carried by these tracts are generated in response to movements of the head. Nerve fibers within the tectospinal tracts conduct motor impulses that control reflex movement of the head and eyes, primarily in response to auditory and visual stimuli. Skeletal muscle tone is also affected by the neurons within the olivospinal tracts, which are largely under conrol of the cerebellum.

The Spinal Reflex Arc

Not all of the sensory impulses carried to the spinal cord over sensory neurons enter one of the ascending tracts of the spinal cord and thus reach a higher center of the central nervous system. Some sensory neurons synapse directly or through an interneuron with motor neurons in the anterior horn of the gray matter of the spinal cord at the same level at which they enter the cord. Other neurons travel only a few segments up or down the cord before synapsing with a motor neuron. The neural pathways by which sensory impulses from receptors reach effectors without traveling to the brain are called **spinal reflex arcs.**

The presence of spinal reflex arcs makes possible automatic, stereotyped reactions to stimuli. These reactions, which are called **reflexes,** are performed without conscious thought. Consequently, a particular stimulus always elicits a particular response. Since spinal reflexes occur at the level of the spinal cord, without involving the pyramidal tracts from the cortex of the brain, they are involuntary responses, even though they often involve skeletal muscles. At the same time that a reflex is occurring, however, information about the stimulus that initiated the reflex may also be transmitted over one of the ascending tracts of the spinal cord to the brain, where a conscious sensation is elicited. For example, when you touch something hot with your hand, you become aware of it through impulses carried over the lateral spinothalamic tracts. By the time you feel the burning sensation, however, you have already withdrawn your hand from the hot object as a result of reflex action. If the spinal cord is severed in the cervical region, making it impossible for impulses to reach the brain from the spinal cord, the hand would still be withdrawn as a result of reflex activity that occurs in the spinal cord itself. In this instance, however, there would be no sensation of pain.

A spinal reflex arc is composed of at least five components (Figure 14.27): **F14.27**

1. A **receptor,** which can be a peripheral ending of a sensory neuron or a specialized cell associated with a peripheral ending of a sensory neuron.

2. A **sensory (afferent) neuron,** which carries impulses through a spinal nerve from the receptor to the spinal cord.

3. A **synapse** between sensory neurons and motor neurons in the spinal cord. If a sensory neuron synapses directly with a motor neuron, the arc formed is known as a **monosynaptic reflex arc.** If there are one or more interneurons located between a sensory and a motor neuron, requiring more than one synapse, the arc is a **polysynaptic reflex arc.** Most CNS reflexes are polysynaptic.

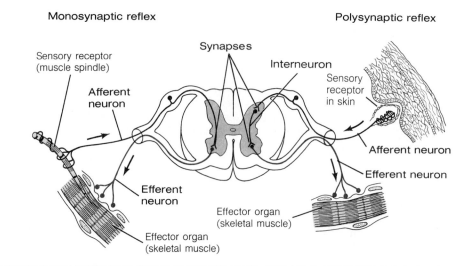

Monosynaptic reflex   Polysynaptic reflex

Figure 14.27
Components of a spinal reflex arc. The left side of the diagram illustrates a monosynaptic reflex; the right side shows a polysynaptic reflex.

4. A **motor (efferent) neuron,** which transmits the nerve impulse from the anterior horn of the gray matter to an effector.

5. An **effector,** which responds to the efferent nerve impulse. Muscle tissue (skeletal, smooth, or cardiac) and glands are the only structures capable of serving as effectors.

Two important spinal reflexes—the stretch reflex and the tendon reflex—influence the contractions of skeletal muscles.

Stretch Reflex

The muscle **stretch reflex** is initiated by encapsulated skeletal muscle receptors called **muscle spindles** (Figure 14.28). The muscle spindles provide information about the lengths of skeletal muscles. Within a muscle spindle there are several specialized muscle cells called *intrafusal fibers.* The intrafusal fibers are contractile only at their ends, where they are supplied by efferent neurons. The noncontractile, central portions of the intrafusal fibers are supplied by afferent neurons that spiral around the intrafusal fibers. The frequency of nerve impulses in the afferent neurons increases in response to a change in the stretch of the central areas of the intrafusal fibers and decreases when central areas of the fibers are compressed. The muscle spindles are oriented parallel to the muscle fibers so that stretching the muscle stretches the muscle spindle (and its intrafusal fibers) and contracting the muscle compresses the muscle spindle.

In the stretch reflex, the stretching of a muscle results in an increased frequency of nerve impulses in the afferent neurons associated with the muscle spindles (Figure 14.29). Within the spinal cord, the afferent neurons synapse with efferent neurons that supply the cells of the muscle. As a result, the increased frequency of nerve impulses in the afferent neurons increases the stimulation of the motor neurons, causing the muscle to contract and resist the stretch. The afferent neurons also synapse with inhibitory neurons in the spinal cord that, in turn, synapse with efferent neurons controlling muscles whose activities oppose the contraction of the stretched muscle (that is, antagonistic muscles). Thus, when the stretch reflex stimulates the stretched muscle to contract, antagonistic muscles that oppose the contraction are inhibited.

The *patellar reflex* (knee jerk) is a common example of a stretch reflex. Striking the patellar tendon at the knee pulls on the tendon and thus stretches the quadriceps muscles of the anterior thigh. If all the components of the reflex arc are intact, the muscle spindles within the quadriceps muscles initiate a stretch reflex that causes the quadriceps muscles to contract and swing the leg forward.

In addition to their synapses with neurons involved in the stretch reflex, the afferent neurons from muscle spindles also synapse with neurons that relay impulses to the brain. As a result, nerve impulses from muscle spindles

F14.28

F14.29

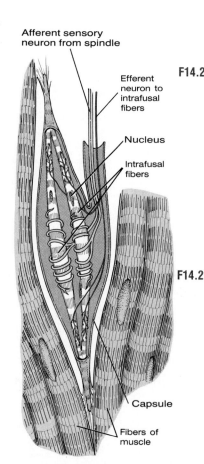

Afferent sensory neuron from spindle

Efferent neuron to intrafusal fibers

Nucleus

Intrafusal fibers

Capsule

Fibers of muscle

Figure 14.28
A muscle spindle.

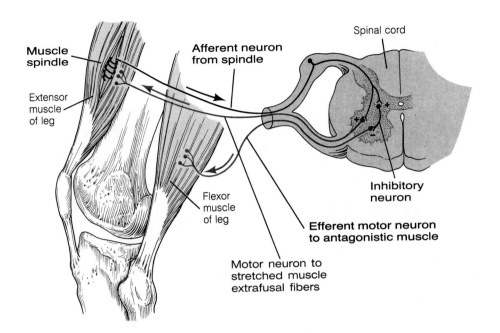

Figure 14.29
Pathways involved in a stretch reflex. Excitation is indicated by a +; inhibition is indicated by a −.

Labels in figure:
Muscle spindle
Afferent neuron from spindle
Spinal cord
Extensor muscle of leg
Inhibitory neuron
Flexor muscle of leg
Efferent motor neuron to antagonistic muscle
Motor neuron to stretched muscle extrafusal fibers

provide information about the state of stretch or contraction of skeletal muscles that is important in maintaining body posture and in coordinating muscular activity.

Tendon Reflex

The **tendon reflex** helps protect tendons and their associated muscles from the damage that could result from excessive tension. The receptors for this reflex are **neurotendinous organs,** which are encapsulated structures located within tendons near the junction of a tendon with a muscle. Unlike muscle spindles, which are sensitive to muscle length, neurotendinous organs are sensitive to tension. Within a neurotendinous organ are afferent neuron endings wrapped around small bundles of collagen fibers. When the tension applied to a tendon increases, so does the frequency of nerve impulses in the afferent neurons; when the tension decreases, so does the afferent impulse frequency. Because of their locations within tendons, the neurotendinous organs are in series with the muscles themselves, and they respond to changes in tension that result from either passive stretch or muscular contraction.

In the tendon reflex, an increase in the tension applied to a tendon—most commonly as a result of muscular contraction—increases the frequency of nerve impulses in the afferent neurons associated with the neurotendinous organs (Figure 14.30). Within the spinal cord, the afferent neurons synapse with inhibitory neurons that, in turn, synapse with motor neurons that supply the muscle associated with the tendon. As a result, the increased frequency of nerve impulses in the afferent neurons from the neurotendinous organs ultimately diminishes the stimulation of the motor neurons to the muscle, and may even completely inhibit muscular contraction. Thus, the tendon reflex protects the muscle and tendon from excessive tension. The afferent neurons from the neurotendinous organs also synapse with stimulatory neurons that, in turn, synapse with motor neurons controlling antagonistic muscles. As a result, the tendon reflex is generally accompanied by a reciprocal stimulation of the antagonistic muscles.

In addition to synapsing with neurons involved in the tendon reflex, the afferent neurons from the neurotendinous organs also synapse with neurons that relay impulses to the brain. Thus, just as nerve impulses from muscle spindles provide the brain with information about the length of a skeletal muscle, nerve impulses from the tendon organs provide information about the tension developed by the muscle.

F14.30

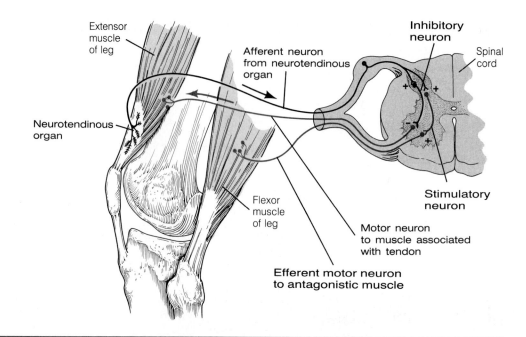

Figure 14.30
Pathways involved in a tendon reflex. Excitation is indicated by a +; inhibition is indicated by a −.

NEURON POOLS

Within the central nervous system, neurons are organized into functional groups called **neuron pools,** and information transmitted from a receptor is often received by a particular neuron pool. Within the pool, this information is processed and integrated with information received from other sources. It is then transmitted from the pool to various destinations.

Neuron pools occur at all levels of the central nervous system, including the cerebral cortex, and there are numerous interconnections among the various pools. Although the neural pathways involved in particular activities are commonly depicted as simple neuron-to-neuron chains, the performance of even the simplest actions generally requires the participation of many neurons whose activities are coordinated in complex circuits in neuron pools. For example, the motor neurons to skeletal muscle are influenced by nerve impulses arriving from muscle spindles, neurotendinous organs, and higher brain centers. As a result, the degree of activity of the motor neurons at any one moment is determined by a combination of influences.

CONDITIONS OF CLINICAL SIGNIFICANCE

The Central Nervous System

Because of the complex structure of the brain and spinal cord, innumerable abnormalities can occur in the central nervous system as the result of either trauma or disease. Central nervous system dysfunctions are particularly serious because the neurons in the CNS are not capable of effective regeneration once they have been damaged. Consequently, complete recovery from CNS dysfunction is not often achieved.

In the following sections we discuss some of the more common and more illustrative abnormalities of the CNS. However, we will begin by discussing the sensa-

tion of pain. Although pain is not generally considered to be an abnormality itself, it is associated with many abnormalities and dysfunctions, and it is certainly of great clinical significance.

Pain

Stimuli strong enough to cause tissue damage commonly give rise to a sensation of pain. Stimuli such as excessive mechanical stress, extremes of heat or cold, and various chemical substances stimulate pain receptors. This stimulation gives rise to neural signals that are

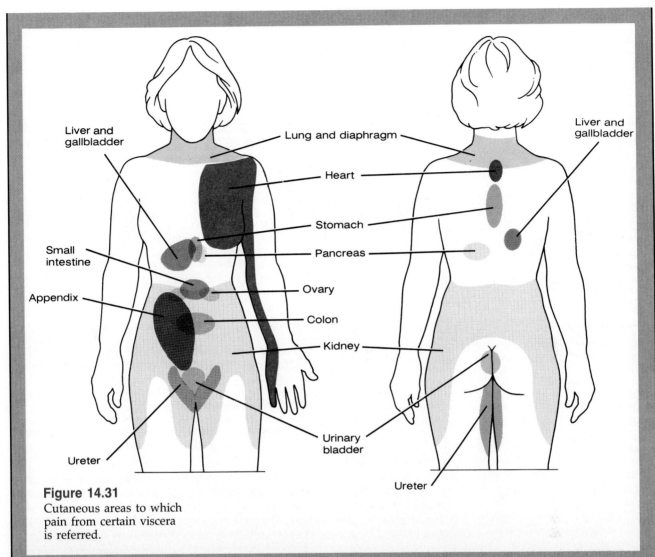

Figure 14.31
Cutaneous areas to which
pain from certain viscera
is referred.

transmitted to brain areas that include the reticular formation, the thalamus, and the somatic sensory areas of
the cerebral cortex. In addition to eliciting a sensation of
pain, the transmission of pain signals to the brain leads
to emotional reactions (for example, crying, anxiety,
fear) and behavioral responses (such as withdrawal or
defensive responses).

Referred Pain

A person can often identify the location of a stimulus
that produces pain because the brain usually projects the
sensation of pain to the site of the stimulus. In some
instances, however, signals from pain receptors in internal organs are incorrectly interpreted by the brain as
coming from areas quite distant from the actual sites of
stimulation—particularly as coming from sites on the
body surface (Figure 14.31). Thus, visceral pain can be
confused with somatic pain. This phenomenon is called
referred pain. The locations of some referred pains are
so consistent that they are used by physicians in diag-
nosing visceral dysfunction. For example, because visceral sensory nerve fibers from the heart enter the spinal
cord through the same thoracic spinal nerves as do somatic sensory fibers from the skin of the chest over the

heart, the left shoulder, and along the medial surface of
the left arm, a heart attack frequently causes referred
pain from these surface areas. This referred pain is
known as *angina pectoris.* The presence of angina pectoris
indicates possible heart dysfunction.

One attempt to explain the cause of referred pain
suggests that sensory neurons that transmit pain signals
from a particular area of the body surface and sensory
neurons that transmit pain signals from an internal
organ connect within the spinal cord with the same ascending neurons. Thus, these ascending neurons carry
pain signals to the brain from both the particular body
surface area and the internal organ. Because cutaneous
pain is much more common than visceral pain, pain signals carried over the ascending neurons are interpreted
by the brain as having originated in the skin rather than
in the viscera, and the pain sensation is projected to the
skin site.

Phantom Pain

Phantom pain is the phenomenon whereby a person
who has undergone an amputation continues to feel
pain that he or she perceives as coming from the amputated body part. Phantom pain, like referred pain, is a

case of inaccurate projection of the pain sensation by the brain. The neurons that supplied the affected structure are, of course, severed as a result of the amputation. However, the remaining portions of the neurons may continue to send nerve impulses to the same area of the brain as previously. The brain continues for some time to interpret impulses from the severed neurons as originating from the same body region as before the amputation. As a result, the sensations evoked in the brain are projected to that region. In this manner, pain (and other sensations) may still be "felt," for example, in the toes, even after the foot has been amputated.

Spinal Cord Dysfunctions

Damage to tissues or organs that results in impairment or loss of function are referred to as *lesions*. Lesions of the spinal cord, either from injury or from disease, generally involve the motor and sensory fiber tracts of which the cord is composed.

Paralysis

Lesions of the motor spinal tracts cause *paralysis* of the structures supplied by the neurons involved. Paralysis of both lower limbs is called *paraplegia*. The condition in which both the upper limb and the lower limb on one side of the body are paralyzed is *hemiplegia. Quadriplegia* is the paralysis of all four limbs. Paralysis may be *flaccid*, where the affected muscles lose their normal slight contraction (tonus) and reflexes are absent; or it may be *spastic*, with increased tonus and reflexes. Spastic paralysis generally results from lesions of descending motor tracts from the brain. These lesions are often in the extrapyramidal system, which eliminates the normal inhibitory effects of the higher centers on the spinal reflexes. Flaccid paralysis generally results when the pyramidal tracts or the lower motor neurons are affected.

Lesions of Sensory Tracts of the Spinal Cord

Generally, lesions of the spinal cord tracts are not so precisely located that they involve only one tract. Damage to sensory tracts can be located by identifying the combination of sensory losses that are experienced. Damage to the tracts in the posterior funiculi (fasciculi gracilis and cuneatus) results in a loss of proprioceptive ability and touch discrimination. A loss of thermal and pain sensations from a particular region of the body indicates damage to the lateral spinothalamic tracts above the affected region. Damage to the ventral spinothalamic tracts causes a slight decrease in the sense of touch.

Specific Dysfunctions of the Spinal Cord

TABES DORSALIS   *Tabes dorsalis* is a condition caused by the progressive degeneration of the posterior funiculi of the spinal cord and the dorsal roots of the spinal nerves. It is a result of invasion of the CNS by the spirochete bacteria of syphilis. Because of the loss of the proprioceptive pathways in the posterior columns and dorsal roots, tabes dorsalis causes a loss of muscular coordination. Consequently, people affected with tabes dorsalis walk unsteadily and must watch the ground in

order to maintain their balance. There may be severe leg and abdominal pain in the early stages of the disease due to the irritation of the sensory neurons in the dorsal roots. As these dorsal roots degenerate, the pain diminishes along with the sense of touch.

POLIOMYELITIS   The disease *poliomyelitis* is caused by a virus that primarily destroys the motor nerve-cell bodies in the anterior horns of the spinal cord, especially those in the cervical and lumbar enlargements. This degeneration produces fever, severe headache, and a flaccid paralysis of the muscles supplied by these neurons. After several weeks of paralysis, the muscles begin to atrophy, and eventually the muscle tissue may be almost completely replaced by connective tissue and adipose tissue. Death can result from respiratory failure or heart failure if the virus invades the nerve cells in the regulatory centers of the medulla.

SYRINGOMYELIA   The condition *syringomyelia* occurs when small fluid-filled sacs (cysts) form in the gray matter of the spinal cord and brain stem, and the neuroglia of the central canal proliferate. Numerous commissural pathways are interrupted in syringomyelia, producing various sensory dysfunctions and muscular weakness accompanied by muscular atrophy. For example, the lateral spinothalamic tracts are often destroyed as they cross within the gray commissure of the cord. As a result, impulses concerned with the sensations of pain and temperature may not be transmitted to the brain from a particular region of the body.

MULTIPLE SCLEROSIS   *Multiple sclerosis* is a chronic condition that results in widespread destruction of the myelin sheaths of the nerve fibers in the spinal cord and brain. The destroyed myelin sheaths are replaced by hardened plaques that interfere with the normal transmission of nerve impulses. The cause of the destruction is not known; however, a virus is suspected by some investigators. Multiple sclerosis causes a great variety of symptoms, involving both motor and sensory tracts, depending on the areas of the CNS where plaque formation occurs. Common symptoms include abnormal sensations, spastic paralysis, and exaggerated reflexes. Although multiple sclerosis is chronic and progressive, it is not unusual for its symptoms to disappear for years at a time. However, with each recurrence there is permanent damage to the neurons of the CNS, thus causing progressive incapacitation.

Dysfunctions of the Brain Stem

All the sensory and motor tracts of the central nervous system travel through the brain stem (medulla, pons, and mesencephalon). The brain stem also contains various centers that control the autonomic nervous system, and most cranial nerves originate there. Consequently, lesions in this region of the CNS can produce a complex mixture of motor and sensory symptoms. Moreover, a condition such as tumor, hemorrhage, or trauma can affect the reticular formation in the brain stem, producing a coma.

Dysfunctions of the Cerebellum

Lesions of the cerebellum or its pathways can cause dysfunctions in the smoothly coordinated actions that normally occur between groups of muscles. These dysfunctions may be evident as disturbances of gait and posture, as the inability to perform smoothly a task that involves the movement of several joints, or as an inability to estimate the range of movement and thus to stop a movement at a particular point.

Dysfunctions of the Basal Ganglia

The basal ganglia, which are part of the extrapyramidal system, are small islands of gray matter deep within the cerebral hemispheres. The neurons in the basal ganglia regulate somatic muscle activity, primarily by inhibiting contractions. This inhibition is in contrast to the excitatory effects of the neurons in the pyramidal tracts. For this reason, lesions of the basal ganglia generally result in spastic movements that appear to be voluntary but are actually beyond the individual's control. The person's limbs may be flung wildly about, writhe continuously, or move in a rapid, jerky manner.

Parkinsonism is the most common dysfunction of the basal ganglia. Since its onset is gradual, this disease primarily affects people over 50. A number of factors are thought to produce Parkinson's syndrome, all of which cause the degeneration of cells in the basal ganglia. The condition is characterized by useless contractions of skeletal muscles, causing tremor and rigidity of the muscles, as well as a decrease in normal associated movements, such as swinging the arms while walking and changing facial expressions related to emotions.

There is impressive evidence that parkinsonism may be related to the abnormal metabolism of the neurotransmitter *dopamine* by the cells of the basal ganglia. In Parkinson's disease, the level of dopamine in the basal ganglia is found to be substantially reduced. It is not known how the reduced dopamine level brings about the symptoms of Parkinson's disease. Since dopamine itself is unable to cross the blood-brain barrier effectively, the administration of supplementary dopamine is ineffective in controlling parkinsonism. However, *L-dopa*, a precursor (forerunner) of dopamine, will pass from the blood to the brain cells and has proved effective in treating the condition.

Inflammatory Diseases of the Central Nervous System

Encephalitis and Myelitis

The invasion of nervous tissue by viruses (and in some instances by bacteria, fungi, and other agents) can produce inflammation of the CNS. Inflammation of the brain is called *encephalitis*. Inflammation of the spinal cord is known as *myelitis*. Both conditions display a variety of possible motor and sensory symptoms, including paralysis, coma, and death.

Meningitis

Meningitis is an inflammation of the meninges that cover the brain and spinal cord. It is caused by a number of microorganisms, the most common being the meningococci, streptococci, pneumococci, and tubercle bacilli. The organisms are thought to enter the body through the nose and throat. Meningitis produces a high fever with a severe headache and stiffness of the neck. In severe cases, meningitis can cause coma and death.

Tumors of the Central Nervous System

Most tumors of the CNS develop from glial cells, including the oligodendrocytes that form the neurilemma in the CNS, but it is not uncommon for cells of the meninges to form tumors also. Only rarely do the neurons themselves give rise to tumors. The presence of tumors in the brain and spinal cord can produce a variety of dysfunctions, depending on their location. The symptoms of a tumor can include headache, convulsions, a change in behavior patterns, pain, and paralysis. These dysfunctions result from the destruction of nervous tissue by the tumor, increased pressure in the skull, or edema of the nervous tissue. Tumors may be treated by surgical removal or destroyed by chemical and radiation therapy.

STUDY OUTLINE

THE BRAIN pp. 373–393

TELENCEPHALON cerebrum in adult; contains two lateral ventricles.

> *Cerebral Cortex* outer surface of gray matter; unmyelinated neurons.
>
> *Basal Ganglia* gray matter structures deep inside cerebral hemisphere.
>
> *White Matter* myelinated nerve-fiber tracts.
>
> *MYELINATED NERVE-FIBER TRACTS*
> 1. *Projection tracts:* carry motor or sensory impulses from one level of brain or spinal cord to another.
> 2. *Association tracts:* connect areas in cortex of same cerebral hemisphere.
> 3. *Commissural tracts:* connect left and right hemispheres.
>
> *GYRI, FISSURES, SULCI, AND LOBES OF CEREBRUM*
>
> *Gyri* convolutions on the surface of the cerebrum.
>
> *Fissures* deeper furrows between gyri.
>
> *Sulci* shallower furrows between gyri.
>
> *Lobes of Cerebrum* division by fissures and sulci into frontal, parietal, temporal, and occipital lobes; insula located deep in the lateral fissure.

FUNCTIONAL AREAS OF THE CEREBRAL COR-TEX

Primary Motor (Primary Somatic Motor) Area in precentral gyrus of frontal lobe; controls precise voluntary contractions of skeletal muscle.

Premotor Area anterior to the primary motor area; causes groups of muscles to contract, producing stereotyped movements.

Primary Sensory (Primary Somatic Sensory) Area in postcentral gyrus of parietal lobe; senses general stimuli: temperature, touch, pain, proprioception.

Special Senses Areas
1. *Primary visual area:* posterior occipital lobe.
2. *Primary auditory area:* upper margin of temporal lobe.
3. *Primary olfactory area:* medial surface of temporal lobe.
4. *Primary taste area:* parietal lobe, near bottom of postcentral gyrus.

Association Areas interconnect motor and sensory areas.
1. *Frontal association area:* site of intellectual activities.
2. *Somatic association area:* integration and interpretation center.
3. *Visual and auditory association areas:* contribute to interpretation of visual and auditory experiences.

BASAL GANGLIA somatic motor functions; part of extrapyramidal system; inhibit motor function.

OLFACTORY BULBS on ventral surface of cerebral hemisphere; smell.

DIENCEPHALON

THALAMUS two oval masses of gray matter forming walls of third ventricles; sensory relay and integrating center of brain; some motor involvement.

HYPOTHALAMUS located below thalamus; regulates:
1. *Sympathetic and parasympathetic activity* (body temperature, water balance, appetite, gastrointestinal activity, sexual activity, for example).
2. *Emotions* (fear and rage, for example).
3. *Pituitary hormone release.*

Mammillary Bodies olfactory reflexes.

Tuber Cinereum contains neurons that transport regulatory hormones (or factors) from hypothalamus to pituitary gland.

Infundibulum hormone transport to posterior lobe of pituitary gland.

Optic Chiasma point of crossing of some optic nerve fibers.

EPITHALAMUS forms roof of third ventricle; includes pineal body that has possible neuroendocrine function.

LIMBIC SYSTEM includes structures that affect emotional responses.

MESENCEPHALON between forebrain and hindbrain. *Cerebral aqueduct* connects third and fourth ventricles.

CEREBRAL PEDUNCLES nerve-fiber tracts from primary motor area of cerebral cortex to pons and spinal cord, and sensory nerve fibers from spinal cord to thalamus. *Red nucleus* coordinates cerebellum and cerebral hemispheres.

CORPORA QUADRIGEMINA visual and auditory functions.

METENCEPHALON superior portion of hindbrain; contains fourth ventricle.

CEREBELLUM concerned with functions below conscious level; directs precise, smooth movements and maintains equilibrium.

PONS consists of nerve-fiber tracts that connect cerebellum with brain stem; connects cerebellum with various levels of central nervous system; also affects respiration rate.

MYELENCEPHALON also known as *medulla oblongata.* Includes motor nerve tracts called *pyramids* on ventral surface.

RETICULAR FORMATION receives input from several centers in brain; sensory input from spinal tracts; activates cerebral cortex; maintains wakefulness. Includes *medullary centers* that control vital functions such as heart rate, respiration, blood vessel dilation and constriction, coughing, swallowing, and vomiting.

VENTRICLES OF THE BRAIN four interconnected spaces filled with cerebrospinal fluid.

CHOROID PLEXUSES in each ventricle; form cerebrospinal fluid.

THE MENINGES three layers of connective tissue that cover the entire CNS.

DURA MATER

ARACHNOID

PIA MATER

CEREBROSPINAL FLUID cushions CNS; formed from blood via choroid plexuses; circulates in and around CNS, and returns to blood via arachnoid villi.

BRAIN FUNCTIONING no area or structure of brain seems to act entirely on its own.

THE SPINAL CORD   pp. 393–405

GENERAL STRUCTURE OF THE SPINAL CORD gives rise to 31 pairs of spinal nerves.

CERVICAL ENLARGEMENT forms brachial plexus; supplies upper limbs.

LUMBAR ENLARGEMENT forms lumbosacral plexus; supplies lower limbs.

MENINGES OF THE SPINAL CORD same as those surrounding brain; dura mater, arachnoid, pia mater.

COMPOSITION OF THE SPINAL CORD white myelinated areas surround gray unmyelinated central area.

GRAY MATTER OF THE SPINAL CORD H-shaped.

Posterior Horns axons of sensory neurons from spinal nerves; develop from alar plate.

Anterior Horns cell bodies of somatic motor neurons whose axons leave cord for a spinal nerve; develop from basal plate.

Lateral Horns only in thoracic and upper lumbar regions; cell bodies of visceral motor neurons; develop from basal plate.

DORSAL AND VENTRAL ROOTS OF SPINAL NERVES dorsal and ventral roots unite on each side of each spinal segment, forming a spinal nerve.

Dorsal Roots enter spinal cord at tips of posterior gray horns; sensory nerves; contain dorsal root ganglia.

Ventral roots leave spinal cord at tips of anterior gray horns; somatic and visceral motor nerves.

WHITE MATTER OF THE SPINAL CORD surrounds gray matter.

Funiculi posterior, lateral, and anterior regions of white matter in each half of spinal cord.

Tracts small bundles of neuron processes within funiculi:
1. *Ascending (sensory) spinal tracts:* carry afferent sensory impulses from peripheral sensory receptors to brain centers; all cross over in CNS.
   a. *Fasciculus gracilis* and *fasciculus cuneatus:* awareness of joint sense, muscle position sense, and fine-touch localization.
   b. *Spinothalamic tracts:* pain, temperature, touch, pressure.
   c. *Spinocerebellar tracts:* unconscious proprioception.
2. *Descending (motor) spinal tracts:* Carry impulses from brain to lower motor neurons that regulate skeletal muscle:
   a. *Pyramidal tracts* (corticospinal tracts): originate from pyramidal cells in cortex of precentral gyrus and travel through medulla pyramids; voluntary control of skeletal muscles, especially fine movements.
   b. *Extrapyramidal tracts:* originate from various nuclei in brain stem; modify muscular contractions for posture and balance.

THE SPINAL REFLEX ARC neural pathway by which sensory impulses from receptors reach effectors without traveling to brain.

RECEPTOR → SENSORY NEURON → SYNAPSE → MOTOR NEURON → EFFECTOR

STRETCH REFLEX initiated by muscle spindles that respond to stretch.

TENDON REFLEX neurotendinous organs respond to increased tension.

NEURON POOLS functional groups of neurons within central nervous system; even simple actions require participation of many neurons whose activities are coordinated in complex circuits in neuron pools.

p. 406

CONDITIONS OF CLINICAL SIGNIFICANCE: The Central Nervous System pp. 406–409

PAIN signals to brain from pain receptors elicit sensations of pain and also lead to such activities as emotional reactions and behavioral responses.

REFERRED PAIN inaccurate projection of pain sensation to site other than site of stimulus.

PHANTOM PAIN person feels pain from amputated body part.

SPINAL CORD DYSFUNCTIONS

PARALYSIS caused by lesions of motor spinal tracts.

LESIONS OF SENSORY TRACTS OF THE SPINAL CORD various sensory losses, depending on site of lesion.

SPECIFIC DYSFUNCTIONS OF THE SPINAL CORD

Tabes Dorsalis progressive degeneration of posterior funiculi and dorsal roots of spinal nerves caused by spirochete bacteria of syphilis.

Poliomyelitis caused by virus that destroys nerve-cell bodies in anterior horns of spinal cord.

Syringomyelia cyst formation in gray matter of cord and brain stem, with proliferation of neuroglia of central canal.

Multiple Sclerosis chronic, widespread destruction of myelin sheaths of neurons in spinal cord and brain.

DYSFUNCTIONS OF THE BRAIN STEM lesions, tumors, hemorrhage, or trauma can produce a variety of motor and sensory symptoms; damage to reticular formation can produce coma.

DYSFUNCTIONS OF THE CEREBELLUM lesions cause dysfunction in coordinated actions between muscle groups.

DYSFUNCTIONS OF THE BASAL GANGLIA lesions result in spastic movements. *Parkinsonism* may be related to abnormal metabolism of dopamine by basal ganglia.

INFLAMMATORY DISEASES OF THE CENTRAL NERVOUS SYSTEM

ENCEPHALITIS brain inflammation.

MYELITIS spinal cord inflammation.

MENINGITIS inflammation of meninges that cover brain and spinal cord.

TUMORS OF THE CENTRAL NERVOUS SYSTEM Symptoms include headaches, convulsions, behavior change, pain, paralysis; tumors destroy nervous tissue, increase pressure within skull, or cause edema of nervous tissue; treatments include surgical removal or chemical and radiation therapy.

SELF-QUIZ

1. The fibers that connect the left and right cerebral hemispheres form the: (a) projection tracts; (b) association tracts; (c) commissural tracts.

2. The cerebrum is divided into left and right hemispheres by the: (a) longitudinal fissure; (b) occipital lobe; (c) parietal lobe.

3. Each hemisphere of the cerebrum is divided by the central sulcus into: (a) frontal and parietal lobes; (b) parietal and occipital lobes; (c) occipital and frontal lobes.

4. The taste area is located in the parietal lobe, deep in the longitudinal fissure. True or false?

5. The control of emotions is influenced largely by the: (a) thalamus; (b) hypothalamus; (c) epithalamus.

6. "Pleasure centers" and "punishment centers" of the brain are located in the: (a) mammillary bodies; (b) limbic system; (c) metencephalon.

7. The cerebellum is part of the: (a) mesencephalon; (b) telencephalon; (c) metencephalon.

8. The part of the brain that directs precise, smooth movements and the maintenance of equilibrium is the: (a) cerebrum; (b) cerebellum; (c) medulla oblongata.

9. The myelencephalon is the most inferior division of the brain. True or false?

10. The control of heart rate, coughing, and swallowing is a function of the: (a) reticular activating system; (b) pons; (c) medullary centers.

11. The vagus cranial nerve is number (a) IX; (b) X; (c) XI.

12. Each ventricle contains a plexus of capillaries. True or false?

13. Cushioning the CNS against jolts and blows is a function of the: (a) meninges; (b) cerebrospinal fluid; (c) ventricles of the brain.

14. The portion of the spinal cord that gives rise to the spinal nerves supplying the upper limbs is the: (a) cervical enlargement; (b) lumbar enlargement; (c) lumbosacral plexus.

15. The white matter of the brain and spinal cord is composed of: (a) meninges; (b) myelinated processes of neurons; (c) nerve-cell bodies.

16. Match the terms associated with the brain with the appropriate description.

Cerebrum	(a)	This structure is divided into frontal, parietal, temporal, and occipital lobes.
Projection tracts		
Basal ganglia		
Cerebral hemisphere	(b)	Located in the posterior portion of the occipital lobe.
Insula		
Primary motor area	(c)	Acts as the sensory relay and integration center of the brain.
Visual area		
Olfactory area		
Thalamus	(d)	Connects the cerebel-

Cerebral peduncles		lum with the brain stem.
Metencephalon	(e)	Carry motor and/or sensory nerve impulses from one level of the CNS to another.
Pons		
Hypothalamus		
Ventricles	(f)	Contains gyri, fissures, and sulci.
Telencephalon		
Reticular formation	(g)	Controls many vital processes including appetite and sexual activity.
	(h)	A fluid-filled system of cavities in the brain.
	(i)	This structure envelops the diencephalon and obscures much of the rhombencephalon.
	(j)	This network of interlacing nerve fibers exerts control over the cerebral cortex.
	(k)	A portion of the cerebral cortex located deep in the lateral fissure.
	(l)	Located on the medial surface of the temporal lobe.
	(m)	Gray-matter structures located deep in the cerebral hemispheres.
	(n)	Two cylindrical nerve tracts on the ventral surface of the mesencephalon.
	(o)	Contains the cerebellum.
	(p)	Located in the precentral gyrus.

17. The ascending tracts of the spinal cord carry afferent (sensory) impulses from peripheral receptors to various centers in the brain. True or false?

18. Sensory information received by receptors on the right side of the body is interpreted in the: (a) left cerebral cortex; (b) right cerebral cortex; (c) sensory information from each side of the body is received in both cerebral cortices.

19. Match the items associated with the spinal cord with the appropriate description.

Cauda equina	(a)	Contains myelinated axons that travel up or down the spinal cord to higher or lower levels in the CNS.
Meninges		
White matter		
Gray matter		
Ventral roots		
Fasciculus gracilis	(b)	Convey impulses concerned with pain and temperature.
Spinothalamic tracts		
Spinocerebellar tracts	(c)	These structures contribute processes of both somatic motor neurons and visceral motor neurons to the spinal nerves.
Pyramidal tracts		
Extrapyramidal tracts		

(d) Dura mater, arachnoid, pia mater.

(e) Modify muscular contractions related to posture and balance.

(f) Major structures involved in the voluntary control of skeletal muscles.

(g) A mass of descending nerve roots below the end of the spinal cord.

(h) Carry information concerning unconscious proprioception from neuromuscular receptors.

(i) Enables you to locate where an object is touching your body.

(j) The anterior horns of this substance contain the cell bodies of somatic motor neurons.

20. Most spinal reflexes are: (a) polysynaptic; (b) monosynaptic; (c) asynaptic.

21. Parkinsonism is a dysfunction of the: (a) basal ganglia; (b) brain stem; (c) cerebellum.

LEARNING OBJECTIVES

After completing this chapter, you should be able to:

- State the function of the peripheral nervous system, and list its two divisions.

- Name the 12 pairs of cranial nerves, and specify the functions of each.

- Describe the distribution patterns of the spinal nerves.

- Distinguish between dorsal rami and dorsal roots and between ventral rami and ventral roots.

- List the three major nerve plexuses, and state the body regions they supply.

- Name and describe the specific distribution of each nerve that is formed by the major nerve plexuses.

- Distinguish between the somatic and autonomic nervous systems by describing the function of each.

- Cite the most common disorders of the peripheral nervous system.

CHAPTER CONTENTS

The Peripheral Nervous System

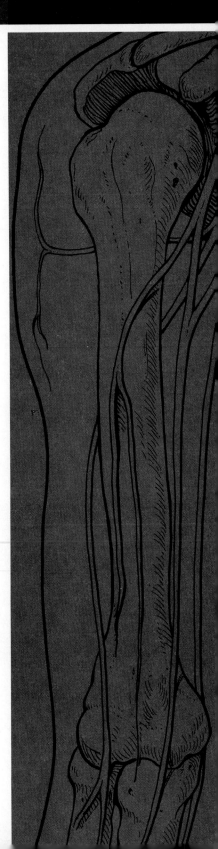

The **peripheral nervous system (PNS)** includes all neurons except those restricted to the brain and spinal cord—that is, all except the neurons in the central nervous system. It consists of pathways of nerve fibers between the central nervous system and all outlying structures of the body. Included in the peripheral nervous system are 12 pairs of **cranial nerves** and 31 pairs of **spinal nerves.** In terms of function, the peripheral nervous system is divided into:

1. the **afferent (sensory) division,** whose nerve fibers relay impulses from all areas of the body, including the viscera, to the central nervous system.

2. the **efferent (motor) division,** which is subdivided into:
   a. the **somatic nervous system,** whose fibers carry motor impulses between the central nervous system and the skeletal muscles.
   b. the **autonomic nervous system,** which connects motor fibers from the central nervous system with smooth muscles, cardiac muscle, and glands.

CRANIAL NERVES

Twelve pairs of cranial nerves arise from the brain (Figure 15.1). Most of the cranial nerves are mixed nerves composed of both motor and sensory neurons, although some cranial nerves carry only sensory impulses. Previously it was thought that some cranial nerves transmitted only motor impulses. We now know, however, that the cranial nerves that were considered to be entirely motor nerves actually contain some sensory neurons from proprioceptors in the muscles they innervate. Thus they convey information to the central nervous system concerning the tension in specific muscles.

In the cranial nerves, as in the spinal nerves, the cell bodies of the sensory neurons are in ganglia located *outside* the central nervous system. Some motor neurons in the cranial nerves supply skeletal muscles. These neurons, which are under conscious control, are called **somatic motor neurons.** Other motor neurons in the cranial nerves are not under conscious control. These supply smooth muscle, cardiac muscle, or glands, and are called **visceral motor neurons.** The visceral motor neurons in the cranial nerves are part of the parasympathetic division of the autonomic nervous system, which is discussed in Chapter 16. With the exception of the vagus nerve, the cranial nerves supply only structures in the head and neck. The cranial nerves are numbered (using Roman numerals) from superior to inferior in the order in which they leave the brain. Table 15.1 summarizes the cranial nerves and their functions.

I: Olfactory Nerves

The first pair of cranial nerves, the **olfactory nerves,** arise from receptor cells in the nasal mucosa (Figure 15.2, Table 15.1). Processes of these receptor cells pass through the perforations of the *cribriform plate* of the ethmoid bone and enter the olfactory bulbs of the telencephalon portion of the brain. In the olfactory bulbs, the nerve fibers synapse with neurons that pass posteriorly in

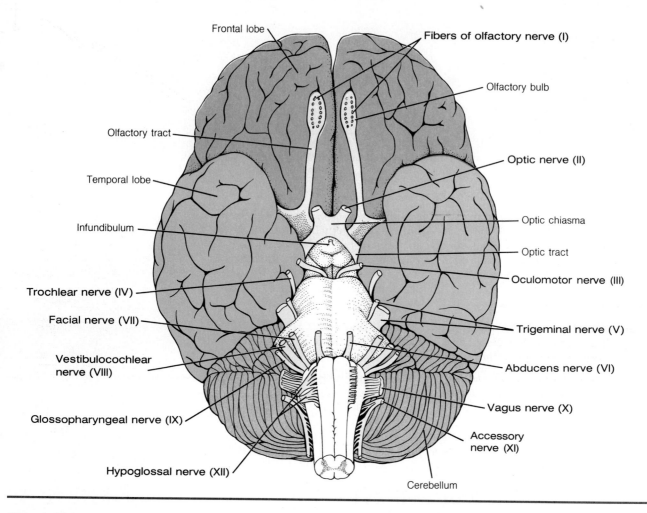

Frontal lobe

Fibers of olfactory nerve (I)

Olfactory bulb

Olfactory tract

Optic nerve (II)

Temporal lobe

Optic chiasma

Infundibulum

Optic tract

Oculomotor nerve (III)

Trochlear nerve (IV)

Trigeminal nerve (V)

Facial nerve (VII)

Vestibulocochlear nerve (VIII)

Abducens nerve (VI)

Vagus nerve (X)

Glossopharyngeal nerve (IX)

Accessory nerve (XI)

Hypoglossal nerve (XII)

Cerebellum

Figure 15.1
The ventral surface of the brain showing the cranial nerves.

the olfactory tracts. The fibers of the olfactory tracts enter the brain, and many of them travel to the cerebral cortex of the medial sides of the temporal lobes. The olfactory nerves are entirely sensory, carrying impulses associated with the sense of smell.

II: Optic Nerves

The **optic nerves** carry impulses associated with vision. Like the olfactory nerves, they are entirely sensory. The optic nerves are actually brain tracts rather than true nerves, since they are formed from outgrowths of the embryonic diencephalon.

The optic nerves arise from the retina of the eyes, on which images are focused. After leaving the posterior surface of the eyeball, each optic nerve exits from the orbital cavity and enters the cranial cavity through an *optic foramen* in the sphenoid bone. Shortly after entering the cranium, the two **F15.3,** optic nerves meet in the *optic chiasma* just anterior to the pituitary gland (Fig-**Table 15.1** ure 15.3, Table 15.1). In the optic chiasma, nerve fibers from the medial half of each retina cross to the opposite side, and those from the lateral half of each retina remain on the same side. The fibers then continue to the brain as the *optic tracts*. Because of the crossing of nerve fibers at the optic chiasma, each optic tract consists of fibers from the retinas of both eyes.

Some nerve fibers in the optic tracts terminate in the superior colliculi of the mesencephalon, where they function in subconscious visual reflexes. However, most of the fibers in the optic tracts travel to the lateral geniculate bodies of the thalamus, where they synapse with neurons that form pathways called **optic radiations.** The neurons of the optic radiations, which are third-order neurons, pass through the internal capsule and terminate in the visual cortex of the occipital lobes.

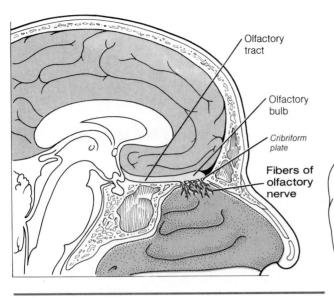

Figure 15.2
Olfactory nerve (cranial nerve I). Sagittal section of the face showing the position of the olfactory bulbs and tracts just above the cribriform plate of the ethmoid bone. The fibers of the olfactory nerve pass through the openings in the cribriform plate to enter the nasal cavity.

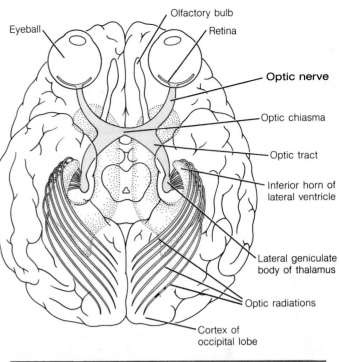

Figure 15.3
Optic nerve (cranial nerve II). Ventral view of the brain showing the optic nerves and the visual pathways to the cortex of the occipital lobes.

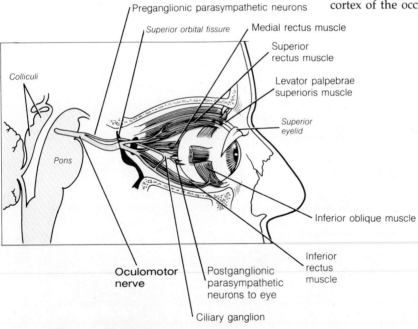

Figure 15.4
Oculomotor nerve (cranial nerve III). Lateral view of the right eye. Note the somatic motor neurons to muscles that surround the right eye and parasympathetic fibers that enter the eye. (The lateral rectus muscle has been cut.)

III: Oculomotor Nerves

The **oculomotor nerves** emerge from the midbrain, just superior to the pons, and enter the orbits through the *superior orbital fissures*, which are located between the small wings and great wings of the sphenoid bone. The oculomotor nerves consist of somatic motor neurons traveling to, and proprioceptive sensory neurons traveling from, four of the six extrinsic muscles that move the eyeball (Figure 15.4, Table 15.1). Specifically, the oculomotor nerves innervate the superior rectus, medial rectus, inferior rectus, and inferior

F15.4,
Table 15.1

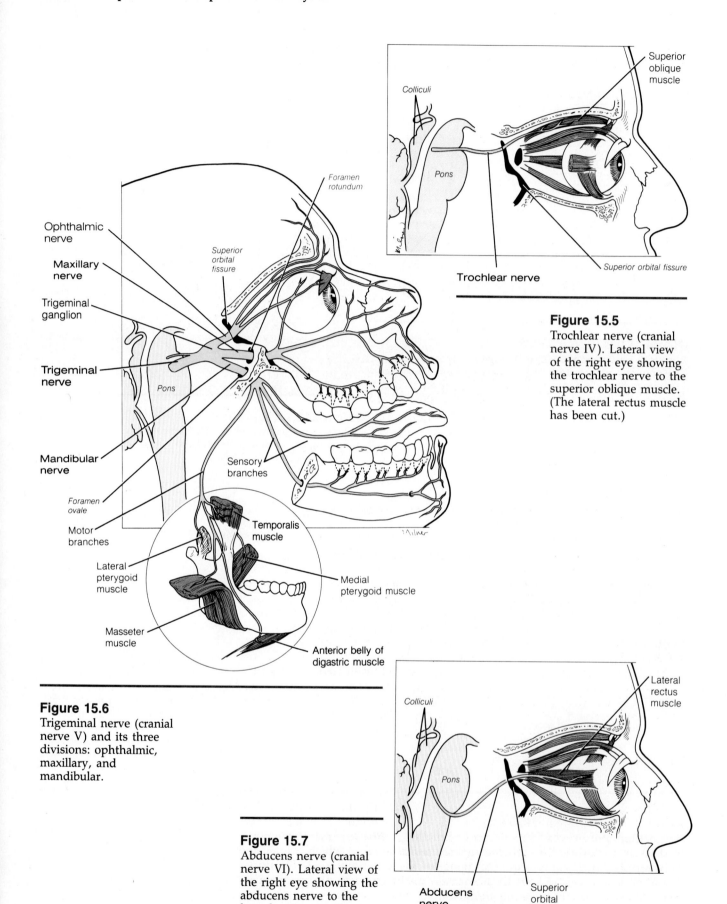

Ophthalmic nerve

Maxillary nerve

Trigeminal ganglion

Trigeminal nerve

Mandibular nerve

Foramen ovale

Motor branches

Lateral pterygoid muscle

Masseter muscle

Foramen rotundum

Superior orbital fissure

Pons

Sensory branches

Temporalis muscle

Medial pterygoid muscle

Anterior belly of digastric muscle

Colliculi

Pons

Superior oblique muscle

Superior orbital fissure

Trochlear nerve

Figure 15.5
Trochlear nerve (cranial nerve IV). Lateral view of the right eye showing the trochlear nerve to the superior oblique muscle. (The lateral rectus muscle has been cut.)

Figure 15.6
Trigeminal nerve (cranial nerve V) and its three divisions: ophthalmic, maxillary, and mandibular.

Colliculi

Pons

Lateral rectus muscle

Abducens nerve

Superior orbital fissure

Figure 15.7
Abducens nerve (cranial nerve VI). Lateral view of the right eye showing the abducens nerve to the lateral rectus muscle.

oblique muscles of the eyeball. In addition, the oculomotor nerves supply the levator palpebrae superioris muscles, which function to elevate the upper eyelids.

The oculomotor nerves also contain neurons of the parasympathetic division of the autonomic nervous system (Chapter 16). The efferent pathways of the autonomic nervous system have two neurons between the central nervous system and the innervated structures. The first neuron, called a **preganglionic (presynaptic) neuron,** always travels from the central nervous system to a ganglion located outside the central nervous system. Within the ganglion, the preganglionic neuron synapses with second neurons, called **postganglionic (postsynaptic) neurons,** which leave the ganglion and travel to the innervated structures. Preganglionic parasympathetic neurons in the oculomotor nerves synapse in ganglia called **ciliary ganglia,** which are located behind the eyeballs. From a ciliary ganglion, postganglionic neurons enter the eye and innervate the intrinsic smooth muscles that regulate the size of the pupil and shape of the lens. Therefore, apart from regulating the voluntary movement of the eyeballs, the oculomotor nerves are also involved in the reflex adjustments of the eyes to varying intensities of light and in focusing the eyes for near and far vision.

IV: Trochlear Nerves

The **trochlear nerves** are small nerves that arise below the inferior colliculi on the dorsal surface of the midbrain. These nerves, which are the only cranial nerves to exit from the dorsal surface of the brain, curve around the lateral sides of the brain and enter the orbits through the *superior orbital fissures* along with the oculomotor nerves (Figure 15.5, Table 15.1). The trochlear nerves F15.5, carry somatic motor neurons to, and proprioceptive neurons from, one of the Table 15.1 extrinsic muscles of the eye—the superior oblique muscle—thus aiding in the voluntary movements of the eyeball.

V: Trigeminal Nerves

The large **trigeminal nerves** emerge from the lateral sides of the pons. As their name indicates, each trigeminal nerve has three divisions: the **ophthalmic, maxillary,** and **mandibular nerves** (Figure 15.6, Table 15.1). The three F15.6, divisions exit the skull through openings in the sphenoid bone: the ophthal- Table 15.1 mic division leaves the skull through the *superior orbital fissure;* the maxillary division passes through the *foramen rotundum;* and the mandibular division exits the skull through the *foramen ovale.* The trigeminal nerves are the major sensory nerves of the face. They contain sensory neurons that originate from the skin of the face and anterior scalp and from the mucous membranes of the nasal cavity and mouth. The cell bodies of these sensory neurons are located in large **trigeminal (semilunar) ganglia** located at the points where the ophthalmic, maxillary, and mandibular nerves join before entering the brain.

The trigeminal nerves also contain motor neurons that travel within the mandibular nerve to the muscles of mastication (medial and lateral pterygoids, masseter, and temporalis) and to the mylohyoid muscle and the anterior belly of the digastric muscle.

VI: Abducens Nerves

The sixth cranial nerves originate from the metencephalon and exit the brain stem just below the pons (Figure 15.7, Table 15.1). The **abducens nerves** enter F15.7, the orbits through the *superior orbital fissures* along with the oculomotor (III) Table 15.1 and trochlear (IV) cranial nerves. The abducens nerves carry somatic motor neurons to, and proprioceptive sensory neurons from, the remaining extrinsic muscles of each eye—the lateral rectus muscles.

Eye movements involve the coordinated contraction of the extrinsic muscles of both eyes. This coordination, in turn, requires a synchronization of the nerve impulses carried by the oculomotor (III), trochlear (IV), and abducens (VI) cranial nerves.

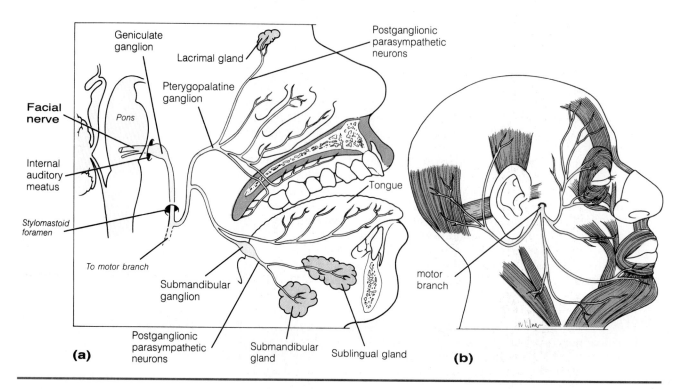

Figure 15.8
Facial nerve (cranial nerve VII). **(a)** Sensory and parasympathetic neurons. **(b)** Voluntary motor branches.

VII: Facial Nerves

The **facial nerves** leave the metencephalon at the lower border of the pons, just lateral to the abducens nerves, and enter the petrous portions of the temporal bones through the *internal auditory meatus*. After traveling through the temporal bones, the facial nerves leave the skull by way of the *stylomastoid foramina*. They then pass forward through the parotid glands and divide into numerous branches that supply somatic motor neurons to the muscles of the face and scalp (Figure 15.8, Table 15.1).

F15.8, Table 15.1

The facial nerves also contain sensory neurons that originate from the taste buds on the anterior two-thirds of the tongue. The cell bodies of these sensory neurons are located in the **geniculate ganglia,** which lie within the petrous portions of the temporal bones.

Parasympathetic neurons to the lacrimal (tear) glands, to mucous glands in the nasal cavity, and to the submandibular and sublingual salivary glands are also carried in the facial nerves. Preganglionic parasympathetic neurons to the lacrimal glands synapse in the **pterygopalatine ganglia** with postganglionic parasympathetic neurons that then pass to the lacrimal glands. The preganglionic parasympathetic neurons to the submandibular and sublingual glands synapse in the **submandibular ganglia** with postganglionic parasympathetic neurons that travel to the glands.

VIII: Vestibulocochlear Nerves

F15.9, Table 15.1

The eighth cranial nerves have two separate divisions: the **cochlear** and **vestibular nerves** (Figure 15.9, Table 15.1). Both of these divisions are entirely sensory and originate from inner-ear receptors located in the petrous portions of the temporal bones. The two divisions join to form a common trunk, the **vestibulocochlear nerve,** which leaves the temporal bones through the *internal auditory meatus* and enters the brain stem just below the pons.

The cochlear divisions transmit impulses related to hearing from the spiral organ, located in the cochlea of the ear (Chapter 17). The cell bodies of the cochlear nerves lie in **spiral ganglia** located within the cochlea.

The vestibular divisions are concerned with equilibrium. Their receptors are located in the ampullae of the semicircular canals and in the saccule and

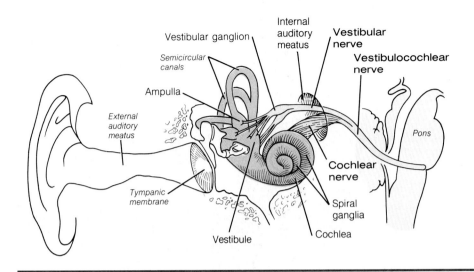

Figure 15.9
Vestibulocochlear nerve (cranial nerve VIII). Notice the vestibular nerve that supplies the vestibule and ampullae and the cochlear nerve that supplies the cochlea. All these structures are within the inner ear.

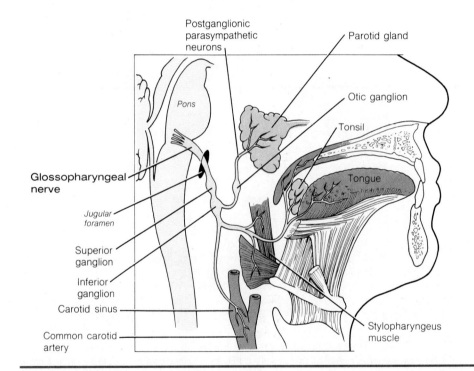

Figure 15.10
Glossopharyngeal nerve (cranial nerve IX) to the tongue, throat, and parotid gland.

utricle of the vestibule of the ear (Chapter 17). The cell bodies of the vestibular nerves are located in the **vestibular ganglia.**

IX: Glossopharyngeal Nerves

The **glossopharyngeal nerves** are mixed nerves, carrying both motor and sensory impulses. As the name indicates, these nerves supply the tongue and the pharynx. They emerge from the medulla and leave the skull through the *jugular foramina* of the temporal bone (Figure 15.10, Table 15.1).

F15.10, Table 15.1

Sensory neurons in the glossopharyngeal nerves carry impulses from the taste buds of the posterior third of the tongue; from the mucous membranes of the pharynx and tonsils; from receptors that are sensitive to changes in blood levels of oxygen and carbon dioxide in the carotid body; and from receptors that monitor the blood pressure in the carotid sinus (Chapter 11). The cell bodies of these sensory neurons are located in the **superior** and **inferior ganglia** of the nerves.

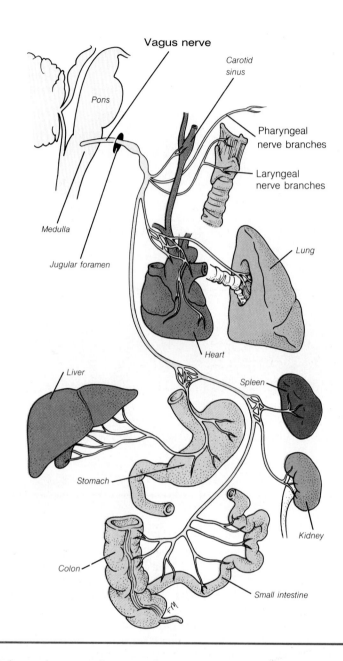

Figure 15.11
Vagus nerve (cranial nerve X), showing its distribution to the neck, thoracic cavity, and abdominal cavity.

The glossopharyngeal nerves also contain motor neurons that supply the stylopharyngeus muscles of the pharynx, which are involved in swallowing. In addition, some motor neurons of the glossopharyngeal nerves intermix with the vagus (X) and accessory (XI) cranial nerves to innervate several other pharyngeal muscles.

Preganglionic parasympathetic neurons that travel in the glossopharyngeal nerves synapse in the **otic ganglia,** from which postganglionic parasympathetic neurons travel to the parotid salivary glands.

X: Vagus Nerves

The **vagus nerves** are the only cranial nerves that are not restricted to the head and neck regions. They leave the sides of the medulla by several rootlets, pass through the *jugular foramina* of the temporal bones, and descend along the pharynx close to the common carotid arteries and the internal jugular veins (Figure 15.11, Table 15.1). After leaving the neck, the vagus nerves enter the thorax and abdomen.

F15.11,
Table 15.1

The vagus nerves carry motor impulses to the voluntary muscles of the pharynx and larynx, and sensory impulses from taste receptors located to-

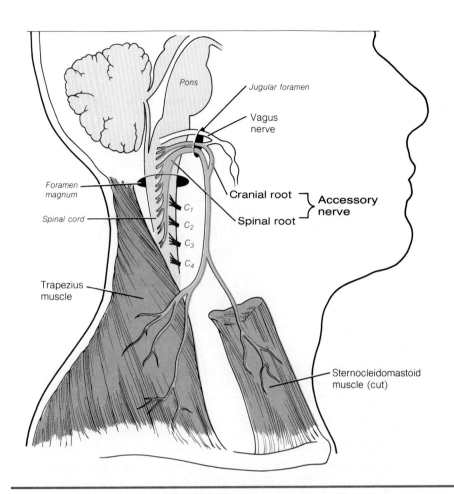

Figure 15.12
Accessory nerve (cranial nerve XI), with cranial and spinal portions separated.

ward the base of the tongue and from the skin of the external ear. Moreover, they have a broad parasympathetic distribution, carrying preganglionic neurons to the involuntary muscles of the thoracic and abdominal viscera as far caudally as the transverse colon of the large intestine. Located close to, or in, the walls of the innervated structures are small ganglia from which short postganglionic parasympathetic neurons supply the structures. The vagus nerves also carry sensory fibers from the viscera. The sensory input from the viscera generally does not reach the conscious level; rather, it automatically regulates heart rate, depth of respiration, blood pressure, digestive processes, and so forth. Under certain conditions these visceral sensations can reach conscious levels, however, as evidenced by sensations of distension and nausea.

XI: Accessory Nerves

The **accessory nerves** are composed of motor neurons to voluntary muscles and some sensory neurons from proprioceptors. Each accessory nerve is actually formed by two nerves. One arises from the medulla and is thus a true cranial nerve, whereas the other arises from the cervical region of the spinal cord and is actually a spinal nerve (Figure 15.12, Table 15.1). The spinal portions pass upward along the sides of the spinal cord and enter the skull through the *foramen magnum* of the occipital bone. In the cranial cavity, the spinal and cranial portions join and leave the skull through the *jugular foramina* along with the glossopharyngeal (IX) and vagus (X) cranial nerves.

The fibers of the cranial portions intermix with the vagus nerves to supply the muscles of the larynx and pharynx. The fibers of the spinal portions supply the trapezius and sternocleidomastoid muscles.

F15.12,
Table 15.1

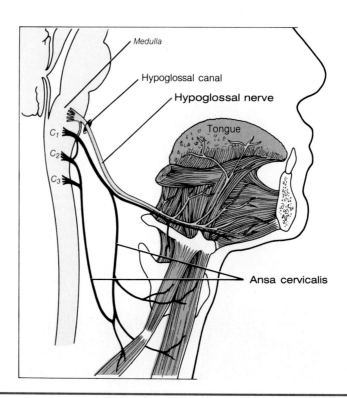

Figure 15.13
Hypoglossal nerve (cranial nerve XII), which supplies the muscles of the tongue. Branches of the first three cervical nerves (shown in black) interconnect with the hypoglossal nerve to supply certain muscles of the throat.

XII: Hypoglossal Nerves

The **hypoglossal nerves** leave the anterior surface of the medulla as a series of rootlets and pass out of the skull through the *hypoglossal canals* of the occipital bone. As their name indicates, they are located beneath the tongue, where they travel anteriorly in close relationship with the first cervical spinal nerve (Figure 15.13, Table 15.1). The hypoglossal nerves consist of motor neurons that supply the intrinsic and extrinsic muscles of the tongue, as well as some sensory neurons from proprioceptors. The first three cervical nerves send motor fibers to some of the muscles of the neck through branches called the *ansa cervicalis*, which appear to arise from the hypoglossal nerves.

F15.13, Table 15.1

SPINAL NERVES

There are 31 pairs of **spinal nerves,** including 8 cervical, 12 thoracic, 5 lumbar, 5 sacral, and 1 coccygeal. With the exception of the first pair of cervical nerves, the spinal nerves leave the vertebral canal by passing through the *intervertebral foramina.* The first pair of cervical nerves exit between the occipital bone and the atlas. The second through seventh pairs of cervical nerves emerge *above* the vertebra for which they are named. The eighth pair of cervical nerves emerge between the seventh cervical and the first thoracic vertebrae. All the remaining pairs of spinal nerves pass *below* the vertebrae for which they are named.

Formation of the Spinal Nerves

The spinal nerves are formed from the union of **ventral** and **dorsal roots** that leave or enter the spinal cord (Figure 15.14). The ventral roots contain axons of motor neurons that leave the anterior and lateral gray horns of the spinal cord. The cell bodies of these motor neurons are located in the spinal cord. The dorsal roots contain axons of sensory neurons that enter the posterior horns of the gray matter. The cell bodies of these sensory neurons lie outside the spinal cord in **dorsal root ganglia (spinal ganglia)** on each dorsal root.

F15.14

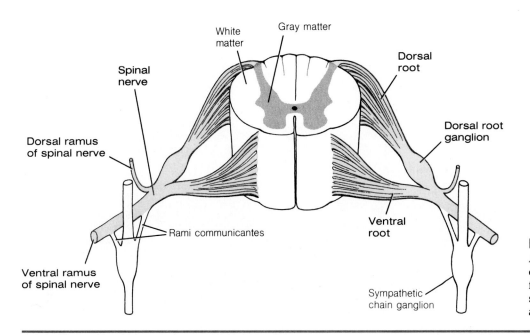

Figure 15.14
A segment of the spinal cord showing the formation of a pair of spinal nerves from dorsal and ventral roots.

Table 15.1 Summary of the Cranial Nerves

Nerves	Site of Exit from Brain	Site of Exit from Skull	Functions
I: Olfactory [F15.2] *sensory*	Telencephalon (cerebral hemisphere)	Cribriform plate of ethmoid	*Sensory* olfaction (sense of smell)
II: Optic [F15.3] *sensory*	Diencephalon	Optic foramen	*Sensory* vision
III: Oculomotor [F15.4] *motor and proprioception*	Mesencephalon (midbrain)	Superior orbital fissure	*Motor* levator palpebrae superioris and the external eye muscles, except superior oblique and lateral rectus *Proprioception* from the innervated muscles *Parasympathetic* ciliary muscle of the lens and the sphincter of pupil
IV: Trochlear [F15.5] *motor and proprioception*	Mesencephalon (midbrain)	Superior orbital fissure	*Motor* superior oblique muscle of the eye *Proprioception* from the superior oblique muscle
V: Trigeminal [F15.6] *mixed*	Metencephalon (pons)		
Ophthalmic division		Superior orbital fissure	*Sensory* cornea; skin of nose, forehead, and scalp
Maxillary division		Foramen rotundum	*Sensory* nasal cavity, palate, upper teeth, skin of cheek, and upper lip
Mandibular division		Foramen ovale	*Sensory* tongue, lower teeth, skin of chin, lower jaw, and temporal regions *Motor* muscles of mastication *Proprioception* from muscles of mastication

Table 15.1   Summary of the Cranial Nerves (continued)

Nerves	Site of Exit from Brain	Site of Exit from Skull	Functions
VI: Abducens [F15.7] *motor and proprioception*	Metencephalon (pons)	Superior orbital fissure	*Motor* lateral rectus muscle of the eye *Proprioception* from lateral rectus muscle
VII: Facial [F15.8] *mixed*	Metencephalon (pons)	Stylomastoid foramen	*Motor* muscles of facial expression *Proprioception* from muscles of facial expression *Sensory* taste from the anterior two-thirds of tongue *Parasympathetic* sublingual and submandibular salivary glands; lacrimal glands; mucous glands of nasal cavity
VIII: Vestibulocochlear [F15.9] *sensory*	Metencephalon (pons)		
Vestibular division		Internal auditory meatus	*Sensory* equilibrium
Cochlear division		Internal auditory meatus	*Sensory* hearing
IX: Glossopharyngeal [F15.10] *mixed*	Myelencephalon (medulla)	Jugular foramen	*Motor* stylopharyngeus muscle; other pharyngeal muscles via cranial nerves X and XI *Proprioception* from innervated muscles *Parasympathetic* parotid salivary glands *Sensory* taste and general sensation from the posterior one-third of the tongue; pharynx, middle-ear cavity, carotid sinus
X: Vagus [F15.11] *mixed*	Myelencephalon (medulla)	Jugular foramen	*Motor* muscles of the pharynx and the larynx *Proprioception* from innervated muscles *Sensory* skin of the external ear; taste from the rear of the tongue; visceral sensory from the thoracic and the abdominal organs *Parasympathetic* organs of the thoracic and abdominal cavities
XI: Accessory [F15.12] *motor and proprioception*	Myelencephalon (medulla)	Jugular foramen	*Motor* trapezius and sternocleidomastoid muscles *Proprioception* from innervated muscles
XII: Hypoglossal [F15.13] *motor and proprioception*	Myelencephalon (medulla)	Hypoglossal canal	*Motor* intrinsic and extrinsic muscles of the tongue *Proprioception* from innervated muscles

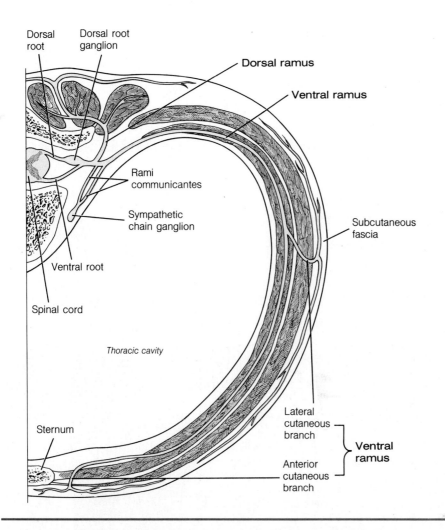

Figure 15.15
Pathways of the ventral and dorsal rami of a thoracic spinal nerve.

After the roots join, all spinal nerves are mixed nerves containing the processes of motor (somatic and visceral) and sensory neurons.

Branches of the Spinal Nerves

Soon after passing through the intervertebral foramina, each spinal nerve divides into two branches: a **dorsal ramus** and a **ventral ramus** (Figure 15.15). The dorsal rami pass posteriorly to supply the skin and muscles of the back. The ventral rami are longer, and their distribution varies in different body regions. Like the spinal nerves, both rami are mixed, containing motor and sensory fibers. In the thoracic region, the ventral rami travel in the intercostal spaces between the ribs to supply the skin and muscles of the lateral and anterior body walls. In the cervical, lumbar, and sacral regions, the ventral rami of successive spinal nerves unite to form *plexuses* (networks) that give rise to nerves supplying the skin, muscles, and joints of the upper and lower limbs.

Distribution of the Spinal Nerves

The rami of the spinal nerves are distributed throughout the body in a fairly systematic pattern, as evidenced by the distribution of sensory fibers to the skin in uniform regions called **dermatomes** (Figure 15.16). Each dermatome is an area of skin that is supplied by the sensory fibers in a single dorsal root of a spinal nerve. The pattern is particularly regular in the trunk region, where the rami of each spinal nerve supply a horizontal band of skin. The regular pattern of nerve distribution is also present, though somewhat modified, in the skin of the limbs. In the upper limbs, the ventral rami of cervical nerves supply the posterior and anterolateral surfaces of the entire limb, as well as the ventral surfaces of the hands. The ventral ramus of the first thoracic nerve

F15.15

F15.16

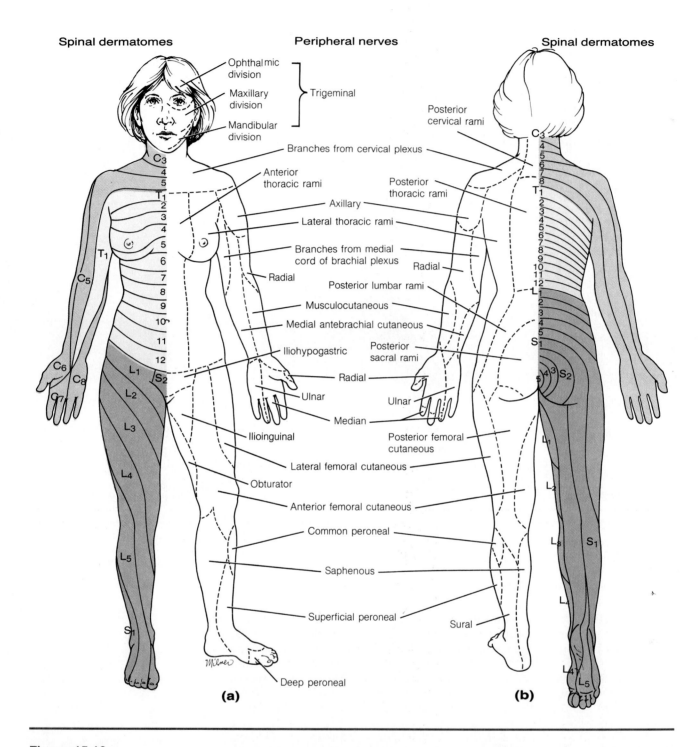

Spinal dermatomes

Peripheral nerves

Spinal dermatomes

Ophthalmic division
Maxillary division
Mandibular division
Trigeminal

Posterior cervical rami

Branches from cervical plexus

Anterior thoracic rami

Posterior thoracic rami

Axillary

Lateral thoracic rami

Branches from medial cord of brachial plexus

Radial

Radial

Posterior lumbar rami

Musculocutaneous

Medial antebrachial cutaneous

Iliohypogastric

Posterior sacral rami

Radial

Ulnar

Ulnar

Median

Ilioinguinal

Posterior femoral cutaneous

Lateral femoral cutaneous

Obturator

Anterior femoral cutaneous

Common peroneal

Saphenous

Superficial peroneal

Sural

Deep peroneal

(a)   **(b)**

Figure 15.16
Dermatome and peripheral distribution of spinal nerve innervations.
(a) Anterior surface of the body. **(b)** Posterior surface of the body.

(T_1) supplies the anteromedial surfaces of the limbs (in the anatomical position). In the lower limbs, the ventral rami of the lumbar nerves supply the anterior surfaces and the ventral rami of the sacral nerves supply the posterior surfaces.

Plexuses and Peripheral Nerves

The peripheral nerves that are formed by the intermixing of the ventral rami of the spinal nerves in plexuses have specific names. These nerves primarily supply the skin and the underlying muscles of the limbs. Since each of these named peripheral nerves is formed of fibers from more than one spinal nerve, their distribution patterns (Figure 15.16) do not duplicate the cutaneous dis-

F15.16

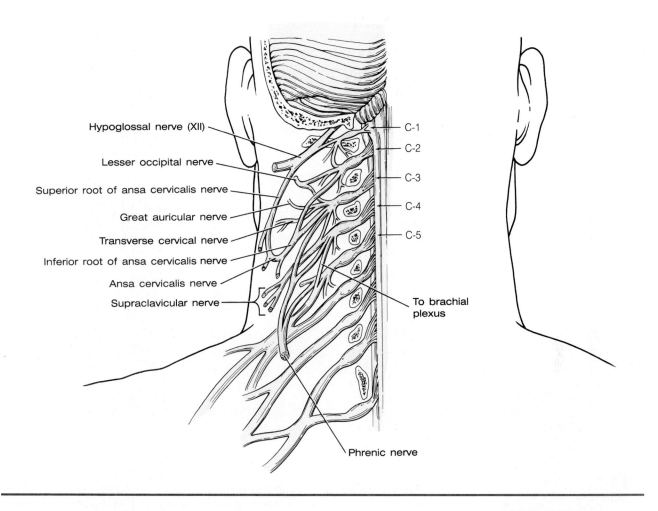

Hypoglossal nerve (XII)

Lesser occipital nerve

Superior root of ansa cervicalis nerve

Great auricular nerve

Transverse cervical nerve

Inferior root of ansa cervicalis nerve

Ansa cervicalis nerve

Supraclavicular nerve

C-1
C-2
C-3
C-4
C-5

To brachial plexus

Phrenic nerve

Figure 15.17
Major branches of the cervical plexus.

tribution patterns of the spinal nerves. The main nerve plexuses, all of which are paired, are the *cervical, brachial,* and *lumbosacral* plexuses.

Cervical Plexus

Each *cervical plexus* is formed by the ventral rami of the first four cervical nerves (Figure 15.17). Branches from the plexus supply the muscles and skin **F15.17** of the head, neck, shoulder, and chest. Branches of the cervical plexus also interconnect with cranial nerves X (vagus), XI (accessory), and XII (hypoglossal). One branch from each of the cervical plexuses, the **phrenic nerve,** passes through the thorax to supply the diaphragm. Although this innervation may seem unusual for a cervical nerve, during embryonic development the diaphragm arises from cervical myotomes. As it assumes its adult position in the lower portion of the thoracic cavity, the diaphragm retains its embryonic nerve supply. The major branches from the cervical plexus are summarized in Table 15.2.   **Table 15.2**

Brachial Plexus

Each **brachial plexus** is formed by the ventral rami of the last four cervical nerves (C_5 to C_8) and the first thoracic nerve (T_1). The brachial plexus extends downward and laterally, passing behind the clavicle, to enter the axilla (Figure 15.18). The *roots* of the plexus, which are the ventral rami of the spinal **F15.18** nerves, unite to form a *superior trunk* (C_5–C_6), a *middle trunk* (C_7), and an *inferior trunk* (C_8–T_1). A *dorsal scapular nerve* to the levator scapulae and the rhomboid muscles, and a *long thoracic nerve* to the serratus anterior muscle branch off of the roots.

Table 15.2   The Cervical Plexus (F15.16, F15.17)

Nerve	Spinal Nerves Involved (Ventral Rami)	Distribution
SUPERFICIAL BRANCHES		
Lesser occipital	C_2	*Skin* of scalp behind and above ear
Greater auricular	C_2 through C_3	*Skin* over parotid gland, over mastoid process, and on back of ear
Transverse cutaneous	C_2 through C_3	*Skin* over side and front of neck
Supraclavicular	C_3 through C_4	*Skin* over upper region of shoulder and chest
DEEP (MUSCULAR) BRANCHES		
Ansa cervicalis (leaves the hypoglossal nerve)		
Superior root	C_1 through C_2	Deep *muscles* of neck, including geniohyoid and thyrohyoid
Inferior root	C_2 through C_3	Infrahyoid *muscles,* including sternohyoid and sternothyroid
Phrenic	C_3 through C_5	*Diaphragm*
Muscular Branches	C_2 through C_4	*Muscles* sternocleidomastoid, trapezius, levator scapulae, and scalenus medius

Each trunk divides into *anterior* and *posterior divisions.* Branching off of the trunks are a *suprascapular nerve* to the supraspinatus and infraspinatus muscles and a nerve to the subclavius muscle. The trunks, in turn, separate into *posterior, medial,* and *lateral cords.* Five major nerves, as well as several smaller ones, arise from these cords and provide the entire nerve supply to the skin and muscles of the upper limbs (Figure 15.19). The major branches of the brachial plexus are summarized in Table 15.3.

F15.19
Table 15.3

POSTERIOR CORD   The **axillary nerve** passes laterally from the posterior cord to supply the skin and muscles of the shoulder. The **radial nerve,** the main branch of the posterior cord, passes behind the humerus and curves around it (in the radial groove) to supply the skin on the posterior surface of the arm, forearm, and hand, as well as the extensor muscles of the arm and forearm. Smaller branches of the posterior cord include the *superior* and *inferior subscapular nerves* to the subscapularis and teres major muscles and the *thoracodorsal nerve* to the latissimus dorsi muscle.

LATERAL CORD   The **musculocutaneous nerve** from the lateral cord supplies the skin on the lateral surface of the forearm and several anterior muscles of the arm. The *lateral pectoral nerve* branches off of the lateral cord and supplies the pectoralis major muscle.

MEDIAL CORD   The **ulnar nerve** leaves the medial cord and passes behind the medial epicondyle of the humerus to supply the skin on the medial surface of the hand and some flexor muscles of the forearm, as well as several intrinsic muscles of the hand. Other branches from the medial cord include the *medial pectoral nerve* to the pectoralis major and minor muscles, the *medial brachial cutaneous nerve* to the skin of the medial side of the arm, and the *medial antebrachial cutaneous nerve* to the skin on the medial side of the forearm.

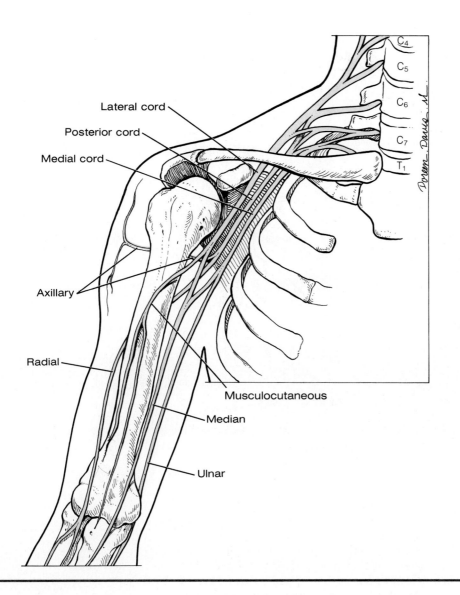

Figure 15.18
Location of the brachial plexus and the formation of the major nerves of the upper limb.

MEDIAN NERVE   The **median nerve** is formed by branches from the medial and lateral cords of the brachial plexus. It supplies the skin and muscles of the lateral portion of the palmar surface of the hand and the flexor muscles of the forearm.

Lumbosacral Plexus

The nerves of each **lumbosacral plexus** supply the skin and muscles of the buttocks, pelvis, lower abdomen, and lower limbs (Figure 15.20). The plexus is divisible into two sections, the *lumbar plexus* and the *sacral plexus*, which are connected by a *lumbosacral trunk* of nerves.   **F15.20**

LUMBAR PLEXUS   Each **lumbar plexus** is formed from the ventral rami of the first four lumbar nerves and some nerve fibers from the twelfth thoracic nerve. The nerves that arise from the lumbar plexus supply the lower abdomen and the anterior and medial portions of the lower limb (Figure 15.16).   **F15.16**
The major nerves that arise from the lumbar plexus are the femoral and the obturator nerves. Smaller branches of this plexus are described in Table 15.4.   **Table 15.4**

The **femoral nerve** passes underneath the inguinal ligament to supply the anterior muscles of the thigh (Figure 15.20*b*). Two superficial branches of the   **F15.20b**
femoral nerve, the **anterior femoral cutaneous nerve** and the **saphenous nerve,** innervate the skin of the anterior medial surface of the thigh, leg, and

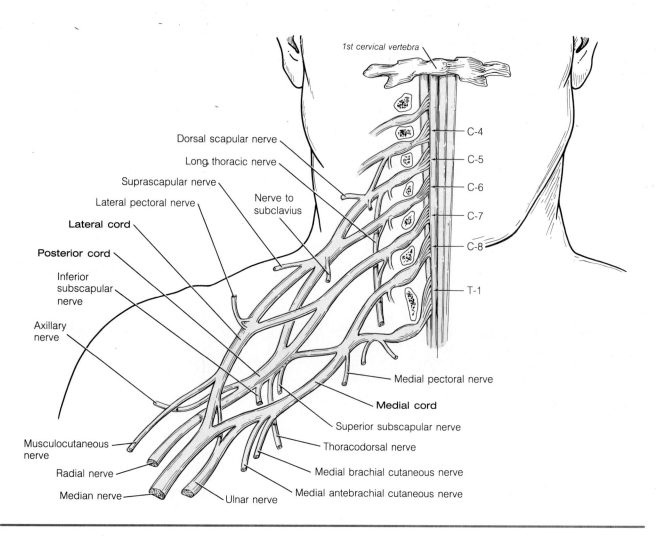

Figure 15.19

Major branches of the
brachial plexus.

Table 15.3 Major Branches of the Brachial Plexus [F15.16, F15.18, 15.19]

Nerve	Spinal Nerves Involved (Ventral Rami)	Distribution
POSTERIOR CORD		
Axillary	C_5 and C_6	*Skin* of the shoulders
		Muscles teres minor and deltoid
Radial	C_5 through C_8; T_1	*Skin* of posterior lateral surface of arm, forearm, and hand
		Muscles triceps brachii, supinator, anconeus, brachioradialis, extensor carpi radialis brevis, extensor carpi radialis longus, extensor carpi ulnaris, and several muscles that move the fingers (see Table 7.14)
LATERAL CORD		
Musculocutaneous	C_5 through C_7	*Skin* of lateral surface of forearm
		Muscles brachialis, biceps brachii, coracobrachialis

notch and travels down the posterior surface of the thigh, innervating the muscles and skin of the region (Figure 15.20c). The sciatic nerve is actually two nerves wrapped in a common sheath. In the lower thigh, these two nerves separate into the common peroneal nerve and the tibial nerve. The **common peroneal nerve** passes obliquely along the lateral side of the popliteal fossa to supply muscles in the anterior and lateral compartments of the leg and the skin on the anterior surface of the leg and the dorsum of the foot. The common peroneal nerve has branches called the **superficial** and **deep peroneal nerves.** The superficial peroneal supplies the muscles of the lateral compartment of the leg; the deep peroneal supplies the muscles of the anterior compartment of the leg. The **tibial nerve** supplies the muscles and skin on the posterior surface of the leg and the sole of the foot. The tibial nerve gives rise to the **sural nerve** (which also receives branches from the common peroneal nerve) and ends on the sole of the foot as the **medial** and **lateral plantar nerves.** The sural nerve supplies the posterior and lateral surfaces of the leg and foot.

F15.20c

Table 15.5   The Sacral Plexus [F15.16, F15.20]

Nerve	Spinal Nerves Involved (Ventral Rami)	Distribution
Superior gluteal	L_4, L_5; S_1	*Muscles* gluteus minimus, gluteus medius, and tensor fasciae latae
Inferior gluteal	L_4, L_5; S_1	*Muscles* gluteus maximus
Posterior femoral cutaneous	S_1 through S_3	*Skin* on posterior surface of the thigh
Pudendal	S_2 through S_4	*Skin and muscles* of perineum External genitalia
Sciatic		
Tibial *Sural*	L_4, L_5; S_1 through S_3	*Skin* of posterior surface of leg and sole of foot
Medial and lateral plantar		*Muscles* biceps femoris, semimembranosus, semitendinosus, flexor digitorum longus, flexor hallucis longus, tibialis posterior, popliteus, and intrinsic muscles of foot (see Table 7.22)
Common peroneal *Superficial and deep peroneal*	L_4, S_5; S_1, S_2	*Skin* of anterior surface of leg and dorsum of foot
		Muscles peroneus brevis, peroneus longus, peroneus tertius, tibialis anterior, extensor hallucis longus, extensor digitorum longus, extensor digitorum brevis

CONDITIONS OF CLINICAL SIGNIFICANCE

The Peripheral Nervous System

Injury and Regeneration of Peripheral Nerves

The most common disorders involving neurons of the peripheral nervous system are those associated with damage or inflammation. When a peripheral nerve is severely damaged or severed, the portion of the nerve that is distal to the injury undergoes degenerative changes (Figure 15.21). Within a few days following the injury, the nerve fibers and their myelin sheaths are broken down by macrophages from the endoneurium. Cells of the sheath of Schwann seem to assist in the degeneration by exerting phagocytic action themselves.

Following the injury, the Schwann cells proliferate and form cords in the endoneurial tubes. In a matter of days, sprouts form on the stumps of the damaged nerve fibers. Some of these sprouts grow into the endoneurial tubes, and if there are no obstructions (such as scar tissue) in the tubes, the fibers may again grow out to the periphery and eventually innervate the structures that were separated from their nerve supply. As the new fibers from the proximal stump of the neuron grow along the endoneurial tubes, they are surrounded by Schwann cells. Because nerve fibers regenerate at a rate of from 1 to 4 mm per day, it is possible to estimate the length of time required for the nerve supply to return to the denervated structure.

The first indication that a peripheral nerve has reached the vicinity of a denervated structure is evidenced by an improved blood supply to the area. This stage is followed by the return of sensory function to the structure. In the case of a paralyzed skeletal muscle, the motor function is the last function to return.

Neuritis

The term *neuritis* means "inflammation of a nerve," but many of the conditions referred to as neuritis are more degenerative than inflammatory. Neuritis is characterized by a range of sensations from mild tingling to sharp, stabbing pains. It can result from a number of conditions, including mechanical damage to the involved nerves, prolonged pressure on the nerves, vas-

cular disorders involving the nerves, and direct invasion of the nerves by pathological organisms.

Bell's palsy is a term used to describe peripheral inflammation of the facial nerve (cranial nerve VII). In Bell's palsy, all the muscles of facial expression are paralyzed on the affected side, causing them to sag. This paralysis causes difficulties in speaking and eating, and it results in the lower portion of the face being pulled to the unaffected side when the person smiles.

Neuralgia

Neuralgia refers to spasms of severe pain along the path of a peripheral nerve. There are many varieties of neuralgia, most of which have unknown causes. In *trigeminal neuralgia (tic douloureux)*, there are short attacks of excruciating pain along the course of the maxillary and mandibular divisions of the trigeminal nerve (cranial nerve V). The attacks of pain are short at first, and they may be followed by periods of several weeks without pain. With time the spasms generally become more frequent and of longer duration. The cause of trigeminal neuralgia is unknown.

Herpes Zoster (Shingles)

The disease *herpes zoster*, also known as *shingles*, is a viral infection of the dorsal root ganglia of the spinal nerves. The infection causes pain and produces fluid-filled vesicles on the skin along the path of the peripheral sensory neurons that are infected.

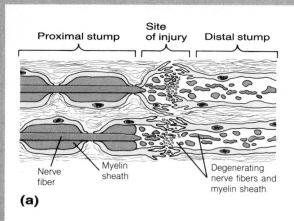

Figure 15.21
Regeneration of a peripheral nerve. **(a)** Nerve fibers degenerate distal (and a short distance proximal) to the site of injury. **(b)** Schwann cells become active, forming cords within the endoneurial tube. **(c)** Nerve fibers then form sprouts that grow into the endoneurial tube alongside the Schwann cell cords. **(d)** Nerve fibers follow the Schwann cell cords. The Schwann cells then wrap around the nerve fibers, forming a myelin sheath around the new fiber sprouts.

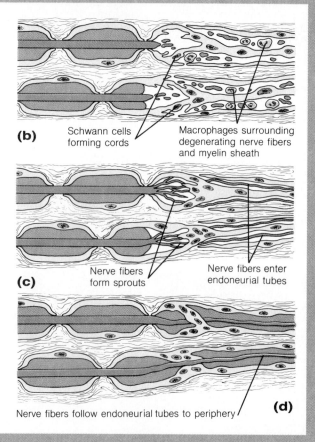

STUDY OUTLINE

PERIPHERAL NERVOUS SYSTEM includes 12 pairs of cranial nerves and 31 pairs of spinal nerves; divided into afferent (sensory) and efferent (motor) divisions; efferent division divided functionally into somatic and autonomic nervous systems. **p. 415**

CRANIAL NERVES 12 pairs. pp. 415–424

I: OLFACTORY NERVES arise from receptor cells in nasal mucosa; entirely sensory; control sense of smell.

II: OPTIC NERVES entirely sensory; carry impulses concerned with vision; arise from retina of eye.

OPTIC CHIASMA where the two optic nerves meet.

OPTIC TRACT beyond optic chiasma; each optic tract consists of fibers from retinas of both eyes.

OPTIC RADIATIONS neurons from thalamus to visual cortex of occipital lobe.

III: OCULOMOTOR NERVES
1. Emerge from midbrain, superior to pons.
2. Innervate four extrinsic muscles of eyeball that assist in voluntary movements of eyeball and levator palpebrae superioris muscle, which elevates upper eyelids.
3. Also contain neurons of parasympathetic nervous system that cause reflex adjustments of eyes to varying light intensities and focus eyes for near and far vision.

IV: TROCHLEAR NERVES
1. Arise below inferior colliculi on dorsal surface of midbrain.
2. Somatic motor neurons to and proprioceptive neurons from superior oblique eye muscle.
3. Assists in voluntary movements of eyeball by controlling contraction of superior oblique muscle.

V: TRIGEMINAL NERVES ophthalmic, maxillary, and mandibular nerves.
1. Emerge from lateral sides of pons.
2. Major sensory nerves of face.
3. Motor neurons to muscles of mastication.

VI: ABDUCENS NERVES
1. Originate from metencephalon below pons.
2. Somatic motor neurons to and proprioceptive neurons from lateral rectus eye muscle.
3. Coordinate with oculomotor and trochlear nerves to cause contraction of extrinsic muscles of both eyes.

VII: FACIAL NERVES
1. Arise from metencephalon at lower border of pons.
2. Somatic motor neurons to face and scalp muscles.
3. Carry sensory neurons from taste buds on anterior two-thirds of tongue.
4. Carry parasympathetic neurons to lacrimal glands and submandibular and sublingual salivary glands.

VIII: VESTIBULOCOCHLEAR NERVES cochlear and vestibular nerves.
1. From inner-ear receptors; entirely sensory.
2. Cochlear division transmits impulses related to hearing.
3. Vestibular division concerned with equilibrium.

IX: GLOSSOPHARYNGEAL NERVES
1. Emerge from medulla.
2. Mixed nerves—motor and sensory.
3. Sensory—impulses from taste buds of posterior third of tongue, from membranes of pharynx and tonsils, from pressure receptors in carotid sinus.
4. Motor—innervate pharyngeal muscles.
5. Parasympathetic neurons to parotid salivary glands.

X: VAGUS NERVES only cranial nerve not restricted to head and neck regions.
1. Arise from sides of medulla.
2. Somatic motor neurons to and sensory neurons from pharynx and larynx.
3. Sensory input from viscera regulates heart rate, respiration depth, blood pressure, digestion, and so forth.
4. Parasympathetic neurons to thoracic and abdominal viscera and blood vessels.

XI: ACCESSORY NERVES
1. Composed of a true cranial nerve (which arises from medulla) and a cervical spinal nerve.
2. Cranial portions supply pharynx and larynx muscles.
3. Spinal portions supply trapezius and sternocleidomastoid muscles.

XII: HYPOGLOSSAL NERVES
1. Arise from anterior surface of medulla.
2. Supply intrinsic and extrinsic tongue muscles.
3. Sensory neurons from proprioceptors.

SPINAL NERVES 31 pairs. pp. 424–435

FORMATION, BRANCHES, AND DISTRIBUTION
1. Union of ventral and dorsal roots.
2. All are mixed (motor and sensory) nerves.
3. Each spinal nerve divides into dorsal and ventral rami.
4. Sensory nerves in rami are distributed in bands called *dermatomes.*

PLEXUSES AND PERIPHERAL NERVES

CERVICAL PLEXUS formed by ventral rami of first four cervical nerves; supply muscles and skin of head, neck, shoulder, and chest; phrenic nerve supplies diaphragm.

BRACHIAL PLEXUS formed by ventral rami of last four cervical nerves and first thoracic nerve. Dorsal scapular and long thoracic nerves branch from roots; suprascapular nerve and nerve to subclavius muscle branch from the trunks of the plexus.

Posterior Cord
1. *Axillary nerve:* skin and muscles of shoulder.
2. *Radial nerve:* skin on posterior surface of arm, forearm, and hand; extensor muscle of arm and forearm.
3. *Superior and inferior subscapular nerves* and *thoracodorsal nerves* branch from posterior cord.

Lateral Cord
1. *Musculocutaneous nerve:* skin of lateral surface of forearm; anterior muscles of arm.
2. *Lateral pectoral nerve.*

Medial Cord
1. *Ulnar nerve:* skin of medial surface of hand; forearm; intrinsic hand muscles.
2. *Medial pectoral nerve; medial brachial cutaneous nerve; medial antebrachial cutaneous nerve.*

Median Nerve from branches of *medial* and *lateral* cords; flexor muscles of forearm; skin and muscles of lateral palmar portion of hand.

LUMBOSACRAL PLEXUS skin and muscles of buttocks, pelvis, lower abdomen, lower limbs.

Lumbar Plexus ventral rami L$_1$ through L$_4$.
1. *Femoral nerve:* anterior thigh muscles; skin of anterior medial thigh, leg, and foot; has anterior femoral cutaneous and saphenous branches.
2. *Obturator nerve:* skin of medial thigh; adductor muscles of thigh.

Sacral Plexus ventral rami L$_4$, L$_5$, and S$_1$ through S$_4$.
1. *Sciatic nerve:* muscles and skin of posterior surface of thigh.
2. *Common peroneal nerve* (through superficial and deep peroneal nerves): muscles of anterior and lateral compartments of leg; skin on anterior surface of leg and dorsum of foot. Has superficial and deep peroneal branches.

3. *Tibial nerve:* muscles on skin on posterior surface of leg and sole of foot. Has sural as well as medial and lateral plantar branches.

CONDITIONS OF CLINICAL SIGNIFICANCE: THE PERIPHERAL NERVOUS SYSTEM pp. 435–436

INJURY AND REGENERATION OF PERIPHERAL NERVES
1. Macrophages break down nerve fibers and myelin sheath distal to site of injury.
2. Schwann cells proliferate, forming cords.
3. Buds form on stumps of nerves.
4. New fibers develop from buds and grow toward periphery.
5. Schwann cells surround new fibers.

NEURITIS
1. Degenerative or inflammatory condition.
2. Tingling to sharp pain.
3. Results from mechanical damage, prolonged pressure, vascular disorders, pathological organism invasion.
4. Bell's palsy is neuritis of facial nerve.

NEURALGIA spasms of severe pain along peripheral nerve path.

TRIGEMINAL NEURALGIA (TIC DOULOUREUX) short attacks of severe pain along maxillary and mandibular divisions of trigeminal cranial nerve.

HERPES ZOSTER (SHINGLES) viral infection of dorsal root ganglia of spinal nerves.

SELF-QUIZ

1. The peripheral nervous system includes the:
(a) brain; (b) spinal cord; (c) ganglia.

2. Which motor neurons are not under conscious control, are within the cranial nerves, and supply smooth or cardiac muscle? (a) visceral motor neurons; (b) somatic motor neurons; (c) sympathetic neurons.

3. The olfactory nerves are: (a) motor and leave the mesencephalon; (b) sensory and enter the pons; (c) sensory and enter the telencephalon.

4. The optic nerves are: (a) sensory and leave the skull through the optic foramen; (b) motor and leave the telencephalon; (c) motor and leave the skull through the superior orbital fissure.

5. The oculomotor nerves: (a) include proprioceptor fibers; (b) are located in the diencephalon; (c) leave the skull through the foramen rotundum.

6. The trochlear nerves: (a) arise in the diencephalon; (b) aid in voluntary movements of the eyeball; (c) are large nerves.

7. The trigeminal nerves are the: (a) major sensory nerves of the face; (b) principal nerves that control eyeball movement; (c) nerves that control the tongue.

8. The abducens nerves: (a) originate from the telencephalon; (b) are sensory nerves; (c) leave the skull through the superior orbital fissure.

9. The facial nerves: (a) leave the skull through the foramen magnum; (b) also supply the viscera of the thorax; (c) are mixed nerves.

10. The vestibulocochlear nerves: (a) are concerned with equilibrium; (b) are motor nerves; (c) arise in the telencephalon.

11. The glossopharyngeal nerves: (a) leave the skull through the stylomastoid foramen; (b) carry motor and sensory impulses; (c) regulate contractions of the abdominal viscera.

12. The vagus nerves: (a) supply most organs of the thorax and abdomen; (b) are restricted to the head and neck region; (c) carry sensory impulses to the pharynx.

13. The accessory nerves are composed of: (a) somatic motor neurons; (b) sensory neurons from proprioceptors; (c) parasympathetic neurons; (d) both a and b.

14. The hypoglossal nerves: (a) supply the intrinsic and extrinsic muscles of the pharynx; (b) arise in the telencephalon; (c) consist of somatic motor neurons that supply the tongue.

15. The spinal nerves are formed from the union of ventral and dorsal roots that leave or enter the spinal cord. True or false?

16. After the ventral and dorsal roots join, all the spinal nerves are mixed nerves containing the processes of motor neurons (somatic and visceral) and sensory neurons. True or false?

17. The cell bodies of the motor neurons are located outside the spinal cord, whereas those of the sensory neurons lie within the spinal cord. True or false?

18. From which plexus does the phrenic nerve pass through the thorax to supply the diaphragm? (a) lumbosacral; (b) brachial; (c) cervical.

19. The skin on the posterior surface of the arm, forearm, and hand, and the extensor muscles of the arm and forearm, are supplied by the: (a) radial nerve; (b) phrenic nerve; (c) musculocutaneous nerve.

20. The palmaris longus and flexor carpi radialis muscles are supplied by the: (a) ulnar nerve; (b) radial nerve; (c) median nerve.

21. The sciatic nerve, which is the largest nerve in the body, is the main branch of the sacral plexus. True or false?

22. It is possible for a peripheral nerve that has been severed to regenerate and innervate the same structures it supplied before it was damaged. True or false?

LEARNING OBJECTIVES

After completing this chapter, you should be able to:

- State the function of the autonomic nervous system.

- Distinguish between the sympathetic and parasympathetic divisions of the autonomic nervous system in terms of structure and function.

- List the paths that the preganglionic sympathetic axons might follow upon entering the chain ganglia.

- Describe biofeedback as it relates to the autonomic nervous system.

CHAPTER CONTENTS

ANATOMY OF THE AUTONOMIC NERVOUS SYSTEM

FUNCTIONS OF THE AUTONOMIC NERVOUS SYSTEM

BIOFEEDBACK

CONDITIONS OF CLINICAL SIGNIFICANCE: THE AUTONOMIC NERVOUS SYSTEM

The Autonomic Nervous System

The **autonomic nervous system (ANS)** is a part of the efferent division of the peripheral nervous system. It is composed entirely of visceral motor (efferent) neurons that innervate and thus regulate the activity of cardiac muscle, smooth muscle, and the glands of the body. This system is normally an involuntary system that functions below the conscious level.

We will consider the autonomic nervous system separately from the peripheral nervous system, but keep in mind that the ANS is structurally and functionally an integral part of the body's single nervous system. In fact, many nerve fibers of the autonomic nervous system travel in the spinal nerves and certain cranial nerves. Even though the autonomic nervous system has only motor functions, visceral sensory fibers of the afferent division of the peripheral nervous system travel along the same pathways as the visceral motor fibers of the autonomic nervous system. The cell bodies of these visceral sensory neurons are located in the dorsal root ganglia of the spinal nerves or in one of the outlying ganglia of certain cranial nerves.

ANATOMY OF THE AUTONOMIC NERVOUS SYSTEM

The efferent pathways of the autonomic nervous system that run from the central nervous system to the effectors are composed of two neurons. One of these neurons, called a **preganglionic (presynaptic) neuron,** has its cell body in the central nervous system. The axon of the preganglionic neuron travels to an **autonomic ganglion** located outside the central nervous system, where it synapses with other neurons called **postganglionic (postsynaptic) neurons.** The axons of postganglionic neurons generally form into branching networks known as **autonomic plexuses,** and then travel to the various effectors. This two-neuron chain is in contrast to the somatic nervous system, where a single motor neuron travels from the central nervous system to the structure innervated.

The autonomic nervous system can be separated structurally as well as functionally into two divisions: the *sympathetic division* and the *parasympathetic division* (Figure 16.1).

Sympathetic Division

The cell bodies of the preganglionic neurons of the **sympathetic division** of the autonomic nervous system are located in the lateral horns of the gray matter of the spinal cord from the first thoracic segment (T_1) through the second lumbar segment (L_2). For this reason the sympathetic division is also called the **thoracolumbar division.** The axons of these visceral motor neurons leave the spinal cord through the ventral roots, along with somatic motor axons, and enter the dorsal and ventral rami of the spinal nerves of the various segments. After going only a short distance in the ventral rami of the spinal nerves, all the preganglionic sympathetic nerve fibers leave the ventral rami and enter one of a series of interconnected **chain (paravertebral, or sympathetic) ganglia** (Figure 16.2). The chain ganglia form longitudinal pathways, called **sympathetic trunks,** on each side of the bodies of the vertebrae

441

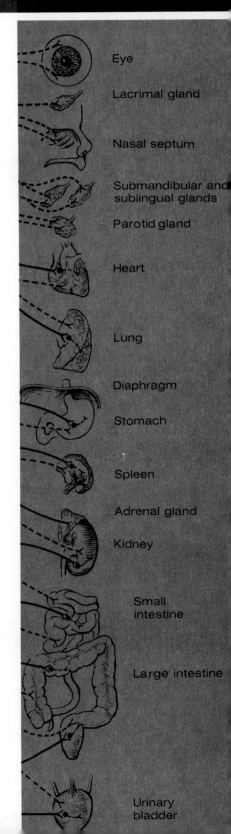

Eye

Lacrimal gland

Nasal septum

Submandibular and sublingual glands

Parotid gland

Heart

Lung

Diaphragm

Stomach

Spleen

Adrenal gland

Kidney

Small intestine

Large intestine

Urinary bladder

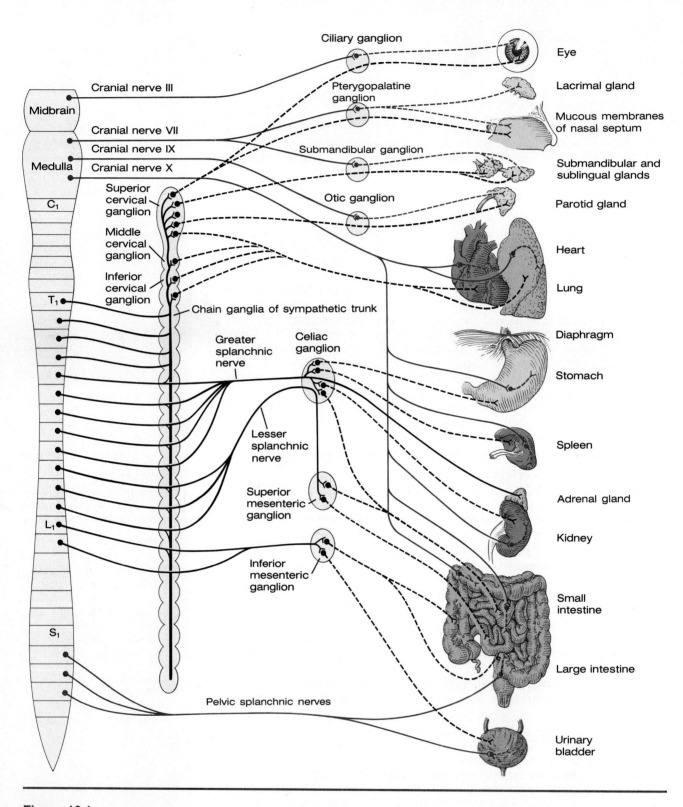

Figure 16.1
The autonomic nervous system. The
parasympathetic division is shown in rust;
the sympathetic division is shown in black.
The solid lines indicate preganglionic nerve
fibers; the dotted lines indicate postganglionic
nerve fibers.

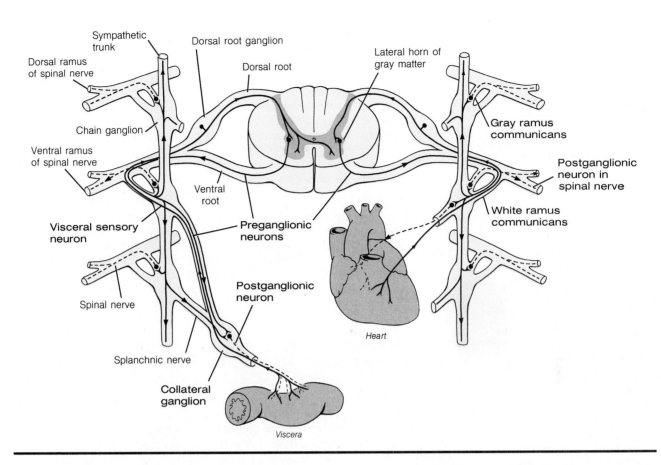

Symptathetic trunk

Dorsal ramus of spinal nerve

Dorsal root ganglion

Dorsal root

Lateral horn of gray matter

Gray ramus communicans

Chain ganglion

Postganglionic neuron in spinal nerve

Ventral ramus of spinal nerve

Ventral root

White ramus communicans

Visceral sensory neuron

Preganglionic neurons

Spinal nerve

Postganglionic neuron

Heart

Splanchnic nerve

Collateral ganglion

Viscera

Figure 16.2
The sympathetic division of the autonomic nervous system. The central connections of the sympathetic neurons are shown as seen in one segment of the thoracic spinal cord. Solid black lines indicate sympathetic preganglionic neurons; dotted black lines indicate sympathetic postganglionic neurons; colored lines indicate sensory neurons. The arrows show the direction of the nerve impulse.

along the entire length of the vertebral column, including the cervical and sacral regions. Except in the cervical region, where several chain ganglia fuse together to form three or four larger ganglia, the ventral ramus of each spinal nerve generally has a chain ganglion associated with it. Since most of the preganglionic sympathetic nerve fibers are myelinated, the short pathways they form in passing from the ventral ramus of the spinal nerve to the sympathetic trunks appear white and are called **white rami communicantes.** There are 14 pairs of white rami communicantes connecting the first thoracic through the second lumbar spinal nerves to the chain ganglia.

Upon entering the chain ganglia, the preganglionic sympathetic axons follow one of three courses:

1. *Preganglionic sympathetic axons may synapse with the cell bodies of postganglionic neurons in the chain ganglion located at the same level at which the preganglionic fibers entered the chain.* The axons of the postganglionic neurons return directly to the spinal nerve and travel to the periphery in the dorsal and ventral rami of the spinal nerve. These axons innervate effectors in the skin, including smooth muscles in the walls of the blood vessels of the skin, sweat glands, and the arrector pili muscles of the hairs. Because most postganglionic axons are unmyelinated, the pathways they form as they pass from the chain ganglia to the spinal nerves appear gray and are called **gray rami communicantes.**

2. *Preganglionic sympathetic axons may travel up or down within the sympathetic trunk before synapsing with postganglionic neurons at a higher or lower level.* The axons of some of these postganglionic neurons enter the cervical and sacral spinal nerves through gray rami communicantes, and ultimately innervate the skin (blood vessels, sweat glands, arrector pili muscles) of the re-

gions supplied by those nerves. Notice that, although preganglionic axons enter the chain ganglia through white rami from only the first thoracic through the second lumbar segments, every chain ganglion is connected to a spinal nerve by a gray ramus. Thus, because preganglionic axons may travel to higher or lower levels in the sympathetic trunk, the rami of every spinal nerve receive axons of postganglionic sympathetic neurons. However, some postganglionic axons in the thoracic and cervical ganglia pass directly from the ganglia to innervate the thoracic viscera and structures in the head, rather than entering the rami of a spinal nerve.

3. *Preganglionic sympathetic axons may pass through the sympathetic trunk without synapsing.* Some preganglionic axons from the thoracic region form pathways, called **greater** and **lesser splanchnic nerves,** that pass through the diaphragm and lead to **collateral (prevertebral) ganglia** located on the front of the abdominal aorta. The major collateral ganglia, which are named for the vascular branches of the aorta nearest them, are the **celiac, superior mesenteric,** and **inferior mesenteric.** Within the collateral ganglia, the preganglionic axons synapse with postganglionic neurons. The axons of the postganglionic neurons leave the collateral ganglia, interconnect to form an autonomic plexus, and supply the viscera of the abdominopelvic cavity.

Preganglionic sympathetic neurons that innervate the adrenal medulla travel in the splanchnic nerves and do not synapse before reaching the gland. Thus there are no postganglionic sympathetic neurons innervating the adrenal medulla—the only exception to the usual two-neuron chain in autonomic efferent pathways. The reason for this exception is explained later in the chapter when we discuss the differences between the divisions of the autonomic nervous system.

Accompanying the sympathetic motor fibers that supply the viscera are sensory fibers of the afferent division of the peripheral nervous system returning from the viscera. These visceral sensory fibers travel without synapsing from the innervated structure, through the chain ganglia and the white rami communicantes, to their cell bodies in the dorsal root ganglia. Therefore, the cell bodies of both visceral and somatic sensory neurons are located in the same ganglia.

Parasympathetic Division

The cell bodies of the preganglionic neurons of the **parasympathetic division** of the autonomic nervous system are located either within nuclei in the brain or within the lateral portions of the gray matter of the spinal cord in the second, third, and fourth sacral segments (Figure 16.1). Because of these origins, the parasympathetic division of the autonomic nervous system is also referred to as the **craniosacral division.** The distribution of the parasympathetic division differs from that of the sympathetic division in that its fibers do not travel throughout the rami of the spinal nerves. Consequently, the sweat glands, arrector pili muscles, and cutaneous blood vessels do not receive parasympathetic innervation. In fact, with few exceptions, the parasympathetic division does not innervate blood vessels anywhere in the body.

Parasympathetic axons whose cell bodies are located in nuclei of the brain travel to the viscera of the head, thorax, and abdomen within the cranial nerves—specifically, the oculomotor, facial, glossopharyngeal, and vagus nerves. (The specific distribution of the parasympathetic axons in these cranial nerves is discussed in Chapter 15 and summarized in Table 15.1). Preganglionic parasympathetic axons in the four cranial nerves synapse with postganglionic neurons in ganglia (ciliary, pterygopalatine, otic, submandibular, and terminal ganglia) that are located close to the structures innervated by the postganglionic neurons.

F16.1

Preganglionic parasympathetic axons whose cell bodies are located in the sacral region of the spinal cord leave the cord in the ventral roots of the sacral spinal nerves. The parasympathetic axons then leave the ventral roots and join together to form the **pelvic splanchnic nerves,** which interconnect in the hypogastric plexus and supply the viscera of the pelvic cavity. The preganglionic axons of the sacral parasympathetic neurons synapse with postganglionic neurons in terminal ganglia located close to the organs they innervate.

Anatomical Differences Between the Divisions

The sympathetic and parasympathetic divisions of the autonomic nervous system differ not only in the locations of the cell bodies of their preganglionic neurons but also in the lengths of their fibers. In the sympathetic division, most preganglionic axons are relatively short, synapsing in the chain ganglia close to the vertebral column. The postganglionic axons are long, extending from the chain ganglia to the structures they innervate. In contrast, the parasympathetic preganglionic axons are relatively long, passing uninterrupted from their origins in the central nervous system to terminal ganglia located on or close to the walls of the organs they supply. The postganglionic axons of the parasympathetic division are short, extending from the terminal ganglia to the organs innervated.

FUNCTIONS OF THE AUTONOMIC NERVOUS SYSTEM

Parasympathetic preganglionic and postganglionic fibers, as well as sympathetic preganglionic fibers, release *acetylcholine*—the same neurotransmitter substance that is secreted by somatic motor neurons (Figure 16.3). Therefore, these nerve fibers are called **cholinergic fibers.** In contrast, sympathetic postganglionic fibers secrete *norepinephrine (noradrenaline)*. Consequently, these nerve fibers are called **adrenergic fibers.** An exception exists in the sympathetic postganglionic fibers that travel through the rami of the spinal nerves to innervate the sweat glands. Since these fibers secrete acetylcholine, they are cholinergic.

F16.3

Norepinephrine (as well as the closely related substance epinephrine) is also secreted by the **adrenal medulla,** an endocrine gland. The adrenal medulla develops in the embryo from neural crest cells. Neural crest cells also give rise to postganglionic neurons of the sympathetic division of the ANS. It is understandable, then, that the adrenal medulla possesses physiological and biochemical properties similar to those of the sympathetic nervous system. The adrenal medulla is innervated by cholinergic preganglionic sympathetic neurons that do not synapse before reaching the gland. Consequently, the adrenal medulla is often considered to be a modified sympathetic ganglion. Norepinephrine has the same effects whether released into the bloodstream by the adrenal medulla or secreted directly onto an organ by a sympathetic nerve fiber. However, when it is released into the bloodstream, norepinephrine is carried to all parts of the body, and its effects may be more widespread.

With few exceptions (such as sweat glands, arrector pili muscles of the hairs, the adrenal medulla, and blood vessels), most body organs are innervated by neurons of both the sympathetic and the parasympathetic divisions of the autonomic nervous system. Because the postganglionic fibers of the parasympathetic division are cholinergic, whereas those of the sympathetic division are mainly adrenergic, the two divisions generally cause opposite responses. For example, if one division increases the activity of an organ, the other may decrease it. Although most organs are predominantly controlled by one division or the other, the dual innervation of an organ by both divisions of the autonomic nervous system contributes to the precise control of the organ's activity. The effects of sympathetic and parasympathetic stimulation on a number of body organs are summarized in Table 16.1.

Table 16.1

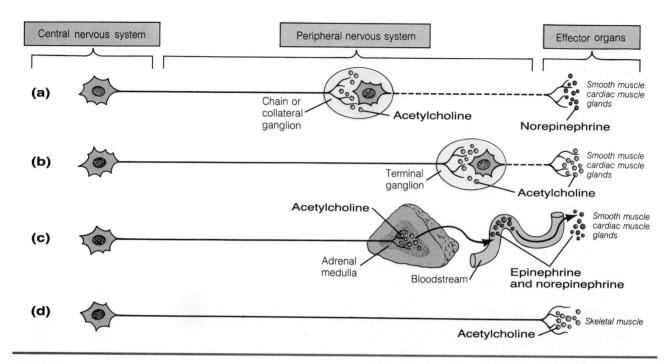

Figure 16.3

Neurotransmitters of the efferent division of the peripheral nervous system. Solid lines indicate preganglionic neurons; dashed lines indicate postganglionic neurons. (a) Preganglionic neurons of the sympathetic division of the autonomic nervous system release acetylcholine at their synapses with postganglionic neurons. Most sympathetic postganglionic neurons release norepinephrine at their junctions with effectors. (b) Preganglionic neurons of the parasympathetic division of the autonomic nervous system release acetylcholine at their synapses with postganglionic neurons, and the postganglionic neurons also release acetylcholine at their junctions with effectors. (c) The adrenal medulla is supplied by preganglionic sympathetic neurons that release acetylcholine. The adrenal medulla releases epinephrine and norepinephrine into the bloodstream, which carries the secretions to effectors. (d) Somatic efferent neurons release acetylcholine at their junctions with effectors.

There is no generalization that indicates whether sympathetic or parasympathetic stimulation will excite or inhibit a particular organ. However, when viewed in broad terms, parasympathetic stimulation tends to produce responses that are primarily concerned with maintaining bodily functions under relatively quiet conditions. For example, parasympathetic stimulation decreases the heart rate and promotes digestive activities. Sympathetic stimulation, in contrast, tends to produce responses that prepare a person for strenuous physical activity, such as may be required in an emergency or in situations that lead to aggressive or defensive behavior. In fact, emotional states such as rage or fear are generally accompanied by a widespread activation of the sympathetic division of the autonomic nervous system. This broad sympathetic activity produces a group of responses—such as increased heart rate and dilation of the bronchii of the lungs—that increase the capability of the body to perform vigorous muscular activity. These responses are particularly beneficial to a person who must defend against or flee from a physical threat or challenge; consequently, they are frequently called "fight or flight" responses.

BIOFEEDBACK

The fact that the autonomic nervous system normally functions below the conscious level implies that a person has no control over the activities governed by this system. However, this is not entirely true.

Table 16.1   Effects of the Autonomic Nervous System

Structure	Effects of Sympathetic Stimulation	Effects of Parasympathetic Stimulation
Heart	Increase rate	Decrease rate
Lungs		
Bronchioles	Dilation	Constriction
Bronchial glands	Possible inhibition of secretion	Stimulation of secretion
Salivary glands	Secretion of viscous fluid	Secretion of watery fluid
Stomach		
Motility	Decreased	Increased
Secretion	Possible inhibition	Stimulation
Intestine		
Motility	Decreased peristalsis	Increased peristalsis
Secretion	Possible inhibition	Stimulation
Pancreas (Exocrine Portion)		Stimulation of secretion
Liver	Increased release of glucose	
Eye		
Iris	Dilation of pupil (contraction of radial muscles)	Constriction of pupil (contraction of sphincter muscles)
Ciliary muscle	Slight relaxation	Contraction (accommodates for near vision)
Sweat glands	Stimulation of secretion (cholinergic)	
Adrenal medulla	Stimulation of secretion (cholinergic preganglionic neurons)	
Urinary bladder	Relaxation	Contraction
Blood vessels of:		
Skin	Constriction	
Salivary glands	Constriction	
Abdominal viscera	Constriction	
External genitalia	Constriction	Dilation

Normally, we receive only limited information at the conscious level about what is occurring within our bodies. For example, blood pressure may fluctuate, or brain-wave patterns change, without our awareness, and therefore we make no conscious effort to react to or control such changes. However, through the use of electronic instruments, researchers using a technique referred to as **biofeedback** have made it possible to monitor some of the subconscious feedback that occurs over the sensory nerve fibers that travel in the autonomic nervous system. Moreover, it has become possible to raise this feedback to the conscious level. The instruments provide information concerning such events as temperature changes and variation in nerve impulse patterns. With this conscious knowledge of feedback that had previously been subconscious, it has been possible in some cases for people to learn to control the responses of their autonomic nervous system. The evidence seems to indicate that visceral responses can be learned in the same way that somatic responses are learned—provided we are made aware of them. For example, using biofeedback techniques, people have learned to lower their heart rate, lower their blood pressure, increase the circulation of blood through their limbs, relieve migraine headaches by reducing the blood pressure within the vessels of the head, and control epileptic seizures. Biofeedback, then, shows considerable promise as a self-administered therapeutic technique with broad applications.

CONDITIONS OF CLINICAL SIGNIFICANCE

The Autonomic Nervous System

Raynaud's Disease

Raynaud's disease is characterized by episodes of pallor or cyanosis of the extremities—especially the fingers and toes and, less frequently, the tip of the nose and the ears. It is thought to be the result of exaggerated vasomotor responses, both central and local, by the sympathetic division of the autonomic nervous system. These responses cause episodes of vasoconstriction of the blood vessels of the affected regions.

The episodes are generally first noticed in cold weather and may be infrequent. The course of Raynaud's disease is variable; it often remains as nothing more than a nuisance for years and in some cases subsides spontaneously. However, occasionally the condition becomes progressive and produces ulcerations and areas of gangrene of the fingertips.

Achalasia

Achalasia, or *cardiospasm*, is characterized by a difficulty in swallowing accompanied by a feeling that food is sticking in the esophagus. It is the result of uncoordi-

nated and ineffectual peristalsis of the esophagus and persistent contraction of the esophagus where it enters the cardiac region of the stomach. These conditions produce a functional obstruction of the esophagus. Achalasia is probably caused by a number of factors. Emotions and a hypersensitivity to the hormone gastrin are implicated, but there may also be structural or functional disorders of the portion of the parasympathetic nervous system that innervates the esophagus.

Hirschsprung's Disease

Hirschsprung's disease, or *megacolon*, is somewhat similar to achalasia except that the functional obstruction occurs in the distal portion of the colon and the rectum. In response to this obstruction the colon above the level of the obstruction dilates greatly (megacolon). This disorder is thought to be caused by a reduction in the parasympathetic innervation to the affected structures. The reduction of parasympathetic innervation allows the sympathetic neurons to inhibit peristalsis and maintain a chronic contraction in the affected region.

STUDY OUTLINE

AUTONOMIC NERVOUS SYSTEM p. 441

VISCERAL MOTOR (EFFERENT) NEURONS innervate and regulate cardiac muscle, smooth muscle, and glands; control involuntary body processes.

ENTIRELY MOTOR FUNCTIONS part of efferent division of peripheral nervous system.

ANATOMY OF THE AUTONOMIC NERVOUS SYSTEM pp. 441–445

EFFERENT PATHWAYS composed of two neurons.

PREGANGLIONIC (PRESYNAPTIC) NEURON cell body in CNS; axon synapses with postganglionic neurons.

POSTGANGLIONIC (POSTSYNAPTIC) NEURON located outside CNS; axons travel to effectors.

SYMPATHETIC DIVISION (THORACOLUMBAR DIVISION) preganglionic neuron cell bodies in lateral horns of spinal cord gray matter from T_1 through L_2. Fibers leave ventral rami and enter a series of chain

ganglia that form longitudinal pathways along each side of vertebral column. *White rami communicantes* (14 pairs) are short pathways formed by myelinated preganglionic nerve fibers as they pass from ventral ramus to chain ganglia. Preganglionic sympathetic neurons follow one of three courses upon entering chain ganglia:

1. May synapse with postganglionic neurons in chain ganglia at same level. Postganglionic axons return to spinal nerve to innervate effectors located in skin. *Gray rami communicantes* are pathways formed by unmyelinated postganglionic axons as they pass from chain ganglia to spinal nerves.
2. May travel up or down within the sympathetic trunks before synapsing with postganglionic neurons that supply effectors in skin, head, or thorax.
3. May pass through chain ganglia without synapsing and synapse with postganglionic neurons in collateral ganglia; postganglionic neurons from collateral ganglia supply viscera of abdominopelvic cavity.

PARASYMPATHETIC DIVISION (CRANIOSACRAL DIVISION)

1. Preganglionic neuron cell bodies are located within nuclei in brain or lateral portions of gray matter of spinal cord at S_2 through S_4.
2. Fibers do not travel through rami of spinal nerves.
3. Preganglionic parasympathetic axons in four cranial nerves (III, VII, IX, X) synapse with postganglionic neurons in ganglia close to structures innervated.
4. Sacral preganglionic parasympathetic axons leave ventral roots of spinal nerves and form pelvic nerve that supplies viscera of pelvic cavity.

ANATOMICAL DIFFERENCES BETWEEN THE DIVISIONS

LOCATION OF PREGANGLIONIC NEURON CELL BODIES

Sympathetic lateral horns of spinal cord gray matter from T_1 through L_2.

Parasympathetic brain and lateral horns of spinal cord gray matter from S_1 through S_4.

FIBER LENGTH
short sympathetic preganglionic axons; long sympathetic postganglionic axons. Long parasympathetic preganglionic axons; short parasympathetic postganglionic axons.

FUNCTIONS OF THE AUTONOMIC NERVOUS SYSTEM   pp. 445–446

CHOLINERGIC FIBERS   release acetylcholine; include parasympathetic preganglionic fibers, parasympathetic postganglionic fibers, and sympathetic preganglionic fibers.

ADRENERGIC FIBERS   release norepinephrine; include sympathetic postganglionic fibers. Exception: postganglionic sympathetic fibers to sweat glands are cholinergic.

1. Most organs are innervated by both sympathetic and parasympathetic neurons.
2. Sympathetic and parasympathetic neurons generally cause opposite responses.
3. Parasympathetic stimulation tends to produce responses primarily concerned with maintaining bodily functions under relatively quiet conditions.
4. Sympathetic stimulation tends to produce responses that prepare a person for strenuous physical activity.

BIOFEEDBACK   technique by which a person is made aware of normally subconscious body activities and learns to exert some voluntary control over them.   pp. 446–448

CONDITIONS OF CLINICAL SIGNIFICANCE: THE AUTONOMIC NERVOUS SYSTEM   p. 448

RAYNAUD'S DISEASE   cyanosis of extremities

ACHALASIA   also called *cardiospasm*.

HIRSCHSPRUNG'S DISEASE   also called *megacolon*.

SELF-QUIZ

1. The autonomic nervous system regulates the activity of: (a) cardiac muscle; (b) skeletal muscles; (c) smooth muscle; (d) both a and c.

2. The autonomic nervous system is normally a voluntary system that functions at the conscious level. . True or false?

3. The autonomic nervous system has both motor and sensory functions. True or false?

4. The efferent pathways of the autonomic nervous system, which run from the central nervous system to the effectors, are composed of how many neurons? (a) 1; (b) 2; (c) 4.

5. The ventral ramus of each spinal nerve generally has associated with it: (a) a collateral ganglion; (b) a chain ganglion; (c) a dorsal root ganglion.

6. Preganglionic sympathetic axons may synapse with the cell bodies of postganglionic neurons in the chain ganglion located at the same level at which the preganglionic axons entered the ganglion. True or false?

7. Most postganglionic axons are myelinated. True or false?

8. Preganglionic sympathetic axons may pass through the chain ganglia without synapsing. True or false?

9. Preganglionic sympathetic axons may travel up or down in the sympathetic trunk before synapsing with: (a) other preganglionic neurons; (b) postganglionic neurons; (c) sensory neurons.

10. The distribution of the sympathetic division differs from that of the parasympathetic division in that its fibers do not travel through the rami of the spinal nerves and, therefore, the cutaneous structures do not receive sympathetic innervation. True or false?

11. The parasympathetic division of the autonomic nervous system is also referred to as the: (a) thoracolumbar division; (b) craniosacral division; (c) neither

12. The sympathetic and parasympathetic divisions of the autonomic nervous system differ in the: (a) location of the cell bodies of their preganglionic neurons; (b) length of their preganglionic fibers; (c) both a and b.

13. Most sympathetic postganglionic fibers secrete: (a) norepinephrine; (b) acetylcholine; (c) neither a nor b.

14. The adrenal medulla causes reactions that are similar to those caused by the sympathetic nervous system. True or false?

15. Most of the organs of the body are innervated by neurons of the: (a) sympathetic division; (b) parasympathetic division; (c) both divisions.

16. The "fight or flight" responses are produced by widespread activation of the: (a) parasympathetic division; (b) somatic nervous system; (c) sympathetic division.

LEARNING OBJECTIVES

After completing this chapter, you should be able to:

- Describe the embryonic development of the eye.

- Name the basic layers of the eye and their main components.

- Describe the microscopic anatomy of the retina.

- Distinguish between presbyopia, astigmatism, and emmetropia.

- Explain how colors are perceived.

- Describe the nerve pathways over which impulses from the retina travel to higher centers in the brain.

- Describe the embryonic development of the ear.

- Name the basic structural compartments of the ear and describe their main components.

- Name and describe the receptor organs for equilibrium.

- Explain how a sound wave is converted into a sound.

- Explain how substances are tasted.

- Explain how odors are detected.

CHAPTER CONTENTS

The Organs of the Special Senses

The special senses of the body are those of *vision* (sight), *hearing* (audition), *balance* or *equilibrium* (the labyrinthine sensations), *smell* (olfaction), and *taste* (gustation). The body possesses specialized receptors for each of these senses. These receptors are located in specific structures. The receptors for vision are located in the eyes; those for hearing and for balance (head position and movement) are located in the ears; the receptors for smell are located in the nose; and the receptors for taste are located in the mouth and throat.

THE EYE—VISION

The **eye** is the sense organ that contains the receptors for vision. Each eye is situated in a bony cavity in the skull called the *orbit*, which protects much of the eye from physical injury. Six extrinsic skeletal muscles hold the eyes in the orbits and move them freely, allowing for an expanded visual field. Most portions of the eye are concerned with gathering and focusing light rays from the external environment. Only one portion, the retina, contains receptors that are sensitive to light rays.

Embryonic Development of the Eye

By the fourth week of embryonic development, lateral outgrowths from either side of the diencephalon have formed structures called **optic vesicles.** As the optic vesicles enlarge, the distal portion of each invaginates, forming a double-layered **optic cup** (Figure 17.1). The optic cup is destined to become **F17.1** the retina of the eye. The internal layer of the cup thickens to form the nervous-tissue layer of the retina, which contains receptor cells and neural elements. The external layer develops into the pigmented layer of the retina. The proximal portion of each optic vesicle narrows into an **optic stalk,** which will become incorporated into the optic nerve.

As the optic cups approach the underside of the surface ectoderm of the embryo, the ectoderm thickens into a **lens placode** directly over each cup. While the cavity of each optic cup deepens, the lens placodes invaginate to form **lens vesicles,** each of which is surrounded by one of the optic cups. With further development, the lens vesicles separate from the surface ectoderm and form a rounded body within the opening of each optic cup. The lens of each adult eye develops from these lens vesicles. Ectodermal cells superficial to the lens vesicles develop into the **cornea** of each eye.

On the undersurface of each optic cup and optic stalk is a groove called the **optic fissure.** The optic fissure serves as a path that directs the fibers of the optic nerve from the inner layer of the retina to the brain and provides a means by which blood vessels are able to supply the interior of the eye. By the time of birth, the optic fissures have closed, completely encompassing the optic nerves and blood vessels.

Loose mesenchymal (mesoderm) cells accumulate around the outside of each optic cup. These cells differentiate into two connective-tissue layers of the eye: the fibrous tunic and the vascular tunic.

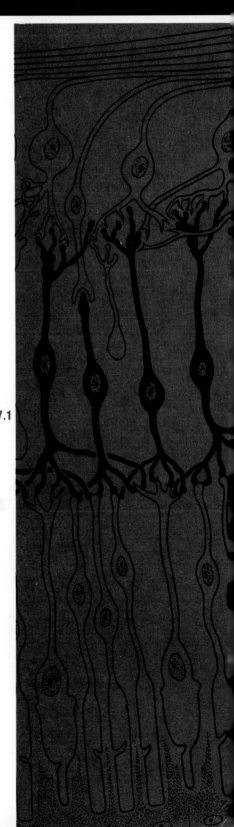

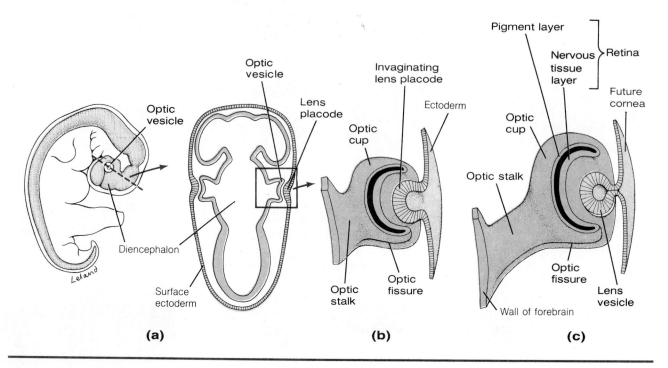

(a) **(b)** **(c)**

Figure 17.1

Embryonic formation of the eye. **(a)** Section through the diencephalon and optic vesicle. (The dotted line indicates the plane of section.) The optic vesicle grows out from the diencephalon and contacts the surface ectoderm. The surface ectoderm then thickens to form a lens placode. **(b)** The lens placode invaginates, forming a hollow lens vesicle. The optic vesicle surrounds the lens vesicle, forming an optic cup that develops into both layers of the retina. **(c)** Sagittal section of the optic cup and lens vesicle. The outer tunics of the eye are formed from mesodermal cells that migrate into the area and surround the optic cup. The optic stalk becomes part of the optic nerve.

Structure of the Eye

The eye is essentially a spherical structure composed of three basic coats, or layers: the *fibrous tunic*, the *vascular tunic*, and the internal tunic, or *retina* (Figure 17.2).

F17.2

Fibrous Tunic

The outermost layer of the eye is called the **fibrous tunic.** The posterior five-sixths of the fibrous tunic, the **sclera,** is white and opaque. The sclera, composed of dense connective tissue, aids in protecting the inner structures of the eye and helps to maintain the shape of the eye. The anterior one-sixth of the fibrous tunic is clear and is called the **cornea.** The cornea is composed primarily of dense connective tissue, with an outer layer of stratified squamous epithelium. It has a greater curvature than the sclera, which causes it to protrude somewhat from the sclera. As light enters the eye it passes through the cornea.

Vascular Tunic

Internal to the fibrous tunic is a layer called the **vascular tunic,** which is well supplied with blood vessels. The vascular tunic is composed of the *choroid,* the *ciliary body,* and the *iris.*

The **choroid,** which lines most of the internal region of the sclera, is darkly pigmented and contains many blood vessels. Around the edge of the cornea the choroid forms the **ciliary body,** which contains smooth muscles called **ciliary muscles.** Three groups of muscle fibers are identifiable in the ciliary muscle, according to the orientation of the fibers: meridional (longitudinal), radial, and circular.

The anterior portion of the vascular tunic is the **iris.** The iris, which is largely a continuation of the choroid, is a thin, muscular diaphragm whose pigmentation is responsible for eye color. In the center of the iris is a rounded opening, the **pupil,** through which light enters the interior regions of the eye. The contraction of circular smooth muscles of the iris constricts the pupil (diminishes its diameter), and the contraction of radially arranged smooth muscles of the iris dilates the pupil (increases its diameter). Constriction and dilation of the pupil occur primarily as reflex responses and help to regulate the amount of light entering the eye.

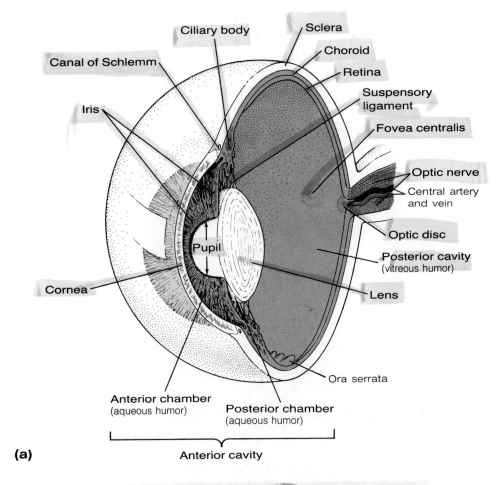

Ciliary body

Sclera

Choroid

Canal of Schlemm

Retina

Suspensory ligament

Iris

Fovea centralis

Optic nerve

Central artery and vein

Pupil

Optic disc

Posterior cavity (vitreous humor)

Cornea

Lens

Ora serrata

Anterior chamber (aqueous humor)

Posterior chamber (aqueous humor)

(a)

Anterior cavity

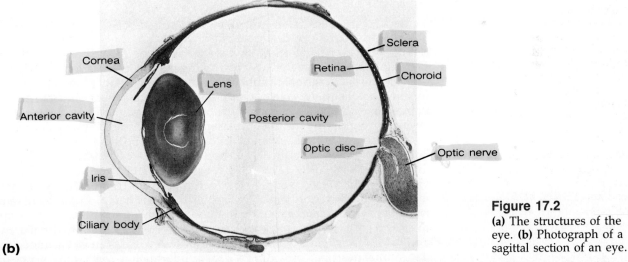

Cornea

Sclera

Lens

Retina

Choroid

Anterior cavity

Posterior cavity

Optic disc

Iris

Optic nerve

Ciliary body

(b)

Figure 17.2
(a) The structures of the eye. **(b)** Photograph of a sagittal section of an eye.

Retina

The innermost layer of the eye is the internal tunic, or **retina** (see Box 17.1). The retina consists of an outer **pigmented layer** and inner **nervous-tissue layers.** The outer pigmented layer is composed of a single layer of heavily pigmented cuboidal epithelial cells that lie in contact with, and are tightly adherent to, the choroid. Both the pigmented layer of the retina and the choroid contain the black-brown pigment *melanin.* The dark pigmentation of these structures reduces the reflection of light that enters the eye. The nervous-tissue layer appears to end anteriorly, near the ciliary body, in a scalloped margin called the **ora serrata.** However, it actually continues forward as

Box 17.1

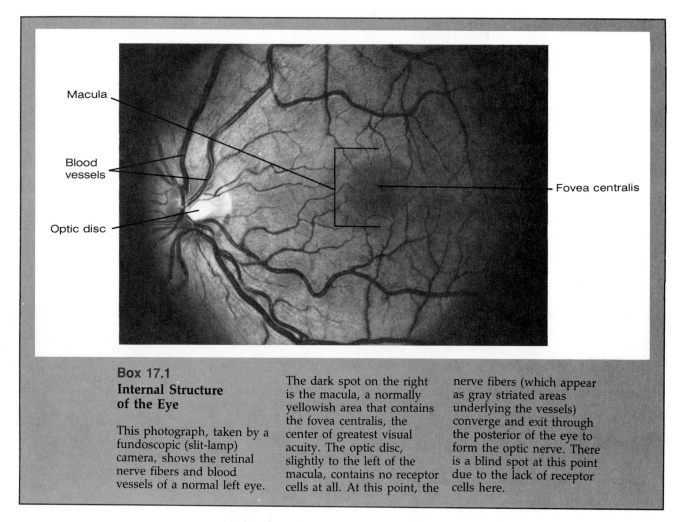

Macula

Blood vessels

Optic disc

Fovea centralis

Box 17.1
Internal Structure of the Eye

This photograph, taken by a fundoscopic (slit-lamp) camera, shows the retinal nerve fibers and blood vessels of a normal left eye.

The dark spot on the right is the macula, a normally yellowish area that contains the fovea centralis, the center of greatest visual acuity. The optic disc, slightly to the left of the macula, contains no receptor cells at all. At this point, the

nerve fibers (which appear as gray striated areas underlying the vessels) converge and exit through the posterior of the eye to form the optic nerve. There is a blind spot at this point due to the lack of receptor cells here.

a thin pigmented layer on the inner surface of the ciliary body and the iris. The nervous-tissue layer is attached to the pigmented layer only around the optic nerve and at the ora serrata. Because the connection is loose it is possible for the nervous-tissue layers of the retina to become detached from the pigmented layer. Such a separation can be repaired surgically, or the layers can be fused together by means of a laser.

The nervous-tissue layers of the retina contain the actual receptors for light—photoreceptors called **rods** and **cones**—as well as numerous neural interconnections. After several synapses in the retina, nerve fibers carrying visual information converge and exit through the rear of the eye, slightly medial to the posterior pole of the eye, as the **optic nerve.** At the point of exit no photoreceptors are present, and this area is called the **blind spot,** or **optic disc.** The *central retinal artery* and *vein* that supply and drain the retina enter and leave the eye through the optic disc. These vessels can be examined with an ophthalmoscope, providing valuable information concerning a person's health. Lateral to the optic disc is a slightly yellow region known as the **macula.** The macula is located almost exactly at the posterior pole of the eye. In the center of the macula is a depression called the **fovea centralis** (central depression). The fovea, which is the most sensitive portion of the retina, does not contain any rod-type photoreceptors, but it does contain very densely packed cones. When you look directly at an object, the image of the object is generally focused on the fovea.

Lens

Behind the pupil is a clear **lens,** which is held in position by a fibrous **suspensory ligament** that attaches to the ciliary body. The lens is a biconvex struc-

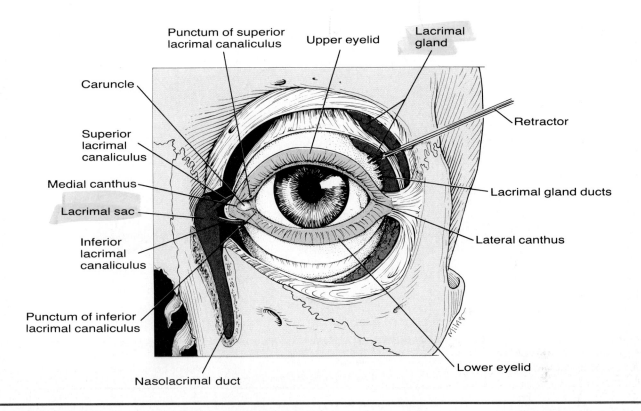

Punctum of superior lacrimal canaliculus
Upper eyelid
Lacrimal gland
Caruncle
Superior lacrimal canaliculus
Medial canthus
Lacrimal sac
Inferior lacrimal canaliculus
Punctum of inferior lacrimal canaliculus
Nasolacrimal duct
Retractor
Lacrimal gland ducts
Lateral canthus
Lower eyelid

ture that is relatively elastic. By changing shape the lens aids in properly focusing on the retina the light that enters the eye. The lens has no blood supply. It receives its nourishment from substances within the eye called the aqueous and vitreous humors. The lens is composed of elongated cells called *lens fibers* and is surrounded by a capsule that is elastic and highly refractive.

Cavities and Humors

The lens separates the interior of the eye into two cavities. In front of the lens is an **anterior cavity;** behind the lens is a **posterior cavity.** The anterior cavity can be subdivided into two chambers. The **anterior chamber** is located in front of the lens and iris and behind the cornea. The **posterior chamber** is located between the iris and the suspensory ligament. The anterior cavity contains a clear fluid known as the **aqueous humor,** and the posterior cavity contains a transparent, jellylike substance called the **vitreous humor.** The aqueous humor is produced by epithelial folds called *ciliary processes,* which project from the ciliary body. After passing into the anterior chamber through the pupil, the aqueous humor drains into a venous sinus called the **canal of Schlemm** and eventually reaches the bloodstream. The canal of Schlemm is located near the junction of the cornea and the iris. The rates of production and removal of aqueous humor are such that a relatively constant pressure is maintained in the eye.

Accessory Structures of the Eye

Several structures associated with the eye contribute in various ways to its functioning. These include the *eyelids,* the *conjunctiva,* the *lacrimal apparatus,* and the *extrinsic eye muscles.*

Eyelids

Skin-covered structures called upper and lower **eyelids (palpebrae)** can be drawn over the anterior surface of the eye to protect the exposed portion (Figure 17.3). When closed they prevent about 99% of the incident light from entering the eye. The movements of the eyelids are controlled by skeletal muscles. Fibers of the **orbicularis oculi** muscle extend into both eyelids and

Figure 17.3
Accessory structures of the eye (anterior view). Portions of the eyelids have been removed to expose the deeper structures, and the lacrimal gland has been retracted to show the ducts.

F17.3

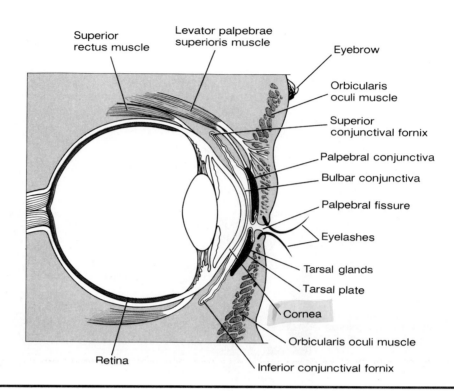

Figure 17.4
Accessory structures of
the eye (sagittal section).

serve to close the lids. The **levator palpebrae superioris** muscle is inserted
F17.4 into the upper lid and functions to raise it (Figure 17.4). Much of the activity
of these muscles occurs as the result of reflexes that produce an eye-blink
response. This response occurs spontaneously about 25 times per minute
when a person is awake, and is particularly evident if a foreign object comes
into contact with the surface of the eye or with the hairs **(eyelashes)** projecting
from the borders of the eyelids.

Within each eyelid is dense connective tissue that forms a structure called
a **tarsal plate.** The tarsal plates are important in maintaining the shape of the
eyelids. Sebaceous glands called **tarsal** *(meibomian)* **glands** are located close to
the inner surfaces of the eyelids and are embedded in the tarsal plates. The
ducts of the tarsal glands open onto the margins of the eyelids, and their
secretions onto these margins help prevent tears from overflowing between
the eyelids. An infection of the tarsal glands can produce a *cyst* on the eyelid.
The eyelids also possess modified sweat glands called **ciliary glands** as well as
sebaceous glands located at the bases of the hair follicles of the eyelashes.
Occasionally, these sebaceous glands may become infected, thereby producing a *sty.* Between the upper and lower eyelid of each eye is a gap through
which the eye is exposed—the **palpebral fissure.** The angle formed by the
lateral junction of the upper and lower eyelids is called the **lateral canthus;**
the angle formed by the medial junction of the upper and lower eyelids is
F17.3 called the **medial canthus** (Figure 17.3). A small reddish mound of tissue at
the medial canthus of each eye, the **caruncle,** contains a few sebaceous
glands.

Conjunctiva

Both the insides of the eyelids and the anterior surface of the eye itself are
covered with a thin protective layer of epithelium that forms a mucous membrane called the **conjunctiva.** The portion of the conjunctiva associated with
the inner surface of the eyelids is called the **palpebral conjunctiva,** and the
portion associated with the surface of the eye is called the **bulbar conjunctiva**
F17.4 (Figure 17.4). The palpebral conjunctiva of the upper eyelid is reflected at its
upper margin onto the anterior surface of the eye as the bulbar conjunctiva.
The angle formed by the palpebral and bulbar portions of the conjunctiva

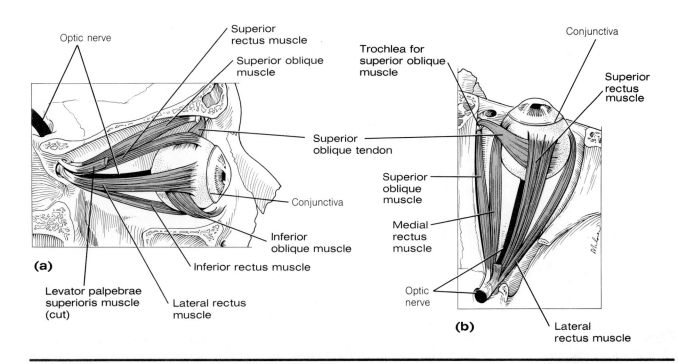

Optic nerve

Superior rectus muscle

Superior oblique muscle

Trochlea for superior oblique muscle

Superior oblique tendon

Conjunctiva

Conjunctiva

Superior rectus muscle

Inferior oblique muscle

Inferior rectus muscle

(a)

Levator palpebrae superioris muscle (cut)

Lateral rectus muscle

Superior oblique muscle

Medial rectus muscle

Optic nerve

(b)

Lateral rectus muscle

Figure 17.5
Extrinsic muscles of the eye viewed **(a)** from the side and **(b)** from above.

at the line of reflection is called the **superior conjunctival fornix.** Similarly, the palpebral conjunctiva of the lower eyelid is reflected at its lower margin onto the anterior surface of the eye and an **inferior conjunctival fornix** is located at the line of reflection. An inflammation of the conjunctiva is called *conjunctivitis.*

Lacrimal Apparatus

In the superior lateral region of the orbit is a gland called the **lacrimal gland** (Figure 17.3). Six to 12 ducts arise from the lacrimal gland and carry its secre- **F17.3** tions onto the surface of the conjunctiva, primarily near the lateral portion of the superior conjunctival fornix. The lacrimal gland produces a watery secre- tion **(tears)** that continually bathes the surface of the eye. Tears—which con- tain salts, some mucin, and a bacteriocidal enzyme called lysozyme—flow over the bulbar conjunctiva from their area of secretion to the medial canthus. Most tears simply evaporate, but excess fluid can drain into the nasal cavity by ducts or overflow from the eye through the palpebral fissure.

Toward the medial end of each eyelid is a small projection called a **lacri- mal papilla.** The lacrimal papilla contains the opening, or **punctum,** of a **lacrimal canaliculus.** The lacrimal canaliculi of the upper and lower eyelids converge medially from their openings and connect to a **lacrimal sac.** The inferior portion of the lacrimal sac opens into a **nasolacrimal duct,** which in turn opens into the nasal cavity beneath the inferior nasal conchae (turbi- nates). Normally, the secretions of a lacrimal gland amount to less than 1 ml per day, but these secretions increase during certain emotional states (such as crying) and when foreign objects or other irritants contact the eye. This is why it is generally necessary to blow one's nose when crying.

Extrinsic Eye Muscles

Six straplike skeletal muscles that originate from the orbit and insert on the connective tissue of the eye are responsible for eye movement (Figure 17.5). **F17.5** Originating from the rear of the orbit and inserting on the lateral surface of the eye is the **lateral rectus** muscle, whose contraction principally turns the eye laterally (outward). The **medial rectus** muscle, which originates from the rear of the orbit and inserts on the medial surface of the eye, primarily turns the eye medially (inward). Originating from the rear of the orbit and inserting on the superior surface of the eye is the **superior rectus** muscle, which principally turns the eye upward and somewhat medially. Originating from the rear of the orbit and inserting on the inferior surface of the eye is the **inferior rectus** muscle, which principally turns the eye downward and somewhat medially. The **inferior oblique** muscle originates from the medial surface of the front of the orbit and inserts on the inferior surface of the eye; it primarily turns the eye upward and laterally. The **superior oblique** muscle originates from the posterior portion of the orbit with the rectus muscles and runs along the medial surface of the orbit. Anteriorly, its tendon passes through a pulleylike loop, the *trochlea*, and turns to insert on the superior surface of the eye. The superior oblique muscle principally turns the eye downward and laterally.

The extrinsic eye muscles are the most rapidly acting and among the most precisely controlled skeletal muscles in the body. They receive impulses over the oculomotor (III), trochlear (IV), and abducens (VI) cranial nerves, as described in Chapter 15.

Refracting Media of the Eye

Light rays traveling in a straight line through a uniform medium of a given optical density (for instance, air) travel at a uniform speed. If these rays encounter a second medium with a different optical density (for instance, glass), their speed may be altered and the light rays bent (*refracted*). If the light rays strike the surface of the second medium at an angle that is not perpendicular to the medium's surface, they are bent or refracted. The greater the deviation from perpendicular at which the light rays strike the second medium, the greater the bending of light.

Substances that bend light are called **refracting media.** In order to reach the rods and cones of the retina, light must pass through four refracting media: the *cornea*, the *aqueous humor*, the *lens*, and the *vitreous humor*.

Focusing Images on the Retina

Four processes are involved in focusing images on the retina: (a) refraction of light rays; (b) accommodation by the lens; (c) constriction of the pupil; and (d) convergence of the eyes.

Refraction

Light rays are bent every time they pass from one refracting medium to another—if each medium has a different optical density. Therefore, light reaching the retina is refracted (1) as it enters the cornea from the air, (2) as it leaves the posterior surface of the cornea and enters the aqueous humor, (3) as it enters the lens, and (4) as it leaves the lens and enters the vitreous humor. The greatest bending of light rays occurs as light enters the cornea.

Accommodation

Normally, the fluid in the eye exerts a pressure (the *intraocular pressure*) that tends to force the walls of the eye apart. The lens is attached to the walls by ciliary processes of the ciliary body that connect to the suspensory ligament of the lens. Thus, the elastic lens is normally under tension and appears relatively flat, with a less pronounced curvature than it might otherwise assume. In this condition, the normal *(emmetropic)* eye focuses parallel light rays from distant points sharply on the retina (light rays from points more than 20 feet away are considered parallel). The more divergent light rays that enter the eye from points closer than 20 feet are focused behind the retina.

Adjustments in the focusing system of the eye must be made if objects closer than about 20 feet are to be focused clearly on the retina. This adjustment process, called *accommodation,* is accomplished by smooth ciliary muscles of the ciliary body. The contraction of the ciliary muscles pulls the ciliary body slightly forward and inward, narrowing the ring of the ciliary body—an action somewhat similar to that of a sphincter muscle. This lessens the tension on the suspensory ligament, thus permitting the elastic lens to assume a more pronounced curvature. This greater curvature of the lens causes the

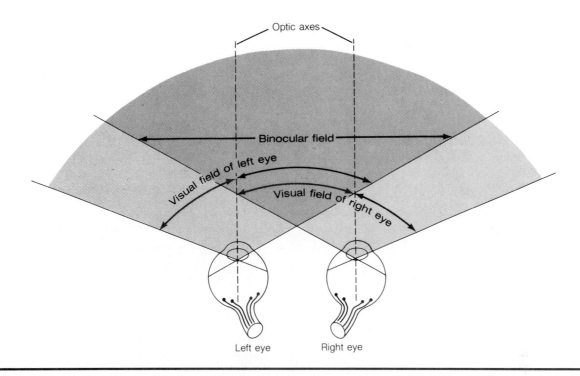

Figure 17.6
Visual fields of the eye. The colored arrows indicate the left portions of the visual field of each eye. The black arrows indicate the right portions. Binocular vision occurs where these visual fields overlap. Light rays from objects located in the left portion of each visual field strike the neurons in the retina shown in color. Those from objects in the right portion of each visual field strike the neurons shown in black.

more divergent light rays that come from points closer than 20 feet to be bent more and thus to be focused on the retina. (Under such conditions, light rays from distant points are focused in front of the retina.)

Constriction

Accommodation also involves constriction of the pupil. This constriction is beneficial because it eliminates the most divergent light rays, which would otherwise pass through the most peripheral portions of the lens. Even a good lens may not be perfect, and light rays that pass through the peripheral regions of the lens may not focus at exactly the same point as those that pass through the more central areas of the lens. Thus the constriction of the pupil aids in the formation of a sharp image on the retina (and also reduces the amount of light that enters the eye).

Convergence

Another event associated with accommodation is the convergence of the eyes. When viewing close objects the eyes turn inward so that the image falls on the fovea, the most sensitive part of the retina. In fact, the eyes are crossed when viewing very close objects, such as the tip of the nose.

Binocular Vision

A person's eyes view portions of the external world that overlap considerably with one another. That is, the visual field of the left eye is quite similar to, but does not completely overlap, that of the right eye (Figure 17.6). Both eyes do   F17.6 not form exactly identical visual images, however, because they occupy slightly different locations. Thus, humans possess what is called **binocular vision.** Binocular vision is basic to accurate depth perception. With only one eye, depth perception depends to a large degree on the use of learned cues such as the fact that the relative sizes of objects in the environment diminish with distance.

   Light rays that come from an object in the environment strike the retinas of both eyes. Normally a person perceives only one object and not two, however, because the retinas possess corresponding points that, when stimulated, result in the perception of a single image. In conditions such as *strabismus,* in which the extrinsic muscles of one eye may be so weak that the eye

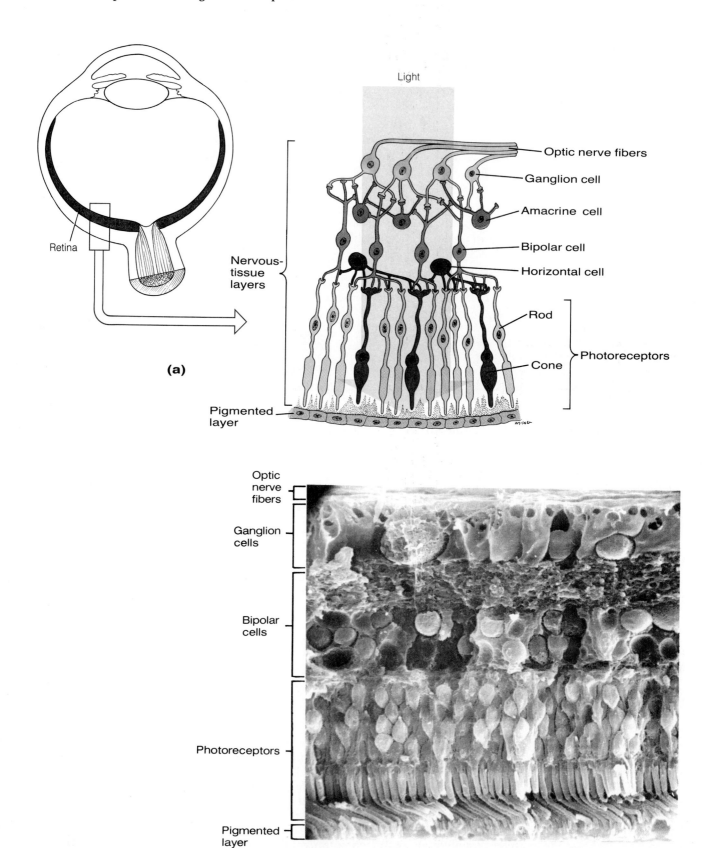

Figure 17.7
Structure of the retina. **(a)** Schematic view. **(b)** Scanning electron micrograph (×1500). (From *Tissues and Organs: A Text-Atlas of Scanning Electron Microscopy* by Richard G. Kessel and Randy H. Kardon. W.H. Freeman and Company. Copyright © 1979.)

does not adequately coordinate its movements with those of the other eye, light rays from an object may not fall on corresponding points on both retinas. As a result, one may see two objects—a condition referred to as double vision, or *diplopia*. In some cases of double vision, the aberrant image may eventually become suppressed; if the condition is not corrected, the suppressed eye can become functionally blind.

Photoreceptors of the Retina

The nervous-tissue portion of the retina is a complex structure composed of a number of layers of nerve cells (Figure 17.7). The outermost portion of the nervous-tissue layers (the portion closest to the pigmented layer of the retina and the choroid) contains the light-sensitive photoreceptors—the **rods** and **cones** (so named because of their microscopic appearance). The rods and cones contain photopigments whose chemical configurations are altered when they are struck by and absorb light. These alterations lead to changes in the polarity of the membranes of the photoreceptors, which ultimately result in the transmission of neural signals from the retina to the brain, where they are interpreted as visual events. Note that light must pass through all of the nervous-tissue layers of the retina in order to reach the photoreceptors, which are located in the deepest layer.

F17.7

There are four different photopigments, each consisting of a protein called an *opsin* to which a chromophore molecule (a variant of vitamin A) called *retinal* is attached. The opsins differ from pigment to pigment and confer specific light-sensitive properties on each pigment. When light of the proper wavelength strikes a photopigment, the retinal breaks away from the opsin. After the light-induced breakdown of the photopigment, the altered retinal is rearranged and rejoined to the opsin to restore the photopigment.

Rods

The photopigment contained in the rods is called *rhodopsin*. Since rhodopsin is very sensitive to light, the rods can respond to very low levels of illumination, such as those present at night or in dimly lit areas. The responses of the rods indicate degrees of brightness, but they do not indicate color. Consequently their responses are interpreted only in shades of gray.

Cones

There are three different types of cones. Each type contains a different photopigment and each is especially sensitive to particular wavelengths of light. Red cones respond more intensely than the others to those wavelengths of light that the brain interprets as red. Green cones respond more intensely than the others to those wavelengths of light that the brain interprets as green. Blue cones respond most intensely to those wavelengths of light that the brain interprets as blue or violet.

The cones are responsible for color vision. The ability to perceive many different colors rather than only three (red, green, and blue) is due to the fact that different wavelengths of light striking the retina evoke different ratios of response from the three basic cone types. These varied responses are interpreted by the brain as a wide variety of different colors.

Because the cones operate only at relatively high levels of illumination, they are the principal photoreceptors during daylight vision or in brightly lit areas. The cones are concentrated in the center of the retina and most highly in the fovea, whereas the rods are more numerous in the peripheral retina.

Neural Elements of the Retina

The various layers in the nervous-tissue portion of the retina are composed of neural elements (Figure 17.7). The photoreceptor cells (rods and cones) that are located in the outermost layer synapse with **bipolar cells** that form a middle layer. The bipolar cells, in turn, synapse with **ganglion cells** that form an inner layer. The axons of the ganglion cells pass along the inner surface of the retina and leave the eye through the optic disc, forming the **optic nerve,** which passes to the brain.

F17.7

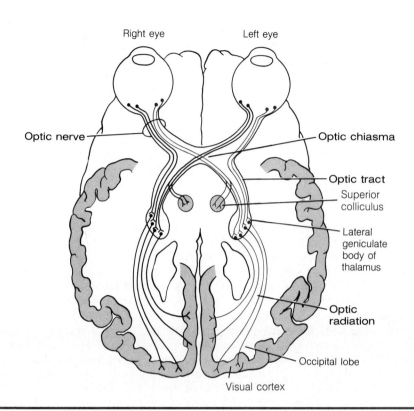

Right eye   Left eye

Optic nerve — Optic chiasma

Optic tract

Superior colliculus

Lateral geniculate body of thalamus

Optic radiation

Occipital lobe

Visual cortex

Figure 17.8
Neural pathways for vision (ventral view of brain).

In general, the neural pathways that transmit the responses of the rods display greater convergence than the pathways that transmit the responses of the cones. For example, in the peripheral regions of the retina many rods synapse with a single bipolar cell, and many bipolar cells synapse with a single ganglion cell. In contrast, within the fovea there are approximately equal numbers of cones, bipolar cells, and ganglion cells, and relatively little convergence occurs. Consequently, the cones in the fovea provide more precise information about the area of the retina stimulated than do the rods in the peripheral retina. As a result, vision is more precise in the fovea than in the peripheral retina. On the other hand, the convergence that occurs along the visual pathways of the rods provides for summation of the afferent impulses, and this, together with the differences in light sensitivity of the rods and cones themselves, contributes to the fact that rod vision is more effective in dim light than cone vision.

Visual Pathways

The two optic nerves, which are formed of the axons of ganglion cells, meet at the **optic chiasma,** which is located just anterior to the pituitary gland (Figure

F17.8 17.8). Within the optic chiasma, ganglion cell axons from the medial half of each retina cross to the opposite side. From the optic chiasma the axons continue as the **optic tracts.** As a consequence of the crossing of the axons in the chiasma, the *left* optic tract consists of ganglion cell axons from the lateral half of the retina of the left eye and the medial half of the retina of the right eye. Both of these fiber groups carry visual information from the right visual field. Conversely, the *right* optic tract consists of axons from the lateral half of the retina of the right eye and the medial half of the retina of the left eye. Both of these fiber groups carry visual information from the left visual field.

Most of the ganglion cell axons in the optic tracts travel to the lateral geniculate bodies of the thalamus. There they synapse with neurons that form pathways called **optic radiations,** which terminate in the *visual cortex* of the occipital lobe. It is in this region that nerve impulses from the retina are interpreted visually by the brain. Some of the axons in the optic tracts terminate in the superior colliculi of the brain, where they function in visual reflexes important for such things as eye-hand coordination.

CONDITIONS OF CLINICAL SIGNIFICANCE

The Eye

Myopia

Myopia is nearsightedness. In this condition, light rays from distant objects are focused in front of the retina and only light rays from close objects can be focused accurately on the retina (Figure 17.9). Myopia is caused by a focusing system that has too great a refractive power with respect to the position occupied by the retina. Most commonly this condition is due to an elongated eyeball, which can result from a weakness of the coats of the eye. Myopia can be corrected by eyeglasses (or contact lenses) with *concave* lenses. These lenses cause light rays from distant points to diverge slightly as they enter the eye so that the refractive system of the eye can focus them on the retina.

Hypermetropia

Hypermetropia (hyperopia) is farsightedness. In this condition, light rays from distant objects are focused behind the retina when the eye is at rest, and accommodation is necessary to focus the rays on the retina. Since the eyes can accommodate only so much, a hypermetropic individual has trouble viewing close objects. Moreover, accommodation is required to view any object, close or far. Hypermetropia is most commonly due to a shortening of the eyeball (Figure 17.10). The condition can usually be corrected with eyeglasses with *convex* lenses. These lenses cause light rays to converge as they enter the eye, and thus assist the refractive system of the eye in focusing on the rays on the retina.

Astigmatism

Previous considerations of the refractive system of the eye have assumed that all its elements possess uniformly curved surfaces (much like a marble). In many instances, however, the surface of one or more elements of the system may not be uniformly curved in all planes. The surface of the cornea, for instance, may have a different curvature in the horizontal plane than in the vertical plane (much like a chicken's egg). As a result, light rays that enter the eye in different planes are focused at different points, causing an out-of-focus image. The condition that results from unequal curvature of portions of the refractive system of the eye is called *astigmatism*. If an astigmatic individual examines a series of radiating lines, some may be sharply focused on the retina and seen clearly while others may be focused either in front or in back of the retina and be seen indistinctly. Astigmatism can be corrected by an eyeglass lens that has the same degree of astigmatism as the eye but is rotated 90° to the eye's astigmatism. Thus the astigmatism of the lens cancels that of the eye.

Cataract

In some cases, the lens of the eye or a portion of it becomes cloudy or opaque and thus impairs vision. This condition is known as a *cataract*. Often such lenses are removed surgically, and effective vision can be restored by the use of special eyeglasses that compensate for the loss of the lens.

Glaucoma

Glaucoma is an abnormal elevation in the intraocular pressure that may result from deficient drainage of the aqueous humor. When severe, the high intraocular pressure of glaucoma can squeeze shut blood vessels that supply the eye, leading to the degeneration of the retina and to blindness.

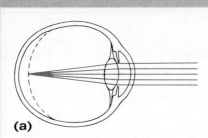

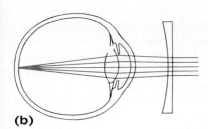

Figure 17.9

Myopia. **(a)** In myopia, light rays from distant objects focus in front of the retina. **(b)** Myopia can be corrected by a concave lens, which causes light rays to diverge as they enter the eye.

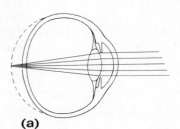

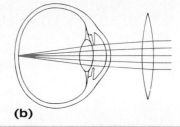

Figure 17.10

Hypermetropia. **(a)** In hypermetropia, light rays from distant objects focus behind the retina. **(b)** Hypermetropia can be corrected by a convex lens, which causes light rays to converge as they enter the eye.

Color Blindness

Color blindness is a deficiency in color perception that can range from an inability to distinguish certain shades of color to a complete lack of color perception. Difficulty in distinguishing reds and greens is the most common form of color blindness. The condition is caused by an absence of one or more of the three different cone types. Color blindness is thought to result from the lack of specific genes on the X chromosomes. Because color genes are missing from the X chromosomes, color blindness is sex-linked—about 8% of males but less than 1.5% of females are red-green color blind.

Effects of Aging on the Eye

With aging, the lens of the eye becomes yellowed because of the effects of ultraviolet rays from such sources as sunlight. In addition, the lens is one of the few body structures that exhibits increased cellular growth with age, which may contribute to the development of cataracts.

As an individual ages, the pupil of the eye is no longer able to dilate fully, and the amount of light that reaches the photoreceptors of the retina by age 70 may be only 50% of the amount that reaches the retina during youth. The inability of the pupil to dilate fully can also lead to poor drainage of the aqueous humor, resulting in increased intraocular pressure and a greater likelihood of glaucoma.

The continuous exposure of the rods and cones to light damages their membranes. This damage results in the accumulation of cellular debris, which can be removed by cells of the pigmented layer of the retina. With aging, however, these cells can become congested, accumulating debris from the rods and cones, and thus contributing to a loss of visual acuity. This loss is particularly striking if it occurs in the fovea centralis or macula region, and the accumulation of cellular debris may contribute to macular degeneration.

With aging, the lens gradually loses its elasticity, and thereby loses some of its ability to change shape during accommodation for viewing near objects. Although changes in lens elasticity begin early in life, the greatest loss of elasticity occurs after age 40 or so. As the ability to accommodate for viewing near objects diminishes, it becomes necessary to hold reading materials farther and farther from the eye in order to focus the printed letters on the retina. Ultimately, books and papers may need to be held so far from the eye that the images of the letters on the retina are too small to be recognized. This condition, called *presbyopia*, can be corrected by the use of eyeglasses with *convex* lenses. The convex lens increases the convergence of the light rays so that the eye's refractive system can focus them on the retina when the printed matter is held reasonably close to the eye.

THE EAR—HEARING AND BALANCE

F17.11 The ear contains receptors for hearing as well as receptors that detect the position and movement of the head. Information from the latter receptors is used to maintain balance and equilibrium. The ear is divided into *external*, *middle*, and *inner* regions (Figure 17.11). The external ear is essentially a funnel-shaped structure used for collecting sound waves. The middle ear contains three small bones, called ossicles, that transmit sound waves from the external ear to the inner ear. The inner ear is composed of a system of fluid-filled semicircular canals and chambers that contain the receptors for the perception of sound, as well as those concerned with the position and movement of the head.

Embryonic Development of the Ear

In the embryo, the development of the ear begins with the formation of a thickened ectodermal plate called an **otic placode,** which is located on the side **F17.12a** of the head in the vicinity of the hindbrain (Figure 17.12a). Each otic placode **F17.12b** invaginates to form an **otic pit** (Figure 17.12b). By the fourth week of development, the otic pit separates from the surface ectoderm and develops into a **F17.12c** closed sac called an **otic vesicle** (Figure 17.12c).

While the otic vesicles are forming, lateral pouches develop from the side of the pharynx. As these **pharyngeal pouches** expand, the surface ectoderm indents toward them, forming the **branchial grooves.**

With further development, the otic vesicle forms the membranous labyrinth of the inner ear. While this is occurring, fibers from the vestibulocochlear nerve (cranial nerve VIII) grow toward the otic vesicle and innervate it. The distal end of the first pharyngeal pouch forms the cavity of the middle ear. The proximal portion of the first pharyngeal pouch forms a tube called an *auditory (eustachian) tube,* which connects the middle-ear cavity with the phar-

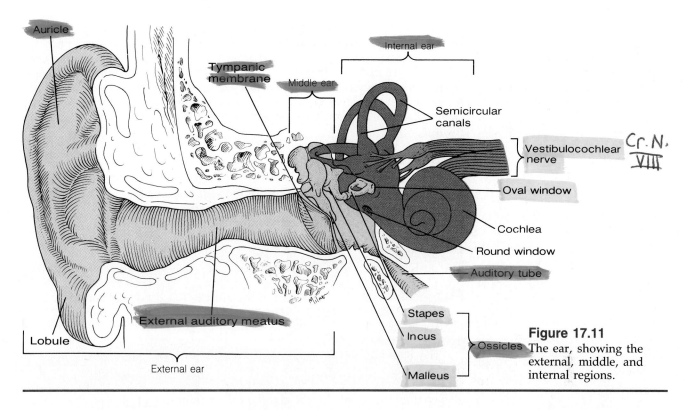

Figure 17.11
The ear, showing the external, middle, and internal regions.

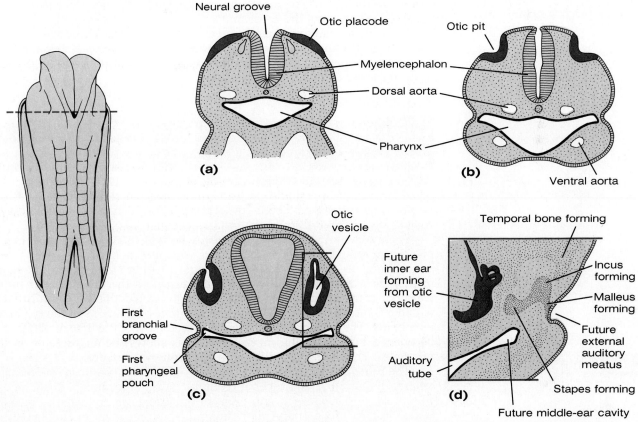

Figure 17.12
Embryonic development of the ear. **(a)** Formation of the otic placodes and the first pharyngeal pouches. **(b)** Invagination of the otic placodes to form the otic pits, and beginning of the formation of the first branchial grooves. **(c)** Otic vesicles form from the otic pits, and the branchial grooves deepen. **(d)** Inner-ear structures form from the otic vesicles. Each first branchial groove develops into an external auditory meatus, and each first pharyngeal pouch becomes an auditory tube and a middle-ear cavity.

F17.12d ynx (Figure 17.12*d*). The three ossicles form from condensations of mesenchymal cells in the middle-ear cavity. On the surface of the embryo, the branchial groove associated with the first pharyngeal pouch deepens and forms a canal called the *external auditory meatus.* The *tympanic membrane (eardrum)* develops from membrane that separates the first pharyngeal pouch from the floor of the branchial groove. The *external ear (auricle)* forms from the coalescence of a series of elevations that develop around the external auditory meatus.

Structure of the Ear

The structures of the various regions of the ear are uniquely suited to collect sound waves, to convert vibrations of the sound waves into mechanical vibrations—first by small bones in the middle ear, then by fluid in the inner ear—and finally to convert the fluid vibrations into nerve impulses.

External Ear

F17.11 The **auricle,** or **pinna,** is the most prominent portion of the external ear (Figure 17.11). It consists of an irregularly shaped framework of elastic cartilage covered with skin. The only part of the auricle that is not supported by cartilage is the **lobule,** a flap of skin-covered connective tissue that extends from the lower margin of the auricle. The auricle directs sound waves into the external auditory meatus.

The **external auditory meatus** (canal) is a curved passageway approximately 2.5 cm long that extends from the auricle to the eardrum. The meatus is lined with skin, and near its entrance are fine hairs and sebaceous glands. It also contains modified sweat glands, called **ceruminous glands,** that secrete *cerumen (earwax).* The hairs and the cerumen help to prevent small foreign objects from reaching the eardrum.

The external auditory meatus serves as a resonator for the range of sound waves typical of human speech (2500 to 5000 cycles per second). Because of its resonating properties, the meatus increases the sound pressure on the eardrum for tones in this frequency range.

Middle Ear

The middle ear is a small air-filled chamber in the temporal bone. It is separated from the external auditory meatus by the **tympanic membrane (eardrum)** and separated from the inner ear by a bony wall in which there are two small membrane-covered openings—the **oval window (fenestra vestibuli)**
F17.11 and the **round window (fenestra cochleae)** (Figure 17.11). An opening in the posterior wall of the middle ear leads to the mastoid sinuses in the mastoid portion of the temporal bone. Another opening connects the middle-ear chamber with the **auditory tube,** which leads to the nasopharynx.

The auditory tube provides a means by which the air pressure in the middle-ear chamber remains equalized with atmospheric pressure. When atmospheric pressure is reduced, as occurs at higher altitudes, the tympanic membrane would bulge outward if the pressure in the middle-ear chamber were not correspondingly reduced. Not only would this be painful, but it would also impair hearing by interfering with the vibrations of the tympanic membrane. The auditory tube, which is closed most of the time in adults, can be opened by swallowing or yawning. This allows the air pressure in the middle-ear chamber to equalize with atmospheric pressure.

The mucous membranes that line the middle-ear chamber, the mastoid sinuses, and the auditory tubes are continuous with the mucous membrane of the throat. For this reason, infections of the throat can readily spread to the middle-ear chamber—particularly in young children, whose auditory tubes are straighter than an adult's and tend to remain open most of the time. Because of the opening between the middle-ear cavity and the mastoid sinuses, infections of the middle-ear cavity can spread to the mucous membrane that lines the mastoid sinuses, producing a condition called *mastoiditis.* The mastoid sinuses are separated from the brain only by thin, bony parti-

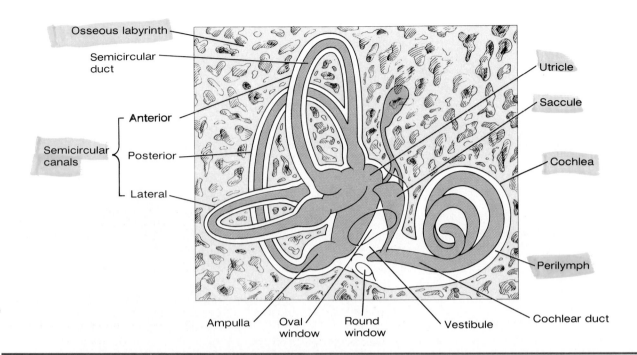

Osseous labyrinth

Semicircular duct

Semicircular canals { Anterior, Posterior, Lateral }

Utricle

Saccule

Cochlea

Perilymph

Ampulla   Oval window   Round window   Vestibule   Cochlear duct

Figure 17.13
Structures of the inner ear. The membranous labyrinth (shown in color) is filled with endolymph and is separated from the osseous labyrinth by perilymph (shown in white).

tions. Therefore, it is also possible for infection to spread from the mastoid sinuses to the meninges of the brain.

The three ear **ossicles,** or middle-ear bones, form a flexible bridge across the middle-ear chamber (Figure 17.11). The handle-shaped portion of the ossicle called the **malleus** *(hammer)* attaches to the tympanic membrane. The foot-plate portion of the ossicle called the **stapes** *(stirrup)* fits against the oval window on the medial wall of the middle-ear chamber. The third ossicle, the **incus** *(anvil)*, lies between the malleus and the stapes and articulates with them. Thus the three ossicles form a bridge between the tympanic membrane and the oval window. The articulations between the ear ossicles are freely movable synovial joints. The ossicles form a lever system that picks up vibrations of the tympanic membrane and transmits them to the oval window, which in turn leads into the inner ear. Two small muscles attach to the ear ossicles: the **stapedius muscle** attaches to the stapes; the **tensor tympani muscle** attaches to the handle of the malleus. Since these muscles contract reflexively in response to sudden loud noises, they dampen the vibrations of the ossicles and thus protect the receptors of the inner ear from damage.

F17.11

Inner Ear

The inner ear is located medially to the middle ear, in the petrous portion of the temporal bone (Figure 17.13). It consists of a series of canals called the **osseous labyrinth,** which are hollowed out of the bone. Within the osseous labyrinth, and following its course, is a membranous labyrinth. The membranous labyrinth is filled with a fluid called **endolymph** and is suspended in a fluid called **perilymph.** Perilymph therefore separates the walls of the membranous labyrinth from the osseous labyrinth. The osseous labyrinth is divided into three areas: the *vestibule,* the *semicircular canals,* and the *cochlea.*

F17.13

VESTIBULE   The **vestibule** is a chamber just medial to the middle-ear chamber. Since the oval window forms a membranous partition between the middle-ear chamber and the vestibule of the inner ear, vibrations of the oval window induced by the stapes are transmitted to the perilymph of the vestibule. Within the vestibule are two enlargements of the membranous labyrinth: the **saccule** and the **utricle.** These structures contain receptor cells that detect the position of the head. Thus, information from the receptor cells contributes to the sense of balance. The utricle is connected to the portion of

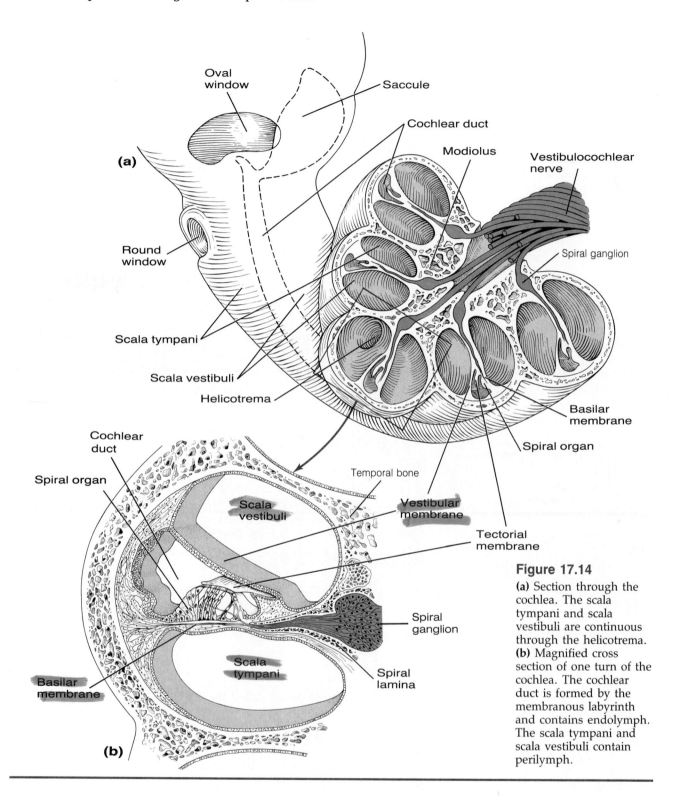

Figure 17.14

(a) Section through the cochlea. The scala tympani and scala vestibuli are continuous through the helicotrema. **(b)** Magnified cross section of one turn of the cochlea. The cochlear duct is formed by the membranous labyrinth and contains endolymph. The scala tympani and scala vestibuli contain perilymph.

the membranous labyrinth that is located in the osseous semicircular canals; the saccule is connected with the membranous labyrinth that is in the cochlea. The membranous labyrinth, therefore, is a continuous series of ducts that are filled with endolymph.

SEMICIRCULAR CANALS   Within the inner ear are three bony **semicircular canals** that contain three membranous **semicircular ducts.** The semicircular canals are arranged at right angles to each other, forming anterior, lateral, and

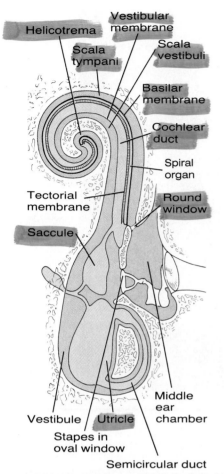

Figure 17.15
Diagrammatic representation of the structures of the inner ear. The red areas are filled with endolymph; the gray areas contain perilymph.

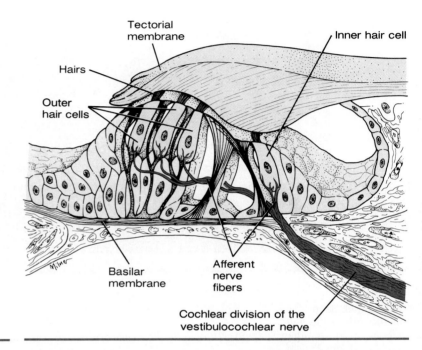

Figure 17.16
Section of the spiral organ.

posterior canals. The anterior and posterior canals are vertical; the lateral canal is horizontal. Each membranous semicircular duct possesses an enlargement, called an **ampulla,** which contains receptor cells that detect certain movements of the head and thereby provide information concerning equilibrium.

COCHLEA The portion of the inner ear associated with hearing is the **cochlea.** It resembles a snail shell, spiraling 2½ turns around a central bony core called the **modiolus** (Figure 17.14a). A bony shelf called the **spiral lamina** extends into the cochlea from the modiolus (Figure 17.14b). Two membranes, the **vestibular membrane** and the **basilar membrane,** extend across the cochlea from the spiral lamina, dividing the cochlea into three longitudinal tunnels.

The central tunnel between the vestibular and basilar membranes is called the **cochlear duct,** or **scala media.** The cochlear duct is the membranous labyrinth of the cochlea and is filled with endolymph. One tunnel, the **scala vestibuli,** is separated from the cochlear duct by the vestibular membrane. The other tunnel, the **scala tympani,** is separated from the cochlear duct by the basilar membrane. The scala vestibuli and scala tympani both contain perilymph and are continuous with one another at the apex of the cochlea through an opening called the **helicotrema.** The scala tympani terminates at the round window, whereas the scala vestibuli ends at the oval window. The relationships of these structures are more easily understood if the cochlea is visualized as being uncoiled, as shown in Figure 17.15.

The **spiral organ** *(organ of Corti),* which contains the receptors for hearing, is located on the basilar membrane. It consists of a series of sensory hair cells and supporting cells (Figure 17.16). The hair cells are arranged in a single row

F17.14a
F17.14b

F17.15

F17.16

Outer hair cells

Inner hair cells

Figure 17.17
Scanning electron
micrograph of the hair
cells of the spiral organ
(×2420). (From *Tissues
and Organs: A Text-Atlas
of Scanning Electron
Microscopy* by Richard G.
Kessel and Randy H.
Kardon. W.H. Freeman
and Company. Copyright
© 1979.)

F17.17 of *inner hair cells* and three rows of *outer hair cells* (Figure 17.17). The hair cells
are innervated by sensory fibers from the cochlear division of the vestibulocochlear nerve (cranial nerve VIII). These sensory fibers travel through the
modiolus and spiral lamina to reach the basilar membrane. Overhanging the
spiral organ is a flexible **tectorial membrane** that is anchored to the spiral
lamina. The hairs of the sensory cells of the spiral organ are in contact with
the tectorial membrane.

Mechanisms of Hearing

Hearing is the perception of sound. Sound is produced when molecules of air
(or other media) are compressed in a regular rhythm, producing a sound
wave. The amplitude (height) of the sound wave determines the intensity, or
loudness, of the sound. The greater the amplitude, the louder the sound. The
frequency of the sound wave (in vibrations per second) determines the
sound's pitch, or tone. The higher the frequency, the higher the pitch. The
human ear is capable of detecting sound waves with frequencies between 20
and 20,000 cycles per second. It is most sensitive, however, to sound waves
with frequencies between 1000 and 4000 cycles per second.

Transmission of Sound Waves to the Inner Ear

Sound waves enter the external auditory meatus and push against the tympanic membrane, causing it to oscillate at the same frequency as the sound
wave. The vibrations of the tympanic membrane are then transmitted across
the middle-ear cavity by the malleus, incus, and stapes to the oval window of
the cochlea. The surface area of the tympanic membrane is about 20 times
greater than the surface area of the foot-plate of the stapes. Because all the
energy of the sound waves that push against the tympanic membrane is
transmitted by the ossicles, the pressure against the oval window is about 20
times greater than the pressure of the sound wave that caused the movement.

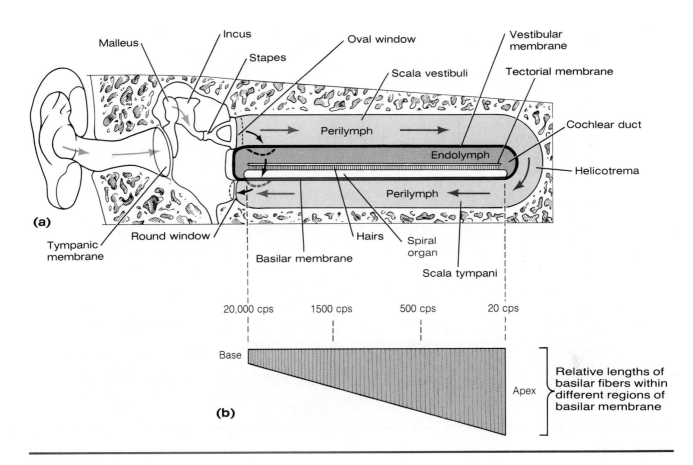

Figure 17.18

(a) Schematic representation of the transmission of pressure waves in the cochlea. Rust-colored arrows indicate the transmission pathway when the movement of the stapes is slow. Black arrows indicate the transmission pathway of higher-frequency pressure waves of the sort associated with sound perception. (b) Graphic representation of the basilar membrane showing that the end of the membrane toward the middle ear contains shorter basilar fibers than the end toward the helicotrema. Maximum vibrations of the basilar membrane occur in the region of the basilar membrane that has the same resonant frequency as the sound wave causing the pressure wave. Low-frequency sound waves cause maximum vibration toward the apex of the basilar membrane. High-frequency waves have their maximal effect toward the base of the basilar membrane.

Function of the Cochlea

If the inward movement of the oval window as a result of the movement of the stapes is very slow, the pressure exerted on the perilymph pushes the perilymph along the scala vestibuli, through the helicotrema, and into the scala tympani, causing the round window to bulge outward into the middle ear (Figure 17.18*a*). Thus, the inward movement of the oval window is com- F17.18a pensated for by the outward bulging of the round window into the middle-ear cavity. This compensatory movement prevents any significant increase in pressure within the inner ear, and thus, very-low-frequency pressure waves have little effect on the basilar membrane. However, higher-frequency pressure waves of the sort associated with sound perception follow a different path. When a sound is perceived, the movement of the stapes is so rapid that an increase in pressure within the perilymph of the scala vestibuli does occur in the base of the cochlea, close to the oval window. This increased pressure of the perilymph is transmitted through the flexible vestibular membrane to the endolymph of the cochlear duct. From the endolymph the pressure waves are transmitted through the basilar membrane to the perilymph of the scala tympani, and finally to the round window. The transmission of the pressure waves through the basilar membrane at the base of the cochlea near the oval

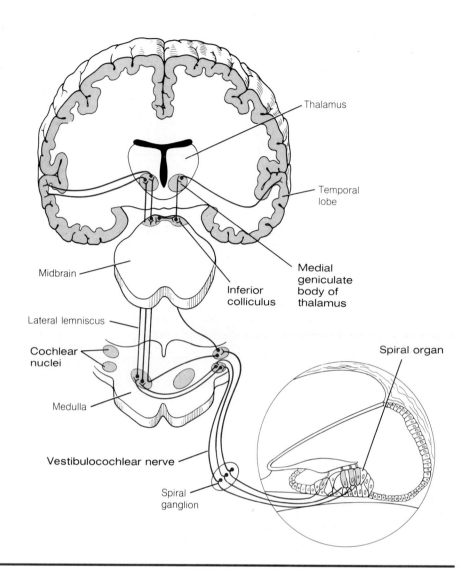

Figure 17.19
Auditory pathways from the spiral organ, through the central nervous system, to the cortex of the temporal lobe of the brain.

and round windows causes this area of the membrane to vibrate. The vibration of the basilar membrane near the base of the cochlea initiates a wave that travels along the basilar membrane toward the helicotrema.

The basilar membrane contains thousands of basilar fibers that project from the spiral lamina toward the outer wall of the cochlea. Although the basilar membrane extends completely across the cochlea from the spiral lamina, its fibers do not. Rather, the distal ends of the fibers are embedded in the basilar membrane. The basilar fibers increase in length and decrease in thickness and rigidity from the base of the cochlea, where the oval window is located, to the apex, near the helicotrema (Figure 17.18b). Because of this, certain regions of the basilar membrane respond more readily than others to the vibrations produced by sound waves of particular frequencies. The amplitude of a sound wave increases as it travels along the basilar membrane, attaining a maximum intensity when it reaches the region of the membrane that has the same maximum resonant frequency as the sound wave. Beyond this region of maximum intensity, the amplitude of the wave drops quickly, traveling no farther along the basilar membrane. Thus, high-frequency sound waves cause maximal movement of the basilar membrane near the base of the cochlea, where the basilar fibers are shorter. The vibrations produced by low-frequency sound waves travel farther along the basilar membrane, causing maximal movement of the membrane toward its apical end, where the basilar fibers are longer.

F17.18b

The hairs of the receptor cells of the spiral organ, which is located on the basilar membrane, are embedded in the tectorial membrane that overhangs them (Figure 17.14). As a result, movement of the basilar membrane causes **F17.14** the hairs to be displaced. The movement of the hairs causes the development of a generator potential in the hair cells. When the basilar membrane moves upward, toward the scala vestibuli, the hair cells of the affected region depolarize, generating afferent impulses in neurons of the cochlear division of the vestibulocochlear nerve that are associated with the hair cells. When the basilar membrane moves downward, the cell membranes of the affected hair cells hyperpolarize, thereby halting the generation of afferent impulses in the neurons of the cochlear division of the vestibulocochlear nerve.

Sound Perception in the Brain

Because sound waves of a given pitch cause maximum vibrations in a particular region of the basilar membrane, the hair cells of that region generate the greatest number of afferent nerve impulses. The impulses that travel from the various hair cells of the spiral organ reach the cortex of the temporal lobe of the brain in a definite spatial arrangement. Thus, when specific hair cells generate nerve impulses, these impulses always travel to a particular site in the cortex. This regular spatial arrangement allows the cortex to localize the region of the basilar membrane that is vibrating maximally, and thus to discriminate between various frequencies and interpret them as different tones or pitches. The neural pathways between the spiral organ and the auditory portion of the cerebral cortex involve synapses in the medulla, the inferior colliculi of the midbrain, and the medial geniculate body of the thalamus (Figure 17.19). **F17.19**

Equilibrium and Balance

In addition to its role in hearing, the inner ear also provides information about the position and movement of the head. This information is utilized in the coordination of movements that maintain body equilibrium and balance. The portion of the inner ear involved in these activities is called the **vestibular apparatus.** Receptor cells of the vestibular apparatus are located in the utricle and saccule and within the ampullae of the semicircular ducts. Changes in the position of the head affect *static balance;* movement of the head affects *dynamic balance.*

Static Balance

The receptors that provide information concerning the position of the head are located in the utricle and the saccule. The utricle serves as the dominant receptor organ; the role of the saccule in humans is unclear.

The receptors in the utricle and saccule are called **maculae.** The maculae are composed of groups of hair cells whose protruding hairs are embedded in a gelatinous substance. Within the gelatinous substance are tiny particles of calcium carbonate called **otoliths** (ear stones). The otoliths make the gelatinous substance heavier than the endolymph that fills the membranous labyrinth. As a result, when the head is in the upright position the force of gravity causes the gelatinous substance to bear down on the hair cells (Figure 17.20). **F17.20** When the position of the head changes, the direction of the force of the gelatinous substance on the hairs changes also, causing the hairs to bend from their normal position. The movement of the hair cells alters the stimulation of the neurons that innervate the hair cells, and a different pattern of nerve impulses is transmitted to the brain. The nerve impulses are carried over the vestibular division of the vestibulocochlear nerve to various motor areas in the central nervous system, primarily in the medulla and the cerebellum. The motor centers, in turn, initiate muscular actions that coordinate various body movements with the position of the head.

Dynamic Balance

The body is able to respond automatically to rotational acceleration or deceleration of the head because of information received by the central nervous

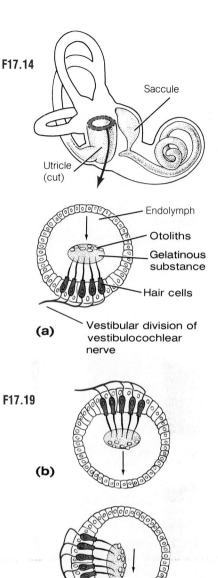

(a)

(b)

(c)

Figure 17.20
Stimulation of the balance receptors (maculae) within the utricle and saccule. The arrows indicate the direction of the force of gravity. **(a)** Shows a macula when the head is in an upright position. **(b)** Illustrates a macula when the head is inverted. **(c)** Shows a macula when the head is in a horizontal position.

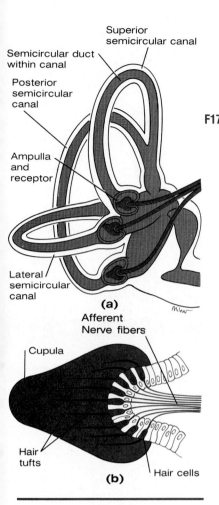

Superior
semicircular canal

Semicircular duct
within canal

Posterior
semicircular
canal

Ampulla
and
receptor

Lateral
semicircular
canal

(a)

Afferent
Nerve fibers

Cupula

Hair
tufts

(b)

Hair cells

Figure 17.21
(a) The semicircular canals and ducts. The ampullae of the ducts have been cut open to show the locations of the balance receptors (cristae) within them.
(b) Enlargement of a crista.

system from receptors located in the ampullae of the membranous semicircular ducts.

Because the three semicircular ducts of each ear are arranged at right angles to one another, movement in any plane affects the endolymph in at least one of the ducts. The ampulla of each duct contains a group of transversely oriented sensory hair cells called a **crista**. The hairs of a crista are F17.21 embedded in a gelatinous mass called the **cupula** (Figure 17.21).

The hair cells of the crista in a semicircular duct are displaced by rotary movements that occur in a plane corresponding to the plane of its semicircular canal. When the head begins to rotate, the inertia of the endolymph causes a momentary delay in the movement of the endolymph. In effect, this delay produces a backward flow of the endolymph in the duct and displaces the cupula and the hairs of that duct's hair cells from their normal positions. The displacement of the hairs alters the stimulation of neurons associated with the hair cells, and a different pattern of afferent nerve impulses is transmitted to the central nervous system by way of the vestibular division of the vestibulocochlear nerve. The endolymph begins to move, after its inertia is overcome, and if rotation continues at a constant velocity, the endolymph is eventually moving at the same rate as the semicircular duct. Under these conditions, the pattern of afferent nerve impulses again resembles the normal pattern and the sensation of movement is decreased.

With deceleration, the momentum of the endolymph causes it to take longer to stop than does the semicircular duct itself. Thus, the endolymph, in effect, flows forward in the semicircular duct. This forward movement displaces the crista in a direction opposite to that which took place during acceleration and causes an alteration in the pattern of afferent nerve impulses. Consequently, the cristae provide information concerning changes in the rate of rotation (either acceleration or deceleration), but little information during periods of rotation at a constant velocity.

The cristae also provide information concerning the *direction* of rotation. Movement in one direction increases the pressure on the cupula of a particular semicircular duct in one ear and thus displaces the hair cells in the ampulla of that duct. At the same time, the pressure on the cupula in the corresponding semicircular duct of the other ear is reduced. Therefore, the central nervous system receives different messages from the two affected cristae, making it possible to determine the direction of rotation.

DISRUPTIONS OF DYNAMIC BALANCE   The continued forward movement of endolymph in the semicircular ducts after rotation of the head has ceased can produce several disorienting sensations, including vertigo, nystagmus, and errors of movement.

In *vertigo* a person experiences an illusion of movement, such as a spinning sensation or a feeling that the external environment is revolving. Vertigo differs somewhat from dizziness, which is accompanied by a feeling of movement within the head, although the two terms are commonly used synonymously. Both vertigo and dizziness can be caused by many factors other than rotation of the body.

Nystagmus is a characteristic movement of the eyes that occurs during rotational acceleration of the body. Since the eyes tend to fix on an object during rotational acceleration, they may drift in a direction opposite to that in which the head is moving. This drift is followed by a rapid and jerky return of the eyes to their original forward position. The eyes then fix on another object and the nystagmus phenomenon is rapidly repeated as long as rotational acceleration continues. If the acceleration is abruptly halted, nystagmus still occurs for a few more seconds, with the eyes jerking in a direction opposite to the direction of rotation. Because of the momentum of the endolymph in the semicircular ducts, the hairs of the cristae are quickly displaced in the opposite direction from which they were displaced during acceleration. Until the fluid stops moving, the brain interprets the pattern of afferent nerve impulses resulting from this displacement as a signal of the reversal of the direction of rotation.

FRONTIERS IN HEALTH

The Bionic Ear

More than 2 million Americans suffer from profound deafness. Brought on by the destruction of hair cells as a result of loud noises or infections such as meningitis, or resulting from birth defects, profound deafness has been considered virtually untreatable. For the unfortunate victims, police sirens, automobile horns, and all other sounds go unperceived. Living such a life is much like watching television with the sound off, 24 hours a day for an entire lifetime.

Profound deafness can seriously affect the development of children born deaf or deafened before they begin to speak. Banished to a silent world, they often fail to mature emotionally. Learning to communicate is difficult, and reading comprehension is generally impaired. In fact, some never advance beyond the elementary reading level.

Hearing aids are usually of no value to those who are born deaf or to those who have completely lost their hearing from some other cause. For this large group of people, however, there is now some hope. It comes in the form of a cochlear implant—a device that simulates the function of the ear.

Pioneered by Dr. William House of the House Ear Institute in Los Angeles, the cochlear implant comprises several parts. Mounted at ear level is a small microphone, which picks up sound and transmits it to a stimulator unit worn on a belt or on the chest. The stimulator, which is about the size of a deck of cards, processes the signals it receives from the microphone and converts them into electric signals, which are sent to an induction coil attached to the outside of the body over the mastoid bone. Smaller than a quarter, the induction coil beams the signal through the skull to an implanted receiver. The receiver, in turn, picks up the faint signal and generates an electric current. This tiny current travels to the inner ear along an extremely fine implanted electrode, and from there to a ground electrode implanted a short distance away. The current produces a sensation in the nerve cells of the auditory nerve. In some models the electrodes are imbedded in the auditory nerve fibers themselves.

This complex method makes it possible for the deaf to hear, but the hearing is very different from normal hearing. Patients with cochlear implants can detect speech and environmental sounds and can distinguish some words from others, but the fine discriminations that allow people to communicate freely through complex language systems are impossible.

However, even this rudimentary form of hearing is tremendously exciting to people who have been closed off from the sounds of the world. For them, the cochlear implant provides valuable outside stimuli and a connection with the world. It also makes living a little safer. The sounds of horns and sirens, which may be annoying to most of us, serve an important purpose. The implant also helps adults monitor and regulate their own voices. Speech reading (reading lips and matching with sound) becomes easier, for the cochlear implant allows the listener to follow normal conversation at a comfortable

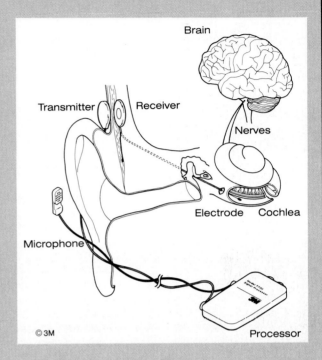

Diagram of the 3M cochlear implant.

level and allows recipients to distinguish some words by their sounds. Thus, the minimal gains achieved by cochlear implants can significantly enhance the lives of profoundly deaf adults. For children, minimal hearing afforded by the cochlear implant can mean the difference between learning to speak and read or remaining mute and illiterate—the difference between communication and complete isolation.

The 3M Company of St. Paul, Minnesota manufactures a single-electrode cochlear implant that is being used by hundreds of adults and children. Some patients who were involved in the earliest experiments have been wearing their bionic ears for more than a decade with no significant problems. In 1984, the Food and Drug Administration approved the single-electrode implant for general use.

However, the single-electrode cochlear implant may soon become obsolete as medical researchers continue to search for ways to introduce more electrodes into the inner ear, thereby bringing a wider range of sounds to the organ of hearing. Dr. Robert L. White and his colleagues at Stanford University, among others, are working on multiple-channel cochlear implants. Recipients of these devices may be able to perceive many distinct words, not just the sound of a telephone ringing or a horn blowing. With this development, normal hearing may be a bit closer for those suffering from profound deafness.

CONDITIONS OF CLINICAL SIGNIFICANCE

The Ear

Middle-Ear Infections (Otitis Media)

Infections of the mucous membrane of the throat may travel through the auditory tubes and cause an inflammation of the mucous membrane that lines the middle-ear cavity, including the inner surface of the tympanic membrane. The inflammation can cause fluid to collect in the middle ear, temporarily interfering with the ability to hear.

Impairment of Hearing

Disease or injury to any portion of the ear can lead to some loss in the ability to perceive sounds, causing partial or total deafness. Such hearing losses can be classified as either conductive deafness or nerve deafness.

Conductive deafness involves interference with the transmission of sound waves through the external or middle ear. The cause of the interference can include physical blockage of the external auditory meatus by a foreign object or impacted cerumen (earwax), inflammation of the eardrum, adhesions between the ossicles, or thickening of the oval window. In conductive deafness there is no damage to the receptor cells of the spiral organ or to the nerve pathways. Hearing aids, which transmit sound waves to the inner ear through the temporal bone rather than through the middle ear, can therefore aid a person suffering from this condition.

In *nerve deafness*, the loss of hearing results from disorders that affect the sound receptors in the inner ear,

the vestibulocochlear nerve, or nerve pathways in the central nervous system. Such disorders can be caused by a number of factors, including infection, tumors, and trauma. Until recently, hearing aids have not been helpful in cases of nerve deafness because the hearing loss involves damage to receptor cells or nerve pathways. However, recent technological advances have resulted in the development of devices that can assist hearing even in cases of nerve deafness.

Effects of Aging on the Ear

Several age-related changes can contribute to the impairment of hearing. As the activity of sweat glands within the meatus diminishes with age, the wax secreted by the ceruminous glands becomes drier and tends to accumulate. This build-up of earwax, by itself, is often a cause of hearing loss in older persons, especially in the low frequency tones.

Most hearing loss in older persons, however, is due to changes that occur in inner ear structures. Due in part to a thickening of the walls of capillaries supplying the cochlea, there is often an age-related degeneration of spiral-organ cells and sensory ganglia cells. There has also been reported to be a decrease in the number of nerve fibers within both divisions of the vestibulocochlear nerve in persons over 45 years of age. Thus, it is not uncommon for older persons to experience problems with balance and equilibrium as well as hearing.

Following rotation, a person commonly has difficulty moving forward to a particular point. As long as the endolymph in the semicircular ducts continues to move, the person receives information indicating that rotation is still occurring. As a consequence, the person attempts to compensate for the rotation by moving forward at an angle opposite to the direction of the imagined rotation. Hence, the person leans and travels toward one side and misses the intended destination.

Continued periodic motion of the endolymph, particularly during repeated vertical movements, produces unpleasant sensations of vertigo and nausea in some people—in other words, *motion sickness.* Most people, however, quickly adapt to the sensations caused by motion.

OLFACTORY ORGANS

F17.22

The receptors for smell *(olfaction)* are specialized neurons located in the epithelium of the nasal mucosa in the upper portion of the nasal cavity, on either side of the nasal septum (Figure 17.22). The receptor neurons are scattered among the columnar cells of the mucous membrane that serve as *supporting cells.*

The olfactory receptor cells have two processes. One of the processes bears many fine hairs, or cilia, which extend into the nasal cavity and are embedded in the mucous lining of the cavity. The second process of the receptor neurons, together with similar processes from other olfactory receptor cells, passes through the cribriform plate of the ethmoid bone and enters the **olfactory bulbs** in the brain as a component of the *olfactory nerves* (Cranial

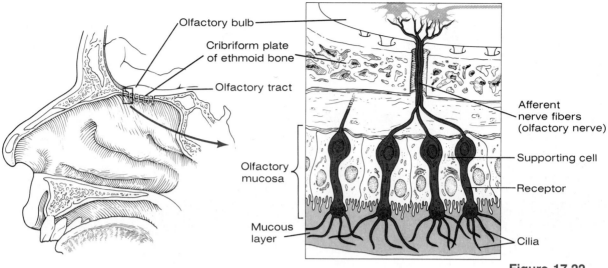

Olfactory bulb

Cribriform plate of ethmoid bone

Olfactory tract

Olfactory mucosa

Mucous layer

Afferent nerve fibers (olfactory nerve)

Supporting cell

Receptor

Cilia

Figure 17.22
Location and structure of the olfactory receptors.

nerve I). Within the olfactory bulbs, the processes of the receptor cells synapse with neurons that leave the bulbs posteriorly as the **olfactory tracts.** From the olfactory tracts, nerve fibers travel to the cerebral cortex on the medial sides of the temporal lobes.

For an odorous substance to be detected, it must reach the olfactory receptors. Normally, air moving through the nose does not pass through the upper portion of the nasal cavity, where the olfactory receptors are located, so odorous substances must diffuse to that region. This process can be augmented by sniffing. Once in the region of the receptors, an odorous substance must dissolve in the mucus layer covering the receptors and interact with them in some way. This interaction is believed to depolarize the receptor, resulting in a generator potential that can initiate nerve impulses that are conveyed to the brain, where they are interpreted as a particular smell.

TASTE ORGANS

The receptors for taste are called **taste buds.** Most taste buds are located on the surface and the papillae of the tongue, but some are found on the roof of the mouth, on the pharynx, and on the larynx. Three types of cells have been identified within a taste bud. These cells have been called *receptor cells, supporting cells*, and *basal cells*. While these names imply a specific functional role for each cell type, it has not been possible to confirm the functional differences. In fact, there is evidence that the cell types may actually be different stages in the development of a single cell type. Microvilli from the upper portions of the receptor cells extend into a small pore (the **taste pore**) at the surface of the taste bud (Figure 17.23). This position allows them to be bathed F17.23 by the fluids of the mouth.

Sensory nerve fibers contact the receptor cells of the taste buds. A single nerve fiber may contact several different receptor cells, and a single receptor cell may have several different nerve fibers contacting it. These fibers run to the brain stem via the facial nerve (from the anterior two-thirds of the tongue), the glossopharyngeal nerve (from the posterior third of the tongue), and the vagus nerve (a few fibers from the pharyngeal region). The axons in these cranial nerves synapse in the brain stem with interneurons that cross to the opposite side and travel through the medial lemnisci to the thalamus. In the thalamus the interneurons synapse with neurons that are conveyed to the tongue area of the somatosensory cortex in the postcentral gyrus of the parietal lobe.

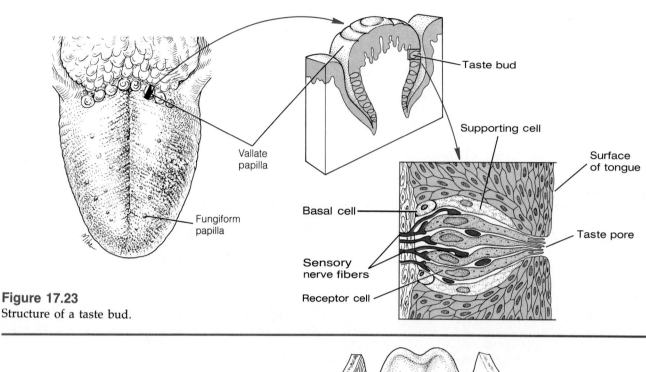

Figure 17.23
Structure of a taste bud.

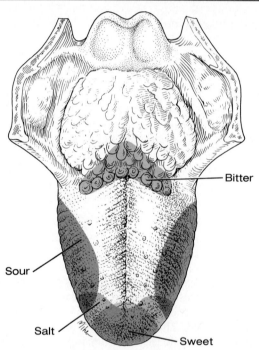

Figure 17.24
Areas of the tongue that are most sensitive to particular taste sensations.

If a substance is to evoke a taste sensation, it must dissolve in the fluids that bathe the tongue and then interact with the receptor cells of the taste buds. Four specific tastes have traditionally been identified—*sweet, salty, bitter,* and *sour*—with each taste being detected best in certain specific regions of the tongue (Figure 17.24). It appears, however, that there is no corresponding specificity of taste receptor cell types. That is, a taste receptor cell can respond to a variety of different substances that belong to more than one specific taste category.

Taste receptor cells have been postulated to possess several different types of receptor sites that can form loose combinations with different types of molecules. The formation of these combinations depolarizes the receptor cell, establishing a generator potential that can ultimately result in action po-

F17.24

tentials in the sensory nerve fibers that contact the receptor cell. Thus, the sensation of a specific taste is a complex phenomenon involving the relative activities and firing patterns of a number of different sensory neurons from a variety of different taste receptor cells.

The sense of taste is closely associated with the sense of smell, and much of what is commonly considered to be taste is in reality related to the stimulation of olfactory receptors. This fact is particularly evident when a person has a cold that produces a congestion of the nasal mucosa that blocks olfactory stimulation. Under such conditions, the person often notices an inability to "taste" food, even though there is no impairment of the taste buds.

STUDY OUTLINE

THE EYE—VISION pp. 451–462

EMBRYONIC DEVELOPMENT OF THE EYE
1. Outgrowths from lateral diencephalon form optic vesicles; each invaginates, forming optic cup, which becomes retina.
2. Optic stalk is incorporated into optic nerve.
3. Ectoderm over optic cup thickens into lens placode; placodes develop into lens vesicles, which later separate from ectoderm, eventually forming lens of adult eye.
4. Cornea forms from cells superficial to lens vesicles.
5. Loose mesenchymal cells form fibrous tunic and vascular tunic of eye.

STRUCTURE OF THE EYE

FIBROUS TUNIC outermost layer.

Sclera posterior; white; dense connective tissue; protects eye and maintains its shape.

Cornea anterior; clear; allows for passage of light.

VASCULAR TUNIC beneath fibrous tunic.

Choroid posterior; dark; contains many blood vessels.

Ciliary Body around edge of cornea; contains smooth ciliary muscles oriented in three directions; produces aqueous humor.

Iris anterior; thin muscular diaphragm; pigmentation responsible for eye color.

Pupil opening in center of iris through which light passes.

RETINA innermost layer of eye (internal tunic); consists of outer pigmented layer and inner nervous-tissue layer.

Pigmented Layer composed of epithelial cells in contact with choroid.

Nervous-Tissue Layer contains photoreceptors—rods and cones.

Optic Nerve area of its exit from eye is blind spot (optic disc).

Macula located at posterior pole of eye; at its center is fovea centralis, with high concentration of cone photoreceptors.

LENS aids in focusing light on retina; relatively elastic, biconvex structure.

CAVITIES AND HUMORS lens separates eye into two cavities: anterior cavity (contains aqueous humor) and posterior cavity (contains vitreous humor).

ACCESSORY STRUCTURES OF THE EYE

EYELIDS anterior, skin-covered structures; provide protection and light screening; controlled by skeletal muscles (orbicularis oculi and levator palpebrae superioris).

CONJUNCTIVA mucous membrane that lines inside of eyelids and covers anterior surface of eye.

LACRIMAL APPARATUS includes lacrimal gland and associated ducts; gland produces tears.

EXTRINSIC EYE MUSCLES lateral, medial, superior, and inferior rectus muscles; inferior and superior oblique muscles. Rapid-acting; among the most precisely controlled skeletal muscles.

REFRACTING MEDIA OF THE EYE light rays bent (refracted) when passing through media of different optical densities.

FOCUSING IMAGES ON THE RETINA involves four processes:
1. *Refraction:* light rays bent when they enter and leave cornea and lens.
2. *Accommodation:* objects closer than 20 feet are focused on retina by accommodation; accomplished by ciliary muscles allowing lens to change shape.
3. *Constriction:* narrowing of pupil prevents divergent light rays from entering eye.
4. *Convergence:* eyes turn inward when viewing close objects; causes image to fall on fovea of retina.

BINOCULAR VISION overlapping visual fields of both eyes; basic to depth perception.

PHOTORECEPTORS OF THE RETINA

RODS responses indicate degrees of brightness but not color; contain rhodopsin; numerous in peripheral retina.

CONES perceive color; some sensitive to red, others to green, others to blue; concentrated in center of retina and fovea.

NEURAL ELEMENTS OF THE RETINA middle portion of retina contains bipolar neurons; inner portion of retina contains ganglion cells; processes from ganglion cells form optic nerve.

VISUAL PATHWAYS

OPTIC CHIASMA neurons from medial half of each retina cross over.

OPTIC TRACTS from optic chiasma to superior colliculi or lateral geniculate bodies.

OPTIC RADIATIONS from lateral geniculate bodies to visual cortex of brain.

CONDITIONS OF CLINICAL SIGNIFICANCE: THE EYE   pp. 463-464

MYOPIA nearsightedness.
1. Light rays from distant objects focused in front of retina.
2. Corrected by concave lenses.

HYPERMETROPIA farsightedness.
1. Distant objects focused behind retina.
2. Corrected by convex lenses.

ASTIGMATISM unequal curvature of portions of refractive system of eye.

CATARACT opaqueness of lens of eye.

GLAUCOMA increase in intraocular pressure.

COLOR BLINDNESS absence or inadequate amount of cone photopigments.

EFFECTS OF AGING ON THE EYE
1. Lens becomes yellow due to exposure to ultraviolet rays.
2. Cataracts can develop.
3. Pupil is less able to dilate, resulting in less light reaching photoreceptors.
4. Poor drainage of aqueous humor increases likelihood of glaucoma.
5. Debris from rods and cones can cause loss of visual acuity.
6. Lens loses elasticity, leading to *presbyopia*.

THE EAR—HEARING AND BALANCE
pp. 464-474

EMBRYONIC DEVELOPMENT OF THE EAR
1. Plate of thickened ectoderm (otic placode) invaginates, forming otic pit; otic pit separates from surface ectoderm to become otic vesicle; eventually forms membranous labyrinth of inner ear.
2. Part of first pharyngeal pouch forms auditory tube.
3. Mesenchymal cells in middle-ear cavity form three ossicles.
4. Branchial groove external to first pharyngeal pouch forms external auditory meatus.
5. Tympanic membrane develops from membrane separating first pharyngeal pouch and branchial groove.

STRUCTURE OF THE EAR

EXTERNAL EAR

Auricle (Pinna) directs sound waves into external auditory meatus.

External Auditory Meatus (Canal) contains hairs and cerumen; serves as resonator.

MIDDLE EAR air chamber in temporal bone; extends from tympanic membrane to oval window.

Auditory Tube regulates pressure in middle-ear chamber.

Ossicles (Malleus, Incus, Stapes) transmit and amplify vibrations of tympanic membrane to oval window.

INNER EAR series of fluid-filled (perilymph) canals (osseous labyrinth) containing membranous labyrinth, which is filled with endolymph.

Vestibule saccule and utricle; associated with static balance (head position).

Semicircular Canals associated with dynamic balance (equilibrium).

Cochlea portion of inner ear for hearing; spiral organ.

MECHANISMS OF HEARING

TRANSMISSION OF SOUND WAVES TO THE INNER EAR
1. Sound waves enter external auditory meatus and cause tympanic membrane to vibrate.
2. Ossicles carry vibrations from tympanic membrane to oval window of cochlea.

FUNCTION OF THE COCHLEA
1. Vibrations of tympanic membrane are transmitted through endolymph of cochlear duct, which vibrates specific regions of basilar membrane.
2. Movement of hairs on spiral organ generates afferent nerve impulses in cochlear division of cranial nerve VIII.

SOUND PERCEPTION IN THE BRAIN nerves from hair cells of spiral organ carry impulses to cortex of temporal lobe of brain.

EQUILIBRIUM AND BALANCE

STATIC BALANCE (POSITION OF HEAD) receptors located in utricle and saccule; change in direction of force of gravity on maculae signals change in position of head.

DYNAMIC BALANCE (MOVEMENT OF HEAD) receptors located in ampullae of semicircular canals; because of inertia of endolymph, hair cells (crista) are displaced by acceleration or deceleration of head.

CONDITIONS OF CLINICAL SIGNIFICANCE: THE EAR   p. 476

MIDDLE-EAR INFECTIONS (OTITIS MEDIA) inflammation of mucous membrane of middle ear.

IMPAIRMENT OF HEARING

CONDUCTIVE DEAFNESS interference of sound-wave transmission by physical blockage, inflammation of eardrum, ossicle adhesions, or oval-window thickening.

NERVE DEAFNESS loss of hearing involving damage to sensory cells or nerve pathways.

EFFECTS OF AGING ON THE EAR
1. Build-up of earwax.
2. Degeneration of cells in sprial organ and sensory ganglia.
3. Decrease in number of nerve fibers in both divisions of vestibulocochlear nerve.

OLFACTORY ORGANS **pp. 476–477**
1. Receptors are neurons in epithelium of nasal mucosa in upper nasal cavity.
2. Odorous substances must dissolve in mucus layer covering receptors and then interact with receptors.
3. Depolarization of receptors initiates nerve impulses that are conveyed to brain (by cranial nerve I), where they are interpreted as smell.

TASTE ORGANS **pp. 477–479**
1. Receptors are in taste buds on tongue, roof of mouth, pharynx, larynx.
2. Taste closely associated with sense of smell.

SELF-QUIZ

1. The retina develops from the embryonic: (a) telencephalon; (b) diencephalon; (c) mesencephalon.

2. The iris belongs to the: (a) fibrous tunic; (b) internal tunic; (c) vascular tunic.

3. The constriction and dilation of the pupil occur primarily as voluntary responses and help regulate the amount of light that enters the eye. True or false?

4. The anterior cavity of the eye is separated from the posterior cavity by the: (a) lens; (b) iris; (c) cornea.

5. The aqueous humor is produced by the: (a) canal of Schlemm; (b) ciliary body; (c) vitreous body.

6. The extrinsic eye muscles are the least rapid-acting but among the most precisely controlled skeletal muscles in the body. True or false?

7. The conjunctiva is a mucous membrane that lines the eyelids and covers the anterior surface of the eye. True or false?

8. The cornea, the lens, and both humors of the eye are all refracting media. True or false?

9. During accommodation for viewing near objects: (a) the pupils dilate; (b) the ciliary muscles relax; (c) neither.

10. The loss of lens elasticity that occurs with aging is called: (a) presbyopia; (b) myopia; (c) hypermetropia.

11. An abnormal increase in intraocular pressure is: (a) strabismus; (b) astigmatism; (c) glaucoma.

12. The eye receptors that enable a person to perceive color are: (a) cones; (b) rods; (c) neither.

13. The right optic tract is formed from neurons from the right portion of the visual field of each eye. True or false?

14. The middle ear contains the: (a) cochlea; (b) semicircular canals; (c) incus.

15. A membranous partition between the middle-ear chamber and the scala vestibuli of the inner ear is formed by the: (a) oval window; (b) round window; (c) fenestra cochleae.

16. The cochlea is the portion of the inner ear that is associated with hearing. True or false?

17. The spiral organ is located on the: (a) tympanic membrane; (b) basilar membrane; (c) vestibular membrane.

18. The loudness of a sound is determined primarily by which aspect of a sound wave? (a) amplitude; (b) frequency; (c) pitch.

19. Receptor cells of the vestibule and the semicircular canals of the inner ear detect changes in: (a) the position of the head; (b) the movement of the head; (c) both.

20. Since the three semicircular ducts of each ear are arranged at right angles to one another, the endolymph in at least one of the ducts is affected by movement in any plane. True or false?

21. A characteristic movement of the eyes that occurs while rotational acceleration of the body is taking place is termed: (a) dizziness; (b) nystagmus; (c) vertigo.

22. Sensory nerve fibers associated with taste travel to the brain via the: (a) glossopharyngeal nerve; (b) vagus nerve; (c) facial nerve; (d) all of these.

LEARNING OBJECTIVES

After completing this chapter, you should be able to:

- Describe how an endocrine gland differs from an exocrine gland.

- List the major endocrine glands, and name the hormones secreted by each.

- Describe the embryonic development of the major endocrine glands.

- Describe the cell types found in the adenohypophysis, and relate these cell types to the production of specific hormones.

- Describe the functional relationship between the pars nervosa of the pituitary gland and the hypothalamus.

- Describe the structure of the thyroid gland.

- Describe the structure of the adrenal gland, and compare the actions of the hormones produced by each region of the gland.

- Describe the structure of the endocrine portion of the pancreas.

- Explain the functions of each hormone.

CHAPTER CONTENTS

The Endocrine System

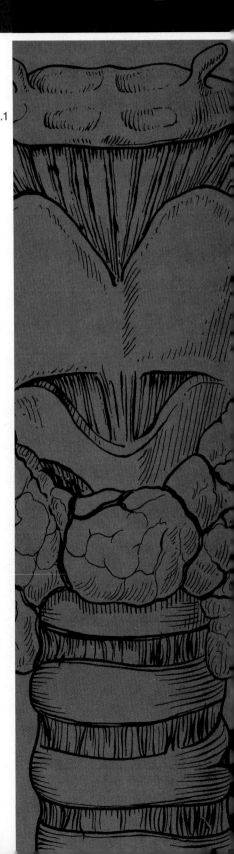

Rather than being a clearly defined anatomical system, the endocrine system is composed of various glands that are located in diverse places throughout the body (Figure 18.1). The **endocrine glands** aid in the regulation of body **F18.1** activities by producing molecules called **hormones.** Because of their influence on the organs of the body, hormones are often considered to be *chemical messengers.* The regulation by hormones may be general, affecting the cells of a large number of organs, or it may be quite specific, affecting the cells of only a particular organ or group of organs. The organs that react to a specific hormone are called **target organs.**

In contrast with exocrine glands, whose secretions are transported to their site of action through ducts, endocrine glands lack ducts—that is, they are *ductless glands.* The cells of endocrine glands, which are derived from epithelium, release their secretions (hormones) directly into the bloodstream. The bloodstream then distributes the hormones throughout the body, and each hormone affects only specific target organs. The principal endocrine glands are the *pituitary, thyroid, parathyroids, adrenals, pancreas,* and *gonads.* These glands and their hormones are summarized in Table 18.2 (pp. 504–505). In addition to these major endocrine glands there are other structures, such as the thymus gland, the gastrointestinal tract, and the placenta, that contain cells that produce hormones and therefore exhibit endocrine activity.

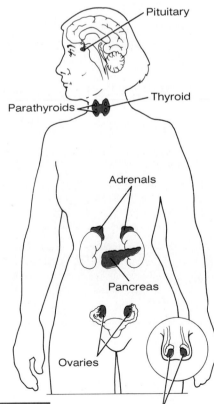

Figure 18.1
The major endocrine glands.

483

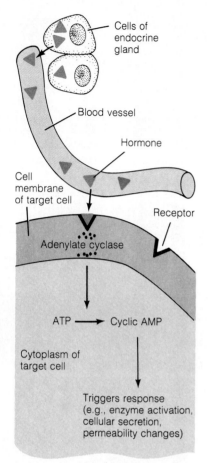

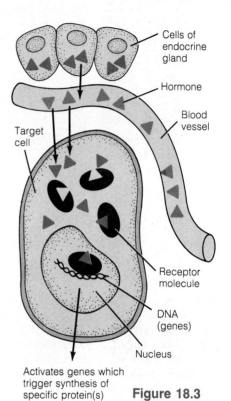

Figure 18.2
Many hormones bind to
receptors on plasma
membranes and influence
cellular function by way
of intracellular mediators
such as cyclic AMP.

Figure 18.3
Mechanism by which
some hormones,
particularly the steroid
hormones, activate genes
within a cell.

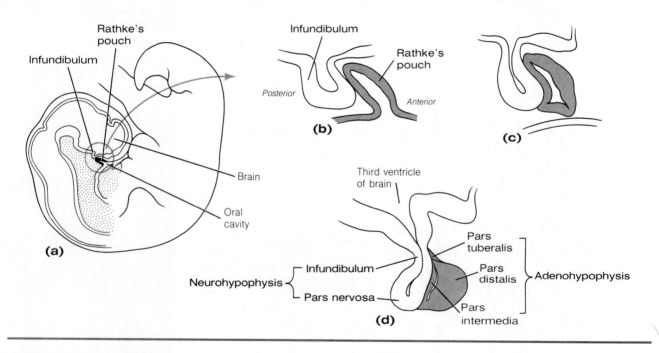

Figure 18.4
Stages in the embryonic
development of the
pituitary gland. See the
text for a detailed
discussion.

CHEMICAL STRUCTURE OF HORMONES

The products of the endocrine structures, the hormones, do not fall into any easily defined class of chemical substances. Some are steroids (for example, cortisol), others are proteins (for example, prolactin), and still others are smaller polypeptides (for example, oxytocin) or are closely related to amino acids (for example, epinephrine). Regardless of their chemical nature, however, hormones aid in the regulation and integration of body processes and in the maintenance of homeostasis. This regulatory role is similar to that played by the nervous system, and the endocrine and nervous system are in fact very closely related.

MECHANISMS OF HORMONE ACTIONS

Different hormones affect the activities of their target cells by different mechanisms. However, two important general mechanisms of hormonal action are: (1) by *utilizing intracellular mediators* and (2) by *activating genes within cells.*

Those hormones that influence cellular function by way of intracellular mediators are believed to bind with receptors on the plasma membrane (Figure 18.2). This binding of hormone with receptor is thought to cause the **F18.2** release of the intracellular mediator, which is often referred to as a *second messenger.* One second messenger is a compound known as *3',5'-cyclic adenosine monophosphate (cyclic AMP).* Cyclic AMP is formed from *adenosine triphosphate* (ATP) by an enzyme called adenylate cyclase. When molecules of a hormone that utilizes cyclic AMP as an intracellular mediator attach to receptors on the plasma membrane, the activity of adenylate cyclase is altered. The alteration of adenylate cyclase activity leads to changes in the level of cyclic AMP within the cell, which, in turn, can affect various cellular functions, such as enzyme activities, secretory activities, and plasma membrane permeability. In this reaction, the hormone is viewed as a first messenger and the cyclic AMP as a second messenger. The ultimate effects of cyclic AMP depend on the type of target cell stimulated. Some hormones are believed to exert their effects by reducing, rather than increasing, the activity of adenylate cyclase.

A different mechanism of action is used by the steroid hormones, which exert at least some of their effects by activating genes within a cell (Figure 18.3). In this method, the steroid hormones enter a cell and combine with **F18.3** receptor proteins in the cytoplasm of the cell. The hormone–receptor complex is transported to the nucleus of the cell, where it interacts with the genetic material and activates certain genes. This gene activation leads to synthesis of messenger-RNA and ultimately to the production of proteins (for example, enzymes) that influence cellular reactions or processes.

PITUITARY GLAND

Because its hormones regulate several other endocrine glands and affect a number of diverse body activities, the **pituitary gland** or **hypophysis,** has been referred to as the "master gland."

Embryonic Development and Structure

The close relationship between the nervous and endocrine systems is especially evident in the development and function of the pituitary gland. The pituitary gland is located beneath the brain and is surrounded by the *sella turcica* of the sphenoid bone. The opening of the sella turcica is covered over by a ring-shaped fold of dura mater called the *diaphragma sellae.* The pituitary develops embryonically from two different ectodermal regions: the floor of the brain and the roof of the mouth (Figure 18.4).   **F18.4**

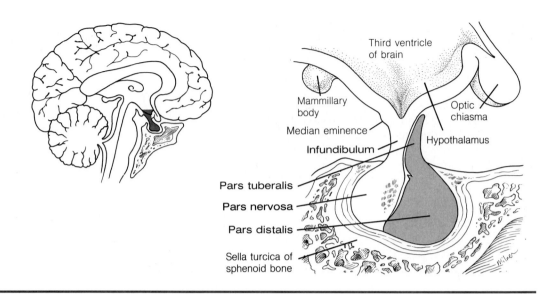

Third ventricle
of brain

Mammillary
body

Median eminence

Infundibulum

Optic
chiasma

Hypothalamus

Pars tuberalis

Pars nervosa

Pars distalis

Sella turcica of
sphenoid bone

Figure 18.5
The pituitary gland.

Neurohypophysis

The portion of the pituitary known as the **neurohypophysis** is an outgrowth
of nervous tissue from the floor of the brain in the region of the hypothalamus. The fully developed neurohypophysis consists of a stalklike **infundibulum** connected to a structure called the **pars nervosa.** Where the infundibulum connects the pars nervosa to the brain there is a small elevation called the
median eminence (Figures 18.4 and 18.5).

F18.4,
F18.5

Adenohypophysis

The second portion of the pituitary arises from ectodermal tissue from the
roof of the mouth (Figure 18.4*b*). An outpouching from the mouth, called
Rathke's pouch, grows toward the developing neurohypophysis. As it meets
the neurohypophysis, Rathke's pouch loses its connection with the mouth
(Figure 18.4*c*). The portion of the pituitary derived from Rathke's pouch becomes the **adenohypophysis** of the fully developed pituitary and includes the
pars distalis, pars tuberalis, and **pars intermedia** (Figure 18.4*d*). The pars
intermedia, however, is virtually nonexistent in the human pituitary.

F18.4b

F18.4c

F18.4d

The pars distalis is the most important region of the adenohypophysis,
producing six hormones. Three types of glandular cells have been identified
in the pars distalis. These cells are classified as *acidophils, basophils,* or *chromophobes,* depending on how they react to laboratory staining techniques. Recent work indicates that these three types of cells can be further subdivided,
as it is now generally thought that each hormone of the adenohypophysis is
made by a separate and distinct type of cell. For our purposes, however, it is
useful to follow the general classification and to consider that acidophils produce two hormones (*growth hormone* and *prolactin*) and that the basophils produce four hormones (*thyrotropin, follicle-stimulating hormone, luteinizing hormone,* and *adrenocorticotropic hormone*). The chromophobes are believed to be
cells that are in nonsecretory stages of development.

Table 18.1

Table 18.1 summarizes the terminology used to identify the divisions and
subdivisions of the pituitary gland. Note that the pituitary gland may also be
divided into an *anterior lobe,* which includes the pars distalis and pars tuberalis, and a *posterior lobe,* which includes the pars nervosa and, when present,
the pars intermedia.

Relationship to Brain

The neurohypophysis of the pituitary remains directly connected to the brain
by way of the infundibulum. The adenohypophysis is also closely associated
with the brain, although it is not directly connected to it. The association
between the brain and the adenohypophysis is maintained by way of the

Table 18.1   Terminology of the Pituitary Gland

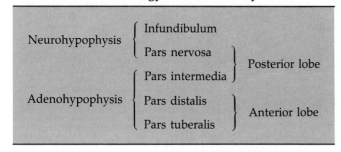

Neurohypophysis	Infundibulum	
	Pars nervosa	Posterior lobe
	Pars intermedia	
Adenohypophysis	Pars distalis	Anterior lobe
	Pars tuberalis	

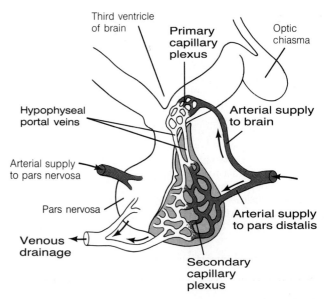

circulatory system. Within the median eminence of the lower hypothalamic region of the brain is a capillary bed known as the **primary capillary plexus** (Figure 18.6). This plexus receives blood from the internal carotid arteries and the cerebral arterial circle (see Chapter 11). Blood in the primary capillary plexus flows downward along the stalk of the pituitary by way of **hypophyseal portal veins** directly to a **secondary capillary plexus** in the adenohypophysis. Thus, the brain and the adenohypophysis are connected by a direct circulatory link.

F18.6

Figure 18.6
Diagrammatic representation of the circulatory linkage between the brain and the adenohypophysis of the pituitary gland.

Neurohypophyseal Hormones and Their Effects

The neurohypophysis is essentially a nerve tract, composed of axons that terminate in the pars nervosa. Because there are no gland cells present, the neurohypophysis does not produce any hormones. However, two peptide hormones that are produced in the hypothalamus and travel along the axons are stored in and released from the pars nervosa.

Antidiuretic Hormone

One of the neurohypophyseal hormones is **antidiuretic hormone (ADH),** which is also called **vasopressin.** An antidiuretic is a substance that reduces the amount of urine formed. ADH promotes the reabsorption of water from the urine-forming structures of the kidneys, and thus it aids in the retention of fluid within the body. At relatively high concentrations ADH has been found to be active in promoting the constriction of small blood vessels called arterioles—an effect that can elevate the blood pressure (hence the name vasopressin).

Oxytocin

Another neurohypophyseal hormone is called **oxytocin.** When released into the circulation, oxytocin stimulates the smooth muscles of the reproductive organs, including the uterus in the female and possibly the ductus deferens in the male. It also promotes the contraction of myoepithelial cells surrounding the saclike alveoli of the mammary glands, resulting in the "letdown" of milk during lactation.

Release of Neurohypophyseal Hormones

As indicated previously, although both ADH and oxytocin are released from the pars nervosa, neither hormone is manufactured there. Rather, both are manufactured in certain hypothalamic nuclei by specialized neural cells called

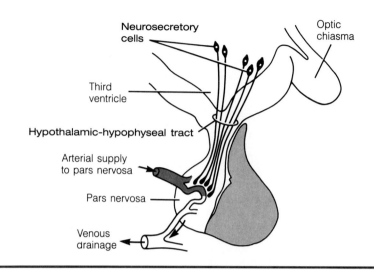

Figure 18.7
Neurosecretory cells in the hypothalamic region of the brain. These cells manufacture antidiuretic hormone and oxytocin, which are transported along the axons of the neurosecretory cells within the hypothalamic-hypophyseal tract to the pars nervosa of the pituitary. From the pars nervosa, the hormones are released into the circulation.

neurosecretory cells. These cells possess long axons that extend from their cell bodies in the hypothalamic nuclei, along the infundibulum, to the pars nervosa. The axons form the **hypothalamic-hypophyseal tract** in the infundibulum (Figure 18.7). ADH and oxytocin synthesized in the hypothalamus by these specialized nerve cells move down the axons to the pars nervosa, where they pass through the walls of capillaries and are thereby released into the circulation. Both hormones are transported from the hypothalamus to the pars nervosa attached to molecules of a carrier protein called *neurophysin.* The direct connection between the brain and the pars nervosa of the pituitary gland clearly demonstrates the close relationship between the nervous and endocrine systems.

Adenohypophyseal Hormones and Their Effects

Six hormones are produced in, and released by, the pars distalis of the adenohypophysis: *follicle-stimulating hormone (FSH), luteinizing hormone (LH), thyrotropin (TSH), adrenocorticotropin (ACTH), growth hormone (GH),* and *prolactin.*

Gonadotropins

Two of the hormones produced by the adenohypophysis are called **gonadotropins** because they particularly affect the gonads (ovaries and testes). In females, **follicle-stimulating hormone (FSH)** stimulates the development of structures called follicles in the ovaries and induces the secretion of the estrogenic female sex hormones. In this activity, FSH probably works in conjunction with the other adenohypophyseal gonadotropin: **luteinizing hormone (LH).** Surging levels of LH together with FSH lead to ovulation and the formation of a structure called the corpus luteum from an ovarian follicle. The corpus luteum produces estrogenic hormones, as well as the female sex hormone progesterone.

Both FSH and LH are also present in males. FSH is involved in the development and maturation of sperm. LH, which in males is called **interstitial-cell-stimulating hormone (ICSH),** stimulates the interstitial cells of the testes to produce the male sex hormones (that is, androgens such as testosterone). Since the androgens are themselves involved in sperm production, LH can be regarded as important in this process as well, and FSH may also play some role in androgen production. Both FSH and LH are glycoproteins (proteins combined with carbohydrate molecules).

Thyrotropin

A third hormone produced by the adenohypophysis is **thyrotropin,** which is also called **thyroid-stimulating hormone (TSH).** TSH stimulates the synthesis and release of thyroid hormones from the thyroid gland. Like FSH and LH, thyrotropin is a glycoprotein.

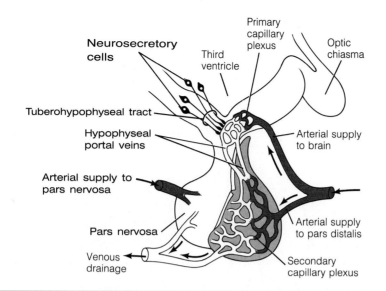

Figure 18.8
Neurosecretory cells in the tuber cinereum of the hypothalamus form the tuberohypophyseal tract and produce releasing substances that enter the primary capillary plexus and travel through the circulation to the pars distalis of the pituitary.

Adrenocorticotropin

Another hormone that is synthesized and released by the adenohypophysis is **adrenocorticotropin (adrenocorticotropic hormone, or ACTH)**. ACTH acts to stimulate the release of hormones from the cortical regions of the adrenal glands, particularly cortisol and other glucocorticoid hormones that are active in carbohydrate metabolism. ACTH is a polypeptide hormone composed of 39 amino acids.

Growth Hormone

The adenohypophysis also produces **growth hormone (GH),** which is sometimes called **somatotropin.** Growth hormone stimulates growth in general, and the growth of the skeletal system in particular. GH enhances the entrance of amino acids into cells and favors their incorporation into protein. In addition GH increases the release of fatty acids from adipose tissue into the blood. The fatty acids can be used as energy sources by most cells. Growth hormone has also been shown to promote the formation of glucose (from liver glycogen) and its release into the blood. In many instances, growth hormone acts synergistically with other hormones to enhance their effects. GH is a protein hormone that consists of 191 amino acids.

Prolactin

Prolactin is still another adenohypophyseal hormone. It is involved in the initiation and maintenance of milk secretion in females. Prolactin plays no clearly recognized role in males. It is a protein hormone, with a structure similar to that of growth hormone.

Release of Adenohypophyseal Hormones

The brain exercises a good deal of control over the synthesis and release of the adenohypophyseal hormones, even though the adenohypophysis is not directly connected to the brain.

Within the brain, specialized neurosecretory cells manufacture a group of compounds known collectively as *releasing* or *inhibiting substances*. If the chemical identity of a releasing or inhibiting substance has been established, the substance is termed a *hormone*. Otherwise the substance is referred to as a releasing or inhibiting *factor*. The neurosecretory cells in the brain that manufacture releasing or inhibiting substances form the **tuberohypophyseal tract** as they pass through the tuber cinereum of the hypophysis to the infundibulum. Within the median eminence the neurosecretory cells release their products into the primary capillary plexus (Figure 18.8). The releasing or inhibiting substances travel through the circulation by way of the hypophyseal portal

F18.8

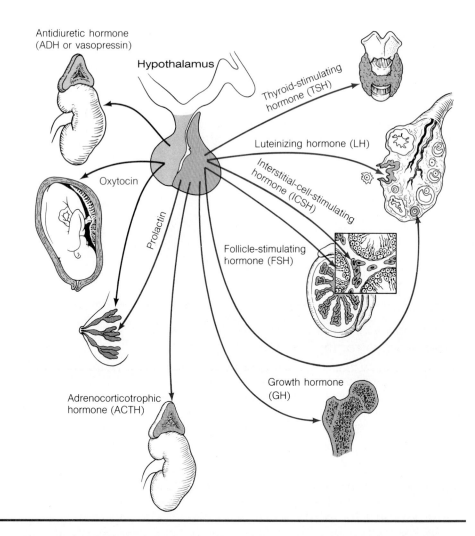

Antidiuretic hormone
(ADH or vasopressin)

Hypothalamus

Thyroid-stimulating
hormone (TSH)

Luteinizing hormone (LH)

Interstitial-cell-stimulating
hormone (ICSH)

Oxytocin

Prolactin

Follicle-stimulating
hormone (FSH)

Adrenocorticotrophic
hormone (ACTH)

Growth hormone
(GH)

Figure 18.9
Summary of the actions
of the pituitary gland
hormones. The neuro-
hypophysis is in gray;
the adenohypophysis is
in color.

CONDITIONS OF CLINICAL SIGNIFICANCE

Pituitary Disorders

Disorders of the glands of the endocrine system are generally the result of either an underproduction of hormones—a condition referred to as *hyposecretion*—or an overproduction (*hypersecretion*) of hormones. Pituitary malfunctions may be the result either of disorders of the gland itself or of difficulties involving the releasing or inhibiting substances from the hypothalamus of the brain. Pituitary disorders may involve the functioning of a major segment of the gland, or they may be limited to disorders in the release of a single pituitary hormone. Occasionally, for example, total deficiencies of the adenohypophysis are encountered. Among the more evident symptoms of such deficiencies of the adenohypophysis are those associated with adrenal cortical insufficiency due to lack of ACTH (see Table 18.2).

Individual hormone disorders produce a number of observable conditions. Deficiencies of the gonadotropins will upset normal gonadal function. These deficiencies can lead to *hypogonadotropic eunuchoidism* in males, which is characterized by gonadal inactivity, and to a failure to menstruate (*amenorrhea*) in females. Deficiencies of prolactin can result in a failure of lactation after delivery, whereas prolactin overproduction can lead to lactation in a woman who has not recently given birth. TSH deficiency can result in underactivity of the thyroid and *hypothyroidism*, whereas excess TSH can contribute to *hyperthyroidism*. As previously mentioned, ACTH deficiency can lead to adrenal cortical insufficiency, whereas excess ACTH can produce symptoms of adrenal cortical hyperfunction. In the young, GH deficiency can result in impaired growth, producing *pituitary dwarfs*. In excess, growth hormone can lead to *gigantism* or, in adults, to *acromegaly*. Acromegaly is characterized by an increase in the total mass of connective tissue in the body, by increased thickening of bones such as those of the hands and jaw, and by a coarsening of the facial features. A deficiency of antidiuretic hormone can result in *diabetes insipidus*, which is characterized by the excretion of large volumes of dilute urine.

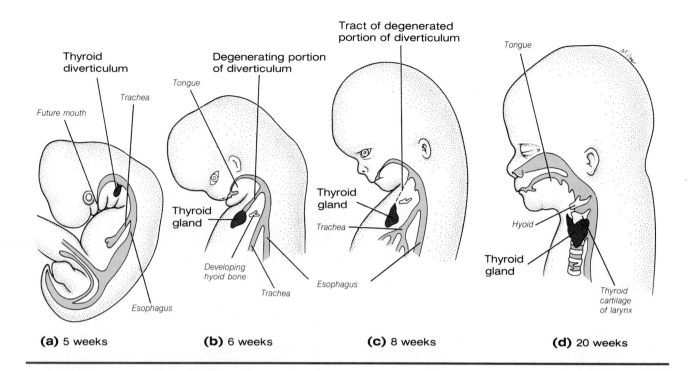

(a) 5 weeks   **(b)** 6 weeks   **(c)** 8 weeks   **(d)** 20 weeks

veins directly to the adenohypophysis, where they induce or inhibit the re-lease of the adenohypophyseal hormones.

The following releasing or inhibiting substances have been either identi-fied or postulated: **gonadotropin-releasing hormone (GnRH),** which stimu-lates the release of both FSH and LH; **thyrotropin-releasing hormone (TRH); corticotropin-releasing factor (CRF); prolactin-releasing factor (PRF);** and **growth-hormone-releasing factor (GH-RF).** In addition, there appears to be a **prolactin-inhibiting factor (PIF)** as well as an inhibiting factor for growth hormone called **somatostatin.**

This brief consideration of the hormones of the pituitary gland and their functions should serve to emphasize the importance of the gland in regulat-ing the activities of many diverse cells throughout the body. These functions are illustrated in Figure 18.9.

F18.9

Figure 18.10
Embryonic development of the thyroid gland.

THYROID GLAND

The fully developed thyroid gland is among the largest of the body's endo-crine organs. It is located anterior to the upper part of the trachea, near its junction with the larynx.

Embryonic Development and Structure

The **thyroid gland** originates as an epithelial thickening in the floor of the pharynx. The thickening grows outward from the pharynx, eventually loses its connection with the gastrointestinal tract, and comes to occupy a position around the trachea just below the larynx (Figure 18.10).

F18.10

The thyroid gland is divided into right and left lobes that are joined across the trachea by a thin band called the **isthmus** (Figure 18.11). The gland is well **F18.11** vascularized. It receives blood from the superior thyroid artery (a branch of the external carotid artery) and the inferior thyroid artery (a branch of the subclavian artery). The veins draining the gland empty into the internal jugu-lar and brachiocephalic veins.

The basic internal structure of the thyroid consists of hollow balls of cells called **follicles** that are bound together with connective tissue (Figure 18.12). **F18.12** The follicular cells are basically cuboidal in shape, but their height varies with

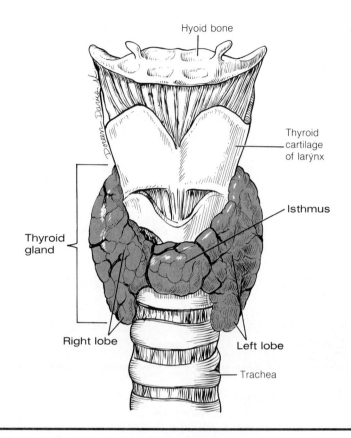

Figure 18.11
Ventral view of the
thyroid gland.

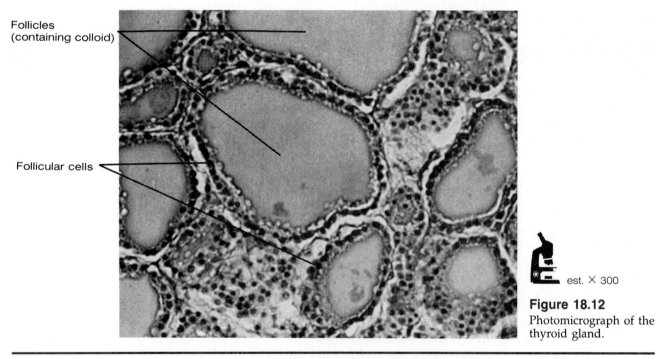

est. × 300

Figure 18.12
Photomicrograph of the
thyroid gland.

their activity—that is, they become taller and more columnar when active and
flatter when inactive. The central region of each follicle contains a protein
substance, the *colloid*, which is a stored form of thyroid hormones. The thyroid gland is different from all other endocrine glands in having extracellular
storage sites—the follicles—for its hormones. The other endocrine glands
store their hormones inside their cells.

CONDITIONS OF CLINICAL SIGNIFICANCE
Thyroid Disorders

In the young, *hypothyroidism* (lowered thyroid function) can lead to abnormal skeletal ossification and connective-tissue growth. The reproductive structures do not develop properly, the skin is dry, and the hair is scant. Among the most significant features of hypothyroidism in the young is a faulty development of the nervous system that can lead to mental retardation. The condition caused by severe hypothyroidism beginning in infancy is called *cretinism*. If the hypothyroid state is recognized early and thyroid hormone replacement therapy begun, the condition can be markedly improved. If treatment is delayed, the mental retardation and other changes can become permanent.

Severe hypothyroidism in the adult is called *myxedema*. This condition, which is considerably more common in women than in men, is characterized by a puffiness of the face and eyelids and a swelling of the tongue and larynx. The skin becomes dry and rough, the hair is scant, and the individual has both a low basal metabolic rate and a low body temperature. There is poor muscle tone, a lack of strength, and a tendency to fatigue easily. Mental activity is generally sluggish and retarded. This condition can be alleviated by the administration of thyroid hormones.

An excess of thyroid hormones acting on the tissues is called *thyrotoxicosis*. One form of thyrotoxicosis is *hyperthyroidism* (thyroid overfunction). Hyperthyroidism is characterized by an elevated basal metabolism and body temperature and a rapid heartbeat. Despite an increased appetite, there may be a large weight loss. The affected person perspires freely and may be nervous, emotionally unstable, and unable to sleep. Hyperthyroidism can be treated surgically, or with drugs that impair thyroid function, or with radioactive iodine, which is taken up by the gland and destroys some of the thyroid cells.

A *goiter* is simply an enlargement of the thyroid gland. It may or may not be associated with hypothyroidism or hyperthyroidism. In a form of thyrotoxicosis known as *exophthalmic goiter*, or *Graves' disease*, the thyroid may exhibit a diffuse enlargement (goiter); in addition, the eyeballs tend to protrude from their sockets (exophthalmos). *Simple goiter* is glandular enlargement without either hypothyroidism or thyrotoxicosis.

Thyroid Hormones and Their Effects

The thyroid gland secretes three hormones. Two of these hormones are concerned with regulating the rate of metabolism in the cells of the body; the other hormone regulates the levels of calcium and phosphate in the blood.

Triiodothyronine and Thyroxine

Two of the thyroid hormones contain *iodine*. Most of the body's iodine is obtained from the diet, and the vast majority of this iodine is actively taken up by the follicles of the thyroid gland. In the follicles, the iodine is attached to the amino acid *tyrosine* to form one of two hormones. If four iodine molecules are attached, the hormone **tetraiodothyronine (T_4, or thyroxine)** is formed; if only three iodine molecules are attached, the hormone **triiodothyronine (T_3)** is formed. There is evidence that these synthetic reactions take place in the colloid of the thyroid follicles, where the tyrosines are part of a glycoprotein molecule known as *thyroglobulin*, which is manufactured by the follicular cells. Thyroglobulin molecules containing tetraiodothyronine and triiodothyronine are therefore produced and stored in the colloid areas of the follicles.

When the thyroid is actively secreting, thyroglobulin from the follicles enters the follicle cells by endocytosis. Enzymes within the cells break down the thyroglobin to T_3 and T_4, which pass out of the follicle cells and enter capillaries in the connective tissue surrounding the follicles. However, the thyroid hormones do not generally remain in a free form in the bloodstream; rather, they become bound to plasma proteins such as *thyroid-binding globulin*. In the peripheral tissues, T_3 and T_4 are again freed from the binding proteins and leave the circulation to enter their target cells. The total levels of T_4 in the circulation are considerably higher than those of T_3.

Among the most evident effects of the thyroid hormones T_3 and T_4 are those associated with metabolism. By affecting the metabolism of carbohydrates, proteins, and lipids, the thyroid hormones increase the body's oxygen consumption and heat production. Most body tissues are responsive to this influence.

Calcitonin

The thyroid releases another hormone—**calcitonin.** Calcitonin lowers blood calcium and blood phosphate levels—probably by accelerating the absorption of calcium by the bones. Calcitonin is a polypeptide and it is produced by cells called *parafollicular cells* ("C" cells), which are located in the walls of the follicles as well as between the follicles. Although calcitonin significantly lowers blood calcium and phosphorus levels in children, it has only a very weak effect on these levels in adults.

Release of Thyroid Hormones

The hormones of the thyroid gland are released from the gland under the control of two different mechanisms:

1. The synthesis and release of triiodothyronine and thyroxine are under the control of thyrotropin (TSH) from the pars distalis of the pituitary. TSH is itself regulated by thyrotropin-releasing hormone (TRH) from the brain.

2. The release of calcitonin is triggered by an increase in the level of calcium in the blood.

PARATHYROID GLANDS

Embedded on the posterior surface of the lobes of the thyroid gland are four small **parathyroid glands** (Figure 18.13). The blood supply of these glands is F18.13 the same as that of the thyroid gland, although the parathyroid and thyroid glands differ in their embryonic development, their structure, and their function.

Embryonic Development and Structure

The parathyroid glands develop embryonically from the dorsal halves of the third and fourth pairs of pharyngeal pouches. With continued development the glands lose their attachments to the pouches and migrate to the neck, where they assume their adult positions on the posterior surfaces of the lobes of the thyroid gland. There are usually two masses of parathyroid tissue (*superior* and *inferior*) on each of the two thyroid lobes.

The parathyroid glands are composed of densely packed masses or cords of cells. At least two types of epithelial cells have been identified in these masses or cords. Most parathyroid cells are of a type called **principal cells** (or **chief cells**), which produce the hormone of the parathyroid glands. Larger, but much less numerous, are the **oxyphilic cells,** which are scattered in clumps throughout the glands. Their function has not been determined.

Parathyroid Hormone and Its Effects

Parathyroid hormone (PTH, or **parathormone**) is a polypeptide, and it is currently believed that two and possibly three forms of the hormone may appear in the bloodstream. Parathyroid hormone is a principal controller of calcium and phosphorus metabolism. Specifically, it increases the plasma calcium concentration, and decreases the plasma phosphate concentration. It does this in part by increasing the kidneys' excretion of phosphate in the urine and by decreasing calcium excretion. Parathyroid hormone also increases bone reabsorption—causing the release of calcium from bone into the blood—by stimulating the activity of osteoclasts. Moreover, parathyroid hormone promotes the metabolic transformation of vitamin D in the kidneys and the absorption of calcium from the gastrointestinal tract. This latter effect may be due to the high levels of vitamin D that result from its metabolic transformation by parathyroid hormone. In fact, adequate levels of vitamin D appear to

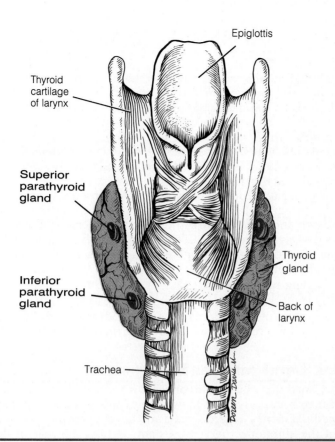

Epiglottis

Thyroid
cartilage
of larynx

**Superior
parathyroid
gland**

**Inferior
parathyroid
gland**

Thyroid
gland

Back of
larynx

Trachea

Figure 18.13
The parathyroid glands
(posterior view).

be necessary for parathyroid hormone to exert its effects on bone and kidneys as well as on the gastrointestinal tract. In general, parathyroid hormone and vitamin D act synergistically to maintain calcium levels.

Because calcium concentrations in the body fluids directly affect the contraction of muscle tissue and the generation of a nerve impulse, parathyroid hormone is a vital homeostatic mechanism of the body. Note that parathyroid hormone and calcitonin from the thyroid have opposite effects on the blood calcium level. As a result of their antagonistic actions, the calcium level in the blood is normally maintained within rather narrow limits.

Release of Parathyroid Hormone

The major controller of parathyroid activity is the calcium level in the blood plasma. When the plasma calcium level falls, the secretion of parathormone increases.

CONDITIONS OF CLINICAL SIGNIFICANCE

Parathyroid Disorders

Hyperparathyroidism, which results in an excess of parathyroid hormone, causes extensive bone decalcification and can lead to deformities and fractures. The plasma calcium level rises and soft tissues—especially the kidneys—may calcify.

Hypoparathyroidism, which results in a deficiency of parathyroid hormone, leads to a lowered plasma calcium level that greatly increases the excitability of the nervous system. The result can be muscular twitching and spasms (tetany).

ADRENAL GLANDS

F18.14

The two **adrenal glands,** or **suprarenal glands,** are pyramid-shaped organs located behind the peritoneum close to the superior border of each kidney (Figure 18.14). Each gland is surrounded by a connective-tissue capsule and is embedded in fat. The adrenal glands are well supplied with blood vessels. Arterial branches from the aorta, the inferior phrenic, and the renal arteries supply the glands. Venous blood from the right adrenal gland drains into the inferior vena cava; blood from the left adrenal gland empties into the left renal vein.

F18.15

Table 18.2

Each adrenal gland consists of two separate portions: an inner *medulla* and an outer *cortex* (Figure 18.15). The medulla and the cortex have different embryonic origins and different structures, and the actions of their hormones differ considerably (Table 18.2). Consequently, each adrenal gland is, in effect, actually two distinct endocrine organs. Let us examine these two portions in greater detail.

Adrenal Medulla

The central portion of each adrenal gland, the **adrenal medulla,** is composed of cells arranged in groups or short cords surrounding blood capillaries and venules.

Embryonic Development and Structure

The adrenal medulla arises embryonically from neural crest cells. Since the neural crest cells also give rise to postganglionic sympathetic neurons, the adrenal medulla can be regarded as a modified portion of the sympathetic division of the autonomic nervous system. In fact, the cells of the adrenal medulla function in a manner similar to postganglionic sympathetic cells. Because they stain with chromium salts, the cells of the medulla are called **chromaffin cells.**

Adrenal Medullary Hormones and Their Effects

The chromaffin cells of the medulla produce two hormones: *epinephrine (adrenaline)* and *norepinephrine (noradrenaline).* Recall that norepinephrine is also released from the end terminals of the postganglionic neurons of the sympathetic nervous system; thus, it is present in the blood from sources other than the adrenal medulla. Epinephrine and norepinephrine are structurally similar molecules and function cooperatively to prepare the body for emergencies or stress in a manner similar to the sympathetic nervous system. Considering the embryonic origin of the adrenal medulla, this similarity to the sympathetic nervous system is not surprising. The two hormones are not identical in function, however. The primary effect of norepinephrine is on the vascular system, whereas epinephrine exerts metabolic as well as cardiovascular effects. The hormones of the adrenal medulla are not essential to life. In their absence the sympathetic nervous system alone does an adequate job of preparing the body for emergencies or stress.

EPINEPHRINE   The hormone **epinephrine** elevates blood-glucose levels and stimulates the release of ACTH from the pituitary. ACTH in turn causes the release of glucocorticoid hormones from the adrenal cortex.

Epinephrine also increases the rate, force, and amplitude of the heartbeat. It constricts blood vessels in a number of body areas, including the skin, mucous membranes, and kidneys, but it may induce vessels to dilate in some areas, such as in skeletal muscles, the coronary arteries to the heart, and the pulmonary arteries to the lungs.

NOREPINEPHRINE   **Norepinephrine** increases the heart rate and the force of contraction of cardiac muscle. It also constricts blood vessels in most areas of the body.

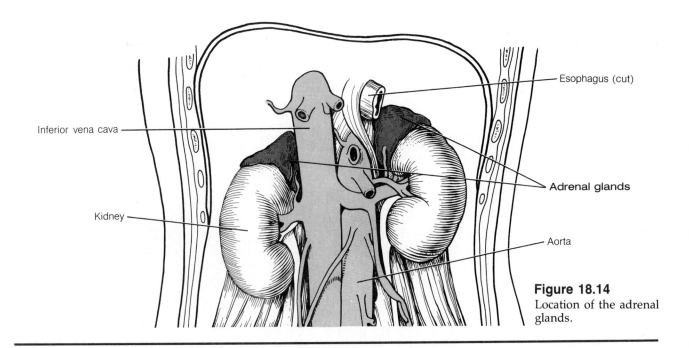

Figure 18.14
Location of the adrenal glands.

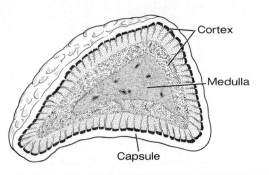

Figure 18.15
An adrenal gland (sectioned). The gland is a dual structure consisting of a central medulla and a surrounding cortex, all enclosed in a fibrous capsule.

RELEASE OF ADRENAL MEDULLARY HORMONES The release of the hormones of the adrenal medulla is controlled by preganglionic neurons to the chromaffin cells of the medulla from the sympathetic division of the autonomic nervous system. A variety of conditions can lead to the release of adrenal medullary hormones, including emotional excitement, injury, exercise, and low blood-glucose levels. The hormones of the adrenal medulla together with the sympathetic nervous system maintain blood pressure and help regulate carbohydrate metabolism.

Adrenal Cortex

The outer portion of each adrenal gland, the **adrenal cortex,** makes up approximately 80% of the total weight of each gland when fully developed.

Embryonic Development and Structure

The adrenal cortex is derived embryonically from mesoderm from the same region that gives rise to the gonads. The outer portion of the adrenal cortex is surrounded by a connective-tissue capsule. Trabeculae (strands of connective tissue) extend from the capsule into the gland itself. The endocrine cells of the adrenal cortex are organized into three layers (Figure 18.16): the *zona glomerulosa*, the *zona fasciculata*, and the *zona reticularis*. F18.16

The **zona glomerulosa** is a relatively thin region in the cortex. It is located directly beneath the capsule and is composed of clusters of cells. Beneath the

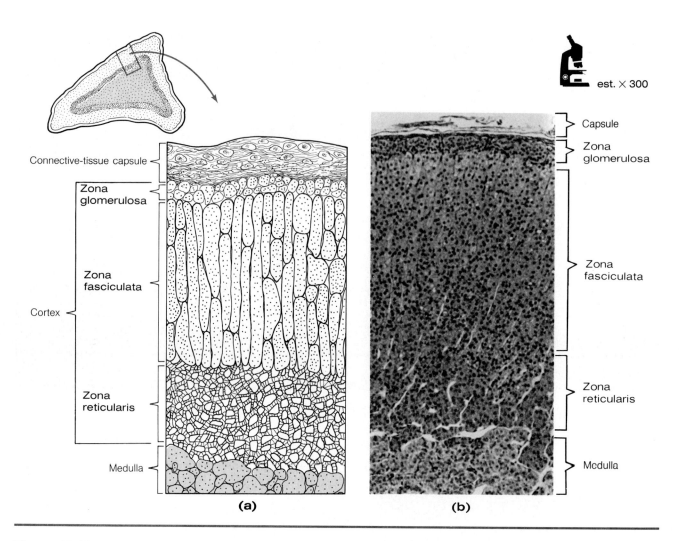

est. × 300

Connective-tissue capsule

Zona glomerulosa

Zona fasciculata

Cortex

Zona reticularis

Medulla

Capsule

Zona glomerulosa

Zona fasciculata

Zona reticularis

Medulla

(a)   (b)

Figure 18.16
The layers of the adrenal cortex. **(a)** Schematic representation. **(b)** Photomicrograph.

zona glomerulosa is a thick region called the **zona fasciculata.** The cells of the zona fasciculata are arranged in parallel columns that run at right angles to the surface of the gland. Occupying the deepest region of the adrenal cortex and lying adjacent to the adrenal medulla is the **zona reticularis,** in which the cells are arranged in a network of interconnecting cords.

Adrenal Cortical Hormones and Their Effects

MINERALOCORTICOIDS   The cells of the *zona glomerulosa* produce the adrenal cortical hormones associated with sodium and potassium metabolism—the **mineralocorticoids,** such as **aldosterone.** Aldosterone promotes the reabsorption of sodium and the excretion of potassium by urine-forming structures in the kidneys called distal tubules.

GLUCOCORTICOIDS   The cells of the *zona fasciculata* produce adrenal cortical hormones that affect carbohydrate metabolism. These are the **glucocorticoids,** such as **cortisol** and **cortisone.** The glucocorticoids supplement and conserve the energy derived from circulating glucose. In response to the glucocorticoids, glucose utilization in peripheral tissues is inhibited, fatty acids from adipose tissue are mobilized, and the source of metabolic energy for muscle tissue shifts from glucose to fatty acids.

SEX HORMONES   In addition to producing mineralocorticoids and glucocorticoids, the adrenal cortex produces a number of other substances. Among these are some androgenic compounds that resemble male sex hormones and

CONDITIONS OF CLINICAL SIGNIFICANCE

Adrenal Disorders

Adrenal cortical *hypofunction* in humans is called *Addison's disease.* Inadequate amounts of aldosterone impair the body's ability to conserve sodium and excrete potassium. This condition can lead to a decreased extracellular fluid volume, weight loss, decreased plasma volume, low blood pressure, decreased cardiac size and output, general weakness, and possibly shock. A deficiency of glucocorticoids in Addison's disease can result in loss of appetite, hypoglycemia, apathy, weakness, and a diminished ability to withstand various types of physiological stress. Hormone administration may alleviate the symptoms of Addison's disease.

Adrenal cortical *hyperfunction* can result in a number of disorders, depending on which hormones are produced in excessive amounts. *Hypercortisolism* can produce *Cushing's syndrome,* which is characterized by adipose-tissue accumulation and weight gain, osteoporosis (softening of bones), weakness, hypertension, and abnormal hairiness (hirsutism). *Hyperaldosteronism* is characterized by potassium depletion and by the retention of sodium and water, which causes the expansion of the extracellular fluid compartment and may produce edema or hypertension. Certain adrenal tumors can affect the cells that produce sex hormones and cause the production of excessive androgenic substances. This produces masculizing effects that are particularly evident in females.

Disorders of the adrenal medulla are relatively rare. Significant overproduction of the medullary hormones is generally caused by the presence of a tumor. The most common type is a benign tumor called *pheochromocytoma,* but a malignant tumor called a *neuroblastoma* is sometimes present. The tumors cause a hypersecretion of epinephrine and norepinephrine. Consequently, they produce symptoms indicating sympathetic overactivity, including high blood pressure, rapid heartbeat, sweating, and nervousness.

a much smaller quantity of estrogenic materials that resemble female sex hormones. These sex hormones are thought to be produced by the cells of the *zona fasciculata,* and perhaps also by those of the *zona reticularis.*

RELEASE OF ADRENAL CORTICAL HORMONES   The release of the *mineralocorticoids* (such as aldosterone) from the adrenal cortex is principally regulated by an enzyme called **renin.** Renin is released by specialized cells in the kidneys in response to sympathetic nervous system activity, lowered sodium levels, or decreased blood pressure or volume. Renin acts on a precursor substance called **angiotensinogen,** which is manufactured by the liver and is normally present in an inactive form in the blood. After undergoing modifications by renin and at least one additional enzyme, angiotensinogen is converted into **angiotensin,** which stimulates the release of aldosterone from the adrenal cortex.

The release of the *glucocorticoids* (such as cortisol) is controlled primarily by ACTH from the pituitary, and the release of ACTH, in turn, is influenced by corticotropin releasing factor (CRF) from the hypothalamus of the brain. A variety of stressful situations (such as trauma or emotional stress) can increase the release of ACTH and consequently increase glucocorticoid secretion. This effect is probably mediated neurally by way of CRF. Once released, cortisol is bound to a plasma glycoprotein called *transcortin* for transport through the circulation.

PANCREAS

The **pancreas,** which is located behind the stomach between the spleen and the duodenum of the small intestine, is both an endocrine gland that produces hormones and an exocrine gland that produces digestive enzymes. The endocrine portion of the pancreas produces hormones that have a role in regulating the metabolic activities of the body, particularly those associated with carbohydrate metabolism.

Embryonic Development and Structure

The pancreas arises embryonically through a fusion of two outgrowths from the duodenum of the small intestine. The exact origin of the cells that make

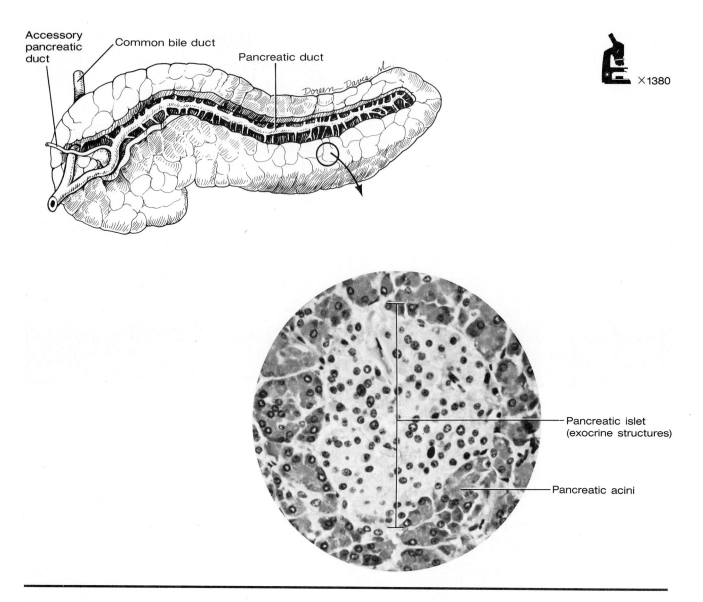

×1380

Figure 18.17

The pancreas. The enlargement shows details of the pancreatic cellular structure.

F18.17

up the endocrine portion of the pancreas remains uncertain. At least some of the endocrine cells are believed to bud off from the lining of the pancreatic ductules—either from endodermal cells that may have arisen at the site or from neuroectodermal cells that migrated into the intestinal mucosa at an earlier developmental stage.

The endocrine portion of the fully developed pancreas consists of aggregations of cells that are clustered into groups called **pancreatic islets** (*of Langerhans*) (Figure 18.17). The islet cells are arranged in irregular cords separated by a rich vascular system of capillary vessels or sinusoids. The main blood supply to the pancreas is through branches of the splenic and superior mesenteric arteries. Nerves from both the sympathetic and parasympathetic divisions of the autonomic nervous system supply the pancreas and play important roles in controlling the synthesis and release of islet hormones.

Pancreatic Hormones and Their Effects

The pancreatic islets contain at least three functionally different cell types. **Alpha cells** produce the hormone *glucagon*, and **beta cells** produce the hormone *insulin*. A third cell type, the **delta cells** produce a substance called *somatostatin*. There is some evidence that somatostatin is a neurotransmitter substance that inhibits the release of both glucagon and insulin.

FRONTIERS IN HEALTH

Insulin Pumps: A Step Toward an Artificial Pancreas

Lauren is one of those unfortunate people whose lives are ruled by an incurable disease, diabetes mellitus, or sugar diabetes. Three times a day she must inject insulin into her body to maintain blood sugar (glucose) levels.

For the most part, Lauren's daily injections work, but occasionally things go awry. Too much insulin or too much exercise can cause her blood glucose levels to fall, producing hypoglycemia and causing her to feel dizzy and weak. At other times, blood glucose rises to excessive level, producing hyperglycemia and with it depression, fatigue, irritability, and weakness. Teetering between hyperglycemia and hypoglycemia, Lauren, like many diabetics, lives a life dependent on her daily shots of insulin.

Diabetes, which afflicts an estimated 6 to 8 million Americans, exists in two forms. Type I diabetes results from an insulin deficiency. For reasons unknown, the pancreas fails to produce enough insulin to regulate blood glucose properly. Some researchers believe that Type I diabetes may be genetically predetermined, with something being required to trigger it—perhaps a viral infection or a chemical imbalance of some sort. Type I diabetes can be treated successfully with insulin. About 1 million Americans now rely on daily injections of insulin to maintain their blood glucose levels within tolerable limits.

The second type of diabetes, called Type II diabetes, mainly strikes overweight people who are middle-aged or older. Victims of this form of diabetes exhibit an unexplained phenomenon: cells of the body, especially muscle and liver cells, become progressively less sensitive to insulin. Extra insulin is produced to compensate for this loss of sensitivity, but over time the extra activity of the pancreas causes it to become exhausted.

Besides worrying about insulin doses and having to be continually concerned with her blood glucose levels, Lauren, like her fellow diabetics, has another worry: long-term complications. Many diabetics develop complications 20 or more years after the onset of the disease. The complications are probably the result of long-term high blood glucose levels, which can damage blood vessels and nerves. Blindness and a host of disabling diseases of the kidney, nervous system, and cardiovascular system are the more common late-appearing complications of diabetes.

To prevent these long-term complications and treat the immediate symptoms as well, physicians have long tried to find ways to mimic the secretion of insulin by the pancreatic islet cells. But simulating the body's intricate and precise glucose-regulating system is not an easy task. The body constantly maintains glucose levels, minute by minute. Even a conscientious diabetic can measure blood glucose only three or four times a day. Adjustments can be made to take into account large meals or extra exercise, but such changes are crude in comparison to the body's elaborate system.

New advances offer some hope for the diabetic. One of the most promising is the insulin infusion pump. Worn outside the body, this pump delivers tiny amounts of insulin to the body day and night, providing the baseline insulin levels needed to maintain proper blood glucose concentrations. The pump also delivers a surge of insulin at meal times to take care of the sudden rise in blood glucose that accompanies the digestion of a meal. The diabetic simply presses a button on the pump 30 minutes before a meal to deliver the necessary preprogrammed surge of insulin. If the meal will be bigger than usual, an additional small amount of insulin can be delivered to protect against hyperglycemia.

Insulin infusion pumps are being used by roughly 8000 Americans. According to Dr. Joseph Loewenstein, an endocrinologist from the Louisiana State University School of Medicine, most of the pumps introduce insulin into the bloodstream by slowly depressing a plunger on a disposable plastic syringe. Insulin is delivered through a small plastic tube connected on one end to the pump and on the other to a needle inserted under the skin in the thigh or the abdomen. The needle and tubing are periodically replaced, usually every 3 or 4 days.

New designs for infusion pumps are now under development. Some of them capitalize on computer technology to regulate insulin flow continuously. According to their developers, the new models are simple to operate, yet offer great flexibility by accommodating exercise and meals of varying size. They even have memories that make it possible to store information on the exact doses given over a period of time. Physicians can use this information to reprogram the pump. Moreover, built-in alarms are designed to detect any discrepancies between the pump's dosage and the preprogrammed levels that were supposed to be administered.

However, the constant presence of tubes, needles, and computerized pumps is cumbersome. Many diabetics give the current pumps low marks for comfort and aesthetics. But the devices win high praise for the level of control over blood glucose they make possible.

For many, the simplicity of daily injections may outweigh the benefits of the pump. However, for diabetic women who have become pregnant, the infusion pump may mean the difference between a normal child and a defective one. Even mild hyperglycemia can seriously affect the fetus. Although an insulin pump may not simplify a diabetic's life, and may well complicate it, it can provide more freedom from constant fluctuations in blood glucose levels.

Insulin

Insulin is a protein composed of two polypeptide chains. One chain consists of 21 amino acids; the other chain consists of 30 amino acids. Circulating insulin molecules are, for the most part, bound to protein carriers in the plasma.

The major effects of insulin are to facilitate the uptake and utilization of glucose by cells and to prevent the excessive breakdown of glycogen in liver and muscle. As a result, insulin is a powerful *hypoglycemic agent*—that is, it decreases the blood sugar level. Insulin also influences lipid and protein metabolism. It favors lipid formation, inhibits the breakdown and mobilization of stored fat, and favors the synthesis of proteins by facilitating the movement of amino acids into cells.

Glucagon

Glucagon is a polypeptide molecule composed of 29 amino acids. The activities of glucagon are generally opposite to those of insulin. Glucagon decreases glucose oxidation and promotes *hyperglycemia*—that is, it increases the blood sugar level. Its main action seems to be to stimulate the breakdown of glycogen in the liver. Glucagon also stimulates the formation of carbohydrates in the liver from noncarbohydrate precursors and the breakdown of lipids in the liver and adipose tissue. Glucagon may also have a mildly stimulating effect on protein breakdown.

Release of Pancreatic Hormones

The release of the pancreatic hormones is under chemical, hormonal, and neural control. Blood glucose level appears to be the major factor governing insulin release. The higher the blood glucose level, the greater the insulin release. Blood glucose level also influences the release of glucagon, but the effect is opposite to that seen for insulin—that is, the lower the blood glucose

CONDITIONS OF CLINICAL SIGNIFICANCE

Pancreatic Disorders

The most common pancreatic endocrine disorder is the condition known as *diabetes mellitus*. This condition is caused by a relative deficiency of insulin that can be due to a deficient secretion of insulin by the pancreas or to a decreased sensitivity to insulin by the target cells. For example, in obese diabetics, insulin levels are normal or even above normal, but insulin sensitivity is decreased.

Diabetes mellitus is characterized by increased levels of blood glucose, coupled with a reduced entry of glucose into cells and an impairment of cellular ability to use glucose. As blood glucose rises it exceeds the capacity of the kidneys to resorb it and glucose appears in the urine. Extra water is osmotically required to excrete the glucose, and a large volume of urine is produced. This condition can lead to dehydration and ultimately to circulatory difficulties.

Free fatty acids are mobilized and released from adipose tissue in diabetes mellitus, and they provide energy sources for the cells. Such energy sources are necessary because of the deficient cellular uptake and utilization of glucose. The oxidation of fatty acids in the liver produces substances known as *ketone bodies*. Ketone bodies may be produced faster than other body tissues can metabolize them, and thus they may appear in the urine. In severe cases of diabetes mellitus, the body's buffering mechanisms may not be able to cope with these acidic substances, in which case the pH of the body fluids falls, resulting in acidosis. Acidosis, in turn, can cause altered respiration, nervous system depression, coma, and death.

Protein synthesis decreases and protein breakdown increases in diabetes mellitus, and this can impair the body's ability to combat infection and repair injured tissue. In the case of a diabetic whose pancreas remains functional but is unable to produce sufficient insulin, drugs taken orally have been used to stimulate insulin production. However, the effectiveness and safety of these drugs remains questionable, and they are not used as commonly as they once were. If the individual's pancreas cannot secrete insulin, the insulin deficiency must be made up by injection.

level, the greater the glucagon release. Amino acids stimulate the simultaneous secretion of both insulin and glucagon. Glucagon and several hormones from the gastrointestinal tract have also been reported to promote insulin release.

Regulation of pancreatic endocrine activity by the nervous system also occurs. Acetylcholine, the chemical released at the endings of the parasympathetic division of the autonomic nervous system, stimulates insulin release. Epinephrine and norepinephrine from the adrenal medulla and terminal endings of the neurons of the sympathetic division of the autonomic nervous system inhibit insulin release. Conversely, glucagon output appears to be increased by the activation of the sympathetic division of the autonomic nervous system. These neural controls are thought to be continuously active in adjusting the basal (that is, minimal) secretion of pancreatic islet hormones.

GONADS

The male gonads (*testes*) and female gonads (*ovaries*) produce hormones as well as gametes (reproductive cells: sperm or ova). The testes produce the male sex hormones—the **androgens**—and the ovaries produce the female sex hormones—the **estrogens** and **progesterone.** These hormones influence body growth, and they control the onset of puberty. The endocrine as well as the reproductive roles of the gonads are discussed further in Chapter 22.

OTHER ENDOCRINE TISSUES AND HORMONES

The **pineal gland** of the brain is a small conical structure that lies above the posterior end of the third ventricle. It reaches its maximum size by seven years of age and gradually decreases in size after that. While its physiology is not clearly understood, the pineal gland is considered to be an endocrine structure, with its main secretion being the hormone **melatonin.** In a variety of experimental animals the pineal has been shown to have an antigonadal activity. Removing the pineal from young animals leads to early puberty and enlargement of the reproductive organs. Melatonin is thought to affect gonadal activity by reducing the amount of gonadotropin-releasing hormone (GnRH) released by the hypothalamus. Pineal activity appears to be influenced by the light cycle to which an animal is exposed. When light strikes the retinas of the animal's eyes, it gives rise to nerve impulses that are ultimately transmitted to the pineal gland. These impulses are believed to inhibit the secretion of melatonin. Thus, the concentrations of melatonin in the blood are greater during the night and lower during the daylight hours. In this manner the pineal gland is thought to influence gonadal activity in a cyclic manner. Whether melatonin has a significant antigonadal effect in humans remains uncertain however. Melatonin also inhibits the production of melanin by the melanocytes of the skin.

The **thymus gland,** which is a lymphoid organ located in the anterior portion of the mediastinum, is also regarded as a potential source of hormonal material. The thymus reaches its maximum size by puberty, after which time it gradually decreases in size. In the adult it is largely replaced by adipose tissue. A polypeptide hormone called *thymosin,* which appears to be involved in the development of immunologically competent lymphocytes called T cells, has been isolated from the thymus.

The **digestive tract** is also the source of a number of hormonal substances (such as gastrin, secretin, cholecystokinin), as is the **placenta** (estrogens, progesterone, and human chorionic gonadotrophic hormone).

Table 18.2   Major Endocrine Glands and Their Hormones

Gland	Hormones	Representative Effects	Selected Disorders
PITUITARY [F18.5]			
Neurohypophysis (hormones are actually manufactured in hypothalamus of the brain)	Antidiuretic Hormone (ADH)	Promotes reabsorption of water from urine-forming structures of kidneys.	Undersecretion leads to diabetes insipidus.
	Oxytocin	Stimulates contraction of uterine smooth muscle and myoepithelial cells around alveoli of mammary glands. As such, it is involved in birth processes and milk "letdown" during nursing.	
Adenohypophysis	Follicle-Stimulating Hormone (FSH) and Luteinizing Hormone (LH)	Stimulates gonads to produce gametes and sex hormones.	Undersecretion causes gonadal inactivity in males, and menstrual failure in females.
	Thyrotropin (TSH)	Stimulates thyroid gland to secrete thyroid hormones.	Undersecretion leads to symptoms of hypothyroidism. Oversecretion may contribute to hyperthyroidism.
	Adrenocorticotropin (ACTH)	Stimulates adrenal cortex to secrete glucocorticoids (such as cortisol).	Undersecretion leads to symptoms of adrenal cortical insufficiency. Oversecretion leads to symptoms of adrenal cortial hyperfunction.
	Growth Hormone (GH)	Stimulates growth in general, and growth of skeletal system in particular. It also affects metabolic functions.	Undersecretion produces pituitary dwarfs. Oversecretion causes gigantism or, in adults, acromegaly.
	Prolactin	Involved in milk secretion in females.	Undersecretion may cause failure to lactate after giving birth. Oversecretion may lead to lactation without recently having given birth.
THYROID [F18.11]	Thyroxine (T$_4$ or tetraiodothyronine) and Triiodothyronine (T$_3$)	Increases oxygen consumption and heat production (calorigenic effect). Important for normal growth and development. These hormones affect many metabolic processes.	Undersecretion leads to symptoms of hypothyroidism possibly causing cretinism in children, or myxedema in adults. Oversecretion leads to hyperthyroidism possibly causing Graves' disease.
	Calcitonin	Lowers blood calcium and phosphate levels.	

Table 18.2 **Major Endocrine Glands and Their Hormones (continued)**

Gland	Hormones	Representative Effects	Selected Disorders
PARATHYROIDS [F18.13]	Parathyroid Hormone	Affects calcium and phosphate metabolism to raise plasma calcium levels and decrease plasma phosphate levels.	Undersecretion leads to nervous excitability and tetanus. Oversecretion leads to bone decalcification, and calcification of soft tissues such as the kidneys may occur.
ADRENALS [F18.14]			
Medulla	Epinephrine	Affects carbohydrate metabolism generally, leading to hyperglycemia. Constricts vessels in skin, mucous membranes, and kidneys, but dilates vessels in skeletal muscle.	
	Norepinephrine	Increases heart rate and force of contraction of cardiac muscle, constricts blood vessels in almost all areas of the body.	
Cortex	Mineralocorticoids (such as aldosterone)	Promote reabsorption of sodium, and excretion of potassium from urine-forming structures of kidneys.	Undersecretion may lead to decreased fluid volume and circulatory difficulties, and contribute to Addison's disease. Oversecretion may cause increased fluid volume, edema, and hypertension.
	Glucocorticoids (such as cortisol)	Affect many aspects of carbohydrate metabolism; generally lead to mobilization and hyperglycemia.	Undersecretion contributes to Addison's disease. Oversecretion leads to Cushing's syndrome.
PANCREAS [F18.17]	Insulin	Affects many aspects of carbohydrate metabolism; generally causes hypoglycemia.	Relative deficiency leads to hyperglycemia and diabetes mellitus.
	Glucagon	Affects metabolism in fashion generally opposite of insulin; generally causes hyperglycemia.	
GONADS			
Ovaries	Estrogens Progesterone	The gonadal hormones are involved in the processes of reproduction. Their functions are discussed in Chapter 22.	
Testes	Androgens (such as testosterone)		

STUDY OUTLINE

ENDOCRINE GLANDS ductless glands that release secretions directly into bloodstream. Hormones help regulate body processes by acting on target organs. **pp. 483–484**

CHEMICAL STRUCTURE OF HORMONES
some hormones are steroids; others are proteins, polypeptides, or related to amino acids. **p. 485**

MECHANISMS OF HORMONE ACTIONS
p. 485

INTRACELLULAR MEDIATORS (SECOND MESSENGER) hormone is first messenger; cyclic AMP is second messenger for cell activation.

GENE ACTIVATION some hormones, particularly steroid hormones, combine with receptor proteins and activate certain genes.

PITUITARY GLAND has been referred to as the "master gland." **pp. 485–491**

EMBRYONIC DEVELOPMENT AND STRUCTURE

NEUROHYPOPHYSIS from ectodermal tissue of brain floor; directly attached to brain.

ADENOHYPOPHYSIS from ectodermal tissue of roof of mouth; associated with brain by circulatory link; contains acidophil, basophil, and chromophobe cells.

NEUROHYPOPHYSEAL HORMONES AND THEIR EFFECTS synthesized in brain by neurosecretory cells; peptide hormones.

ANTIDIURETIC HORMONE (ADH), OR VASOPRESSIN promotes reabsorption of water from urine-forming structures of kidneys; may elevate blood pressure.

OXYTOCIN stimulates smooth muscles of uterus; promotes contraction of myoepithelial cells that surround alveoli of mammary glands.

RELEASE OF NEUROHYPOPHYSEAL HORMONES transported to pars nervosa from hypothalamus by neurosecretory cells that form hypothalamic-hypophyseal tract.

ADENOHYPOPHYSEAL HORMONES AND THEIR EFFECTS

GONADOTROPINS follicle-stimulating hormone (FSH) and luteinizing hormone (LH); LH called ICSH in male.
1. Ovarian follicle development and estrogen production.
2. Ovulation and formation of corpus luteum.
3. Spermatogenesis and androgen production.

THYROTROPIN (TSH) synthesis and release of thyroid hormones.

ADRENOCORTICOTROPIN (ACTH) stimulates cortical region of adrenal gland (glucocorticoids).

GROWTH HORMONE (GH), OR SOMATOTROPIN affects growth of skeletal system; involved in retention of amino acids and incorporation into proteins; increases fatty-acid release into blood.

PROLACTIN initiation and maintenance of milk secretion in females.

RELEASE OF ADENOHYPOPHYSEAL HORMONES hormone releasing and inhibiting substances manufactured in brain by neurosecretory cells; travel through circulation by way of hypophyseal portal veins to adenohypophysis, where they influence release of specific adenohypophyseal hormones.

CONDITIONS OF CLINICAL SIGNIFICANCE: PITUITARY DISORDERS disorders of gland itself or difficulty involving releasing substances from brain.
1. Adrenocortical insufficiency due to lack of ACTH.
2. Adrenocortical hyperfunction due to excess ACTH.
3. Pituitary dwarf due to growth hormone deficiency.
4. Gigantism and acromegaly due to excess growth hormone.

THYROID GLAND largest endocrine gland.
pp. 491–494

EMBRYONIC DEVELOPMENT AND STRUCTURE originates as epithelial thickening in pharynx floor.

THYROID HORMONES iodine attached to tyrosine molecules; triiodothyronine (T_3) and thyroxine (T_4); stored as thyroglobulin.

EFFECTS OF T_3 AND T_4
1. Involved in wide range of metabolic activities.
2. Increase carbohydrate oxidation; accelerate energy liberation.
3. Stimulate many aspects of lipid metabolism.

CALCITONIN hormone from thyroid that lowers blood calcium and blood phosphorus levels.

RELEASE OF THYROID HORMONES degradation of thyroglobulin and release of T_3 and T_4 into bloodstream. Calcitonin released in response to increased blood calcium level. TSH from pituitary controls synthesis and release of T_3 and T_4; thyrotropin releasing factor involved in control. Extraneous factors such as low temperature and stress affect thyroid activity.

CONDITIONS OF CLINICAL SIGNIFICANCE: THYROID DISORDERS

CRETINISM severe hypothyroidism beginning in infancy.

MYXEDEMA severe hypothyroidism in adult.

THYROTOXICOSIS excess of thyroid hormones acting on tissues; hyperthyroidism, for example.

GOITER enlarged thyroid gland.

PARATHYROID GLANDS embedded on posterior surface of thyroid gland. **pp. 494–495**

EMBRYONIC DEVELOPMENT AND STRUCTURE from dorsal halves of third and fourth pairs of pharyngeal pouches. Usually two masses of parathyroid tissue on posterior surface of each lateral lobe of thyroid. Composed of masses or cords of principal cells.

PARATHYROID HORMONE AND ITS EFFECTS two or three forms of parathyroid hormone may exist; controls calcium and phosphorus metabolism.

RELEASE OF PARATHYROID HORMONE released in response to decrease in plasma calcium level.

CONDITIONS OF CLINICAL SIGNIFICANCE: PARATHYROID DISORDERS

HYPERPARATHYROIDISM bone decalcification; calcification of soft tissues.

HYPOPARATHYROIDISM lowered plasma calcium level; increased nervous system excitability.

ADRENAL GLANDS located along superior border of each kidney; composed of inner medulla and outer cortex. **pp. 496–499**

ADRENAL MEDULLA

EMBRYONIC DEVELOPMENT AND STRUCTURE arises from neural crest cells; composed of chromaffin cells.

ADRENAL MEDULLARY HORMONES AND THEIR EFFECTS

Epinephrine (Adrenaline) has primarily metabolic and cardiovascular effects.

Norepinephrine (Noradrenaline) primarily affects vascular system; also released by sympathetic postganglionic neurons.

Release of Adrenal Medullary Hormones may be caused by emotional excitement, injury, exercise, low blood glucose, certain kinds of stress.

ADRENAL CORTEX

EMBRYONIC DEVELOPMENT AND STRUCTURE arises from mesoderm from same region that gives rise to gonads. Arranged in three layers: zona glomerulosa, zona fasciculata, and zona reticularis.

ADRENAL CORTICAL HORMONES AND THEIR EFFECTS

Mineralocorticoids (for example, aldosterone) promote reabsorption of sodium and excretion of potassium.

Glucocorticoids (for example, cortisol) supplement and conserve energy derived from circulating glucose.

Release of Adrenal Cortical Hormones renin from kidney controls release of mineralocorticoids; ACTH controls release of glucocorticoids.

CONDITIONS OF CLINICAL SIGNIFICANCE: ADRENAL DISORDERS

ADDISON'S DISEASE hypofunction of adrenal cortex; inadequate aldosterone; impaired ability to conserve sodium and excrete potassium.

CUSHING'S SYNDROME adipose-tissue accumulation; weight gain.

HYPERALDOSTERONISM potassium depletion; edema.

ADRENAL TUMORS adrenal cortical tumors may produce masculizing effects in females; medullary tumors produce symptoms of sympathetic overactivity.

PANCREAS both an endocrine and an exocrine gland. **pp. 499–503**

EMBRYONIC DEVELOPMENT AND STRUCTURE arises through fusion of two outgrowths from duodenum of small intestine; endocrine portion composed of cell clusters called pancreatic islets.

PANCREATIC HORMONES AND THEIR EFFECTS pancreatic islets contain three different cell types; alpha cells produce glucagon, beta cells produce insulin, and delta cells produce somatostatin.

INSULIN facilitates uptake and utilization of glucose by cells; prevents excessive glycogen breakdown in liver and muscle; influences lipid and protein metabolism.

GLUCAGON activities are generally opposite to those of insulin; stimulates breakdown of liver glycogen.

RELEASE OF PANCREATIC HORMONES chemical, hormonal, and neural control.
1. Increased plasma glucose levels increase insulin release and decrease glucagon release.
2. Amino acids stimulate secretion of insulin and glucagon.
3. Some gastrointestinal hormones promote insulin release.
4. Acetylcholine stimulates insulin release; epinephrine and norepinephrine inhibit insulin release.

CONDITIONS OF CLINICAL SIGNIFICANCE: PANCREATIC DISORDERS diabetes mellitus caused by relative deficiency of insulin.

GONADS testes and ovaries produce hormones as well as gametes. Testes produce androgens; ovaries produce estrogens and progesterone. **p. 503**

OTHER ENDOCRINE TISSUES AND HORMONES *digestive tract* and *placenta* are sources of hormonal substances; *pineal gland* produces hormone melatonin, which has antigonadal effects; *thymus gland* produces thymosin. **pp. 503–505**

SELF-QUIZ

1. Endocrine glands are generally ductless glands composed of epithelial cells that release their secretions directly into the target organ. True or false?

2. The pars nervosa: (a) retains direct connection with the brain in the adult; (b) synthesizes releasing factors; (c) is embryonically derived from Rathke's pouch.

3. ADH and oxytocin are manufactured in the: (a) pars nervosa of the pituitary gland; (b) hypothalamus; (c) pars distalis.

4. Gonadotropins are: (a) produced by the ovaries and testes; (b) neurohypophyseal hormones; (c) adenohypophyseal hormones.

5. Surging levels of LH together with FSH are important in: (a) stimulation of the thyroid gland; (b) ovulation (c) stimulation of the cortical region of the adrenal glands.

6. The acidophil cells of the adenohypophysis are thought to produce which of the following hormones? (a) growth hormone; (b) ACTH; (c) FSH; (d) all of these hormones.

7. Which hormone moves along the nerve fibers of the hypothalamic-hypophyseal tract? (a) ADH; (b) oxytocin; (c) both a and b.

8. Match the hormones with their appropriate sources and functions.

ADH	(a) An adenohypophyseal hormone.
Oxytocin	
FSH	(b) A neurohypophyseal hormone.
TSH	
ACTH	(c) Stimulates the cortical region of the adrenal gland.
GH	
	(d) Stimulates the reabsorption of water from the urine-forming structures of the kidneys.
	(e) Stimulates growth.
	(f) Stimulates the smooth muscle of the uterus.
	(g) Stimulates the synthesis of thyroid hormones.
	(h) Stimulates production of gametes and sex hormones by gonads.

9. Specialized neurosecretory cells in the brain manufacture a group of compounds known collectively as: (a) follicle-stimulating hormone; (b) luteinizing hormone; (c) releasing and inhibiting substances.

10. The structure of the thyroid gland is unique in having storage follicles for the hormone. True or false?

11. The thyroid gland develops embryonically from the floor of the brain. True or false?

12. Hormones that contain iodine are produced by which gland? (a) pituitary; (b) adrenal medulla; (c) thyroid.

13. The storage form of thyroid hormone is: (a) thyroglobulin; (b) thyroid-binding globulin; (c) free thyroxine.

14. The thyroid hormones generally travel in their active forms in the bloodstream. True or false?

15. In the young, hypothyroidism may lead to: (a) elevated body temperature; (b) elevated basal metabolism; (c) abnormal skeletal ossification.

16. Match the pituitary gland disorders with their possible symptom.

Lack of ACTH	(a) Underactivity of thyroid.
Gonadotropin deficiency	
TSH deficiency	(b) Cushing's syndrome.
TSH overproduction	(c) Failure to menstruate.
ACTH overproduction	(d) Pituitary dwarfism.
Growth hormone deficiency	(e) Acromegaly.
Growth hormone overproduction	(f) Adrenal cortical insufficiency.
Antidiuretic hormone deficiency	(g) Diabetes insipidus.
	(h) Hyperthyroidism.

17. Enlargement of the thyroid gland is known as: (a) Addison's disease; (b) folliculitis; (c) goiter.

18. This hormone is secreted by the adenohypophysis of the pituitary gland: (a) thyrotropin; (b) antidiuretic hormone; (c) oxytocin.

19. The same blood vessels supply the parathyroid glands and the thyroid gland. True or false?

20. When the plasma calcium level falls, parathyroid hormone secretion: (a) decreases; (b) remains constant; (c) increases.

21. Hypoparathyroidism: (a) increases the excitability of the nervous system; (b) raises serum calcium levels; (c) leads to calcification of the kidneys.

22. The adrenal medulla and the adrenal cortex both develop from the same type of embryonic tissue. True or false?

23. The release of epinephrine is controlled by nerves to the adrenal medulla from the sympathetic division of the autonomic nervous system. True or false?

24. Epinephrine is able to counteract the hypoglycemic effects of insulin by: (a) elevating blood sugar levels; (b) lowering renal potassium levels; (c) raising renal sodium levels.

25. The production site of adrenocortical hormones associated with sodium and potassium metabolism is the: (a) zona reticularis; (b) zona fasciculata; (c) zona glomerulosa.

26. ACTH from the pituitary gland controls the release of: (a) aldosterone; (b) cortisol; (c) FSH.

27. Two major effects of insulin are to facilitate the uptake and utilization of glucose by cells and to prevent the excessive breakdown of glycogen in liver and muscle. True or false?

28. The alpha cells of the pancreas produce: (a) insulin; (b) thymosin; (c) glucagon.

LEARNING OBJECTIVES

After completing this chapter, you should be able to:

- Trace the route of a given volume of air from the time it is inhaled through the nose until it enters the lungs, and name the various structures it passes through.

- Distinguish between the regions of the pharynx, and describe their respiratory functions.

- Describe the framework of the larynx.

- Describe the subdivisions of the bronchi.

- Describe the gross anatomical structure of the lungs.

- Describe the composition of the respiratory membrane.

- Distinguish between inspiration and expiration by describing the mechanics of each.

- Explain the various lung volumes that are combined into the total lung capacity.

- Discuss some of the common diseases of the respiratory system.

CHAPTER CONTENTS

The Respiratory System

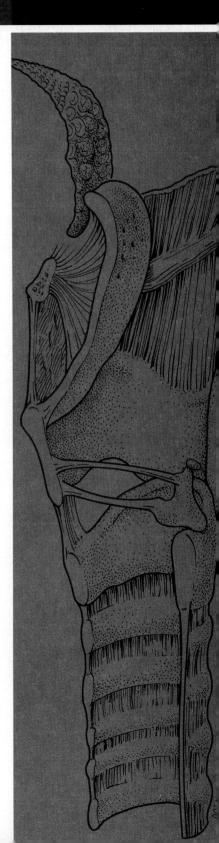

In order for the cells of the body to carry on their metabolic activities under aerobic conditions, they require a constant supply of oxygen and an efficient means of removing the carbon dioxide that their activities produce. Oxygen is supplied and carbon dioxide is removed by the **respiratory system,** with the assistance of the circulatory system. The respiratory system also makes vocalization possible. We are able to speak, sing, and laugh by varying the tension of the vocal folds as exhaled air passes over them.

The exchange of oxygen and carbon dioxide between the air and the blood occurs in the lungs. To reach the exchange sites in the lungs, the air flows through a series of conducting passageways that branch from one another much like tree branches. In fact, the passageways associated with breathing are often referred to as the *respiratory tree.* Air that enters the nose or mouth passes into the **pharynx** and is conveyed to the lungs by the **trachea,** which provides a branch—a **primary bronchus**—to each lung. In the lung each primary bronchus divides into smaller bronchi which, in turn, branch into tiny tubules called **bronchioles.** The bronchioles themselves divide and terminate in small air sacs called **alveoli.** Gaseous exchange occurs in the alveoli.

EMBRYONIC DEVELOPMENT OF THE RESPIRATORY SYSTEM

The first indication of the development of the respiratory system is the formation of an outpouch (*diverticulum*) from the ventral surface of the endoderm of the digestive tract, just behind the pharynx (Figure 19.1). This diverticulum, which appears in the four-week-old embryo, is called the **laryngotracheal bud.** As the bud elongates, the proximal portion develops into the *trachea* and its distal end bifurcates, forming two buds that will develop into the *bronchi.* The bronchial buds continue to grow and rebranch, giving rise to many small tubes called *bronchioles.* The closed terminal portions of the bronchiole buds become dilated as they develop into the *alveoli* of the lungs.

It is clear from the pattern of embryonic development that the epithelium lining the entire respiratory tract is derived from endoderm. In the adult, this lining is called the **respiratory epithelium.** With the exception of the lining of the smallest bronchioles and the alveoli, the respiratory epithelium is composed of pseudostratified ciliated columnar cells and scattered mucus-secreting goblet cells. The cartilage, muscles, and connective tissues of the trachea and the connective tissues of the lungs develop from embryonic mesoderm that becomes massed around the laryngotracheal bud.

ANATOMY OF THE RESPIRATORY SYSTEM

The respiratory system consists of the nose, nasal cavity, pharynx, larynx,

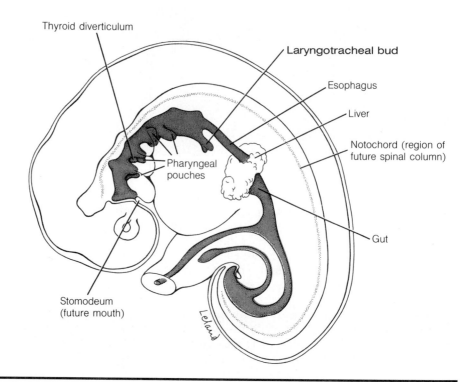

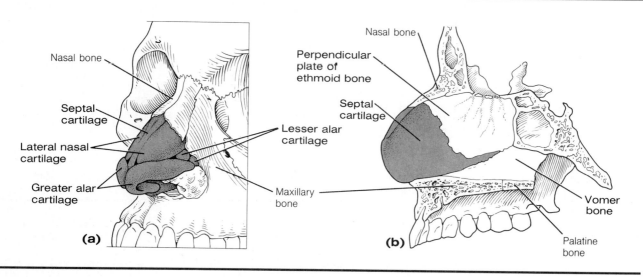

Figure 19.1
Lateral view of an embryo. The pharyngeal pouches, the thyroid diverticulum, and the laryngotracheal bud develop as outpouches of the pharynx.

Figure 19.2
(a) Cartilage plates of the nose. **(b)** Midsagittal view of the skull showing the nasal septum.

trachea, bronchi, and lungs. The following sections describe the structures of these organs. Notice each organ's structure is uniquely suited to its function.

Nose and Nasal Cavity

Air enters the respiratory system through the **external nares (nostrils),** which lead to the **vestibule** of the nose. The lower part of the vestibule contains hairs that serve to trap the largest particles that might be drawn into the respiratory system during inspiration. The bridge of the nose is formed by the nasal bones. The rest of the framework of the nose consists of several plates of

F19.2a cartilage held together by fibrous connective tissue (Figure 19.2a). To form the **nasal septum,** one of the cartilages—the **septal cartilage**—joins with the nasal bones above, the vomer bone below, the perpendicular plate of the

F19.2b ethmoid posteriorly, and the maxillae inferiorly (Figure 19.2b). The nasal septum divides the **nasal cavity** into right and left chambers. A deviation or deflection of the septum can interfere with the free flow of air through the nasal cavity, but this condition can be surgically corrected.

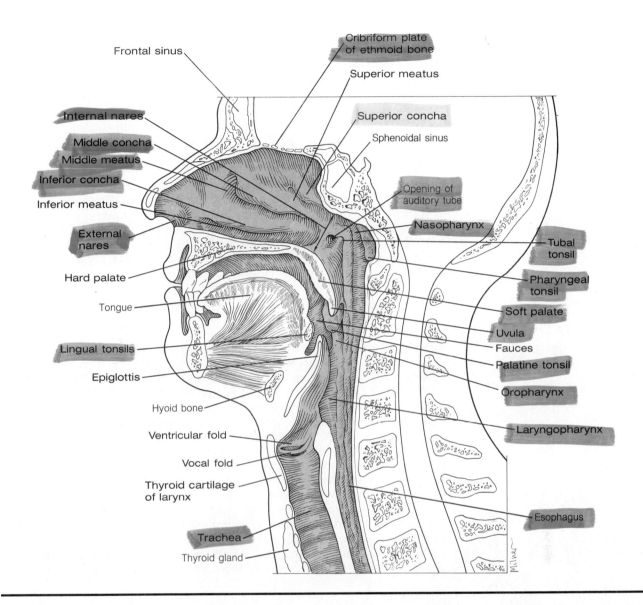

Frontal sinus

Cribriform plate
of ethmoid bone

Superior meatus

Internal nares

Middle concha

Middle meatus

Inferior concha

Inferior meatus

External
nares

Hard palate

Tongue

Lingual tonsils

Epiglottis

Hyoid bone

Ventricular fold

Vocal fold

Thyroid cartilage
of larynx

Trachea

Thyroid gland

Superior concha

Sphenoidal sinus

Opening of
auditory tube

Nasopharynx

Tubal
tonsil

Pharyngeal
tonsil

Soft palate

Uvula

Fauces

Palatine tonsil

Oropharynx

Laryngopharynx

Esophagus

Figure 19.3
Sagittal view of the head
and neck showing the
mouth, the pharynx, and
the lateral wall of the
nasal cavity.

The bony roof of the nasal cavity is formed by the cribriform plate of the ethmoid bone (Figure 19.3). The lateral walls, which are irregular, are formed by the **superior** and **middle conchae** of the ethmoid bone and the separate **inferior concha** bones. Beneath the shelves formed by the conchae are recesses called the **superior, middle,** and **inferior meatuses.** The floor of the nasal cavity is formed by the bony **hard palate** (horizontal plates of the palatine bones and palatine processes of the maxillary bones) and the more posterior, muscular **soft palate.** The palate separates the nasal cavity from the oral cavity. The nasal cavity opens posteriorly into the nasopharynx through the **internal nares (choanae).**

F19.3

Passageways from the **paranasal sinuses** drain into the nasal cavities. Most of them open into the meatuses formed by the conchae. The paranasal sinuses are air spaces located in the frontal, maxillary, ethmoid, and sphenoid bones. The *nasolacrimal ducts,* which drain fluid (tears) from the surface of the eyes, also empty into the nasal cavity.

Lining of the Nasal Cavity and the Paranasal Sinuses

The vestibule of the nose is lined with stratified squamous epithelium that is continuous with the skin. The rest of the nose, the nasal cavity, and the paranasal sinuses are lined with a continuous mucous membrane of pseudostratified ciliated columnar epithelium that contains many mucus-secreting goblet cells. The part of the mucous membrane that is located at the top of the nasal cavity, just beneath the cribriform plate of the ethmoid bone, is a specialized epithelium called the **olfactory epithelium.** This region is supplied by the **olfactory nerve** (cranial nerve I), which passes through holes in the cribriform plate to reach the brain. Although the olfactory epithelium is sensitive to smells, it does not lie in the direct path of air flow, so it is helpful to sniff air in order to detect odors.

The mucous membrane has an extensive blood supply that warms the air as it is inhaled. Moreover, the mucous membrane saturates with water the air passing over it. The membranes of the nasal cavity and the delicate portions of the lungs are thus protected from becoming frozen or dried out. A sheet of mucus covering the mucous membrane further protects the respiratory system by trapping any small particles that get past the hairs guarding the external nares. The cilia of the membrane move in such a manner as to carry the particle-filled mucus toward the pharynx, where it can be removed by coughing or swallowing.

Infections of the Mucous Membranes

The mucous membranes of the nasal cavity may become inflamed because of infections (such as the common cold) or allergies. When inflamed, the blood vessels dilate, the membranes swell, and mucus secretion increases. The resulting congestion interferes with breathing and often causes a "runny nose." Such infection can spread into the mucous membranes of the paranasal sinuses, blocking their connections with the nasal cavity and causing them to fill with mucus. Since the sinuses act as resonance chambers, congestion changes the sound of the voice and can also cause a pressure increase in the sinuses such that a severe headache results. Infections of the mucous membrane of the nasal cavity can also extend through the nasolacrimal duct to the covering (conjunctiva) of the eye. Thus, it is not uncommon to have red, watery eyes along with a cold. It is also possible for infection to travel from the mucous membrane of the nasal cavity into the pharynx, thus producing a "sore throat." From the pharynx, the infection can spread into the bronchi of the lungs, causing coughing and possibly bronchitis, or through the auditory tubes into the middle ear. This occurs particularly in young children, in whom the lumina of the auditory tubes are relatively larger than in adults.

Pharynx

The **pharynx** is a tube that is used by the digestive system as well as the respiratory system. It communicates with the nasal cavity (through the internal nares), the oral cavity (through the fauces), the middle-ear cavity (through the auditory tubes), the larynx (through the glottis), and the esophagus. The pharynx is a muscular structure lined with a mucous membrane that is continuous with the mucous membrane of the structures with which it communicates. For descriptive purposes the pharynx is divided into three parts: *nasopharynx, oropharynx,* and *laryngopharynx.*

Nasopharynx

F19.3

The **nasopharynx** is located immediately behind the nasal cavity and is continuous with it through the internal nares (Figure 19.3). The mucous membrane of the nasopharynx, like that of the nasal cavity, is formed of pseudostratified columnar epithelium. On its lateral walls, the nasopharynx receives the **auditory (eustachian) tubes** that connect the nasopharynx with the cavity of the middle ear. Located near the openings of the auditory tubes are small masses of lymphoid tissue called **tubal tonsils.** On the posterior wall are the larger **pharyngeal tonsils.** When these tonsils become enlarged in response to an infection they are called *adenoids.* Such enlargement can be chronic and can interfere with breathing through the nose, making it necessary to breathe through the mouth. The **soft palate** and **uvula** form the floor of the nasopharynx.

Oropharynx

The **oropharynx** is a continuation of the nasopharynx, extending from the soft palate to the beginning of the laryngopharynx (Figure 19.3). It communicates with the oral cavity through the **fauces.** The oropharynx therefore receives food from the oral cavity and air from the nasopharynx. During exercise, air is also drawn into the oropharynx through the mouth, thus increasing the ventilation rate of the lungs. The mucous membrane lining of the oropharynx is stratified squamous epithelium, which is the epithelium typically found in the oral cavity and the upper portion of the digestive tract. This type of epithelium serves to protect the region from the abrasiveness of swallowed food. On the side walls are two **palatine tonsils.** Embedded in the base of the tongue is an aggregate of **lingual tonsils.** The tonsils are formed of lymphoid tissues and are therefore part of the body's immune system.

F19.3

Laryngopharynx

The **laryngopharynx** extends from the oropharynx above to the esophagus below (Figure 19.3). It communicates anteriorly with the larynx. Like the oropharynx, the laryngopharynx serves as a passageway for food and air and consequently is lined with a protective stratified squamous epithelium.

F19.3

Larynx

The **larynx** connects the laryngopharynx with the trachea, which continues below the larynx. Air traveling to or from the lungs passes through the larynx. Any solid substance that enters the larynx, such as food, is generally expelled by violent coughing. The larynx forms a **laryngeal prominence** (*Adam's apple*) on the anterior surface of the neck. This prominence is particularly noticeable in males after puberty, when the larynx becomes larger than in females and the anterior region of its framework forms a more acute angle.

Framework of the Larynx

The larynx is formed by nine cartilages—three unpaired and three paired (Figure 19.4). These cartilages are held together, and attached to the hyoid bone above and the trachea below, by ligaments and muscles. The **thyroid cartilage** is the largest of the unpaired cartilages. It is formed by the midline junction of two broad plates anteriorly, producing the laryngeal prominence. The plates remain separated posteriorly, which leaves a wide opening in the laryngopharynx. Just below the thyroid cartilage is the ring-shaped **cricoid cartilage,** which is anchored to the thyroid cartilage above and the trachea below. The posterior region of the cricoid cartilage is taller than the anterior region. The third unpaired cartilage is the leaf-shaped **epiglottis.** The epiglottis is attached by its narrow end to the inner surface of the anterior region of the thyroid cartilage. When air is flowing into or out of the larynx its free upper portion projects like a flap behind the base of the tongue, and the entrance to the trachea is exposed. During swallowing, the larynx is pulled upward, tipping the epiglottis so that it tends to block the opening of the larynx and deflect solids and fluids into the esophagus.

F19.4

The **arytenoid cartilages** are the most important of the paired cartilages. Each arytenoid cartilage is shaped like a small pyramid and rests on the superior-posterior border of the cricoid cartilage. The posterior ends of the vocal folds are attached to the arytenoid cartilages, and movement of the cartilages is responsible for varying the tension on the folds. The other paired cartilages, the **cuneiform** and **corniculate cartilages,** are small and closely related to the arytenoid cartilages.

Mucous Membrane of the Larynx

The mucous membrane that covers the epiglottis and the upper parts of the larynx, where they are in direct communication with the laryngopharynx, is lined with stratified squamous epithelium. The remainder of the larynx is lined with pseudostratified ciliated columnar epithelium. The mucous membrane near the entrance to the larynx forms two pairs of horizontal folds that

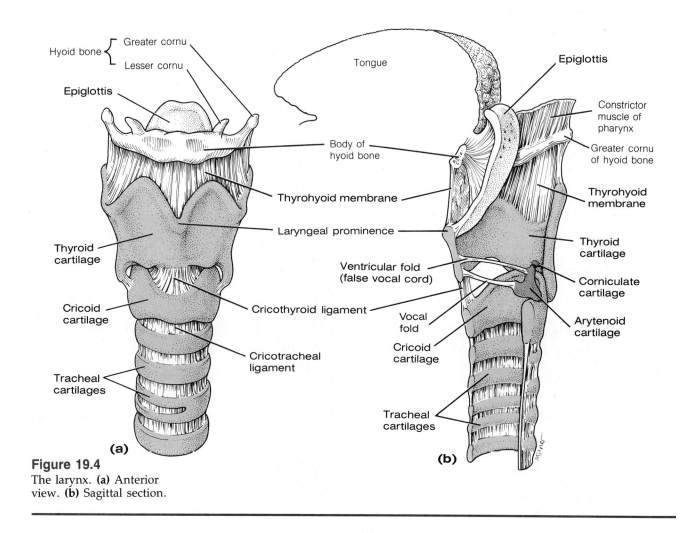

Figure 19.4
The larynx. **(a)** Anterior view. **(b)** Sagittal section.

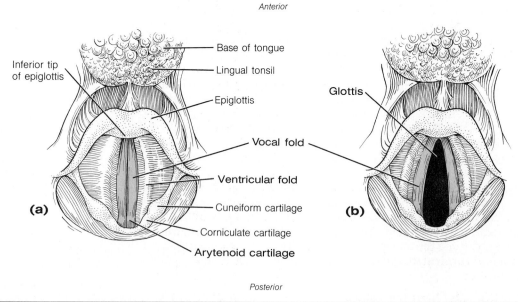

Figure 19.5
The larynx. Superior view showing **(a)** the glottis closed and **(b)** the glottis open.

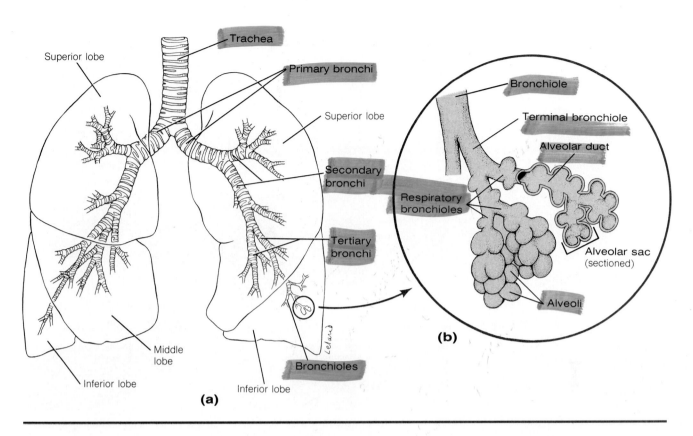

(a)

Figure 19.6
(a) The trachea, the bronchi, and the lungs. (b) Respiratory bronchioles divide into alveolar ducts, which lead to alveoli.

extend posteriorly on each side from the thyroid cartilage to the arytenoid cartilages. The upper pair of folds are called the **ventricular folds** (*false vocal cords*). The lower pair are the **vocal folds** (*true vocal cords*). The opening between the vocal folds through which air enters the larynx is the **glottis** (Figure 19.5). Within the vocal folds are bands of elastic ligaments that connect to the thyroid, cricoid, and arytenoid cartilages. The elastic vocal ligaments can be tightened or slackened by the actions of certain intrinsic muscles of the larynx. These muscles also rotate the arytenoid cartilages, thus further varying the amount of stretch on the vocal folds. As a result of the actions of the intrinsic muscles, the glottis can be narrowed or widened. Air passing through the glottis causes the vocal folds to vibrate and produce a sound. The frequency of the vibrations, and therefore the pitch of the sound produced, depends upon the tension on the vocal folds.

The mucous membrane that covers the vocal folds becomes swollen when irritated or inflamed. This swelling interferes with the ability of the folds to vibrate, and hoarseness may result—a condition referred to as *laryngitis.*

Trachea

The **trachea** (windpipe) is a tube approximately 2.5 cm in diameter and 11 cm long. It extends from the larynx to the level of the sixth thoracic vertebra, where it divides into *left* and *right primary bronchi* (Figure 19.6a). The trachea lies against the anterior surface of the esophagus. The air passageway of the trachea is surrounded by a series of C-shaped hyaline cartilages that function to keep the trachea from collapsing, in the same manner as the wire rings in the hose of a vacuum cleaner. The cartilages are enclosed in an elastic fibrous membrane, and elastic fibers remain an important layer in the walls of all subsequent parts of the respiratory tree. Smooth muscle fibers, called the **trachealis muscle,** and elastic connective tissue join the rings together posteriorly and close the opening of the C. When the passage of air through the upper airways is impeded, a direct airway into the trachea is sometimes created surgically through the anterior surface of the neck, between the second and third cartilaginous rings. This procedure is called a *tracheotomy.*

F19.5

F19.6a

Cilia

Goblet cell

Figure 19.7
Photomicrograph of the
inner lining of the
trachea (×927). (From
*Tissues and Organs: A
Text-Atlas of Scanning
Electron Microscopy* by
Richard G. Kessel and
Randy H. Kardon. W.H.
Freeman and Company.
Copyright © 1979.)

F19.7   The trachea is lined with a mucous membrane of pseudostratified ciliated
columnar epithelium that contains numerous goblet cells (Figure 19.7). Since
the cilia beat upward, they carry foreign particles and excessive mucus secretions away from the lungs to the pharynx, where they can be swallowed.

Bronchi, Bronchioles, and Alveoli

As the trachea passes behind the arch of the aorta, it divides into two smaller
branches: the **left** and **right primary bronchi.** Each primary bronchus divides
into still smaller **secondary (lobar) bronchi,** one for each lobe of the lung. The
secondary bronchi, in turn, branch into many **tertiary (segmental) bronchi**
that further branch repeatedly, ultimately giving rise to tiny **bronchioles.** The
bronchioles themselves subdivide many times, forming **terminal bronchioles,** each of which gives rise to several **respiratory bronchioles.** Each respiratory bronchiole subdivides into several **alveolar ducts** that end in clusters of
F19.6b   small, thin-walled air sacs called **alveoli** (Figure 19.6b). Often, several alveoli
open into a common chamber called an **alveolar sac.** A few scattered alveoli
are also present in the walls of the respiratory bronchioles. The bronchial tree
is lined with pseudostratified ciliated columnar epithelium. However, in the
respiratory bronchioles the epithelial lining loses its cilia and changes to cuboidal and then to squamous cells as the bronchioles extend distally. The
many tiny alveolar sacs and the bronchial system of tubes through which the
air travels to enter or leave the alveoli give the lungs a spongy texture (Figure
F19.8   19.8).
    The walls of the primary bronchi, like the trachea, are supported by in-
complete cartilage rings. In the lungs, the rings are replaced by small plates of
cartilage that completely encircle the bronchus. Smooth muscle also encircles
the bronchus. With further branching, the cartilage plates gradually become
fewer and smaller, forming incomplete rings, and the smooth muscle that
surrounds the air passageways becomes more prevalent. The walls of the
bronchioles contain no cartilage and are completely surrounded by smooth

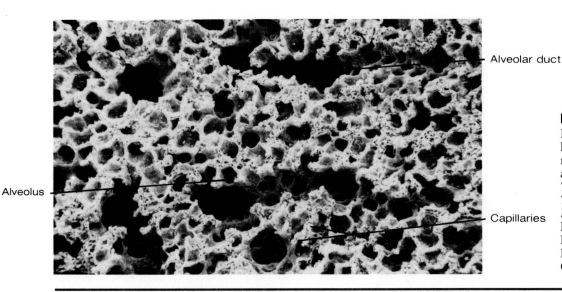

Figure 19.8
Photomicrograph of a lung section (×113). The numerous tiny openings are capillaries. (From *Tissues and Organs: A Text-Atlas of Scanning Electron Microscopy* by Richard G. Kessel and Randy H. Kardon. W.H. Freeman and Company. Copyright © 1979.)

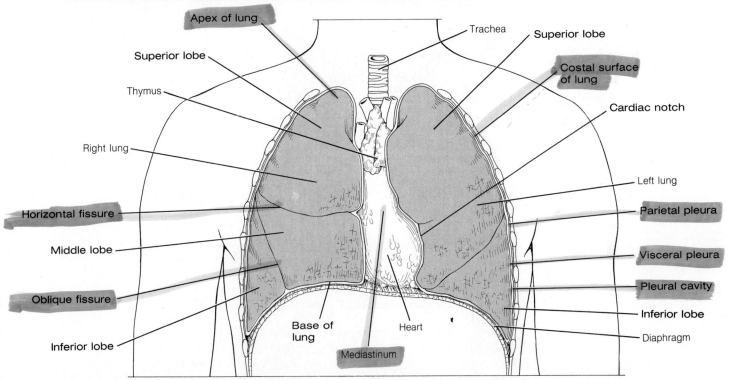

Figure 19.9
The lungs and the thoracic cavity.

muscle. The fact that bronchioles have walls of smooth muscle explains one of the most disturbing symptoms of an asthma attack. In response to various allergens, the muscles of the bronchioles undergo spasms, and since there is no supporting cartilage the air passageways are squeezed shut, making breathing very difficult.

Lungs

The **lungs** are shaped somewhat like cones, with the pointed **apex** of each lung fitting into the narrow space at the top of the thoracic cavity behind the clavicle (Figure 19.9). The **base** of each lung is broad and concave and rests on the convex surface of the diaphragm. A depression called the **hilus** is on the mediastinal surface of the lung. The hilus is the region where the structures that form the *root* of the lung—that is, the bronchus, blood vessels, lymphat-

F19.9

Lung bronchopulmonary segments (anterior view)

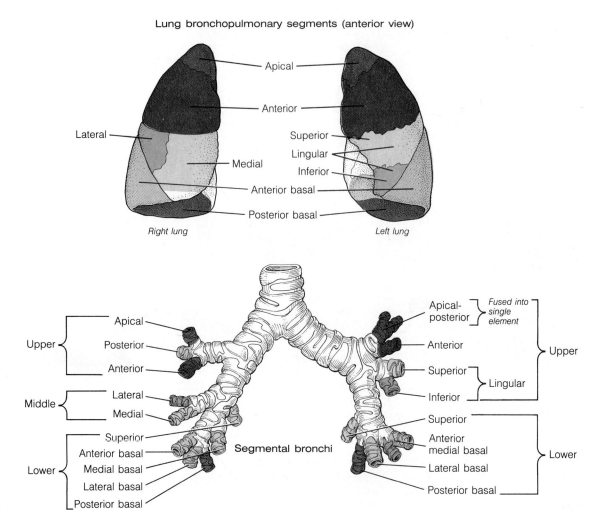

Segmental bronchi

Lung bronchopulmonary segments (posterior view)

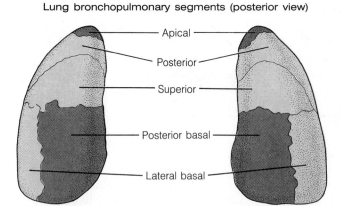

Figure 19.10

The bronchopulmonary segments of the lungs, and the branching of the bronchi into the segments.

ics, and nerves—enter or leave the lung. The **costal surface,** which lies against the ribs, is rounded to match the curvature of the ribs. The left lung has a concavity for the heart, called the **cardiac notch,** on its medial surface. Each lung is divided into **superior** and **inferior lobes** by an **oblique fissure.** The right lung is further divided by a **horizontal fissure,** which bounds a **middle lobe.** The right lung therefore has three lobes, (and receives three secondary bronchi), whereas the left has only two (and receives only two secondary bronchi). In addition to these five lobes, which are visible externally, each lung is subdivided by connective tissue partitions into smaller units called **bronchopulmonary segments** (Figure 19.10). Each bronchopul- **F19.10** monary segment represents the portion of the lung that is supplied by a specific tertiary bronchus. The superior lobe of the right lung has three segments; the middle lobe has two segments; the inferior lobe has five segments. In the left lung, the superior lobe and the inferior lobe each have five segments, although some of them are not distinctly separated. The bronchopulmonary segments are important surgically because a diseased segment can be removed without having to remove an entire lobe or the entire lung. Also, disease does not spread so easily across the partitions that separate the segments, so pathology tends to be confined to one or several segments rather than spreading freely throughout the lung.

The two lungs are separated by a space called the **mediastinum.** Important structures are located in the mediastinum, including the heart, aorta, venae cavae, pulmonary vessels, esophagus, part of the trachea and bronchi, and the thymus gland.

The Pleura

Each lung is enclosed in a double-walled sac called the **pleura** (Figure 19.9). **F19.9** Both layers of the pleura are formed of serous membrane. The portion of the pleura that adheres firmly to the lungs is the **visceral pleura.** The portion that lines the walls of the thoracic cavity is the **parietal pleura.** The visceral and parietal layers are continuous at the hilus of the lung. Between the two pleural layers is a very narrow **pleural cavity,** which is filled with **pleural fluid.** The pleural fluid is secreted by the pleura, and acts as a lubricant to reduce the friction between the two layers during respiratory movements.

Blood Supply of the Lungs

There is an important difference in the blood supply to the alveoli and to the bronchi. The alveoli are supplied by branches of the pulmonary artery (which contains oxygen-poor blood), whereas small bronchial arteries from the descending aorta (which contains oxygen-rich blood) supply the bronchi. Thus, the pulmonary arteries carry blood that will be aerated within the alveoli, whereas the blood within the bronchial arteries provides for the nutrition of the lung tissue.

The Respiratory Membrane

The air in the alveoli is separated from the blood by a very thin **respiratory membrane** formed by the alveolar epithelium and its basal lamina and the endothelium of the capillary and its basal lamina. In some places there is a small amount of connective tissue in the form of reticular fibers and elastic fibers between the two basal laminae (Figure 19.11). It is obviously advanta- **F19.11** geous to have a thin respiratory membrane—since oxygen and carbon dioxide diffuse across it during respiration—while still maintaining the integrity of the blood vascular system. For the efficient diffusion of oxygen and carbon dioxide, the respiratory membrane must be moist; consequently, the alveolar surfaces exposed to the air are coated with a thin layer of fluid.

The epithelium that lines the alveoli contains two types of cells. Most of the cells are of the simple squamous type through which gases diffuse readily. These cells are called *alveolar type I cells.* Scattered among the simple squamous cells are rounded or cuboidal cells—called *alveolar type II cells*—that may bulge into the lumen of the alveolus. The type II cells secrete a substance called **surfactant,** which serves to lower the surface tension of the fluid that coats the alveoli, thus preventing the collapse of the alveoli and decreasing the muscular effort required to expand the lungs.

Phagocytic cells (macrophages) are also commonly present within the respiratory membrane of the alveolar walls, and they may be found free in the alveolar space. These phagocytes are sometimes called *dust cells,* since they often contain carbon particles obtained from inhaled air.

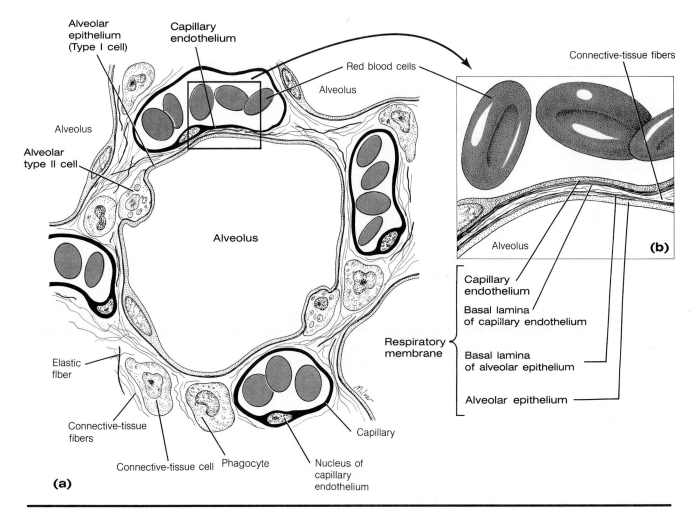

Figure 19.11
(a) Cross section of an alveolus surrounded by capillaries, showing the respiratory membrane that separates the air in the alveolus from the blood in the capillaries. (b) The respiratory membrane of the lungs.

MECHANICS OF BREATHING

In order to maintain a concentration of oxygen and carbon dioxide within the alveolar air that is favorable to their diffusion across the alveolar membrane, it is necessary to constantly bring in fresh air and remove the air already in the lungs. At rest, about 500 ml of air enters and leaves the lungs with each respiration. Since respiration is repeated about 16 times a minute, the oxygen from 8000 ml (8 liters) of air enters the respiratory system every minute.

The thoracic cavity is airtight, with flexible side walls (rib cage) and flexible floor (diaphragm). The lungs, enclosed in the double-layered pleura, are suspended in this airtight cavity. The alveoli and passageways within the lungs communicate with the atmosphere through the trachea.

In the newborn child the lungs fill the entire pleural cavity without being stretched. During childhood, however, the thoracic cage grows faster than the lungs. In fact, if it were not for the presence of a small quantity of pleural fluid in the pleural cavity, between the surface of the lung and the inner wall of the thorax, the thoracic wall would pull away from the lung. As the growing thoracic cage tends to pull away from the lungs, however, it causes the pressure in the pleural cavity to drop below the air pressure (atmospheric pressure) in the lungs. As a consequence, the lungs are caused to expand and stretch as the thoracic cage grows.

Because of the negative pressure in the pleural cavity plus the surface tension caused by the small amount of pleural fluid present in the cavity, the visceral pleura—and as a consequence the surface of the lung—is held tightly to the parietal pleura. These membranes can slide against one another, but

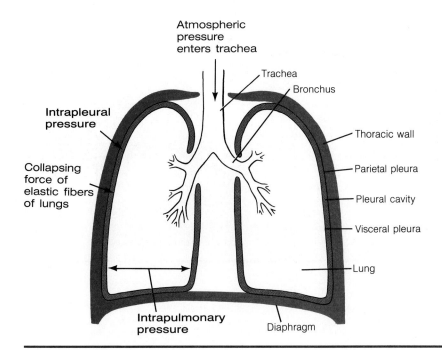

Atmospheric
pressure
enters trachea

Trachea

Bronchus

Intrapleural
pressure

Thoracic wall

Parietal pleura

Collapsing
force of
elastic fibers
of lungs

Pleural cavity

Visceral pleura

Lung

Intrapulmonary
pressure

Diaphragm

Figure 19.12
Diagram of the thorax
indicating the pressures
involved in respiration.

they cannot be easily separated. Therefore, the lungs do not collapse in the body, in spite of their tendency to do so because of the natural recoil of the elastic connective tissue in their walls and partitions.

Air moves through the trachea and into the lungs when the pressure in the lungs (*intrapulmonary pressure*) is less than atmospheric pressure (Figure **F19.12** 19.12). Air moves out of the lungs when the pressure in the lungs is greater than atmospheric pressure. Between respirations the pressure in the lungs is equal to atmospheric pressure. Because the rigid thoracic wall does not follow the lungs while the elastic fibers within them tend to retract the lungs away from the walls, a partial vacuum is created in the pleural cavity. The pressure in the pleural cavity (*intrapleural*, or *intrathoracic*, *pressure*) is thus maintained at slightly less than atmospheric pressure. This reduced intrapleural pressure represents the "collapsing" force of the lungs. During fetal development, however, the pressure in the pleural cavity is not less than atmospheric pressure; as a consequence, the lungs of the fetus are collapsed and contain no air. As breathing commences immediately following birth, the thoracic cage expands and pulls the lungs out with it. Once stretched and filled with air, the lungs normally remain that way throughout life.

Inspiration

Inspiration refers to the movement of air *into* the lungs. As previously indicated, such movement occurs when the pressure in the lungs drops below atmospheric pressure. Air then moves into the lungs and reestablishes the pressure equilibrium. The slight drop in pressure within the lungs is achieved by increasing the volume of the thoracic cavity.

There are two ways in which the volume of the thoracic cavity can be increased. One way is by contracting the diaphragm. When contracted, the diaphragm flattens, lowering its dome. This action increases the longitudinal dimension of the thoracic cavity. The second way is by elevating the ribs. In the resting position, the ribs slant downward and forward from the vertebral column. The contraction of muscles such as the subcostals and the intercostals pulls the ribs upward, thereby increasing the internal dimensions of the thoracic cavity and increasing its volume.

During normal, quiet inspiration, the contraction of the diaphragm is the dominant means of increasing the volume of the thoracic cavity and lowering the pressure within the lungs. The elevation of the ribs is more evident during forced inspiration.

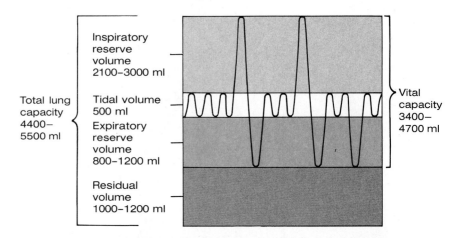

Figure 19.13
Lung volumes.

Expiration

Expiration refers to the movement of air *out of* the lungs, back into the atmosphere. It occurs when the volume of the thoracic cavity decreases, thus increasing the pressure in the lungs to greater than atmospheric pressure. During quiet breathing, the volume of the thoracic cavity is decreased by passive processes that do not involve muscular contractions. When the muscles involved in inspiration relax, the elastic recoil of the lungs, chest wall, and abdominal structures returns the ribs and diaphragm to their resting positions. This activity reduces the volume of the thoracic cavity and raises the pressure within the lungs until it is slightly greater than atmospheric pressure. The increased pressure in the lungs forces air out of the lungs until intrapulmonary and atmospheric pressure are again in equilibrium.

During forced expiration, as occurs in exercise, muscles are involved in the further reduction of the thoracic cavity's volume. The intercostals, the transversus thoracis, the quadratus lumborum, and the serratus posterior inferior muscles all assist in reducing the volume of the thoracic cavity by depressing the rib cage. Moreover, the muscles of the anterior abdominal wall aid in forced expiration by exerting pressure on the abdominal viscera, thus forcing the diaphragm upward.

Traditionally, it has been thought that the roles of the intercostal muscles in respiratory movements could be distinguished, with the external intercostals assisting in forced inspiration and the internal intercostals assisting in forced expiration. However, recent studies have shown that both sets of intercostals are active in both forced inspiration and forced expiration, serving to control the distances between ribs.

Rhythmic Breathing Movements

The rhythmic pattern of inspiration and expiration that is characteristic of normal respiration depends on the cyclic activity of the neurons that supply the respiratory muscles—primarily neurons whose cell bodies are located in the medulla oblongata of the brain. The area in which these nerve-cell bodies are located is called the **respiratory center.**

Lung Volumes

In young males, the lungs have a total capacity of approximately 5900 ml. Once the lungs are filled with air following birth, however, it is impossible to exhale all the air from them. Therefore the total 5900 ml is not available for respiratory purposes and about 1200 ml always remains in the lungs no matter how forced the expiration. This remaining volume is called the **residual volume** (Figure 19.13).

F19.13

Table 19.1 Average Lung Volumes in Young Males and Females (Age 20–30 Years)

	Volume (in ml)	
	Male	Female
Total lung capacity	5900	4400
Vital capacity	4700	3400
Inspiratory reserve	3000	2100
Tidal volume	500	500
Expiratory reserve	1200	800
Residual volume	1200	1000

During normal, quiet respiration, about 500 ml of air moves into and out of the lungs—this is the **tidal volume. Inspiratory reserve** is the extra volume of air (approximately 3000 ml) that can be inspired in addition to the normal tidal air. Following a normal, passive expiration of tidal air, additional air can be forced out. This **expiratory reserve** is about 1200 ml. The **vital capacity** is the maximum amount of air that can be moved into and out of the lungs— from the deepest possible inspiration to the most forced expiration. Vital capacity thus represents the sum of the inspiratory reserve, the tidal air, and the expiratory reserve. Note that it does not include residual air. In healthy young men the vital capacity is about 4700 ml. It is somewhat less in women because they tend to have smaller thoracic cages and smaller lung capacities (see Table 19.1). Table 19.1

CONDITIONS OF CLINICAL SIGNIFICANCE
The Respiratory System

Common Cold

The _common cold_, or _acute coryza_, is an inflammation of the mucosa of the upper respiratory tract, one that is familiar to most people. The initial inflammation is caused by various viruses, and is often followed by bacterial infection of the sinuses, ears, or bronchi. When inflamed, the mucosa becomes engorged and swollen, causing discomfort and difficulty in breathing. Later, the mucosa discharges a watery fluid that makes its presence known in the form of a runny nose. Such discharges from the mucosa lining the paranasal sinuses can irritate the larynx and trachea, producing a cough.

Hay Fever

Hay fever is an allergic (antigen-antibody) response to certain allergens (antigens) that are present in plant pollens and dust. This allergic response causes a localized inflammation of the respiratory mucous membranes, producing symptoms similar to those of the common cold.

Bronchial Asthma

Like hay fever, _asthma_ is an allergic response to foreign substances that generally enter the body via air breathed or food eaten. It is characterized by episodes of wheezing and difficult breathing. In response to the allergens, the mucous membranes of the respiratory system secrete excessive amounts of mucus, sometimes producing mucus plugs that completely block the bronchioles. Moreover, the smooth muscles that surround the smaller bronchi and bronchioles go into spasms. These spasms narrow the passageways, making it difficult for air to move into or out of the alveoli.

Bronchitis

Bronchitis is an acute or chronic inflammation of the bronchial tree. It is caused by bacterial infection or by irritants in the inhaled air (such as smoke or chemical fumes). The mucous membranes of the respiratory system produce a sticky secretion that inhibits the normal protective function of the macrophages of the respiratory tract and hinders the self-cleaning actions of the cilia of the cells that line the bronchi. As the secretions accumulate in the bronchi, they are removed by coughing. Coughing is annoying, but it serves the useful purpose of helping to keep the lungs clear.

Tuberculosis

Tuberculosis, an infection caused by the tubercle bacillus (*Mycobacterium tuberculosis*), can affect many parts of the body. Because the bacterium most commonly enters the body by inhalation, tuberculosis of the lungs is the most prevalent form. Even if not inhaled, the bacilli can enter the lymphatic system or the bloodstream and thereby reach the lungs. When the bacilli reach the lungs, the lung tissue reacts by forming small clumps (tubercles) around them. Many of the bacilli are then engulfed by protective cells (phagocytes), and fibrous walls form around them. If the bacilli are not successfully walled off, the lung tissue is destroyed and the site of infection spreads. The process may continue until both lungs are extensively destroyed. Even if such a massive involvement of the lungs does not result, the fibrosis in the affected portions interferes with the diffusion of gases, and causes the lungs to lose their elasticity, thereby reducing the vital capacity.

Emphysema

Emphysema is a condition that develops slowly as a secondary response to other respiratory problems, such as chronic bronchitis and tuberculosis, or to environmental irritants, such as cigarette smoke and industrial pollutants. (For a discussion of the effects of smoke on the lungs, see Box 19.1.) In persons suffering from emphysema the alveoli become overdistended, and the walls of the alveoli break down and are often replaced by fibrous tissue. This destruction of alveoli greatly reduces the surface area across which gaseous exchange can occur. As a consequence, the amount of oxygen in the blood is reduced, and even mild exercise increases the breathing rate. In addition, the elastic tissues of the overexpanded lungs are reduced, making expiration difficult. A reduced expiratory volume is an early symptom of emphysema.

Thus, two basic problems face the victim of emphysema: the lungs are "fixed" in inspiration; and the respiratory surfaces of the lungs have deteriorated so much that they can no longer accomplish normal gas exchange. Unfortunately, the disease is progressive and irreversible.

Pneumonia

The inflammation of *pneumonia* causes a fibrinous exudate to be produced in the alveoli. The lung, or a part of it, becomes solid and airless, which makes it very difficult for gaseous exchange to occur in the alveoli. Most cases of pneumonia are probably caused by one of several viruses. Another common cause is the *pneumococcus bacterium*; however, it has been suggested that most cases of bacterial pneumonia are preceded by a virus infection. Pneumonia can also result from the inhalation of food or other foreign bodies that cause obstruction of a bronchus. Such an obstruction can lead to collapse of the lung, accumulation of fluids in the lung, and subsequent infection.

Pleurisy

Pleurisy is an inflammation of the pleural membranes that is commonly caused by bacteria—usually *pneumococci*, *streptococci*, or the *tubercle bacillus*. In the early stage, the inflamed pleural membranes are "dry" and covered with fibrous material. This condition causes a sharp pain in the chest during respiration. Adhesions between the layers of the pleura can develop as a result of pleurisy, and in severe cases surgery may be necessary to remove them. In later stages of pleurisy, there is often excessive secretion of pleural fluid into the pleural cavity.

Pneumothorax

Pneumothorax refers to any condition that allows air to enter the pleural cavity—that is, the space between the parietal pleura and the visceral pleura. Normally, the lung is kept in an expanded condition by the negative pressure in this space. If the pleural cavity is opened to the outside air—either by perforation of the chest wall and the parietal pleura or by rupture of the visceral pleura and the underlying lung tissue (*spontaneous pneumothorax*)—air enters the pleural cavity and causes the lung to collapse (*atelectasis*). In spontaneous pneumothorax, the collapse of the lung generally seals off the perforation and, with rest, the ruptured visceral pleura heals. When the chest wall has been perforated, it is necessary to seal the puncture with a dressing that allows no air to pass through. In either case the air that enters the pleural cavity is slowly reabsorbed, and eventually, the negative pressure is reestablished.

Effects of Aging on the Respiratory System

Tissue changes that occur in the respiratory system with increased age cause the chest wall to become more rigid and the lungs to become less elastic. Thus, although the total lung volume does not change with age, the increased rigidity of the chest wall and the loss of elasticity in the lungs result in a diminished ventilating capacity. Because of these changes, the vital capacity of the lungs in males decreases from about 4700 ml at age 20 to about 4000 ml at age 70. Accompanying these changes is a decrease in the amount of oxygen carried in arterial blood. This decrease in arterial oxygen is quite pronounced when a person is lying in a supine position, which makes breathing more difficult. For this reason there is a tendency for elderly people to experience difficulty getting adequate oxygen during sleep and to be more comfortable if supported by pillows.

With increasing age, there is also a decrease in the phagocytic activity of macrophages and in the activity of the cilia of the epithelial linings of the respiratory tract. As a consequence, the cleaning of the respiratory tract lining by cilia is less efficient. The rigidity of the chest wall, the loss of elasticity in the lungs, and the diminished phagocytic activity and ciliary action all make elderly people more susceptible to infections of the respiratory system, such as pneumonia.

Box 19.1
Smoker's and Nonsmoker's Lungs

The ill-effects of tobacco smoke are clearly evident in this comparison between a smoker's lungs (top) and a nonsmoker's lungs (bottom). In the smoker's lungs, area A depicts a cancerous growth and the resultant narrowing of the lumen of the bronchus. The tumor has extended beyond the walls of the bronchus into the surrounding tissue (B). Lung cancer is typified by neoplastic growth—the presence of an abnormal mass (malignant tumor). The cancerous cells multiply and spread, destroying healthy tissue. Lung cancer is also characterized by metastatic growth, in which cancer cells spread throughout the body through the blood or lymphatic systems. Lung cancer can be treated with radiation, chemotherapy, or surgery to remove part or all of the lung.

Smoking is a contributing factor to heart disease, emphysema, and chronic bronchitis, as well as lung disease. Just one cigarette speeds up the heart rate, increases blood pressure, upsets the flow of blood and air in the lungs, and causes a drop in the skin temperature of fingers and toes. Less well known, but equally important, is the deleterious effect of secondhand smoke on the nonsmoker. It has been estimated that when a nonsmoker leaves a smoky environment, it takes four to five hours for 50% of the carbon monoxide to leave his or her body.

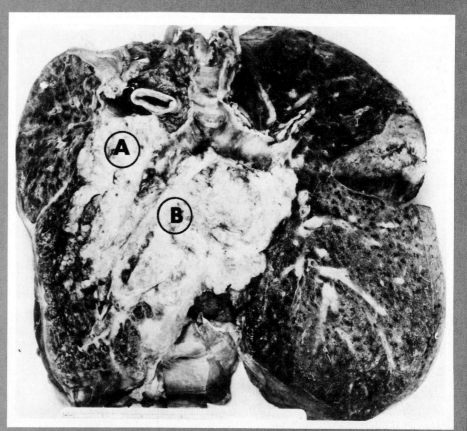

Lungs of a smoker

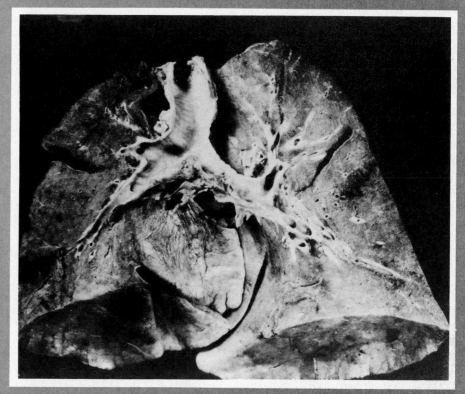

Lungs of a nonsmoker

STUDY OUTLINE

EMBRYONIC DEVELOPMENT OF THE RESPIRATORY SYSTEM p. 511

LARYNGOTRACHEAL BUD diverticulum from ventral endoderm of digestive tract.

TRACHEA from proximal part of laryngotracheal bud.

BRONCHI from bifurcation of distal end of laryngotracheal bud.

BRONCHIOLES AND ALVEOLI from bronchi.

RESPIRATORY EPITHELIUM derived from endoderm; lines entire respiratory tract.

ANATOMY OF THE RESPIRATORY SYSTEM
respiratory system consists of nose, pharynx, larynx, trachea, bronchi, and lungs. pp. 511–521

NOSE AND NASAL CAVITY air passes through the external nares to the vestibule of the nose and enters the nasal cavity.

SEPTUM formed from septal cartilage, vomer bone, and perpendicular plate of ethmoid bone.

ROOF formed by cribriform plate of ethmoid.

LATERAL WALLS formed by superior, middle, and inferior conchae.

FLOOR formed by hard and soft palates.

VESTIBULE stratified squamous epithelial lining.

NASAL CAVITY, SINUSES mucous membrane of pseudostratified ciliated columnar epithelium. Blood supply warms air and saturates it with water; cilia trap small particles; internal nares connect nasal cavity with nasopharynx.

PHARYNX muscular structure lined with mucous membrane.

NASOPHARYNX receives internal nares and auditory tubes; contains tubal and pharyngeal tonsils.

OROPHARYNX food and air passageway; contains palatine and lingual tonsils; communicates with oral cavity through fauces.

LARYNGOPHARYNX food and air passageway; between oropharynx and esophagus and larynx.

LARYNX air conduction between laryngopharynx and lungs.
1. Framework of larynx consists of nine cartilages.
2. Mucous membranes of larynx are stratified squamous and pseudostratified ciliated columnar epithelium; form pairs of ventricular folds and vocal folds.
3. Glottis is opening between vocal folds.

TRACHEA (WINDPIPE)
1. Kept open by C-shaped hyaline cartilages.
2. Ciliated cells carry foreign particles away from lungs to pharynx.

BRONCHI, BRONCHIOLES, AND ALVEOLI
1. Primary, secondary, and tertiary bronchi → terminal and respiratory bronchioles → alveolar ducts → alveoli.
2. With branching, supportive cartilage gradually replaced by smooth muscle.

LUNGS
1. Three-lobed right lung; two-lobed left lung; each lobe further divided into bronchopulmonary segments.
2. *Hilus*—site where bronchi, blood vessels, lymphatics, and nerves pass into or out of lungs.
3. *Mediastinum*—space that separates the lungs.

THE PLEURA
1. Visceral and parietal pleura—double-walled serous membrane sac that encloses each lung.
2. Pleural cavity between layers—filled with pleural fluid.

BLOOD SUPPLY OF THE LUNGS pulmonary artery (oxygen-poor blood) supplies alveoli; bronchial arteries from descending aorta (oxygen-rich blood) supply bronchi.

RESPIRATORY MEMBRANE thin membrane that separates alveolar air from blood; O_2 and CO_2 diffuse across membrane. Consists of alveolar epithelium and its basal lamina and capillary endothelium and its basal lamina. Contains two types of cells: (1) *alveolar type I cells*—simple squamous cells that allow for gas exchange; (2) *alveolar type II cells*—rounded or cuboidal cells that produce surfactant.

MECHANICS OF BREATHING pp. 522–525
1. Surface tension and negative pressure in pleural cavity hold visceral pleura and surface of lung against parietal pleura.
2. Air enters lungs when pressure in the lungs is slightly less than atmospheric pressure.
3. Air leaves lungs when pressure in the lungs is slightly greater than atmospheric pressure.
4. Pressure in pleural cavity is slightly less than atmospheric pressure due to collapsing force of elastic fibers in lungs.

INSPIRATION air movement into lungs.
1. Drop in pressure in the lungs.
2. Increased volume of thoracic cavity due to contraction of diaphragm and elevation of ribs.

EXPIRATION air movement out of lungs back into atmosphere.
1. Pressure in the lungs is greater than atmospheric pressure.
2. Decreased volume of thoracic cavity due to relaxation of diaphragm and return of ribs to resting position.

RHYTHMIC BREATHING MOVEMENTS rhythmic pattern of inspiration and expiration under control of respiratory center in medulla oblongata.

LUNG VOLUMES

RESIDUAL VOLUME about 1000–1200 ml of air always in lungs.

TIDAL VOLUME 500 ml of air moving into and out of lungs during quiet respiration.

INSPIRATORY RESERVE extra volume of air (about 2100–3000 ml) that can be inspired in addition to normal tidal air.

EXPIRATORY RESERVE 800–1200 ml that can be forced out after normal passive expiration of tidal air.

VITAL CAPACITY sum of inspiratory reserve, tidal air, and expiratory reserve (3400–4700 ml).

CONDITIONS OF CLINICAL SIGNIFICANCE: THE RESPIRATORY SYSTEM
pp. 525–526

COMMON COLD viral inflammation and sometimes subsequent bacterial infection of mucosa of upper respiratory tract.

HAY FEVER allergic response to allergens in plant pollens and dust.

BRONCHIAL ASTHMA allergic response characterized by wheezing; caused by excessive mucus and spasms of smooth muscles of bronchioles.

BRONCHITIS acute or chronic inflammation of bronchial tree; caused by bacterial infection or irritants.

TUBERCULOSIS infection caused by *Mycobacterium tuberculosis*, which most commonly affects the lungs.

EMPHYSEMA progressive condition that is secondary response to other respiratory problems; overdistension and destruction of alveoli and reduced elastic tissue of lungs produce reduced oxygen in blood and difficult expiration.

PNEUMONIA fibrinous exudate produced in alveoli causes part of lung to become solid and airless.

PLEURISY inflammation of pleural membranes.

PNEUMOTHORAX air allowed to enter pleural cavity, causing lung collapse.

EFFECTS OF AGING ON THE RESPIRATORY SYSTEM
1. Chest wall rigidity.
2. Loss of lung elasticity.
3. Decreased ciliary activity on epithelial lining of respiratory tract.
4. Decreased phagocytic activity, which increases susceptibility to infection.

SELF-QUIZ

1. In the embryo the respiratory epithelium develops from: (a) endoderm; (b) ectoderm; (c) mesoderm.

2. The nasal septum is formed from the: (a) vomer bone; (b) ethmoid bone; (c) septal cartilage; (d) all of these contribute to the nasal septum.

3. The floor of the nasal cavity is formed by the: (a) bony hard palate; (b) membranous soft palate; (c) inferior meatus; (d) both a and b.

4. The paranasal sinuses, like the nasal cavity, are lined with stratified squamous epithelium. True or false?

5. The pharynx communicates with the nasal cavity via the: (a) fauces; (b) glottis; (c) middle meatus; (d) internal nares.

6. Which of the following is located in the oropharynx? (a) pharyngeal tonsils; (b) palatine tonsils; (c) lingual tonsils; (d) both b and c; (e) all of these.

7. The largest of the unpaired cartilages of the larynx is the: (a) cricoid cartilage; (b) arytenoid cartilage; (c) thyroid cartilage; (d) cuneiform cartilage.

8. The vocal folds are attached to: (a) the thyroid cartilage; (b) the arytenoid cartilage; (c) the cricoid cartilage; (d) both a and b.

9. The trachea divides directly into: (a) tertiary bronchi; (b) primary bronchi; (c) respiratory bronchi; (d) terminal bronchi; (e) alveolar ducts.

10. Alveoli are associated with the: (a) trachea; (b) pharynx; (c) lungs; (d) nose.

11. The region where the bronchi and blood vessels enter and leave the lung is called the: (a) costal surface; (b) oblique fissure; (c) hilus; (d) cardiac notch.

12. Air enters the lungs when the intrapulmonary pressure is less than atmospheric pressure. True or false?

13. Contraction of the diaphragm is the dominant means of raising the pressure in the lungs during normal, quiet inspiration. True or false?

14. Match the terms associated with the lung with the appropriate description.

Costal surface	(a) The space that separates the two lungs.
Cardiac notch	
Oblique fissure	(b) A double-walled sac that encloses each lung.
Mediastinum	
Pleura	
Visceral pleura	(c) The portion of the pleura that adheres firmly to the lungs.
Parietal pleura	
Alveoli	(d) Lies against the ribs.
	(e) Clusters of small, thin-walled air sacs.
	(f) The portion of the pleura that lines the walls of the thoracic cavity.
	(g) Divides each lung into superior and inferior lobes.
	(h) A concavity in the left lung.

15. The volume of gas that remains in the lungs after the most forceful exhalation is the: (a) tidal volume; (b) dead space; (c) residual volume; (d) vital capacity.

16. During quiet inspiration: (a) the pressure in the thoracic cavity exceeds that in the lungs; (b) the lungs actively expand; (c) the diaphragm contracts.

17. The lungs are located in the: (a) pleural cavities; (b) abdominal cavity; (c) mediastinum.

18. Which has the greatest volume? (a) tidal air; (b) residual air; (c) vital capacity; (d) inspiratory reserve.

19. When air enters the pleural cavity, the condition is known as: (a) pleurisy; (b) pneumothorax; (c) pulmonary edema.

20. Emphysema generally develops as a secondary response to other respiratory problems. True or false?

LEARNING OBJECTIVES

After completing this chapter, you should be able to:

- List the principal components of the digestive system, and cite the chief functions of each.

- Describe the microscopic and gross anatomy of the components of the digestive system.

- List the types of teeth, according to shape, and cite the specific function of each type.

- List the basic layers of the digestive tube, and cite the function of each.

- Describe the modifications of the basic layers found in various regions of the gastrointestinal tract.

- Distinguish between peristalsis and segmentation as major digestive activities.

- Describe the structure of the pancreas, liver, and gallbladder, and cite the functions of each.

- Name the structural features of the small intestine that enable it to digest and then absorb food materials.

- Describe the mechanical processes involved in digestion.

CHAPTER CONTENTS

The Digestive System

Every cell in the body requires a constant source of energy in order to perform its particular functions—whether these functions be contraction, secretion, synthesis, or any other. Ingested food provides the basic materials from which this energy is produced and new molecules are synthesized. Most food, however, cannot enter the bloodstream and be used by the cells of the body until it is broken down into simpler molecules. The digestive system alters the ingested food by mechanical and chemical processes so that it can ultimately cross the wall of the gastrointestinal tract and enter the blood vascular and lymphatic systems. The vascular system then carries these food molecules through the hepatic portal vein to the liver before distributing them to cells throughout the body. After entering the cells, the digested food molecules may be reassembled into proteins, carbohydrates, and lipids, or they may be used in the production of energy to support body activity.

The digestive system consists of a tube—called the **gastrointestinal tract** or **alimentary canal**—that extends from the mouth to the anus (Figure 20.1). F20.1

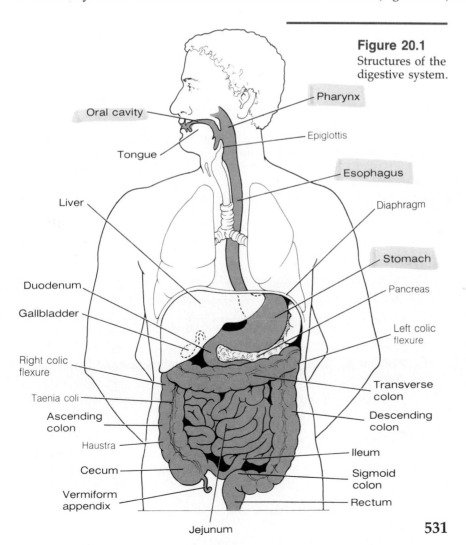

Figure 20.1
Structures of the digestive system.

Oral cavity

Tongue

Liver

Duodenum

Gallbladder

Right colic flexure

Taenia coli

Ascending colon

Haustra

Cecum

Vermiform appendix

Jejunum

Pharynx

Epiglottis

Esophagus

Diaphragm

Stomach

Pancreas

Left colic flexure

Transverse colon

Descending colon

Ileum

Sigmoid colon

Rectum

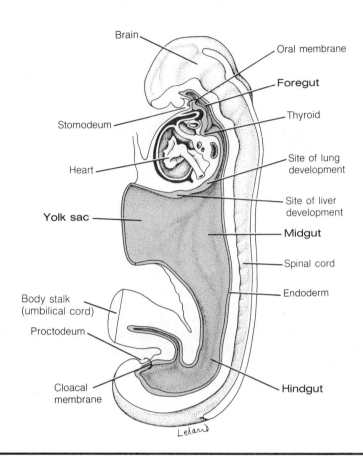

Figure 20.2
Sagittal section of an
embryo (approximately
22 days old) showing the
future foregut and
hindgut forming. Note
that the midgut remains
open to the yolk sac.

As long as food remains in the gastrointestinal tract, it is actually still outside
the body. To "enter" the body it must cross the epithelium that lines the wall
of the digestive tract. Emptying into the digestive tube are the secretions of
the salivary glands, gastric glands, intestinal glands, liver, and pancreas, all
of which assist in the digestion of food. Although the gastrointestinal tract is
a continuous tube, it is divisible into specialized regions each of which performs specific functions in the digestion of food. These regions include the
mouth, pharynx, esophagus, stomach, small intestine, and *large intestine.*

The activities of the digestive system can be divided into six basic processes:

1. Ingestion of food into the mouth

2. Movement of food along the digestive tract

3. Mechanical preparation of food for digestion

4. Chemical digestion of food

5. Absorption of digested food into the circulatory and lymphatic
   systems

6. Elimination of indigestible substances and waste products from
   the body by defecation

In this book we are primarily interested in the anatomy of the digestive
system and the mechanical processes involved in preparing food for digestion. However, the chemical processes of digestion are discussed briefly with
each digestive organ, and summarized toward the end of the chapter.

EMBRYONIC DEVELOPMENT OF THE DIGESTIVE SYSTEM

Early in development, the embryo is a hollow cylinder covered on the outside with ectoderm. Its internal cavity, which is lined with endoderm, is the developing digestive tract (Figure 20.2). The portion of the tract that extends anteri- **F20.2** orly into the head region is the **foregut;** the portion that extends posteriorly is the **hindgut;** and the central region of the tract is the **midgut.** Until the fifth week of development, the midgut opens into a pouch called the yolk sac. After the fifth week, the attached portion of the yolk sac constricts, sealing the midgut.

In the early stages of development, the anterior region of the foregut expands to form the pouches of the pharynx. With continued growth, the foregut contacts the surface ectoderm at the point where the ectoderm has formed a depression called the **stomodeum** (the future mouth). The **oral membrane** that separates the foregut from the stomodeum breaks through during the fourth week of development, and the foregut is then continuous with the outside of the embryo through the mouth. In a similar manner, the hindgut contacts the surface ectoderm at a depression called the **proctodeum** (the future anus). With the rupture of the **cloacal membrane** that separates the hindgut from the proctodeum, the digestive tract forms a continuous tube from the mouth to the anus.

With further development, hollow buds from the endoderm form at various places along the foregut. These buds will grow into the mesoderm that surrounds the digestive tract and give rise to the thyroid gland, parathyroid glands, salivary glands, liver, gallbladder, and pancreas. The thyroid and parathyroid glands later lose their connections with the digestive tract and function as endocrine glands. The salivary glands, liver, gallbladder, and pancreas retain their connections with the digestive tube by means of ducts. Consequently, they serve as accessory glands to the digestive system.

ANATOMY OF THE DIGESTIVE SYSTEM

It is important to keep in mind that the entire digestive tract is lined with **mucous membrane.** This membrane protects the underlying tissues, and at the same time, it allows for the absorption of digested food in the intestine.

In order to function in absorption, the membrane must be thin and moist. The secretion of *mucus* by cells of the mucous membrane keeps the membrane moist; because the mucus is viscous, it also serves as a protective mechanism. Thus, the thin membrane that lines the absorptive regions of the digestive tract provides adequate protection as long as it is coated with mucus.

Mouth

The **mouth** is the first part of the digestive tract. It extends from the lips to the oropharynx (Figure 20.3). The outer surfaces of the lips are covered with skin **F20.3** that has a relatively transparent surface layer of cells, which allows the underlying blood capillaries to show through. For this reason the lips appear red. Since the surface layer of the lips is not keratinized (horny), evaporation occurs from the lips. Consequently, the lips must be moistened frequently to prevent them from becoming dry and cracked. The inside surfaces of the lips and the rest of the mouth are lined with a mucous membrane that has a stratified squamous surface layer of nonkeratinized cells. No food is absorbed in the mouth, so the cells lining it need not be capable of absorption; and the stratified surface layer affords an extra degree of protection from abrasive food particles.

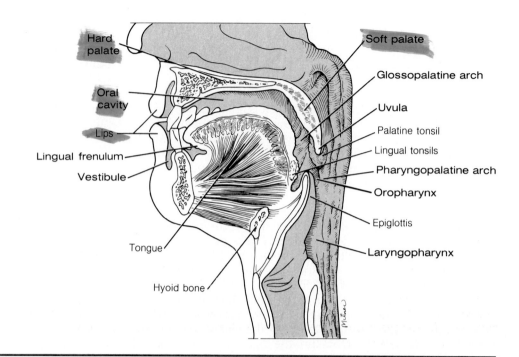

Figure 20.3
Sagittal section of the oral and pharyngeal cavities.

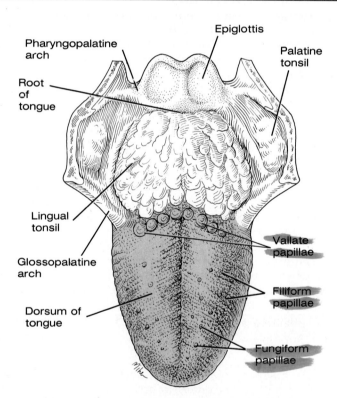

Figure 20.4
Dorsal surface of the tongue.

For descriptive purposes, the mouth can be divided into the *oral cavity*, which is the large space internal to the teeth that houses the tongue, and the *oral vestibule*, which is the small space separating the lips and the cheeks from the teeth and gums. The lips and cheeks aid in moving food between the upper and lower teeth during *mastication* (chewing). They also aid in speech.

The roof of the mouth is formed anteriorly by the **hard palate** and posteriorly by the **soft palate**. The hard palate is formed by the maxillary and palatine bones. The soft palate, which extends posteriorly from the hard palate, separates the oral cavity from the nasopharynx. It is composed primarily of muscles. The soft palate is pushed upward during swallowing, blocking the entrance to the nasal cavity from the pharynx and thus serving to prevent food and drink from entering the nasal cavity. The **uvula** is a small muscular flap that hangs down from the posterior margin of the soft palate. It serves as a friction pad to prevent the soft palate from being pushed into the nasal cavity during swallowing. The soft palate is attached laterally to the tongue by the **glossopalatine arches;** it is connected to the wall of the oropharynx by the **pharyngopalatine arches.** The palatine tonsils, which are composed mainly of lymphoid tissue, are located in the fossae between the two arches, a region called the **fauces.**

Tongue

The tongue forms the floor of the mouth (Figure 20.3). It is composed of **F20.3** interwoven bundles of skeletal muscles covered with mucous membrane. The extrinsic muscles of the tongue originate from the hyoid bone, the mandible, and the styloid processes of the temporal bones. These muscles protrude the tongue, retract it, and move it sideways. The intrinsic muscles originate and insert in the tongue. Their fibers run in various directions and modify the shape of the tongue in many different ways. Because of these two sets of muscles, the tongue is quite versatile and is used in manipulating food, swallowing, and speaking.

The mucous membrane that covers the dorsum of the tongue is modified by the presence of numerous small projections called **papillae** (Figure 20.4). **F20.4** The papillae vary in shape; each type tends to be more common in certain regions of the tongue surface. The **filiform papillae,** which are small cones, are distributed in V-shaped rows over the entire dorsal surface of the tongue. Interspersed among the filiform papillae are flattened, mushroom-shaped **fungiform papillae** (Figure 20.5a). About a dozen large **vallate papillae** form **F20.5a,** an inverted V toward the back of the tongue (Figure 20.5b). **Taste buds** are **F20.5b** found on the fungiform and vallate papillae. The filiform papillae make the surface of the tongue rough, allowing it to move food around during chewing. The posterior dorsal surface of the tongue contains an aggregate of small lymph nodules called the **lingual tonsil.**

The tongue is connected ventrally to the floor of the mouth by a fold of mucous membrane called the **lingual frenulum.** If the frenulum is so short that it hinders the movement of the tongue, it interferes with speech. Such a condition, which can be corrected surgically, is called "tongue-tied."

Teeth

The teeth extend into the mouth from **alveoli** (sockets) located along the alveolar processes of the mandible and maxillary bones. **Gums** (*gingivae*), composed of stratified squamous epithelium and dense fibrous connective tissue, cover the alveolar processes. The sockets are lined with a fibrous membrane called the **periodontal membrane** (Figure 20.6). Destruction of the peri- **F20.6** odontal membrane by bacteria can lead to infection within a socket and loss of its tooth.

The portion of each tooth that extends from the gum into the mouth is called the **crown.** One or more **roots** anchor the tooth to the alveolus. Between the crown and the root is a slightly constricted **neck.**

Each tooth is composed mostly of a hard, calcified substance called **dentin.** The dentin of the crown is covered by **enamel,** which is even harder than dentin. A bonelike substance called **cementum** covers the dentin of the roots and anchors the tooth to the periodontal membrane that lines the alveoli. The central region of the tooth contains the **pulp cavity,** in which are found blood vessels, nerves, and a connective tissue called **pulp.** The pulp cavity extends down into each root as a **root canal.** At the end of each root canal is an **apical foramen,** through which blood vessels and nerves enter the pulp cavity.

Certain bacteria found in the mouth can produce enzymes and acids that are capable of breaking down tooth enamel. The points of enamel destruction

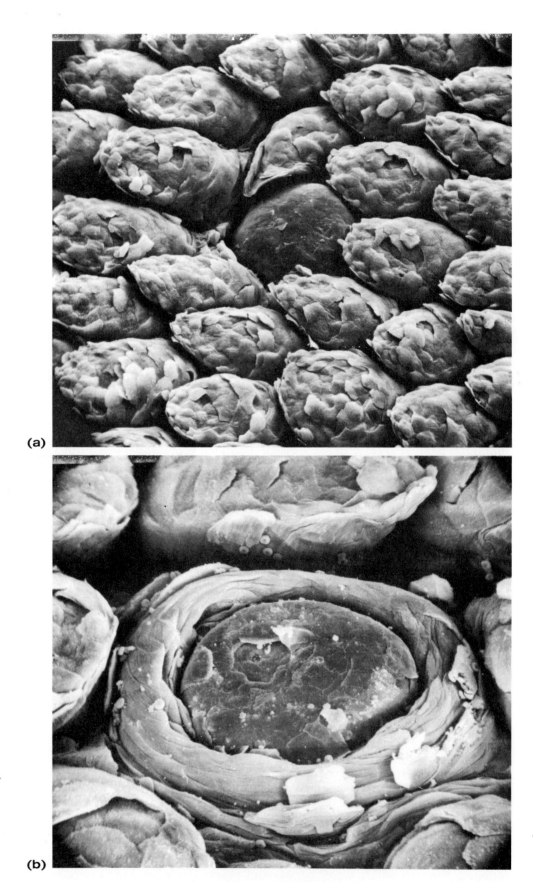

(a)

(b)

Figure 20.5
Papillae of the tongue.
(a) Fungiform papillae
(center) surrounded by
filiform papillae (×195).
(b) Vallate papilla (×606).
(From *Tissues and Organs:
A Text-Atlas of Scanning
Electron Microscopy* by
Richard G. Kessel and
Randy H. Kardon. W.H.
Freeman and Company.
Copyright © 1979.)

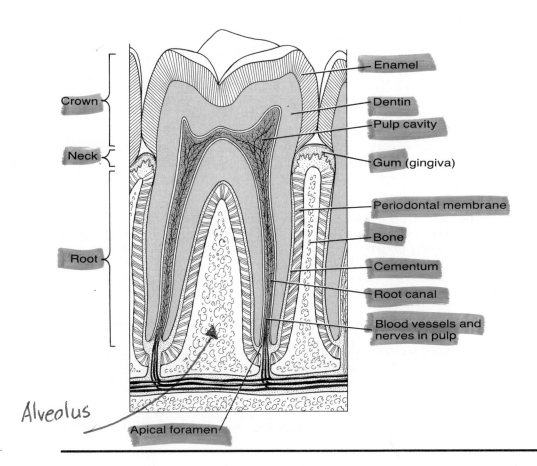

Crown

Neck

Root

Alveolus

Enamel

Dentin

Pulp cavity

Gum (gingiva)

Periodontal membrane

Bone

Cementum

Root canal

Blood vessels and nerves in pulp

Apical foramen

Figure 20.6
Longitudinal section of a molar tooth within an alveolus.

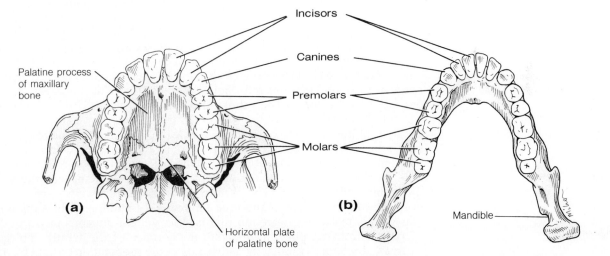

Incisors

Canines

Premolars

Molars

Palatine process of maxillary bone

Horizontal plate of palatine bone

Mandible

(a)   **(b)**

Figure 20.7
Permanent teeth of
(a) the upper jaw and
(b) the lower jaw.

are called *dental caries,* or *cavities.* Once the enamel has been penetrated, the dentin may also be destroyed by the enzymes and acids. If the decay reaches the pulp, it can irritate nerves in the pulp cavity, causing a toothache. The application of fluoride, either directly to the teeth or in drinking water, seems to make the enamel more resistant to bacterial enzymes and acids.

There are four types of teeth, named according to function or shape (Figure 20.7). Each type of tooth performs a specific function in preparing food for digestion. The front chisel-shaped teeth, called **incisors,** are especially adapted for cutting. On each side of the incisors are conical **canines** *(cuspids),* which serve to tear food. The **premolars** *(bicuspids)* and **molars** have broad crowns with rounded cusps, which aid in crushing and grinding food.

F20.7

Table 20.1 Numbers of Specific Types of Teeth in Each Jaw

	Deciduous	Permanent
Incisors	4	4
Canines (cuspids)	2	2
Premolars (bicuspids)	0	4
Molars	4	6
TOTAL IN BOTH JAWS	20	32

Table 20.1

Normally, two sets of teeth develop during a person's lifetime. There are 20 **deciduous,** or **milk teeth,** that erupt through the gums at regular intervals, beginning with the incisors at about six months of age. All the deciduous teeth are usually present by age 2½. There are 32 teeth in the **permanent** set. The permanent teeth begin to replace the deciduous teeth at about six years of age and continue to do so until about age 17, when all the temporary teeth generally have been replaced. The **third molars** (*wisdom teeth*) are the last to erupt, generally between the ages of 17 and 25. It is not unusual for the wisdom teeth to fail to erupt or to become wedged (impacted) in the jaw so that they are unable to erupt. Table 20.1 summarizes the number of each type of tooth that are present in each jaw, in both the deciduous and permanent sets.

Salivary Glands

F20.8

About 1000 to 2000 ml of saliva is secreted daily into the mouth. Some of this is produced by many small **buccal glands** located throughout the mucous membrane of the oral cavity. These glands secrete continuously and keep the mucous membranes moist. Most of the saliva, however, is secreted by three pairs of **salivary glands** that are activated primarily by stimuli associated with food (Figure 20.8). These large salivary glands are of the compound tubuloalveolar type, being composed of blindly ending tubules and alveoli. The largest of the paired salivary glands, the **parotid glands,** are located below and anterior to the ear, on the posterior surface of the masseter muscle. Each parotid gland has a duct that crosses the masseter muscle, pierces the buccinator muscle, and empties into the vestibule of the mouth opposite the second upper molar tooth. The **submandibular glands** lie medial to the angle of the mandible. The ducts of the submandibular glands travel anteriorly under the floor of the mouth to open at the base of the frenulum of the tongue. The **sublingual glands** are located on the floor of the mouth within a fold of mucous membrane. Each sublingual gland has several small ducts that open onto the floor of the mouth and, often, a larger duct that joins with the duct of the submandibular gland. The parotid glands produce a watery serous secretion, the sublingual glands produce a mucous secretion, and the submandibular glands produce a secretion that is a serous and mucous mixture.

Apart from keeping the mucous membranes of the mouth moist and cleansing the mouth and teeth, saliva aids in the preparation of food by moistening it—thus enhancing its formation into a mass (*bolus*) and making it easier to chew and swallow. Saliva also dissolves some of the food molecules, thus allowing the food to be tasted. Moreover, the secretions of the paired salivary glands contain the enzyme *salivary amylase,* which initiates carbohydrate digestion.

Pharynx

Food that is swallowed passes from the mouth into the oropharynx and then on into the laryngopharynx. These two portions of the **pharynx,** which serve as common passageways for the respiratory and digestive systems, are described in more detail in Chapter 19. A major function of the pharynx is to serve as the site of muscular contractions involved in swallowing. Upon leaving the laryngopharynx, food enters the esophagus.

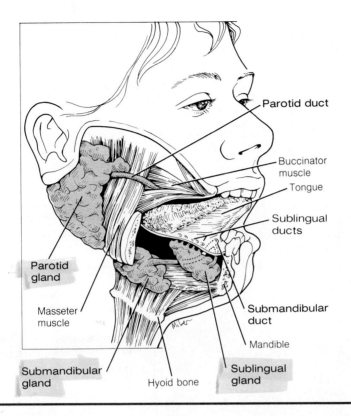

Figure 20.8
The major salivary glands. The right side of the mandible has been removed to expose the submandibular and sublingual glands.

Labels in figure: Parotid duct · Buccinator muscle · Tongue · Sublingual ducts · Parotid gland · Submandibular duct · Masseter muscle · Mandible · Submandibular gland · Hyoid bone · Sublingual gland

Gastrointestinal Tract Wall

Before beginning a detailed study of the gastrointestinal tract wall it is advisable for you to review the membrane relationships described in Chapter 1. Recall that the organs in the abdominopelvic cavity are covered by the *visceral peritoneum,* which is a continuation of the *parietal peritoneum* that lines the wall of the cavity. Also recall that most organs in the abdominopelvic cavity are suspended in the cavity by a membrane referred to as a *mesentery.* The mesenteries are double-layered extensions of the parietal peritoneum. In the following discussions we will at times be using specific names for the mesenteries that support particular organs or structures. Some of these mesenteries are illustrated in Figure 20.11.

F20.11

Beginning in the esophagus and continuing all the way to the anus, the wall of the digestive tube has the same basic arrangement of four layers **(tunics)**, with complex networks of nerves interconnecting the tunics (Figure 20.9). Although the structure of the wall is modified in various regions of the digestive tract, the four basic layers present are, from the lumen (cavity) of the gut outward: the *tunica mucosa,* the *tunica submucosa,* the *tunica muscularis,* and the *tunica serosa* or *adventitia.*

F20.9

Tunica Mucosa

The **tunica mucosa** is the mucous membrane that lines the digestive tract. It consists of three layers:

1. An **epithelial layer** borders on the lumen. In the esophagus, it is stratified squamous epithelium. In the remainder of the tract, it is simple columnar epithelium.

2. The **lamina propria** is composed of loose connective tissue to which the epithelial cells are attached. Blood vessels, lymph nodules, and small glands are generally located in this layer.

3. Outside the lamina propria is a thin layer of smooth muscle fibers called the **muscularis mucosae.**

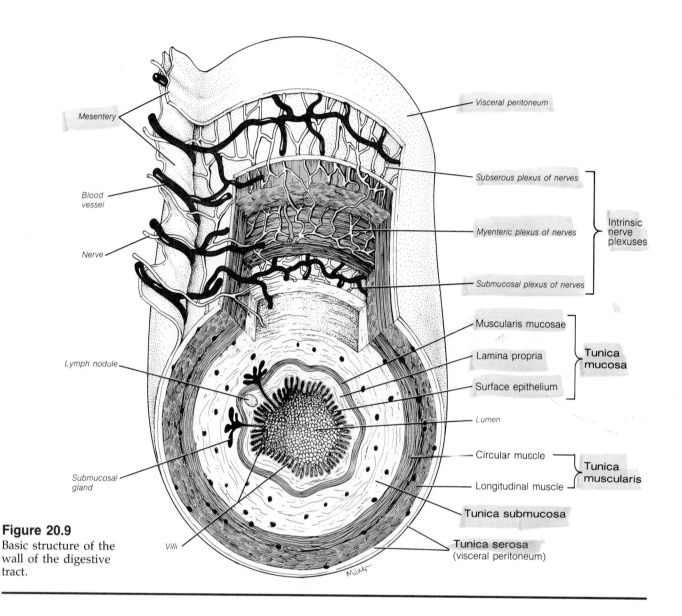

Mesentery

Blood vessel

Nerve

Lymph nodule

Submucosal gland

Villi

Visceral peritoneum

Subserous plexus of nerves

Myenteric plexus of nerves

Submucosal plexus of nerves

Intrinsic nerve plexuses

Muscularis mucosae

Lamina propria

Surface epithelium

Tunica mucosa

Lumen

Circular muscle

Longitudinal muscle

Tunica muscularis

Tunica submucosa

Tunica serosa (visceral peritoneum)

Figure 20.9
Basic structure of the wall of the digestive tract.

Tunica Submucosa

The **tunica submucosa** is a thick layer of either dense or loose connective tissue located deep to the mucosa. It contains blood vessels, lymphatic vessels, nerves, and in some regions, glands.

Tunica Muscularis

In most regions of the digestive tract, the **tunica muscularis** is a double layer of muscle tissue. The muscle fibers of the inner layer are arranged circularly around the tube, whereas the outer fibers are oriented longitudinally along its long axis. At several points along the tract, the fibers of the circular layer are thickened, forming sphincters that control the movement of food from one region of the digestive tract to another. In the upper part of the esophagus, the tunica muscularis is composed of skeletal (voluntary) muscle fibers. Throughout the rest of the tract, it is smooth (involuntary) muscle. Rhythmic contractions of these muscles produce peristalsis, which moves food toward the anus.

Tunica Serosa or Adventitia

The outermost tunic of the digestive tube is composed primarily of a layer of connective tissue. In the esophagus, this connective-tissue layer merges into the connective tissue of surrounding structures and is called the **adventitia.**

Along the rest of the digestive tract, the connective tissue is covered with a serous membrane consisting of a single layer of squamous epithelial cells, forming the visceral peritoneum. In this case, the outer tunic is called the **serosa.**

Intrinsic Nerve Plexuses

Between the four tunics that form the wall of the digestive tract, there are complex interconnections of neurons organized into intrinsic nerve plexuses. One plexus, called the **subserous plexus,** is distributed throughout the tunica serosa. Another plexus, the **myenteric plexus,** is located between the two layers of smooth muscle forming the tunica muscularis. A third nerve plexus, the **submucosal plexus,** is in the tunica submucosa. The intrinsic nerve plexuses coordinate much of the activity of the digestive tract.

Esophagus

The **esophagus** is a muscular tube that connects the pharynx with the stomach. It is located behind the trachea. The esophagus travels through the mediastinum of the thorax and passes through the diaphragm by means of an opening called the *esophageal hiatus.* Food is moved through the esophagus by waves of contractions *(peristalsis)* of the muscles in its wall. In the upper portion, near the pharynx, the wall of the esophagus contains skeletal muscles. In the lower portions of the esophagus the muscles in the wall are smooth.

Stomach

Shortly after passing through the diaphragm, the esophagus empties into the **stomach** (Figure 20.10). The stomach, which functions to prepare ingested **F20.10** food by mechanical and chemical means for digestion, lies to the left of the midplane just beneath the diaphragm. The opening from the esophagus into the stomach is called the **cardiac orifice.** The exit from the stomach, where it joins with the small intestine, is guarded by the **pyloric sphincter.**

The right border of the stomach, which is concave, is called the **lesser curvature.** The convex left border is called the **greater curvature.** The lesser curvature is attached to the undersurface of the liver by a mesentery consisting of a double layer of visceral peritoneum. This mesentery is called the **lesser omentum** (Figure 20.11). The two layers of the lesser omentum sepa- **F20.11** rate at the lesser curvature and form the serosa on the surfaces of the stomach. The membrane layers rejoin at the greater curvature to form the **greater omentum.** The greater omentum is a folded mesentery forming an apron that covers the anterior surface of the transverse colon and the coils of the small intestine. It generally has deposits of fat attached to it.

The main portion of the stomach is called the **body.** The **fundus** is the portion that bulges above the entrance of the esophagus. The body of the stomach tapers inferiorly to form a region called the **pylorus,** which joins the duodenum of the small intestine.

The wall of the stomach is composed of the four basic layers (tunics) that are typical of the digestive tract (Figure 20.12). In the stomach, however, the **F20.12** epithelial cells of the mucosa are simple columnar rather than stratified squamous as in the esophagus. The epithelium of the mucosa remains simple columnar throughout the intestines. When the stomach is empty, the mucosa and submucosa form longitudinal folds called **rugae.** These folds flatten as the stomach fills.

A modification of the mucosa in the stomach is the presence of many **gastric glands** that occupy the lamina propria. These glands, which secrete gastric juice, empty onto the surface of the mucosa through small invaginations called **gastric pits** (Figure 20.13). The gastric glands located in the fun- **F20.13** dus and body of the stomach (**gastric glands proper,** or **fundic glands**) contain several types of cells:

1. **Mucous neck cells** are located near the gastric pits.

2. **Zymogenic (chief) cells** secrete *pepsinogen,* a precursor of the digestive enzyme *pepsin.*

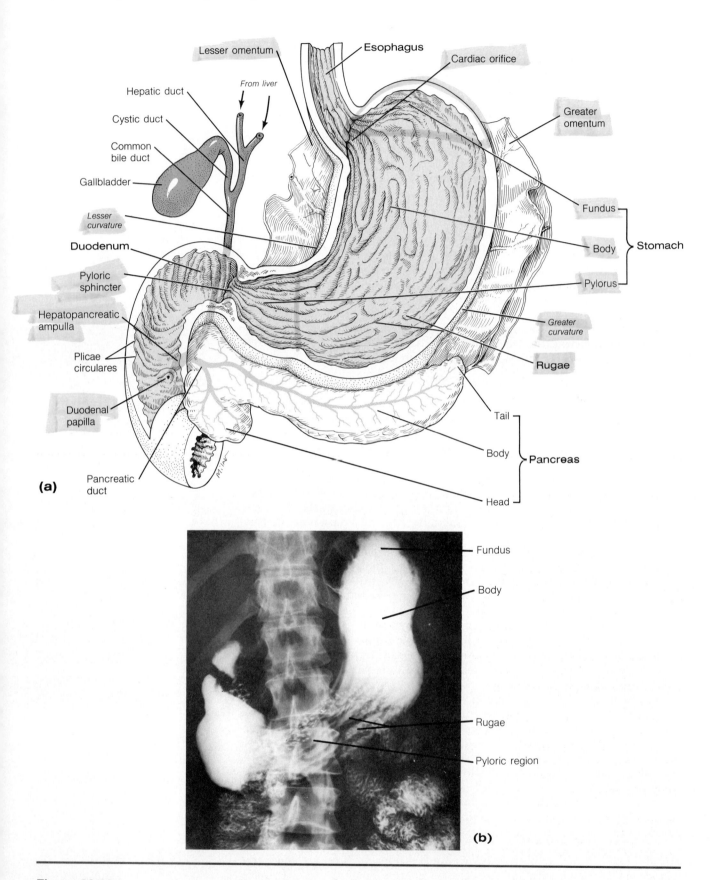

Lesser omentum
Esophagus
Cardiac orifice
Hepatic duct
From liver
Cystic duct
Greater omentum
Common bile duct
Gallbladder
Fundus
Lesser curvature
Body
Stomach
Duodenum
Pyloric sphincter
Pylorus
Hepatopancreatic ampulla
Greater curvature
Plicae circulares
Rugae
Duodenal papilla
Tail
Body
Pancreas
Pancreatic duct
Head
(a)

Fundus
Body
Rugae
Pyloric region
(b)

Figure 20.10

(a) Frontal section of the stomach and duodenum showing the pancreatic and bile ducts.
(b) X ray of the upper gastrointestinal tract following a barium (contrast medium) meal. The lower stomach is in contraction, separating the rugae into visible folds.

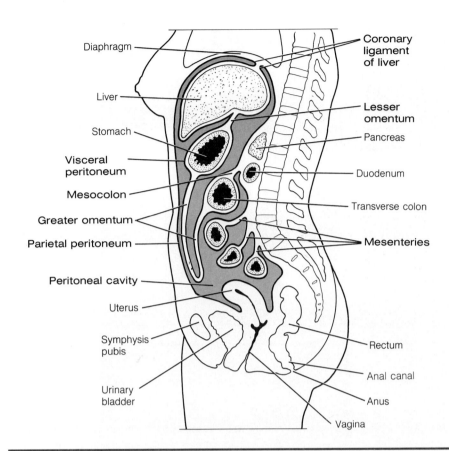

Figure 20.11
Sagittal section of the female abdominopelvic cavity showing the peritoneum, the omenta, and the mesenteries.

3. **Parietal cells** produce *hydrochloric acid* and *intrinsic factor*.

4. **Enteroendocrine (argentaffin) cells** produce the hormone *gastrin*.

The glands of the cardiac region **(cardiac glands)** and the pyloric region **(pyloric glands)** secrete mainly mucus. The many mucous cells of the stomach produce a layer of mucus that adheres to the stomach lining and protects the gastric mucosa. Hydrochloric acid kills many of the bacteria that enter the digestive tract along with food. Moreover, it aids in the digestion of protein and converts pepsinogen, which is inactive, to pepsin. Pepsin initiates the digestion of proteins by breaking certain chemical bonds that link amino acids together, thereby forming smaller chains of amino acids called *polypeptides*. Gastrin stimulates the secretion of hydrochloric acid by the parietal cells. Intrinsic factor must be present in order for vitamin B_{12} to be absorbed in the small intestine.

Another modification of the tunics of the stomach is found in the tunica muscularis. Apart from the circular and longitudinal coats, the stomach wall contains an oblique layer of muscle between the circular layer and the submucosa. The additional layer of muscle in the wall makes extra strong contractions possible in the stomach and aids in its major functions—mashing the food and mixing it with digestive juices.

Small Intestine

The stomach empties into the **small intestine**—the longest (about 6 m) and most convoluted portion of the digestive tract. The small intestine joins the large intestine at the ileocecal valve. The small intestine is lined by simple columnar epithelium that contains cells specialized to absorb nutrients, which is a major function of the small intestine. On the basis of differences in microscopic structure, the small intestine can be divided into three regions that are not otherwise distinct from each other:

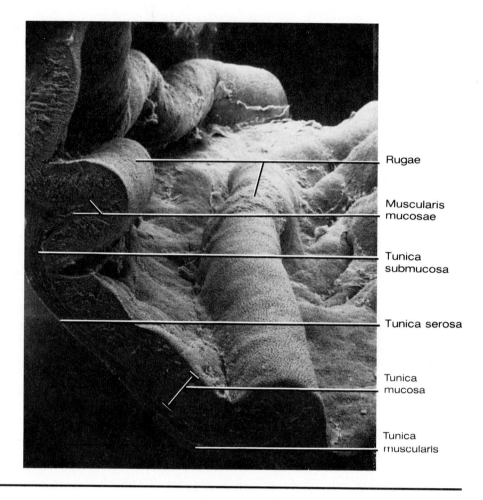

Rugae

Muscularis mucosae

Tunica submucosa

Tunica serosa

Tunica mucosa

Tunica muscularis

Figure 20.12
The tunics of the stomach wall (×28). (From *Tissues and Organs: A Text-Atlas of Scanning Electron Microscopy* by Richard G. Kessel and Randy H. Kardon. W.H. Freeman and Company. Copyright © 1979.)

F20.10a

1. The **duodenum,** which is the first 25 cm or so of the small intestine, curves around the head of the pancreas (Figure 20.10*a*). The **common bile duct** from the liver and the **pancreatic duct** from the pancreas join together to form the **hepatopancreatic ampulla** (*ampulla of Vater*), which empties into the duodenum at the **duodenal papilla.** This opening is surrounded by a sphincter muscle called the **hepatopancreatic sphincter** (*sphincter of Oddi*). The common bile duct carries bile, and digestive enzymes are transported through the pancreatic duct. The duodenum is retroperitoneal—that is, situated behind the peritoneum—and is attached tightly to the posterior body wall (Figure 20.11).

F20.11

2. The next 2.5 m or so of the small intestine is the **jejunum.** This portion is suspended in the abdominal cavity by a mesentery.

3. The **ileum** is the remaining 3.5 m or so of the small intestine. The entrance of the ileum into the cecum of the large intestine is guarded by the **ileocecal valve.** Rather than being a true sphincter, this valve is composed of two folds of tissue. The ileum, like the jejunum, is suspended from the posterior body wall by a mesentery. The mesentery allows the small intestines to move during the contractions of peristalsis, and it also provides support for blood vessels, lymphatic vessels, and nerves that supply the intestines.

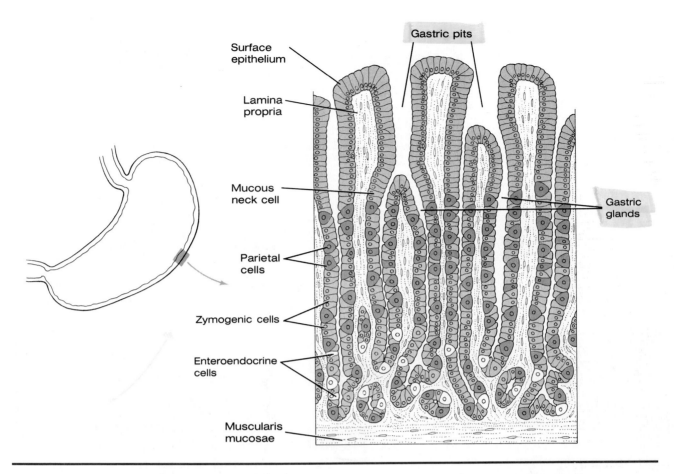

Figure 20.13
Structure of the gastric glands.

The modifications in the wall of the small intestine are within the tunica mucosa and the submucosa. These layers form permanent, circular, shelflike folds **(plicae circulares)** that extend into the lumen of the small intestine and increase the surface area of the mucosa (Figure 20.10). The surface area is **F20.10** increased further by **villi,** which are tiny fingerlike projections of the mucosa into the lumen (Figure 20.14). The villi are so numerous and so close together **F20.14** that the mucosa of the small intestine has a velvety appearance. Each villus has a surface coat consisting of a single layer of epithelial cells, and contains a lymphatic capillary called a **lacteal.** The lacteal is surrounded by a network of blood capillaries. Two main types of epithelial cells cover the surface of the villi: *goblet cells,* which secrete mucus, and more numerous *absorptive cells,* which participate in the absorption and digestion of food materials. The free surfaces of the absorptive cells are folded into very small projections called **microvilli.** Microvilli greatly increase the total surface area of the intestinal mucosa. The increased surface area provided by the plica circulares, villi, and microvilli aids in the absorption of digested foods, which must pass through the mucosa before entering blood capillaries or lacteals.

The mucosa contains many tubular **intestinal glands** (*crypts of Lieberkuhn*) located between the bases of the villi. In the submucosa of the duodenum are mucous glands called **duodenal (Brunner's) glands,** or **submucosal glands.** These glands generally empty into the intestinal glands.

The intestinal glands secrete *intestinal juice* that contains almost no digestive enzymes. Rather, the glands serve as the source of enzyme-rich epithelial cells that migrate up into the villi, where they replace the surface cells that are constantly being lost into the lumen of the intestine. The enzymes are thought to remain attached to the plasma membranes of the microvilli of the absorptive cells, therefore they digest foods within the cells rather than within the lumen of the small intestine. The intestinal glands produce

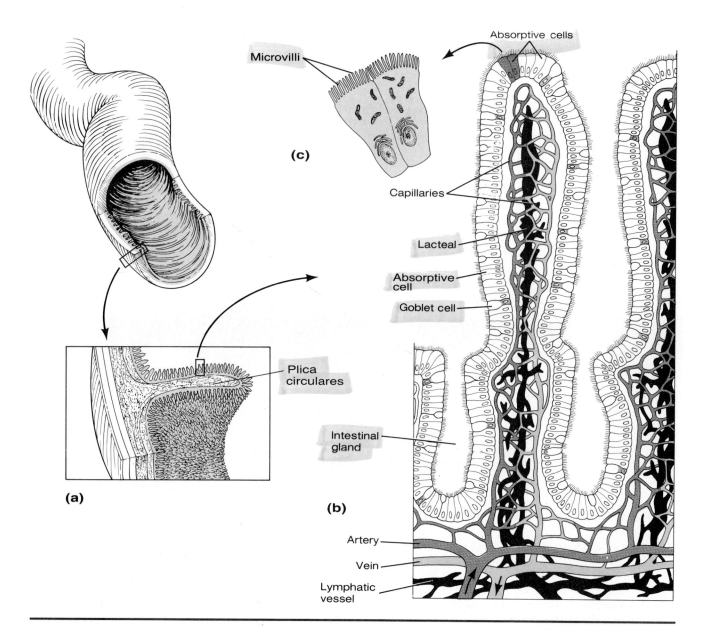

(c) Absorptive cells

Microvilli

Capillaries

Lacteal

Absorptive cell

Goblet cell

Plica circulares

Intestinal gland

Artery

Vein

Lymphatic vessel

(a)

(b)

Figure 20.14

Microscopic structure of the small intestine.
(a) Enlargement showing villi extending from a plica circulares.
(b) Higher magnification of a section through two villi, showing blood vessels and lacteals.
(c) Enlargement of two absorptive cells, showing microvilli on their free surfaces.

Table 20.2

cells containing enzymes that digest carbohydrates, proteins, and lipids (Table 20.2).

Most digestive enzymes that act in the small intestine originate in the pancreas, being transported to the small intestine through the pancreatic duct. The pancreatic enzymes also act on carbohydrates, proteins, and lipids. Bile from the liver is carried to the small intestine through the common bile duct. Bile aids in the digestion of lipids. Because of the presence of bile as well as intestinal and pancreatic enzymes that act on all types of food, most digestion occurs in the small intestine. And because of the large surface area created by the villi and microvilli, most absorption of digested food also occurs there.

Large Intestine

The **large intestine,** which is about 1.5 m long, extends from the ileocecal valve to the anus (Figure 20.15, Figure 20.16). It is so named because its diameter in most regions is greater than that of the small intestine. Like the small intestine, the large intestine is lined with simple columnar epithelium having absorptive cells and goblet (mucous) cells, the latter being most abundant. Very few enzymes, if any, are produced by the epithelial cells. The large

F20.15, F20.16

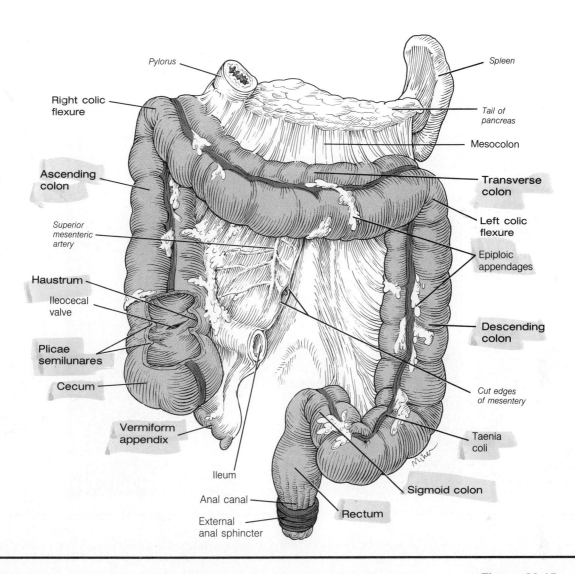

Pylorus

Right colic
flexure

Ascending
colon

*Superior
mesenteric
artery*

Haustrum

Ileocecal
valve

Plicae
semilunares

Cecum

Vermiform
appendix

Ileum

Anal canal

External
anal sphincter

Spleen

*Tail of
pancreas*

Mesocolon

Transverse
colon

Left colic
flexure

Epiploic
appendages

Descending
colon

*Cut edges
of mesentery*

Taenia
coli

Sigmoid colon

Rectum

Figure 20.15
The large intestine. The
cecum has been opened
to expose the ileocecal
valve.

intestine begins as a blind pouch called the **cecum,** which receives the ileum of the small intestine. The **vermiform appendix** is a narrow, blind tube that extends downward from the cecum. The wall of the appendix contains numerous lymphatic nodules. The large intestine extends upward from the cecum as the **ascending colon.** The ascending colon is not supported by a mesentery; instead, it lies tightly against the posterior wall of the abdomen. Just beneath the liver, the ascending colon bends sharply to the left (**right colic flexure**) and crosses the abdominal cavity as the **transverse colon.** This portion of the colon is suspended by a mesentery called the **mesocolon.** In the vicinity of the spleen, the transverse colon bends downward (**left colic flexure**) and forms the **descending colon.** The descending colon, like the ascending colon, is retroperitoneal. Where the descending colon reaches the left pelvic brim, it curves to the midplane via an S-shaped **sigmoid colon.**

Several variations occur in the four tunics of the colon. The tunica mucosa, for example, contains intestinal glands and a large number of mucous cells, but lacks villi. Moreover, while the longitudinal layer of the muscles of the tunica muscularis is continuous as a thin layer over the entire surface of the colon, most of the muscles of this layer are in the form of three bands—called **taeniae coli**—that run the length of the colon. The taeniae are shorter than the colon and therefore cause the colon to form small pouches called **haustra** (singular: *haustrum*). The mucosa between the haustra is thrown into **plica semilunares** (semilunar folds) that extend into the lumen of the gut. Unlike the plicae circulares of the small intestine, the folds of the colon extend

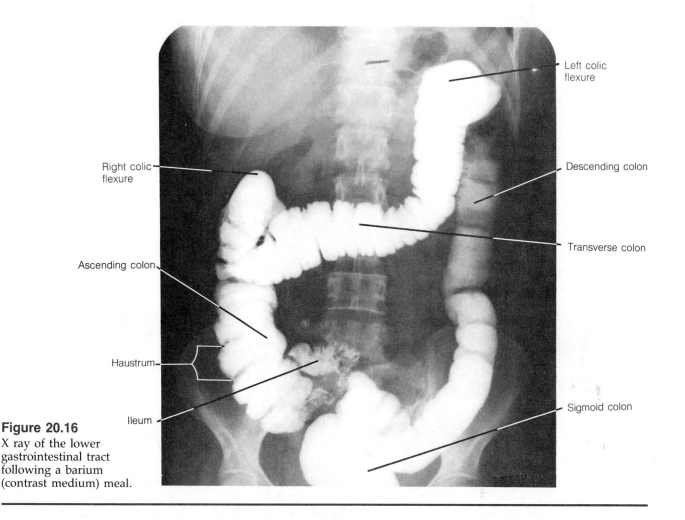

Figure 20.16
X ray of the lower gastrointestinal tract following a barium (contrast medium) meal.

Labels: Left colic flexure, Descending colon, Transverse colon, Sigmoid colon, Right colic flexure, Ascending colon, Haustrum, Ileum

only partway around the gut tube. Another unique characteristic of the colon is the presence on its external surface of small tabs called **epiploic append-ages.** These tabs are composed of folds of peritoneum filled with fat. Although no significant digestion occurs in the large intestine, it does serve as a major site of absorption of sodium, chloride, and water.

Rectum and Anal Canal

F20.15 Beyond the sigmoid colon, the large intestine passes downward in front of the sacrum. This portion is called the **rectum** (Figure 20.15). The rectum has the same structure as the colon except that taeniae coli are not present, the longitudinal layer of muscles being again spread evenly over the entire surface.

F20.17 The terminal 3 to 4 cm of the large intestine is called the **anal canal** (Figure 20.17). This region is located below the pelvic diaphragm and thus is outside the pelvis.

The mucosa of the anal canal forms a series of longitudinal folds known as **anal columns.** The anal columns are separated from each other by furrows called **anal sinuses,** which end distally in membraneous **anal valves.** The anal valves unite the lower ends of the anal columns. Within the anal canal, the epithelium changes from simple columnar, which is characteristic of the stomach and small and large intestines, to stratified squamous, which is typical of the mouth and esophagus.

The anal canal opens to the exterior through the **anus.** The anal canal is surrounded by **internal** and **external anal sphincter muscles.** The internal sphincter is formed from a thickening of the circular layer of the smooth muscles of the tunica muscularis and is therefore not under voluntary control.

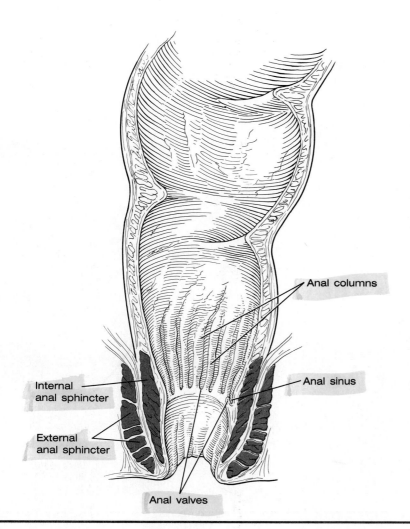

Anal columns

Anal sinus

Internal anal sphincter

External anal sphincter

Anal valves

Figure 20.17
Longitudinal section of
the anal canal.

The external sphincter is formed by skeletal muscle and is under voluntary control. Normally, therefore, a person can voluntarily control bowel movements.

The circulation of blood through veins that travel the length of the anal columns is sometimes interfered with, causing them to become enlarged. This condition is called *hemorrhoids.* Hemorrhoids can be irritated by bowel movements, producing discomfort and bleeding, particularly if the feces are too dry, as in constipation.

ACCESSORY DIGESTIVE ORGANS

In addition to the many small glands situated in the wall of the digestive tract, there are large glands located outside the tract. The secretions of these glands, which are very important in the digestion of food, are carried to the digestive tract by way of ducts. The ducts and the secretory portions of the accessory digestive glands are derived from the endodermal lining of the gut tube of the embryo. These glands include the *salivary glands,* which were discussed earlier in this chapter and whose ducts empty into the mouth, and the *pancreas* and *liver,* both of which empty their secretions into the duodenum of the small intestine.

Pancreas

The pancreas is located behind the peritoneum and beneath the stomach (Figure 20.10*a*). The **head** of the pancreas is nestled in the curve of the duode- F20.10a

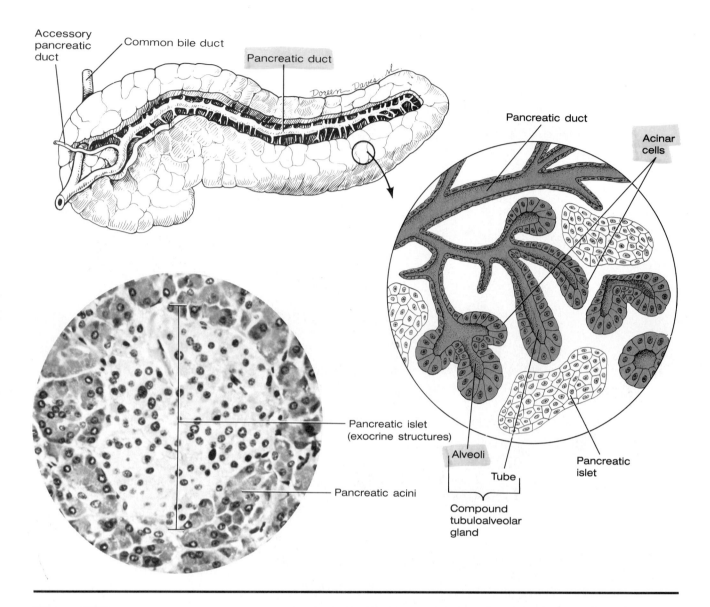

Figure 20.18
The pancreas. One inset shows the form of its compound tubuloalveolar glands, and the other is a photomicrograph of the acini and islets.

F20.18

num, with the **neck, body,** and **tail** regions extending to the left. The tail reaches the vicinity of the spleen.

Microscopically, the pancreas resembles the salivary glands, containing many compound tubuloalveolar glands, with their secretory cells arranged in short tubules or small sacs called **acini** (Figure 20.18). The acini consist of a single row of pyramid-shaped epithelial cells, with their apices converging toward the lumen. In acinar cells that are actively secreting, vesicles called *zymogen granules* accumulate toward the apex of the cells. The acinar cells secrete **pancreatic juice,** which contains digestive enzymes. The digestive enzymes of the pancreas are thought to be synthesized in the cytoplasm at the base of the acinar cells, where they enter the endoplasmic reticulum and are transported to the Golgi apparatus in the apical region of the cell. The Golgi apparatus concentrates the enzymes into zymogen granules, which are released as pancreatic juice.

Pancreatic juice is transported to the duodenum by the **pancreatic duct** *(duct of Wirsung).* The pancreatic duct usually joins with the common bile duct, and they enter the duodenum together. An **accessory pancreatic duct** *(duct of Santorini)* often branches from the pancreatic duct and empties into the duodenum independently.

The pancreas not only produces digestive enzymes, however, but also functions as an endocrine gland. Scattered among the acinar cells are clusters

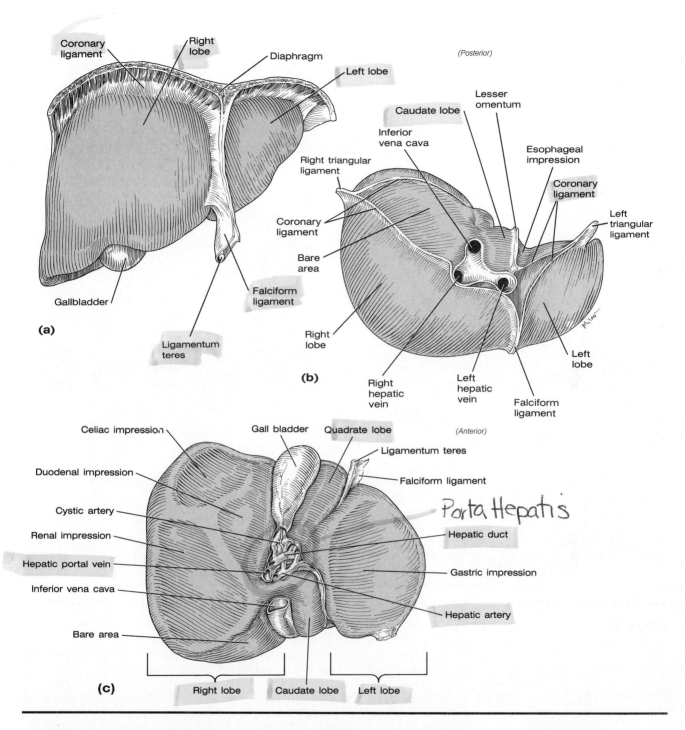

(a)

(b)

(c)

(Posterior)

(Anterior)

Porta Hepatis

Figure 20.19
The liver and its supporting ligaments.
(a) Anterior surface.
(b) Superior (diaphragmatic) surface.
(c) Posterior-inferior surface.

of endocrine cells called **pancreatic islets** *(islets of Langerhans).* The secretions of the cells of the pancreatic islets are not transported in ducts. Rather, they enter the bloodstream from the pancreas and travel throughout the body. The endocrine secretions of the pancreas are discussed along with the endocrine glands in Chapter 18.

Liver

The **liver** is a large organ that lies under the right side of the diaphragm. It is divided into two major regions: the **right** and **left lobes** (Figure 20.19*a*). On the inferior surface of the right lobe are smaller **caudate** and **quadrate lobes.** The right and left lobes are separated by a fold of parietal peritoneum called the **falciform ligament,** which attaches the liver to the anterior abdominal wall. In the free margin of the falciform ligament is a fibrous cord called the

F20.19a

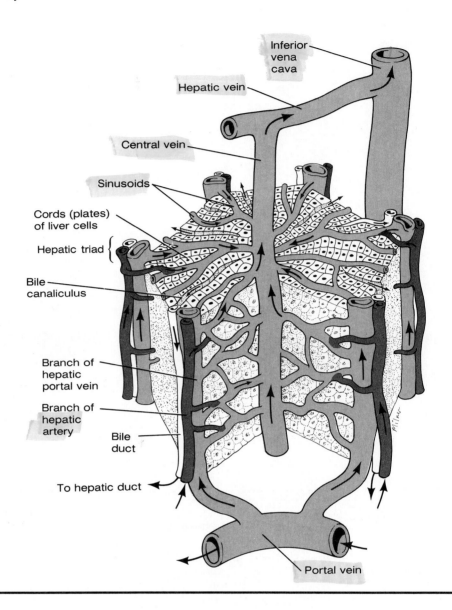

Figure 20.20
Microscopic anatomy of a liver lobule. The arrows show the direction of the flow of blood and bile.

ligamentum teres (round ligament). The ligamentum teres is the remnant of the fetal umbilical vein that transported blood from the placenta to the liver. In postnatal life, it extends from the umbilicus to the undersurface of the liver. The falciform ligament is continuous on the superior surface of the liver with the **coronary ligament,** a fold of the parietal peritoneum that attaches the liver to the undersurface of the diaphragm (Figure 20.19*b*). The coronary ligament consists of anterior and posterior layers that are joined at their lateral margins by **right** and **left triangular ligaments.** Between the two layers of the coronary ligament is a region called the **bare area** of the liver. The bare area, which rests against the diaphragm, is the only portion of the liver that is not covered with visceral peritoneum. The undersurface of the liver is anchored to the lesser curvature of the stomach by the **lesser omentum,** through which the hepatic artery, the hepatic portal vein, and the common bile duct travel. The inferior vena cava is partially embedded on the posterior surface of the liver (Figure 20.19*c*).

The liver receives blood from two sources: the **hepatic artery,** which supplies it with oxygenated blood from the aorta, and the **hepatic portal vein,** which carries venous blood from the digestive tract, pancreas, and spleen. The oxygen saturation of the blood in the hepatic portal vein is relatively low, but the blood contains a high concentration of dissolved nutrients absorbed from the intestines as a result of digestion. These nutrients are acted on in

F20.19b

F20.19c

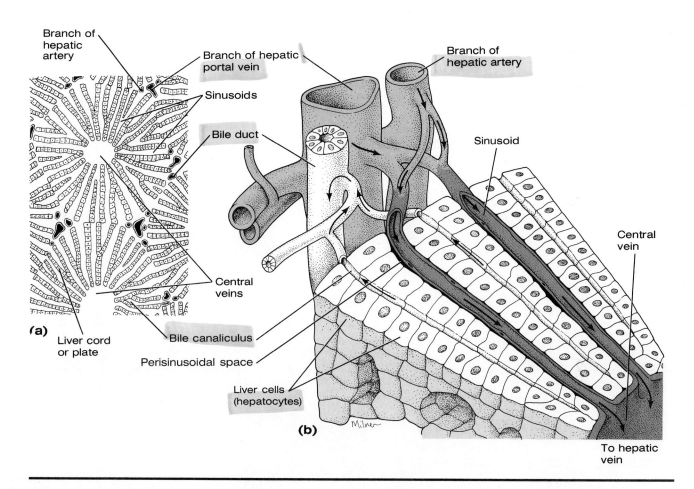

(a)

Branch of hepatic artery

Branch of hepatic portal vein

Sinusoids

Bile duct

Central veins

Liver cord or plate

Bile canaliculus

Perisinusoidal space

Liver cells (hepatocytes)

Branch of hepatic artery

Sinusoid

Central vein

(b)

Milner

To hepatic vein

Figure 20.21
Microscopic structure of the liver. **(a)** Enlargement showing parts of several lobules; two central veins are visible. **(b)** A higher magnification of two sinusoids showing their relationship with the branches of the hepatic portal vein and the hepatic artery. The colored arrows indicate the flow of blood through the sinusoids and into the central vein; the black arrows trace the flow of bile through the bile canaliculi and into the bile duct.

various ways as the blood passes through the liver. Approximately 1500 ml of blood flows through the liver every minute, of which 1100 ml arrives from the hepatic portal vein and 400 ml from the hepatic artery.

The liver is composed of many tiny hexagonal compartments called **liver lobules,** with the corners of the compartments occupied by a **portal canal** (*porta hepatis*). Branches of the hepatic artery and the hepatic portal vein, as well as a **bile duct,** travel in the portal canal. These three structures in close approximation within a portal canal are referred to as a **hepatic triad** (Figure 20.20). The most unusual aspect of the hepatic circulation is that the branches of both the hepatic artery and the hepatic portal vein empty into the same **sinusoids,** which therefore contain a mixture of arterial and venous blood (Figure 20.21). The sinusoids of each lobule pass between the liver cells **(hepatocytes),** which are arranged in rows or sheets called **liver cords,** or **plates,** and empty into a common **central vein** that passes through the center of each lobule, at right angles to the sinusoids. The central vein of each lobule drains into a **hepatic vein.** There are three hepatic veins, all of which empty into the inferior vena cava, which transports the blood to the heart. The pathway of blood through the liver is summarized in Figure 20.22.

F20.20

F20.21

F20.22

The liver sinusoids are lined with an endothelium that is highly permeable; it even allows proteins to diffuse out of the blood as it passes through. Because of this permeability, substances absorbed into the blood from the intestines can leave the blood freely and enter a space **(perisinusoidal space** or **space of Disse)** which separates the lining of the sinusoids from the hepatic cells. Microvilli from the hepatic cells extend into the space, greatly enhancing the movement of substances from the blood into the hepatic cells, where they are metabolized or otherwise altered. Moreover, interspersed between the hepatocytes that line the sinusoids are star-shaped cells called **stellate macrophages** *(Kupffer cells).* Projections of the macrophages extend into the lumen

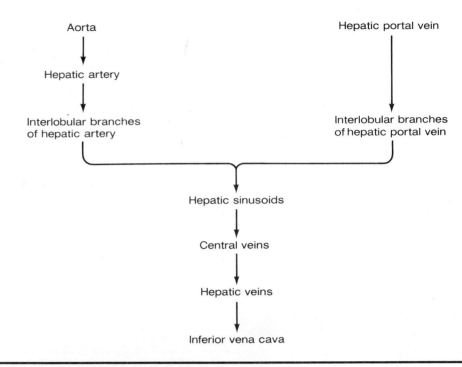

Figure 20.22
The pathway of blood through the liver.

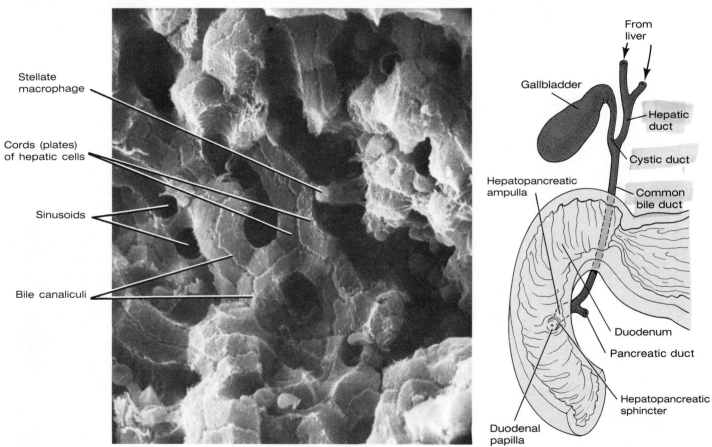

Figure 20.23
Microscopic anatomy of the liver (×1716). (From *Tissues and Organs: A Text-Atlas of Scanning Electron Microscopy* by Richard G. Kessel and Randy H. Kardon. W.H. Freeman and Company. Copyright © 1979.)

Figure 20.24
The gallbladder and bile ducts.

of the sinusoid. The macrophages are active phagocytes that remove bacteria and other foreign materials from the blood as it passes through the liver.

Another function performed by hepatic cells is the secretion of bile. Between the hepatic cells are tiny canals, called **bile canaliculi** (Figure 20.23), which carry the bile to bile ducts at the periphery of each lobule. Within the canaliculi, the bile travels in the opposite direction from the bloodflow in the sinusoids. The many bile ducts join together to form two main trunks that leave the liver and unite to form the **hepatic duct.**

Gallbladder and Bile Ducts

The **gallbladder,** a small sac on the inferior surface of the liver, is lined with columnar epithelium. It serves as a storage site for bile, which it receives from the liver. It also concentrates the bile by reabsorbing water from it. The gallbladder is drained by the **cystic duct,** which joins with the hepatic duct from the liver to form the **common bile duct** (Figure 20.24). The pancreatic duct joins with the common bile duct, and the two ducts share a common entrance into the duodenum. This entrance is surrounded by the hepatopancreatic sphincter. When the sphincter relaxes and the smooth muscle in the wall of the gallbladder contracts, bile is propelled into the intestine. When the sphincter contracts, bile from the liver fills the common bile duct and then enters the gallbladder via the cystic duct.

MECHANICAL PROCESSES OF THE DIGESTIVE SYSTEM

In order for digestion to occur, ingested food must be continually moved along the gastrointestinal tract so that it can be acted on by the digestive enzymes that are secreted into various regions of the tract. Moreover, to ensure that all food particles come into contact with these enzymes, the contents of the digestive tract must be constantly churned and mixed. This churning also brings the food into contact with the wall of the tract, which allows for absorption of the digested food into the bloodstream.

The moving and mixing of food in the digestive tract is accomplished by the rhythmic contraction and relaxation of the muscles associated with the tract. The muscles of the mouth, pharynx, upper portion of the esophagus, and external anal sphincter are skeletal muscles and are therefore under voluntary control. The muscles in the remainder of the gastrointestinal tract are visceral muscles that contract rhythmically and automatically, generally not under voluntary control.

Two basic forms of movement occur within the gastrointestinal tract—mixing and propulsion. One major mixing movement is **segmentation,** in which stationary muscular contractions occur at intervals along a portion of the digestive tract, thus dividing the tract into constricted and unconstricted regions (Figure 20.25). As segmentation proceeds, constricted regions relax and unconstricted regions constrict, thoroughly mixing the contents of the tract with digestive juices.

One important propulsive movement is **peristalsis,** in which the muscles surrounding a portion of the digestive tract undergo a wavelike contraction (Figure 20.26). This contraction produces a ring of constriction that moves along the tract, forcing materials within the tract to move ahead of it. Peristaltic waves normally travel toward the anus, probably because of the organization of the intrinsic nerve plexuses in the wall of the digestive tract.

Chewing

The first mechanical process associated with the digestive tract is chewing *(mastication).* It is accomplished by opening, closing, and lateral movements of the jaws, which are accompanied by the continual positioning of food between the teeth by the tongue and the muscles of the cheeks. Chewing converts the ingested materials into smaller, more digestible pieces and mixes

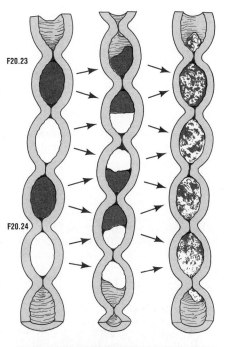

Figure 20.25
The mixing action of segmentation in the small intestine.

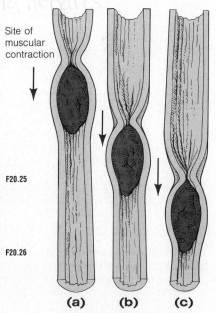

Site of muscular contraction

(a)   (b)   (c)

Figure 20.26
The movement of food by peristalsis, as in the esophagus and intestine.

FRONTIERS IN HEALTH
Removing Gallstones the Easy Way

Jennifer has an odd, yellowish color to her skin, a condition called jaundice. She also experiences increasingly frequent episodes of severe, steady pain in the upper abdomen, accompanied by chills and nausea. Like many other people, Jennifer is suffering from gallstones, small nodules made of cholesterol or bile pigments that develop within the gallbladder.

Approximately 25 million Americans have gallstones. Each year a million Americans learn that they have them; about half of them will require surgery. Removal of the gallbladder—an operation called a cholecystectomy—is one of the most common operations performed in the United States. It is estimated that gallstones cost more than a billion dollars a year in medical expenses and lost work time.

The most common gallstones are composed of cholesterol, with small amounts of calcium. Physicians have long been seeking a way to dissolve stones before they start causing pain by migrating down into the cystic duct or the common bile duct, where they may obstruct the flow of bile. Such obstruction can cause severe infections in the gallbladder and the pancreas.

Medical researchers may have a partial cure for gallstones. They have found a naturally occurring bile acid, chenodeoxycholic acid (CDCA) which, when taken orally, dissolves cholesterol stones. CDCA is used in some 40 countries and has been approved for regular use in the United States by the Food and Drug Administration. It is sold under the name chenodiol. However, chenodiol must be taken for months, or sometimes for as long as 2 years, to erode away even small gallstones. For patients like Jennifer, whose conditions are acute, CDCA is too slow to provide relief from pain and possible infection. CDCA is also ineffective in individuals whose gallbladders no longer function properly. Nor does it break up larger stones, or stones made of pigments, or stones containing more than 4% calcium. Regardless of these disadvantages, thousands of people who cannot risk an operation, such as the elderly and those with heart disease, may find relief using this drug.

In addition to being very slow acting, chenodiol has some side effects. Liver changes and diarrhea are the most common, but they are usually mild and disappear once treatment stops. In the search for a replacement for CDCA, medical researchers have begun testing another bile acid that has fewer side effects and seems to work faster than CDCA. The new drug is ursodeoxycholic acid (UDCA), and is called ursodiol.

Medical experts cautiously warn the public not to expect miracles from CDCA or UDCA. These drugs completely dissolve stones in only one of every five patients. Moreover, gallstones reform in up to half of all patients whose stones have been completely dissolved.

Suppose that a gallstone does slide down into the common bile duct, as it did in Jennifer. Can surgery still be avoided? Thanks to some pioneering research, there is a relatively painless way to dislodge stones that have become stuck in the common bile duct. This new method uses an instrument called a flexible endoscope. The endoscope is a flexible tube, which is swallowed by

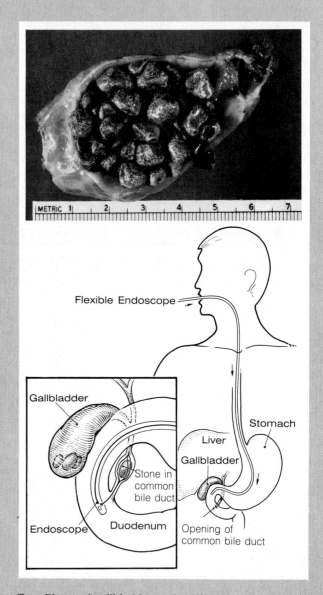

Top: Photo of gallbladder with gallstones. Bottom: Insertion of endoscope to the common bile duct.

the patient and guided by the surgeon through the stomach and into the small intestine. From there it is snaked into the common bile duct. Inserting the tube into the bile duct widens the duct, and this often suffices to flush the stone into the duodenum. In other cases, a small wire snare inserted through the tube is used to pull the stone out.

The endoscope contains thousands of tiny glass fibers that project light and convey a sharp image of the area being examined back to the endoscopist. Unfortunately, the endoscope can only be used to remove stones from the ducts draining the gallbladder. If there are stones within the gallbladder, surgery is often required.

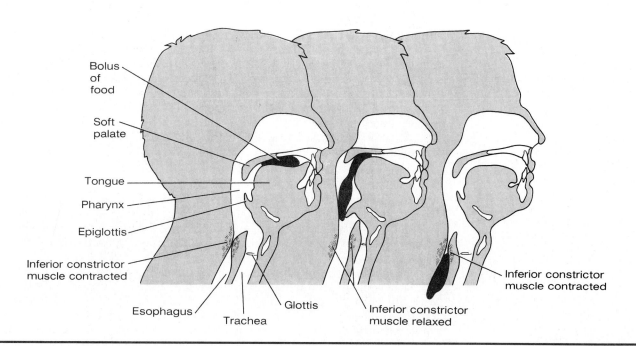

them thoroughly with saliva, thus forming a more manageable mass called a *bolus*.

Swallowing

The bolus of food is moved along the digestive tract of the process of swallowing *(deglutition)*. Swallowing is initiated by placing the tip of the tongue against the hard palate, thus forcing the bolus into the oropharynx. From there, instead of entering the esophagus, the food could go upward into the nasopharynx or downward into the trachea. But food is generally prevented from entering the nasopharynx by the elevation of the soft palate and uvula against the rear wall of the pharynx. And the entrance to the trachea is protected during swallowing by closing the glottis and moving the larynx upward. In this position, the movement of the bolus into the esophagus tips the epiglottis backward, thus covering the closed glottis (Figure 20.27).

All these events are coordinated through cranial nerves V, IX, XI, and XII by interconnected nerve centers—referred to as the *swallowing center*—in the brain stem. Generally, these events occur so effortlessly that they are below the level of awareness. Sometimes, however, the swallowing center miscoordinates these events, and food then enters the nasal cavity or the larynx. The usual response to food that enters the larynx is a violent cough, which expels the food back into the oropharynx.

The bolus is moved toward the esophagus from the pharynx by the contraction of the pharyngeal constrictor muscles. Before the bolus can enter the esophagus, it is necessary for the lowest fibers of the inferior constrictor muscle of the pharynx, which surrounds the entrance of the esophagus, to relax. In contrast to the other constrictor muscles of the pharynx, the lower fibers of the inferior constrictor muscle maintain a constant tonic contraction. This muscle thus acts as a sphincter between the pharynx and the esophagus. In this way, it normally prevents the regurgitation of substances from the esophagus back into the pharynx.

Once in the esophagus, the bolus is propelled by peristaltic contractions. Peristalsis in the esophagus is caused by the contraction of the circular layer of muscle in the tunica muscularis of the esophagus, and is relatively weak. In response to stretching of the wall by the bolus, a ring of circular muscles behind the food contracts, thereby narrowing the tube and forcing the food downward.

Just above the cardiac orifice, the smooth muscle of the esophagus is normally in a state of tonic contraction, and it acts as a physiological

F20.27

Figure 20.27
Swallowing, showing the movement of a bolus of food through the mouth, the oropharynx, the laryngopharynx, and the upper esophagus.

sphincter—referred to as the *lower esophageal sphincter*—that keeps the lower portion of the esophagus closed. As the peristaltic wave moves along the esophagus during swallowing, the lower esophageal sphincter relaxes, allowing the bolus to enter the stomach. The presence of a sufficient amount of gas in the stomach can produce enough pressure to open the lower esophageal sphincter. The gas is then able to escape from the stomach by means of belching. At times, some of the contents of the stomach are forced past the sphincter into the esophagus. The acidic gastric contents may irritate the esophagus, producing *heartburn*.

Gastric Motility

The mechanical activities of the stomach include: (1) storing ingested food until it can be utilized by the remainder of the gastrointestinal tract; (2) mixing the food with gastric secretions; and (3) moving the food into the duodenum of the small intestine at a rate consistent with efficient intestinal digestion and absorption.

As food enters the stomach, it forces already-present food against the stomach wall. As additional food enters the stomach, it forms into layers, with the most recently swallowed food remaining near the cardiac orifice. Thus, newly swallowed food remains in a central position for a period of time. Since the acidity of gastric juice inactivates salivary amylase from the salivary glands, the layering of food in the stomach has the advantage of delaying the mixing of gastric juice with the most recently swallowed food, thus allowing additional time for salivary amylase to function.

The stomach enlarges as more food enters it from the esophagus. As the stomach wall is distended, receptors in the wall are stimulated. The stimulation of these receptors initiates nerve reflexes that increase gastric motility. Many of the nerve impulses that increase gastric motility reach the stomach through the vagus nerve. Others arrive from the intrinsic nerve plexuses in the wall of the stomach. In addition, the hormone *gastrin* is released by cells in the stomach lining in response to distension. Gastrin enters the bloodstream and ultimately returns to the stomach, where it stimulates gastric motility.

The muscles of the fundus region of the stomach contract slightly but not in peristaltic waves. The fundus serves as a holding and mixing area for food until it is moved into the lower regions of the stomach. As the volume of food in the stomach increases, peristaltic contractions begin in the body of the stomach and move toward the pylorus. The pylorus, which leads into the duodenum, is surrounded by the **pyloric sphincter.** The pyloric sphincter is generally partially open and therefore offers only limited resistance to the movement of the stomach contents into the duodenum. Nevertheless, despite the strong contractions of the stomach, only a small amount of the most finely emulsified particles of food is forced past the pyloric sphincter with each peristaltic wave. Most of the food is forced back into the body of the stomach, where it is subjected to further churning. As the food is repeatedly squeezed and mixed with gastric juice, it is converted to a semifluid consistency and is referred to as **chyme.**

The amount of chyme in the duodenum and its chemical composition are important factors controlling both gastric motility and emptying. As the duodenum becomes filled with chyme, its wall is stretched, which reflexly inhibits the parasympathetic motor neurons that travel to the wall of the stomach through the vagus nerve. Ordinarily, motor neurons in the vagus nerve transmit nerve impulses that cause the muscles in the wall of the stomach to contract. The presence of chyme in the duodenum initiates sensory impulses that travel to the medulla, where they inhibit the motor neurons that pass to the stomach in the vagus nerve and also stimulate the sympathetic neurons to the stomach. As a result of these reflex activities, the effect of chyme in the duodenum is to slow the emptying of the stomach. Moreover, the acidity of the chyme in the duodenum, as well as the presence of certain amino acids and fatty acids, triggers the release of hormones such as *secretin, cholecystokinin,* and *gastric inhibitory peptide* from cells of the duodenum and jejunum. These hormones travel through the bloodstream to the stomach, where they exert an inhibitory effect on gastric motility.

Intestinal Motility

Most absorption of digested food occurs in the intestines. It is not surprising, therefore, that intestinal motility provides an efficient means for moving food through the intestines in a fairly regular manner—in contrast to the repeated mixing and churning it undergoes in the stomach.

Small Intestine

Both segmentation and peristalsis take place in the small intestine. Segmentation may cause some movement of materials along the intestine, but its primary contributions are to mix the chyme thoroughly with the digestive juices and to expose different portions of the chyme to the intestinal mucosa. Peristaltic contractions are primarily responsible for moving the chyme along the small intestine. In general, the peristaltic waves of the small intestine are weak, and they die out after traveling only a short distance. Therefore, the movement of the chyme through the small intestine tends to be relatively slow, allowing ample time for digestion and absorption.

The motility of the small intestine, like that of the stomach, is affected by the amount of chyme present as well as by the acidity and osmotic concentration of the chyme. In addition, intestinal motility is under the control of the autonomic division of the nervous system. Parasympathetic activity (via the vagus nerves) increases intestinal motility, and sympathetic activity (via the splanchnic nerves) decreases it. Because of this autonomic control, a person's emotional state can also affect intestinal (and gastric) motility.

The **ileocecal sphincter,** which surrounds the entrance of the last portion of the small intestine (ileum) into the cecum of the large intestine, is usually closed. Thus, the contents of the large intestine cannot be regurgitated back into the small intestine and the chyme in the small intestine is retained there long enough for it to be thoroughly mixed.

Large Intestine

The passage of food into and out of the large intestine is regulated by the ileocecal sphincter and the anal sphincter, respectively. Segmentation and peristalsis, the same two movements that are typical of the small intestine, occur within the colon of the large intestine. Movement through the colon is much slower than in the small intestine, however, requiring 18–24 hours or more to reach the rectum. This slow movement allows bacteria to grow in the colon. The action of these bacteria on the contents of the colon sometimes produces gas, causing uncomfortable distension.

About three or four times each day, long, slow waves of peristalsis, called *mass peristalsis*, move the contents of the colon toward the rectum. Mass peristalsis often occurs following a meal—which indicates that the presence of food in the stomach activates a reflex that increases the motility of the colon.

Since the rectum is generally empty, the mass movement of fecal material into it stimulates receptors in its wall that initiate the *defecation reflex*, which tends to move material out of the lower colon and rectum. The neural center for this reflex is located in the sacral region of the spinal cord. The defecation reflex stimulates peristaltic contraction of the descending colon, the sigmoid colon, and the rectum. It also causes the relaxation of the internal anal sphincter, which surrounds the anal opening. This combination of contraction and relaxation propels the feces through the anus. However, a second sphincter—the external anal sphincter—also surrounds the anal opening. The external anal sphincter is composed of skeletal muscle, and thus it can be voluntarily controlled by higher centers in the brain. If the external anal sphincter is voluntarily relaxed, defecation will occur. If the external anal sphincter is voluntarily contracted, defecation can be prevented. Thus, evacuation of the rectum can be delayed voluntarily for a period of time. If defecation is delayed, the muscles in the wall of the rectum relax and the urge to defecate diminishes temporarily. When the next mass peristalsis moves more feces into the rectum, the defecation reflex is again activated.

The defecation reflex is usually assisted through voluntary efforts when a person has a bowel movement. This is accomplished by taking a deep inspiration followed by closing the glottis and contracting the abdominal wall muscles. These actions raise the pressure in the abdomen, which aids defecation by squeezing the rectum.

CHEMICAL PROCESSES OF THE DIGESTIVE SYSTEM

The chemical breakdown of large molecules of foodstuffs into smaller molecules that can be absorbed from the gastrointestinal tract into the bloodstream is accomplished by digestive enzymes. Although we are primarily concerned in this book with the anatomy of the digestive system, it will be instructive to survey the chemical digestive processes that occur in various regions of the gastrointestinal tract. Table 20.2 summarizes the digestive enzymes and their actions.

Table 20.2

Digestion in the Mouth

Saliva, which is produced by the salivary glands, contains only one digestive enzyme: **salivary amylase.** Salivary amylase begins the digestion of carbohydrates by splitting large starch molecules into smaller fragments such as dextrins and maltose. Since food does not remain in the mouth very long, much of the activity of salivary amylase occurs in the esophagus as the food travels to the stomach.

Table 20.2   Digestive Enzymes—Their Locations and Functions

Location	Enzyme	Function
Mouth	Salivary Amylase	*Carbohydrate digestion:* converts starch into dextins and maltose
Stomach	Pepsin	*Protein digestion:* converts proteins into polypeptides
Small Intestine *Pancreatic juice*	Pancreatic amylase	*Carbohydrate digestion:* breaks starch into the disaccharides maltose and isomaltose
	Trypsin Chymotrypsin	*Protein digestion:* break proteins into peptones, proteoses, and dipeptides
	Carboxypeptidase	*Protein digestion:* reduces proteins or protein derivatives to free amino acids
	Pancreatic lipase	*Lipid digestion:* reduces fats into monoglycerides and free fatty acids
Microvilli of intestinal absorptive cells	Maltase Isomaltase Sucrase Lactase	*Carbohydrate digestion:* complete digestion into glucose, galactose, lactose, and/or fructose
	Aminopeptidase Dipeptidase	*Protein digestion:* reduce small polypeptides and dipeptides into amino acids
	Intestinal lipase	*Lipid digestion:* intracellular digestion of monoglycerides into fatty acids and glycerol

Digestion in the Stomach

The zymogenic cells of the gastric glands of the stomach produce a number of different digestive enzymes, but since these enzymes all have basically similar functions they are considered together as **pepsin.** Pepsin begins the digestion of proteins by breaking them into smaller chains of amino acids called *polypeptides.* Assisting in the digestion of proteins in the stomach is hydrochloric acid, which is secreted by parietal cells of the gastric glands. Hydrochloric acid converts the inactive precursor *pepsinogen,* which is secreted by the cells of the gastric glands, into active pepsin. The parietal cells tend to be located higher up in the gastric glands than are the zymogenic cells. This arrangement ensures that pepsinogen, which is strongly corrosive, remains inactive until it is almost out of the gland and into the stomach, whose lining is well protected by a coat of mucus.

Digestion in the Small Intestine

The small intestine is the most important portion of the alimentary tract as far as digestion and absorption are concerned. Not only do the cells of the small intestine produce digestive enzymes that remain on the plasma membrane of their villi, but the small intestine also receives digestive enzymes from the pancreas and bile from the liver.

The pancreatic juice contains an enzyme **(pancreatic amylase)** that breaks starch and dextrins into the disaccharides (12-carbon sugars) maltose and isomaltose. Pancreatic juice also contains three enzymes—**trypsin, chymotrypsin,** and **carboxypeptidase**—that reduce proteins to smaller molecules (such as peptones, proteoses, and dipeptides) or digest them completely into their component amino acids. Fats are acted on for the first time in the digestive tract by **pancreatic lipase.** The action of lipase in breaking fat into monoglycerides (a fatty acid attached to a glycerol molecule) and free fatty acids is accelerated by the presence of bile salts from the liver.

The enzymes that remain attached to the plasma membrane of the absorptive cell microvilli act on carbohydrates, proteins, and lipids. The digestion of carbohydrates, which is begun in the mouth and continued in the presence of pancreatic amylase, is completed by the intestinal enzymes **maltase, isomaltase, sucrase,** and **lactase.** These enzymes break down the disaccharides into monosaccharides (glucose, galactose, lactose, and fructose). **Aminopeptidase** and **dipeptidase** enzymes of the intestinal juice digest small polypeptides and dipeptides into free amino acids. Although most fat digestion occurs as the result of the activity of pancreatic lipase, there is also an **intestinal lipase** produced by the cells of the intestinal mucosa. Much of the intestinal lipase is thought to act in the intestinal cells, where it reduces the absorbed monoglycerides into fatty acids and glycerol.

CONDITIONS OF CLINICAL SIGNIFICANCE

The Digestive System

Peptic Ulcer

A *peptic ulcer* is an erosion of the wall of the gastrointestinal tract in an area of the tract exposed to gastric juice containing acid and pepsin. Peptic ulcers are most commonly found in the stomach (gastric ulcers) and the duodenum (duodenal ulcers). They may be caused by an excessive acid-pepsin secretion or they may result from an insufficient secretion of mucus, which normally protects the gastrointestinal mucosa from being digested by the gastric juice.

The most common symptom of peptic ulcer is pain. In some cases, the pain can be temporarily relieved by the ingestion of food, apparently because the food provides some protection by coating the ulcer.

If the erosion due to a gastric ulcer is sufficiently severe, blood vessels in the stomach wall are damaged and bleeding occurs into the stomach itself (a *bleeding ulcer*). In extreme cases, a peptic ulcer can lead to perforation—that is, a hole through the wall of the gastrointestinal tract. A perforation allows the contents of the gastrointestinal tract to pass into the abdominal cavity.

Such an occurrence is very serious because it can cause *peritonitis*, an inflammation of the lining of the abdominopelvic cavity. Perforated ulcers often require immediate surgery.

Gastroenteritis

Gastroenteritis is an acute or chronic inflammation of the mucosa of the stomach and intestine. It is often caused by irritants such as excessive alcohol or cathartics (medicines that cause bowel movements), but can have any of a wide variety of other causes, including viral infections, food allergies, and overeating.

Gallstones

Gallstones are particles of cholesterol and bile salts that sometimes precipitate in bile. The stones can become lodged in such a position as to block the cystic duct from the gall bladder or the common bile duct to the duodenum. Gallstones are often painful, and those that block the cystic duct are especially so because the contractions of the gallbladder exert force on them.

In the case of cystic gallstones, bile is still able to reach the duodenum directly from the liver through the common bile duct. Blockage of the common bile duct, on the other hand, prevents bile from reaching the intestine and thus interferes with the proper absorption of fat. In addition, the bile pigments cannot reach the intestine to be excreted. These pigments accumulate in the blood and are eventually deposited in the skin. This produces a yellow color in the skin, a condition called *jaundice*.

Pancreatitis

Pancreatitis is an inflammation of the pancreas that is often caused by the digestion of parts of the organ by pancreatic enzymes normally carried to the small intestine within the pancreatic ducts. In pancreatitis, the enzymes become activated within the ducts and they can destroy the ducts and the pancreatic cells. Since pancreatic enzymes are very important in the digestion of carbohydrates, proteins, and fats, pancreatitis can produce severe nutritional problems.

Hepatitis

Hepatitis is an infection of the liver. Most commonly, it is a viral infection transmitted either by virus-infected blood or by contaminated food or water. The infected liver becomes enlarged, and its functioning is impaired, which can lead to jaundice. In some cases, liver function is depressed for a year or more.

Cirrhosis

Cirrhosis is a chronic inflammation of the liver that is progressive and diffuse. In the affected areas of the liver, some cells are replaced by fibrous connective tissue, thereby interfering with liver function. In this way, cirrhosis can cause a reduction in the production of bile, in the excretion of bile pigments, and in the production of blood-clotting factors and plasma albumin. Cirrhosis also reduces the liver's ability to detoxify the blood, and toxins accumulate.

Appendicitis

The appendix is a blind pouch extending from the cecum. If it becomes obstructed (for example, with hardened fecal material), its venous circulation may be interfered with. Such interference reduces the oxygen supply to the area and permits bacteria to flourish. The appendix then becomes inflamed and filled with pus, a condition called *appendicitis*.

If an inflamed appendix is not surgically removed soon enough, the accumulating pus can exert so much pressure that it ruptures, releasing its contents into the abdominal cavity. This occurrence can cause *peritonitis*, which is an inflammation of the lining of the abdominopelvic cavity.

Effects of Aging on the Digestive System

With aging, the digestive secretions tend to diminish somewhat and the muscles in the walls of the tract become weaker, reducing the strength of peristaltic contractions.

The loss of teeth and a gradual atrophy of the salivary glands interfere with the chewing of food in older persons. The reduction in the volume of saliva not only causes the mouth to be dry, but also diminishes the ability to taste and provides less cleansing for the mouth and teeth. It is not uncommon for older persons to experience difficulty swallowing because of incomplete relaxation of the lower esophageal sphincter.

Gradual atrophy of the mucous glands and the gastric glands in the stomach reduce the secretion of mucus, hydrochloric acid, and digestive enzymes. This may particularly interfere with the digestion of proteins and cause a chronic inflammation of the stomach lining.

Atrophy of the cells lining the small intestine alters the structure of the villi, thereby reducing the surface area across which absorption may occur. The cells in the wall of the large intestine also undergo gradual atrophy. This atrophy may so weaken the walls that outpockets develop, a condition referred to as diverticulosis.

STUDY OUTLINE

EMBRYONIC DEVELOPMENT OF THE DIGESTIVE SYSTEM   p. 533

1. Formed from endoderm.
2. Divided into foregut, midgut, hindgut.
3. Oral membrane breaks through to stomodeum, forming mouth.
4. Cloacal membrane ruptures through to proctodeum, forming anus.
5. Endodermal buds along foregut, together with surrounding mesoderm, give rise to salivary glands, liver, gallbladder, pancreas.

ANATOMY OF THE DIGESTIVE SYSTEM
pp. 533–549

MUCOUS MEMBRANE LINING
1. Secretes mucus for protection and moisture.
2. Allows for absorption of digested food.

MOUTH from lips to oropharynx.
1. Lined with stratified squamous layer of nonkeratinized cells (except outer lips).
2. Lips and cheeks aid chewing and speech.
3. Roof formed anteriorly by hard palate; posteriorly by soft palate.

TONGUE forms floor of mouth.
1. Protruded, retracted, and moved sideways by extrinsic muscles.
2. Shape modified by intrinsic muscles.
3. Assists in food manipulation, swallowing, speech.
4. Papillae present on dorsal surface; taste buds present on some papillae.
5. Connected ventrally to mouth floor by frenulum, which can cause tongue-tied condition.

TEETH
1. Enamel-covered crown anchored by roots into alveolus (socket).
2. Pulp cavity, root canal, and apical foramen enclose vessels and nerves.
3. Incisors, canines, premolars, molars.
4. 20 deciduous teeth; 32 permanent teeth.

SALIVARY GLANDS

PAROTID GLANDS below, and anterior to ear.

SUBMANDIBULAR GLANDS medial to mandibular angle.

SUBLINGUAL GLANDS on floor of mouth.

SALIVA moistens mucous membranes; moistens food to form bolus; dissolves some food molecules; contains salivary amylase to initiate carbohydrate digestion.

PHARYNX from oropharynx through laryngopharynx.

GASTROINTESTINAL TRACT WALL four tunics in wall of digestive tube; this basic structure modified in some regions of digestive tract.

TUNICA MUCOSA mucous-membrane lining; consists of epithelial layer, lamina propria, and muscularis mucosa.

TUNICA SUBMUCOSA dense or loose connective tissue; contains blood vessels, lymphatic vessels, nerves, and glands.

TUNICA MUSCULARIS double layer of muscle tissue; inner layer circular; outer layer longitudinal.

TUNICA SEROSA (ADVENTITIA) outer layer of connective tissue.

ESOPHAGUS muscular tube that connects pharynx to stomach; upper portion contains skeletal muscles; lower portion has smooth muscles.

STOMACH from cardiac orifice to pyloric sphincter; body, fundus, and pyloric regions; greater and lesser curvatures.

TUNICA MUCOSA MODIFICATIONS
1. Columnar epithelium.
2. Rugae.

3. Presence of gastric glands.
   a. Mucous neck cells secrete mucus.
   b. Chief cells secrete pepsinogen.
   c. Parietal cells produce hydrochloric acid.
   d. Enteroendocrine cells produce gastrin.
   e. Cardiac and pyloric glands secrete mucus.

TUNICA MUSCULARIS MODIFICATION includes an oblique muscle layer, as well as circular and longitudinal layers.

SMALL INTESTINE
1. 6 m long; site of most digestion and absorption.
2. Three regions:

DUODENUM: retroperitoneal; receives common bile duct and pancreatic duct at duodenal papilla.

JEJUNUM: suspended by mesentery.

ILEUM: suspended by mesentery; entrance into large intestine surrounded by ileocecal valve.

3. Modifications in wall: Plicae circulares, villi, and microvilli increase surface area. Contains goblet cells, intestinal glands, and duodenal glands.
4. Plasma membrane of microvilli of absorptive cells have attached enzymes that digest proteins, carbohydrates, and lipids; small intestine also receives digestive enzymes from pancreas and bile from liver.

LARGE INTESTINE
1. Composed of cecum; ascending, transverse, descending, and sigmoid colons; rectum; anal canal.
2. Modifications in tunics:
   a. Tunica mucosa has many mucous cells; lacks villi.
   b. Tunica muscularis: longitudinal muscles form short bands (taeniae coli) that cause colon to form pouches (haustra).
   c. Epiploic appendages on external surface.

RECTUM AND ANAL CANAL
1. Rectum lies anterior to sacrum.
2. Anal canal mucosa forms longitudinal rectal columns.
3. Involuntary internal anal sphincter; voluntary external anal sphincter.

ACCESSORY DIGESTIVE ORGANS
pp. 549–555

PANCREAS
1. Cells in sacs called acini secrete pancreatic juice, which is transported to duodenum by pancreatic duct.
2. Endocrine secretions of pancreatic islet cells enter bloodstream.

LIVER
1. Right and left lobes separated by falciform ligament.
2. Ligamentum teres is remnant of fetal umbilical vein.
3. Coronary ligament attaches liver to undersurface of diaphragm.
4. Hexagonal lobules consist of rows of cuboidal cells radiating outward from a central vein that empties into a hepatic vein.
5. Bile ducts collect bile from canaliculi that transport bile in opposite direction from bloodflow.

6. Hepatic circulation: Hepatic artery supplies oxygenated blood to liver from aorta. Hepatic portal vein carries venous blood from digestive tract, pancreas, and spleen. Liver sinusoids contain mixture of arterial and venous blood. Sinusoids are lined with highly permeable endothelium that has attached phagocytic stellate macrophages. Hepatic veins empty blood from sinusoids into inferior vena cava.

GALLBLADDER AND BILE DUCTS
1. Gallbladder is small sac on inferior surface of liver; serves as bile storage site.
2. Cystic duct drains gallbladder; joins hepatic duct to form common bile duct, which joins with pancreatic duct to enter duodenum.

MECHANICAL PROCESSES OF THE DIGESTIVE SYSTEM pp. 555–560
1. Segmentation is major mixing movement.
2. Peristalsis is major propulsive movement.

CHEWING mixes food with saliva and reduces size of pieces, forming bolus.

SWALLOWING
1. Swallowing moves bolus to pharynx.
2. Pharyngeal constrictor muscles move bolus toward esophagus.
3. Peristalsis propels bolus in esophagus.

GASTRIC MOTILITY
1. Fundus is holding and mixing area.
2. Peristaltic contractions churn food and mix it with gastric juice to convert it to semifluid chyme.
3. Amount and chemical composition of chyme in duodenum help control gastric motility and emptying.
4. Certain hormones also affect gastric motility.

INTESTINAL MOTILITY

SMALL INTESTINE
1. Peristaltic waves are weaker than those of stomach.
2. Segmentation occurs.
3. Motility affected by amount, acidity, and osmotic concentration of chyme.

LARGE INTESTINE
1. Peristalsis and segmentation—very slow movement.
2. Distension of rectum initiates defecation reflex.

CHEMICAL PROCESSES OF THE DIGESTIVE SYSTEM pp. 560–561

DIGESTION IN THE MOUTH salivary amylase begins digestion of carbohydrates.

DIGESTION IN THE STOMACH
1. Pepsinogen converted into active pepsin by HCL.
2. Pepsin breaks proteins into polypeptides.

DIGESTION IN SMALL INTESTINE
1. Pancreatic juice contains pancreatic amylase (acts on carbohydrates); trypsin, chymotrypsin, and carboxypeptidase (act on proteins); and lipase (acts on fats).
2. Microvilli of absorptive cells contain maltase, isomaltase, sucrase, and lactase (act on carbohydrates); aminopeptidase and dipeptidase (act on protein derivatives); and lipase (acts on fats).

CONDITIONS OF CLINICAL SIGNIFICANCE: THE DIGESTIVE SYSTEM pp. 561–562

PEPTIC ULCER Erosion of gastrointestinal tract wall in an area of tract exposed to gastric juice containing acid and pepsin.

GASTROENTERITIS Acute or chronic inflammation of mucosa of stomach and intestine; caused by irritants, viral infection, or food allergy.

GALLSTONES Particles containing cholesterol and bile salts; can block cystic or common bile ducts.

PANCREATITIS Inflammation of pancreas; often caused by digestion of parts of pancreas by pancreatic enzymes.

HEPATITIS Infection of liver, usually viral; causes enlargement of liver and impaired liver function.

CIRRHOSIS Chronic liver inflammation; may result in reduced bile production, reduced bile pigment excretion, reduced blood-clotting factor production, and blood toxin accumulation.

APPENDICITIS Inflammation of appendix; can be due to obstruction that interferes with venous circulation.

EFFECTS OF AGING ON THE DIGESTIVE SYSTEM Digestive secretions diminished; peristaltic contractions weakened; loss of teeth and reduced saliva. Atrophy of mucous glands and digestive glands in stomach and small intestine. Wall of large intestine weakened.

SELF-QUIZ

1. The entire digestive tract is lined with: (a) cloacal membrane; (b) proctodeum; (c) mucous membrane.

2. The hard palate extends posteriorly from the soft palate, separating the oral cavity from the nasopharynx. True or false?

3. The hardest part of the tooth is the: (a) dentin; (b) enamel; (c) cementum.

4. The teeth specialized to tear food are the: (a) canines; (b) incisors; (c) bicuspids.

5. The largest of the salivary glands are the: (a) parotids; (b) sublinguals; (c) buccal.

6. Which one of the four layers of the digestive tube is responsible for peristalsis? (a) tunica mucosa; (b) tunica submucosa; (c) tunica muscularis; (d) tunica serosa.

7. Chief cells of the stomach produce: (a) bicarbonate ions; (b) pepsinogen; (c) hydrochloric acid.

8. The point of exit from the stomach where it joins with the small intestine is guarded by a: (a) pyloric sphincter; (b) greater omentum; (c) lesser omentum.

9. Most digestion and absorption occur in the small intestine. True or false?

10. The ducts from the pancreas and liver empty into the alimentary canal at the: (a) stomach; (b) duodenum; (c) ileum.

11. Pancreatic enzymes are produced by cells called: (a) pancreatic islets; (b) acinar cells; (c) goblet cells; (d) parietal cells.

12. Duodenal glands produce: (a) amylase; (b) pepsinogen; (c) mucus.

13. As the descending colon reaches the left pelvic brim, it curves to the midline via an S-shaped sigmoid colon. True or false?

14. The internal anal sphincter is formed by a thickening of the circular layer of the smooth muscles of the tunica muscularis and is therefore under voluntary control. True or false?

15. The pancreatic islets are endocrine cells. True or false?

16. Bile is produced in the: (a) liver; (b) gallbladder; (c) pancreas.

17. When the hepatopancreatic sphincter relaxes, bile from the liver fills the common bile duct and then enters the gallbladder via the cystic duct. True or false?

18. When you swallow, food first enters the: (a) oropharynx; (b) nasopharynx; (c) trachea.

19. The strength and velocity of peristaltic contractions are increased by impulses carried over: (a) cranial nerve XI; (b) vagus nerve; (c) phrenic nerve.

20. The presence of chyme in the duodenum: (a) inhibits peristaltic contractions in the stomach; (b) stimulates the production of more chyme; (c) inhibits swallowing.

21. Both segmentation and peristalsis occur in the small intestine. True or false?

22. Which movements occur in the colon? (a) peristalsis; (b) segmentation; (c) mass peristalsis; (d) a and c only; (e) all of these.

23. Protein digestion occurs in the: (a) mouth; (b) stomach; (c) small intestine; (d) all of these; (e) both b and c.

24. No significant digestion occurs in the large intestine. True or false?

25. Blood from the liver sinusoids enters the: (a) central vein; (b) bile canaliculi; (c) hepatic duct; (d) gallbladder.

LEARNING OBJECTIVES

After completing this chapter, you should be able to:

- State the general functions of the kidneys.
- Name the components of the urinary system.
- Describe the embryonic development of the kidneys.
- Describe the gross anatomical structure of the kidneys.
- Describe the microscopic anatomy of a renal tubule.

- Describe the filtration barrier that separates the blood in the glomerulus from the capsular space.
- Describe the pathway of blood through the kidney.
- Explain the unique features of bloodflow through the kidney.
- Distinguish between the ureter and urethra by citing the structure and function of each.
- Describe the structure of the urinary bladder.

CHAPTER CONTENTS

The Urinary System

If the cells of the body are to survive and carry out their functions effectively, they must be surrounded by a stable environment. As we have noted previously, the relatively constant state of the body's internal environment is called *homeostasis*. To maintain homeostasis the concentrations of such substances as water, sodium, potassium, calcium, and hydrogen ions must remain relatively constant, as must the concentrations of a wide variety of cellular nutrients and products. Cellular metabolism constantly tends to upset the body's internal environmental balance by consuming some substances (such as oxygen and glucose) and producing wastes and toxins (such as carbon dioxide and urea). In addition, substances may be added to the internal environment as a result of ingestion and removed from the internal environment by excretion.

Maintaining homeostasis involves most of the body systems. The digestive system, for example, supplies nutrients and also serves as a means of excreting some waste products. The lungs supply oxygen to the body and eliminate carbon dioxide and some water. The skin also plays a minor role in excretion—sweat, for example, contains small amounts of urea and ammonia. The **kidneys,** however, as the main excretory organs, are critically important in maintaining the balance of substances required for internal constancy. The kidneys eliminate from the body a variety of metabolic products, such as urea, uric acid, and creatinine. Further, the kidneys conserve or excrete water and electrolytes as required so that the internal balance of these substances will be maintained. In fact, kidney malfunction can cause severe and even fatal problems as a result of upsets in fluid and electrolyte balance. The kidneys also act as endocrine structures by producing a hormone called *erythropoietin*, which stimulates the production of red blood cells.

Because they are selective in what they excrete, the kidneys are able to maintain the internal environment within a range that is optimal for the survival of the cells. As a result of the selectivity of the kidneys, substances that are vital to the cells may not be excreted at all. Other substances are excreted in varying amounts that depend largely on the needs of the body. Therefore, the kidneys have a major role in regulating the composition and pH of the interstitial fluid. In a similar manner, the amount of water removed by the kidneys from the blood, and thus from the interstitial fluid, varies with the needs of the body.

If the kidneys fail, there is no way for many of the substances they normally excrete to be removed from the blood. As a consequence, these substances accumulate in the blood and eventually in the interstitial fluid. Within a matter of days following kidney failure, the internal environment of the body can change so much that the body's cells no longer function. To prevent death, it is necessary to replace the nonfunctioning kidneys with healthy ones by means of a kidney transplant or to remove potentially harmful substances regularly from the blood with an artificial kidney machine (see Box 21.1). **Box 21.1**

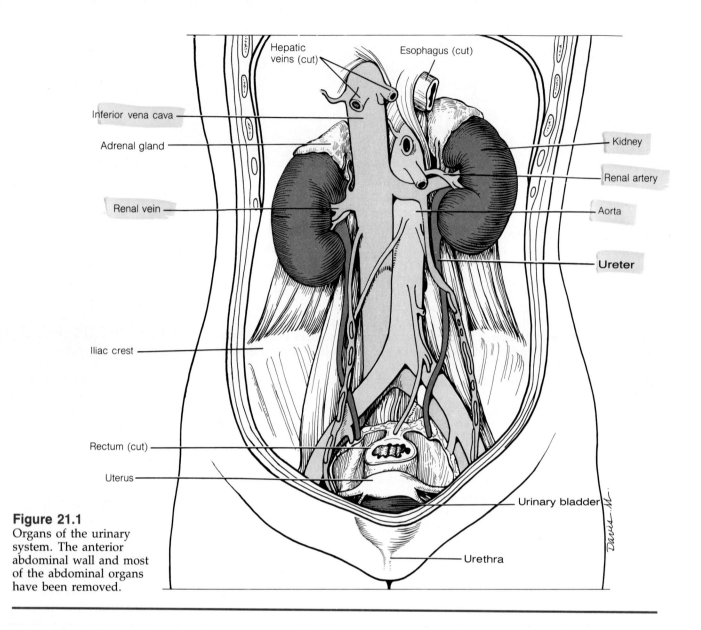

Figure 21.1
Organs of the urinary system. The anterior abdominal wall and most of the abdominal organs have been removed.

COMPONENTS OF THE URINARY SYSTEM

F21.1 The urinary system consists of the *kidneys*, which produce urine; the *ureters*, which carry urine to the *urinary bladder*, where it is temporarily stored; and the *urethra*, which transports urine to the outside of the body (Figure 21.1).

EMBRYONIC DEVELOPMENT OF THE KIDNEYS

The kidneys develop in the embryo from columns of mesodermal cells, called *intermediate mesoderm*, that form just lateral to the somites. The somites give rise to the vertebrae that will form the vertebral column and to the trunk muscles. The columns of intermediate mesoderm begin to form in the superior trunk region of the three-week-old embryo. As the more inferior regions

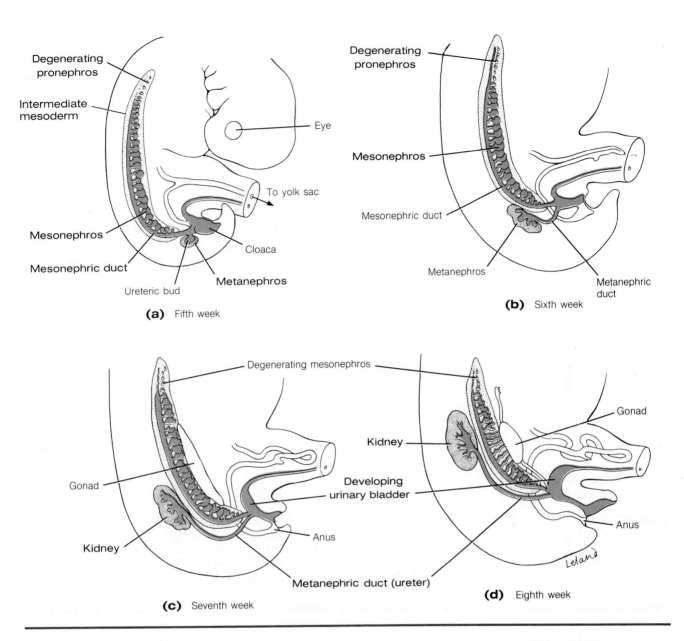

Degenerating pronephros

Intermediate mesoderm

Eye

To yolk sac

Mesonephros

Mesonephric duct

Ureteric bud

Cloaca

Metanephros

(a) Fifth week

Degenerating pronephros

Mesonephros

Mesonephric duct

Metanephros

Metanephric duct

(b) Sixth week

Degenerating mesonephros

Gonad

Kidney

Gonad

Kidney

Developing urinary bladder

Anus

Metanephric duct (ureter)

Anus

(c) Seventh week

(d) Eighth week

Figure 21.2
Embryonic development of the kidneys. The illustration shows the sequential development of the pronephros, the mesonephros, and the metanephros on one side of the embryo.

of the intermediate mesoderm develop into kidneys, the older, more superior regions degenerate.

During embryonic development, three pairs of kidneys form in the intermediate mesoderm (Figure 21.2). The first and most superior kidney to form **F21.2** is called the **pronephros.** Even though the pronephros is never functional in the human, a **pronephric duct** develops and connects each pronephros with the exterior of the body through the embryonic cloaca. The pronephros begins to degenerate in the fourth week of embryonic development and is completely gone by about the sixth week. The pronephric ducts remain, however, and are utilized by the second pair of kidneys, the **mesonephros.**

As the pronephros is degenerating, the mesonephros develops from the columns of intermediate mesoderm inferior to the region of the pronephros. Tubules from the mesonephros join with the pronephric ducts, which are then called the **mesonephric ducts.** By the sixth week, when the development of the mesonephros has reached its inferior limit, its superior portions begin to degenerate. By about the eighth week, only the most inferior regions of the

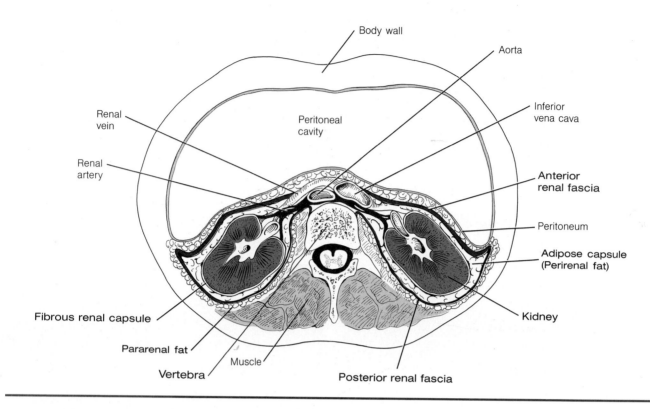

Figure 21.3

Transverse section of the body trunk showing the retroperitoneal location of the kidneys and the renal fascia that surrounds them.

mesonephros remain. Although there is no direct evidence that the mesonephros functions in the human embryo, it appears to be structurally capable of producing urine.

The **metanephros** is the third pair of kidneys to develop from the embryonic intermediate mesoderm. This pair develops into the adult kidneys. In about the fifth week of embryonic development, hollow outgrowths called **ureteric buds** arise from the distal end of each mesonephric duct, near its junction with the embryonic cloaca. The ureteric buds grow dorsally and cranially into the most inferior regions of the intermediate mesoderm. The upper ends of the ureteric buds, which come into contact with the intermediate mesoderm, enlarge. Each ureteric bud then elongates and carries its cap of mesoderm cranially, eventually reaching what will be the permanent positions of the kidneys in the upper lumbar region. With further development, the *nephrons*, which are actively involved in the production of urine, form from the cap of intermediate mesoderm that covers each ureteric bud. The ureteric buds themselves develop into the *collecting tubules*, the *calyces*, the *renal pelvis*, and the *ureters*—all of which transport the urine that is formed in the nephrons to the urinary bladder. These structures remain in the adult; they are described in the next section.

ANATOMY OF THE KIDNEYS

The kidneys are paired reddish-brown organs situated on the posterior wall of the abdominal cavity, one on each side of the vertebral column (Figure **F21.1** 21.1). Each kidney is capped by an endocrine gland called the **adrenal gland**. The kidneys are approximately 11 cm long, extending from the level of the eleventh or twelfth thoracic vertebra to the third lumbar vertebra. Because of the presence of the liver, the right kidney is generally slightly lower than the left. The kidneys are located between muscles of the back and the peritoneal **F21.3** cavity (Figure 21.3). This retroperitoneal location makes it possible for the kidneys to be exposed surgically through the posterior body wall, without opening the peritoneal cavity.

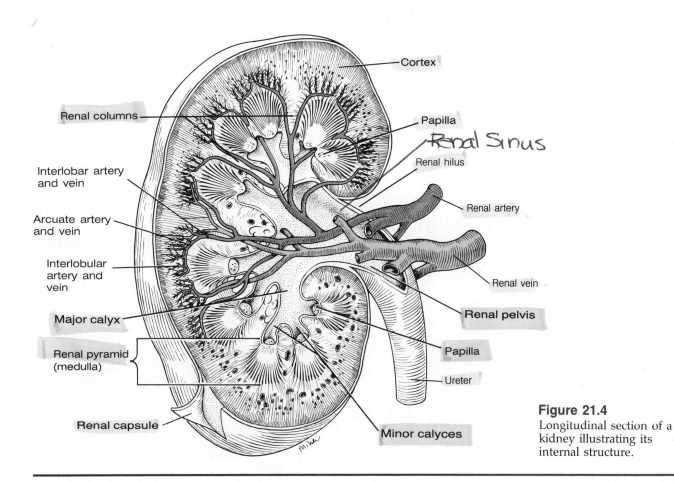

Renal Sinus

- Cortex
- Renal columns
- Papilla
- Renal hilus
- Interlobar artery and vein
- Renal artery
- Arcuate artery and vein
- Interlobular artery and vein
- Renal vein
- Major calyx
- **Renal pelvis**
- Renal pyramid (medulla)
- Papilla
- Ureter
- Renal capsule
- Minor calyces

Figure 21.4
Longitudinal section of a kidney illustrating its internal structure.

Tissue Layers Surrounding the Kidney

Each kidney is surrounded by three layers of tissues (Figure 21.3). The inner- **F21.3**
most layer, which covers the surface of the kidney, is the fibrous **renal capsule.** Surrounding the renal capsule is a mass of perirenal fat called the **adipose capsule.** The third tissue layer that covers the kidney is a double layer of
fascia called the **renal fascia.** The renal fascia surrounds the kidney and the
adipose capsule, completely enclosing them and anchoring the kidney to the
posterior abdominal wall. There is an additional accumulation of fat (*pararenal
fat*) outside of the renal fascia.

External Structure of the Kidney

The kidneys are bean-shaped, with convex lateral borders and concave medial borders. The medial border contains an indentation, called the **renal
hilus,** through which the renal arteries enter the kidney and the renal veins
and the ureter leave. The hilus opens into a space, called the **renal sinus,** in
which the renal vessels and the renal pelvis are located.

Internal Structure of the Kidney

Three general regions can be distinguished in each kidney: the *cortex,* the
medulla, and the *pelvis* (Figure 21.4).   **F21.4**

The **cortex** is the outer layer of the kidney, just deep to the renal capsule.
Extensions of the cortex, called **renal columns,** pass into the medulla of the
kidneys.

The **medulla** is located deep to the cortex and consists of several (up to 18)
triangular **renal pyramids.** The pyramids are oriented so that their broad
bases are covered by the cortex and their tips **(papillae)** project toward the
renal pelvis. The pyramids are separated from one another by the cortical

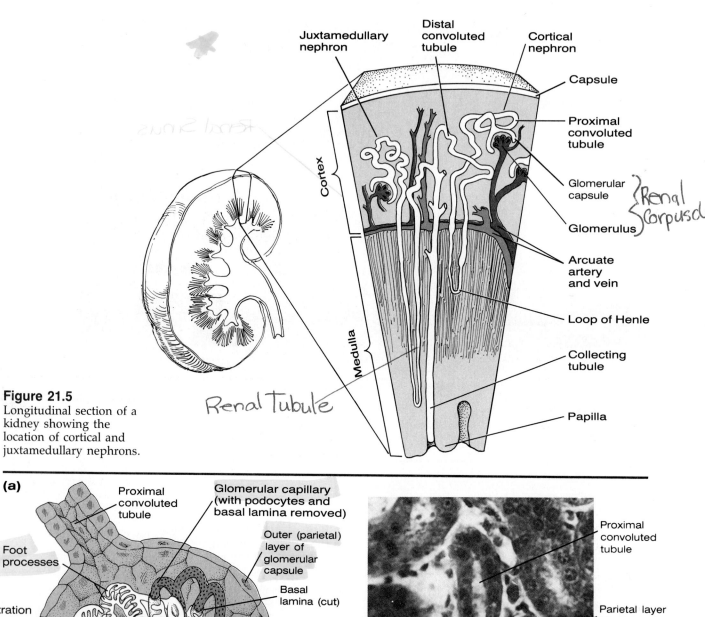

Figure 21.5
Longitudinal section of a
kidney showing the
location of cortical and
juxtamedullary nephrons.

Labels: Juxtamedullary nephron; Distal convoluted tubule; Cortical nephron; Capsule; Proximal convoluted tubule; Glomerular capsule; Glomerulus; Arcuate artery and vein; Loop of Henle; Collecting tubule; Papilla; Cortex; Medulla

Handwritten notes: Renal Sinus; Renal Corpuscle; Renal Tubule

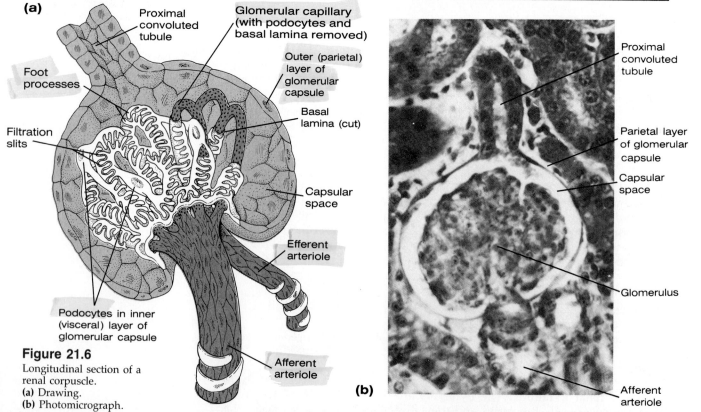

(a)

Figure 21.6
Longitudinal section of a
renal corpuscle.
(a) Drawing.
(b) Photomicrograph.

Labels (a): Proximal convoluted tubule; Glomerular capillary (with podocytes and basal lamina removed); Outer (parietal) layer of glomerular capsule; Basal lamina (cut); Capsular space; Efferent arteriole; Afferent arteriole; Foot processes; Filtration slits; Podocytes in inner (visceral) layer of glomerular capsule

(b)

Labels (b): Proximal convoluted tubule; Parietal layer of glomerular capsule; Capsular space; Glomerulus; Afferent arteriole

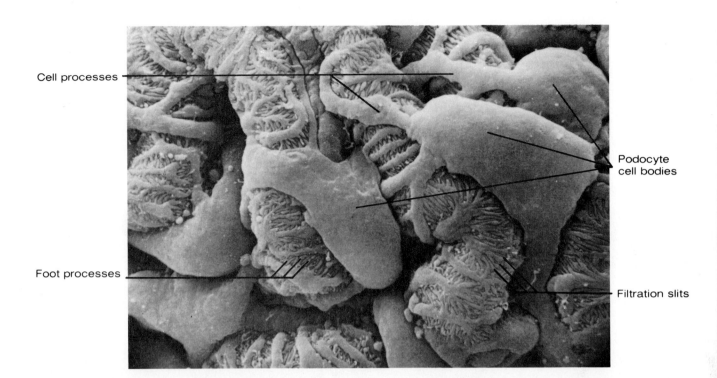

Cell processes

Podocyte
cell bodies

Foot processes

Filtration slits

Figure 21.7
Scanning electron
micrograph of podocytes
surrounding glomerular
capillaries (×3725). (*From
Tissues and Organs: A Text-
Atlas of Scanning Electron
Microscopy* by Richard G.
Kessel and Randy H.
Kardon. W.H. Freeman
and Company. Copyright
© 1979.)

renal columns. Blood vessels that supply the cortex and medulla pass through the renal columns.

The papilla of each pyramid projects into a funnel-shaped chamber called a **minor calyx.** Several minor calyces join together to form a **major calyx.** There are generally 2 or 3 major calyces and 8 to 13 minor calyces in each kidney. The major calyces join with one another to form the **renal pelvis,** which is the expanded upper end of the ureter. Urine passes as droplets from tiny pores in the papillae into the minor calyces. From there it travels into the major calyces, the renal pelvis, and finally into the ureter, which carries it to the urinary bladder.

The Renal Tubules

The **renal tubules,** which are the functional units of the kidneys where urine is formed, consist of **nephrons** and **collecting tubules.** There are estimated to be over 1 million nephrons in each kidney. Some nephrons—called **juxta-medullary nephrons**—extend deep into the medulla. Other nephrons—called **cortical nephrons**—do not penetrate as deeply into the medulla (Figure 21.5). Each nephron consists of two parts: (1) a network of parallel capillaries **F21.5** called a **glomerulus** and (2) a **tubule.** Various regions of each tubule differ from one another anatomically. Epithelial variations along the length of a tubule are related to variations in function. The proximal end forms a double-walled cup known as the **glomerular capsule** (or *Bowman's capsule*), which surrounds the glomerulus. The capsule and the glomerulus together are called the **renal corpuscle** (Figure 21.6). The renal corpuscles are located in **F21.6** the cortical region of the kidney.

The outer (parietal) layer of the glomerular capsule is composed of simple squamous epithelium that rests on a thin *basal lamina.* The inner (visceral) layer of the capsule is composed of specialized cells called **podocytes.** The podocytes have several processes that radiate from a central cell body and adhere to the basal lamina covering the squamous cells of the endothelium of the capillaries that form the glomerulus (Figure 21.7). These processes, in **F21.7** turn, branch into secondary and tertiary processes. The tertiary processes of the podocytes are called **foot processes,** or **pedicles.** The foot processes of one podocyte interdigitate with those of an adjacent podocyte, leaving an elabo-

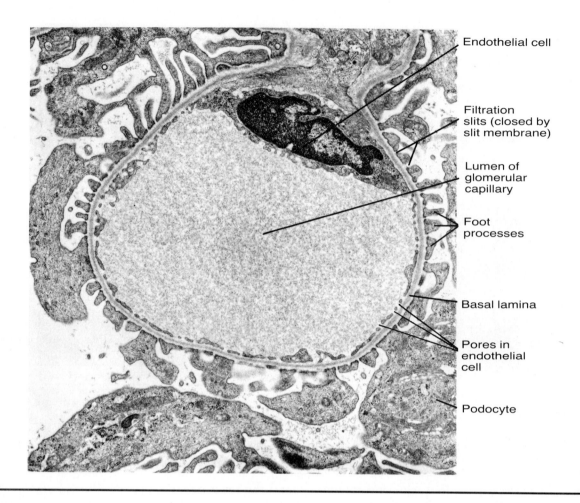

Endothelial cell

Filtration slits (closed by slit membrane)

Lumen of glomerular capillary

Foot processes

Basal lamina

Pores in endothelial cell

Podocyte

Figure 21.8

Transmission electron micrograph of a cross section of a glomerular capillary. The filtration barrier is composed of fenestrated endothelium, basal lamina, and filtration slits.

rate network of small clefts between them. These intercellular clefts are called **filtration slits,** or **slit pores.** A very thin **slit membrane** extends between the foot processes of adjacent cells, forming a diaphragm-like barrier that restricts the passage of molecules through the filtration slits. The capillaries that form the glomeruli are similar to other capillaries in that they consist of endothelium formed of a single layer of squamous cells; but they are a bit different due to the presence of many small (600–900-Å diameter), open pores that perforate the endothelial cells. Because of these pores, the endothelium of the glomeruli is referred to as *fenestrated endothelium* (*fenestra* = window).

The *filtration barrier* that separates the blood in the glomerular capillaries from the space in a glomerular capsule **(capsular space)** consists only of (1) the fenestrated endothelium, (2) the basal lamina, and (3) the slit membrane that

F21.8 covers the filtration slits (Figure 21.8). Consequently, many substances are able to pass through the barrier, and the renal corpuscle is the site where most substances leave the blood and enter a nephron. Not all molecules are able to pass through the filtration barrier, however. The endothelial pores restrict the movement of blood cells and molecules larger than 160 Å in diameter, and the basal lamina and the slit membrane act as barriers to smaller molecules, allowing only those smaller than the plasma proteins (70 Å in diameter) to pass through. As a result, the **glomerular filtrate** that enters the glomerular cap-sule includes most of the substances present in the blood plasma except for

F21.9 blood cells and plasma proteins (Figure 21.9). The composition of the filtrate is similar to that of interstitial fluid.

Beyond its glomerular capsule each nephron forms a tightly looping tu-bule whose lumen is continuous with the capsular space. This region of the

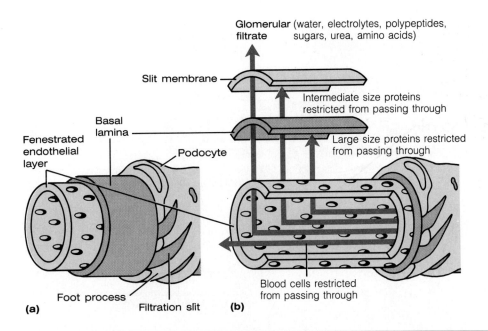

Glomerular (water, electrolytes, polypeptides,
filtrate   sugars, urea, amino acids)

Slit membrane

Intermediate size proteins
restricted from passing through

Basal
lamina

Large size proteins restricted
from passing through

Fenestrated
endothelial
layer

Podocyte

Foot process

Filtration slit

Blood cells restricted
from passing through

(a)

(b)

Figure 21.9
(a) Structure of a glomerular capillary with a surrounding podocyte. **(b)** Restrictions imposed by the various layers of the filtration barrier.

nephron is referred to as the **proximal convoluted tubule** because it is located in the cortex, close to the capsule, and is twisted (Figure 21.10). The wall of each proximal convoluted tubule consists of a single layer of thick cuboidal or pyramidal cells. The free surfaces of these cells, facing the lumen of the tubule, have many microvilli.

Beyond its proximal convoluted tubule, each nephron first has a straight portion, the **proximal straight tubule,** and then forms a **loop of Henle (ansa nephroni)** (Figure 21.10), which passes into a pyramid of the medulla of the kidney. The loops of Henle of juxtamedullary glomeruli are longer than those of the cortical glomeruli. As a consequence, the loops of Henle of juxtamedullary nephrons extend deeper into the medulla than do the loops of Henle of the more superficially located cortical nephrons. The portion of each loop of Henle that descends into the medulla is called the **descending limb** of the loop of Henle. Because the epithelium of the tubule wall changes to thin squamous cells in the descending limb, this region is also referred to as the **thin segment** of the loop of Henle. Within the medulla the tubule makes a hairpin turn and ascends toward the cortex as the **ascending limb** of the loop of Henle, which passes out of the medulla and back into the cortex. The wall of the ascending limb is composed primarily of cuboidal cells and this portion of the nephron is therefore also referred to as the **thick segment** of the loop of Henle. The thick segment of the loop of Henle is also known as the **distal straight tubule.** In juxtamedullary nephrons the thin segment often extends around the hairpin turn and into the ascending limb. In these nephrons, therefore, the thick segment does not begin until the upper portion of the ascending limb.

Beyond its distal straight tubule, each nephron again becomes highly coiled. This portion is called the **distal convoluted tubule.** The wall of each distal tubule is composed of a single layer of cuboidal cells. Within the cortex of the kidney the distal tubule contacts the blood vessel (the *afferent arteriole*) that carries blood to the glomerulus. At this point of contact, the cells of both the blood vessel and the tubule are modified. Among the smooth-muscle cells of the tunica media of the afferent arterioles are cells that contain prominent granules in their cytoplasm. These cells are called *juxtaglomerular cells.* Where the distal convoluted tubule contacts the juxtaglomerular cells, there is a concentration of nuclei and the cells appear higher than in other parts of the distal tubule. This region is the *macula densa.*

F21.10

F21.10

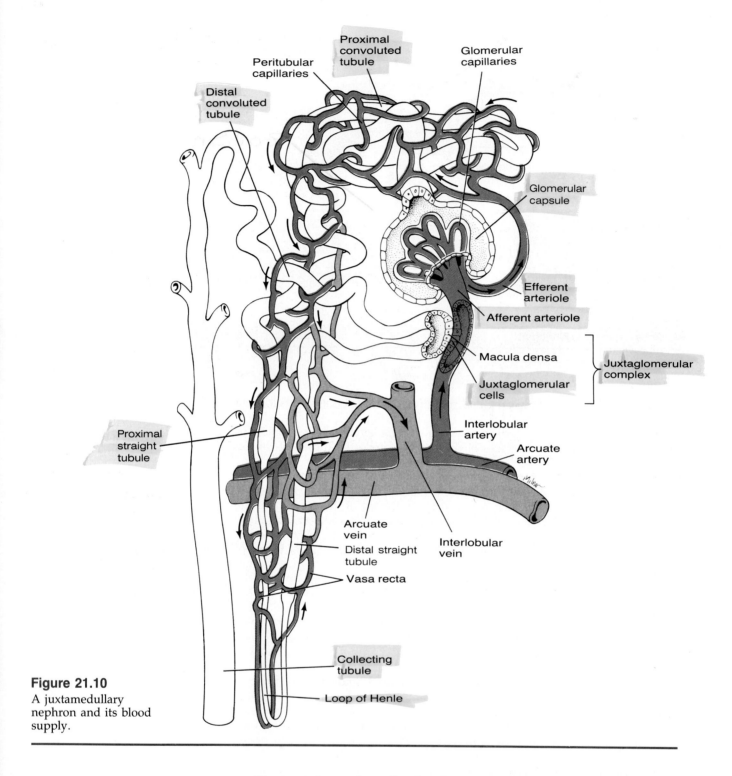

Figure 21.10
A juxtamedullary nephron and its blood supply.

F21.10 The juxtaglomerular cells plus the macula densa form a structure called the **juxtaglomerular complex,** (or **juxtaglomerular apparatus**) (Figure 21.10). The juxtaglomerular cells are thought to secrete an enzyme *(renin)* in response to lowered blood pressure. Through its effects on angiotensinogen, a plasma protein, renin causes arterioles to constrict and promotes the release of the hormone aldosterone from the adrenal cortex. Aldosterone acts on the kidney tubules, causing them to reabsorb more sodium, and the increased sodium reabsorption causes an increase in the reabsorption of water. The increased reabsorption of water into the blood increases the volume of the blood. As a consequence of both the constriction of arterioles and the increased blood

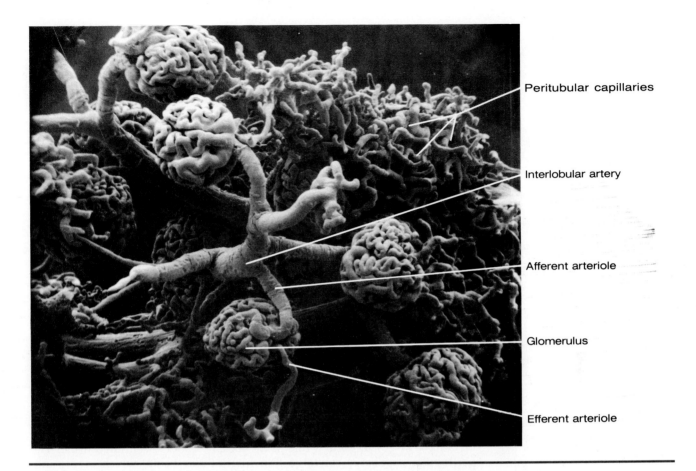

Peritubular capillaries

Interlobular artery

Afferent arteriole

Glomerulus

Efferent arteriole

Figure 21.11

Scanning electron micrograph of the blood vessels associated with glomeruli (×206). (*From Tissues and Organs: A Text-Atlas of Scanning Electron Microscopy* by Richard G. Kessel and Randy H. Kardon. W.H. Freeman and Company. Copyright © 1979.)

volume, renin indirectly raises the blood pressure in the kidney. This elevated kidney blood pressure plays an important role in kidney functioning, since the blood pressure in the glomerulus must be maintained at a level high enough to cause filtration of substances into a glomerular capsule.

The distal convoluted tubules of several nephrons empty into a common **collecting tubule** that transports urine back into the renal pyramids of the medulla. From 10 to 25 collecting tubules open on the papilla of each pyramid and empty into a minor calyx. The cells that make up the walls of the collecting tubules vary from cuboidal to columnar.

Blood Vessels of the Kidney

The importance of the relationship between the kidneys and the blood vascular system becomes apparent when one considers the large size of the **renal arteries** that supply the organ. It has been estimated that, at rest, these vessels carry to the kidneys about 20% of the total cardiac output. In young adults, approximately 1100 ml of blood passes through the two kidneys every minute. Very little of this blood supplies the kidneys' nutritive needs. Rather, the large bloodflow is related to the fact that the kidneys can maintain the homeostasis of the blood only if a considerable amount of blood passes through them.

Shortly after entering the hilus of the kidney, the renal artery divides into ventral and dorsal sets of vessels that pass on either side of the renal pelvis (Figure 21.4). These vessels travel between the pyramids and through the renal columns as the **interlobar arteries.** At the bases of the pyramids, which represent the junction between the medulla and the cortex, the interlobar arteries form arching branches that run parallel to the surface of the kidney. These are the **arcuate arteries.** At intervals, the arcuate arteries give off small **interlobular arteries** that travel through the cortex toward the surface of the

F21.4

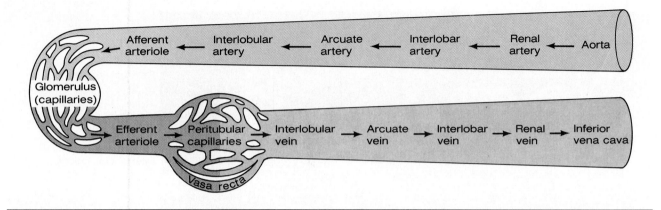

Figure 21.12
Summary of the pathway of blood through the kidney.

kidney. The interlobular arteries divide into several **afferent arterioles,** each of which supplies a renal corpuscle and forms a capillary network called the **glomerulus** (Figure 21.11). As mentioned earlier, it is at the glomerulus that blood comes in close contact with the cells of the glomerular capsule, and the glomerular filtrate is formed.

F21.11

Blood leaves each glomerulus through an **efferent arteriole** that passes out of the renal corpuscle and divides into a network of capillaries surrounding the proximal and distal convoluted tubules of the nephron. These capillaries are called the **peritubular capillaries** (Figure 21.10). Peritubular capillaries also surround the loops of Henle of the cortical nephrons. Thin-walled vessels called **vasa recta** extend from the efferent arterioles of the juxtamedullary nephrons to supply their loops of Henle and the collecting ducts. The vasa recta have a very important role in the formation of concentrated urine. The peritubular capillaries converge into **interlobular veins,** which empty into **arcuate veins** and **interlobar veins** before leaving the kidney through the **renal veins.** The veins follow the same general course as the arteries of the same names. The pathway of blood through the kidney is summarized in Figure 21.12.

F21.10

F21.12

There are several unusual and important features concerning bloodflow through the kidney. As blood moves through the kidney, it flows through two sequential series of capillary beds—the glomerulus and the peritubular capillaries. The kidney is one of the few places in the body where this arrangement occurs. Generally, when blood leaves capillaries it enters venules. In the kidney, however, efferent arterioles transport the blood from the first capillary bed (glomerulus) to the second capillary bed (peritubular capillaries). The presence of arterioles leading to and from the glomeruli has great functional significance. Like all arterioles, these vessels have smooth muscles in their walls that permit them either to constrict or to dilate in response to various types of stimulation. Thus, it is possible for a fairly constant blood pressure to be maintained in the glomeruli even if the systemic pressure fluctuates. This constant blood pressure contributes to the efficient functioning of the kidney in spite of varying systemic conditions.

PHYSIOLOGY OF THE KIDNEYS

The excretory and regulatory activities of the kidneys depend in large measure on the efficient functioning of the glomeruli and the renal tubules (nephrons and collecting tubules). The glomeruli and renal tubules participate in several different activities, including **glomerular filtration, tubular reabsorption,** and **tubular secretion** (Figure 21.13). As blood flows through the kidneys, some of the plasma is *filtered* out of the vascular system through the fenestrated endothelium, the basal lamina, and the slit membranes into the glomerular capsules of the renal tubules. This process, which is primarily due

F21.13

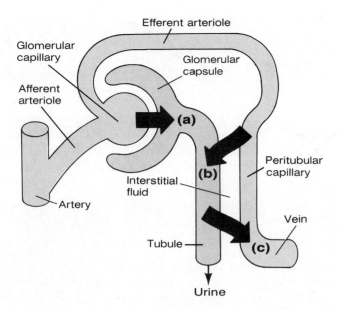

Figure 21.13
Substances can enter the renal tubules either (a) by filtration from the glomerular capillaries into a glomerular capsule or (b) by secretion into the tubules from the peritubular capillaries. Substances can leave the tubules and enter the interstitial fluid and return to the blood by (c) tubular reabsorption.

to the blood pressure within the glomeruli, is not very selective, and almost all plasma components—with the exception of large protein molecules—can enter the renal tubules. As the filtrate flows along the tubules, water, electrolytes, glucose, amino acids, and other substances required by the body are *reabsorbed* from them and returned to the blood by both active and passive transport processes. Moreover, materials within the blood in the peritubular capillaries and the vasa recta that were unable to pass through the filtration barrier into the renal tubules can be *secreted* into the tubules. The reabsorption and secretion processes are selective, and hormonal mechanisms enable the kidneys to exert a high degree of control over the amounts of many materials— particularly water and electrolytes—that are reabsorbed and returned to the blood or secreted into the tubular fluid.

When the reabsorption and secretion processes are completed, the fluid remaining within the renal tubules is transported to other components of the urinary system to be excreted as urine. Thus, urine consists of water and materials that were filtered or secreted into the renal tubules but not reabsorbed.

As a result of filtration, reabsorption, and secretion, the renal tubules maintain the composition, volume, and pH of the blood plasma within narrow limits. Through their effects on the blood plasma, the renal tubules also regulate the composition, volume, and pH of the tissue fluid.

URETERS

The urine drips from the collecting tubules at the tips of the papillae and enters the minor calyces. The minor calyces join with the major calyces, which, in turn, join with the renal pelvis. From the renal pelvis, urine is transported to the urinary bladder by **ureters,** one from each kidney (Figure 21.14). The ureters descend between the parietal peritoneum and the body wall to the pelvic cavity, where they turn medially and enter the urinary bladder on its posterior lateral surfaces. Before opening into the bladder, the ureters travel obliquely through the bladder wall. As a result, contraction of the muscles of the bladder wall can compress the ureters and help prevent urine from flowing back into the ureters from the bladder. This occurs during bladder emptying *(micturition).* In effect, the muscles of the bladder wall act as

F21.14

FRONTIERS IN HEALTH

New Surgery for Kidney Stones

Kidney stones are one of the most painful maladies that afflict humans. The sharp, often excruciating pain stabs the back and side or lower abdomen with such force that even powerful pain-killers are sometimes ineffective.

One in every ten men and one in every forty women will require medical attention because of kidney stones. Hundreds of thousands of Americans enter hospitals for treatment of kidney stones each year. Luckily, for the majority of them no surgery is required. The stones are allowed to pass through the urinary ducts and are excreted in the urine. But for some 20,000 to 50,000 people each year, surgery to remove the stones becomes necessary.

Until recently, kidney surgery was a major undertaking, and recovery was slow and painful. A 6- to 8-inch incision in the patient's side was required to allow surgeons to remove the stones, and a hospital stay of about 10 days was usual. After release from the hospital, another 8 weeks of recuperation at home were necessary before a patient could return to work.

Thanks to research that began in West Germany, many patients can now be up and walking the day after their surgery and can be back on the job in a week. The only visible sign of the surgery is a ¼-inch scar.

Instead of making a large incision to expose the kidney, surgeons now use a small hollow metal device, called a nephroscope. The nephroscope is inserted into the abdominal cavity through a tiny incision in the patient's side and, with the aid of fiber optics, guided into the kidney, from which it is possible to view the entire upper urinary tract. After locating the kidney stone the surgeon aspirates it if the stone is small. For larger stones, a tiny basket-like grabbing device is introduced through the nephroscope. The stone is clasped in the claws of this device and physically crushed. The crushed stone can then be sucked out through the hollow nephroscope. For still larger stones, however, a new method called percutaneous ultrasound lithotripsy is often used.

Dr. P. Alken and his colleagues at the University of Mainz Medical School in Mainz, West Germany were the first to use percutaneous ultrasound lithotripsy to pulverize kidney stones prior to their removal. This technique was introduced into the United States in 1981, and today it is used routinely in several major medical centers.

Lithotripsy is one of the many new medical uses of ultrasound—very high frequency sound waves inaudible to the human ear. To break up kidney stones, surgeons insert a hollow metal ultrasound probe through the nephroscope. The probe is then pushed into contact with the stone and turned on. Once the ultrasonic vibrations cause the hard outer "shell" of the kidney stone to crack, the remainder of the stone generally falls apart. Surgeons then suck out the fragments with a vacuum aspirator.

Clinical trials of lithotripsy have been very encouraging. A 1984 paper in the medical journal *Radiology* reports that surgeons were successful in removing all frag-

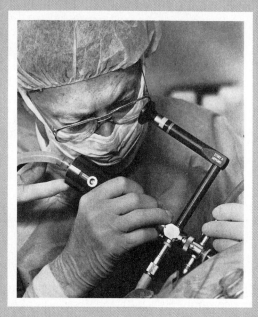

Surgeon using ultrasound through a nephroscope to break up kidney stones.

ments of the kidney stone in 80% of the patients on the first attempt. The remainder of the cases required a second try. Overall, the success rate was 97%.

Although complications from ultrasound lithotripsy are about the same as for conventional kidney surgery, this new technique has several advantages for the patients. Discomfort and pain are drastically reduced, and recovery time and medical costs are cut to a fraction of what they were prior to the introduction of this technique.

As successful as ultrasound lithotripsy has been, however, the West German researchers have continued to look for new ways to remove kidney stones. A new method they are now working with may someday replace ultrasound lithotripsy.

Imagine a surgical technique that requires no surgery. A patient with a kidney stone is mildly sedated, given a local anesthetic, and placed in a warm water bath. Sitting perfectly still, he can listen to his favorite music while ultrasound waves, focused on the kidney stone, penetrate his skin and shatter the stone that is lodged in his kidney. When the procedure is completed, he returns home to pass the tiny fragments in the urine.

Sound like fantasy? At least 1000 West Germans have been treated with the technique and are now free of their painful stones. Six medical centers in the United States began testing the technique in the summer of 1984, and although the studies are not complete, early results are quite promising. However, despite an outstanding success rate and only minor side effects, this new technique may not become available to everyone. The machine alone costs $2 million, and only the largest medical centers will be able to afford it.

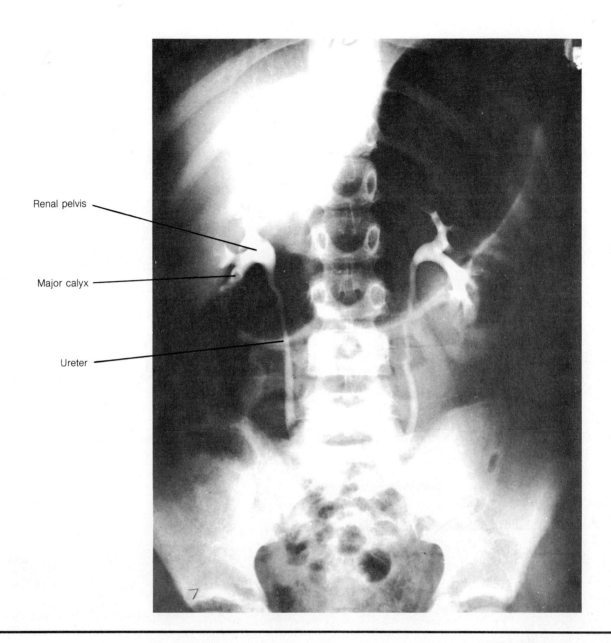

Renal pelvis

Major calyx

Ureter

Figure 21.14
X ray of the urinary tract
by retrograde pyelogram.
The lower halves of the
ureters are not visible,
since they do not contain
contrast medium.

sphincters on the ureters. Valvelike folds of the mucous membrane lining of the bladder cover the orifices of the ureters and assist in preventing urine from flowing back into the ureters during micturition.

The walls of the ureters are composed of three layers: an inner *mucosal layer*, a middle *muscular layer*, and an outer *fibrous layer*. The mucosa has a surface layer of transitional epithelium, which is also typical of the urinary bladder and the urethra. The muscular layer of the ureter is capable of undergoing peristaltic contractions, thus propelling urine to the bladder.

URINARY BLADDER

The **urinary bladder** is a hollow muscular organ used to store urine (Figure 21.15). The bladder rests on the floor of the pelvic cavity; like other urinary **F21.15** structures, it is retroperitoneal. The anterior surface of the bladder lies just behind the pubic symphysis. In males, the bladder is in front of the rectum. In females, it lies just anterior to the uterus and the superior portion of the

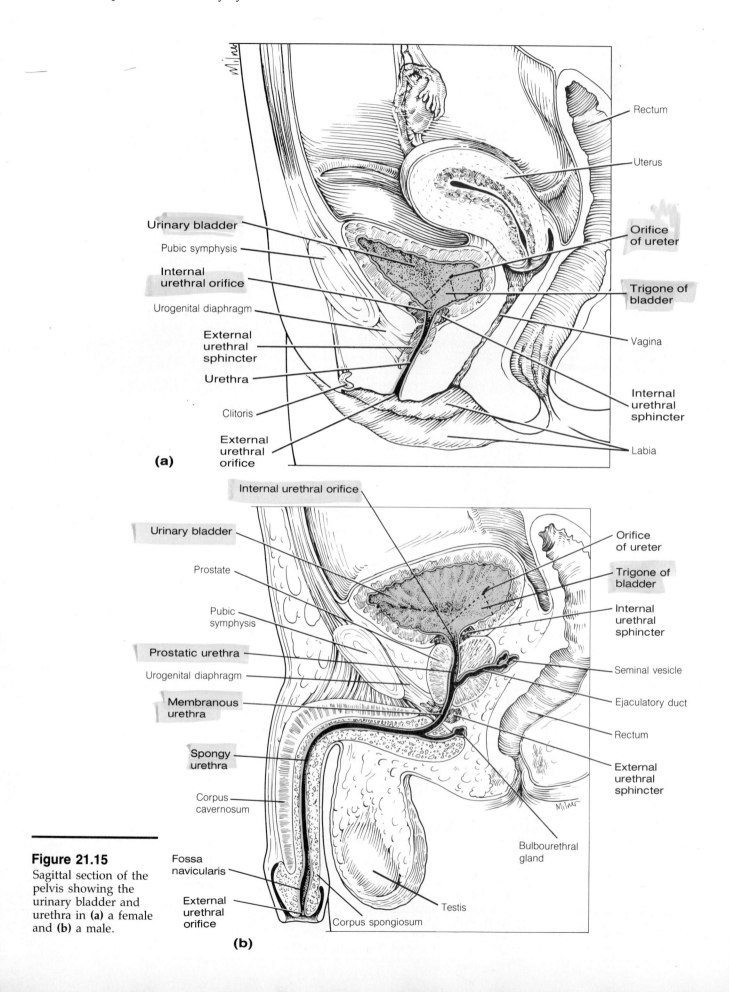

(a)

Rectum

Uterus

Urinary bladder

Pubic symphysis

Internal urethral orifice

Urogenital diaphragm

External urethral sphincter

Urethra

Clitoris

External urethral orifice

Orifice of ureter

Trigone of bladder

Vagina

Internal urethral sphincter

Labia

Internal urethral orifice

Urinary bladder

Prostate

Pubic symphysis

Prostatic urethra

Urogenital diaphragm

Membranous urethra

Spongy urethra

Corpus cavernosum

Fossa navicularis

External urethral orifice

Corpus spongiosum

(b)

Orifice of ureter

Trigone of bladder

Internal urethral sphincter

Seminal vesicle

Ejaculatory duct

Rectum

External urethral sphincter

Bulbourethral gland

Testis

Figure 21.15
Sagittal section of the pelvis showing the urinary bladder and urethra in **(a)** a female and **(b)** a male.

vagina. When full, the bladder is spherical; but when empty, it resembles an inverted pyramid.

The urinary bladder, like the ureters, is lined with a mucous membrane of transitional epithelium. Covering the transitional epithelium is a tunic of muscle called the **detrusor muscle,** which consists of three layers of smooth muscle—inner and outer longitudinal layers on either side of a prominent circular layer. Throughout most of the bladder, the mucous membrane is loosely attached to the underlying muscular coat, and it appears wrinkled when the bladder is contracted. However, the internal opening of the urethra anteriorly and the openings of the two ureters laterally mark the corners of a triangular area in which the mucous membrane is firmly attached to the muscular coat and where it is, therefore, always smooth. This smooth, triangular region is called the **trigone** of the bladder.

The bladder can hold 600 to 800 ml of urine, but it is generally emptied before it reaches this capacity. As the bladder fills with urine, its walls are stretched, stimulating receptors within the walls to transmit increasing numbers of sensory impulses to the sacral region of the spinal cord. These impulses ultimately stimulate parasympathetic neurons that supply the smooth muscles of the bladder wall and inhibit somatic motor neurons that supply the external urethral sphincter muscle. As a consequence, when approximately 300 ml of urine has accumulated within the bladder, the muscles of the bladder wall contract, the external urethral sphincter relaxes, and *micturition* (urination or voiding) occurs.

Although micturition as just described is essentially the result of a local spinal reflex, it is influenced by higher brain centers in the brain stem and cerebral cortex. As sensory impulses are being transmitted to the spinal cord from receptors within the bladder walls, some impulses are also sent to higher brain centers. These impulses can lead to a sensation of a full bladder and to a feeling of a need to urinate. Moreover, impulses sent from the brain can either facilitate or inhibit this reflex emptying of the bladder, and with training it is possible to gain a high degree of voluntary control over micturition. As a result, urination can be either voluntarily induced or postponed until an opportune time. However, until control is developed and training is complete, the reflex response is the dominant factor governing bladder emptying. A baby, therefore, urinates whenever its bladder is sufficiently full to activate the spinal reflex.

URETHRA

The **urethra** is a muscular tube, lined with mucous membrane, that exits from the inferior surface of the urinary bladder. It carries urine from the bladder to the exterior of the body. At the junction of the urethra and bladder, the smooth muscle of the bladder surrounds the urethra and acts as a sphincter—called the **internal urethral sphincter**—that tends to keep the urethra closed. During bladder emptying, the contraction of the bladder and the resulting changes in its shape pull the sphincter open. Thus, no special mechanism is required to relax this sphincter. As the urethra passes through the pelvic floor **(urogenital diaphragm),** it is surrounded by skeletal muscles that form the **external urethral sphincter.** When constricted, this sphincter—which is under voluntary control—is able to hold the urethra closed against strong bladder contractions.

In the female, the urethra is short (approximately 4 cm), and it runs along the anterior surface of the vagina (Figure 21.15*a*). It opens to the exterior at the **F21.15a** **external urethral orifice,** which is located between the clitoris and the vaginal orifice.

The male's urethra is about 20 cm long and extends to the external urethral orifice at the tip of the penis (Figure 21.15*b*). In the male, the urethra is **F21.15b** divisible into three parts—the *prostatic, membranous,* and *spongy urethrae,* which are named according to the regions through which the urethra passes. The **prostatic urethra** passes through the prostate gland and receives the ejaculatory ducts, which carry sperm, on its posterior walls. Beyond the point of junction with the ejaculatory ducts, the male urethra is used for reproduction as well as urine transport. The short portion of the urethra that passes through the urogenital diaphragm is called the **membranous urethra.** The **spongy** (*cavernous*) **urethra** is the longest portion, extending from just below the urogenital diaphragm to the external urethral orifice on the glans penis. Within the glans penis, the urethra dilates into a small chamber called the **fossa navicularis.** A short distance below the urogenital diaphragm, the spongy urethra receives the ducts of the bulbourethral glands of the reproductive system. In the penis, the urethra passes through the **corpus spongiosum,** one of three columns of erectile tissue.

Box 21.1
Artificial Kidney

Since effective kidney function is necessary for survival, kidney disease can often prove fatal. The development of the artificial kidney machine, however, has greatly alleviated a number of the problems that accompany kidney disease. The artificial kidney employs the principle of dialysis to accomplish the removal of waste materials from the blood. Dialysis involves the separation of molecules of different size by passing the solution they are in through a semipermeable membrane.

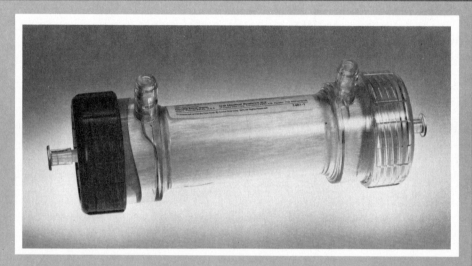

Dialysis unit of an artificial kidney. Blood enters and leaves the unit through the openings on the ends. The dialysis solution enters and leaves through the two openings on the side of the unit.

In the artificial kidney, the patient's blood is passed through a dialysis tubing that allows urea, electrolytes, and other small molecules to move freely across its wall but does not allow the movement of large protein molecules. The dialysis tubing is immersed in a bath containing various substances. If urea or some other small molecule is present in the blood but not in the bath, it diffuses out of the dialysis tubing and into the bath and thus is removed from the blood. If a specific low-molecular-weight substance is present in the bath at the same concentration as within the blood, then it diffuses into the tubing as fast as it diffuses out, and there is no net loss of the material from the blood. Thus, regulating the composition and concentration of materials within the bath acts to regulate the types and amounts of materials that leave the blood by net diffusion out of the dialysis tubing. In this way, waste products are removed from the blood while needed constituents are retained.

CONDITIONS OF CLINICAL SIGNIFICANCE

The Urinary System

Pressure-Related Pathologies

Factors that upset the pressure relationships determining glomerular filtration rates can interfere with normal kidney functioning.

Prostate Hypertrophy

In males, it is not unusual for the prostate gland, which surrounds the urethra just below the bladder, to hypertrophy (enlarge) and compress the urethra so that it becomes difficult for urine to leave the bladder. As urine accumulates within the bladder, the pressure within the bladder rises. If the pressure increases enough, it causes a backing up of urine in the ureters, which produces a dilation of the renal pelvis and calyces and leads to an increased pressure within the glomerular capsules. The increased capsular pressure reduces the glomerular filtration rate and thus interferes with the kidneys' regulation of body fluid composition.

Low Arterial Blood Pressure

Blood pressure changes associated with heart failure, hemorrhage, or shock can cause renal failure. In all of these conditions, there is a drop in arterial blood pressure, which reduces the ability of the kidneys to form glomerular filtrate. In addition, a substantial drop in blood pressure can activate reflexes involved in maintaining normal pressure within the major arteries of the body. These reflexes cause the afferent arterioles of the kidneys to constrict. The constriction of the afferent arterioles helps elevate the general body blood pressure, but it diminishes the blood pressure within the glomeruli and further reduces the formation of glomerular filtrate. The reduction in the formation of glomerular filtrate diminishes the kidneys' ability to excrete wastes and regulate body fluid composition.

Glomerulonephritis

Glomerulonephritis is an inflammatory condition that primarily affects the glomeruli, though there is secondary damage to parts of the nephrons. It is often due to streptococcal infection, although the bacteria do not invade the kidneys themselves. Rather, glomerulonephritis seems to be an allergic (antigen–antibody) reaction to the toxins produced by certain streptococcal bacteria from infections in other parts of the body, such as throat infections and endocarditis. In the acute phase, the glomeruli are so inflamed and swollen that they allow erythrocytes and large amounts of plasma proteins to pass into the glomerular filtrate. The inflammation can become chronic, and many of the glomeruli and renal tubules may be replaced by scar tissue. Chronic glomerulonephritis can lead to elevated blood pressure and renal failure unless a kidney transplant is performed or an artificial kidney is used. Chronic glomerulonephritis generally results in death. However, most individuals afflicted with acute glomerulonephritis have normal kidney function again within a few months.

Pyelonephritis

Pyelonephritis is a bacterial infection of the renal pelvis and the surrounding tissues of the kidney. In contrast to glomerulonephritis, the bacteria in pyelonephritis are present in the kidneys themselves. The bacteria reach the kidneys from other sites of infection by way of the bloodstream or lymphatics, or the infection may spread up the ureters from the bladder. In pyelonephritis, the kidney may become swollen and congested and the renal pelvis may become inflamed and filled with pus. Abscesses often develop on the surface of the kidney. Pyelonephritis generally responds well to treatment with antibiotics. In chronic cases, however, extensive scar tissue is formed in the kidney and renal failure becomes a possibility.

Proteinuria

In a number of renal diseases, the permeability of the glomerular capillaries may be increased to such an extent that large amounts of plasma proteins (mostly albumin) pass into the glomerular filtrate and are excreted in the urine. This condition is called *proteinuria*, or *albuminuria*. In severe cases, the loss of plasma proteins can be so great that the plasma osmotic pressure decreases substantially. As a result there is an increased tendency for fluid to leave the systemic blood vessels and enter the tissue spaces—a condition that produces generalized edema (swelling) of the body.

Uremia

If the metabolic products (such as urea) that are derived from the breakdown of proteins are not excreted properly, they accumulate in the blood and produce a condition called *uremia*. Uremia affects several body systems, including the nervous system (convulsions and coma), digestive system (vomiting and diarrhea), and respiratory system (dyspnea, or labored breathing).

Kidney Stones

Kidney stones (renal calculi) sometimes form in the renal pelvis or the urinary bladder. The stones generally consist of various combinations of uric acid, calcium oxalate, calcium phosphate, and certain other substances. What causes the stones to form is not known. However, there seems to be some correlation between their formation and kidney infections, high concentration of salts in the urine, vitamin A deficiency, and hyperparathyroidism caused by a tumor.

A stone formed in the renal pelvis may remain there, or it may enter the ureters and pass to the bladder. The stone often causes severe, painful contractions of the ureter as it travels through it. A more serious condition results if a stone becomes lodged in the ureter, obstructing the flow of urine to the bladder. In addition to the retention of urine, kidney stones can cause ulcerations in the lining of the urinary tract, which makes the tract more prone to infection.

Cystitis

Cystitis is an inflammation of the urinary bladder accompanied by frequent and burning urination and by blood in the urine. In the acute form of cystitis, the mucous membrane lining of the bladder becomes swollen and some bleeding occurs. In the chronic condition, the wall of the bladder may become thickened and its capacity reduced.

The bladder is generally quite resistant to bacterial infection, but under certain conditions bacteria become established in the bladder lining, thus producing cystitis. Cystitis can also be caused by chemicals or by mechanical irritation, such as catheterization. Women have a higher incidence of cystitis than men, probably due to their short urethra, which makes it easier for bacteria to reach the bladder from outside the body. It is not uncommon for *E. coli* bacteria from a woman's anal region to infect her urethra as a result of improper cleansing of the area.

Effects of Aging on the Kidneys

There is a progressive decline in kidney function with aging. At age 70 the glomerular filtration rate is only about 50% of the rate at age 40. Renal bloodflow decreases from approximately 1100 ml per minute at ages 20 to 45 to only about 475 ml per minute at 80 to 89 years of age. There is a corresponding decrease in the function of the renal tubules and in their ability to concentrate the tubular fluid. The decrease in tubular functioning may slow the removal of drugs administered for medical treatments from the body, causing the medications to accumulate in the blood in higher than optimal levels. For this reason, the dosages of certain medications may need to be adjusted as a person becomes older. However, the kidneys do retain their ability to regulate the acid–base balance of the body, although they respond less quickly to a sudden, large acid load.

The pathological changes that often occur in aging kidneys largely involve degeneration and sclerosis (hardening) of glomeruli and the consequent functional loss of sclerotic glomeruli.

STUDY OUTLINE

COMPONENTS OF THE URINARY SYSTEM
urinary system is composed of kidneys, ureters, urinary bladder, and urethra.   **p. 568**

EMBRYONIC DEVELOPMENT OF THE KIDNEYS   **pp. 568–570**

FROM INTERMEDIATE MESODERM three successive pairs of embryonic kidneys develop:

PRONEPHROS (with pronephric duct)

MESONEPHROS (with mesonephric duct)

METANEPHROS becomes adult kidney.

URETERIC BUDS develop into collecting tubules, calyces, renal pelvis, and ureters.

NEPHRONS develop from cap of intermediate mesoderm that covers ureteric bud.

ANATOMY OF THE KIDNEYS   paired reddish-brown, bean-shaped organs on posterior abdominal wall; retroperitoneal.   **pp. 570–578**

TISSUE LAYERS SURROUNDING THE KIDNEYS three layers: renal capsule (fibrous), adipose capsule (perirenal fat), renal fascia (double-layered).

EXTERNAL STRUCTURE OF THE KIDNEY renal hilus is medial indentation; passageway for blood vessels and ureters. Renal pelvis located in renal sinus.

INTERNAL STRUCTURE OF THE KIDNEY
1. Cortex: outer layer; also forms renal columns that pass into medulla.
2. Medulla: consists of renal pyramids.
3. Renal pelvis: expanded upper end of ureter; formed from joining of major calyces.

THE RENAL TUBULES functional units of kidneys; consist of nephrons and collecting tubules. Each nephron consists of a glomerulus (network of parallel capillaries) and a tubule (proximal end forms glomerular capsule).

RENAL CORPUSCLE capsule and glomerulus; site of transfer between blood and nephron.
1. Visceral layer of glomerular capsule composed of podocytes; foot processes separated by filtration slits, which are covered by slit membranes.
2. Glomeruli have fenestrated endothelium.

FILTRATION BARRIER formed of fenestrated endothelium, basal lamina, and slit membrane.

PROXIMAL CONVOLUTED TUBULE single layer of cuboidal cells with microvilli.

LOOP OF HENLE descending limb (thin; squamous cells) and ascending limb (thick; cuboidal cells).

DISTAL CONVOLUTED TUBULE cuboidal cells; empties into collecting tubule.

JUXTAGLOMERULAR APPARATUS distal tubule contacts afferent arteriole; composed of juxtaglomerular cells of afferent arteriole and macula densa of tubule; secretes renin.

BLOOD VESSELS OF THE KIDNEY renal artery provides substantial bloodflow, which allows kidneys to maintain blood homeostasis. Interlobar artery → arcuate artery → interlobular artery → afferent arterioles → glomerulus → efferent arteriole → peritubular capillaries → interlobular vein → arcuate vein → interlobar vein → renal vein.

SPECIAL FEATURES OF BLOODFLOW THROUGH THE KIDNEY
1. Two capillary beds: glomerulus and peritubular capillaries.
2. Afferent and efferent arterioles maintain constant blood pressure in glomerulus.

PHYSIOLOGY OF THE KIDNEYS   renal tubules maintain composition, volume, and pH of blood and tissue fluid by means of glomerular filtration, tubular reabsorption, and tubular secretion.   **pp. 578–579**

URETERS   **pp. 579–581**

URINE TRANSPORT from renal pelvis to urinary bladder; retroperitoneal.

THREE-LAYERED WALLS

INNER MUCOSAL LAYER transitional epithelium.

MIDDLE MUSCULAR LAYER peristaltic contractions.

OUTER FIBROUS LAYER

URINARY BLADDER   **pp. 581–583**

HOLLOW MUSCULAR ORGAN urine storage.
1. Retroperitoneal.
2. Lined with transitional epithelium.
3. Three layers of smooth muscle.

MICTURITION (URINATION)
1. Full bladder activates spinal reflex that causes muscles of bladder wall to contract, resulting in micturition.
2. Impulses from brain can facilitate or inhibit reflex emptying of bladder; with training, can come under voluntary control.

URETHRA muscular tube lined with mucous membrane; carries urine from bladder to exterior of body; surrounded by external urethral sphincter where it passes through urogenital diaphragm.   **pp. 583–584**

FEMALE short; runs along anterior surface of the vagina.

MALE functions in urine passage and reproduction; long; has three parts.

PROSTATIC URETHRA passes through prostate gland.

MEMBRANOUS URETHRA passes through urogenital diaphragm.

SPONGY URETHRA passes through penis.

CONDITIONS OF CLINICAL SIGNIFICANCE: THE URINARY SYSTEM **pp. 584–586**

PRESSURE-RELATED PATHOLOGIES

PROSTATE HYPERTROPHY can cause fluid backup that interferes with regulation of body fluid composition.

LOW ARTERIAL BLOOD PRESSURE reduces ability of kidneys to form glomerular filtrate.

GLOMERULONEPHRITIS inflammatory condition that affects glomeruli; due to allergic reaction to toxins produced by streptococci bacterial infection in other body parts.

PYELONEPHRITIS bacterial infection of renal pelvis and surrounding tissue.

PROTEINURIA plasma proteins pass into urine.

UREMIA metabolic products, including urea, accumulate in blood due to improper excretion.

KIDNEY STONES (RENAL CALCULI) formed in renal pelvis or urinary bladder by combination of uric acid, calcium oxalate, and calcium phosphate; may cause urine retention, pain, and infection due to blockage of ureters and ulceration of urinary tract lining.

CYSTITIS urinary bladder inflammation that produces frequent, burning urination and blood in urine; more common in women, probably because urethra is short.

EFFECTS OF AGING ON THE KIDNEYS
1. Progressive decline in function: decreased glomerular filtration rate; decreased renal bloodflow; decreased renal tubule function.
2. Acid–base regulation: ability retained by aging kidneys.
3. Pathological changes: degeneration and sclerosis of glomeruli.

SELF-QUIZ

1. The kidneys are supplementary excretory organs that are important but not essential in maintaining the balance of substances required for internal constancy. True or false?

2. The kidneys: (a) excrete urea; (b) produce erythropoietin; (c) do both.

3. The kidneys develop from the embryonic: (a) ectoderm; (b) mesoderm; (c) endoderm.

4. The medulla of the kidney contains: (a) the adipose capsule; (b) glomeruli; (c) renal pyramids.

5. The expanded, funnel-shaped upper end of the ureter within the kidney is the: (a) renal pelvis; (b) renal hilus; (c) urethra; (d) nephron.

6. The capillary found in the glomerular capsule of a nephron is the: (a) glomerulus; (b) proximal convoluted tubule; (c) adipose capsule; (d) juxtaglomerular complex.

7. The space in each kidney that contains the renal vessels and the renal pelvis is called the renal: (a) capsule; (b) sinus; (c) cortex.

8. Beyond the loop of Henle, the distal straight tubule often exits the medulla and runs back into the cortex. True or false?

9. Blood in an interlobular vein in the kidney would next flow into: (a) peritubular capillaries; (b) an arcuate vein; (c) an efferent arteriole; (d) a glomerulus.

10. A large flow of blood through the kidneys is essential for the kidneys to maintain the homeostasis of the blood. True or false?

11. From the collecting tubules urine enters the: (a) renal pelvis; (b) major calyces; (c) minor calyces; (d) distal tubule.

12. Podocytes are responsible for the formation of: (a) filtration slits; (b) fenestrated endothelium; (c) capsular space; (d) pronephros.

13. Renal corpuscles are located in the cortical region of the kidney. True or false?

14. Microvilli are especially numerous in the proximal convoluted tubule. True or false?

15. The juxtaglomerular complex is formed from modification of the cells of the: (a) distal tubule; (b) afferent arteriole; (c) both a and b.

16. Capillary branches that supply the loops of Henle and the collecting ducts are called: (a) glomeruli; (b) vasa recta; (c) macula densa; (d) arcuate veins.

17. The urethra of the female is divisible into three parts—the prostatic, membranous, and spongy urethra, which are named according to the regions through which it passes. True or false?

18. The ureters and the urinary bladder are lined with simple squamous epithelium. True or false?

19. While the internal urethral sphincter is formed of smooth muscle, the external urethral sphincter is composed of striated muscle. True or false?

20. The condition in which large amounts of plasma proteins pass into the glomerular filtrate and become excreted in the urine is called: (a) pyelonephritis; (b) proteinuria; (c) cystitis.

21. In the artificial kidney, the patient's blood is passed through a dialysis tubing that allows the free movement of (a) urea; (b) protein; (c) electrolytes; (d) both a and c.

22. Which one of the following is least affected by aging? (a) glomerular filtration rate; (b) renal bloodflow rate; (c) renal tubule function; (d) renal regulation of body pH.

LEARNING OBJECTIVES

After completing this chapter, you should be able to:

- Describe the embryonic development of the internal and external reproductive structures.

- Describe the structure and function of each part of the male internal reproductive structures and external genitalia.

- Describe the structure and function of each part of the female internal reproductive structures and external genitalia.

- Describe a spermatozoon and explain spermatogenesis.

- Explain the functional value of the testes' location in the scrotum outside the abdominopelvic cavity.

- Describe oogenesis and relate it to the ovarian cycle.

- Describe the changes that the endometrium undergoes during the menstrual cycle.

- Discuss some common pathologies of the reproductive system.

CHAPTER CONTENTS

The Reproductive System

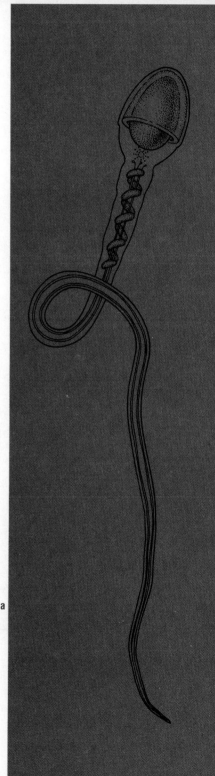

The organs of the male and female reproductive systems ensure the continuance of the species. They do this by producing **gametes,** or **germ cells,** and by providing a method by which the gametes of the male **(spermatozoa,** or **sperm)** can be introduced into the body of the female, where one of them joins with a gamete **(ovum)** of the female. This penetration of an ovum by a spermatozoon is called *fertilization.* The organs of the female reproductive system provide a suitable environment in which the fertilized ovum *(zygote)* can develop to a stage at which it is capable of surviving outside the mother's body.

The organs that produce the gametes are referred to as the **primary sex organs,** or **essential sex organs.** These are the **gonads**—the **testes** in the male and the **ovaries** in the female. In addition to producing the gametes, the primary sex organs also produce hormones that influence the development of male or female secondary sex characteristics and regulate the reproductive cycle. In the male, specialized cells in the testes produce a group of hormones called *androgens.* The most active androgen is *testosterone.* In the female, the ovaries produce *estrogens* and *progesterone.*

The structures that transport, protect, and nourish the gametes after they leave the gonads are called **accessory sex organs.** In the male the accessory sex organs include the *epididymis, ductus deferens, seminal vesicles, prostate gland, bulbourethral glands, scrotum,* and *penis.* Female accessory sex organs include the *uterine tubes, uterus, vagina,* and *vulva.*

EMBRYONIC DEVELOPMENT OF THE REPRODUCTIVE SYSTEM

The gonads are formed within retroperitoneal elevations called **genital ridges,** which protrude into the coelom (body cavity) just medial to the developing kidneys (mesonephros). Although the sex of the embryo is determined at the time of fertilization, the genital ridges do not differentiate into testes or ovaries until the end of the eighth week of embryonic development. Germ cells, which originate in the yolk sac, migrate to the genital ridges in the sixth week of development, prior to the differentiation of the ridges. During the undifferentiated period, the gonads develop in close approximation to the **mesonephric ducts,** which drain the embryonic kidneys. A second pair of ducts—the **paramesonephric ducts**—also develops. The paramesonephric ducts run parallel to the mesonephric ducts and empty into the urogenital sinus at the posterior end of the embryo (Figure 22.1*a*). Males retain the meso- F22.1a nephric ducts for the transport of sperm, while females retain the paramesonephric ducts for the transport of ova and the nourishment of the fetus.

Development of the Internal Reproductive Structures

Following the undifferentiated period, during which all embryos have the same structure, different developmental paths are followed by male and female embryos. As a result the internal reproductive structures that are unique to each sex are formed.

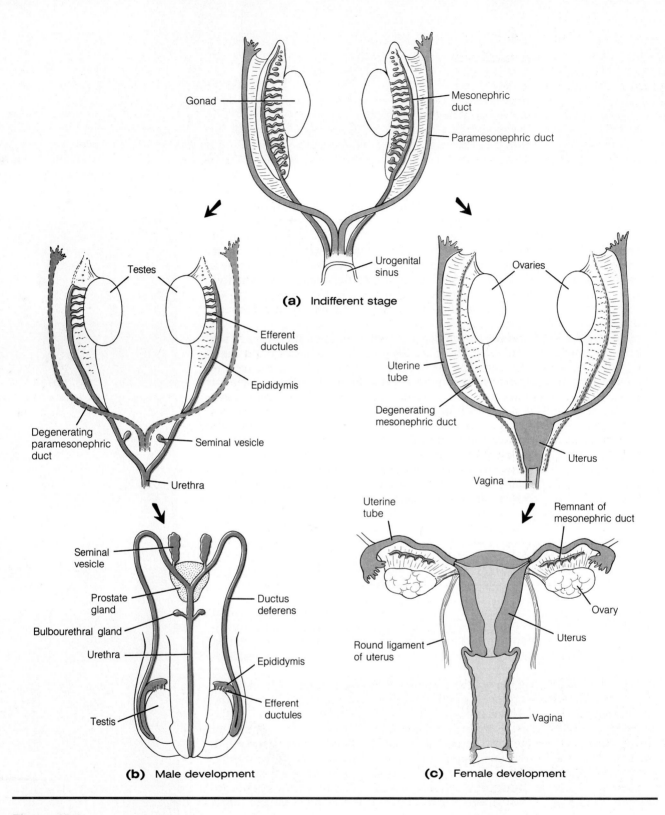

Figure 22.1

Embryonic development of male and female internal reproductive structures. Notice that the testes descend out of the abdominopelvic cavity into the scrotum, but the ovaries, which also descend, remain within the pelvic cavity.

Male Embryo

In the male embryo, the inner portion **(medulla)** of the undifferentiated gonads develops tubules that join with the mesonephric duct. These tubules become the **seminiferous tubules** of each testis, in which spermatozoa eventually are produced. In the male, the mesonephric duct is utilized for the transport of sperm from the testes to the exterior of the body. With further development, each mesonephric duct forms **efferent ductules,** an **epididymis,** and a **ductus deferens** of the adult (Figure 22.1*b*). Shortly after the gonads begin to differentiate into testes, the paramesonephric ducts degenerate. **F22.1b**

Female Embryo

In the female embryo, the outer portion **(cortex)** of the undifferentiated gonads undergoes the greater development and forms follicles, in which ova develop. Coinciding with the differentiation of the gonads into ovaries, the distal ends of the paramesonephric ducts join together to form the **uterus** and **vagina.** The portion of the paramesonephric ducts between the ovaries and the uterus forms the **uterine tubes** (*oviducts,* or *fallopian tubes),* through which ova travel to the uterus (Figure 22.1*c*). In the female, the mesonephric ducts degenerate without contributing any functional structures to the reproductive system. **F22.1c**

Effects of Hormones on Genital Duct Development

The development of the genital ducts into either male or female structures appears to be under the control of hormones produced by the testes. In the presence of the androgens that are produced after the development of the genital ridges into testes, the mesonephric ducts remain and the paramesonephric ducts degenerate. If the gonads (testes) are removed from an embryo that is genetically a male (XY), the embryo develops the reproductive structures of a female. That is, the paramesonephric ducts undergo further development and the mesonephric ducts degenerate. Similarly, the injection of androgens into a genetic female (XX) causes the embryo to develop according to the male pattern. These results indicate that in the absence of the male hormones, all embryos, regardless of their genetic makeup, would develop into females. Under normal conditions, of course, embryos that are genetically male produce androgens and therefore develop male reproductive structures.

Development of the External Reproductive Structures

The embryonic development of the external genitalia is also controlled by hormones produced by the gonads. The external genitalia, like the internal reproductive organs, remain in an undifferentiated state until about the eighth week. Before the production of hormones begins, all embryos develop a conical elevation—called the **genital tubercle**—at the point where the mesonephric and paramesonephric ducts open to the exterior of the body (Figure 22.2*a*). On the inferior surface of this tubercle is a shallow depression called **F22.2a** the **urethral groove,** which opens into the urogenital sinus of the embryo. On either side of the urethral groove are slight elevations called the **urethral folds.** Still further laterally from the folds are rounded elevations called the **labioscrotal swellings.**

Male Embryo

If the individual is a male, the genital tubercle elongates to form the **penis.** The urethral folds fuse, leaving an opening only at the distal end of the penis. The tube thus formed becomes the spongy portion of the urethra. The labioscrotal swellings develop into a pouch **(scrotum)** that eventually will receive the testes (Figure 22.2*b*). **F22.2b**

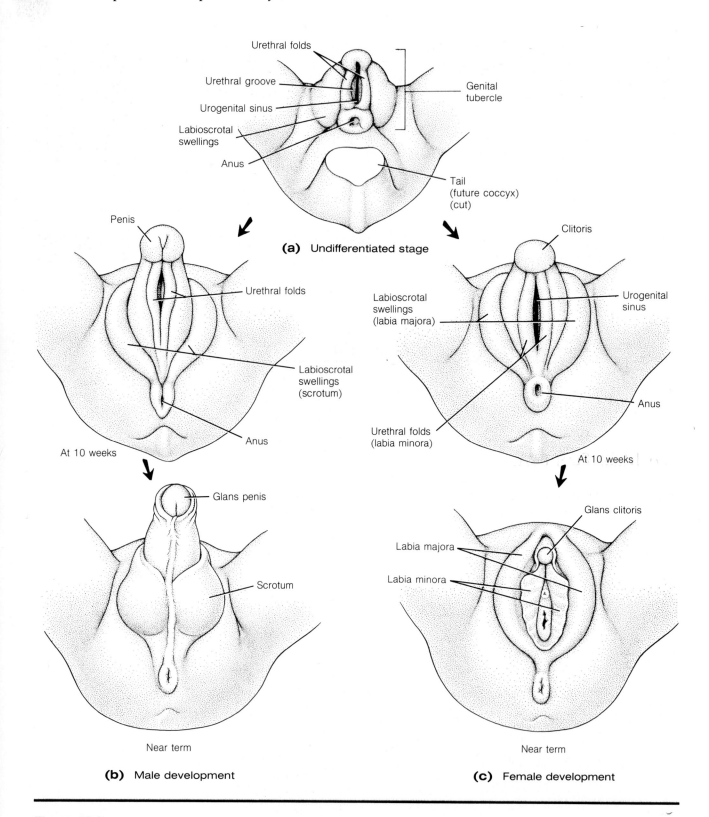

(a) Undifferentiated stage

At 10 weeks

At 10 weeks

Near term

(b) Male development

Near term

(c) Female development

Figure 22.2
Embryonic development of male and female external reproductive structures.

F22.2c

Female Embryo

In the female embryo, the genital tubercle does not elongate as much as in the male and it becomes the **clitoris**. Nor do the urethral folds fuse in the female. Rather, they remain as the **labia minora**, surrounding the entrance into the vagina. The labioscrotal folds are not destined to receive gonads in the female. They remain as elevations called the **labia majora**, which flank the labia minora (Figure 22.2c).

CONDITIONS OF CLINICAL SIGNIFICANCE
Sexual Malformations During Embryonic Development

An understanding of the patterns of development of the reproductive structures makes it apparent how even slight malformations during their embryonic development can cause any number of abnormalities in the adult. The most extreme abnormality is that of the *hermaphrodite*, where the individual possesses gonads and internal and external reproductive structures of both sexes.

It is estimated that some degree of sexual abnormality occurs in one out of every thousand persons. In the male, for example, the urethral folds may fail to join completely, which leaves openings into the urethra on the undersurface of the penis. In the female, an ovary may descend into the labia majora in a manner similar to the testes entering the scrotum. Many other sexual abnormalities occur, but it is beyond the scope of this book to consider them.

ANATOMY OF THE MALE REPRODUCTIVE SYSTEM

The male reproductive system includes the *testes*, which produce spermatozoa; a number of *ducts* that store, transport, and nourish the spermatozoa; several accessory *glands* that contribute to the formation of semen; and the *penis*, through which semen is conveyed outside the body.

The Male Perineum

The **perineum** includes all the structures that are located between the pubic symphysis anteriorly, the coccyx posteriorly, and the ischiopubic rami and the sacrotuberous ligaments laterally. The portion of the sacrotuberous ligament that bounds the perineum runs from the lateral margin of the sacrum and coccyx to the ischial tuberosity. The muscles of the perineum are described in Chapter 7. Here, we are interested only in the surface anatomy of the region in males.

The perineum can be divided into two triangles by a transverse line that passes between the ischial tuberosities (Figure 22.3). The anterior triangle, **F22.3** which contains the external genitalia, is called the **urogenital triangle.** The posterior triangle is called the **anal triangle** because it contains the anus.

Testes and Scrotum

The **testes** are the organs in which the production of spermatozoa **(spermatogenesis)** occurs. The testes are located in a skin-covered pouch called the **scrotum.** The scrotum consists of an outer layer of skin covering a thin layer of smooth muscle called the **dartos tunic.** Contraction of the muscles of the dartos tunic gives the scrotum a wrinkled appearance.

Each testis is an oval organ that is surrounded by a connective-tissue capsule called the **tunica albuginea.** Inward extensions **(septa)** of the tunica divide the testes into compartments (Figure 22.4). Each compartment holds **F22.4** several highly convoluted **seminiferous tubules** that contain germ cells in various stages of development. Through the process of spermatogenesis that occurs inside the seminiferous tubules, the germ cells eventually develop into spermatozoa.

The seminiferous tubules in each compartment of the testis join together to form a short, straight tube called the **tubulus rectus** (*rectus* = straight). Toward the posterior portion of the testis, the tubuli recti from all the compartments form a network of tubes termed the **rete testis.** The tubes of the rete testis, in turn, empty into the **efferent ductules** that leave the testis and enter the epididymis.

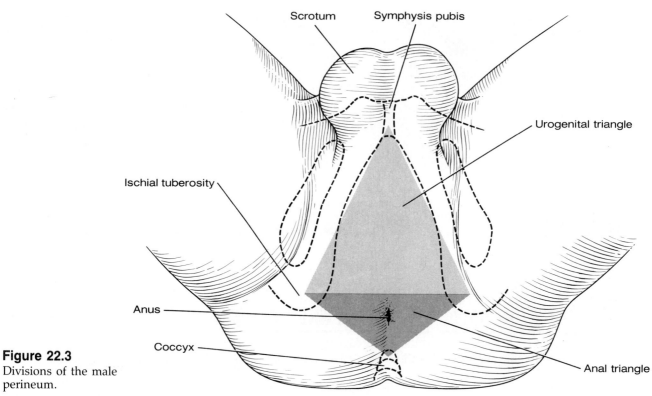

Figure 22.3
Divisions of the male perineum.

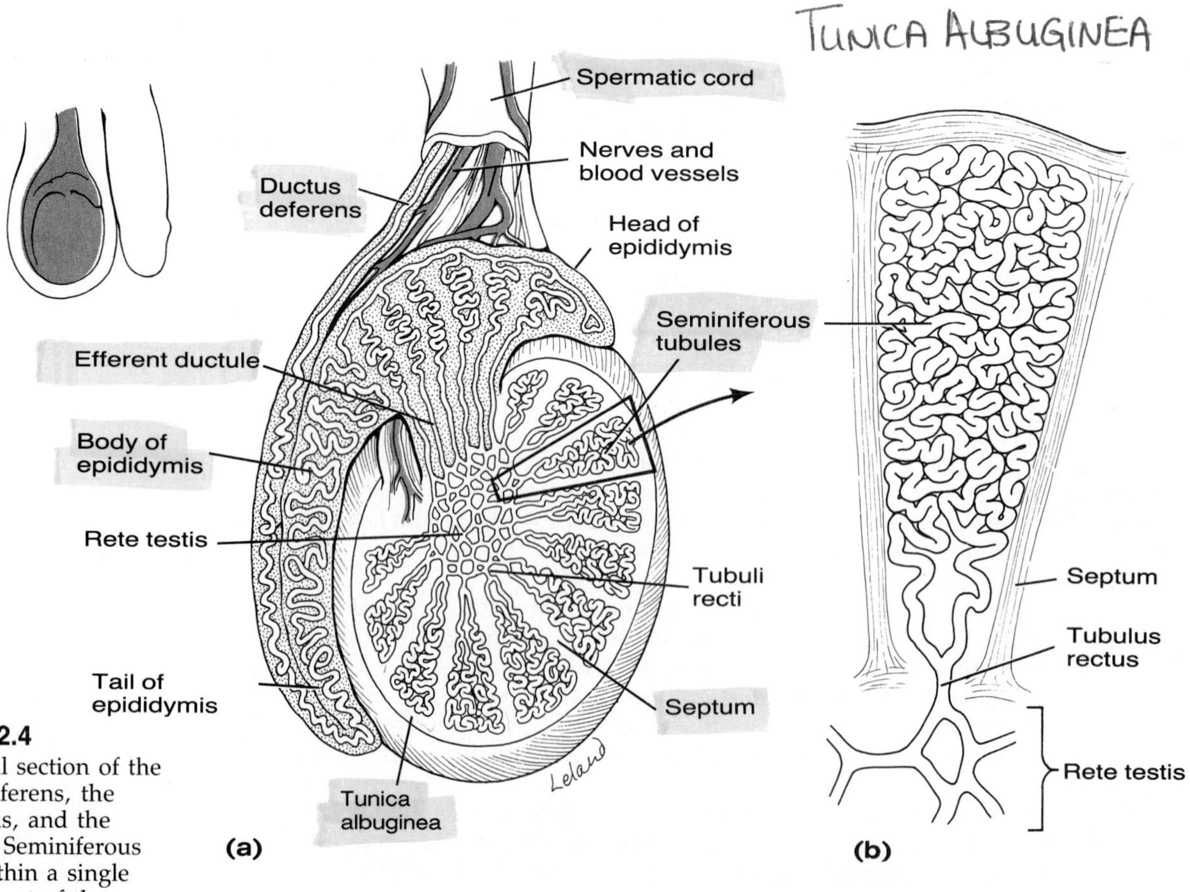

TUNICA ALBUGINEA

Figure 22.4
(a) Sagittal section of the ductus deferens, the epididymis, and the testis. (b) Seminiferous tubule within a single compartment of the testis.

(a)

(b)

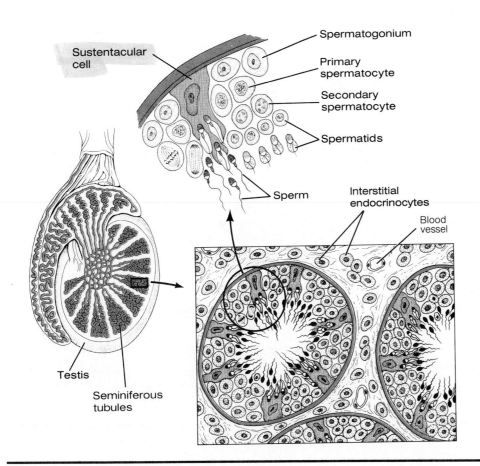

Figure 22.5
Section of a testis showing spermatogenesis within the seminiferous tubules.

Clusters of cells called **interstitial endocrinocytes** (*interstitial cells of Leydig*) (Figure 22.5) are located in the loose connective tissue between the seminifer- **F22.5** ous tubules. These interstitial cells secrete the male hormones, the androgens.

Spermatogenesis

The production of mature gametes—spermatozoa and ova—is called *gametogenesis.* This phenomenon involves a unique type of cell division: *meiosis.* (Meiosis is described in Chapter 2, and the specific events of this process should be reviewed at this time.) The purpose of meiosis is to reduce the number of chromosomes in each gamete by half. The cells of the human body are *diploid:* that is, they contain 23 chromosomes from each parent, for a total of 46. Cells that have undergone meiosis have lost half of their total number of chromosomes, retaining 23 chromosomes. Such cells are said to be *haploid.* The formation of haploid gametes ensures that, following fertilization, when a male and a female gamete unite, the resulting cell will be a diploid cell that has 46 chromosomes.

The production of male gametes, **spermatozoa (sperm),** occurs in the seminiferous tubules of the testes. The process by which sperm are produced is called **spermatogenesis** (Figure 22.6). Located within the outer portion of **F22.6** each tubule are cells called **spermatogonia,** which have the diploid number of chromosomes (44 autosomes plus one X and one Y sex chromosome). The spermatogonia divide mitotically, thereby providing a continuous source of new cells that are used for the production of spermatozoa. Some spermatogonia move toward the lumen of the tubule and undergo a period of growth. These enlarged cells are called **primary spermatocytes.** Each primary spermatocyte undergoes a first meiotic division, which results in the formation of two **secondary spermatocytes.** One of these secondary spermatocytes contains 22 autosomes plus the X sex chromosome; the other has the remaining 22 autosomes and the Y sex chromosome. Recall that during cell division the

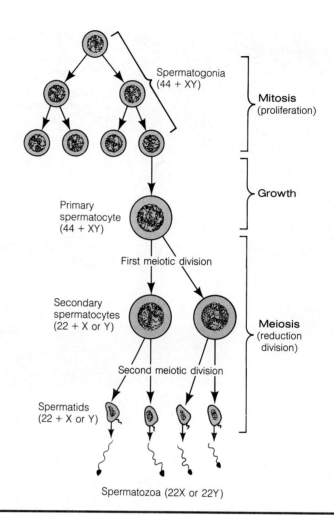

Figure 22.6
Meiosis in the testis.
Note that during meiosis
each spermatogonium
produces four haploid
spermatozoa.

chromosomes duplicate, and consist of two connected chromatids. In the first meiotic division the double-stranded chromosomes do not separate.

During the second meiotic division the double-stranded chromosomes split and each secondary spermatocyte divides into two small spherical cells called **spermatids.** Thus, meiosis results in the formation of four haploid spermatids from each diploid spermatogonium. Each spermatid undergoes a complex series of structural changes by which it develops into a mature **spermatozoon.** During these changes, much of the cytoplasm of the cell is discarded and a tail containing a group of contractile proteins forms. At this stage spermatozoa are still functionally immature. The final stages of maturation of the spermatozoa occur within the epididymis.

During spermatogenesis, spermatids as well as spermatocytes are observed to be deeply indented into the surface of columnar cells that extend **F22.5** from the basal lamina of the seminiferous tubule to the lumen (Figure 22.5). These cells, which are called **sustentacular cells** *(Sertoli cells)*, are attached to each other by tight junctions. As the spermatocytes develop into spermatids the tight junctions open and reform, allowing the developing spermatids to gradually move from the peripheral region of the sustentacular cells toward the region that borders the lumen of the seminiferous tubules. Because they are firmly connected by tight junctions the sustentacular cells form a barrier that hinders the movement of substances between the seminiferous tubules and the blood. This **blood-testis barrier** provides a stable microenvironment that is favorable for the development of sperm and is thought to prevent antigens from the developing sperm from entering the blood. This barrier is necessary because sperm are not produced for the first 12 years-or-so of life, and by this time the immune system has already been programmed to recog-

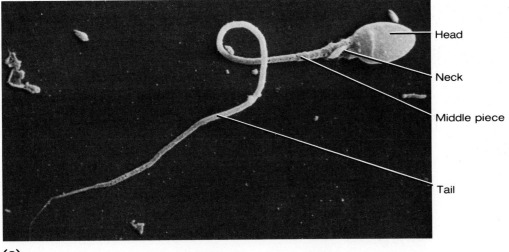

(a)

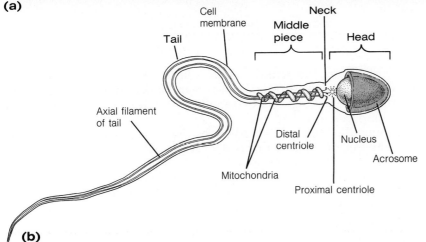

(b)

Figure 22.7
Structure of a spermatozoon.
(a) Scanning electron micrograph (×4859).
(b) Drawing. (Micrograph from *Tissues and Organs: A Text-Atlas of Scanning Electron Microscopy* by Richard G. Kessel and Randy H. Kardon. W.H. Freeman and Company. Copyright © 1979.)

nize a person's own body tissues. If the antigens of mature sperm entered the blood they would be recognized as foreign bodies and attacked by the body's immune system. Moreover, since spermatozoa do not contain much cytoplasm and therefore have only limited metabolic capabilities, the sustentacular cells are thought to provide nourishment to them. The sustentacular cells also secrete a hormone called *inhibin* which reduces the rate of spermatogenesis in the seminiferous tubules by inhibiting the secretion of follicle-stimulating hormone by the pituitary gland.

Although spermatogenesis is triggered by the *follicle-stimulating hormone* (FSH), which is released from the adenohypophysis, the ensuing events proceed in a regular sequence under some unknown control. At any one time, all the primary spermatocytes in one portion of a seminiferous tubule may be dividing while in an adjacent segment the secondary spermatocytes may be dividing. In humans, the entire process—from spermatogonia to spermatozoa—requires about 72 days and occurs continuously, beginning at puberty. It has been estimated that several hundred million sperm reach maturity each day in the human male.

The Spermatozoon

The **spermatozoon** is a flagellated cell that is divisible into **head, neck, middle piece,** and **tail** regions (Figure 22.7). The head consists almost entirely of the condensed nucleus capped with a fluid-filled vesicle (the **acrosome**) at its tip. The acrosome develops from the Golgi apparatus of the spermatid. A number of enzymes have been identified in the acrosomal fluid. In order for a sperm   **F22.7**

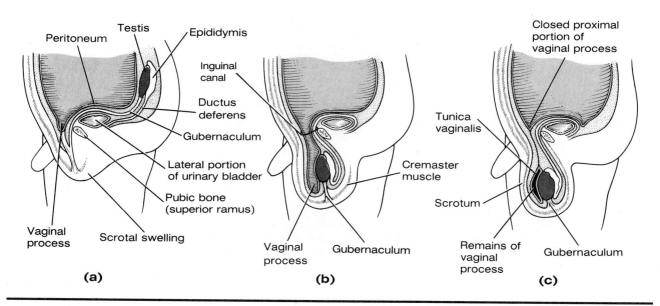

Figure 22.8
Descent of a testis.
(a) Formation of the vaginal process within the scrotal swellings. **(b)** Descent almost complete. **(c)** Upper portion of the vaginal process is obliterated, while the lower portion remains as the double-layered tunica vaginalis.

to fertilize an ovum, the acrosome must release its enzymes, which break down several layers that surround the ovum. Located in the neck region are two centrioles at right angles to one another. The distal centriole is oriented along the long axis of the sperm, and its microtubules extend into the tail, where it forms the flagellum (which gives the sperm motility). The middle piece is composed of mitochondria that meet end to end and spiral around the distal centriole. Most of the small amount of cytoplasm present in a mature sperm is located around the mitochondria of the middle piece. Because there is so little cytoplasm present, spermatozoa cannot survive very long on their own. Instead, they must derive nourishment from the semen in which they are suspended.

Descent of the Testes

As discussed earlier, the testes begin their development as retroperitoneal structures in the abdominopelvic cavity, just below the kidneys. As development proceeds, the testes move caudally toward swellings of the abdominal wall called *labioscrotal swellings*. These labioscrotal swellings, which are located just behind the penis in the anterior portion of the urogenital triangle of

F22.8 the perineum, develop into the scrotum (Figure 22.8). As the testes move toward the scrotal swellings, an outpouching of the peritoneum called the **vaginal process** protrudes over the superior ramus of the pubic bone and extends through an **inguinal canal** into the two chambers of the scrotum. The testes, which remain behind the peritoneum, follow the vaginal process out of the abdominopelvic cavity through the inguinal canals and into the scrotum. As the testes descend into the scrotum from their original position in the abdominopelvic cavity, their blood supply travels with them. Therefore, the testicular arteries leave the aorta—and the testicular veins join the inferior vena cava (the left testicular vein via the left renal vein)—in the region of the kidneys, near the original site of the testes, and follow their paths of descent into the scrotum.

After the testes have entered the scrotum, the inguinal canal narrows, constricting the upper portion of the vaginal process. The lower portion of each vaginal process forms a double-layered sac that covers the testis. This portion is then called the **tunica vaginalis.** If the upper portion of the vaginal process does not close off completely, it is possible for small loops of intestine to protrude into the inguinal canal. This condition is known as an *inguinal hernia*. Even if the vaginal process closes completely, this is an area of weakness in males, and thus a potential site of hernia. Because in females the gonads do not pass through the body wall and thus do not stretch and weaken the abdominal muscles surrounding the inguinal canals, inguinal hernias are less common in females than in males.

The actual cause of the descent of the testes into the scrotum is not known. However, it seems to be initiated by testosterone from the testes and certain hormones from the pituitary gland. A fibromuscular band called the **gubernaculum** is also thought to assist in the descent, but the precise manner in which this happens is not certain. The gubernaculum extends from the caudal surface of each testis, passes through the body wall, and is attached to the floor of the scrotum. As the embryo grows, the gubernaculum shortens; but it is questionable whether it is strong enough to actually pull each testis toward the scrotum.

The location of the testes outside the abdominopelvic cavity in the scrotum is necessary for the normal development of sperm. Occasionally, one or both of the testes fail to leave the abdominopelvic cavity. This condition is called *cryptorchidism*. Eventually, the undescended testis will atrophy. Since sperm are not produced in cryptorchid testes, the person is sterile if both testes fail to descend. The problem in cryptorchidism is that normal spermatogenesis cannot occur at body-core temperature. The scrotum provides a means by which the testes can be maintained at a temperature lower than body-core temperature. Actually, by means of the **cremaster muscle** the scrotum is able to regulate the temperature of the testes to some extent. The cremaster muscle, which is a continuation of the internal abdominal oblique muscle, extends down over the spermatic cord and the testes as a series of loops. When the environmental temperature is low, the cremaster muscle contracts, which brings the testes closer to the body and thus warms them. At higher environmental temperatures, the cremaster muscle relaxes and the scrotum becomes flaccid and elongates, thus removing the testes still further from the warmth of the body.

Epididymis

The **epididymis,** which is located in the scrotum, is the first portion of the duct system that transports sperm from the testes to the exterior of the body (Figure 22.4). Each epididymis is an elongated structure that fits tightly **F22.4** against the posterior surface of the testis. It consists of a tightly coiled tube that receives the sperm from the testis through the efferent ductules. The tube is lined with pseudostratified columnar epithelium. The free surfaces of some of the columnar cells have long, nonmotile microvilli called *stereocilia*. The stereocilia serve to facilitate the passage of nutrients from the epithelium to the sperm. If uncoiled, the epididymal tube would be 4 to 6 m in length. Thus it is well suited to serve as a storage site for sperm. The efferent ductules enter the upper region (*head*) of the epididymis. In its lower region (the *tail*), the tube becomes continuous with the ductus deferens. There are smooth muscles in the wall of the epididymal tube that contract during ejaculation. These contractions move the sperm into the ductus deferens. During their slow passage through the epididymis, the sperm undergo a maturation process without which they would be nonmotile and infertile when they entered the female reproductive tract.

Ductus Deferens

The **ductus (vas) deferens** is a continuation of the duct of the epididymis (Figure 22.9). Each ductus deferens is a straight tube that passes along the **F22.9** posterior surface of the testis, medial to the epididymis, and ascends through the scrotum. The ductus deferens passes through the body wall in the region of the inguinal canal and enters the abdominopelvic cavity.

As it passes from the epididymis to the point where it enters the abdominopelvic cavity, the ductus deferens lies next to the vessels and nerves supplying the testis. All of these structures are enclosed in a sheath of fascia called the **spermatic cord.** Included in the spermatic cord along with the ductus deferens are the testicular artery, testicular vein, lymphatic vessels, and nerves. The veins returning from the testes form a network of connecting

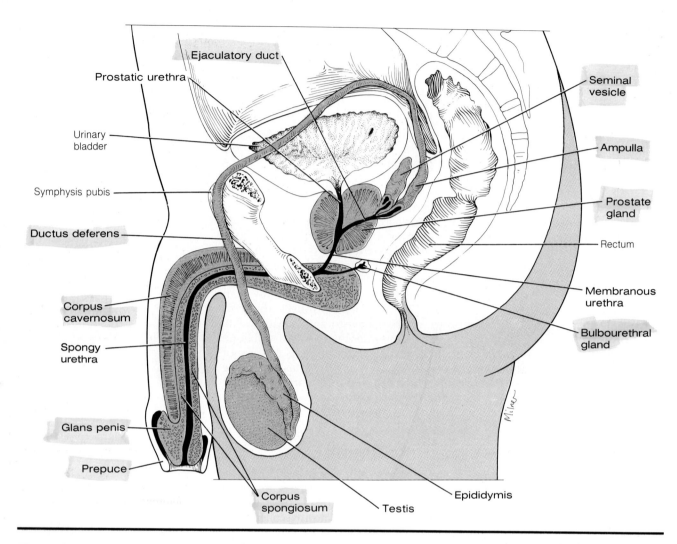

Figure 22.9
Sagittal section of the male pelvis with a portion of the left pubic bone attached to illustrate the path of the left ductus deferens.

branches (the *pampiniform plexus*) around the testicular artery. This plexus is thought to absorb heat from the blood in the testicular artery, thereby lowering the temperature of the arterial blood to the testes and helping to maintain the temperature of the testes slightly below body-core temperature. This lower temperature is essential for normal spermatogenesis.

Because the portion of the ductus deferens in the spermatic cord is so easy to reach surgically, cutting the two ducts has become a rather common means of birth control. This procedure is called *vasectomy* (vessel removal). Vasectomy involves making a small incision on each side of the scrotum, opening the two spermatic cords, and tying each ductus deferens in two places. The portion of each ductus deferens between the ties is then cut. A vasectomy does not interfere with the production of hormones or sperm by the testes, but it does prevent the passage of sperm from the testes into the urethra.

Inside the abdominopelvic cavity, the ductus travels beneath the peritoneum along the lateral wall of the cavity, crosses over the top of the ureter, and then descends along the posterior surface of the urinary bladder, where it

F22.10 enlarges to form an **ampulla** (Figure 22.10). On reaching the inferior surface of the urinary bladder, each ductus deferens is joined by the duct of a seminal vesicle to form a short **ejaculatory duct** that passes through the prostate gland to join the prostatic portion of the urethra. The ductus deferens, like the epididymis, is lined with pseudostratified columnar epithelium that has stereocilia on its free surface. The muscular coat of the wall of the ductus deferens

F22.11 is quite thick, consisting of three layers of smooth muscle (Figure 22.11). The muscle of the wall of the duct undergoes peristaltic contractions during ejaculation, propelling sperm into the ejaculatory duct.

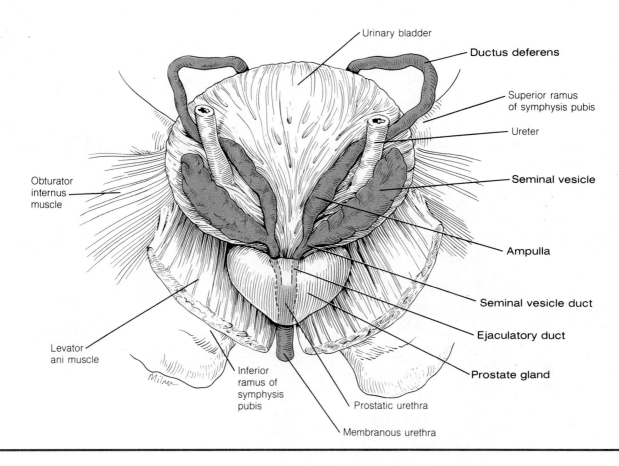

Urinary bladder

Ductus deferens

Superior ramus
of symphysis pubis

Ureter

Seminal vesicle

Obturator
internus
muscle

Ampulla

Seminal vesicle duct

Ejaculatory duct

Levator
ani muscle

Prostate gland

Inferior
ramus of
symphysis
pubis

Prostatic urethra

Membranous urethra

Figure 22.10
Posterior view of the male urinary bladder illustrating its
relationship to the ductus deferens, the ureters, the
seminal vesicles, and the prostate gland.

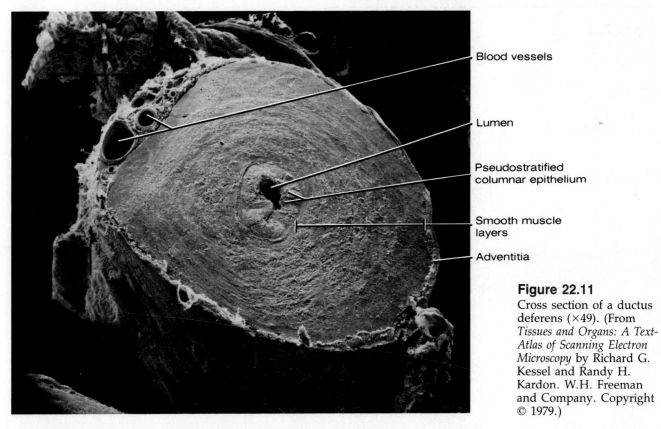

Blood vessels

Lumen

Pseudostratified
columnar epithelium

Smooth muscle
layers

Adventitia

Figure 22.11
Cross section of a ductus
deferens (×49). (From
*Tissues and Organs: A Text-
Atlas of Scanning Electron
Microscopy* by Richard G.
Kessel and Randy H.
Kardon. W.H. Freeman
and Company. Copyright
© 1979.)

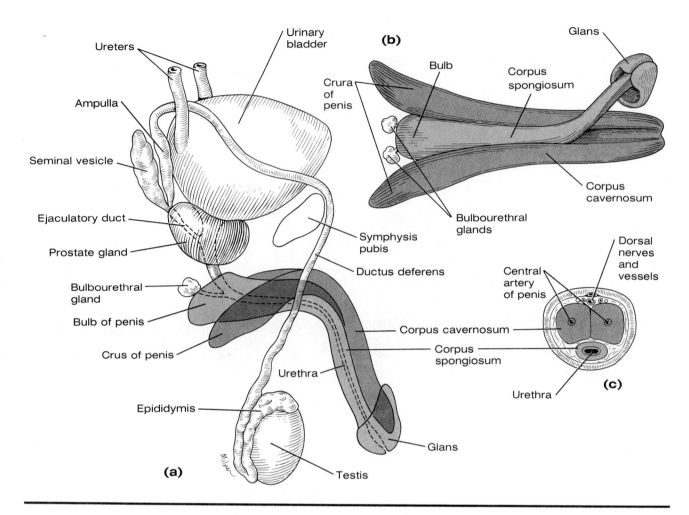

(a) Schematic representation of the male reproductive system. **(b)** Inferior view of a dissected penis with the distal portion of the corpus spongiosum displaced to one side. **(c)** Cross section of a penis.

Figure 22.12

Seminal Vesicles

F22.10, F22.12

The **seminal vesicles** are two membranous pouches located lateral to the ductus deferens on the posterior surface of the urinary bladder (Figure 22.10, Figure 22.12). The duct of each seminal vesicle joins with a ductus deferens to form an **ejaculatory duct.** The ejaculatory ducts penetrate the prostate gland and enter the urethra just below its point of exit from the urinary bladder. Contraction of the ejaculatory ducts propels spermatozoa from the ductus deferens and secretions from the seminal vesicles into the urethra. The seminal vesicles secrete a viscid fluid that contributes to the formation of semen.

Prostate Gland

F22.9, F22.12

The **prostate gland** (Figure 22.9, Figure 22.12) is a single organ that encompasses the urethra just below the bladder. Because it is located directly in front of the rectum, the prostate can be manually examined by means of a rectal examination in which a finger is placed in the anal canal and the gland palpated through the anterior wall of the rectum. The gland is composed of about 30 small tubuloalveolar glands from which a number of excretory ducts open independently into the urethra. It is enclosed in a capsule of fibrous connective tissue and smooth muscle fibers that extend into the gland and divide it into indistinct lobes. The prostate secretes a thin, alkaline, milky fluid that contributes to semen. The prostate tends to enlarge in most older men and may cause difficulty in urination by squeezing the prostatic portion of the urethra.

Bulbourethral Glands

F22.12

The **bulbourethral glands** (Figure 22.12) are a pair of small glands located below the prostate on either side of the membranous portion of the urethra.

Their secretion, which is a thick, alkaline mucus, is carried to the urethra by a duct from each gland. The secretion of the bulbourethral glands is released early in sexual stimulation, prior to ejaculation. It serves as a lubricant for sexual intercourse and is believed to protect the sperm by neutralizing any acidic urine that may remain in the urethra.

Penis

The **penis** is the copulatory organ by which spermatozoa are placed in the female reproductive tract. It consists of a shaft covered by loosely attached skin and an expanded tip, the **glans** (Figure 22.12). The skin continues over **F22.12** the glans as the **prepuce (foreskin).** In order to make it easier to keep the glans clean, the prepuce is often surgically removed shortly after birth in a procedure called *circumcision*.

The penis contains three cylindrical bodies, each of which is surrounded by a fibrous capsule. The three cylindrical bodies are held together by a connective-tissue sheath that is covered by skin. Each cylindrical body is composed of highly vascular connective tissue called *erectile tissue*, which contains numerous spongelike spaces that fill with blood during sexual arousal, causing the penis to enlarge and become firm. This phenomenon is referred to as an *erection*. The two dorsal cylindrical bodies are called **corpora cavernosa penis.** The single ventral body is the **corpus spongiosum penis.** The urethra passes through the corpus spongiosum. The expanded distal end of the corpus spongiosum forms the glans of the penis. The proximal end of the corpus spongiosum is slightly enlarged, forming the **bulb** of the penis. The bulb is attached to the urogenital diaphragm, which forms the floor of the pelvic cavity. The two corpora cavernosa separate at their proximal ends and form the **crura** of the penis, which anchor it to the rami of the ischial and pubic bones.

Semen

Semen is a mixture of spermatozoa from the testes and fluids from the seminal vesicles, prostate gland, and bulbourethral glands. The secretion of the seminal vesicles contributes about 60% of the bulk of semen. Semen serves as a nutritive source for the spermatozoa and activates them so that they become motile.

Each ejaculation has a volume of about 2 ml and contains about 300 million sperm. Although an ovum is fertilized by a single sperm only, many sperm must be present in order for fertilization to occur. When the number of sperm per ejaculation is less than 60 million, the male is generally incapable of fertilizing an ovum, although fertilization has been accomplished by males with sperm counts lower than that.

Semen is slightly alkaline (pH 7.5). This alkalinity, which protects the sperm from the acid pH of the vagina, is due in large part to the fluid secreted by the prostate gland. Semen contains prostaglandins, fructose, choline, citric acid, lipids, creatine, the enzyme hyaluronidase, and several other substances. The roles of some of these substances in the semen are not clear; however, fructose has been shown to be the major energy source available to the ejaculated spermatozoa. Since spermatozoa contain very little cytoplasm, they have only limited intracellular glycogen available from which to derive energy. Instead, they depend on extracellular fructose as an energy source. The prostaglandins are thought to facilitate the process of fertilization by reacting with mucus at the cervix of the female uterus to make it more receptive to sperm and by stimulating reverse peristaltic contractions that enhance the movement of sperm along the uterus and uterine tubes.

ANATOMY OF THE FEMALE REPRODUCTIVE SYSTEM

The female reproductive system includes: the *ovaries*, which produce ova; the *uterine tubes*, which transport, protect, and nourish the ova; the *uterus*, which

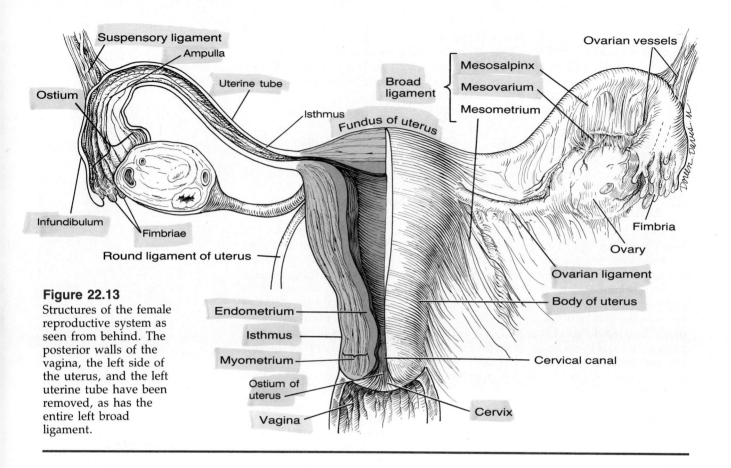

Figure 22.13
Structures of the female reproductive system as seen from behind. The posterior walls of the vagina, the left side of the uterus, and the left uterine tube have been removed, as has the entire left broad ligament.

provides a suitable environment for the development of an embryo; and the *vagina*, which serves as a receptacle for sperm.

Ovaries

The female gonads, or primary sex organs, are the **ovaries,** in which the female gametes **(ova)** are produced. Moreover, the *estrogens* and *progesterone*—hormones that influence the development of the female accessory sex organs and secondary sex characteristics—are secreted by the ovaries.

Each of the paired ovaries is oval and slightly smaller than a testis. The ovaries do not migrate as extensively during embryonic development as do the testes. Rather, they descend only as far as the pelvis and lie against the lateral wall on either side of the uterus (Figure 22.13, Figure 22.14). The ovaries are held in this position by several ligaments. The largest is the **broad ligament,** which also supports the uterine tubes, the uterus, and the vagina. The ovaries are suspended from the posterior surface of the broad ligament by a short fold of peritoneum called the **mesovarium.** A fibrous band called the **ovarian ligament** lies within the broad ligament and extends from the superior-lateral margins of the uterus to the ovary. The lateral margins of the broad ligament form a fold that attaches the ovary to the pelvic wall. This fold, through which the ovarian vessels reach the ovary, is known as the **suspensory ligament.**

The peritoneum that covers the surface of the ovary is composed of relatively short, simple cuboidal cells rather than squamous cells as is typical of the rest of the peritoneum. This outer epithelial covering of the ovary is called the **germinal epithelium** (Figure 22.15). In spite of what the name implies, the germinal epithelium does not give rise to germ cells. Beneath the germinal epithelium is a fibrous connective-tissue layer, the **tunica albuginea.** The ovary itself can be divided into an outer **cortex** surrounding a central **medulla.** The medulla is composed of connective tissue and contains blood vessels, lymphatic vessels, and nerves that enter and leave the ovary where it attaches

F22.13,
F22.14

F22.15

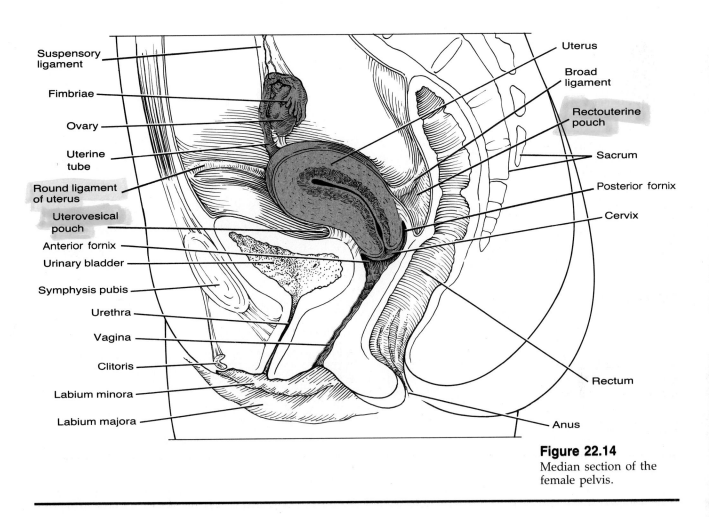

Figure 22.14
Median section of the female pelvis.

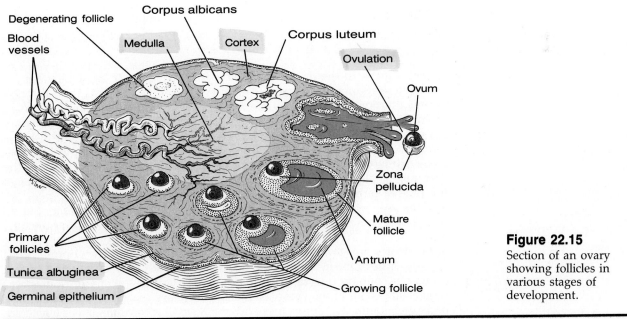

Figure 22.15
Section of an ovary showing follicles in various stages of development.

to the mesovarium. At birth the cortex of each ovary contains hundreds of thousands of immature ova within small individual spheres that are composed of a single layer of cells. Each of these structures is called a **primary follicle,** and the enveloping cells are called *follicular cells.* Most follicles remain

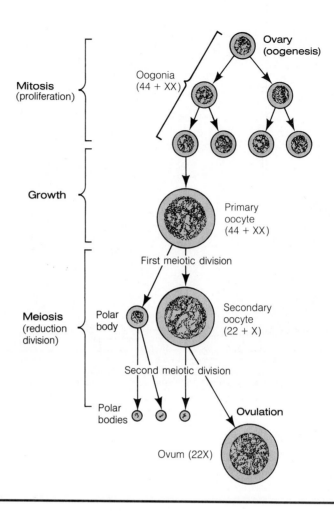

Figure 22.16
Meiosis in the ovary.
Note that during meiosis
each oogonium produces
a single haploid ovum.

as primary follicles until after puberty, which generally occurs between the ages of 12 and 15.

Oogenesis

F22.16 Gametogenesis in the female is called *oogenesis* (Figure 22.16). Oogenesis involves the production of **ova** in the ovary. The precursor cells that, through mitotic division, provide a source for the cells that will develop into ova are diploid cells called **oogonia.** Each oogonium contains 44 autosomes and two X sex chromosomes. By the time embryonic development is completed, a few hundred thousand oogonia have undergone a growth phase and have become **primary oocytes.** The primary oocytes enter the prophase stage of the first meiotic division, but they do not complete it and are in first prophase at the time of birth.

It is not until puberty—some 12 to 15 years later—that a primary oocyte completes the first meiotic division and produces two *haploid* cells of unequal size. After puberty, the first meiotic division generally occurs in one primary oocyte every month. The small cell, which receives half of the chromosomes but very little of the cytoplasm, is the **first polar body.** The large cell, which is called a **secondary oocyte,** contains the other half of the chromosomes and retains almost all of the cytoplasm. The secondary oocyte and the first polar body each contain 22 autosomes and one X sex chromosome, all of which are composed of two connected chromatids. The secondary oocyte immediately begins a second meiotic division; however, it halts at metaphase and is in this stage when it is ovulated. Second meiosis is not completed by the secondary oocyte unless it is fertilized. However, if fertilization does occur, meiosis is then quickly completed.

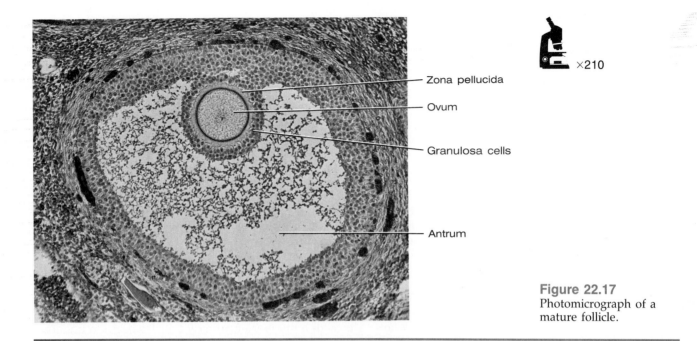

×210

Zona pellucida

Ovum

Granulosa cells

Antrum

Figure 22.17
Photomicrograph of a mature follicle.

When second meiosis does occur, it produces, as does the first meiotic division, one large cell—the mature **ovum**—and a small **second polar body.** The first polar body often undergoes second meiosis, which produces two additional polar bodies; thus, oogenesis produces only one ovum and three polar bodies. The polar bodies, which ultimately degenerate, serve as a means of discarding half of the chromosomes while allowing the ovum to retain almost all of the cytoplasm. Retaining the cytoplasm becomes important after fertilization, since until it becomes implanted in the endometrium of the uterus, the fertilized ovum must depend on its own cytoplasm for a supply of the materials that produce the energy needed during early embryonic development.

Ovarian Cycle

The first *ovarian cycle* occurs at puberty. The ovarian cycle consists of a series of changes within an ovary, including the development of follicles, the release of an ovum from a mature follicle at ovulation, and the formation of a structure called the *corpus luteum.* The length of the ovarian cycle, which varies from about 20 days to 40 days, averages about 28 days. Therefore, the ovarian cycle is commonly considered to be a 28-day cycle. The ovarian cycle is closely correlated with another cycle, the *menstrual cycle,* which involves a repeating series of changes within the lining of the uterus (and to a lesser degree of the vagina).

DEVELOPMENT OF FOLLICLES   Under the influence of *follicle-stimulating hormone* (FSH) from the pituitary gland, some primary follicles undergo further development (Figure 22.15). The single layer of follicle cells that forms F22.15 the wall of each of these primary follicles proliferates, forming a stratified layer of cells, which are now called *granulosa cells.* These follicles are then referred to as **growing follicles,** or **secondary follicles.** Some cells in the connective tissue outside the follicles condense into a layer around the follicle. This layer is called the *theca.* Cells in the inner part of the theca produce the *estrogens,* which are instrumental in the cyclic changes that occur in a female after puberty.

With continued growth, a clear, noncellular region called the **zona pellucida** forms within some of the growing follicles (Figure 22.17). The zona pellu- F22.17 cida separates the developing ovum within each follicle from the surrounding granulosa cells. As the solid, growing follicle becomes larger, a fluid-filled cavity called the **antrum** forms in the layers of the granulosa cells. The antrum displaces the ovum and a few layers of granulosa cells that surround it to one side of the follicle. As more fluid accumulates within the antrum, the follicle continues to enlarge as it moves toward the surface of the ovary, where it produces a bulge. Such a follicle is ready for ovulation, and it is called a **mature** (*Graafian*) **follicle.** Several follicles begin this series of changes each month, but generally only one reaches the mature phase. The rest degenerate. The process of degeneration of a follicle—and the ovum within it—is called *atresia,* and the follicles undergoing atresia are referred to as *atretic follicles.*

F22.15 OVULATION Under the proper hormonal conditions, a mature follicle ruptures and releases its ovum into the abdominopelvic cavity. This event is called *ovulation* (Figure 22.15). During ovulation, those follicle cells that directly surround the ovum remain with it. Therefore, the ovulated ovum is surrounded by a zona pellucida and a sphere of follicle cells that is now called the **corona radiata.** It takes from 10 to 14 days for a primary follicle to develop into a mature follicle. During this time, the developing ovum completes the first meiotic division and reaches metaphase of the second meiotic division, as described previously. The ovum is in this stage when it is released from the ovary at ovulation. Although a number of primary follicles begin to undergo further development during an ovarian cycle, only one usually matures fully and releases its ovum during each cycle.

There are thought to be approximately 400,000 primary follicles present in the two ovaries at birth. Since, during the 30 to 40 years in which a female is capable of reproduction, only one follicle generally matures fully and ruptures each month, a total of only about 400 ova mature during a woman's lifetime.

FORMATION OF THE CORPUS LUTEUM Following ovulation and the accompanying loss of the follicular fluid, the ruptured mature follicle collapses. At this time the adenohypophysis of the pituitary gland is producing increasing amounts of *luteinizing hormone* (LH). Shortly thereafter the cells of the follicle increase in size and take on a yellowish color, due in part to the accumulation of lipid granules. The resulting structure is called a **corpus luteum** (that is, yellow body). The future of the corpus luteum depends on the fate of the ovum. If the ovum is not fertilized and pregnancy does not occur, the corpus luteum reaches its maximum development about eight to ten days after ovulation and then begins to degenerate. Eventually all that remains is a white connective-tissue scar called the **corpus albicans** (that is, white body). However, if the ovum is fertilized and pregnancy does occur, hormones produced by the placenta cause the corpus luteum to continue to develop and to remain for several months into the pregnancy before it degenerates. The corpus luteum serves as an important source of *progesterone* and *estrogens,* which maintain the mucosal lining of the uterus in a condition favorable for implantation and development of the embryo.

Uterine Tubes

F22.13 An ovum released at ovulation is carried to the uterus by a **uterine tube** (Figure 22.13). Each uterine tube, which extends from the vicinity of the ovary to the superior lateral angle of the uterus, lies between the layers of the broad ligament. The portion of the broad ligament that anchors each uterine tube is called the **mesosalpinx.** The medial constricted portion of the uterine tube is the **isthmus.** It opens into the uterus. The uterine tube is expanded somewhat in the region where it curves around the ovary. This region is called the **ampulla.** Fertilization generally occurs within the ampulla. The distal end of each uterine tube is the **infundibulum.** The infundibulum opens into the abdominopelvic cavity, very close to an ovary. This opening, called the **ostium,** is surrounded by small, ciliated, fingerlike projections called **fimbriae.** One of the fimbriae is generally attached to the ovary. The movements of the fimbriae and their cilia are thought to cause currents of peritoneal fluid to enter the uterine tube and thus carry the ovum that has been released from its follicle into the tube. Since the reproductive tract of the female opens into the peritoneal cavity through the two uterine tubes (one tube to each ovary) and to the exterior of the body through the uterus and vagina, infections of the lower reproductive tract can spread into the body cavity and cause peritonitis. (In males, there is no opening of the reproductive tract into the body cavity.)

The wall of the uterine tube is covered by the peritoneum. The *muscular layer* of the uterine tube, which is deep to the serosa, contains circular and longitudinal layers of smooth muscles. The inner *mucosa* is thick and its epithelium is composed of simple columnar cells. The mucosa forms numerous branched folds that extend into the lumen of the uterine tube (Figure 22.18). **F22.18** Two types of cells are present in the epithelium of the mucosa: those with cilia and those without. Most of the cells have cilia that beat rhythmically toward the uterus. Thus, once within the uterine tube, the ovum is carried to the uterus by a weak fluid current caused by the beating of the cilia and possibly aided by peristaltic contractions of the smooth muscles in the walls of the uterine tube. Interspersed among the ciliated cells are cells that lack cilia. These cells are thought to be secretory cells that maintain a moist environment in the uterine tube and may serve as a source of nourishment for the ovum.

Uterus

The **uterus** is a single, hollow, pear-shaped organ that receives the uterine tubes at its upper lateral angles and is continuous below with the vagina (Figure 22.13). The upper portion of the uterus is called the **body.** Below its **F22.13** body, the uterus narrows into the **isthmus,** and where it joins with the vagina it forms the cylindrical **cervix.** The opening of the uterus into the vagina is the **ostium** of the uterus. The dome-shaped region of the uterine body above and between the points of entrance of the uterine tubes is the **fundus.**

The uterus lies in the pelvis, behind the urinary bladder and in front of the sigmoid colon and rectum. The peritoneum that covers the superior and posterior surfaces of the urinary bladder is folded back up the anterior surface of the uterus from the floor of the pelvic cavity. The space thus formed is called the **uterovesical pouch** (Figure 22.14). In a similar manner, the **recto- F22.14 uterine pouch** is formed where the peritoneum that continues over the top of the uterus and down its posterior surface is folded back up the anterior surface of the rectum from the pelvic cavity floor.

The layers of peritoneum that cover the uterus anteriorly and posteriorly fuse along its lateral margins and extend to the lateral walls and floor of the pelvic cavity as the **broad ligament** (Figure 22.13). The portion of the broad **F22.13** ligament that is below the mesovarium and anchors the uterus is called the **mesometrium.** Blood vessels and nerves reach the uterus and uterine tubes by traveling between the peritoneal layers of the broad ligament. The *uterine arteries*, which are branches of the internal iliac arteries, travel within the broad ligament to reach the cervical portion of the uterus. They then follow its lateral margin up to the isthmus of the uterine tube, where they give off tubal and ovarian branches.

Assisting the broad ligament in holding the uterus in place are three pairs of suspensory ligaments: the round ligaments, the uterosacral ligaments, and the cardinal ligaments. The **round ligaments** are fibrous bands that run within the broad ligaments from the lateral margins of the uterus near the junctions with the uterine tubes through the inguinal canals and into the labia majora. The ovarian and round ligaments form a continuum that is *homologous* to—that is, formed from the same embryonic tissue as—the gubernaculum of the male. Because of this homology it is possible for the ovarian and round ligaments to cause an ovary to descend into a labium during embryonic development in a manner similar to the descent of the testes into the scrotum. Recall that the labia majora and the scrotum are also homologous structures. The **uterosacral ligaments** pass from the lateral surfaces of the uterus to the anterior surface of the sacrum. The **cardinal (lateral cervical) ligaments** pass

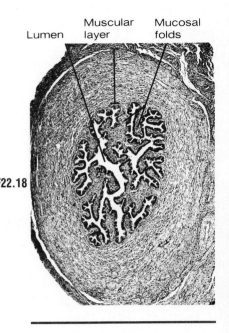

Lumen   Muscular layer   Mucosal folds

Figure 22.18
Photomicrograph of uterine tube (×30).

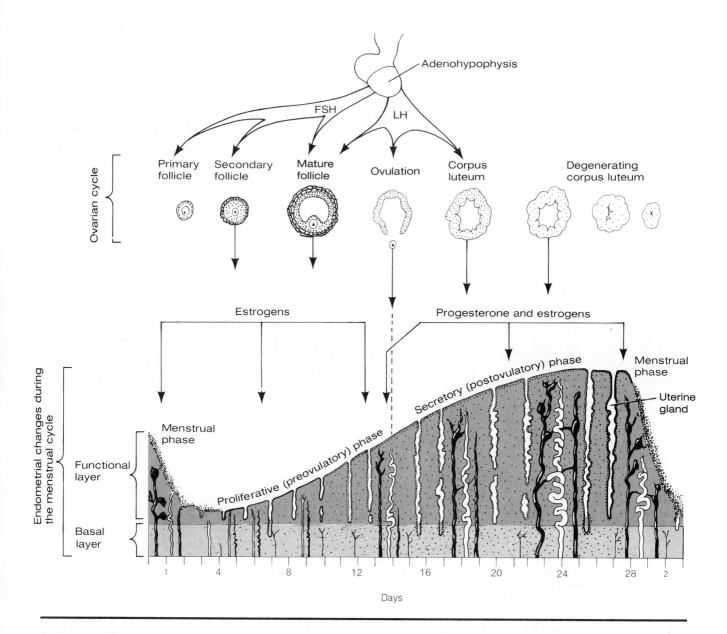

Figure 22.19
The ovarian and menstrual cycles, illustrating cyclic changes in the ovarian follicles and the endometrium of the uterus.

from the cervix of the uterus and the upper region of the vagina to the lateral walls of the pelvis. The uterosacral and cardinal ligaments tend to prevent the uterus from moving downward.

The ligaments do not hold the uterus tightly in place. Rather, they allow it to have limited movement. Normally, the uterus is tipped forward, its longitudinal axis forming an angle of about 90° with the longitudinal axis of the vagina. This angle varies, however, depending on how full the urinary bladder or the rectum is. Abnormal degrees of tipping, either forward or backward, can interfere with blood circulation in the uterus, causing painful menstruation. The ligaments do not in fact provide much support for the uterus. Its chief support is provided from below, by the muscles and membranes that form the floor of the pelvic cavity—that is, the **pelvic diaphragm** and the **urogenital diaphragm.** These muscles all attach to a small, circular tendon located just posterior to the vaginal opening. This tendon is called the *central tendon* of the perineum. If the central tendon is weakened, which sometimes occurs as a result of childbirth, the uterus can move downward into the vagina. In this condition, the uterus is said to be *prolapsed.*

The wall of the uterus consists of the same three layers as the wall of the uterine tubes. It is covered by a **serosa** formed by the peritoneum that is folded back over the uterus from the broad ligament. Beneath the serosa is a thick muscular layer called the **myometrium** (Figure 22.13). The myometrium, **F22.13** which makes up most of the thickness of the wall of the uterus, is composed of bundles of smooth muscles running in various directions. The cavity in the uterus is lined with an epithelium of ciliated columnar cells called the **endometrium.** The surface epithelium invaginates to form numerous tubular *uterine glands*, which extend down into a thick lamina propria referred to as the *endometrial stroma.* The endometrium consists of a thick superficial layer called the **functional layer** and a thin, deep **basal layer.** The functional layer undergoes marked developmental changes during the menstrual cycle. Each month it thickens and becomes engorged with blood in preparation for receiving a fertilized ovum. If the ovum is not fertilized, the functional layer of the endometrium is sloughed off as the menstrual flow. The structural changes that the uterus undergoes during menstruation are described in more detail in the next section.

Uterine Changes During the Menstrual Cycle

The combined effects of the estrogens and progesterone that are produced by the ovarian follicle and the corpus luteum during an ovarian cycle bring about cyclic changes in the female reproductive tract that result in the **menstrual cycle.** Therefore, it is the ovarian cycle that normally controls the menstrual cycle (Figure 22.19). The menstrual cycle is a series of changes that occur in **F22.19** the endometrium of the uterus each month.

The menstrual cycle is divisible into three phases: (1) *menstrual phase,* (2) *proliferative (preovulatory,* or *follicular) phase,* and (3) *secretory (postovulatory,* or *luteal)* phase. For purposes of description, the first day of menstrual flow is considered to be the beginning of the cycle, although physiologically it actually marks the end of the cycle.

MENSTRUAL PHASE The *menstrual phase* of the menstrual cycle is the period during which the superficial functional layer of the endometrium, which had thickened and become vascular during the other two phases of the cycle, breaks down and is lost along with the blood from the damaged vessels of the endometrium. The menstrual flow, or **menses,** therefore, consists of blood, disintegrated endometrial tissue, secretions of the uterine glands, and mucus. Menses lasts for various lengths of time in different women, most generally for three to six days. During this phase of the cycle, the levels of estrogens and progesterone in the blood are generally low, although they begin to rise slowly. At the end of the menstrual phase, the endometrium is relatively thin, and torn blood vessels and glands are exposed on its surface.

The low levels of gonadal hormones present during the menstrual phase and late in the secretory phase of the preceding cycle permits the hypothalamus to release FSH-releasing factor (FSH-RF). The FSH-RF causes FSH to be released by the adenohypophysis, and a small amount of LH is also released and works synergistically with FSH. FSH triggers the ovarian cycle by stimulating at least one primary follicle in an ovary to begin further development.

PROLIFERATIVE PHASE By the end of the first week of the menstrual cycle, in response to FSH, the maturing follicle is producing a sufficient amount of estrogens to cause the functional layer of the endometrium of the uterus to thicken and undergo the repair necessary following menstruation. Straight tubular uterine glands lengthen, and blood vessels invade the endometrium. This marks the beginning of the *proliferative phase* of the menstrual cycle. The length of this phase varies, but it averages seven to nine days. During the proliferative phase of the menstrual cycle, the ovarian cycle progresses to the point of development of a mature follicle, and the level of estrogens in the blood increases. Toward the latter part of this phase there is a slow rise in LH secretion by the adenohypophysis and some production of progesterone by the mature follicle.

The level of estrogens in the blood peaks on about the twelfth day of the

FRONTIERS IN HEALTH

Premenstrual Syndrome: Everyone Has a Cure, But Is There One?

After years of telling women that their premenstrual irritability, depression, pain, and tension—symptoms of premenstrual syndrome—were "all psychological," medical scientists have changed their minds. Premenstrual syndrome (PMS), physicians now agree, is a legitimate medical problem.

No one knows quite how to classify this condition, which strikes four of every ten women of reproductive age. Complaints of those who suffer from PMS cover a wide range of physical and psychological symptoms. According to a recent report in the Harvard Medical School Health Letter, "PMS is easier to define by the timing of the symptoms than by the symptoms themselves." Almost without exception, the symptoms emerge just before menstruation begins. Luckily, most women suffer only a few of the discomforts, the most common being mood swings, nervous tension, and uncontrollable crying spells. For others there are fatigue, depression, headaches, bloating, swelling and tenderness of the breasts, and joint pain.

As difficult as it is to classify this disease, it is even more difficult to determine the cause of the complaints. Moreover, medical science is at a loss to describe how the psychological and physical symptoms are related to one another.

The cause of PMS is now under investigation at the National Institutes of Mental Health. Dr. David Rubinow, a psychiatrist who studies the effects of hormones on the brain, has begun a long-term study of the roles hormones and neurotransmitters may play in bringing about the many symptoms of PMS. Dr. Rubinow is studying the effects of estrogen and progesterone (ovarian hormones), luteinizing hormone and follicle-stimulating hormone (pituitary hormones that regulate the development of ovarian follicles and orchestrate the menstrual cycle), norepinephrine (an adrenal hormone), beta endorphin (the brain's own natural pain killer), and aldosterone (an adrenal hormone that regulates salt and water balance). Because PMS is such a complex condition, it will probably be a long time before medical scientists can truly pinpoint its cause or causes.

Ignorance of the cause of PMS has not deterred the health care industry from providing remedies. Dozens of "proven cures" now exist, ranging from massive doses of progesterone to vitamin B_6. Clinics specializing in PMS have opened, offering cures to suffering women. Because they are desperate, women will often pay large sums of money for a promised cure.

Let the buyer beware, though. Despite claims of success, there is very little scientific evidence to indicate that any of the "cures" really work. Consider the home remedies touted by popular magazines and health-food advocates. Exercise, stress management, and reductions in caffeine and sugar are among the most popular, but they have never been tested in the systematic way that is

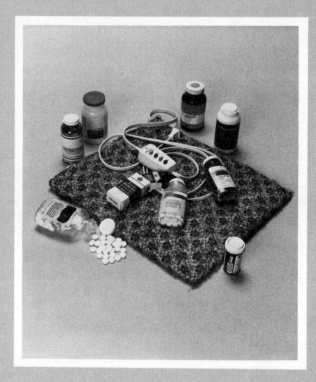

A number of home and medical remedies for PMS.

required for their acceptance by the medical profession. According to the Harvard Medical School Health Letter, "In all cases, these home remedies for PMS have received support from testimonials, but no solid research backs them up."

The medical remedies that physicians prescribe are, quite surprisingly, no more firmly established than the plethora of home remedies. Perhaps the most popular of all medically accepted treatments is progesterone, which physicians are now prescribing in large doses to patients suffering from PMS, despite the absence of any significant clinical evidence of its effectiveness. Prostaglandin, a hormone with broad effects, has also recently begun to be used. Reports of success are encouraging but not yet sufficiently substantiated.

Work is presently under way to test the efficacy of these various treatments, but medical researchers will need several years to assemble a list of proven remedies for PMS. In the meantime, those who suffer from this condition should see a doctor to make certain that the symptoms are not the result of some other condition that has a proven cure.

menstrual cycle, after which it declines relatively quickly. Immediately following the peaking of the production of the estrogens—and perhaps caused by it—large amounts of LH (and to a lesser extent FSH) are released by the pituitary gland. The effect of LH on the estrogen-stimulated mature follicle is to cause it to rupture, and *ovulation* occurs. Ovulation most frequently occurs 14 days after the onset of menstruation, which is the fourteenth day of the 28-day cycle.

SECRETORY PHASE Following ovulation, the menstrual cycle enters the *secretory phase*. This phase lasts approximately 13 days, from day 15 to day 28, and then menstruation occurs, marking the beginning of a new 28-day cycle. The dominant structure of the ovarian cycle during the secretory phase of the menstrual cycle is the corpus luteum, the formation of which is initiated by LH. The corpus luteum produces both progesterone and estrogens, and the levels of these hormones increase sharply during the early part of the final two weeks of the menstrual cycle. As the levels of progesterone and estrogens rise, they inhibit the release of LH and FSH from the pituitary gland.

Under the influence of estrogens and progesterone, the glands of the endometrium of the uterus continue to grow and begin to secrete small amounts of fluid rich in glycogen. The arteries in the endometrium become enlarged and spiraled. In this condition, the endometrium is prepared to provide nourishment for the embryo if fertilization should occur. Moreover, blood vessels are readily available for the formation of a placenta. Progesterone also diminishes the spontaneous contractions of the myometrium. Presumably this effect would assist a fertilized ovum to implant in the endometrium. In essence, the uterus prepares for pregnancy every 28 days.

In the fourth week of the menstrual cycle, for reasons that are not completely clear, the corpus luteum begins to regress and ceases to produce estrogens and progesterone. Recall that the high levels of these hormones are responsible for the prolific development of the endometrium during the secretory phase of the menstrual cycle. As the levels of these hormones decline and their stimulatory effects are withdrawn, some blood vessels of the endometrium undergo prolonged spasms (contractions). These spasms reduce the bloodflow to the areas of the endometrium supplied by those vessels. The resultant lack of blood causes the tissues in the affected regions of the endometrium to degenerate. Eventually the vessels relax, which allows blood to flow through them again. However, capillaries in the area have become so weakened that blood leaks through their walls. The blood that leaks from the vessels takes some of the deteriorating endometrial tissue with it, producing the menstrual flow.

The deepest portion of the endometrium, the basal layer, is not greatly affected during the menstrual cycle, and remains intact. The troughs of the uterine glands extend into the basal layer and provide new cells for growth in the next proliferative phase. When the levels of estrogens and progesterone have dropped low enough, the pituitary gland again begins to produce FSH and follicular development is once again stimulated. Thus, the ovarian cycle and the menstrual cycle repeat themselves.

Vagina

The **vagina** is the canal that leads from the cervix of the uterus to the exterior of the body (Figure 22.14). The smooth muscle tunic in the wall of the vagina **F22.14** is much thinner than the muscular tunic of the uterus. The mucosa that lines the vagina has a protective surface layer of stratified squamous epithelium, as is typical of all the canals that open onto the body surface. The vaginal mucosa proliferates during the menstrual cycle in a manner similar to the endometrial changes of the uterus, but there is considerably less development in the mucosa of the vagina than in the mucosa of the uterus. The mucosa of the vagina contains numerous transverse ridges, or rugae. Near the entrance to the vagina the mucosa usually forms a vascular fold called the **hymen.** Gener-

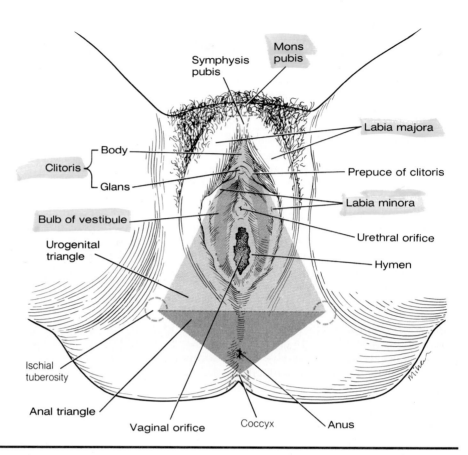

Figure 22.20
External genital organs of the female showing divisions of the perineum.

ally, the hymen only partially blocks the vaginal opening, but in some cases it completely closes the orifice. The hymen, which is stretched and often torn during sexual intercourse, may also be torn by other means. Since the hymen may persist after sexual intercourse, its presence or absence is not a reliable method by which to prove or disprove virginity.

F22.14   The lumen of the vagina is generally quite small, and the walls that surround it are usually in contact with each other. The canal is capable of considerable stretching, however, as when the vagina receives the male penis during sexual intercourse or when it serves as the birth canal. The upper end of the vaginal canal surrounds the cervix of the uterus, forming a recess (the **vaginal fornix**) around the cervix (Figure 22.14). The largest portion of the recess lies dorsal to the cervix and is called the **posterior fornix**. The **anterior fornix** and the two **lateral fornices** are not as deep.

Female External Genital Organs

F22.20   When considered collectively, the external genital organs of the female are known as the **vulva**, or **pudendum** (Figure 22.20).

Under the influence of estrogens, there is a tendency in the female for fatty tissue to be deposited over the pubic symphysis. This deposition produces a mound called the **mons pubis**. Because of the lack of estrogens, this fat deposit is lacking in males. The skin over the mons becomes covered with hair after puberty.

Two rounded folds—the **labia majora**—extend backward from the mons. The outer surfaces of the labia majora are pigmented and covered with hair. Their inner surfaces are smooth, lack hair, and are moist because of the presence of large sebaceous glands.

The **labia minora** are two smaller folds located medial to the labia majora. Anteriorly they surround the clitoris. The labia minora are highly vascular and lack hair. They surround a space (the **vestibule**) into which the vagina and urethra open. The vaginal opening into the vestibule may be partially blocked by the hymen. A number of glands open into the vestibule and keep its walls moist. The ducts of the **greater vestibular glands,** which are homologous to the bulbourethral glands of the male, open into the vestibule on either side of the vagina. On each side of the external urethral orifice are the openings for the **lesser vestibular glands.** The vestibular glands lubricate the vestibule and thus facilitate sexual intercourse.

The **clitoris** is a small elongated structure located at the anterior junction of the labia minora. Most of the body of the clitoris is enclosed by a **prepuce** (foreskin) formed by the labia minora. The free exposed portion of the clitoris is called the **glans.** The clitoris is homologous to the dorsal portion of the male penis and, like the penis, is composed of erectile tissue. The clitoris contains two **corpora cavernosa clitoris,** but it does not contain a corpus spongiosum (which, in the male, surrounds the urethra). The female's urethra is not within the clitoris. Rather, it opens separately posterior to the clitoris. The crura of the corpora cavernosa attach the clitoris to the rami of the pubic and ischial bones. The clitoris, which is very sensitive to touch, becomes engorged with blood and erect when stimulated, contributing to the sexual arousal of the woman.

Just deep to the labia are two elongated masses of erectile tissue called the **bulb of the vestibule.** The bulb, which is homologous to the corpus spongiosum penis and bulb of the penis, runs on either side of the vaginal orifice. During sexual arousal, the bulb becomes engorged with blood, thus narrowing the opening into the vagina and squeezing against the male penis during sexual intercourse.

The Female Perineum

As in the male, the female perineum can be divided by a transverse line between the ischial tuberosities into an anterior **urogenital triangle,** which contains the external genitalia, and a posterior **anal triangle,** which contains the anus (Figure 22.20). The region between the vagina and the anus is re- **F22.20** ferred to as the *clinical perineum* because this area is sometimes torn by the stretching that occurs during childbirth. Sometimes the tears even damage the anal sphincter. To prevent tearing, an incision called an *episiotomy* is often made in the perineum during delivery. This operation enlarges the vestibule and makes delivery easier. Moreover, a surgical incision is easier to repair than a tear.

Mammary Glands

Even though **mammary glands (breasts)** are also present in males, we discuss their anatomy under the female reproductive system because it is only in the female that the breasts are at all involved in reproduction. Even in the female they are not involved in the actual reproductive process, however, but serve to produce milk for the nourishment of the infant.

Each mammary gland is a skin-covered hemispherical elevation located superficial to the pectoral muscles (Figure 22.21). Just below the center of each **F22.21** mammary gland is a protruding **nipple** surrounded by a circular **areola.** The areola has many small elevations due to the presence of numerous large sebaceous glands called **areolar glands.** The areolar glands produce a waxy secretion that prevents chafing of the nipple during nursing. Both the nipple and the areola are pigmented and have capillary beds located close to their surfaces. The pigmentation becomes darker during pregnancy and diminishes somewhat afterward. Smooth muscles in the areola and nipple cause the nipple to become erect as a consequence of stimulation.

Internally, the periphery of each mammary gland is composed of adipose tissue held in place by a connective-tissue stroma. Bands of connective tissue extend from the anterior aspect of the stroma and attach into the dermis. These connective tissue septa are called the **suspensory ligaments** of the breast. The septa subdivide the fat that lies superficial to the glandular tissue, and give a smooth contour to the breast. Centrally, there are 15 to 25 lobes,

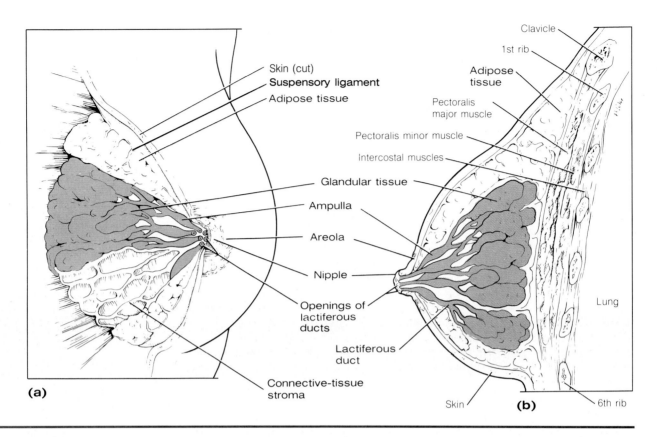

(a)

Skin (cut)
Suspensory ligament
Adipose tissue
Glandular tissue
Ampulla
Areola
Nipple
Openings of lactiferous ducts
Lactiferous duct
Connective-tissue stroma

Clavicle
1st rib
Adipose tissue
Pectoralis major muscle
Pectoralis minor muscle
Intercostal muscles
Lung
Skin
(b)
6th rib

Figure 22.21
The mammary gland.
(a) Anterior view of a partially dissected left breast. **(b)** Sagittal section.

each consisting of a separate compound tubuloalveolar gland. Each lobe is drained by a single **lactiferous duct,** which opens onto the nipple. The nipple, therefore, is perforated by numerous openings. Just before reaching the nipple, each lactiferous duct expands into a small milk reservoir called an **ampulla.** Glandular tissue also extends beyond the breast, into the axilla. Milk is released from the glands by a modified form of apocrine secretion.

The mammary glands begin to develop after puberty, when they are exposed to cyclic stimulation by estrogens and progesterone. They further increase in size during pregnancy and reach their maximum development in a nursing mother. The role of prolactin and oxytocin in milk production and secretion is discussed in Chapter 18.

CONDITIONS OF CLINICAL SIGNIFICANCE

The Reproductive System

Genetic Disorders

In the human, each cell contains 23 pairs of chromosomes that are inherited from the individual's parents. One of these pairs of chromosomes differs in males and females, and therefore are referred to as *sex chromosomes*. In females, both sex chromosomes have the same appearance and are therefore designated as XX. In males, one sex chromosome appears identical to the X chromosome of females but the other is considerably smaller. This small chromosome is called a Y chromosome. Males are therefore genetically XY.

Because of abnormal separation of chromosomes during meiosis, some embryos are genetically XXY. Such

embryos develop normal male genitalia and male secondary sex characteristics. The development of the seminiferous tubules is abnormal, however, and the adult with this genetic condition is sterile because of atrophied testes. Some apparently normal males have been found to be XYY. Some females have been found to be genetically XXX. In some cases, this configuration does not seem to cause any obvious abnormalities, but mental retardation and menstrual irregularities have been reported in others. A more serious condition is the inheritance of only one sex chromosome. If this produces a person who is genetically XO, female internal and external reproductive structures develop but the gonads either remain rudimentary or are completely absent. The

YO condition, where no X chromosome is inherited, results in the death of the embryo.

Venereal Diseases

The most prevalent diseases of the reproductive systems of both male and female are the *venereal diseases*. These infectious diseases are spread through sexual contact.

Gonorrhea

Gonorrhea is an inflammation of the mucous membranes of the urogenital tract or the rectum caused by the bacterium *Neisseria gonorrhoeae*. This inflammation can cause both an infectious discharge of pus from the urethra and painful urination. In females, the inflammation can spread through the cervix of the uterus and the uterine tubes and cause pelvic inflammatory disease. However, it is not unusual for a woman who has gonorrhea to transmit it through sexual contacts and yet show no apparent symptoms herself. In males, the spread of the gonococcus within the reproductive tract can lead to inflammation of the prostate, the seminal vesicles, and the epididymis.

Syphilis

Syphilis is an infectious venereal disease caused by the bacterium *Treponema pallidum*. The bacterium is transmitted by sexual contact and usually produces a lesion (*chancre*) at the site where it enters the body. Most commonly the chancre is on the penis or in the vagina. The chancre soon heals and there may be no symptoms for several weeks. However, during that time the infection spreads throughout the body by way of the bloodstream. After about six weeks, a skin rash accompanied by fever and aching joints may occur. These secondary symptoms may then disappear, and the disease may enter a latent (inactive) period that can last for many years. During this latent period, the body may develop an immunity to the bacterium and destroy it, or the bacterium may spread to many different sites, including the nervous and vascular systems, as well as various organs. The bacterium causes degeneration in these structures and produces severe and varied symptoms that depend on the location of the infection. Syphilis can be detected by several different blood tests, one of which is called the *Wassermann reaction*.

Genital Herpes

Genital herpes is an increasingly common venereal disease. It is the venereal counterpart of fever blisters (see Chapter 4), since both are caused by the *Herpes simplex* virus. The lesions of genital herpes are small vesicles surrounded by swollen, inflamed areas. In men the lesions may appear on the inner surface of the prepuce, on the glans penis, or on the skin in the vicinity of the penis. In women they are generally on the cervix, but lesions may also appear in the vagina or on the vulva. Genital herpes may be *recurrent*, with the vesicles reappearing several times a year and disappearing within 10 days, or it may be *chronic*. The virus can cause serious malformations of children born to infected mothers, and some researchers believe that women with genital herpes have a higher-than-usual risk of developing cancer of the cervix.

Chlamydia

Chlamydia is a sexually transmitted disease that has been recognized only since 1974. Prior to this time any inflammation of the urethra that was not caused by gonorrhea was called nongonococcal urethritis. In 1974 it was determined that most of these inflammations were caused by the bacterium *Chlamydia trachomatis*, and the condition is now more commonly referred to simply as chlamydia. Despite the fact that chlamydia responds well to antibiotic therapy, it is now the most common sexually transmitted disease in the United States.

In males the symptoms of chlamydia are often similar to those of gonorrhea, including discharge from the penis, and painful and frequent urination. Women frequently show no symptoms for extended periods of time, during which they may pass the infection to sexual partners. When symptoms are present they include vaginal discharge, urethritis, and painful intercourse. The infection may spread from the urethra throughout the woman's reproductive tract. If a woman gives birth while infected with chlamydia the child may have a chronic infection of the eyes.

Male Disorders

Prostate Conditions

Although disorders of the male reproductive system are not restricted to the prostate gland, it is the structure most commonly affected. Bacterial inflammation of the prostate produces a condition called *prostatitis*. In this condition, the gland is swollen and tender. In severe cases, abscesses of the gland can form. The prostate gland is also the most common site of cancerous *tumors* in males. Moreover, enlargement of the prostate gland reduces the diameter of the prostatic portion of the urethra, making urination difficult.

Impotence

Impotence is a fairly common male disorder. It is the condition in which the male is unable to attain an erection of the penis, or to retain an erection long enough for sexual intercourse to be completed. Impotence can result from physical disorders of the vascular or nervous systems, but it is often associated with psychological and emotional problems.

Infertility

Male *infertility* is the inability of the male to fertilize an ovum. This condition can be caused by inadequate production of spermatozoa by the testes, the production of abnormal or nonmotile spermatozoa, or an obstruction that prevents the delivery of spermatozoa from the testes to the female vagina. The production of normal sperm can be interfered with by such things as exposure to X rays, malnutrition, and certain diseases, including mumps.

Female Disorders

Abnormal Menstruation

Abnormalities of the menstrual cycle, which are among

the most common disorders of the female reproductive system, can be caused by infections of the reproductive organs or malfunctioning of the ovaries or the pituitary gland. Emotional and psychological factors may also be involved.

Amenorrhea, which is the complete absence of menstrual periods, is most commonly a result of pregnancy. However, it can also be caused by endocrine disorders resulting from abnormal functioning of the ovaries, the pituitary, or the hypothalamus.

Dysmenorrhea, or painful menstruation, is the result of strong contractions of the uterus. It has generally been thought to be associated with low levels of progesterone in the blood, but recent evidence indicates that the presence of prostaglandins is also a major factor. Dysmenorrhea may also be caused by conditions such as pelvic tumors, displacement of the uterus, and pelvic inflammatory disease.

Endometriosis

Some cases of severe dysmenorrhea and pelvic pain are caused by a condition called *endometriosis*. In this condition endometrial tissue occurs in abnormal locations, often outside the uterus. The most common site for endometriosis is in the ovaries, but aberrant endometrial tissue has also been reported in the uterine ligaments, the pelvic peritoneum, and occasionally in various other locations. The tissue is thought either to enter the pelvic cavity from the uterus by way of the uterine tubes during menstruation or else to originate in the pelvic cavity as the result of abnormal embryonic differentiation of the epithelial cells that line the cavity.

Endometriosis becomes clinically important during menstruation, when the aberrant endometrial tissue, like the endometrium of the uterus, apparently reacts to hormone levels in the body and undergoes cyclic bleeding. Unlike the endometrial tissue in the uterus, blood from the aberrant endometrial tissue generally has no means by which it can drain to the outside of the body. As a consequence, the blood collects in the aberrant tissue, causing pain and, in some cases, more serious complications. The condition is most common during a woman's reproductive life and declines in frequency after the age of 40.

Pelvic Inflammatory Disease

Pelvic inflammatory disease is an inflammation involving the uterine tubes, the ovaries, or the peritoneum of the pelvic cavity. It is caused by bacteria—usually gonococci, streptococci, or staphylocci—that generally reach the pelvic cavity by traveling through the vagina, the uterus, and the uterine tubes. In some cases, the bacteria reach the pelvic structures after traveling in the bloodstream from distant sites of infection.

Tumors

Tumors can develop in several locations in the female reproductive system. Such tumors can be cancerous (malignant), and thus capable of spreading to other locations in the body, or they can be noncancerous (benign).

Ovarian tumors can be cystic (containing fluid) or solid. Ovarian cysts are the more common form and are usually noncancerous. A high percentage of solid ovarian tumors are malignant.

The most common tumors are those of the uterus and cervix. These tumors can be diagnosed early in their development by means of a simple technique called a *Pap smear*. This test involves the removal (with a swab) of cells from the cervix and the surrounding area. These cells are then examined microscopically for signs of malignancy. With this procedure, cancerous cells can often be detected before any symptoms have appeared. If identified early, uterine tumors can be removed surgically or destroyed with radiation treatments before they spread to other parts of the body. Since many women now have regular gynecological examinations that include Pap smears, the death rate from uterine cancer has declined steadily.

Tumors are also common in the breasts, particularly after age 30. Breast tumors, whether malignant or benign, can be best detected by regular manual self-examination. Generally, tumors of the breast are removed surgically. If a biopsy shows the tumor to be malignant, the breast—and possibly the underlying pectoral muscles and axillary lymph nodes—is removed in an attempt to prevent the cancer from spreading through the blood or lymphatics to other regions of the body. This procedure is known as a *mastectomy*. Radiotherapy and chemotherapy are often used to destroy the tumor instead of removing it surgically, or following surgery, to destroy any cancerous cells the operation might have missed.

Effects of Aging on the Reproductive System

In males there is a slight, but gradual, age-related decline in the secretion of testosterone—which may reduce muscle strength, contribute to decreased sexual desire, and affect sperm production. However, even in quite elderly men, abundant viable sperm may be produced. The prostate gland begins to atrophy by about 50 years of age, and the volume of its secretion contributed to semen and the force behind its contraction during orgasm are both reduced. The penis undergoes some atrophy with age, causing changes in the walls of the penile blood vessels that may affect the ability of an older man to attain an erection.

In the reproductive tract of women age-related changes occur first in the ovaries and result in reduced secretion of estrogens and progesterone by them. The declining levels of these hormones, in turn, cause degenerative changes in the uterus, vagina, and external genitalia. One of the surest signs of aging in the female is menopause, when she no longer experiences menstrual cycles. After menopause, glands that normally lubricate the vagina gradually atrophy, and the vagina becomes dry and sexual intercourse may be painful.

STUDY OUTLINE

EMBRYONIC DEVELOPMENT OF THE REPRODUCTIVE SYSTEM pp. 591–594

EARLY DEVELOPMENT genital ridges, paramesonephric ducts, and mesonephric ducts.

DEVELOPMENT OF INTERNAL REPRODUCTIVE STRUCTURES

MALE
1. Medulla of gonads gives rise to seminiferous tubules.
2. Mesonephric ducts form efferent ductules, epididymis, and ductus deferens; paramesonephric ducts degenerate.

FEMALE
1. Cortex of gonads becomes site of ova production.
2. Mesonephric ducts degenerate; paramesonephric ducts form uterus, vagina, and uterine tubes.

ANATOMY OF THE MALE REPRODUCTIVE SYSTEM pp. 595–605

THE MALE PERINEUM divided into anterior urogenital triangle and posterior anal triangle.

TESTES AND SCROTUM

TESTES spermatogenesis occurs in seminiferous tubules.

INTERSTITIAL ENDOCRINOCYTES secrete androgens.

SPERMATOGENESIS
1. Spermatogonia (diploid) develop into primary spermatocytes (diploid).
2. Primary spermatocytes undergo first meiotic division, forming secondary spermatocytes (haploid)
3. Secondary spermatocytes undergo second meiotic division, producing spermatids.
4. Spermatids develop into mature spermatozoa.

THE SPERMATOZOON flagellated; contains acrosome, centrioles, mitochondria, and little cytoplasm.

DESCENT OF TESTES testes follow vaginal process through inguinal canal into scrotum.
1. Location of testes in scrotum necessary for normal sperm development.
2. Cremaster muscle assists in regulating temperature of testes.

EPIDIDYMIS first portion of duct system; storage of mature nonmotile sperm.

DUCTUS (VAS) DEFERENS continuation of duct of epididymis; heavy wall of smooth muscle; located in spermatic cord; site of vasectomy.

SEMINAL VESICLES two membranous pouches lateral to ductus deferens.

PROSTATE GLAND encompasses urethra below bladder; secretes thin, milky, alkaline fluid that contributes to semen.

BULBOURETHRAL GLANDS pair of glands below prostate; secretion is a thick, alkaline mucus.

PENIS copulatory organ; contains three cylindrical cavernous bodies; deposits spermatozoa in female reproductive tract.

SEMEN mixture of spermatozoa from testes and fluids from seminal vesicles, prostate, and bulbourethral glands; nutritive source; activates spermatozoa to become motile; alkaline; fructose provides energy source for spermatozoa.

ANATOMY OF THE FEMALE REPRODUCTIVE SYSTEM pp. 605–618

OVARIES production of ova; estrogen and progesterone production; cortex contains primary follicles.

OOGENESIS
1. Oogonia (diploid) develop into primary oocytes (diploid).
2. Primary oocytes undergo first meiotic division, forming secondary oocytes (haploid) and polar bodies.
3. Secondary oocyte stops in second meiotic division; completes it to become fully developed ovum only if fertilized.
4. Beginning at puberty, one secondary oocyte (ovum) usually released at ovulation each month.

OVARIAN CYCLE begins at puberty.

Development of Follicles
1. Under influence of follicle-stimulating hormone (FSH).
2. Growing follicles develop zona pellucida and fluid-filled antrum.
3. Ultimately develops into mature follicle.

Ovulation rupture of mature follicle releases ovum.

Formation of the Corpus Luteum cells of ruptured mature follicle increase in size and become yellow.
1. Degenerates in absence of fertilization.
2. Remains as source of progesterone and estrogens after fertilization.

UTERINE TUBES carry ovum to uterus.
1. Fimbriae surround opening.
2. Thick mucosa of single columnar cells; some ciliated.
3. Cilia and peristaltic contractions carry ovum to uterus.

UTERUS composed of body, isthmus, and cervix.
1. Supported by central tendon of pelvic and urogenital diaphragms.

2. Endometrium thickens and becomes engorged to prepare for fertilized ovum each month.

UTERINE CHANGES DURING THE MENSTRUAL CYCLE Caused by combined effects of estrogens and progesterone from ovarian follicle and corpus luteum.

Menstrual Phase (3–6 days)
1. Thick, vascular endometrium breaks down.
2. Menses composed of blood, endometrium, secretions of uterine glands, and mucus.
3. Low level of gonadal hormones.

Proliferative Phase (7–9 days)
1. Mature follicle development.
2. Maturing follicle produces estrogens.
3. Blood estrogens increase.
4. Endometrial thickness and vascularity increase.
5. Ovulation occurs toward end of cycle (day 14).

Secretory Phase (approximately 13 days)
1. Corpus luteum formed.
2. Sharp increase in estrogens and progesterone.
3. Endometrium prepared for implantation.
4. Decline in LH and FSH levels toward end of phase begins endometrial degeneration.

VAGINA canal from cervix to exterior of body; mucosa of stratified squamous epithelium; entrance partially covered by hymen.

FEMALE EXTERNAL GENITAL ORGANS (VULVA, OR PUDENDUM)
1. Labia majora and minora.
2. Greater and lesser vestibular glands lubricate vestibule.
3. Clitoris homologous to male penis.

THE FEMALE PERINEUM divided into anterior urogenital triangle and posterior anal triangle. *Clinical perineum* is located between vagina and anus.

MAMMARY GLANDS
1. Nipple and areola pigmented.
2. Composed of adipose tissue and lobes of compound tubuloalveolar glands.
3. Ampullae serve as milk reservoirs.

CONDITIONS OF CLINICAL SIGNIFICANCE: THE REPRODUCTIVE SYSTEM pp. 618–620

GENETIC DISORDERS result from abnormal separation of X and Y chromosomes during meiosis.

VENEREAL DISEASES infectious diseases spread through sexual contact.

GONORRHEA inflammation of mucous membranes of genital tract or rectum caused by bacterium *Neisseria gonorrhoeae*.

SYPHILIS caused by bacterium *Treponema pallidum*.
1. Produces chancre where it enters body; usually on penis or vagina.
2. Rash, fever, and aching joints may appear weeks later.
3. Latent period may be followed by degeneration of nervous and vascular structures.

GENITAL HERPES caused by *Herpes simplex* virus. Produces small vesicles surrounded by inflamed area; may be recurrent or chronic.

CHLAMYDIA urethritis caused by bacterium *Chlamydia trachomatis*. Infection may spread throughout woman's reproductive tract.

MALE DISORDERS

PROSTATE CONDITIONS prostatitis, tumors, gland enlargement.

IMPOTENCE inability to attain or retain erection long enough to complete sexual intercourse.

INFERTILITY caused by inadequate or abnormal spermatozoa production or obstruction in duct.

FEMALE DISORDERS

ABNORMAL MENSTRUATION caused by infection, malfunction, or emotional stress.

AMENORRHEA absence of menstrual periods.

DYSMENORRHEA painful menstruation.

ENDOMETRIOSIS endometrial tissue in aberrant locations.

PELVIC INFLAMMATORY DISEASE caused by bacterial infection.

TUMORS can occur in ovaries, uterus, or breasts.

EFFECTS OF AGING ON THE REPRODUCTIVE SYSTEM

MALES decline in secretion of testosterone; atrophy of prostate gland and penis.

FEMALES reduced secretion of estrogens and progesterone by ovaries causes degenerative changes in uterus, vagina, and external genitalia.

SELF-QUIZ

1. The male gonads produce: (a) estrogens; (b) testosterone; (c) progesterone; (d) all of these.

2. In the male embryo, which of these structures degenerate shortly after the gonads begin to differentiate into testes? (a) the paramesonephric ducts; (b) the seminiferous tubules; (c) the mesonephric duct.

3. In the absence of the male hormones, all embryos, regardless of their genetic makeup, will develop into females. True or false?

4. Spermatogenesis occurs in the: (a) tunica albuginea; (b) seminiferous tubules; (c) sustenticular cells; (d) prostate gland.

5. The male hormones, the androgens, are produced in the: (a) seminal vesicles; (b) tubulus rectus; (c) interstitial endocrinocytes; (d) mature follicle.

6. Occasionally, one or both of the testes fail to leave the abdominopelvic cavity, a condition called cryptorchidism, which results in sterility. True or false?

7. The scrotum regulates the temperature of the testes through the action of the: (a) gubernaculum; (b) cremaster muscle; (c) epididymis.

8. The four haploid structures formed in the male during spermatogenesis are called: (a) spermatids; (b) primary spermatocytes; (c) spermatogonia.

9. Match the structures of the male reproductive system with the appropriate description.

   Epididymis
   Ductus deferens
   Seminal vesicles
   Ejaculatory ducts
   Prostate gland
   Bulbourethral glands

   (a) These structures penetrate the prostate gland and enter the urethra just below its exit from the bladder.
   (b) Encompasses the urethra just below the bladder.
   (c) This structure transports sperm to the urethra.
   (d) The first portion of the duct system that transports mature sperm from the testes to the outside of the body.
   (e) Located below the prostate on either side of the membranous portion of the urethra.
   (f) These structures join with the ductus deferens to form the ejaculatory duct.

10. A patient's semen is analyzed and found to contain 8 million sperm per ejaculation. In all likelihood, he is: (a) impotent; (b) sterile; (c) a eunuch.

11. The alkalinity of semen protects the sperm from the acid pH of the vagina. True or false?

12. The developing ovarian follicle produces: (a) FSH; (b) LH; (c) estrogen; (d) both a and b.

13. The ligament that supports the ovary from the posterior surface of the broad ligament is: (a) mesosalpinx; (b) mesovarium; (c) round ligament; (d) ovarian ligament.

14. Each uterine tube opens into the abdominopelvic cavity through an ostium. True or false?

15. Ova mature in the: (a) uterus; (b) corpus luteum; (c) ovarian follicles; (d) cervix.

16. About 4000 of the total number of ova mature during a woman's lifetime. True or false?

17. Following ovulation: (a) what remains of the follicle develops into a corpus luteum; (b) the entire follicle migrates to the uterus and degenerates; (c) the mesosalpinx undergoes transformation.

18. Second meiosis is generally not completed by a secondary oocyte unless it is fertilized. True or false?

19. Which of these structures thickens each month and becomes engorged with blood in preparation for receiving a fertilized ovum? (a) endometrium; (b) pelvic diaphragm; (c) myometrium; (d) labia.

20. The vagina is lined with stratified squamous epithelium. True or false?

21. Collectively, the external genital organs of the female are known as the: (a) mons pubis; (b) labia majora; (c) vulva; (d) pudendum; (e) both c and d.

22. The structure in the female that is homologous to the dorsal portion of the male penis is the: (a) hymen; (b) mons pubis; (c) clitoris; (d) labia majora.

23. An incision called an episiotomy is often made in the clinical perineum to ease delivery of a child. True or false?

24. The nipple of a female breast is perforated by a single opening. True or false?

25. An individual who is genetically XXY: (a) develops abnormal male genitalia; (b) has abnormal male secondary sex characteristics; (c) develops abnormal seminiferous tubules; (d) generally has the abnormalities in a, b, and c.

LEARNING OBJECTIVES

After completing this chapter, you should be able to:

- Describe the events involved in the fertilization of the ovum and the implantation of the blastocyst.

- Describe the formation of the placenta, and indicate the relationship between maternal and fetal blood in the placenta.

- Describe how each embryonic membrane is formed, and state its function.

- Describe how the body coelom is formed.

- Describe the unique features of fetal circulation.

- Describe the circulatory changes that occur at birth.

- Distinguish between ectopic pregnancy, placenta previa, and toxemia.

CHAPTER CONTENTS

Pregnancy and Early Embryonic Development

23

We saw in the previous chapter that two of the main functions of the reproductive system are to provide a means by which fertilization can occur and to furnish an environment that is conducive to the development of the fertilized ovum *(zygote)* to a stage at which it is capable of surviving on its own outside the mother's body. During the time that the zygote is developing within a woman, the woman is said to be **pregnant.** In this chapter we consider how fertilization occurs and discuss the major structural changes that occur in the woman and in the zygote during early embryonic development.

PREGNANCY

For pregnancy to occur, it is necessary for spermatozoa to be deposited in the vagina, generally by sexual intercourse, and to be transported to the upper region of the uterine tube, where fertilization generally occurs. The fertilized ovum must then be transported to the uterus and become implanted within the endometrium, and a placenta must form.

Fertilization

An ovum remains capable of being fertilized for 12 to 24 hours following ovulation; sperm remain viable within the female reproductive tract for 24 to 72 hours after ejaculation. Therefore, for fertilization to occur, sperm should be deposited within the female tract no earlier than 72 hours before ovulation and no later than 24 hours after ovulation.

Once within the vagina, spermatozoa travel through the uterus and reach the upper portion of the uterine tubes rather quickly, often requiring only a few minutes to do so. The rate at which the sperm are transported is too rapid for it to be due solely to the motility of the sperm. It has been suggested that the prostaglandins found in semen might assist sperm transport by stimulating reverse peristaltic contractions of the smooth muscles of the uterus and uterine tubes. The transport problem in the uterine tubes is further complicated by the fact that the tubes must transport sperm in one direction at the same time as they are transporting an ovum in the opposite direction. The semen deposited in the vagina during sexual intercourse usually contains several hundred million sperm. However, sperm mortality is high, and only a few thousand reach the uterine tubes.

During their passage through the uterus and the uterine tubes, sperm undergo changes that make them capable of fertilizing an ovum. This process is referred to as **capacitation.** The process of capacitation is not entirely understood, but it appears to involve an alteration of the sperm membrane that is brought about by secretions of the female reproductive tract. Sperm that have undergone capacitation are able to release enzymes from their acrosome upon contact with the corona radiata surrounding the ovum.

Upon reaching the ovum, several sperm pass through the corona radiata, but only one sperm actually fertilizes the ovum (Figure 23.1*a*). Nevertheless, large numbers of sperm must be present in the vicinity of the ovum for fertilization to occur. This is thought to be due to the need for the enzymes present in the acrosome of each spermatozoon—including *hyaluronidase* and *proteases*. Upon contacting the corona radiata, the acrosome of a capacitated sperm

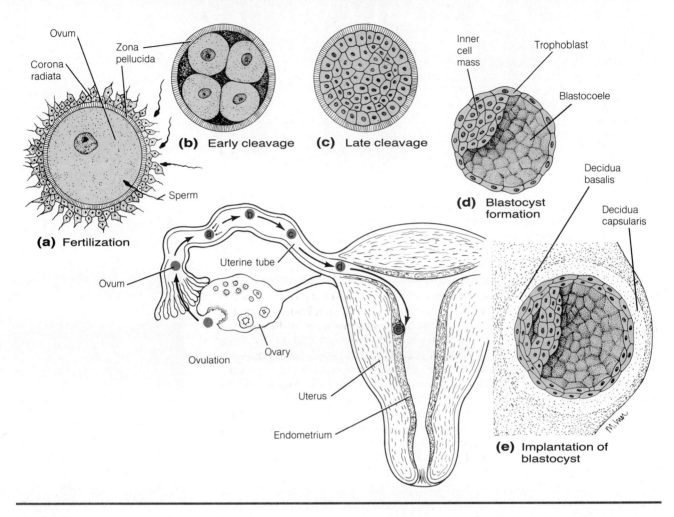

Figure 23.1
Fertilization of an ovum
in the uterine tube;
cleavage and
implantation of the
blastocyst.

opens and releases its enzymes. The hyaluronidase breaks down the zona pellucida and the material that holds the cells of the corona together *(hyaluronic acid),* thus permitting the sperm access to the ovum. It has been suggested that enzymes from several sperm are required in order to remove the barriers from around the ovum.

When one of the sperm contacts the ovum, the cell membranes of the sperm and the ovum fuse and break down, thus allowing the nucleus of the sperm to enter the ovum. The ovum then completes its second meiotic division, and the nuclei (now called *pronuclei)* of the sperm and the ovum join into a single nucleus. Thus, the ovum is fertilized and the fertilized ovum, or **zygote,** contains the full complement of 46 chromosomes. Once the ovum is fertilized, rapid chemical changes are thought to occur in the zona pellucida and in the cell membrane of the ovum that prevent the entry of additional sperm.

Sex Determination

The genetic sex of an individual is determined at the time of fertilization, even though the individual's sex does not become apparent until about the eighth week of development. If the ovum is fertilized by a sperm containing an X sex chromosome, a female (XX) normally develops. If the ovum is fertilized by a sperm containing a Y sex chromosome, a male (XY) develops.

Development and Implantation of the Blastocyst

The zygote immediately begins to divide, that is, undergo *cleavage* (Figure 23.1*b, c*). As the zygote divides it passes along the uterine tube to the uterus—a journey that requires about three to four days. Indeed, if the zygote were to

F23.1b, c

reach the lower end of the uterine tube much sooner, it could not enter the uterus because this region of the uterine tube remains spastically contracted for about three days following ovulation. By the time the uterus is reached, or shortly thereafter, enough cell divisions have occurred for the zygote to have developed into a fluid-filled sphere of cells called a **blastocyst** (Figure 23.1*d*). **F23.1d** Within the blastocyst, the cavity that contains the fluid is called the **blastocoele.** The blastocyst has an accumulation of cells, called the **inner cell mass,** that extend into one side of the blastocoele. The embryo develops from this inner cell mass. The outer sphere of cells that surrounds the blastocoele is called the **trophoblast** (or *trophectoderm*). This layer will contribute to the formation of the placenta.

The blastocyst remains free within the lumen of the uterus for several days before it implants itself in the endometrium. During this period, it obtains nourishment from the uterine fluids and continues to undergo cell division. At the same time, the zona pellucida disintegrates, allowing the trophoblast to expand. About seven days after ovulation, the blastocyst undergoes *implantation* in the endometrium (Figure 23.1*e*). At this time, the woman is in **F23.1e** approximately the twenty-first day of her menstrual cycle, and the endometrium of her uterus has been prepared for implantation by estrogens and progesterone. Upon contact with the endometrium, the trophoblast begins to grow rapidly and its cells release enzymes that digest the cells of the endometrium, thus eroding the endometrial surface. This erosion makes additional fluids and nutrients available to the blastocyst and allows the trophoblast to burrow into the endometrium. As the blastocyst erodes deeper, the endometrium grows over it, and within a few days the blastocyst is completely implanted in the endometrium.

For the first seven days following implantation, the embryo depends entirely on the trophoblast's destruction of endometrial cells for nutrients. Eventually, nutrients are obtained through the placenta, but erosion by the trophoblast continues to supply a significant fraction of the embryo's nutrients for the first two months of development, while the placenta is gradually enlarging and becoming increasingly functional.

Maintenance of the Endometrium

Because of the importance of the endometrium for the nourishment of the embryo, it is vital that the endometrium be maintained in a highly developed state so pregnancy can continue. Thus, the menstrual and ovarian cycles must be interrupted; menses must be prevented from occurring; and the development of additional follicles in the ovary must be suppressed. In order to maintain the endometrium, there must be high levels of estrogens and progesterone within the blood. Such high levels of estrogens and progesterone also suppress the secretion of follicle-stimulating hormone (FSH) and luteinizing hormone (LH) from the pituitary gland, thus preventing the further development of any follicles.

During early pregnancy, the required estrogens and progesterone are produced by the corpus luteum. In the absence of pregnancy, the corpus luteum would regress within two weeks after ovulation. However, when pregnancy occurs, the corpus luteum remains and continues to secrete estrogens and progesterone for approximately three months.

Although it is not completely understood why the corpus luteum persists during early pregnancy, one factor that causes it to continue to secrete its hormones is the presence of another hormone, called *human chorionic gonadotropin* (HCG). This hormone is produced by the **chorion.** The chorion develops from the trophoblast, and it later becomes involved in the formation of the placenta. HCG has properties that are very similar to those of LH; that is, it maintains the activity of the corpus luteum and thus maintains the development of the endometrium. HCG is secreted as early as the second week of pregnancy. Its level reaches a peak during the third month and then sharply declines, remaining low through the remainder of pregnancy.

Another source of estrogens and progesterone during pregnancy is the placenta. By about the end of the second month, the placenta itself is secreting sufficient estrogens and progesterone to maintain the endometrium, and

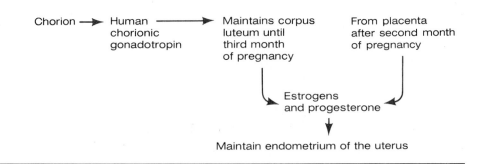

Figure 23.2
The roles of human chorionic gonadotropin, estrogens, and progesterone during pregnancy.

as pregnancy progresses it continues to produce increasing amounts of these hormones. Thus, beyond the second month, the continuance of the pregnancy normally no longer depends on the secretion of estrogens and progesterone by the corpus luteum (Figure 23.2). The third month, therefore, can be a critical period of a pregnancy. During this month, the level of HCG begins to decline, which leads to the degeneration of the corpus luteum. This, in turn, would cause reduced levels of estrogens and progesterone if the placenta were not yet producing sufficient amounts of these hormones to offset their reduction from the corpus luteum. In such a case, the endometrium could not be maintained sufficiently to allow pregnancy to continue, and spontaneous abortion of the embryo might occur.

Development of the Placenta

The **placenta** is an organ that, from the third month to the time of birth some six months later, serves to supply nutrients to, and remove wastes from, the embryo—which after the second month is called a **fetus.** Moreover, the placenta serves as an endocrine gland that produces a number of hormones, including estrogens and progesterone.

Because the superficial layer of the thickened endometrium is discarded by the body, either with the menses or at birth, it is called the **decidua** ("falling off"). The portion of the decidua that lies deep to the blastocyst is called the **decidua basalis** (Figure 23.1e). The decidua basalis is involved in the formation of the maternal portion of the placenta. The portion of the decidua superficial to the blastocyst is the **decidua capsularis.**

As the blastocyst becomes implanted in the endometrium, the trophoblast separates into two layers: an outer **syncytiotrophoblast,** in which the cell boundaries disappear, and an inner **cytotrophoblast,** which is composed of distinctly separate cells (Figure 23.3b). At this time a layer of mesodermal cells develops on the inside of the trophoblast. This layer of cells combines with the trophoblast to form a membrane called the **chorion** (Figure 23.3c). The chorion was mentioned earlier as the source of human chorionic gonadotropin, but it also forms an important part of the placenta. The syncytiotrophoblast proliferates and releases enzymes that erode the endometrium. At the same time, the chorion develops many fingerlike projections called **chorionic villi** (Figure 23.3d). Branches from the umbilical artery and vein of the fetus grow into the villi and form capillary beds. With continued development, the villi become surrounded by pools of the mother's blood, which has collected in sinuses, called *lacunae,* within the endometrium. These blood pools result from the syncytiotrophoblast's erosion of the decidua basalis region of the endometrium and the capillaries in the endometrium.

Thus, while maternal and fetal blood are separated by only a few layers of cells (chorion and endothelium of the fetal capillaries), there is normally no actual mixing of the two blood supplies. Oxygen, carbon dioxide, glucose, amino acids, fatty acids, various ions, and other substances cross the placenta, in one direction or the other, primarily by diffusion. The presence of a large number of villi provides an extensive surface area across which substances can pass. It is estimated that a fully developed placenta has a surface area of about 16 square meters.

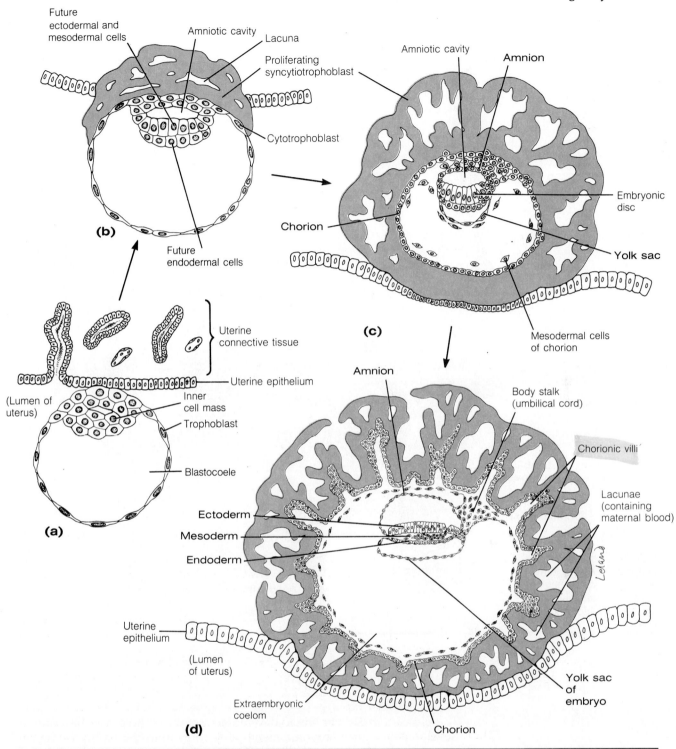

Future ectodermal and mesodermal cells

Amniotic cavity

Lacuna

Proliferating syncytiotrophoblast

Cytotrophoblast

(b)

Future endodermal cells

Uterine connective tissue

(Lumen of uterus)

Uterine epithelium

Inner cell mass

Trophoblast

Blastocoele

(a)

Amniotic cavity

Amnion

Chorion

Embryonic disc

Yolk sac

(c)

Mesodermal cells of chorion

Amnion

Body stalk (umbilical cord)

Chorionic villi

Lacunae (containing maternal blood)

Ectoderm

Mesoderm

Endoderm

Uterine epithelium

(Lumen of uterus)

Extraembryonic coelom

Yolk sac of embryo

Chorion

(d)

It is clear, from the manner in which it is formed, that the placenta is a combination of maternal tissue (decidua basalis) and fetal tissue (chorion) (Figure 23.4). The fetal blood, which remains in the fetal vessels, enters the placenta through the **umbilical arteries** and passes through capillaries in the villi (where the exchange of substances occurs). The fetal blood then leaves the placenta—and returns to the fetus—through the **umbilical vein.** The umbilical vessels travel between the placenta and the fetus in the **umbilical cord.** The mother's blood enters the placenta through the **uterine arteries** (branches of the internal iliac arteries), flows through the blood pools of the endometrium (where the exchange of substances occurs), and leaves the placenta through the **uterine veins.**

F23.4

Figure 23.3
Successive stages of embryonic development showing proliferation of the trophoblast, implantation in the endometrium, and early formation of the amnion, the chorion, the yolk sac, and the three primary germ layers.

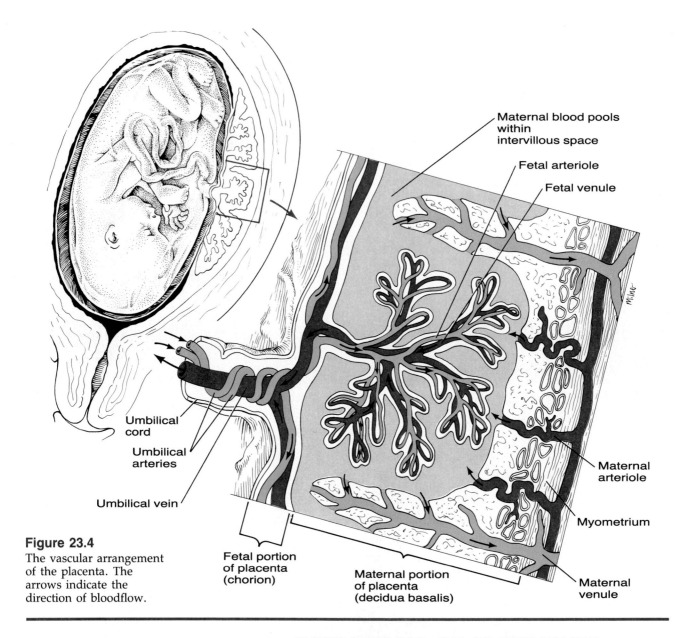

Figure 23.4
The vascular arrangement of the placenta. The arrows indicate the direction of bloodflow.

Maternal blood pools within intervillous space

Fetal arteriole

Fetal venule

Maternal arteriole

Myometrium

Maternal venule

Umbilical cord

Umbilical arteries

Umbilical vein

Fetal portion of placenta (chorion)

Maternal portion of placenta (decidua basalis)

FORMATION OF EMBRYONIC MEMBRANES AND GERM LAYERS

The placenta makes continued development of the embryo possible. Early in development four embryonic membranes form from the embryo—the *amnion, yolk sac, allantois,* and *chorion*. These membranes assist in the embryo's development in various ways, although they contribute very little to the body of the embryo itself.

While the embryonic membranes are forming, the cells of the inner cell mass undergo movements that result in the formation of three germ layers: the *endoderm, ectoderm,* and *mesoderm*. The embryo—including all of its organs and structures—develops from these germ layers.

Embryonic Membranes

Amnion

Following implantation, while the placenta is developing, the inner cell mass proliferates and a cavity forms that separates it from the trophoblast. The membrane that surrounds this cavity is the **amnion.** The amnion is a fluid-

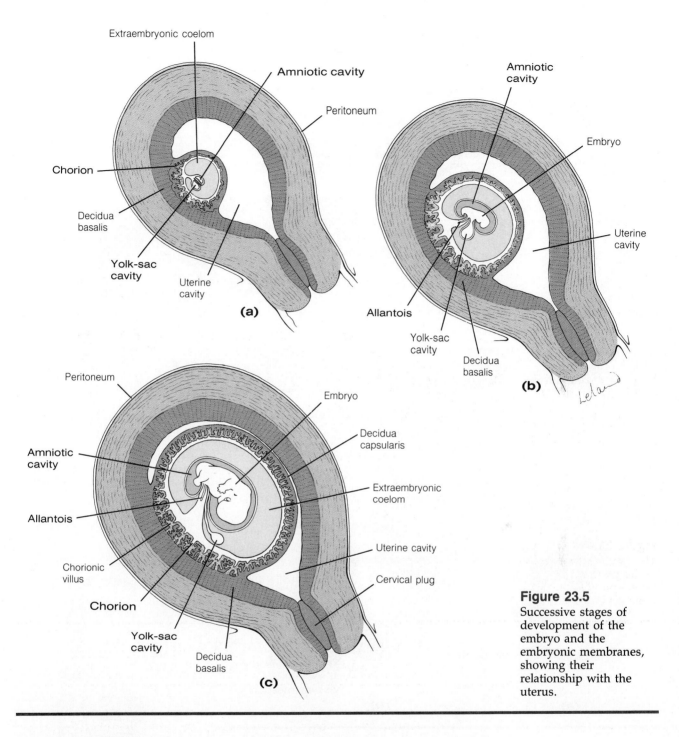

Figure 23.5
Successive stages of development of the embryo and the embryonic membranes, showing their relationship with the uterus.

filled structure that eventually surrounds the fetus (Figure 23.3, Figure 23.5). The amniotic fluid protects the embryo against physical injury, helps to maintain the constancy of its temperature, and allows it to move freely. Those cells of the inner cell mass that form the floor of the amniotic cavity make up the **embryonic disc.** The cells of the embryonic disc, from which the embryo will be formed, separate into two layers: a layer of *ectodermal cells* that face the amniotic cavity and a layer of *endodermal cells* on the side toward the space *(blastocoele)* within the blastocyst. Included in the layer of ectodermal cells are cells that will contribute to the formation of mesoderm.

F23.3, F23.5

Yolk Sac

The endoderm cells of the embryonic disc undergo rapid mitosis and form the second embryonic membrane: the **yolk sac.** The yolk sac develops as a small

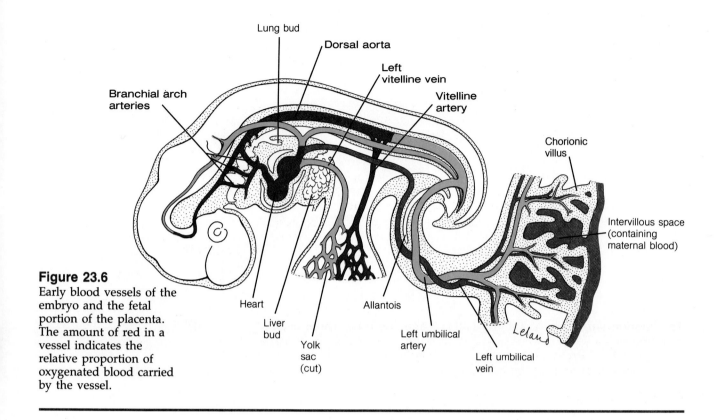

Figure 23.6
Early blood vessels of the embryo and the fetal portion of the placenta. The amount of red in a vessel indicates the relative proportion of oxygenated blood carried by the vessel.

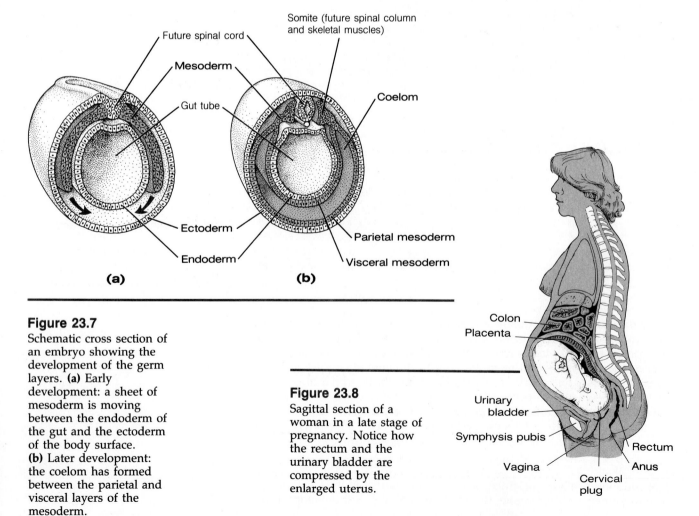

Figure 23.7
Schematic cross section of an embryo showing the development of the germ layers. **(a)** Early development: a sheet of mesoderm is moving between the endoderm of the gut and the ectoderm of the body surface.
(b) Later development: the coelom has formed between the parietal and visceral layers of the mesoderm.

Figure 23.8
Sagittal section of a woman in a late stage of pregnancy. Notice how the rectum and the urinary bladder are compressed by the enlarged uterus.

cavity on the inner surface of the embryonic disc and eventually extends into the umbilical cord. In the human, where very little yolk is present in the embryo, the yolk sac never becomes very large and serves no nutritive function. The yolk sac is, however, the source of the primordial germ cells that migrate to the gonads. With continued development, the endodermal cells on the undersurface of the embryonic mass that form the upper portion of the yolk sac fold together, forming the gut tube of the embryo. The midgut region remains open to the yolk sac.

Allantois

Another diverticulum develops from the endoderm of the hindgut. This pouch is the **allantois,** the third embryonic membrane. The allantois serves as a storage site for embryonic waste products in most animals, though perhaps not in humans. Like the yolk sac, the allantois is in the umbilical cord. The blood vessels that supply the allantois (umbilical arteries and veins) carry the fetal blood to and from the placenta (Figure 23.6).   F23.6

Chorion

The fourth embryonic membrane, the **chorion,** is formed from the trophoblast of the blastocyst plus the mesoderm that lines the inside of the trophoblast. The chorion surrounds the entire embryo and the other three embryonic membranes. As the amnion enlarges, it fuses with the inner layer (mesoderm) of the chorion. In combination with the umbilical blood vessels, the chorion forms the fetal portion of the placenta.

Germ Layers

In the second week of development, some of the surface cells of the embryonic disc settle deeper in the disc and spread between the **ectoderm** and **endoderm** layers that were formed by an earlier separation of the embryonic disc (Figure 23.7). These cells will form the **mesoderm.** As the sheet of meso-   F23.7 derm moves between the ectoderm and endoderm, it splits into two layers: *parietal* and *visceral.* The space between the layers becomes the **coelom,** or body space. Table 23.1 summarizes the main structures formed by the three   Table 23.1 primary germ layers.

GESTATION

The *gestation* (pregnancy) period usually lasts about 280 days from the beginning of the last menstrual period. During this time the developing individual increases in size, and all the organs and structures of the adult are formed (see Box 23.1). During gestation, the total mass of the uterus increases to accom-   Box 23.1 modate the growing fetus. In the later stages of pregnancy, the uterus occupies such a large portion of the abdominopelvic cavity that it can exert pressure on the rectum and the urinary bladder, causing constipation and frequent urination (Figure 23.8).   F23.8

The presence of a growing fetus places extra demands on a woman's body systems. For example, a pregnant woman generally produces somewhat more urine than she did before pregnancy because her kidneys must process the excretory products of the fetus as well as her own excretory products. Diet is especially important during pregnancy, and an expectant mother's diet must include sufficient amounts of vitamins, minerals, proteins, and other substances to supply both her own needs and those of the fetus.

FETAL CIRCULATION

The circulatory system of the fetus contains some unique features that are well adapted to development in the uterus but must change dramatically following birth if the child is to survive.

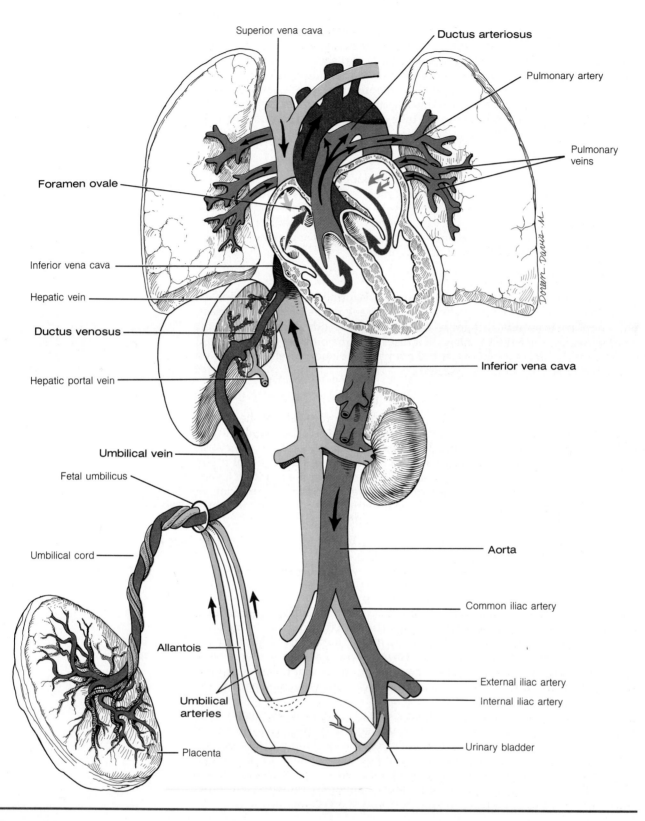

Figure 23.9

Fetal circulation. Blood that is highly oxygenated and rich in nutrients leaves the placenta and enters the fetus through the umbilical vein. After circulating through the fetus, the blood returns to the placenta through the umbilical arteries. The ductus venosus, the foramen ovale, and the ductus arteriosus allow the blood to bypass the fetal liver and lungs. The amount of red in a vessel indicates the relative proportion of oxygenated blood carried by the vessel.

Table 23.1   Structures Formed by the Three Primary Germ Layers

ENDODERM

Epithelium of the digestive tract and its glands (e.g., the liver and the
   pancreas)
Epithelium of the urinary bladder and the urethra
Epithelium of the pharynx, the auditory tube, the larynx, the trachea, the
   bronchi, and the lungs
Epithelium of the tonsils, the thyroid, parathyroid, and thymus glands

MESODERM

Skeletal, smooth, and cardiac muscle
Cartilage, bone, and other connective tissues
Blood, bone marrow, and lymphoid tissue
Epithelium of blood vessels and lymphatics
Epithelium of the coelom and the joint cavities
Epithelium of the kidneys and the ureters
Epithelium of the gonads and reproductive ducts
Epithelium of the adrenal cortex
Dermis of the skin

ECTODERM

Epidermis of the skin
Hair, nails, and glands of the skin
Lens of the eye
Receptor cells of the sense organs
Epithelium of the mouth, the nostrils, the sinuses, and the anal canal
Enamel of the teeth
All nervous tissue
Adrenal medulla

Fetal Blood Pathways

Fetal blood travels to the placenta in the **umbilical arteries,** which are
branches of the internal iliac arteries (Figure 23.9). After circulating through
the capillaries of the chorionic villi, where exchanges of gases, nutrients, and
wastes occur, the blood leaves the placenta in a single **umbilical vein.** The
umbilical vein, therefore, carries blood that is rich in nutrients and oxygen. It
is the only fetal vessel to do so. All other vessels of the fetus carry mixtures of
arterial and venous blood. Between the placenta and the **umbilicus (navel)** of
the fetus the umbilical arteries and the umbilical vein travel within the umbilical cord. Part of the blood in the umbilical vein enters the liver and follows the
usual path of blood through that organ and into the inferior vena cava. The
remainder joins with blood from the hepatic portal vein and enters a bypass
around the liver called the **ductus venosus.** The ductus venosus directly enters the inferior vena cava. Having much of the fetal blood bypassing the fetal
liver causes no functional problems, since the mother's liver serves the needs
of the fetus as the result of the exchange of materials across the placenta.

In the fetus, blood from the inferior vena cava enters the right atrium,
from which it can follow several pathways. It can follow the usual adult route,
for example, entering the right ventricle and traveling to the lungs within the
pulmonary arteries. Since the lungs are collapsed and nonfunctioning, however, they offer considerable resistance to bloodflow. Therefore, most of the
blood that leaves the right ventricle never reaches the lungs. Instead, it flows
through a low-resistant shunt called the **ductus arteriosus,** which connects
the pulmonary trunk to the arch of the aorta.

Another pathway is for the blood in the right atrium of the fetus to enter
the left atrium directly by means of a flap-covered opening, called the **foramen ovale,** in the interatrial septum. The blood then passes into the left ventricle and the aorta. The foramen ovale is located in such a position that most

F23.9

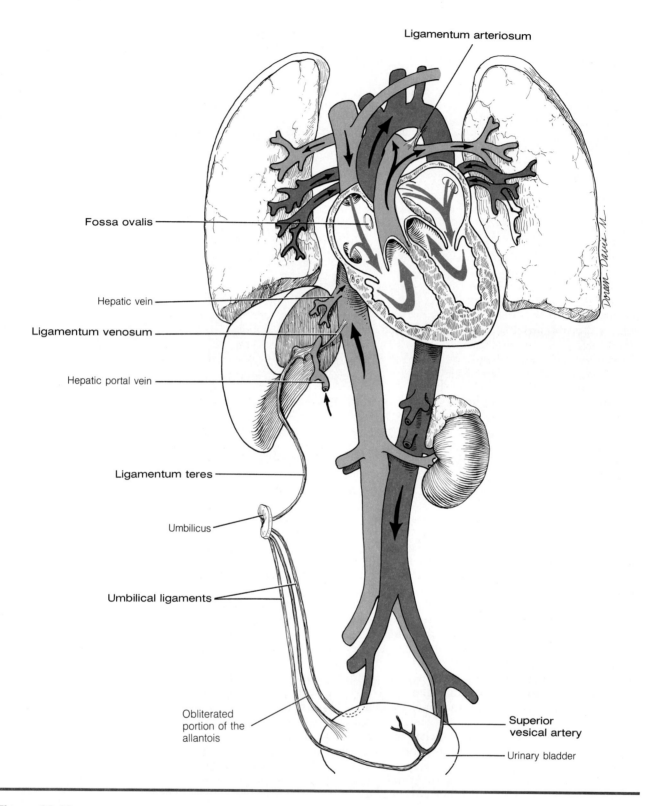

Figure 23.10
Changes in circulatory pathways following birth. These changes result from the closure of the foramen ovale and transformation of the ductus venosus, the ductus arteriosus, the umbilical vein, and the distal portions of the umbilical arteries into ligaments. Red indicates vessels carrying oxygenated blood.

of the blood from the inferior vena cava (which is highly oxygenated) is directed through it. Since the brain is supplied by arteries that branch off the aortic arch, the brain thus receives a good portion of this highly oxygenated blood immediately after it leaves the left ventricle. In contrast, much of the blood that enters the heart from the superior vena cava (venous blood) is directed into the right ventricle and pulmonary trunk. As we have noted, most of this blood passes through the ductus arteriosus, which joins the aorta beyond the arch. This blood is destined for the trunk and lower limbs of the fetus, where the oxygen demand is not so vital as that of the brain tissues.

Circulatory Changes at Birth

Separation of the infant from the placenta, which occurs when the umbilical cord is cut following birth, suddenly deprives the newborn of its oxygen source and its means of eliminating carbon dioxide. The resultant oxygen shortage and buildup of carbon dioxide cause the infant to take its first breath. The expansion of the lungs immediately lowers their resistance to bloodflow, which allows as much as five times more blood to pass through the lungs after birth than before. This action also increases the return of blood from the lungs to the heart, which in turn increases the pressure in the left atrium. The increased pressure in the left atrium forces the flap of the foramen ovale against the interatrial septum, which immediately blocks the opening. All blood that enters the right atrium must then travel to the right ventricle and the pulmonary trunk. Eventually the foramen ovale becomes permanently closed with fibrous connective tissue, leaving an indentation, called the **fossa ovalis,** on the interatrial septum (Figure 23.10). If the infant's foramen ovale **F23.10** fails to close, the higher pressure in the left atrium forces some of the blood that has returned from the lungs into the right atrium, thus recirculating the blood back to the lungs again. If the defect is large enough, the volume of blood that flows from the left atrium into the right atrium can cause pulmonary congestion and failure of the right ventricle. A severe defect in the interatrial septum must be surgically corrected.

   With the lowered resistance in the lungs following birth, the pressure in the pulmonary arteries and the chambers of the right side of the heart is also reduced. As a result, blood is no longer forced through the ductus arteriosus into the aorta, where the pressure is now greater. At the same time, the ductus arteriosus constricts, which prevents the reverse flow of blood from the aorta to the pulmonary trunk. Within a few weeks, the ductus arteriosus is completely obliterated by fibrous connective tissue and is then called the **ligamentum arteriosum.**

   Once the umbilical cord is cut, blood no longer flows through the umbilical vessels or the ductus venosus. The proximal portions of the umbilical arteries continue to supply the urinary bladder—which has developed from the proximal portion of the allantois of the fetus—as **superior vesical arteries.** Beyond the bladder, the umbilical arteries fill in with connective tissue and travel on the inner wall of the abdomen to the umbilicus as the **umbilical ligaments.** In a similar manner, the umbilical vein becomes obliterated with connective tissue and is called the **ligamentum teres** (*round ligament of the liver*). Because blood is no longer carried in the umbilical vein, the ductus venosus gradually constricts, shunting more and more of the newborn's blood from the hepatic portal vein into the liver. Within a few weeks, the ductus venosus is permanently obliterated, remaining as the **ligamentum venosum.** With the closure of the ductus venosus the bypass around the liver is lost, so all the blood from the hepatic portal vein must now pass through the liver. This change is of vital importance because the hepatic portal vein drains the digestive system, and the infant no longer has the mother's liver available to act on the various substances carried in the blood.

FRONTIERS IN HEALTH
Obstetrical Ultrasonics

Imagine having a window in the womb, through which physicians could view the largely secret world of the fetus. Such a window would be useful in detecting congenital birth defects, the position of the fetus just before delivery, and more, and could help save thousands of lives each year.

Is such an idea preposterous? In the strictest sense, perhaps it is. There is no serious thought of developing a porthole through which to observe the growing fetus. Medical scientists, however, now offer the next best thing, using a special technique called ultrasound imaging.

Ultrasound became quite popular in the 1970s as a "safe" substitute for X rays. X rays were once used to determine, among other things, the position of the fetus just before birth, but as more knowledge of the harmful effects of radiation became available, the use of X rays in obstetrics was greatly reduced.

The equipment needed for ultrasonic viewing of the unborn is relatively simple. A probe placed on the woman's abdomen gives off sound waves pitched so high that they escape detection by the human ear. Just as radar waves bounce off aircraft, ultrasound bounces off the tissues and travels back to a special receiving device in the probe. The information received is relayed to a small computer, which interprets the signals and creates a moving picture of the baby. Ultrasound images allow physicians to check for physical malformations such as missing limbs, determine the sex of the fetus, make measurements to determine whether growth is proceeding normally, and so forth.

Ultrasound imaging is also a valuable tool for physicians who specialize in surgery on the fetus. With the aid of this technique, surgeons have guided a needle into a fetus's bladder to drain urine that had accumulated there because of an obstruction in the urethra, and into the brain to drain the fluid-filled ventricles of a hydrocephalic fetus.

Ultrasound is proving to be a valuable tool in the medical laboratory as well. Scientists are using it to gain a better understanding of fetal motor development—determining what reflexes and what movements a fetus has and when they develop. The benefits of research along these lines are far reaching. Perhaps most important, a better understanding of fetal neural development will help physicians detect abnormal central nervous system development, which produces physiological symptoms that are much more subtle than gross physical defects. As an example, normal fetuses will turn away from the sound of a buzzer; a fetus with a hearing problem will not. Ultrasonic imaging allows the observation of such fetal movements. Early detection of deafness and other neurological defects offers hope of effective medical treatment.

Few people doubt the usefulness of ultrasound in medical diagnosis. Many medical scientists, however, warn that heat generated by ultrasound may have subtle, long-term effects on cells, perhaps causing cancer or genetic mutations.

Such concerns raise the question of just how safe ultrasound is. To date, at least 35 studies in animals and a smaller number in humans have failed to answer the risk question to the satisfaction of the medical community. While there is no evidence that ultrasound causes birth defects, retards physical growth, or induces cancer, more work is needed to assess fully the potential long-term effects. A number of laboratory studies which indicate that ultrasound can transform fibroblasts into tumor cells are cause for concern. Ultrasound has also been shown to produce changes in the surface properties and the motility of fibroblasts, and human lymphocytes undergo mutation when exposed to ultrasound in the laboratory. Whether this also happens in the body is not known.

The significance of these alterations must be determined before this promising technique can find a permanent niche in the field of obstetrics.

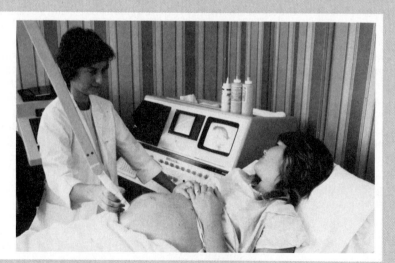

Ultrasound imaging.

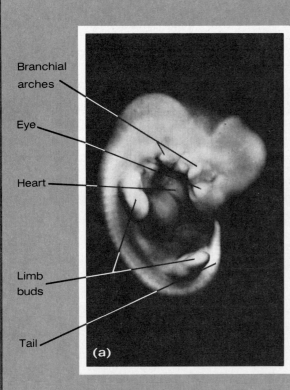

Branchial arches

Eye

Heart

Limb buds

Tail

(a)

Box 23.1
Gestation of the Human Fetus

The developing human at **(a)** 5 weeks, **(b)** 8 weeks, and **(c)** 28 weeks. At 5 weeks, the embryo has developed limb buds, which eventually develop into arms, legs, fingers, and toes. A rudimentary tail is present, as well as primitive branchial arches. At this stage, the brain is covered by only a thin layer of tissue, the eyes are just beginning to form, and the heart is larger in proportion to the body than at any other time in development.

By 8 weeks, the tail has degenerated, and the

branchial arches have evolved into the jaw and various other structures. Gender is now determinable, and most of the major organs are present, including the liver, pancreas, and heart. By this stage, the fetus has acquired reflexes and is thus aware of touch, although it is unaware of sound (except for the mother's heartbeat).

By 28 weeks, eyelids are able to open and close, fingernails are present, and respiratory, circulatory, and nervous systems are established.

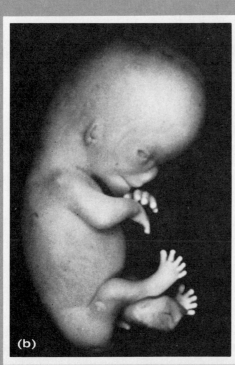

(b)

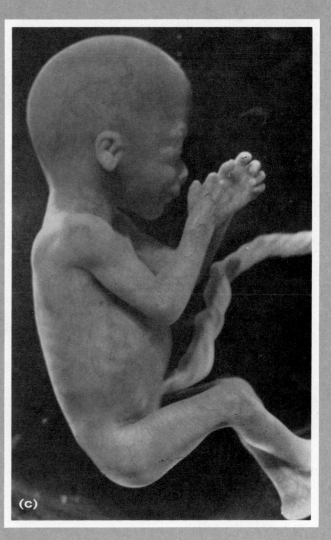

(c)

CONDITIONS OF CLINICAL SIGNIFICANCE
Disorders of Pregnancy

Ectopic Pregnancy

The blastocyst generally becomes implanted in the endometrium of the upper regions of the uterus. However, in some cases, implantation occurs at a site other than the uterus. Such an occurrence is called an *ectopic pregnancy*. The most common ectopic pregnancy is a *tubal pregnancy*, which occurs within a uterine tube.

Ectopic pregnancies can endanger the life of the mother. They frequently cause internal hemorrhaging, since the trophoblast of the blastocyst destroys tissue at the site of implantation. The endometrium, which is the normal site of implantation, can withstand the destructive effects of the trophoblast, but most tissues simply break down. Such destruction may involve the walls of blood vessels, causing an internal hemorrhage. The lack of room for expansion is another danger in ectopic pregnancies. In the case of a tubal pregnancy, for example, the uterine tube may rupture as the embryo grows.

Placenta Previa

Occasionally, when implantation occurs in the lower regions of the uterus, the placenta may cover the inner opening of the cervix. This condition is called *placenta previa*—that is, "placenta leading the way." In this position, the placenta can irritate the cervix, which causes the uterus to contract and can result in a spontaneous

abortion. If the pregnancy goes to term, the placenta is expelled before the baby is born, which causes the mother to hemorrhage severely. The expulsion of the placenta prior to the birth of the fetus can also stimulate the fetus to begin breathing while still in the birth canal, causing respiratory problems due to the inhalation of mucus.

Toxemia of Pregnancy

Toxemia (poison blood) associated with pregnancy continues to be a serious problem. Despite intensive research, the cause of toxemia is not known. It is probable that numerous factors produce toxemia—including malnutrition, metabolic disorders, endocrine imbalances, production of toxins, and a decreased blood supply to the uterus. As a prelude to toxemia, the woman often shows a strong tendency to reabsorb salt from the renal tubules, and with it, water. This excessive reabsorption of salt and water can usually be controlled with a low-salt diet or by the use of diuretic drugs to increase water loss.

If toxemia persists, among other symptoms there may be excessive weight gain; edema of the lungs, kidneys, and brain; visual disturbances; and eventually convulsions. If the convulsive stage is reached, the condition is called *eclampsia*.

STUDY OUTLINE

PREGNANCY pp. 625–629

FERTILIZATION
1. Normally occurs in upper portion of uterine tube.
2. Sperm passing through uterus and uterine tubes undergo changes (called capacitation) that make them capable of fertilizing an ovum.
3. Sperm contact corona radiata, releasing enzymes that break down corona radiata and zona pellucida, thus permitting access to ovum.
4. Cell membrane of one sperm and cell membrane of ovum fuse; sperm nucleus enters ovum.
5. Ovum completes second meiotic division; nuclei of sperm and ovum unite.

SEX DETERMINATION
1. If ovum is fertilized by a sperm containing X sex chromosome, a female develops.
2. If ovum is fertilized by a sperm containing Y sex chromosome, a male develops.

DEVELOPMENT AND IMPLANTATION OF THE BLASTOCYST
1. Blastocyst is fluid-filled sphere of cells; embryo develops from inner cell mass; outer sphere of cells called trophoblast.

2. After several days within lumen of uterus, trophoblast cells release enzymes that digest cells of endometrium; blastocyst is implanted in uterine wall.

MAINTENANCE OF THE ENDOMETRIUM
1. High blood levels of estrogens and progesterone are required.
2. During early pregnancy, estrogens and progesterone produced by corpus luteum.
3. By end of second month, placenta is usually secreting sufficient estrogens and progesterone to maintain endometrium even after corpus luteum has degenerated.

DEVELOPMENT OF THE PLACENTA provides fetal nutrient supply and waste removal; chorionic villi cover extensive surface area; connected to fetus by blood vessels in umbilical cord.

DECIDUA portion of endometrium discarded at birth or with menses.

DECIDUA BASALIS AND FETAL CHORION combine to form placenta.

FORMATION OF EMBRYONIC MEMBRANES AND GERM LAYERS pp. 630–633

EMBRYONIC MEMBRANES

AMNION fluid-filled; surrounds fetus; provides protection; permits movement; maintains temperature.

YOLK SAC develops from endoderm of embryonic disc; provides nourishment.

ALLANTOIS develops from endoderm of embryonic disc; allows for waste storage; allantoic blood vessels carry fetal blood to and from placenta.

CHORION develops from trophoblast plus mesoderm; forms fetal portion of placenta.

GERM LAYERS ectoderm and endoderm formed by separation of embryonic disc into two layers; mesoderm formed from cells of embryonic disc that spread between ectoderm and endoderm. *Coelom* formed when mesoderm sheet splits into parietal and visceral layers.

GESTATION lasts approximately 280 days from beginning of last menstrual period. **p. 633**

FETAL CIRCULATION pp. 633–639

FETAL BLOOD PATHWAYS

BLOOD TO PLACENTA carried by umbilical arteries.

UMBILICAL VEIN carries high-nutrient, high-oxygenated blood from placenta.

FETAL VESSELS most carry mixed arterial and venous blood.

FETAL CIRCUIT

Lungs collapsed; therefore provide much resistance to bloodflow.

Ductus Venosus serves as liver bypass.

Ductus Arteriosus carries blood from pulmonary trunk to arch of aorta.

Foramen Ovale opening in interatrial septum.

CIRCULATORY CHANGES AT BIRTH

1. Pulmonary circuit becomes functional.
2. Foramen ovale closes; forms fossa ovalis.
3. Ductus arteriosus is obliterated; forms ligamentum arteriosum.
4. Umbilical arteries become obliterated; form umbilical ligaments.
5. Ductus venosus is obliterated; forms ligamentum venosum.

CONDITIONS OF CLINICAL SIGNIFICANCE: DISORDERS OF PREGNANCY p. 640

ECTOPIC PREGNANCY implantation occurs at site other than uterus.

PLACENTA PREVIA placenta covers inner opening of cervix; at birth, placenta is expelled first.

TOXEMIA OF PREGNANCY may be caused by numerous factors; often shows tendency for salt and fluid retention; results in lung, kidney, and brain edema.

SELF-QUIZ

1. Fertilization usually occurs within the: (a) upper region of the uterine tube; (b) uterus; (c) ovary.

2. Sperm are not capable of fertilizing an ovum until they have traveled in the female reproductive tract. True or false?

3. During fertilization: (a) several sperm usually enter the ovum; (b) the acrosome enters the ovum; (c) usually only one sperm enters the ovum.

4. The interaction between the sperm that enters the ovum and the cell membrane of the ovum causes changes in the corona radiata and zona pellucida that prevent the entry of additional sperm. True or false?

5. An ovum that is fertilized by a spermatozoon containing an X chromosome normally: (a) develops into a male; (b) develops into a female; (c) has an equal probability of developing into either a male or a female.

6. During the first two months of development the embryo receives most of its nutrients from the: (a) placenta; (b) yolk sac; (c) endometrium; (d) allantois.

7. Cleavage transforms the zygote into a fluid-filled sphere of cells called a: (a) blastocoele; (b) inner cell mass; (c) coelom; (d) blastocyst.

8. The endometrium is eroded and the blastocyst is able to become implanted in the uterine wall due to enzymes that are released by the: (a) inner cell mass; (b) acrosome; (c) blastocoele; (d) trophoblast.

9. Which one of the following is a function of human chorionic gonadotropin (HCG)? (a) maintains the corpus luteum in an active state; (b) helps the blastocyst to implant; (c) breaks down the corona radiata and zona pellucida.

10. The superficial layer of the endometrium is called the: (a) chorion; (b) decidua; (c) embryonic disc; (d) zona pellucida.

11. The placenta is formed from the: (a) decidua basalis; (b) chorion; (c) a and b combine to form the placenta.

12. The membrane that surrounds the cavity formed in the inner cell mass is the: (a) amnion; (b) yolk sac; (c) allantois; (d) chorion.

13. The layer of cells on the side of the embryonic disc toward the yolk sac is composed of endodermal cells. True or false?

14. The chorion forms when mesoderm lines the inside of the trophoblast. True or false?

15. The storage site for embryonic waste products is the: (a) yolk sac; (b) amnion; (c) allantois; (d) chorion.

16. Which of the following are not formed from mesoderm? (a) muscles; (b) bone; (c) blood; (d) all of these are formed from mesoderm.

17. The umbilical arteries of the fetus carry blood that is rich in nutrients and oxygen. True or false?

18. Most of the blood that leaves the right ventricle of the fetus never reaches the lungs. Instead, it enters the aorta through a shunt called the: (a) ductus venosus; (b) foramen ovale; (c) ductus arteriosus; (d) umbilicus.

19. Following birth the umbilical vein becomes obliterated with connective tissue and is called the: (a) fossa ovalis; (b) ligamentum teres; (c) ligamentum venosum; (d) umbilical ligament.

20. A tubal pregnancy is a type of ectopic pregnancy. True or false?

Self-Quiz Answers

CHAPTER 1

1. c *p. 3*
2. F *p. 4*
3. a *p. 5–6*
4. c *p. 6*
5. c *p. 8*
6. Epithelial tissues: c, e
   Muscular tissues: a, g
   Nervous tissue: d, f
   Connective tissues:
   b, h *p. 6*
7. Prone position: k
   Anterior (ventral): g
   Posterior (dorsal): l
   Anatomical position: h
   Superior (cranial): a
   Medial: n
   Supine position: c
   Inferior (caudal): d
   Lateral: b
   Proximal: e
   Cervical: m
   Plantar: f
   Distal: i
   Lumbar: j
   Palmar: o *pp. 8–10*
8. a *p. 8*
9. T *p. 10*
10. c *p. 10*
11. b *p. 11*
12. T *p. 11*
13. T *p. 11*
14. a *p. 12*
15. a *p. 14*
16. c *p. 14*
17. b *p. 14*
18. F *p. 14*
19. c *p. 14*

CHAPTER 2

1. c *p. 22*
2. Active transport: c
   Exocytosis: d
   Facilitated diffusion: a
   Osmosis: b
   Dialysis: e
   Filtration: f *pp. 20–24*
3. b *p. 23*
4. T *p. 27*
5. F *p. 28*
6. a *p. 30*
7. F *p. 30*
8. c *p. 32*
9. b *p. 33*
10. a *p. 34*
11. Ribosomes: d
    Golgi apparatus: e

Peroxisomes: a
Basal bodies: b
Microfilaments: c
pp. 28–35
12. a *p. 33*
13. T *p. 35*
14. F *p. 37*
15. c *p. 37*
16. Autosomes: h
    XY: g
    Homologous
    chromosomes: a
    46: f
    Diploid: c
    23: i
    Haploid: e
    Interkinesis: b
    Crossing over: d
    pp. 37–44

CHAPTER 3

1. b *p. 52*
2. a *p. 53*
3. T *p. 53*
4. a *p. 53*
5. b *p. 55*
6. T *p. 60*
7. Squamous cells: c
   Stratified epithelium: f
   Simple columnar: e
   Columnar cells: a
   Cuboidal cells: d
   Simple squamous
   epithelium: b *pp. 54–59*
8. c *p. 57*
9. Mucous membranes: d
   Lamina propria: c
   Serous membranes: f
   Mucin: g
   Mesothelium: b
   Endothelium: a
   Goblet cells: e *pp. 59–60*
10. b *p. 60*
11. Fibroblasts: g
    Macrophages: f
    Ground substance: d
    Fibrocytes: a
    Fibers: c
    Collagenous: h
    Elastic: e
    Reticular: b *pp. 61–63*
12. F *p. 61*
13. c *p. 61*
14. a *p. 65*

15. c *p. 66*
16. c *p. 67*
17. T *p. 61*
18. F *p. 69*
19. T *p. 71*
20. c *p. 71*

CHAPTER 4

1. b *p. 77*
2. T *p. 79*
3. Stratum basale: e and f
   Keratohyalin: h
   Melanin: j
   Stratum corneum: a
   Eleidin: j
   Stratum granulosum: i
   Carotene: c
   Stratum lucidium: d
   Keratin: g *pp. 78–80*
4. b *p. 80*
5. T *p. 80*
6. a *p. 82*
7. F *p. 83–84*
8. c *p. 84*
9. F *p. 85*
10. a *pp. 85*
11. T *p. 86*
12. b *p. 87*
13. a *p. 87*
14. T *p. 87*
15. c *p. 87*
16. c *p. 87*
17. b *p. 87*
18. Acne: d
    Warts: g
    Psoriasis: f
    Impetigo: h
    Moles: c
    Herpes simplex: a
    Herpes zoster: i
    Cancers: e
    Eczema: b *pp. 87–88*
19. b *p. 89*
20. T *p. 89*

CHAPTER 5

1. c *p. 93*
2. b *p. 95*
3. T *p. 98*
4. F *p. 98*
5. c *p. 99*
6. c *p. 99*

7. a *p. 102*
8. b *p. 104*
9. c *p. 102*
10. Process: i
    Trochanter: f
    Tubercle: j
    Condyle: d
    Sulcus: k
    Crest: h
    Facet: a
    Fossa: g
    Fovea: b
    Foramen: c
    Meatus: e *p. 107*
    (Table 5.2)
11. F *p. 111*
12. a *p. 123*
13. Fontanels: f
    Frontal bone: j
    Sinuses: d
    Parietal bones: a
    Occipital bone: h
    Ethmoid bone: b
    Temporal bones: l
    Maxillary bones: c
    Zygomatic bones: i
    Lacrimal bones: k
    Mandible: e
    Vomer bone: g
    An auditory ossicle: q
    Hyoid bone: o
    Cervical vertebrae: m
    Thoracic vertebrae: r
    Sacral vertebrae: p
    Kyphosis: n *pp. 107–126*
14. a *p. 127*
15. c *p. 127*
16. a *p. 127*
17. d *p. 129*
18. b *p. 134*
19. c *p. 132*
20. a *p. 136*
21. c *p. 138*
22. c *p. 138*
23. F *p. 142*
24. Ilium: h
    Ischium: k
    Pubis: a
    Acetabulum: m
    False pelvis: b
    Pelvic outlet: j
    Femur: e
    Greater trochanter: i
    Condyles: c
    Adductor tubercle: g
    Tibia: d
    Anterior crest: f
    Fibula: l *pp. 136–142*

CHAPTER 6

1. F   *p. 154*
2. b   *p. 153*
3. c   *p. 155*
4. a   *p. 155*
5. T   *p. 155*
6. Synarthrosis: d
   Sutures: g
   Syndesmoses: b
   Synostosis: a
   Amphiarthroses: f
   Synchondroses: h
   Symphyses: c
   Synovial: e   *pp. 153–155*
7. b   *p. 156*
8. F   *p. 156*
9. F   *p. 156*
10. a   *p. 156*
11. Flexion: d
    Extension: f, b
    Abduction: c
    Adduction: a
    Plantar flexion: e
    Dorsiflexion: g   *p. 158*
12. F   *p. 158*
13. c   *p. 158*
14. b   *p. 160*
15. b   *p. 161*
16. Nonaxial: d, g
    Gliding: g
    Uniaxial: f, i
    Hinge: i
    Pivot: e
    Biaxial: a, b, h
    Condyloid: h
    Saddle: b
    Triaxial: c
    Spheroid: c   *p. 161*
17. a   *p. 163 (Table 6.1)*
18. a   *pp. 166 and 170*
19. b   *p. 170*
20. Distal tibia/fibula: a, g
    Pubic/pubic: b
    Ulna/humerus: f
    Sternum/clavicle: d, e
    Radius/carpals: c
    Tarsal/tarsal: d, e
    Occipital bone/atlas:
    c   *pp. 163–164
    (Table 6.1)*
21. Iliofemoral: H
    Arcuate popliteal: K
    Ligamentum flava: V
    Posterior longitudinal: V
    Pubofemoral: H
    Anterior and posterior
    cruciate: K
    Ligamentum teres: H
    Supraspinous: V
    Medial and lateral collateral: K
    Articular capsule: V,
    H, K   *pp. 164–170*

22. T   *p. 172*
23. c   *p. 173*

CHAPTER 7

1. Epimysium: c
   Perimysium: h
   Endomysium: f
   Aponeuroses: d
   Origin: a
   Insertion: g
   Unipennate: b
   Bipennate: i
   Multipennate: e
   pp. 182–184
2. b   *p. 187*
3. c   *p. 186*
4. c   *p. 188*
5. a   *p. 181*
6. Depressor labii inferioris: d
   Epicranius
   occipitalis: c
   Corrugator: b
   Orbicularis oculi: f
   Orbicularis oris: g
   Procerus: e
   Risorius: a   *p. 198;
   Table 7.1*
7. b   *p. 201; Table 7.4*
8. Multifidus: c
   Interspinales: d
   Splenius capitis and
   cervicis: b
   Spinalis thoracis and
   cervicis: a   *pp. 203–
   204; Table 7.5*
9. c   *p. 199; Table 7.2*
10. F   *p. 209*
11. a   *p. 209; Table 7.7*
12. b   *pp. 209–210; Table
    7.8*
13. a   *p. 211*
14. Subclavius: b
    Rhomboideus major: c
    Pectoralis minor: a
    pp. 211
15. Deltoid: a, b, d, e, f
    Latissimus dorsi: b, c,
    e
    Teres minor: f
    Teres major: b, c, e
    Coracobrachialis: a, c
    Infraspinatus: c,
    f   *p. 218; Table 7.12*
16. a, b, e   *pp. 220–221;
    Table 7.13*
17. Flexor carpi radialis: h
    Extensor indicus: d
    Extensor carpi ulnaris: e
    Extensor pollicis longus: f
    Extensor carpi radialis
    longus: b

Extensor digiti minimi: c
Flexor carpi ulnaris: g
Flexor digitorum superficialis: a   *pp. 221–
222; Table 7.14*
18. Gluteus maximus: d
    Adductor magnus: e
    Pectineus: b
    Adductor longus: c
    Gluteus medius: a
    pp. 227–229; Table 7.17
19. c   *p. 231; Table 7.18*
20. b   *pp. 239–240; Table
    7.20*
21. Tibialis anterior: c
    Extensor digitorum
    longus: e
    Peroneus longus: b
    Flexor hallucis longus: a
    Popliteus: d   *pp. 239–
    240; Table 7.20*

CHAPTER 8

1. b   *p. 250*
2. a   *p. 250*
3. c   *p. 250*
4. c   *p. 253*
5. a   *p. 253*
6. b   *p. 256*
7. a   *p. 256*
8. b   *p. 256*
9. c   *p. 256*
10. a   *p. 256*
11. b   *p. 256*
12. c   *p. 262*
13. a   *p. 262*
14. a   *p. 262*
15. d   *p. 262*

CHAPTER 9

1. b   *p. 265*
2. e   *p. 266; Table 9.1*
3. T   *p. 267*
4. Hemoglobin: c
   Heme: d
   Erythropoietin: e
   Reticulocytes: a
   Globin: f
   Macrophages: b
   pp. 269–273
5. e   *p. 273*
6. T   *p. 273*
7. e   *p. 273*
8. c   *p. 274*
9. F   *pp. 274–275*
10. a   *p. 274*
11. Erythrocytes: e
    Platelets: g
    Neutrophils: a
    Eosinophils: d

Basophils: f
Monocytes: c
Lymphocytes: b
pp. 267–275
12. T   *p. 275*
13. b   *p. 276*
14. F   *p. 276*
15. T   *p. 276*

CHAPTER 10

1. b   *p. 283*
2. Pericardial cavity: g
   Atria: d
   Ventricles: i
   Interatrial septum: a
   Inferior vena cava: f
   Aorta: b
   Epicardium: j
   Myocardium: c
   Endocardium: h
   Trabeculae carneae: e   *pp. 283–287*
3. c   *p. 291*
4. T   *p. 291*
5. T   *p. 292*
6. a   *p. 296*
7. c   *p. 295*
8. T   *p. 296*
9. c   *p. 296*
10. F   *p. 287*
11. b   *p. 291*
12. c   *p. 287–288*
13. b   *p. 293*
14. a   *p. 295*
15. T   *p. 295*
16. b   *p. 297*
17. c   *p. 297*
18. F   *p. 297*

CHAPTER 11

1. c   *p. 303*
2. b   *p. 306*
3. c   *p. 303*
4. T   *p. 304*
5. T   *p. 304*
6. c   *p. 306*
7. F   *p. 307*
8. a   *p. 303*
9. d   *p. 306*
10. b   *p. 309*
11. c   *p. 309*
12. F   *p. 309*
13. T   *p. 310*
14. F   *p. 310*
15. Aorta: k
    Coronary arteries: g
    Internal carotid artery: n
    Vertebral artery: a
    External carotid artery: e
    Axillary artery: i

Radial artery: f
Inferior phrenic arteries: b
Left gastroepiploic arteries: l
Superior mesenteric artery: c
Inferior mesenteric artery: m
Common iliac arteries: d
Popliteal artery: j
Anterior tibial artery: h *pp. 310–319*

16. Sinusoids: g
Brachiocephalic vein: e
Superior vena cava: j
Cephalic vein: a
Median cubital vein: i
Azygos vein: h
Inferior vena cava: b
Splenic vein: d
Great saphenous vein: f
Hepatic portal vein: c *pp. 319–329*

17. T *p. 311*
18. F *p. 311*
19. d *p. 311*
20. d *pp. 321–322*

CHAPTER 12

1. F *p. 335*
2. c *p. 335*
3. a *p. 336*
4. b *p. 336*
5. T *p. 338*
6. T *p. 339*
7. d *p. 340*
8. d *p. 340*
9. T *p. 342*
10. F *p. 342*
11. F *pp. 344–346*

CHAPTER 13

1. T *p. 349*
2. c *p. 349*
3. a *p. 350*
4. a *p. 351*
5. F *pp. 351–352*
6. c *p. 352*
7. a *p. 361*
8. F *p. 355*
9. c *p. 355*
10. Nuclei: d
Soma: f
Center: h
Ganglia: a
Dendrite: b
Axon: i
Nerve fiber: c

Sheath of Schwann: e
Myelin: g *pp. 355–356*
11. c *p. 357*
12. T *p. 359*
13. T *p. 359*
14. T *p. 359*
15. b *p. 360*
16. b *p. 367*
17. a *p. 367*
18. c *p. 356*
19. a *p. 357*
20. T *p. 363*
21. Acetylcholine: g
Motor neuron endings: e
Sensory neuron endings: a
Pacinian corpuscles: b
End bulbs of Krause: f
Ruffini's corpuscles: d
Muscle spindles: i
Neurotendinous organs: j
Free nerve endings: c
Meissner's corpuscles: h *pp. 363–364*
22. b *p. 364*
23. a *p. 364*
24. a *p. 365*
25. c *p. 365*
26. c *p. 364*
27. T *p. 365*
28. b *p. 365*
29. c *p. 367*
30. a *p. 367*

CHAPTER 14

1. c *p. 374*
2. a *p. 376*
3. a *p. 376*
4. F *p. 378*
5. b *p. 384*
6. b *p. 385*
7. c *p. 387*
8. b *p. 387*
9. T *p. 388*
10. c *p. 389*
11. b *p. 388*
12. T *p. 389*
13. b *p. 391*
14. a *p. 395*
15. b *p. 397*
16. Cerebrum: i
Projection tracts: e
Basal ganglia: m
Cerebral hemisphere: a
Insula: k
Primary motor area: p
Visual area: b
Olfactory area: l
Thalamus: c
Cerebral peduncles: n
Metencephalon: o

Pons: d
Hypothalamus: g
Ventricles: h
Telencephalon: f
Reticular formation: j *pp. 373–393*
17. T *p. 397*
18. a *p. 398*
19. Cauda equina: g
Meninges: d
White matter: a
Gray matter: j
Ventral roots: c
Fasciculus gracilis: i
Spinothalamic tracts: b
Spinocerebellar tracts: h
Pyramidal tracts: f
Extrapyramidal tracts: e *pp. 393–403*
20. a *p. 403*
21. a *p. 409*

CHAPTER 15

1. c *p. 415*
2. a *p. 415*
3. c *pp. 415–416*
4. a *p. 415*
5. a *p. 417*
6. b *p. 419*
7. a *p. 419*
8. c *p. 419*
9. c *p. 420*
10. a *p. 420*
11. b *p. 421*
12. a *p. 422*
13. d *p. 423*
14. c *p. 424*
15. T *p. 424*
16. T *p. 427*
17. F *p. 424*
18. c *p. 429*
19. a *p. 430*
20. c *p. 431*
21. T *p. 435*
22. T *pp. 435–436*

CHAPTER 16

1. d *p. 441*
2. F *p. 441*
3. F *p. 441*
4. b *p. 441*
5. b *p. 443*
6. T *p. 443*
7. F *p. 444*
8. T *p. 444*
9. b *p. 444*
10. F *p. 444*
11. b *p. 444*
12. c *p. 444*
13. a *p. 445*
14. T *p. 445*

15. c *p. 445*
16. c *p. 446*

CHAPTER 17

1. b *p. 451*
2. c *p. 452*
3. F *p. 452*
4. a *p. 455*
5. b *p. 455*
6. F *p. 458*
7. T *p. 456*
8. T *p. 458*
9. c *p. 458*
10. a *p. 464*
11. c *p. 463*
12. a *p. 461*
13. F *p. 462*
14. c *p. 466*
15. a *p. 466*
16. T *p. 469*
17. b *p. 469*
18. a *p. 470*
19. c *pp. 473–474*
20. T *p. 474*
21. b *p. 474*
22. d *p. 477*

CHAPTER 18

1. F *p. 483*
2. a *p. 486*
3. b *p. 487*
4. c *p. 488*
5. b *p. 488*
6. a *p. 486*
7. c *p. 487*
8. ADH: b, d
Oxytocin: b, f
FSH: a, h
TSH: a, g
ACTH: a, c
GH: a, e *pp. 487–489*
9. c *p. 489*
10. T *p. 492*
11. F *p. 491*
12. c *p. 493*
13. a *p. 493*
14. F *p. 493*
15. c *p. 493*
16. Lack of ACTH: f
Gonadotropin deficiency: c
TSH deficiency: a
TSH overproduction: h
ACTH overproduction: b
Growth hormone deficiency: d
Growth hormone overproduction: e
Antidiuretic hormone deficiency: g *p. 490*
17. c *p. 493*

18. a *p. 488*
19. T *p. 494*
20. c *p. 495*
21. a *p. 495*
22. F *pp. 496–497*
23. T *p. 497*
24. a *p. 496*
25. c *p. 498*
26. b *p. 499*
27. T *p. 502*
28. c *p. 500*

CHAPTER 19

1. a *p. 511*
2. d *p. 512*
3. d *p. 513*
4. F *p. 513*
5. d *p. 514*
6. d *p. 515*
7. c *p. 515*
8. d *p. 517*
9. b *p. 517*
10. c *p. 518*
11. c *pp. 519*
12. T *p. 523*
13. T *p. 523*
14. Costal surface: d
    Cardiac notch: h
    Oblique fissure: g
    Mediastinum: a
    Pleura: b
    Visceral pleura: c
    Parietal pleura: f
    Alveoli: e *pp. 518–521*
15. c *p. 524*
16. c *p. 524*
17. a *p. 521*

18. c *p. 524*
19. b *p. 526*
20. T *p. 526*

CHAPTER 20

1. c *p. 533*
2. F *p. 534*
3. b *p. 535*
4. a *p. 537*
5. a *p. 538*
6. c *p. 540*
7. b *p. 541*
8. a *p. 541*
9. T *p. 546*
10. b *p. 544*
11. b *p. 550*
12. c *p. 545*
13. T *p. 547*
14. F *p. 548*
15. T *p. 551*
16. a *p. 555*
17. F *p. 555*
18. a *p. 557*
19. b *p. 559*
20. a *p. 558*
21. T *p. 559*
22. e *p. 559*
23. e *p. 561*
24. T *p. 561*
25. a *p. 553*

CHAPTER 21

1. F *p. 567*
2. c *p. 567*
3. b *p. 568*

4. c *p. 571*
5. a *p. 573*
6. a *p. 573*
7. b *p. 571*
8. T *p. 575*
9. b *p. 578*
10. T *p. 577*
11. c *p. 577*
12. a *p. 574*
13. T *p. 573*
14. T *p. 575*
15. c *p. 575*
16. b *p. 578*
17. F *p. 583*
18. F *pp. 581–583*
19. T *p. 583*
20. b *p. 585*
21. d *p. 584*
22. d *p. 586*

CHAPTER 22

1. b *p. 591*
2. a *p. 593*
3. T *p. 593*
4. b *p. 598*
5. c *p. 597*
6. T *p. 601*
7. b *p. 601*
8. a *p. 598*
9. Epididymis: d
    Ductus deferens: c
    Seminal vesicles: f
    Ejaculatory ducts: a
    Prostate gland: b
    Bulbourethral
    glands: e *pp. 601–605*
10. b *p. 605*

11. T *p. 605*
12. c *p. 609*
13. b *p. 606*
14. T *p. 610*
15. c *p. 608*
16. F *p. 610*
17. a *p. 610*
18. T *p. 608*
19. a *p. 613*
20. T *p. 615*
21. e *p. 616*
22. c *p. 617*
23. T *p. 617*
24. F *p. 618*
25. c *p. 618*

CHAPTER 23

1. a *p. 625*
2. T *p. 625*
3. c *p. 625*
4. T *p. 626*
5. b *p. 626*
6. c *p. 627*
7. d *p. 627*
8. d *p. 627*
9. a *p. 627*
10. b *p. 628*
11. c *p. 629*
12. a *p. 630*
13. T *p. 631*
14. T *p. 633*
15. c *p. 633*
16. d *p. 635 (Table 23.1)*
17. F *p. 635*
18. c *p. 635*
19. b *p. 637*
20. T *p. 640*

Appendix 2
Word Roots, Prefixes, Suffixes, and Combining Forms

Dear Student:

Your success in the study of anatomy depends, to a great degree, on your ability to master a new vocabulary. This appendix should be a valuable resource for you throughout the course. Good luck!

A. P. Spence

PREFIXES AND COMBINING FORMS

a-, an- *absence or lack* acardia, lack of a heart; anaerobic, in the absence of oxygen

ab- *departing from; away from* abnormal, departing from normal

acou- *hearing* acoustics, the science of sound

acr-, acro- *extreme or extremity; peak* acrodermatitis, inflammation of the skin of the extremities

ad- *to or toward* adorbital, toward the orbit

aden-, adeno- *gland* adeniform, resembling a gland in shape

amphi *on both sides; of both kinds* amphibian, an organism capable of living in water and on land

angi- *vessel* angiitis, inflammation of a lymph vessel or blood vessel

ant-, anti- *opposed to; preventing or inhibiting* anticoagulant, a substance that prevents blood coagulation

ante- *preceding; before* antecubital, in front of the elbow

arthr-, arthro- *joint* arthropathy, any joint disease

aut-, auto- *self* autogenous, self-generated

bi- *two* bicuspid, having two cusps

bio- *life* biology, the study of life and living organisms

blast- *bud or germ* blastocyte, undifferentiated embryonic cell

broncho- *bronchus* bronchospasm, spasmodic contraction of bronchial muscle

bucco- *cheek* buccolabial, pertaining to the cheek and lip

caput- *head* decapitate, remove the head

carcin- *cancer* carcinogen, a cancer-causing agent

cardi-, cardio- *heart* cardiotoxic, harmful to the heart

cephal- *head* cephalometer, an instrument for measuring the head

cerebro- *brain, especially the cerebrum* cerebrospinal, pertaining to the brain and spinal cord

chondr- *cartilage* chondrogenic, giving rise to cartilage

circum- *around* circumnuclear, surrounding the nucleus

co, con- *together* concentric, common center, together in the center

contra- *against* contraceptive, agent preventing conception

cost- *rib* intercostal, between the ribs

crani- *skull* craniotomy, a skull operation

crypt- *hidden* cryptomenorrhea, a condition in which menstrual symptoms are experienced but no external loss of blood occurs

cyt- *cell* cytology, the study of cells

de- *undoing, reversal, loss, removal* deactivation, becoming inactive

di- *twice, double* dimorphism, having two forms

dia- *through, between* diaphragm, the wall through or between two areas

dys- *difficult, faulty, painful* dyspepsia, disturbed digestion

ec-, ex-, ecto- *out, outside, away from* excrete, to remove materials from the body

en-, em- *in, inside* encysted, enclosed in a cyst or capsule

entero- *intestine* enterologist, one who specializes in the study of intestinal disorders

epi- *over, above* epidermis, outer layer of skin

eu- *well* euesthesia, a normal state of the senses

exo- *outside, outer layer* exophthalmos, an abnormal protrusion of the eye from the orbit

extra- *outside, beyond* extracellular, outside the body cells of an organism

gastr- *stomach* gastrin, a hormone that influences gastric acid secretion

glosso- *tongue* glossopathy, any disease of the tongue

hema-, hemato-, hemo- *blood* hematocyst, a cyst containing blood

hemi- *half* hemiglossal, pertaining to one half of the tongue

hepat- *liver* hepatitis, inflammation of the liver

hetero- *different or other* heterosexuality, sexual desire for a person of the opposite sex

hist- *tissue* histology, the study of tissues

hom-, homo- *same* homeoplasia, formation of tissue similar to normal tissue; homocentric, having the same center

hydr-, hydro *water* dehydration, loss of body water

hyper- *excess* hypertension, excessive tension

hypno- *sleep* hypnosis, a sleeplike state
hypo- *below, deficient* hypodermic, beneath the skin; hypokalemia, deficiency of potassium
hyster-, hystero- *uterus or womb* hysterectomy, removal of the uterus; hysterodynia, pain in the womb
im- *not* impermeable, not permitting passage, not permeable
inter- *between* intercellular, between the cells
intra- *within, inside* intracellular, inside the cell
iso- *equal, same* isothermal, equal, or same, temperature
leuko- *white* leukocyte, white blood cell
lip-, lipo- *fat, lipid* lipophage, a cell that has taken up fat in its cytoplasm
macro- *large* macromolecule, large molecule
mal- *bad, abnormal* malfunction, abnormal functioning of an organ
mamm- *breast* mammary gland, breast
mast- *breast* mastectomy, removal of a mammary gland
meningo- *membrane* meningitis, inflammation of the membranes of the brain
meso- *middle* mesoderm, middle germ layer
meta- *beyond, between, transition* metatarsus, the part of the foot between the tarsus and the phalanges
metro- *uterus* metroscope, instrument for examining the uterus
micro- *small* microscope, an instrument used to make small objects appear larger
mito- *thread, filament* mitochondria, small, filamentlike structures located in cells
mono- *single* monospasm, spasm of a single limb
morpho- *form* morphology, the study of form and structure of organisms
multi- *many* multinuclear, having several nuclei
myelo- *spinal cord, marrow* myeloblasts, cells of the bone marrow
myo- *muscle* myocardium, heart muscle
narco- *numbness* narcotic, a drug producing stupor or numbed sensations
nephro- *kidney* nephritis, inflammation of the kidney
neuro- *nerve* neurophysiology, the physiology of the nervous system
ob- *before, against* obstruction, impeding or blocking up
oculo- *eye* monocular, pertaining to one eye
odonto- *teeth* orthodontist, one who specializes in proper positioning of the teeth in relation to each other
ophthalmo- *eye* ophthalmology, the study of the eyes and related disease
ortho- *straight, direct* orthopedic, correction of deformities of the musculoskeletal system
osteo- *bone* osteodermia, bony formations in the skin
oto- *ear* otoscope, a device for examining the ear
oxy- *oxygen* oxygenation, the saturation of a substance with oxygen
pan- *all, universal* panacea, a cure-all
para- *beside, near* paraphrenitis, inflammation of tissues adjacent to the diaphragm
peri- *around* perianal, situated around the anus
phago- *eat* phagocyte, a cell that engulfs and digests particles or cells
phleb- *vein* phlebitis, inflammation of the veins
pod- *foot* podiatry, the treatment of foot disorders
poly- *multiple* polymorphism, multiple forms

post- *after, behind* posterior, places behind (a specific) part
pre-, pro- *before, ahead of* prenatal, before birth
procto- *rectum, anus* proctoscope, an instrument for examining the rectum
pseudo- *false* pseudotumor, a false tumor
psycho- *mind, psyche* psychogram, a chart of personality traits
pyo- *pus* pyocyst, a cyst that contains pus
retro- *backward, behind* retrogression, to move backward in development
sclero- *hard* sclerodermatitis, inflammatory thickening and hardening of the skin
semi- *half* semicircular, having the form of half a circle
steno- *narrow* stenocoriasis, narrowing of the pupil
sub- *beneath, under* sublingual, beneath the tongue
super- *above, upon* superior, quality or state of being above others or a part
supra- *above, upon* supracondylar, above a condyle
sym-, syn- *together, with* synapse, the region of communication between two neurons
tachy- *rapid* tachycardia, abnormally rapid heartbeat
therm- *heat* thermometer, an instrument used to measure heat
tox- *poison* antitoxic, effective against poison
trans- *across, through* transpleural, through the pleura
tri- *three* trifurcation, division into three branches
viscero- *organ, viscera* visceroinhibitory, inhibiting the movements of the viscera

SUFFIXES

-able *able to, capable of* viable, ability to live or exist
-ac *referring to* cardiac, referring to the heart
-algia *pain in a certain part* neuralgia, pain along the course of a nerve
-ary *associated with, relating to* coronary, associated with the heart
-atresia *imperforate* proctatresia, an imperforate condition of the rectum or anus
-cide *destroy or kill* germicide, an agent that kills germs
-ectomy *cutting out, surgical removal* appendectomy, cutting out of the appendix
-emia *condition of the blood* anemia, deficiency of red blood cells
-ferent *carry* efferent nerves, nerves carrying impulses away from the CNS
-fuge *driving out* vermifuge, a substance that expels worms of the intestine
-gen *an agent that initiates* pathogen, any agent that produces disease
-gram *data that are systematically recorded, a record* electrocardiogram, a recording showing action of the heart
-graph *an instrument used for recording data or writing* electrocardiograph, an instrument used to make an electrocardiogram
-ia *condition* insomnia, condition of not being able to sleep
-iatrics *medical specialty* geriatrics, the branch of medicine dealing with disease associated with old age
-itis *inflammation* gastritis, inflammation of the stomach
-logy *the study of* pathology, the study of changes in structure and function brought on by disease
-lysis *loosening or breaking down* hydrolysis, chemical

decomposition of a compound into other compounds as a result of taking up water

-malacia *soft* osteomalacia, a process leading to bone softening

-mania *obsession, compulsion* erotomania, exaggeration of the sexual passions

-odyn *pain* coccygodynia, pain in the region of the coccyx

-oid *like, resembling* cuboid, shaped as a cube

-oma *tumor* lymphoma, a tumor of the lymphatic tissues

-opia *defect of the eye* myopia, nearsightedness

-ory *referring to, of* auditory, referring to hearing

-pathy *disease* psychopathy, any disease of the mind

-phobia *fear* acrophobia, fear of heights

-plasty *reconstruction of a part, plastic surgery* rhinoplasty, reconstruction of the nose through surgery

-plegia *paralysis* paraplegia, paralysis of the lower half of the body or limbs

-rrhagia *abnormal or excessive discharge* metrorrhagia, uterine hemorrhage

-rrhea *flow or discharge* diarrhea, abnormal emptying of the bowels

-scope *instrument used for examination* stethoscope, instrument used to listen to sounds of various parts of the body

-stasis *arrest, fixation* hemostasis, arrest of bleeding

-stomy *establishment of an artificial opening* enterostomy, the formation of an artificial opening into the intestine through the abdominal wall

-tomy *to cut* appendectomy, surgical removal of the appendix

-ty *condition of, state* immunity, condition of being resistant to infection or disease

-uria *urine* polyuria, passage of an excessive amount of urine

Units of the Metric System

Unit	Metric equivalent	Symbol	English equivalent
LINEAR MEASURE			
1 kilometer	= 1000 meters	km	0.62137 mile
1 meter	= 10 decimeters	m	39.37 inches
1 decimeter	= 10 centimeters	dm	3.937 inches
1 centimeter	= 10 millimeters	cm	0.3937 inch
1 millimeter	= 1000 microns	mm	
1 micron	= 1/1000 millimeter or 1000 millimicrons	μ	no English equivalents
1 millimicron	= 10 angstrom units	$m\mu$	
1 angstrom unit	= 1/100,000,000 centimeter	Å	
MEASURES OF CAPACITY			
1 kiloliter	= 1000 liters	kl	35.15 cubic feet or 264.16 gallons
1 liter	= 10 deciliters	l	1.0567 U.S. liquid quarts
1 deciliter	= 100 milliliters	dl	0.03 fluid ounce
1 milliliter	volume of 1 g of water at standard temperature and pressure (STP)	ml	
MEASURES OF MASS			
1 kilogram	= 1000 grams	kg	2.2046 pounds
1 gram	= 100 centigrams	g	15.432 grains
1 centigram	= 10 milligrams	cg	0.1543 grain
1 milligram	= 1/1000 gram	mg	0.01 grain (about)
MEASURES OF VOLUME			
1 cubic meter	= 1000 cubic decimeters	m^3	
1 cubic decimeter	= 1000 cubic centimeters	dm^3	
1 cubic centimeter	= 1000 cubic millimeters	cm^3	
1000 cubic millimeters	= 1 milliliter (ml)	mm^3	

Glossary

Abdomen *ab'-do-men* the portion of the body between the diaphragm and the pelvis.

Abduct *ab-duct'* to move away from the midline of the body.

Abscess *ab'-ses* a localized accumulation of pus and disintegrating tissue.

Absorption *ab-sorp'-shun* passage of a substance into or across a membrane or blood vessel.

Accommodation (1) adaptation or adjustment by an organ or organism in response to differences or needs; (2) adjustment of the eye for seeing objects at various distances.

Acetabulum *as-e-tab'-u-lum* the cuplike cavity where the femur fits on the lateral surface of the pelvic bone.

Acetylcholine *a-see-til-ko'-lene* a chemical transmitter substance released by nerve endings.

Acetylcholinesterase *es'-ter-ase* the enzyme that inactivates acetylcholine; present in muscle and nervous tissue.

Achilles tendon *a-kil'-leze ten'-don* the tendon at the back of the heel (calcaneus) that attaches the calf muscles.

Achondroplasia *a-kon-dro-play'-zee-ah* a condition, sometimes influenced by hereditary factors as well as hormonal levels, that produces a dwarf with short arms and legs but normal trunk and head.

Acid a substance that dissociates into hydrogen ions and anions when in an aqueous solution (compare with *base*).

Acidosis *as-i-do'-sis* a condition in which the blood has a higher hydrogen ion concentration than normal with a decreased pH.

Acne inflammatory disease of the skin.

Acromegaly *ak-ro-meg'-a-lee* an abnormal pattern of bone and connective-tissue growth characterized by enlarged hands, face, and feet and associated with excessive pituitary growth hormone that is secreted after the epiphyseal cartilages have been replaced.

Acromion *ak-ro'-mee-un* the outer projection of the spine of the scapula; forms the highest point of the shoulder.

Acrosome *ak'-ro-some* crescent-shaped structure molded to the nucleus and forming the anterior sperm head.

Actin *ak'-tin* a contractile protein of muscle fiber; myosin is another protein found in muscle.

Action potential an event occurring when a stimulus of sufficient intensity is applied to a neuron, allowing sodium ions to move into the cell and reverse the polarity.

Active transport net movement of a substance across a membrane against a concentration gradient; requires release and use of energy.

Adaptation (1) any change in structure, form, or habits to suit a new environment; (2) decline in the frequency of sensory nerve excitation when a receptor is stimulated continuously.

Addison's disease condition resulting from abnormal, deficient secretion of adrenal cortical hormones.

Adduct *ad-duct'* to move toward the midline of the body.

Adenohypophysis *ad-e-no-high-pof'-i-sis* the part of the pituitary gland that develops from Rathke's pouch.

Adenoids *ad'-e-noidz* enlargement of the pharyngeal tonsils in children.

Adenosine diphosphate (ADP) *a-den'-o-sene di-fos'-fate* the substance formed when ATP is split apart and releases energy.

Adenosine triphosphate (ATP) *a-den'-o-sene tri-fos'-fate* the compound that is an important intracellular energy source.

Adipose *ad'-i-pos* fatty.

Adrenal glands *ad-reen'-al* hormone-producing glands located superior to the kidneys; each consists of medulla and cortex areas.

Adrenalin *ad-ren'-a-lin* trademark name for epinephrine.

Adrenergic fibers *ad-ren-ur'-jick* nerve fibers whose terminals release norepinephrine or epinephrine upon stimulation.

Adrenocorticotropic hormone (ACTH) *a-dree'-no-kort-i-ko-tro'-pik* a hormone that influences the activity of the adrenal cortex and is released by the anterior portion of the pituitary.

Adventitia *ad-ven-tish'-yah* the outermost layer or covering of an organ.

Aerobic *ay-er-o'-bick* requiring oxygen to live or grow.

Afferent *af'-er-ent* carrying to or toward a center.

Afferent neuron *noo'-ron* nerve cell that carries impulses toward the central nervous system.

Agglutinin *ag-gloo'-tin-in* an antibody in blood plasma that causes clumping of corpuscles or bacteria.

Agglutinogen *ag-gloo-tin'-o-jen* (1) an antigen that stimulates the formation of a specific agglutinin; (2) an antigen found on red blood cells that is responsible for determining the ABO blood group classification.

Agonist *ag'-o-nist* a muscle whose contraction opposes the action of another muscle, its antagonist, which at the same time relaxes.

Albumin *al-bu'-min* a protein found in virtually all animal tissue and fluid.

Albuminuria *al-bu-min-oor'-ee-ah* the presence of albumin in the urine.

Aldosterone *al-dos'-ter-own* a hormone produced by the adrenal cortex that is important in sodium retention and reabsorption.

Alimentary *al-im-en'-ta-ree* pertaining to nourishment or the digestive organs.

Alkalosis *al-kal-o'-sis* a condition in which the blood has a lower hydrogen ion concentration than normal with an increased pH.

Allantois *a-lan'-to-is* an embryonic membrane; its blood vessels develop into blood vessels of the umbilical cord.

Allergy *al'-ler-jee* a condition in which there is hypersensitive response to a particular substance or environmental element.

Alveolus *al-ve'-o-lus* (1) a general anatomic term referring to a small cavity or depression; (2) an air sac in the lungs.

Amenorrhea *a-men-or-ree'-ah* absence of menstruation.

Amino acid *a-mee'-no* an organic compound containing nitrogen, carbon, hydrogen, and oxygen; the building blocks of protein.

Amnion *am'-nee-on* the innermost fetal membrane; it forms a fluid-filled sac for the embryo.

Amphiarthroses *am-fi-ar-thro'-seez* a type of articulation in which little motion occurs because of fibrocartilaginous connection of the articulating bony surfaces.

Ampulla *am-pul'-lah* a saclike dilation of a duct or tube.

Anabolism *an-ab'-o-li-zem* the energy-requiring building-up phase of metabolism in which simpler substances are synthesized into more complex substances.

Anaerobic *an-ay-er-o'-bick* requiring no oxygen to live or grow.

Anastomosis *a-nas-to-mo'-sis* a union or joining of blood vessels or other tubular structures.

Anatomy the science of the structure of organs of organic beings.

Androgen *an'-dro-jen* a hormone or other substance that controls male secondary sex characteristics.

Anemia *a-nee'-mee-ah* a decreased number of erythrocytes or decreased percentage of hemoglobin in the blood.

Aneurysm *an'-yu-riz-em* a localized blood-filled sac in an artery wall caused by dilation or weakening of the wall.

Angina pectoris *an-jee'-nah pek'-tor-is* a severe, suffocating chest pain caused by brief lack of oxygen to heart muscle.

Angiotensin *an-jee-o-ten'-sin* a vasoconstrictor substance found in the blood.

Annulus fibrosus *an'-u-lus fi-bro'sus* firm, ring-shaped outer portion of the intervertebral disc.

Anorexia *an-or-ek'-see-ah* loss of appetite or desire for food.

Anorexia nervosa *ner-vo'-sah* a nervous condition in which an extreme loss of appetite leads to emaciation, malnutrition, and possible worse consequences.

Anoxia *an-ok'-see-uh* a deficiency of oxygen.

Antagonist *an-tag'-o-nist* a muscle that acts in opposition to an agonist or prime mover.

Anterior the front of an organ or part; the ventral surface.

Antibody a specialized substance produced by the body that can provide immunity against a specific antigen.

Antidiuretic hormone (ADH) *an-tee-dye-u-re'-tik* a pituitary gland hormone that controls the reabsorption of water by the kidney.

Antigen *an'-ti-jen* any substance—including toxins, foreign proteins, or bacteria—which, when introduced to the body, causes antibody formation.

Antrum *an'-trum* an open space or chamber; a cavity, especially within a bone.

Anus *ay'-nus* the distal end of the digestive tract and the outlet of the rectum.

Aorta *ay-or'-tah* the major system artery; arises from the left ventricle of the heart.

Aortic arch *ay-or'-tik* the curved and most superior portion of the aorta.

Aortic body a receptor in the aortic arch sensitive to changing oxygen, carbon dioxide, and pH levels of the blood.

Aortic hiatus *high-ay'-tus* an opening behind the diaphragm giving passage to the aorta.

Aphasia *a-fay'-zhee-ah* a loss of the power of speech or a defect in speech.

Aprocrine gland *ap'-o-krin* a gland in which the secretions accumulate toward the outer ends of the secreting cells; the cell ends pinch off and are released with the secretions.

Aponeurosis *ap-o-noo-ro'-sis* the fibrous or membranous sheet serving as a connection between muscle and the part it moves.

Appendix *a-pen'-dicks* a wormlike sac attached to the large intestine.

Aqueous humor *ay'-kwee-us hyu'-mer* the watery fluid of the anterior and posterior chambers of the eye.

Arachnoid *a-rak'-noid* weblike; specifically, the weblike, middle layer of the three meninges.

Areola *ah-ree'-o-lah* (1) any minute opening or space in a tissue; (2) the circular, pigmented area surrounding the nipple.

Arrector pili *ar-rek'-tor pih'-lee* tiny, smooth muscles of the skin which, upon contraction, pull the hair follicle, causing the hair to stand up.

Arteriole *ar-te'-ree-ole* a minute artery.

Arteriosclerosis *ar-te-ree-o-skle-ro'-sis* any of a number of proliferative and degenerative changes in the arteries.

Artery a vessel that carries blood away from the heart.

Arthritis *ar-thright'-us* inflammation of the joints.

Articulate *ar-tik'-u-late* to join together in such a way as to allow motion between the parts.

Ascites *as-sigh'-teez* accumulation of serous fluid in the abdominal cavity.

Asphyxia *as-fik'-see-ah* loss of consciousness resulting from a deficiency in the oxygen supply.

Asthma *az'-ma* a disease or allergic response characterized by bronchial spasms and difficult breathing.

Astigmatism *a-stig'-ma-tiz-em* a visual defect resulting from irregularity in the lens or cornea of the eye causing the image to be out of focus.

Ataxia *a-tak'-see-a* lack of muscular coordination.

Atelectasis *at-e-lek'-ta-sis* lung collapse or incomplete expansion of a lung.

Atherosclerosis *ath-er-o-skle-ro'-sis* changes in the walls of large arteries consisting of lipid deposits on the artery walls.

Atlas the first cervical vertebra; articulates with the occipital bone of the skull and the second cervical vertebra (axis).

Atom the smallest particle of an element; composed of electrons, protons, and neutrons; capable of existing individually or in combination with atoms of the same or another element.

Atresia *a-tree'-zhee-ah* (1) the abnormal closure of a body canal or opening; (2) degeneration of the ovarian follicle and the ovum within it.

Atrioventricular bundle (AV bundle) *a-tree-o-ven-trik'-u-lar* the bundle of specialized fibers serving to conduct impulses from the AV node to the right and left ventricles; failure of the AV bundle results in heart block; also called bundle of His.

Atrioventricular node (AV node) a specialized compact mass of conducting cells located at the atrioventricular junction in the heart.

Atrium *ay'-tree-um* a chamber of the heart receiving blood from the veins.

Atrophy *a'-tro-fee* a reduction in size or wasting away of an organ or cell resulting from disease or lack of use.

Auditory *aw'-di-to-ree* pertaining to the sense of hearing.

Auditory ossicles *aw'-di-to-ree oss'-i-kuls* the three tiny bones serving as transmitters of vibrations and located within the middle ear: the malleus, incus, and stapes.

Auditory (eustachian) tube *yoo-stay'-shee-un* the connection between the middle ear and the pharynx.

Auricle *aw'-ri-kel* the external ear.

Auscultation *aws-kul-tay'-shun* the act of examination by listening to body sounds.

Automaticity *aw-to-ma-ti'-si-tee* the ability of a structure, organ, or system to initiate its own activity.

Autonomic *aw-to-nom'-ik* self-directed; self-regulating; regulating; independent.

Autonomic nervous system the division of the nervous system that functions involuntarily and is responsible for innervating cardiac muscle, smooth muscle, and glands.

Axial skeleton *aks'-ee-al* the bones of the head and trunk: the skull, vertebral column, thorax, and sternum.

Axilla *aks-il'-ah* the armpit.

Axis (1) the second cervical vertebra; has a vertical projection called the odontoid process around which the atlas rotates; (2) the imaginary line about which a joint or structure revolves.

Axon *aks'-on* the process of a nerve cell by which impulses are carried away from the cell; efferent process; the conducting portion of a nerve cell.

Bacteria any of a large group of microorganisms, usually non-spore-producing and generally one-celled; found in humans and other animals, plants, soil, air, and water; have a broad range of functions.

Baroreceptor *bar-o-ree-sep'-tor* a receptor that is stimulated by pressure changes.

Basal ganglia *bay'-zel gan'-glee-ah* gray matter structures deep inside each of the cerebral hemispheres.

Basal layer columnar cells in which mitosis takes place; also called stratum basale of the epidermis.

Basal metabolic rate *met-ah-bol'-ik* the rate at which energy is expended (heat produced) by the body per unit time under controlled (basal) conditions: 12 hours after a meal, at rest.

Base a substance that accepts hydrogen ions; capable of uniting with water to form an acid.

Basement membrane a thin layer of substance to which epithelial cells are attached in mucous surfaces.

Basophil *bay'-so-fil* white blood cells that readily take up basic dye; have a relatively pale nucleus and granular-appearing cytoplasm.

Benign *bee-nine'* not malignant.

Biceps *bigh'-seps* two-headed, especially applied to certain muscles.

Bicuspid *bigh-kus'-pid* having two points or cusps.

Bifurcation *bi-fur-ka'-shun* division into two branches.

Bile a greenish-yellow or brownish fluid produced in and secreted by the liver, stored in the gallbladder, and released into the small intestine.

Bilirubin *bil-i-roo'-bin* the red pigment of bile.

Biliverdin *bil-i-ver'-din* the green pigment of bile.

Biofeedback an area of research in which subconscious feedback is raised to the conscious level.

Biopsy *bigh'-op-see* the removal and examination of live tissue.

Bipennate *bigh-pen'-nate* muscles in which the fibers are attached obliquely on both sides of the tendon.

Blastocyst *blas'-to-sist* a stage of early embryonic development.

Blood-brain barrier a mechanism that inhibits passage of materials from the blood into brain tissues and cerebrospinal fluid.

Bolus *bo'-lus* (1) a rounded mass of food prepared by the mouth for swallowing; (2) a concentrated mass of a pharmaceutical preparation, usually given intravenously.

Bowman's capsule *bo'-manz* the double-walled cup at the end of a nephron.

Brachial *bray'-ki-al* pertaining to the arm.

Bradycardia *brad-i-kar'-dee-ah* slowness of the heart rate; below 60 beats per minute.

Brain stem the portion of the brain consisting of the medulla, pons, and midbrain.

Bronchitis *brong-kigh'-tis* inflammation of the bronchi.

Bronchus *brong'-kus* one of the two large branches of the trachea leading to the lungs.

Buccal *bu'-kal* pertaining to the cheek.

Buffer a substance or substances that tend to stabilize the pH of a solution.

Bundle branch block a blocking of heart action resulting from a local lesion of the bundle of His; delayed contraction of one ventricle.

Bursa *bur'-sah* a small sac or cavity filled with fluid and located at friction points, especially joints.

Calcitonin *kal-si-to'-nin* a hormone released by the thyroid that lowers calcium and phosphate levels of the blood.

Calculus *kal'-ku-lus* a stone formed within various body parts.

Callus *kal'-us* (1) a bonelike material that protrudes between ends of a fractured bone; (2) a thickening of the skin caused by rubbing or friction.

Calyx *kay'-liks* (1) a cuplike division of pelvis of the kidney.

Canal a duct or passageway; a tubular structure.

Canaliculi *kan-al-ik'-u-lee* extremely small tubular passages or channels.

Cancer a malignant, invasive cellular tumor that has the capability of spreading throughout the body or body parts.

Capacitation *ka-pass-i-tay'-shun* the process in which sperm undergo changes in the female reproductive tract making them capable of fertilization.

Capillary *kap'-il-lar-ee* a minute blood vessel connecting arterioles with venules.

Carbohydrates *kar-bo-high'-drates* organic compounds composed of carbon, hydrogen, and oxygen with the hydrogen and oxygen present in a 2:1 ratio; include starches, sugars, cellulose.

Carcinogen *kar-sin'-o-jen* cancer-causing agent.

Carcinoma *kar-sin-o'-ma* cancer; a malignant growth.

Cardiac *kar'-dee-ak* pertaining to the heart.

Cardiac muscle specialized muscle of the heart.

Cardiac output the blood volume (in liters) ejected per minute by the left ventricle.

Carotene *kar'-o-teen* a yellow pigment; influences skin color.

Carotid *ka-rot'-id* the main artery in the neck.

Carotid body a receptor at the bifurcations of the common carotid arteries sensitive to changing oxygen, carbon dioxide, and pH levels of the blood.

Carotid sinus *sigh'-nus* a dilation of a common carotid artery at its bifurcation; involved in regulation of systemic blood pressure.

Carpals *kar'-puls* the eight bones of the wrist.

Cartilage *kar'-ti-lej* white, semiopaque, fibrous connective tissue.

Caruncle *kar'-un-kul* a small fleshy protuberance.

Catabolism *ka-tab'-a-liz-em* the process in which living cells break down more complex substances into simpler substances; destructive metabolism.

Cataract *kat'-a-rakt* partial or complete loss of transparency of the crystalline lens of the eye or its capsule.

Catheter *ka'-the-ter* a narrow, hollow tube that can be inserted into a body cavity for withdrawal of fluids.

Caudal *kaw'-dal* in humans, the inferior portion of the anatomy.

Cecum *see'-kum* the blind-end pouch at the beginning of the large intestine.

Cell the basic biological unit of living organisms (except viruses), containing a nucleus and a variety of organelles; usually enclosed in the membrane.

Cell membrane the selectively permeable membrane forming the outer layer of most animal cells; also called plasma membrane or unit membrane.

Cellulose *sel'-u-lose* a fibrous carbohydrate that is the main structural component of plant tissues.

Cementum *se-men'-tum* the bony connective tissue that covers the root of a tooth.

Central nervous system (CNS) the brain and the spinal cord.

Centriole *sen'-tree-ole* a minute body found in the nucleus of the cell; active in cell division.

Cerebellum *ser-e-bel'-lum* part of the hindbrain; controls movement coordination; consists of two hemispheres and a central portion (vermis).

Cerebral aqueduct *ser-ee'-bral a'-kwe-duct* the elongated, slender cavity of the midbrain that connects the third and fourth ventricles; also called the aqueduct of Sylvius.

Cerebral arterial circle a union of arteries at the base of the brain.

Cerebrospinal fluid *ser-e-bro-spy'-nal* the fluid produced in the cerebral ventricles; fills the ventricles and surrounds the central nervous system.

Cerebrum *ser'-i-brum or se-ree'-brum* the largest part of the brain; consists of right and left cerebral hemispheres.

Cerumen *se-roo'-men* earwax.

Cervical *ser'-vi-kal* refers to the neck or the necklike portion of an organ or structure.

Cervix *ser'-vix* (1) the cylindrical, inferior portion of the uterus leading to the vagina; (2) any necklike structure.

Chemoreceptor *kem-o-ree-sep'-tor* receptors sensitive to various chemical stimulations and changes.

Chiasma *kee-az'-ma* a crossing or intersection of two structures, such as the optic nerves.

Cholecystectomy *kol-e-sis-tek'-tom-ee* removal of the gallbladder.

Cholesterol *ko-les'-ter-ol* an organic alcohol found in animal fats and oil as well as in most body tissues, especially bile.

Cholinergic fibers *ko-lin-ur'-jick* nerve endings that, upon stimulation, release acetylcholine at their terminations.

Chondroblast *kon'-dro-blast* a cell that forms the fibers and matrix of cartilage.

Chondrocyte *kon'-dro-site* a mature cartilage cell.

Chorion *kor'-ee-on* the outermost fetal membrane; forms the placenta.

Choroid *ko'-roid* (1) skinlike; (2) the pigmented covering of the eye.

Chromosome *kro'-mo-some* the structures in the nucleus that carry the hereditary factors (genes).

Chyle *kile* a milky fluid consisting of fat globules in lymph formed in the small intestine during digestion.

Chyme *kime* the semifluid contents of the stomach consisting of partially digested food and gastric secretions.

Cilia *sil'-lee-ah* tiny, hairlike projections on cell surfaces that move in a wavelike manner.

Circumcision *sur-kum-si'-zhun* removal of the foreskin of the penis.

Circumduction *sur-kum-duk'-shun* circular movement of a body part.

Cirrhosis *sir-o'-sis* a chronic disease, particularly of the liver, characterized by an overgrowth of connective tissue.

Clitoris *kli'-to-ris* a small, erectile structure in the female, homologous to the penis in the male.

Cochlea *koak'-lee-ah* a cavity of the inner ear resembling a snail shell.

Coenzyme *ko-en'-zime* a nonprotein substance associated with and activating an enzyme.

Coitus *ko'-i-tus* sexual intercourse.

Colloid *kol'-loid* solute particles dispersed in a medium; particles do not settle out readily and do not pass through natural membranes.

Colloidal osmotic pressure (COP) *kol-loi'-dal os-mot'-ick* the pressure exerted on a membrane by the particles in a colloid; usually refers to the osmotic pressure of blood plasma and body fluids resulting from the presence of protein.

Colostrum *kol-os'-trum* the first milk secreted after pregnancy.

Coma *ko'-ma* unconsciousness from which the person cannot be aroused.

Compound a substance composed of two or more different elements.

Concave *kon'-kave* having a curved, depressed surface.

Concha *kon'-kah* a shell-shaped structure.

Condyle *kon'-dile* a rounded projection at the end of a bone that articulates with another bone.

Cones one of the two types of photosensitive cells in the retina of the eye.

Congenital *kon-jen'-i-tal* existing at birth.

Conjunctiva *kon-junk-ti'-va* the thin protective membrane of the insides of the eyelids and the anterior surface of the eye itself.

Conjunctivitis *kon-junk-ti-vigh'-tis* an inflammation of the conjunctiva of the eye.

Connective tissue a primary type of tissue; form varies extensively, as does function, which includes support and storage.

Contraception *kon-tra-sep'-shun* the prevention of conception; birth control.

Contractility *kon-trak-til'-i-tee* a substance's ability to shorten or develop tension upon the application of a stimulus.

Contralateral *kon-tra-lat'-er-al* opposite; acting in unison with a similar part on the opposite side of the body.

Convergence *kon-verj'-ence* turning toward or approaching a common point from different directions.

Convoluted *kon'-vo-lu-ted* rolled, coiled, or twisted.

Copulation *kop-u-lay'-shun* sexual intercourse.

Cornea *kor'-nee-ah* the transparent anterior portion of the eyeball.

Corona radiata *ko-ro'-nah ray-dee-aw'-tah* (1) the arrangement of elongated follicle cells around a mature ovum; (2) the crownlike arrangement of nerve fibers radiating from the inner capsule of the brain to every part of the cerebral cortex.

Corpus *kor'-pus* body; the major portion of an organ.

Cortex *kor'-teks* the outer surface layer of an organ.

Corticoid *kor'-ti-koid* a substance whose function and properties are similar to those of corticosteroids.

Corticosteroids *kor'-ti-ko-ste'-roidz* the steroid hormones released by the adrenal cortex.

Cortisol *kor'-ti-sol* a glucocorticoid produced by the adrenal cortex.

Costal *kos'-tal* pertaining to the ribs.

Cramp a painful, involuntary contraction of a muscle.

Cranial *kray'-nee-al* pertaining to the skull.

Creatine phosphate *kree'-a-tene fos'-fate* a compound that serves as an alternative energy source for muscle tissue.

Crenation *kre-nay'-shun* the shriveling of an erythrocyte resulting from withdrawal of water.

Cretinism *kree'-tin-izm* a severe thyroid deficiency in the young that leads to stunted physical and mental growth.

Crista *kris'-ta* a crest or ridge.

Cryptorchidism *kript-or'-kid-izm* a developmental defect in which the testes fail to descend into the scrotum.

Crystalloid *kris'-tal-oid* a substance having a crystalline structure as opposed to a colloidal composition; particles can pass through a natural membrane.

Cubital *ku'-bi-tal* pertaining to the forearm.

Cupula *ku'-pu-la* a domelike structure.

Cushing's syndrome *ku'-shingz sin'-drome* a disease produced by excess secretion of adrenocortical hormone; characterized by adipose tissue accumulation, weight gain, and osteoporosis, for example.

Cutaneous *ku-tay'-nee-us* pertaining to the skin.

Cyanosis *sigh-a-no'-sis* a bluish coloration of the mucous membranes and skin caused by deficient oxygenation of the blood.

Cyst *sist* a sac with a distinct wall, containing fluid or other material; may be pathological or normal.

Cystitis *sis-tigh'-tis* inflammation of the urinary bladder.

Cytokinesis *sigh-to-ki-nee'-sis* the changes in the cytoplasm during cell division.

Cytology *sigh-tol'-o-jee* the science concerned with the study of cells.

Cytoplasm *sigh'-to-plaz-um* the protoplasm of a cell other than that of the nucleus.

Deamination *dee-am-i-nay'-shun* the removal of an amino group from an organic compound by reduction, hydrolysis, or oxidation.

Deciduous *dee-sid'-u-us* temporary.

Deciduous (milk) teeth the 10 temporary teeth replaced by permanent teeth; "baby" teeth.

Defecation *def-e-kay'-shun* the elimination of the contents of the bowels (feces).

Deglutition *dee-gloo-tish'-un* the act of swallowing.

Dehydration *dee-high-dray'-shun* a condition resulting from excessive loss of water.

Dendrite *den'-drite* branching; the branching neuron process that transmits the nerve impulse to the cell body; the receptive portion of a nerve cell.

Dental caries *den'-tal kar'-eez* tooth cavity.

Dentin *den'-tin* the calcified tissue beneath the enamel forming the major part of a tooth.

Deoxyribonucleic acid (DNA) *dee-ox-i-rye-bo-nu-kle'-ik* a nucleic acid found in all living cells: carries the organism's hereditary information.

Depolarization *dee-po-lar-i-zay'-shun* the neutralization to a state of nonpolarity; the loss of a negative charge inside the cell.

Depressor *dee-pres'-sor* any substance that causes slowing, reduction of activity, or inhibition of another structure, organ, or substance.

Dermatitis *der-ma-tigh'-tis* an inflammation of the skin; nonspecific skin allergies.

Dermis *der'-mis* the deep layer of dense, irregular connective tissue of the skin; also called corium.

Desmosome *des'-mo-some* small, apposed ellipsoidal plates in membranes of adjacent cells; also called macula adherens.

Diabetes insipidus *dye-uh-bee'-teez in-sip'i-dus* a disease characterized by passage of a large quantity of urine of low specific gravity plus intense thirst and dehydration; a hypothalamic disorder is the cause.

Diabetes mellitus *mel'-li-tus* a disease caused by deficient insulin release, leading to failure of the body tissue to oxidize carbohydrates at a normal rate.

Dialysis *dye-al'-i-sis* the separation of substances from one another in a solution by taking advantage of their differing rates of diffusion through a semipermeable membrane.

Diapedesis *dye-a-pe-dee'-sis* the passage of blood cells through unruptured vessel walls into the tissues.

Diaphragm *dye'-a-fram* (1) any partition or wall separating one area from another; (2) a muscle that separates the upper thoracic cavity from the lower abdominopelvic cavity.

Diarthrosis *dye-ar-throw'-sis* a freely movable joint; a synovial joint.

Diastole *dye-as'-to-lee* a period (between contractions) of relaxation and dilation of the heart during which it fills with blood.

Diencephalon *dye-en-cef'-a-lon* that part of the forebrain between the telencephalon and the mesencephalon including the thalami and most of the third ventricle.

Diffusion *di-fu'-zhun* the spreading of particles in a gas or solution with a movement toward uniform distribution of particles.

Digestion *di-jest'-yun* the bodily process of breaking down foods chemically and mechanically into compounds capable of being absorbed by body cells.

Dilate *dye'-late* to stretch; to open; to expand.

Distal *diss'-tal* farthest from the center or midpoint of a limb structure.

Diverticulum *dye-ver-tik'-u-lum* a pouch or sac in the walls of a hollow organ or structure.

Dorsal *dor'-sal* pertaining to the back; posterior.

Duct a canal or passageway.

Duodenum *doo-o-dee'-num* the first part of the small intestine.

Dura mater *doo'-rah may'-ter* the outermost and toughest of the three membranes (meninges) covering the brain and spinal cord.

Dysfunction *dis-funk'-shun* lack of normal function; disorder.

Dysmenorrhea *dis-men-or-ree'-ah* difficult, painful menstruation.

Dyspnea *disp'-nee-ah* labored, difficult breathing.

Dystrophy *dis'-tro-fee* a disorder caused by a defect or dysfunction of nutrition.

Ectopic *ek-to'-pik* not in the normal place; for example, in an ectopic pregnancy the egg is fertilized at a place other than the uterus.

Edema *e-dee'-mah* an abnormal accumulation of fluid in body parts or tissues; causes swelling.

Effector *ef-fek'-tor* a motor or sensory nerve ending in an organ, gland, or muscle.

Efferent *ef'-er-ent* carrying away or away from, especially a nerve fiber that carries impulses away from the central nervous system.

Ejaculation *e-jak-u-lay'-shun* the sudden ejection of a fluid from a duct, especially semen from the penis.

Elastin *e-las'-tin* the main protein in elastic fibers of connective tissues.

Electrocardiogram (ECG) *ee-lek-tro-kar'-dee-o-gram* a graphic record of the electric current associated with heartbeats.

Electroencephalogram (EEG) *ee-lek-tro-en-sef'-a-lo-gram* a graphic record of the activity of nerve cells in the brain.

Electrolyte *ee-lek'-tro-lite* a substance that breaks down into ions when in solution and is capable of conducting an electric current.

Electron *ee-lek'-tron* a negatively charged particle in motion around the nucleus of an atom.

Embolism *em'-bo-liz-em* the obstruction of a blood vessel by a clot floating in the blood; may also be a bubble of air in the vessel (air embolism).

Embryo *em'-bree-oh* an organism in its early stages of development; in humans, the first two months after conception.

Emesis *em'-e-sis* vomiting.

Emphysema *em-fi-see'-muh* a condition caused by overdistension of the pulmonary alveoli or abnormal presence of air or gas in body tissues.

Enamel the hard, calcified substance that covers the crown of a tooth.

Encephalitis *en-sef-a-ligh'-tis* an inflammation of the brain.

Endocardial *en-do-kar'-di-al* pertaining to the inner lining of the heart.

Endocarditis *en-do-kar-di'-tis* an inflammation of the inner lining of the heart.

Endocardium *en-do-kar'-di-um* the membrane lining the interior of the heart; endothelium and connective tissue.

Endochondral *en-do-kon'-dral* pertaining to the development of structures in cartilage.

Endocrine glands *en'-do-krin* ductless glands that empty their secretions directly into the blood.

Endometrium *en-do-me'-tree-um* the mucous membrane lining of the uterus.

Endomysium *en-do-miz'-ee-um* the thin connective tissue between the fibers of a muscle bundle.

Endoplasmic reticulum *en-do-plaz'-mik re-tik'-u-lum* a membranous network of tubular or saclike channels through the cytoplasm of a cell.

Endothelium *en-do-thee'-lee-um* the single layer of simple squamous cells that line the walls of the heart and the vessels that carry blood and lymph.

Energy the capacity to do work.

Enzyme *en'-zime* a substance formed by living cells that acts as a catalyst in bodily chemical reactions.

Eosinophil *ee-o-sin'-o-fil* a granular white blood cell whose granules readily take up a stain called eosin.

Epidermis *e-pi-der'-mis* the outer layer of cells of the skin.

Epididymis *e-pi-did'-i-mis* that portion of the seminal duct in which sperm mature and are transported from testes to body exterior.

Epiglottis *e-pi-glot'-tis* the elastic, membrane-covered cartilage at the back of the throat; guards the glottis during swallowing.

Epimysium *e-pi-miz'-ee-um* the sheath of connective tissue surrounding a muscle.

Epinephrine *e-pi-nef'-rine* the chief hormone of the adrenal medulla.

Epiphysis *e-pif'-i-sis* the extremities of a long bone.

Epithelium *e-pi-thee'-lee-um* one of the primary tissues; covers the surface of the body and lines the body cavities, ducts, and vessels.

Eponychium *ep-o-neech'-ee-um* the fold of skin overlying the root of the nail.

Equilibrium *ee-kwi-lib'-ri-um* balance; a state when opposite reactions or forces counteract each other exactly.

Erythrocyte *e-ree'-throw-site* red blood cell.

Erythropoiesis *e-rith-ro-poi-ee'-sis* the formation process of erythrocytes.

Estrogen *es'-tro-jen* any substance that stimulates female secondary sex characteristics; female sex hormones.

Eupnea *yoop-nee'-ah* easy, normal breathing.

Excretion *eks-kree'-shun* the elimination of waste products from the body.

Exocrine glands *eks'-o-krin* glands that have ducts through which their secretions are carried to a particular site.

Exogenous *eks-og'-en-us* developing or originating outside the organ or part.

Expiration *ex-pi-ray'-shun* the act of expelling air from the lungs.

Exteroceptor *eks'-ter-o-sep-tor* an end organ that responds to stimuli from the external world.

Extracellular *eks-tra-sel'-u-lar* outside a cell.

Extracellular fluid fluid within the body but outside the cells.

Extrinsic *eks-trin'-zik* originating from outside an organ or part.

Exudate *eks'-yu-date* material including fluid, pus, or cells that has escaped from blood vessels and has been deposited in tissues.

Facet *fas'-et* a smooth, nearly flat surface on a bone for articulation.

Fallopian tube *fal-low'-pee-an* the oviduct or uterine tube; the tube through which the ovum is transported to the uterus.

Fascia *fash'-ee-ah* the layers of fibrous tissue under the skin or covering and separating muscles.

Fasciculus *fas-ik'-u-lus* a bundle of nerve, muscle, or tendon fibers separated by connective tissues.

Feces *fee'-seez* material discharged from the bowel composed of food residue, secretions, and bacteria.

Fenestrated *fen'-es-tray-ted* pierced with one or more small openings.

Fetus *fee'-tus* the unborn young; in humans, from the third month in the uterus until birth.

Fibrillation *fib-ri-lay'-shun* irregular, uncoordinated contraction of muscle fibers.

Fibrin *figh'-brin* the fibrous insoluble protein formed during the clotting of blood.

Fibrinogen *figh-brin'-o-jen* a protein that is coverted to fibrin during blood-clotting.

Filtration *fil-tray'-shun* the passage or straining of a solvent and dissolved substances through a membrane or filter.

Fissure *fis'-sure* (1) a groove or cleft; (2) the deepest linear depressions on the brain.

Fistula *fis'-tu-lah* an abnormal passage between organs or between a body cavity and the outside.

Fixator *fiks'-ay-tor* a muscle acting to immobilize a joint or a bone; fixes the origin of prime movers so that muscle action can be exerted at the insertion.

Flaccid *flak'-sid* soft; flabby; relaxed.

Flagella *fla-jel'-ah* long whiplike extensions of the cell membrane of some bacteria; serve as agents for locomotion.

Flexion *flek'-shun* bending; the movement that decreases the angle between bones.

Follicle *fol'-i-kal* a small sac or gland.

Follicle-stimulating hormone (FSH) a hormone produced by the anterior pituitary that stimulates ovarian follicle production in females and sperm production in males.

Fontanels *fon-tan-els'* the fibrous membranes at the body area where bones have not yet formed; babies' "soft spots."

Foramen *fo-ra'men* a hole or opening in a bone or between body cavities.

Forebrain the anterior portion of the brain consisting of the telencephalon and the diencephalon.

Fossa *fos'-ah* a depression; often used as an articular surface.

Fovea *fo-vee'-ah* a pit, generally used for attachment rather than for articulation.

Frontal (coronal) plane a longitudinal section that runs at right angles to sagittal planes dividing the body into anterior and posterior parts.

Fulcrum *ful'-krum* the pivot point of a lever.

Fundus *fun'-dus* the base of an organ; that part farthest from the opening of the organ.

Funiculi *fu-nik'-ye-lee* (1) cordlike structures; (2) anterior and lateral divisions of white matter in the spinal cord.

Gallbladder the sac beneath the right lobe of the liver used for bile storage.

Gallstones particles of cholesterol or calcium carbonate that are occasionally formed in gallbladder and bile ducts.

Gamete *gam'-eet* male or female reproductive cell.

Gametogenesis *gam-e-to-jen'-e-sis* the origin and formation of gametes.

Ganglion *gan'-glee-on* a group of nerve-cell bodies, usually located in the peripheral nervous system.

Gap junction intercellular specialization with the cell membranes of adjacent cells only 20 Å apart; also called nexus.

Gastroenteritis *gas-tro-en-ter-i'-tis* an inflammation of mucosa of stomach and intestine.

Gastroesophageal sphincter *gas-tro-e-soff-a-jee'-al sfink'-ter* narrowing between the esophagus and the stomach.

Gene *jeen* one of the biological units of heredity located on chromosomes; transmits hereditary message.

Genetics *jen-e'-tiks* the science of heredity.

Genitalia *jen-i-tay'-lee-ah* the external sex organs.

Gestation *jes-tay'-shun* the period of pregnancy; about 280 days for humans.

Gingiva *jin-jigh'-vah or jin'-ji-vah* the gums.

Gland an organ specialized to secrete or excrete substances for further use in the body or for elimination.

Glans a small glandlike mass of erectile tissue at the tip of the penis and the clitoris.

Glaucoma *glaw-ko'-mah* an abnormal elevation of the pressure within the eye.

Glia *glee'-a* see neuroglia.

Globin *glow'-bin* the protein component of hemoglobin.

Glomerulus *glom-er'-u-lus* a knot of coiled capillaries in the kidney.

Glottis *glot'-tis* the opening between the vocal cords; entrance to the larynx.

Glucagon *gloo'-ka-gon* a hormone formed by cells of the pancreatic islets; raises the glucose level of blood.

Glucocorticoids *gloo-ko-kor'-ti koidz* the adrenal cortex hormones that affect metabolism of fats and carbohydrates.

Glucose *gloo'-kose* the principal sugar in the blood.

Glycerol *gliss-e-rol* an important alcoholic component of fat.

Glycogen *gligh-ko-jen* an animal starch; the main carbohydrate stored in animal cells.

Glycolysis *gligh-kol'-i-sis* the body's breakdown of glucose into simpler compounds, especially lactic acid.

Goblet cells the individual cells of the respiratory and digestive tracts that function as glands.

Goiter *goi'-ter* an enlargement of the thyroid gland.

Gonad *go'-nad* a gland or organ producing gametes; an ovary or testis.

Gonadotropins *go-nad-o-tro'-pinz* the gonad-stimulating hormones produced by the anterior pituitary.

Graafian follicle *graf'-ee-an fol'-i-kal* a mature ovarian follicle.

Gray matter the gray area of the central nervous system; contains neurons.

Groin the junction of the thigh and the trunk; the inguinal area.

Growth hormone a hormone that stimulates growth in general; produced in the anterior pituitary; also called somatotropin.

Gubernaculum *goo-ber-nak'-u-lum* a guiding, connecting cord structure; the cord between the testis and the scrotal sac, for example.

Gustation *gus-tay'-shun* taste.

Gyrus *jigh'-rus* a convolution on the surface of the cerebral cortex.

Hamstring muscles the posterior thigh muscles: the biceps femoris, semimembranosus, and semitendinosus.

Haustra *haw'-stra* pouches of the colon.

Haversian system or osteon *ha-ver'-zee-an, os'-tee-on* an organized system of interconnecting canals in the microscopic structure of adult compact bone.

Hay fever an acute allergic reaction of conjunctiva and upper air passages due to pollen sensitivity.

Heart block a defective transmission of impulses from atrium to ventricle.

Heart murmur an abnormal heart sound.

Hematocrit *hem-a'-to-krit* the percentage of erythrocytes to total blood volume.

Heme *heem* the iron-containing pigment that is essential to oxygen transport by hemoglobin.

Hemiplegia *hem-i-plee'-jee-ah* paralysis of one side of the body.

Hemocytoblasts *hee-mo-sigh'-to-blasts* stem cells that give rise to all the formed elements of the blood.

Hemoglobin *hee-mo-glo'-bin* the oxygen-transporting component of erythrocytes composed of heme and globin.

Hemolysis *hee-mol'-i-sis* the destruction of erythrocytes.

Hemopoiesis *hem-o-poi'-ee-sis* the formation of blood.

Hemorrhage *hem'-o-ridj* the escape of blood from the vessels by flow through ruptured walls; bleeding.

Heparin *hep'-a-rin* a substance that prevents clotting found in many tissues, especially the liver.

Hepatic portal system *he-pat'-ik* the liver circulatory arrangement where the hepatic portal vein carries dissolved nutrients to the liver tissues.

Hepatitis *hep-at-eye'-tis* an inflammation and/or infection of the liver.

Hernia *her'-nee-ah* the abnormal protrusion of an organ or a body part through the containing wall of its cavity.

Herpes simplex *her'-peez sim'-pleks* a fever blister or cold sore; a virus-caused condition.

Herpes zoster *zos'-ter* an infection of the dorsal root ganglia of spinal nerves by a virus, causing pain and fluid-filled vesicles on the skin; also called shingles.

Heterosexuality *he-ter-o-seks-u-al'-i-tee* sexual interest in or desire for persons of the opposite sex.

Hilum, hilus *high-lum, high-lus* the notched or depressed area where vessels enter and leave an organ.

Histology *his-tol'-o-jee* the branch of anatomy dealing with the microscopic structure of tissues.

Holocrine glands *hol'-o-krin* glands that accumulate their secretions within their cells; secretions are discharged only upon rupture and death of the cell.

Homeostasis *hom-ee-o-stas'-sis* the state when the body organs function together to maintain a stable internal environment for the general well-being of the entire body; a state of body equilibrium.

Homeotherm *ho'-me-o-therm* an organism that produces its own heat and maintains a constant body temperature.

Homologous *ho-mol'-o-gus* parts or organs corresponding in structure but not necessarily in function.

Homosexuality *ho-mo-seks-u-al'-i-tee* sexual interest in or desire for persons of the same sex.

Hormones *hor'-mones* the secretions of endocrine glands; responsible for specific regulatory effects on certain parts or organs.

Hyaline *high'-a-line* glassy; transparent.

Hydrocarbon *high-dro-kar'-bon* a molecule composed of only carbon and hydrogen.

Hydrochloric acid *high-dro-klo'-rik* HCl; facilitates protein digestion in the stomach; produced by parietal cells.

Hydrolysis *high-drol'-i-sis* the process in which a chemical compound unites with water and is then split into smaller molecules.

Hydrostatic pressure *high-dro-stat'-ik* the pressure of fluid in a system.

Hyperopia *high-per-o'-pi-a* farsightedness.

Hypertension *high-per-ten'-shun* high blood pressure.

Hypertonic *high-per-ton'-ik* excessive, above normal, tone or tension.

Hypertrophy *high-per'-tro-fee* an increase in the size of a tissue or organ independent of the body's general growth.

Hypodermis *high-po-der'-mis* the subcutaneous connective tissue; also called superficial fascia.

Hyponychium *high-po-neech'-ee-um* the thickened horny zone of the epidermis beneath the free border of the nail.

Hypothalamus *high-po-thal'-a-mus* the region of the diencephalon forming the floor of the third ventricle of the brain.

Hypotonic *high-po-ton'ik* below normal tone or tension.

Hypoxia *hip-ox'-ee-a* a condition in which a physiologically inadequate amount of oxygen is available to tissues.

Ileum *il'-ee-um* the lower part of the small intestine between the jejunum and the cecum of the large intestine.

Impetigo *im-pe-tee'-go* a highly contagious skin infection common in children.

Impotence *im'-po-tense* (1) lack of power, inability; (2) a male's inability to have sexual intercourse or maintain an erection.

In vitro *in vee'-tro* in a test tube, glass, or artificial environment.

In vivo *in vee'-vo* in the living body.

Infarct *in-farkt'* a region of dead, deteriorating tissue resulting from blood flow interference.

Inferior (caudal) pertaining to a position near the tail end of the long axis of the body.

Inflammation *in-flam-may'-shun* a physiological response of the body to tissue injury; includes dilation of blood vessels and an increase in vessel permeability.

Infundibulum *in-fun-di'-bu-lum* a funnel-shaped body part or passageway.

Inguinal *ing'-gwi-nal* pertaining to the groin region.

Innervation *in-ner-vay'-shun* the supply or distribution of nerves or nerve stimuli to a part.

Insertion *in-ser'-shun* the place or mode of attachment of a muscle; the movable part of a muscle as opposed to origin.

Inspiration *in-spi-ray'-shun* the drawing of air into the lungs.

Insulin *in'-su-lin* the hormone produced in the pancreas affecting carbohydrate and fat metabolism, blood glucose levels, and other systemic processes.

Integumentary system *in-teg-u-men'-tar-ee* the skin and its accessory structures.

Intercellular *in-ter-sel'-u-lar* between the cells of the body or part.

Intercellular matrix *may'-triks* the material between adjoining cells.

Internal capsule the band of white matter between the basal ganglia and the thalamus.

Internal environment the environment within the body.

Internal respiration the exchange of gases between blood and tissue fluid and between tissue fluid and cells.

Interoceptor *in'-ter-o-sep-tor* a nerve ending situated in the viscera sensitive to changes and stimuli within the body's internal environment; also called visceroceptor.

Interstitial fluid *in-ter-stish'-al* the fluid between the cells or body parts.

Intervertebral discs *in-ter-ver'-te-bral* the discs of fibrocartilage between bodies of vertebrae.

Intervertebral foramina *fo-rah'-mi-nah* the openings between the pedicles of adjacent vertebrae through which the spinal nerves pass.

Intracellular *in-tra-sel'-u-lar* within a cell.

Intracellular fluid fluid within a cell.

Intramural pressure *in-tra-mu'-ral* the pressure built up in the walls of an organ.

Intrinsic muscle a muscle that has both its origin and its insertion in an organ.

Invert to turn inward.

Involuntary muscle a muscle not under control of the will; independent muscle.

Ion *eye'-on* an atom with a positive or negative electric charge.

Ipsilateral *ip-si-lat'-er-al* situated on the same side.

Iris *eye'-ris* the pigmented, circular diaphragm in front of the eye's lens.

Ischemia *is-kee'-mee-uh* a local decrease in blood supply resulting from obstruction of arterial inflow.

Isometric *i-so-met'-rik* of the same length.

Isotonic *i-so-ton'-ik* having uniform tension under pressure or stimulation.

Jaundice *jawn'-dis* an accumulation of bile pigments in the blood producing a yellow color of the skin.

Jejunum *je-joo'-num* the part of the small intestine between the duodenum and the ileum.

Joint the junction of two or more bones; an articulation.

Junctional complex the junction between cells in columnar and some cuboidal epithelium.

Juxtamedullary *jux'-ta-med'-u-lair-y* referring to the inner portion of the cortex of the kidney, adjacent to the medulla.

Karyokinesis *kar-ee-o-ken-ee'-sis* see mitosis.

Keratin *ker'-a-tin* a fibrous insoluble protein found in tissues such as hair or nails.

Kinetic energy *ki-net'-ik* the energy of motion.

Kinesthesia *ki-nez-thee'-zhah* the ability to perceive muscle movement.

Krebs cycle the citric acid cycle; the series of reactions during which energy is liberated from metabolism of carbohydrates, fats, and amino acids.

Labia *lay'-bee-ah* lips.

Lacrimal *la'-kri-mal* pertaining to tears.

Lactation *lak-tay'-shun* the production and secretion of milk.

Lacteal *lak'-tee-al* the special lymphatic capillaries of the small intestine that take up chyle.

Lactic acid *lak'-tik* the product of anaerobic glycolysis, especially in muscle.

Lacuna *la-ku'-nah* a little depression or space; in bone or cartilage, lacunae are occupied by cells.

Lamina *lam'-i-nah* (1) a thin layer or flat plate; (2) the portion of a vertebra between the transverse process and the spinous process.

Laryngeal prominence *la-rin'-jul prom'-i-nense* the tubercle of the thyroid cartilage; Adam's apple.

Laryngitis *lar-en-jight'-us* an inflammation of the larynx.

Larynx *lar'-inks* the organ of the voice; located between the trachea and the base of the tongue.

Lateral *lat'-ur-al* away from the midline of the body.

Lateral sacs the reticular sites in muscle from which calcium is released.

Lens the elastic, doubly convex structure behind the pupil of the eye; focuses the light entering the eye on the retina.

Lesion *lee'-zhun* a tissue injury or wound.

Leukemia *loo-kee'-mee-ah* a cancerous condition in which there is an excessive production of leukocytes.

Leukocyte *loo'-ko-site* a white blood cell.

Ligament *lig'-a-ment* a band or sheetlike fibrous tissue connecting bones or parts.

Lingual *ling'-gwal* pertaining to the tongue.

Lipid *lip'-id* a substance that is almost insoluble in water but soluble in fat solvents; fatty acids and fats.

Lobe a curved, rounded structure or projection.

Lordosis *lor-do'-siss* an excessive curve in the anterior lumbar spine; otherwise known as swayback.

Lumbar *lum'-bar* the portion of the back between the thorax and the pelvis.

Lumbar puncture a procedure involving insertion of a needle between the third and fourth lumbar vertebrae and into the subarachnoid space to sample cerebrospinal fluid.

Lumbosacral trunk *lum-bo-sak'-ral* a group of nerves that connects the lumbar plexus and the sacral plexus.

Lumen *loo'-men* the space inside a tube, blood vessel, or intestine.

Luteinizing hormone *lu'-tee-in-eye-zing* an anterior pituitary hormone that stimulates maturation of cells in the ovary and acts on interstitial cells of the male testis.

Lymph *limf* the watery fluid in the lymph vessels collected from the tissue fluids.

Lymph node a mass of lymphatic tissue.

Lymphatic system *lim-fat'-ik* a system of vessels carrying lymph closely related anatomically and functionally to the circulatory system.

Lymphocyte *lim'-fo-site* a granular white blood cell formed in the lymphoid tissue.

Lysosomes *ligh'-so-somes* tiny organelles that originate from the Golgi apparatus and contain strong digestive enzymes.

Macula *mak'-u-la* the slightly yellow region lateral to the optic disc.

Macula adherens *mak'-u-la ad-hear'-uns* see desmosome.

Mammary glands *mam'-mar-ee* milk-producing glands of the breasts.

Malignant *ma-lig'-nant* life threatening; pertains to diseases that spread and lead to death, such as cancer.

Mastication *mas-ti-kay'-shun* the act of chewing.

Matrix *may'-trix* the homogeneous intercellular substance of any tissue.

Meatus *mee-ay'-tus* the external opening of a canal.

Mechanoreceptor *mek-an-o-ree-sep'-tor* a receptor sensitive to mechanical pressures such as touch, sound, or contractions.

Medial *mee'-dee-al* toward the midline of the body.

Mediastinum *mee-dee-as-tigh'-num* the portion of the thoracic cavity between the lungs.

Medulla *med-u'-la* the central portion of certain organs.

Meiosis *my-o'-sis* the last two cell divisions in gamete formation producing nuclei with half the full number of chromosomes (haploid).

Melanin *mel'-a-nin* the dark pigment responsible for skin color.

Melanocyte *mel-an'-o-site* a cell that produces melanin.

Menarche *me'-nar-kee* the onset of menstruation.

Meninges *men-in'-jeez* the membranes that cover the brain and spinal cord.

Meningitis *men-in-ji'-tis* an inflammation of the meninges covering the brain and spinal cord.

Menopause *men'-o-pawz* the physiological termination of menstrual cycles.

Menses *men'-seez* the recurrent monthly discharge of menstruation.

Menstruation *men-stroo-ay'-shun* the periodic, cyclic discharge of blood, secretions, tissue, and mucus from the mature female genital canal in the absence of pregnancy.

Merocrine glands *mer'-o-krin* glands that produce secretions intermittently; secretions do not accumulate in the gland.

Mesencephalon *mes-en-sef'-a-lon* the midbrain.

Mesenteries *mes'-en-ter-eez* the doubled-layered membranes of the peritoneum that support most of the organs in the abdominal cavity.

Metabolic rate *met-a-bol'-ik* the energy expended by the body per unit time.

Metabolism *me-tab'-o-liz-em* the sum total of the chemical reactions that occur in the body.

Metabolize *me-tab'-o-lize* to transform substances into energy or materials the body can use or store by means of anabolism or catabolism.

Metacarpals *met-a-kar'-puls* the five bones of the palm of the hand.

Metastasize *me-tas'-ta-size* the spread of disease from one body part or organ into another not directly connected to it.

Metatarsals *met-a-tar'-suls* the five bones between the instep and the phalanges.

Metencephalon *met-en-sef'-a-lon* the anterior part of the hindbrain composed of the cerebellum and pons.

Microbodies *my-kro-bod'-eez* membrane-bound cytoplasmic structures containing oxidative enzymes; also called peroxisomes.

Microfilament *my-kro-fil'-a-ment* filaments associated with contractile activities of the cell and developmental modifications of cell and organ shape.

Microtubule *my-kro-too'-bule* cytoplasmic structures not bound in membranes having a support function.

Microvilli *my-kro-vil'-lee* the tiny protoplasmic projections formed on the free surfaces of some epithelial cells; appear brushlike when viewed through a microscope.

Mineralocorticoid *min-er-al-o-kor'-ti-koid* an adrenal cortical steroid hormone that regulates mineral metabolism and fluid balance.

Minerals the inorganic chemical compounds found in nature.

Mitochondria *my-to-kon'-dree-ah* the cytoplasmic organelles in the form of granules, rods, filaments responsible for generation of metabolic energy for cellular activities.

Mitosis *my-to'-sis* the division of the cell nucleus; often followed by division of the cytoplasm of a cell; also called karyokinesis.

Mixed nerves the nerves containing the processes of motor and sensory neurons; their impulses travel to and from the central nervous system.

Molar *mo'-lar* a solution concentration determined by mass of solute—one liter of solution contains an amount of solute equal to its molecular weight in grams.

Molecule *mol'-e-kewl* a very small mass of matter composed of atoms held together as a unit.

Moles elevations of the skin that are pigmented.

Monocyte *mon'-o-site* a large, single-nucleus white blood cell.

Mons pubis *monz pu'-bus* the fatty eminence over the pubic symphysis in the female.

Motor nerve cells the nerves that carry impulses leaving the brain and spinal cord.

Motor unit a neuron and the muscle cells it supplies.

Mucus *mew'-kus* a sticky, thick fluid secreted by mucous glands and mucous membranes that keeps the free surface of membranes moist.

Mucous membranes the membranes that form the linings of the digestive, respiratory, urinary, and reproductive tracts.

Multipennate *mul-ti-pen'-nate* the muscles in which the fibers have a complex arrangement involving converging of tendons.

Multiple sclerosis *skler-o'-sis* a chronic condition characterized by destruction of the myelin sheaths of neurons in the spinal cord and the brain.

Multipolar neurons *mul-ti-pol'-ar* neurons that have one long axon and numerous dendrites.

Muscle fibers muscle cells.

Muscle spindles the complex capsules found in skeletal muscles that are sensitive to stretching.

Muscular dystrophy *dis'-tro-fee* a progressive disorder marked by atrophy and stiffness of the muscles.

Mutation *mu-tay'-shun* an alteration in the genetic material.

Myelencephalon *my-e-len-sef'-a-lon* the lower part of the hindbrain, especially the medulla oblongata.

Myelin *my'-e-lin* the white, fatty lipid substance forming a sheath around some nerves.

Myelinated fibers *my'-e-li-nay-ted* axons (projections of a nerve cell) covered with myelin.

Myelitis *my-a-light'-us* an inflammation of the spinal cord.

Myocardial infarction *my-o-kar'-dee-al in-fark'-shun* a condition characterized by dead tissue areas in the myocardium caused by interruption of blood supply to the area.

Myocardium *my-o-kar'-di-um* the cardiac muscle layer of the wall of the heart.

Myofibril *my-o-figh'-bril* a fibril found in the cytoplasm of muscle.

Myofilament *my-o-fil'-a-ment* the filamentous structures making up sarcomere consisting of thick and thin types.

Myogenic *my-o-jen'-ik* having the potential to contract automatically without nervous system stimulation.

Myoglobin *my-o-glo'-bin* muscle hemoglobin.

Myometrium *my-o-me'-tree-um* the thick uterine musculature.

Myopia *my-o'-pee-ah* nearsightedness.

Myosin *my'-o-sin* one of the principal proteins found in muscle.

Myotome *my'-o-tome* that part of a somite that differentiates into skeletal muscle.

Nares *nar'-eez* the nostrils

Necrosis *ne-kro'-sis* the death or disintegration of a cell or tissues caused by disease or injury.

Nephron *nef'-ron* the functional part of the kidney.

Nerve fiber axon (nerve cell projection) together with certain sheaths or coverings.

Neuralgia *noo-ral'-jee-ah* a severe paroxysmal (spasm-producing) pain along the course of a nerve.

Neuritis *noo-righ'-tis* an inflammatory or degenerative condition of the nerves.

Neurofibril *noo-ro-fi'-bril* the fibril of a nerve cell, usually extending from the processes and traversing the cell body.

Neuroglia *noo-rahg'-lee-a* the nonneuronal tissue of the central nervous system that performs supportive and other functions; also called glia.

Neurohypophysis *noo-ro-high-pof'-i-sis* the portion of the pituitary gland derived from the brain.

Neuromuscular junction *noo-ro-mus'-ku-lar* the region where a motor neuron approaches skeletal muscle plasma membrane.

Neurons *noo'-ronz* the nerve cells that transmit messages throughout the body.

Neutron *noo'-tron* an uncharged particle located in the nucleus of an atom.

Neutrophil *noo'-tro-fil* the most abundant of the white blood cells.

Nexus *nex'-us* see gap junction.

Nuchal lines *nu'-kal* the three slight ridges on the external surface of the occipital bone.

Nucleoli *noo-klee'-o-lee* the small spherical bodies in the cell nucleus.

Nucleotide *noo'-klee-o-tide* a component of DNA and RNA consisting of a sugar, a base, and a phosphate group.

Nucleus *noo'-klee-us* (1) a dense central body in most cells containing the genetic apparatus of the cell; (2) the core of an atom.

Nucleus pulposes *noo'-klee-us pul-po'-sus* the central gelatinous part of the intervetebral disc.

Nystagmus *nis-tag'-mus* an oscillatory movement of the eyeballs.

Obesity *o-bee'-sit-ee* a condition of a person being overweight; often leads to other complications.

Occipital *ok-sip'-i-tal* pertaining to the back of the head area.

Occlusion *o-kloo'-zhun* closure or obstruction.

Olfaction *ol-fak'-shun* smell.

Oogenesis *o-o-jen'-e-sis* the process of origin, growth, and formation of the ovum.

Ophthalmic *of-thal'-mik* pertaining to the eye.

Optic *op'-tik* pertaining to the eye.

Optic chiasma *op'-tik kee-az'-ma* the meeting of the optic nerves after entering the cranium.

Oral relating to the mouth.

Ora serrata *o'-ra ser-rat'a* the serrated margin of the retina.

Organ a part of the body combining two or more tissues to perform a specialized function.

Organelle *or-gan-el'* a specialized structure or part of a cell having a definite function to perform.

Organic compound any hydrocarbon or hydrocarbon derivative.

Orgasm *or'-gaz-um* the intense emotional and physical climax associated with sexual stimulation.

Origin the end of attachment of a muscle that remains relatively fixed during muscular contraction.

Osmoreceptor *os-mo-ree-sep'-tor* a structure sensitive to osmotic pressure.

Osmosis *os-mo'-sis* the passage (diffusion) of a solvent through a membrane from a dilute solution into a more concentrated one.

Ossicles *os'-si-kalz* the three bones of the middle ear: malleus, stapes, and incus.

Osteoblasts *os'-tee-o-blasts* the bone-forming cells.

Osteoclasts *os'-tee-o-klasts* the large cells that reabsorb or erode bone substance.

Osteocyte *os'-tee-o-site* a mature bone cell found in each lacuna.

Osteomalacia *os-tee-o-mal-a'-she-ah* the softening of bone resulting from vitamin D deficiency in the adult.

Osteomyelitis *os-tee-o-my-a-light'-us* a disease in which the periosteum, the contents of the marrow cavity, and the bone tissue become inflamed.

Osteoporosis *os-tee-o-por'-o-sis* an increased softening of the bone resulting from a gradual reduction in the rate of bone formation while the rate of bone absorption remains normal; a common condition in older people.

Ostium *os'-tium* a small opening into a hollow structure.

Otic *o'-tik* pertaining to the ear.

Otitis media *o-tigh'-tis mee'-dee-ah* middle-ear infection.

Otolith *o'-to-lith* one of the small calcareous masses in the utricle and saccule of the inner ear.

Ovarian cycle *o-va'-ree-an* the monthly cycle of follicle development, ovulation, and corpus luteum formation in the ovary.

Ovary *o'-va-ree* the female sex organ in which ova (eggs) are produced.

Ovulation *ov-u-lay'-shun* the maturation and release of an ovum.

Ovum *o'-vum* the female gamete (germ cell); an egg cell.

Oxidation *oks-i-day'-shun* the loss of electrons by a molecule; the process of substances combining with oxygen.

Oxygenated *oks'-i-je-nay-ted* the condition in which a substance is saturated with oxygen.

Oxyhemoglobin *oks-i-hee-mo-glo'-bin* oxidized hemoglobin.

Oxytocin *oks-i-to'-sin* a hormone released by the posterior pituitary that stimulates contractility of smooth muscles of the uterus and myoepithelial cells surrounding the alveoli of the mammary glands during labor.

Palate *pal'-et* the roof of the mouth.

Palmar *pal'-mar* the anterior surface of the hands.

Palpation *pal-pay'-shun* examination by touch.

Palpebral fissure *pal'-pe-bral* the gap between the upper and lower eyelids.

Pancreas *pan'-kree-as* the gland located behind the stomach, between the spleen and the duodenum, producing both endocrine and exocrine secretions.

Pancreatic juice *pan-kree-at'-ik* a secretion of the pancreas containing enzymes for digestion of carbohydrates.

Pancreatitis *pan-kree-a-tigh'-tis* an inflammation of the pancreas.

Papilla *pa-pil'-lah* a small elevation, nipple-shaped or cone-shaped.

Papillary muscles *pa'-pil-lar-ee* cone-shaped muscles such as those that project into the cardiac ventricular lumen.

Paralysis *pa-ral'-i-sis* the loss of muscle function or of sensation.

Paraplegia *par-a-plee'-gee-ah* paralysis of the lower limbs.

Parasympathetic division *par-a-sim-pa-thet'-ik* a division of the autonomic nervous system; also referred to as the craniosacral division.

Parathyroid glands *par-a-thigh'-roid* the several small endocrine glands posterior to the capsule of the thyroid gland.

Parathyroid hormone the hormone released by the parathyroid glands that regulate blood calcium level.

Parenchyma *par-en'-ki-mah* the functional components of an organ.

Parenteral *par-en'-ter-al* occurring through some route other than the alimentary canal, such as intravenous.

Parietal *par-eye'-i-tal* pertaining to the walls of a cavity.

Parotid *par-ot'-id* located near the ear.

Parturition *par-toor-i'-shun* the act of giving birth.

Patella *pa-tel'-ah* the kneecap.

Pathogenesis *path-o-jen'-e-sis* the development of a disease.

Pectoral *pek'-to-ral* pertaining to the chest.

Pedicle *ped'-i-kul* the portion of the neural arch between the centrum and the transverse process.

Peduncle *pe-dun'-kal* a stalk of fibers, especially that connecting the cerebellum to the pons, mesencephalon, and medulla oblongata.

Pelvis a basin-shaped structure, especially the lower portion of the body's trunk.

Penis *pee'-nis* the male organ of copulation and urinary excretion.

Pepsin an enzyme capable of digesting proteins in an acid pH.

Peptide bond *pep'-tide* a bond joining the amino group of one amino acid to the acid carboxyl group of a second amino acid with the loss of a water molecule.

Pericardium *per-i-kar'-dee-um* the closed membranous sac enveloping the heart.

Perichondrium *per-i-kon'-dree-um* a fibrous, connective-tissue membrane covering the external surface of cartilaginous structures.

Perimysium *per-i-meez'-ee-um* the connective tissue enveloping bundles of muscle fibers.

Perineum *per-i-nee'-um* that region of the body extending from the anus to the scrotum in males and from the anus to the vulva in females.

Periosteum *per-ee-os'-tee-um* a double-layered connective tissue that covers, invests, and nourishes the bone.

Peripheral nervous system (PNS) *per-if'-er-al* a system of nerves that connect the outlying parts of the body and their receptors with the central nervous system.

Peristalsis *per-is-tal'-sis* the progressive wave of contraction seen in tubes.

Peritoneum *per-i-ton-ee'-um* the membrane lining the interior of the abdominal cavity.

Peritonitis *per-i-ton-eye'-tis* an inflammation of the peritoneum.

Permeability *per-mee-a-bil'-i-tee* that property of membranes which permits transit of molecules and ions.

Peroneal *per-o-nee'-al* pertaining to the outer side of the leg.

Peroxisome *per-ox'-i-some* a membrane-bounded organelle in cells that contains oxidase and peroxidase; also called microbody.

pH the symbol for hydrogen ion concentration; a measure of the relative acid or base level of a substance or solution.

Phagocyte *fag'-o-site* a cell capable of engulfing and digesting particles or cells harmful to the body.

Phagocytosis *fag-o-sigh-to'-sis* the ingestion of foreign solids by cells.

Phalanges *fay-lan'-jeez* the bones of the finger or toe.

Phantom pain a phenomenon whereby a person who has undergone amputation continues to feel pain from the amputated body part.

Pharynx *far'-inks* the muscular, membranous tube extending from the base of the skull to the esophagus.

Phlebitis *fle-by'-tis* an inflammation of a vein.

Photoreceptor *fo-to-ree-sep'-tor* the specialized receptor cells that can convert light energy into a nerve impulse.

Physiology *fiz-ee-ol'-o-jee* the science of the functions of organic beings.

Pinna *pin'-nah* the irregularly shaped elastic cartilage covered with skin forming the most prominent portion of the outer ear.

Pinocytosis *pie-no-sigh-to'-sis* the engulfing of liquid by cells.

Pituitary gland the gland located beneath the brain that serves a variety of functions including regulation of the gonads, thyroid, adrenal cortex, and other endocrine glands.

Placenta *pla-sen'-tah* the organ to which the embryo attaches by the umbilical cord for nourishment and waste removal; has an endocrine function as well.

Placode *pla'-kode* an ectodermal thickening in the early embryo from which a sense organ or structure develops.

Plantar *plan'-tar* pertaining to the sole of the foot.

Plasma the fluid portion of the blood or lymph.

Platelet *plate'-let* one of the disc-shaped components of blood; involved in clotting.

Pleura *ploor'-ah* the membrane covering the lungs.

Plexus *plek'-sus* a network of interlacing nerves or anastomosing blood vessels or lymphatics.

Plica *ply'-ka* a fold.

Pneumothorax *noo-mo-tho'-raks* the presence of air or gas in a pleural cavity.

Podocyte *pod'-o-site* an epithelial cell located on the basement membrane of the glomerulus, spreading thin cytoplasmic projections over the membrane.

Polar body a minute cell given off by the ovum during maturation divisions.

Polarized *po'-lar-ized* the state of an unstimulated neuron in which the inside of the cell is relatively negative in comparison to the outside.

Poliomyelitis *po'-lee-o-my-a-light'-us* the viral destruction of nerve cell bodies within the anterior horns of the spinal cord.

Polycythemia *pol-ee-sigh-theem'-ee-a* the presence of an abnormally large number of erythrocytes in the blood.

Polypeptide *pol-ee-pep'-tide* a small chain of amino acids.

Polyribosome *pol-ee-ribe'-o-some* a multiple structure composed of several ribosomes held together by a molecule of messenger RNA; also called polysome.

Polysome *pol'-ee-some* see polyribosome.

Pons (1) any bridgelike structure or part; (2) the structure connecting the cerebellum with the brain stem providing linkage between upper and lower levels of the central nervous system.

Postganglionic (postsynaptic) neuron *post-gang-lee-on'-ik* a neuron of the autonomic nervous system having its cell body in a ganglion with the axon extending to an organ or tissue.

Preganglionic (presynpatic) neuron *pree-gang-lee-on'-ik* a neuron of the autonomic nervous system having its cell body in the brain or spinal cord and its axon terminating in a ganglion.

Prepuce *pree'-puss* the loose fold of skin that covers the glans penis or glans clitoris.

Pressoreceptor *pre-so-ree-sep'-tor* a nerve ending in the wall of the carotid sinus and aortic arch sensitive to vessel stretching.

Prime movers those muscles whose contractions are primarily responsible for a particular movement.

Process (1) a prominence or projection; (2) a series of actions or method of action for a specific purpose.

Proctologic *prok-to-loj'-ik* pertaining to the rectum or anus.

Progesterone *pro-jes'-ter-own* a hormone responsible for preparing the uterus for the fertilized ovum.

Pronation *pro-nay'-shun* the inward rotation of the forearm causing the radius to cross diagonally over the ulna—palms face posteriorly.

Prone refers to a body lying horizontally with the face downward.

Pronucleus *pro-noo'-clee-us* one of two nuclear bodies (one male and one female) of a newly fertilized ovum, the fusion of which results in formation of a cleavage nucleus.

Proprioceptor *pro-pree-o-sep'-tor* a receptor located in a muscle, tendon, joint, or vestibular apparatus of the internal ear; concerned with locomotion and posture.

Prostaglandin *pros-ta-glan'-din* a substance included in seminal vesicle secretion thought to facilitate fertilization; causes uterine contractions.

Protein *pro'-teen* a complex nitrogenous substance found in various forms in animals and plants as the principal component of protoplasm.

Proteinuria *pro-te-in-oo'-ree-ah* the passage of albumin or other protein in the urine.

Proton *pro'-ton* a particle carrying a positive charge; located in the nucleus of an atom.

Protrude *pro-trood'* to project or assume an abnormally prominent position.

Proximal *proks'-i-mal* toward the attached end of a limb or the origin of a structure.

Psoriasis *so-rye'-uh-sis* a chronic inflammatory skin disease characterized by development of red patches covered with silvery, overlapping scales.

Puberty *pu'-ber-tee* the period at which reproductive organs become functional.

Pulmonary *pul'-mon-ar-ee* pertaining to the lungs.

Pulmonary circuit the circulatory vessels of the lungs.

Pulmonary edema *e-dee'-mah* an effusion of fluid into the air sacs and interstitial tissue of the lungs.

Pulse the rhythmic expansion in arteries resulting from heart contraction; can be felt from the outside of the body.

Pupil an opening in the center of the iris through which light enters the eye.

Purkinje fibers *pur-kin'-jee* the modified cardiac muscle fibers of the conduction system of the heart.

Pus the fluid product of inflammation composed of white blood cells, the debris of dead cells, and a thin fluid.

Pyelonephritis *pie-el-o-nef-rye'-tis* an inflammation of the kidney pelvis and surrounding kidney tissues.

Pyloric glands *pie-lor'-ik* the glands that secrete thin mucus; located in the region of the pylorus of the stomach.

Pyloric region the final portion of the stomach preceding the duodenum.

Pyramid any conical eminence of an organ, especially a body of longitudinal fibers on each side of the anterior median fissure of the medulla oblongata.

Pyrogen *pie'-ro-jen* an agent that induces fever.

Quadriplegia *quad-ri-plee'-jee-ah* the paralysis of all four limbs.

Radiate *ray'-dee-ate* diverging from a central point.

Ramus *ray'mus* a branch of a nerve, artery, vein, or bone, especially a primary division.

Receptor *ree-sep'-tor* a peripheral nerve ending in the skin; specialized for response to particular types of stimuli.

Reduction the gain of electrons by a molecule.

Reflex automatic, stereotyped reactions to stimuli.

Refracted bent.

Refracting media substances that bend light.

Renal *ree'-nal* pertaining to the kidney.

Renal calculus *ree'-nal kal'-ku-lus* a kidney stone formed in the renal pelvis or urinary bladder.

Renin *ren'-in* a substance released by the juxtaglomerular complex of the kidneys; involved with raising blood pressure.

Rete *ree'-tee* a network; often composed of nerve fibers or blood vessels.

Reticulocyte *re-tik'-u-lo-site* a nonnucleated, young erythrocyte.

Reticulum *re-tik'-u-lum* a fine network.

Retract to draw back, shorten, contract.

Rhinencephalon *rye-nen-sef'-a-lon* that portion of the cerebrum concerned with reception and integration of olfactory impulses.

Rhodopsin *ro-dop'-sin* the photopigment contained in the rods of the retina.

Ribonucleic acid (RNA) *rye-bo-nu-kle'-ik* the nucleic acid that contains ribose.

Ribosomes *rye'-bo-somes* the cytoplasmic structures at which proteins are synthesized.

Rickets *rik'-ets* a disease occurring in infants and young children characterized by softening of the bone caused by demineralization from malnutrition.

Rods one of the two types of photosensitive cells in the retina.

Rotate to turn about an axis.

Rugae *ru'-je* elevations or ridges, as in the mucosa of the stomach, uterus, and palate.

Sacral *sa'-kral* the lower portion of the back, just superior to the buttocks.

Sagittal plane *saj'-i-tal* a longitudinal section that divides the body or any of its parts into right and left portions.

Saliva *sa-ligh'-vah* the combined secretions of salivary and mucous glands of the mouth.

Salivary amylase *sal'-i-ver-ee am'-i-laze* a starch-splitting enzyme contained in saliva.

Salt a compound that, when dissolved in water, dissociates into cations other than hydrogen ions and anions other than hydroxide ions.

Sarcoplasmic reticulum *sar-ko-plas'-mik re-tik'-u-lum* the membranous network running through skeletal muscle cells.

Sclera *skle'-rah* the firm, fibrous outer layer of the eyeball; functions for protection and maintenance of eyeball shape.

Scoliosis *sko-lee-o'-siss* a lateral curve in the vertebral column.

Scrotum *skro'-tum* the two-layered sac enclosing the testes.

Sebaceous glands *se-bay'-shus* glands that develop from and empty their sebum secretion into hair follicles.

Sebum *see'-bum* the secretion of sebaceous glands; oily substance rich in lipids.

Secretion *see-kree'-shun* the passage of material formed by a cell from the inside to the outside of its plasma membrane.

Segmentation *seg-men-tay'-shun* the process of cleavage or splitting; the division of an organism into somites.

Semen *see'-men* the fluid produced by male reproductive structures; contains sperm.

Semilunar valves *sem-i-loo'-nar* valves that prevent blood return to the ventricle after contraction.

Seminiferous tubules *sem-i-nif'-er-us* highly convoluted tubes within the testes containing spermatocytes.

Sensory nerve a nerve that contains only the processes of sensory neurons and carries nerve impulses to the central nervous system.

Sensory nerve cell an initiator of nerve impulses following receptor activity.

Serous fluid *ser'-us* a clear, watery fluid secreted by the cells of the mesothelium.

Serum *se'-rum* the amber-colored fluid that exudes from clotted blood as the clot shrinks and then no longer contains fibrinogen.

Sesamoid *ses'-a-moid* denoting a small bone that is embedded in a tendon or in a joint capsule.

Sex chromosome *kro'-mo-some* chromosome that determines sex of the fertilized egg; X and Y chromosomes.

Sinuatrial node *sigh-noo-ay'-tree-al* the mass of specialized myocardial cells in the wall of the right atrium.

Sinus *sigh'-nus* (1) a mucous-membrane-lined, air-filled cavity in certain cranial bones; (2) a dilated channel for the passage of blood or lymph, which lacks the coats of an ordinary vessel.

Smooth (nonstriated) muscle muscle consisting of spindle-shaped, unstriped (nonstriated) muscle cells; involuntary muscles.

Solute *sol'-yoot* the dissolved substance in a solution.

Solution a homogeneous mixture of two or more components.

Somatic nervous system *so-ma'-tik* a division of the peripheral nervous system; also called the voluntary nervous system.

Somite *so'-mite* a segment of the body of an embryo.

Sperm the mature male germ cell; a spermatozoan.

Spermatogenesis *sper-ma-to-jen'-e-sis* the process of meiosis (cell division) in the male to produce mature male germ cells.

Sphenoid *sfen'-oid* wedgelike.

Sphincter *sfink'-ter* a muscle surrounding and enclosing an orifice.

Sprain the wrenching of a joint producing stretching or laceration of the ligaments.

Squamous *skway'-mus* pertaining to flat and thin cells that form the free surface of epithelial tissue.

Stasis *stay'-sis* (1) a decrease or stoppage of flow; (2) a state of equilibrium.

Static balance *stat'-ic* balance concerned with changes in the position of the head.

Stenosis *sten-o'-sis* constriction or narrowing.

Steroids *ster'-oidz* a specific group of chemical substances including certain endocrine secretions and cholesterol.

Stimulus *stim'-u-lus* an excitant or irritant; an alteration in the environment of a living thing producing a response.

Stomodeum *sto-mo-dee'-um* the primitive oral cavity of the embryo.

Strabismus *stra-bis'-mus* a squint; that abnormality of the eyes in which the visual axes do not meet at the desired objective point.

Stratum *stray'-tum* a layer.

Striated muscle *stry-ay'-ted* muscle consisting of cross-striated (cross-striped) muscle fibers.

Stricture *strik'-chur* a contraction or inward pinching of a canal or duct.

Stroke a condition in which a cerebral blood vessel is blocked.

Stroke volume a volume of blood ejected by the left ventricle during a systole.

Stroma *stro'-mah* the supporting framework of an organ including connective tissue, vessels, and nerves.

Sty an inflammation of the connective tissue of the eyelid, near a hair follicle.

Subcutaneous *sub-ku-tay'-nee-us* beneath the skin.

Sublingual *sub-ling'-wal* located beneath the tongue.

Sulcus *sul'-kus* a furrow or linear groove; on the brain, a less deep depression than a fissure.

Summation *sum-may'-shun* the accumulation of effects, especially those of muscular, sensory, or mental stimuli.

Superficial (external) located close to or on the body surface.

Superior refers to the head or upper; higher.

Supination *soo-pi-nay'-shun* the outward rotation of the forearm causing palms to face anteriorly.

Supine refers to a body lying with the face upward.

Suspension *sus-pen'-shun* a dispersing of particles throughout a body of liquid.

Suspensory ligament of eye *sus-pen'-so-ree lig'-a-ment* fibrous ligament that holds the lens in place in the eye.

Sutures *soo'-churz* the immovable joints that connect the bones of the adult skull.

Sweat glands the glands that secrete a watery solution of sodium chloride (salt water).

Sympathetic division a division of the autonomic nervous system; opposes parasympathetic functions.

Synapse *sin'-apse* the region of communication between neurons.

Synaptic cleft *sin-ap'-tik* the space at a synapse between neurons.

Synaptic delay *sin-ap'-tik* the time required for an impulse to cross a synapse between two neurons.

Synarthrosis *sin-ar-throw'-sis* a fibrous joint; two types: sutures and syndesmoses.

Synergist *sin'-er-jist* a muscle cooperating with another to produce a movement neither alone can produce.

Synostosis *sin-os-tow'-sis* a union of originally separate bones by osseous material.

Synovial fluid *sin-o'-vi-al* a fluid secreted by the synovial membrane; lubricates joint surfaces and nourishes articular cartilages.

System a group of organs that function cooperatively to accomplish a common purpose; there are ten major systems in the human body.

Systemic *sis-tem'-ik* general; pertaining to the whole body.

Systemic circuit the circulatory vessels of the body.

Systemic edema *e-dee'-ma* an accumulation of fluid in body organs or tissues.

Systole *sis'-to-lee* the contraction phase of a cardiac cycle.

Systolic pressure *sis-tol'-ik* the pressure generated by the left ventricle during systole.

Tachycardia *tak-ee-kar'-dee-ah* the abnormal, excessive rapidity of heart action; over 100 beats per minute.

Tarsals *tar'-suls* the seven bones that form the ankle and heel.

Taste buds the receptors for taste on the tongue, roof of mouth, pharynx, and larynx.

Telencephalon *tel-en-sef'-a-lon* the anterior subdivision of the primary forebrain that develops into olfactory lobes, cerebral cortex, and corpora striata.

Tendon *ten'-don* a band of dense fibrous tissue forming the termination of a muscle and attaching the muscle to a bone.

Testis *tes'-tis* the male primary sex organ that produces sperm.

Tetanus *tet'-a-nus* (1) the tense, contracted state of a muscle; (2) an infectious disease.

Thalamus *thal'-a-mus* the mass of gray matter at the base of the brain.

Theca *thee'-ka* a sheath.

Thermoreceptor *ther-mo-ree-sep'-tor* a receptor sensitive to temperature changes.

Thoracic *tho-ras'-ik* refers to the chest.

Thorax *tho'-raks* that portion of the trunk above the diaphragm and below the neck.

Threshold the lower limit of stimulus capable of producing an impression on consciousness or evoking a response in an irritable tissue.

Thrombin *throm'-bin* an enzyme that induces clotting by converting fibrinogen to fibrin.

Thrombocyte *throm'-bo-site* a blood platelet thought to be part of the blood-clotting mechanism.

Thrombocytopenia *throm-bo-sigh-to-pee'-nee-ah* a condition in which there is a decrease in the number of blood platelets below normal.

Thrombophlebitis *throm-bo-fle-by'-tis* an inflammation of a vein associated with blood-clot formation.

Thrombus *throm'-bus* a clot that is fixed or stuck to a vessel wall.

Thymus gland *thigh'-mus* a potential source of hormonal material; active in immune response.

Thyroid gland *thigh'-roid* one of the largest of the body's endocrine glands.

Tissue a group of similar cells and fibers forming a distinct structure.

Tonic *ton'-ik* refers to the state of continuous muscular or neuron activity.

Tonofibril *tone-o-figh'-bril* one of the fine fibrils in the cytoplasm of epithelial cells.

Toxemia *tok-see'-mee-ah* a condition in which blood contains poisonous products.

Toxic *tok'-sik* poisonous.

Trabecula *tra-bek'-u-lah* any one of the fibrous bands extending from the capsule into the interior of an organ.

Trachea *tray'-kee-uh* the windpipe; the cartilaginous and membranous tube extending from larynx to bronchi.

Tract a collection of nerve fibers having the same origin, termination, and function.

Transverse processes the projections that extend laterally from each neural arch.

Trauma *traw'-mah* an injury, wound, or shock; usually produced by external forces.

Trochanter *tro-kan'-tur* a large, somewhat blunt process.

Trophic *tro'-fik* pertains to nutrition.

Trypsin *trip'-sin* an active enzyme that splits proteins.

Tubercle *too'-bur-kul* a nodule or small rounded process.

Tuberosity *too-bur-os'-i-tee* a broad process, larger than a tubercle.

Tumor an abnormal growth of cells; a swelling; cancerous at times.

Tunica *too'-ni-kah* a covering or tissue coat; membrane layer.

Twitch a brief contraction of muscle in response to a stimulus.

Tympanic membrane *tim-pan'-ik* the eardrum.

Ulcer *ul'-ser* a lesion or erosion of the mucous membrane, such as gastric ulcer of stomach.

Umbilical cord *um-bil'-i-kal* a structure bearing arteries and veins connecting the placenta and the fetus.

Umbilicus *um-bil'-i-kus* the navel; marks site that gave passage to umbilical vessels in the fetal stage.

Unipennate *yoon-i-pen'-nate* muscles in which all fasciculi insert onto one side of the tendon.

Unipolar neurons *yoon-i-pol'-ar* neurons in which embryological fusion of the two processes leaves only one process extending from the cell body.

Unitary smooth muscle muscle in which the cells contact one another at gap junctions and the impulses spread from cell to cell.

Unmyelinated fibers *un-my'-e-li-nay-ted* nerve axons that are not covered with myelin.

Urea *yoo-ree'-ah* the main nitrogen-containing waste excreted in the urine.

Uremia *yoo-ree'-mee-ah* a toxic accumulation in the blood of substances normally excreted by the kidneys.

Ureter *yoo-ree'-ter* the tube that carries urine from kidney to bladder.

Urethra *yoo-ree'-thrah* the canal through which urine passes from the bladder to the outside of the body.

Uvula *yoo'-vu-la* conical appendix hanging from soft palate.

Vacuole *vak'-yoo-ole* a clear space in a cell.

Valvular insufficiency *val'-vu-lar* a condition in which the cusps of the cardiac valves do not close tightly.

Varicose vein *var-uh-kos'* a dilated, knotted, tortuous blood vessel.

Vas a duct; vessel.

Vasa recta *va'-sa rek'-ta* capillary branches that supply loops of Henle and collecting ducts.

Vascular *vas'-ku-lar* pertaining to channels or vessels.

Vasoconstriction *vaz-o-kon-strik'-shun* the narrowing of blood vessels.

Vasodilatation *vaz-o-die-lay-tay'-shun* the relaxation of the smooth muscles of the vascular system producing dilated vessels.

Vasomotion *vaz-o-mo'-shun* an increase or decrease in caliber of a blood vessel.

Vasomotor center *vaz-o-mo'-tor* an area of brain concerned with regulation of blood vessel resistance.

Vasomotor nerve fibers *vaz-o-mo'-tor* the motor nerve fibers that regulate the constriction or dilation of blood vessels.

Vasopressin *vaz-o-pres'-sin* another name for antidiuretic hormone.

Vein a vessel carrying blood away from the tissues toward the heart.

Ventral *ven'-tral* anterior or front.

Ventricle *ven'-tri-kal* (1) a small cavity or pouch; (2) the blood propulsion chamber of the heart.

Ventricles of the brain *ven'-tri-kalz* the cavities in the interior of the brain.

Venule *ven'-ule* a small vein.

Vertigo *ver'-tee-go* dizziness; the feeling of movement such as a sensation that the external environment is revolving.

Vesicle *ves'-i-kal* a small liquid-filled sac or bladder.

Viscera *vis'-ser-ah* the internal organs.

Visceral *vis'-ser-al* pertaining to the internal part of a structure or the internal organs.

Viscosity *vis-kos'-i-tee* the state of being sticky or thick.

Visual acuity *a-kyoo'-i-ty* the ability of the eye to distinguish detail.

Vital capacity the volume of air that can be expelled from the lungs by forcible expiration after the deepest inspiration.

Vitamins the organic compounds required by the body in minute amounts for physiological maintenance and growth.

Voluntary muscle muscle under control of the will; skeletal muscle.

Vulva *vul'-va* female external genitalia.

White matter the white substance of the central nervous system; the nerve fibers.

Zonula *zon'-u-la* a small zone, usually beltlike.

Zonula adherens that part of a junctional complex of epithelial cells where the cell membranes are not modified and are separated by a gap of 200 Å.

Zonula occludens that part of a junctional complex of epithelial cells where the outer layers of cell membranes of adjacent cells fuse.

Zygote *zy'-goat* the fertilized ovum before splitting (cleavage); produced by union of two gametes.

Photo Credits

Figure 1.1, Box 14.1 p. 367: Dr. J. de Groot, Department of Anatomy and Radiology, University of California, San Francisco.

Figure 1.2: Courtesy of the Mayo Clinic.

Figure 2.2: Photo by M. Weinstock, Department of Anatomy, McGill University. From Ham and Cormack, *Histology* 8th ed., 1979, Lippincott Co., Fig. 5–3, p. 109.

Figures 2.7, 2.14, 2.16: Kimball, J. W. 1978. *Biology* 4th ed. Reading, MA. Addison-Wesley Publishing Co.

Figures 2.12, 2.13: Courtesy of Dr. Barry L. Batzing.

Figure 2.17: T. E. Schroeder, University of Washington/BPS.

Figure 2.18 b, c: Courtesy of E. de Harven, Rockefeller University.

Figure 2.19: Barry King, University of California, Davis/BPS.

Box 2.1: Cancer Research Center/Media Vision.

Figures 3.2, 3.3, 3.4, 3.5, 3.6, 3.8, 3.12, 3.14, 3.16, 3.20, 3.21, 3.22, 3.23, 4.1: Photos by Ed Reschke.

Figures 3.7, 9.5, 13.9, 13.13, 14.6(c), 14.20(b), 22.17: Manfred Kage/Peter Arnold.

Figures 3.13, 3.15, 3.18, 3.19: Biophoto.

Frontiers box p. 43: Wide World Photos.

Box 4.1 p. 81, Frontiers box p. 170: Leonard Kamsler/Medichrome.

Box 5.1: Carolina Biological Supply.

Figures 5.7, 6.7, 6.35: Courtesy of Dr. Henry Jones, Department of Radiology, Stanford Medical Center.

Frontiers box p. 122: Dr. Freiburger/Peter Arnold.

Figures 5.35, 5.36: Lester Bergman & Associates.

Frontiers box p. 143: Illustration by John W. Karapelou. Courtesy of Harold M. Dick, M.D., New York Orthopedic Hospital, Columbia Presbyterian Medical Center, New York.

Figures 5.41, 5.43: Richard Humbert, Stanford University.

Box 6.1, p. 160: Solon I. Finkelstein, M.D.

Figure 6.21: Bobbe Wolf, TIME Magazine.

Box 6.2 p. 176, Box 10.1 p. 284: © Carroll H. Weiss, 1973.

Figure 7.4: Clara Franzini-Armstrong, University of Pennsylvania.

Frontiers box p. 190: Courtesy of Nautilus, Inc.

Frontiers box p. 273: Courtesy of Dr. C. A. Hunt, University of California, San Francisco.

Frontiers boxes pp. 296 and 329, Figures 11.2, 18.17, 20.18: Ed Reschke/Peter Arnold.

Figure 11.3: P. Phelps and J. Luft, "Electron Microscopal Study of Relaxation and Constriction in Frog Arterioles," *Am. J. Anat.* 125: Figs. 2 and 3, p. 409, 1969.

Figure 11.5: R. Bolender, "Stereological Analysis of the Guinea Pig Pancreas." Reproduced from the *Journal of Cell Biology*, 1974, vol. 61, by copyright permission of The Rockefeller University Press.

Frontiers box p. 310: Photo by Rod Cowe.

Figures 12.3, 12.9: Courtesy of Ward's Natural Science Establishment, Inc.

Figure 12.7: Biophoto.

Figure 12.13: Andrejs Liepins, Sloan-Kettering Institute for Cancer Research.

Figure 13.10: P. Schultz/Biology Media.

Figure 13–15 a–d: Photos by Victor B. Eichler.

Figure 14.1(b): Lester Bergman & Associates.

Figures 14.2(b), 14.8(b), 14.9(b): M. Rotker/Taurus Photos, Inc.

Frontiers box p. 362: Courtesy of Dr. Joseph H. Schulman, Neurodyne Corporation, and Pacesetter Systems.

Figure 17.1: Michael Coppinger, Department of Opthamology, Pacific Medical Center.

Figure 17.2(b): Don Fawcett/Photo Researchers/Science Source.

Frontiers box p. 475: 3M.

Box 19.1, p. 527: American Cancer Society, Inc.

Frontiers box p. 556: Illustration by Harriet R. Greenfield. Photo by D. Niccolini/Medichrome.

Figure 21.6(b): Lester Bergman & Associates.

Figure 21.8: G. Tyson and R. Bulger, "Endothelial Detachment Sites in Glomerular Capillaries of Vinblastine Treated Rats," *Anat. Rec.* 172: Fig. 1, p. 671, 1972.

Frontiers box p. 582: Dr. Irving H. Goldman, Crouse Irving Memorial Hospital, Syracuse.

Box 21.1: Biological Photo Service.

Figure 22.18: Bloom and Fawcett, *A Textbook of Histology*, 10th ed. W. B. Saunders Co., © 1985.

Frontiers box p. 616: Wayland Lee.

Box 23.1: Dr. Robert Rugh and Dr. Landrum Shettles.

Frontiers box p. 640: David York/Medichrome.

Index